AF868769

Ruthenium-Containing Polymers

Ulrich S. Schubert • Andreas Winter •
George R. Newkome

Ruthenium-Containing Polymers

Springer

Ulrich S. Schubert
Laboratory for Organic
and Macromolecular Chemistry (IOMC)
Friedrich Schiller University Jena
Jena, Thuringia, Germany

Andreas Winter
Laboratory for Organic
and Macromolecular Chemistry (IOMC)
Friedrich Schiller University Jena
Jena, Thuringia, Germany

George R. Newkome
Center for Molecular Biology
and Biotechnology
Florida Atlantic University
Jupiter, FL, USA

ISBN 978-3-030-75597-3 ISBN 978-3-030-75598-0 (eBook)
https://doi.org/10.1007/978-3-030-75598-0

© The Editor(s) (if applicable) and The Author(s), under exclusive license to Springer Nature Switzerland AG 2021
This work is subject to copyright. All rights are solely and exclusively licensed by the Publisher, whether the whole or part of the material is concerned, specifically the rights of translation, reprinting, reuse of illustrations, recitation, broadcasting, reproduction on microfilms or in any other physical way, and transmission or information storage and retrieval, electronic adaptation, computer software, or by similar or dissimilar methodology now known or hereafter developed.
The use of general descriptive names, registered names, trademarks, service marks, etc. in this publication does not imply, even in the absence of a specific statement, that such names are exempt from the relevant protective laws and regulations and therefore free for general use.
The publisher, the authors and the editors are safe to assume that the advice and information in this book are believed to be true and accurate at the date of publication. Neither the publisher nor the authors or the editors give a warranty, expressed or implied, with respect to the material contained herein or for any errors or omissions that may have been made. The publisher remains neutral with regard to jurisdictional claims in published maps and institutional affiliations.

This Springer imprint is published by the registered company Springer Nature Switzerland AG
The registered company address is: Gewerbestrasse 11, 6330 Cham, Switzerland

Preface

In 1920, Hermann Staudinger proposed that polymers are indeed long chains, which are formed from repeating molecular units *via* covalent bonds. This, then highly controversial, discovery marks the birth of synthetic polymer chemistry. Starting from this epoch-making publication, polymer chemistry has experienced a tremendous evolution in which not only the synthesis of tailor-made architectures became possible—but today, designer polymers can be found in a wide range of utilitarian applications in our daily lives.

When did metal centers come into play? The first metal-containing polymers, based on ferrocene, were reported in the 1950s; however, metallopolymers were sparsely regarded as an useful class of materials due to constraints in both their available synthetic as well as characterization procedures. These limitations were finally overcome in the mid-1990s and, thus, such materials no longer represent laboratory oddities but rather they are employed in a highly diverse spectrum of applications, such as opto-electronics, biomedicine, and catalysis.

Ruthenium is a remarkable metal which, despite its limited natural abundance, plays a prominent role in diverse cutting-edge fields of research. In particular, photovoltaics, anti-cancer treatment, bioimaging, and catalysis have to be noted in this context. The activity of Ru compounds or nanoparticles in these fields has led to the development of Ru-containing polymers ("ruthenopolymers") which today also find a diverse range of applications. The same holds true for other macromolecular Ru-containing materials, such as dendrimers, metal–organic frameworks, and covalent organic frameworks.

Herein, the various Ru-containing compounds are organized according to their classes. In each case, fundamental aspects as well as examples of materials with respect to their synthesis, properties, and potential applications are summarized. We have attempted to cover each field as detailed as possible in order to support future adventures in these rapidly expanding areas; however, the subject matter had to be presented in a compact fashion and contributions might have been overlooked. Therefore, we apologize in advance to those authors whose work was not incorporated.

It is doubtful that Ru-containing materials will ever be used on an industrial scale (owing, in part, to the price and the limited abundance of Ru in Nature); however, substantial progress has already been made in biomedical or catalytic applications, where comparably less Ru is required. Hopefully, this overview will help researchers open new pathways to Ru-containing polymers (and related classes of materials) as well as their applications.

Jena, Germany
Jupiter, FL, USA
February 2021

Ulrich S. Schubert
Andreas Winter
George R. Newkome

Contents

Abbreviations

2-mim	2-methyl-imidazolate
2PA	two-photon absorption
3PA	three-photon absorption
A2780	a human ovarian cancer cell line
A2780cis	a cisplatin-resistant human ovarian cancer cell line
AAS	atomic absorption spectroscopy
acac	acetylacetonate
ADH	alcohol dehydrogenase
ADMET	acyclic diene metathesis
AEM	anion-exchange membrane
AFM	atomic force microscopy
AIBN	azo*bis*(isobutyronitrile)
ALD	atomic-layer deposition
Alq_3	tris(8-hydroxyquinolinato)aluminium
AM	air mass
AMPS	acrylamidomethylpropane sulfonate
ANQ	anthraquinone
aPP	atactic polypropylene
ATRP	atom-transfer radical polymerization
AUC	analytical ultracentrifugation
BCP	2,9-dimethyl-4,7-diphenyl-1,10-phenanthroline
bda	2,2'-bipyridine-6,6'-dicarboxylate
bdc	1,4-benzenedicarboxylate (*i.e.*, terephthalate)
BET	Brunauer–Emmett–Teller
BHJ	bulk heterojunction
BIAN	*bis*(arylimino)acenaphthene
biq	2,2'-bisquinoline
box	2-(2-hydroxyphenyl)benzoxazole
bpdc	4,4'-biphenyldicarboxylate
bpp	2,6-*bis*(1*H*-pyrazol-1-yl)pyridine
bpy	2,2'-bipyridine
BSA	bovin serum albumin
btc	1,3,5-benzenetricarboxylate
btp	2,6-*bis*(1*H*-1,2,3-trizol-4-yl)pyridine

BZ	Belousov–Zhabotinsky
Caco-2	an immortalized cell line of human colorectal adenocarcinoma cells
CAL-72	Centre Antoine Lacassagne-72 (a human osteosarcoma cell line)
CALB	lipase-A from *Candida antarctica*
CD	circular dichroism **or** cyclodextrin
CDI	*N,N'*-carbonyldiimidazole
CDSA	crystallization-driven self-assembly
CE	coloration efficiency
CIE	Commission Internationale de l'Éclairage
CL	ε-caprolactone
CM	cross-metathesis
cmc	critical micellization concentration
CNT	carbon nanotube
cod	η^4-*cis,cis*-1,5-cyclooctadiene
COF	covalent organic framework
CONASH	coordination nanosheet
cot	η^6-*cis,cis,cis*-1,3,5-cyclooctatriene
Cp	cyclopentadienyl
cp*	1,2,3,4,5-pentamethylcyclopentadienyl
CPDT	4*H*-cyclopenta[2,1-*b*:3,4-*b'*]dithiophene
CPL	circularly polarized luminescence
CROP	cationic ring-opening polymerization
CRP	controlled-radical polymerization
CSS	charge-separated state
ct	calf thymus
CTAB	cetyltrimethylammonium bromide
CTF	covalent triazine frameworks
CuAAC	Cu^I-catalyzed azide-alkyne cycloaddition
CV	cyclic voltammetry
Cy	cyclohexyl
Đ	dispersity
DBU	1,8-diazabicyclo[5.4.0]undec-7-ene
dca	dicyanamide
dcbpy	4,4'-dicarboxy-2,2-bipyridine or 5,5'-dicarboxy-2,2-bipyridine
DDQ	2,3-Dichloro-5,6-dicyano-1,4-benzoquinone
DFT	density-functional theory
dip	4,7-diphenyl-1,10-phenanthroline
DLS	dynamic light scattering
dmbpy	4,4'-dimethyl-2,2'-bipyridine
DNA	deoxyribonucleic acid
DOSY	diffusion-ordered NMR spectroscopy
DP	degree of polymerization
DPA	9,10-diphenylantharcene
DPP	differential pulse polarography
dqp	2,6-di(quinoline-8-yl)-pyridine

DSC	differential scanning calorimetry
DSPEC	dye-sensitized photoelectroysnthesis cells
DSSC	dye-sensitized solar cells
DTPA	diethylenetriaminepentaacetic acid
DVB	1,4-divinylbenzene
DVP	differential pulse voltammetry
EC	electrochromism
ECL	electro-generated chemoluminescence or electrochemical luminescence
EDOT	ethylene dioxythiophene
EDTA	ethylenediamine tetraacetate
EDX	energy dispersive X-ray analysis
EGDMA	ethylene glycol dimethacrylate
EL	electroluminescence
ELD	electroluminescent device
ElT	electron transfer
EM	electron microscopy
emim	*N*-ethyl-*N*'-methylimidazolium
EnT	energy transfer
EO	ethylene oxide
EPR	electron paramagnetic resonance or enhanced permeability and retention
EQCM	electrochemical quartz crystal microbalance
EQE	external quantum efficiency
ESI	electrospray ionization
EXAFS	extended X-ray absorption fine structure
Fc	ferrocene
fcc	face-centered cubic
FIB	focused ion beam
FJI	Fujian Institute of Research
FRET	Förster-type resonance energy transfer
FRP	free-radical polymerization
FT	Fourier transformation
FTO	fluorine-doped tin oxide
GC	glassy carbon
$g\text{-}C_3N_4$	graphitic carbon nitride
GDH	glucose dehydrogenase
gMS^2	gradient tandem mass spectrometry
GO	graphite oxide
HAADF	high-angle annular dark-field imaging
hcp	hexagonal close-packed
HCT116	a human colon cancer cell line
HEEDTA	hydroxyethylethylene diaminetriacetic acid sodium salt
hEGF	human epidermal growth factor
HEK	healthy human embryonic kidney cells

HeLa	a human cervical cancer cell line
HER	hydrogen-evolution reaction
HKUST	Hong Kong University of Science and Technology
HMPA	hexamethylphosphoramide
hMSCs	human mesenchymal stem cells
HOMO	highest occupied molecular orbital
HOPG	highly-ordered pyrolytic graphite
HP	hyperbranched polymer
HPC	hierarchically porous carbon
HPLC	high-pressure liquid chromatography
HREM	high-resolution electron microscopy
HRP	horseradish peroxidase
HSA	human serum albumin
HWE	Horner–Wadsworth–Emmons
IC_{50}	half-maximal inhibitory concentration
ICP	inductively coupled plasma
IDA	interdigitated array (electrode)
IDP	isodesmic supramolecular polymerization
IIP	ion-imprinted polymer
IM	ion mobility
IPCE	incident photo-to-electron conversion efficiency
iPP	isotactic polypropylene
IR	infrared
IRMOF	isoreticular MOF
ITC	isothermal titration calorimetry
ITO	indium tin oxide
IVCT	intervalence charge-transfer
K_a	association constant
KP1019	*trans*-[tetrachloro*bis*(1*H*-indazole)ruthenate(III)]
LBL	layer-by-layer
LC	ligand-centered or liquid crystalline
LEC	light-emitting electrochemical cell
LED	light-emitting diode
LF	ligand field
LLDPE	linear low-density polyethylene
LMCT	ligand-to-metal charge-transfer
LNCaP	an androgen-sensitive human prostate adenocarcinoma cell line
LUMO	lowest unoccupied molecular orbital
MAC	macrocyclic building block
MALDI	matrix-assisted laser desorption/ionization
MAPTAC	methacrylamidopropyltrimethylammonium chloride
MAS	magic-angle spinning
MCF7	Michigan Cancer Foundation-7 (a breast cancer cell line)
MCP	metallacycling polymerization
Me_2Ntsi	$C(SiMe_3)_2(SiMe_2NMe_2)$

MIL	matériaux de l'Institut Lavoisier
MIP	molecularly imprinted polymer
MLCT	metal-to-ligand charge-transfer
MMA	methyl methacrylate
M_n	number-averaged molar mass
MOCN	metal–organic coordination network
MOF	metal–organic framework
MOFSC	MOF-based solar cells
MPD	multi-patterned display
MPEG	PEG monomethyl ether
MS	mass spectrometry
MV^{2+}	methylviologen (*i.e.*, *N,N'*-dimethyl-4,4'-bipyridinium or paraquat)
M_w	weight-averaged molar mass
MWCNT	multi-walled carbon nanotubes
NADH	nicotinamide adenine dinucleotide
NAMI-A	nickname for a ruthenodrug
nbd	norbornadiene (*i.e.*, bicyclo[2.2.1]hepta-2,5-diene)
NDI	naphthalenediimide
NHC	*N*-heterocyclic carbene
NIR	near infrared
NLO	non-linear optical
NMP	nitroxide-mediated polymerization
NMR	nuclear magnetic resonance
NP	nanoparticle
NSL	nanosphere lithography
OD	optical density
OEG	oligo(ethylene glycol)
OER	oxygen-evolution reaction
OFET	organic field-effect transistor
OLED	organic light-emitting diode
ORR	oxygen–reduction reaction
Ovcar-3	a human ovarian carcinoma cell line
P2VP	poly(2-vinylpyridine)
P4VP	poly(4-vinylpyridne)
PAA	poly(acrylic acid)
PACT	photoactivated chemotherapy
PAF	porous aromatic framework
PAMAM	poly(amidoamine)
PANI	polyaniline
PBT	poly(butylene terephthalate)
PC	polycarbonate
PCL	poly(ε-caprolactone)
PCN	porous coordination network
PCP	porous cross-linked polymer
PDA	photodiode array

PDDA	poly(diallyldimethylammonium chloride)
PDMS	poly(dimethylsiloxane)
PDT	photodynamic therapy
PE	polyethylene
PEB	poly(ethylene-*co*-butylene)
PEC	photoelectrochemical cell
PEDOT	poly(3,4-ethylenedioxythiophene)
PEG	poly(ethylene glycol)
PEI	poly(ethyleneimine)
PES	polyethersulfone
PET	poly(ethylene terephthalate)
PFDMS	poly(ferrocenyldimethylsilane)
PFGSE	pulsed field gradient spin-echo
phen	1,10-phenanthroline
PI	polyisoprene
PISA	polymerization-induced self-assembly
Piv	pivalate
PLA	poly(L-lactic acid)
PLED	polymer-based LED
PLGA	poly(L-lactic-*co*-glycolic acid)
PMA	polymethacrylate
PMMA	poly(methyl methacrylate)
PNDI	poly(naphthalene diimide)
PNR	polynorbornadiene
POM	polyoxometalate or polarized optical microscopy
POP	porous organic polymers
PPD	polyphenylene dendrimer
PPh_3	triphenylphosphine
PPI	poly(propylene imine)
PPO	poly(propylene oxide)
PPV	poly(*p*-phenylenevinylene)
PPy	polypyrrole
PS	polystyrene
PS2	photosystem-2
PSf	polysulfone
PSS	poly(styrene sulfonate)
PT	polythiophene
PTA	1,3,5-triaza-7-phosphaadamantane
PTARA	poly(triarylamine)
PtBA	poly(*tert*-butyl acrylate)
PTFE	poly(tetrafluoroethylene)
PTFMS	poly(4-trifluoromethylstyrene)
PTZ	phenothiazine
PVA	polyvinylalcohol
PVC	poly(vinyl chloride)

PVDF	poly(vinylidene fluoride)
PVFc	poly(vinylferrocene)
PVI	poly(*N*-vinylimidazole)
PVK	poly(*N*-vinylcarbazole)
PVP	poly(*N*-vinylpyrrolidone)
PVS	poly(vinyl sulfate)
py	pyridine
pydc	3,5-pyridinedicarboxylate
QM/MM	quantum mechanics/molecular mechanics
RAFT	reversible addition–fragmentation chain-transfer
RAPTA-C	Ru(η^6-*para*-cymene)Cl_2(PTA)
RCM	ring-closing metathesis
RDE	rotating-disk electrode
Re	Reynolds number
RGO	reduced graphene oxide
ROM	ring-opening metathesis
ROMP	ring-opening metathesis polymerization
ROP	ring-opening polymerization
RSC	redox-switchable catalysis
Ru	Ruthenium
RWLC	Randy W. Larsen compound (a MOF structure)
salen	*bis*(salicylidene)ethylenediamine
SAXS	small-angle X-ray scattering
SBU	secondary building unit
SCE	saturated calomel electrode
SCILL	solid catalyst with ionic liquid layer
SEC	size-exclusion chromatography
SEM	scanning electron microscopy
SET	single-electron transfer
SK-HEP-1	a human liver hepatic adenocarcinoma cell line
SOP	self-oscillating polymer
STEM	scanning-tunneling electron microscopy
SWCNT	single-walled carbon nanotube
TA	transient absorption
TEM	transmission electron microscopy
TEMPO	2,2,6,6-tetramethylpiperidine-1-oxyl
TEOA	triethanolamine
TFBE	trifluoroborane etherate
TGA	thermogravimetric analysis
TIPNO	2,2,5-trimethyl-4-phenyl-3-azahexane nitroxide
TMEDA	tetramethylethylenediamine
TNT	2,4,6-trinitrotoluene
TOF	time-of-flight or turnover frequency
TON	turnover number
TPA	triphenylamine

TPBi	1,3,5-*tris*(1-phenyl-1*H*-benzimidazol-2-yl)benzene
tpy	2,2‘:6‘,2“-terpyridine
trpy	2-(1*H*-1,2,3-triazol-4-yl)pyridine
TSD	tube-shaped displays
TTA	triplet–triplet annihilation
TTF	tetrathiafulvalene
TWIM	travelling-wave ion mobility
U87MG	Uppsala 87 Malignant Glioma (a human cancer cell line)
UHV	ultrahigh vacuum
UiO	Universitetet i Oslo (*i.e.*, Oslo University)
UPy	4-ureidopyrimidone
USF	University of South Florida
UV	ultraviolet
WAXD	wide-angle X-ray diffraction
WNV	West Nile virus
WOC	water-oxidation catalyst
WOR	water-oxidation reaction
XPS	X-ray photoelectron spectroscopy
XRD	X-ray diffraction
ZIF	zeolitic imidazole framework

Introduction 1

Abstract

In this chapter, the history of ruthenium and its complexes is discussed briefly. A general introduction of the various types metallopolymers, in particular those containing ruthenium centers, has also been presented.

Polymer + Ruthenium = Ruthenopolymer. A simple equation with diverse variables

"Polymers are long chains of high molecular weight which are comprised of covalently linked repeating subunits, the self-styled monomers." Though being highly controversial at the time, Hermann Staudinger's famous proposal—published in 1920—represents the original spark of modern polymer science; he received the 1953 Nobel Prize in Chemistry for his ground-breaking research on macromolecules [1–3]. Not much later, Wallace H. Carothers suggested a classification of synthetic macromolecules according to the incorporated monomers and the technique from which they were polymerized [4]. About 90 years later, this fundamental classification-scheme is still contemporary and commonly used to explain the differences between polymers synthesized by different procedures (e.g., step-/chain-growth or ring-opening polymerization). Since then, a broad range of polymer architectures has been prepared, generally following these three basic mechanisms. In the last few decades, the various controlled/living techniques (i.e., metathesis, ionic, and radical) have widened this synthetic toolbox; nonetheless, all of these polymerization methods obey Carothers' fundamental systematics.

The last century also witnessed a pervasive evolution of polymer and materials science away from the initial core research areas, such as tire technology or

© The Author(s), under exclusive license to Springer Nature Switzerland AG 2021
U. Schubert et al., *Ruthenium-Containing Polymers*,
https://doi.org/10.1007/978-3-030-75598-0_1

packaging means, *inter alia* towards energy- or biomedical-related targets, thereby addressing the special needs of the twenty-first century [5]. Moreover, a broad range of methodologies has been established allowing the preparation of polymers in a highly precise and controlled fashion. Thus, advanced polymer architectures with tailor-made properties can be offered for a broad range of potential applications with particular relevance for today's daily life—these include energy storage [6, 7], energy interconversion [8–10], smart coatings [11, 12], self-healing materials [13], shape-memory materials [14], bioimaging [15], carriers in drug/gene delivery [10, 16, 17], biodegradable package [18], etc. Thus, the introduction of metal centers has paved the way towards new classes of functional materials combining the characteristics of organic polymers, i.e., their mechanical or electronic properties, with that of the introduced metal centers. For example, variation of the involved metal-to-ligand interactions might shift the material's properties from a "static" towards a "dynamic" macromolecular behavior. Moreover, different metal ions feature unique properties on their own that are difficult (or impossible) to realize when only focusing on the "traditional" polymer world: bioactivity [19–22], conductivity [23, 24], ferroelectrics [25, 26], molecular magnetism [27–29], non-linear optics [30, 31] or sensitization [32, 33] represent some areas of interest to be mentioned in this context. Immobilizing catalytically active metal complexes onto polymeric supports, can generate either homogeneous or heterogeneous catalysts for use in organic and macromolecular chemistry, is one further important field of research, since the 1970s. In particular, the various cross-coupling, metathesis or polymerization reactions have to be named here [34, 35].

The first attempts to prepare metal-containing polymers originated back in the 1950s, soon after the discovery of ferrocene, when its vinyl derivative was utilized as a monomer in free radical polymerization reactions (polyvinylferrocene was first reported by Arimoto and Haven Jr. in 1955) [36–38]. However, establishing metallopolymers as a useful materials' class was hampered, for a long time, by simply the constraints in available synthetic methods. This obstacle can be rationalized by the fact that many of the common protocols to synthesize well-defined organic polymers did not (or only in a very limit fashion) tolerate the presence of metal centers. As a consequence, undesirable side reactions might be observed or materials possessing only low molar masses were formed. Another downside, products isolated in the pioneering work were often poorly soluble and, thus, a structural characterization was difficult or impossible when relying on the analytical techniques at that time. These drawbacks called for intense research and, at least since the mid-1990s, a rich toolbox of synthetic methods being well-compatible with the presence of metal ions or their complexes has become available [39]. Thus, the gram-scale synthesis of metallopolymers with high molar masses and great variability with respect to the macromolecular structure has become more straightforward. Improving the solubility of such materials also facilitated their characterization and, if desired, enabled the fabrication of devices by processing the materials from solution (e.g. by spin-coating or inkjet-printing techniques) [40]. As important, the field of polymer analytics has evolved tremendously over the last decades and, today, a wide range of highly developed analytical tools can also be applied to investigate metal-containing polymers [41–46]. As a result, metal-containing materials no longer represent laboratory oddities but have already been employed in a highly diverse spectrum of applications [13, 39, 40, 47–86].

From all the potential metals that might be incorporated into polymeric materials, only a handful of representatives is of particular relevance to polymer science. Two aspects account for this restriction: the materials' stability and potential applications. Besides iron (Fe) [87], iridium (Ir) [76], platinum (Pt) [88], and gallium (Ga) [86], ruthenium (Ru) has attracted widespread attention [63, 85, 88]; however, Ru is an extremely rare metal being only the 74th most abundant element in Earth's crust (0.001 ppm) [89].

At the beginning of the nineteenth century, J. Smithson Tennant and William H. Wollaston developed the first physicochemical method for the processing of Pt ores in practical quantities. These researchers isolated palladium, rhodium, osmium, and iridium from the residues of Pt ores from the Ural [90, 91]. It is speculated that the Polish chemist Jędrzej Śniadecki was possibly the first to isolate the element with the atomic number 44 from South American Pt ores. He disclosed his finding in 1808 and named the metal "vestium" (after the asteroid Vesta, which was discovered previously) [92]. However, his research was heavily opposed and never approved; disappointed by this he later withdrew his claim of discovery [89]. The existence of the metal was also proposed by Gottfried Osann in 1828 when he analyzed the same ores as Tennant and Wollaston [93]; however, not being able to repeat his experiments, he also withdrew his claims after a long controversary with Jöns Jacob Berzelius [94]. As the last member of the Pt elements [95], the metal was finally isolated, though only in little amounts, by Carl Ernst Claus in 1844 from Uralian Pt ores; he named the metal "ruthenium" after the origin and, simultaneously, honoring his motherland Russia and recognizing at least Osann's endeavor (Fig. 1.1) [96].

Fig. 1.1 The Baltic-German chemist Carl Ernst Claus (1796–1864), Kazan State University, who discovered ruthenium: "I named the new body, in honor of my motherland." Figure reproduced with kind permission; © 2020 HFDR [97]

The biggest portion of the annually produced Ru is used in electrical applications, but also the wide field of catalysis represents a considerable consumer (in 2016, *ca.* 30.9 t of Ru were produced with a share of 44.7% and 24.9%, respectively, going into these two fields [98]). Due to its robust coordination chemistry, remarkable redox behavior (oxidation states from 0 to +VIII and −II are accessible) as well as the rich photophysical and biomedical properties, many complexes with ruthenium ions play a significant role in modern science. The 2001 Nobel Prize in Chemistry was awarded *inter alia* to Ryōji Noyori for his contributions to asymmetric hydrogenations; he introduced complex **I** for the enantioselective hydrogenation of polar substrates, such as ketones or imines [99]. Further striking examples in this respect are Grubbs' catalysts for the various olefin metathesis reactions (**II**) [100], Grätzel's photosensitizing black dye (**III**) [101], and the potential anti-cancer drug RAPTA-C (**IV**), which is currently in (pre)clinical studies (Fig. 1.2) [102]. However, the $[Ru(bpy)_3]^{2+}$ complex (bpy: 2,2-bipyridine) experienced the most remarkable evolution from a barely valued laboratory observation in 1936 [103] to a formidable photo-redox catalyst [104]. A spotlight on bpy's history, taking these aspects into account, has been published elsewhere [105, 106]. Besides these organometallic species, stabilized ruthenium nanoparticles are of relevance as heterogeneous catalysts, e.g., in hydrogenation [107].

With an increasing interest in molecular Ru complexes, the synthesis of defined ruthenium-containing metallopolymers moved into the focus-of-interest. A range of basic architectures is conceivable to bring Ru centers and polymer topology together. As pointed out elsewhere [41, 108], one has to be aware that the term "metallopolymer" is used by inorganic and polymer chemists in quite different ways. Thus, the fundamental definition, saying that "any polymer containing metal atoms in the repeating monomer, either as part of the backbone or in the side chains" has to be considered as a metallopolymer is mistakable. For inorganic chemists, coordination compounds of whatever dimensionality are commonly referred to as metallopolymers. In this context, the 2D coordination nanosheets (CONASH) [109] or the 3D metal–organic frameworks (MOFs) [110, 111] represent the most prominent examples. These solid-state assemblies, having a crystalline or an amorphous structure, typically tend to dissociate into their molecular subunits in the presence of strongly coordinating solvents. MOFs containing Ru^{II}

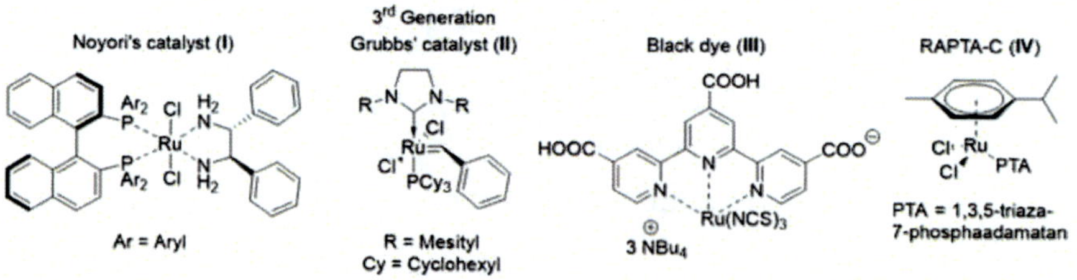

Fig. 1.2 Schematic representation of four important Ru complexes

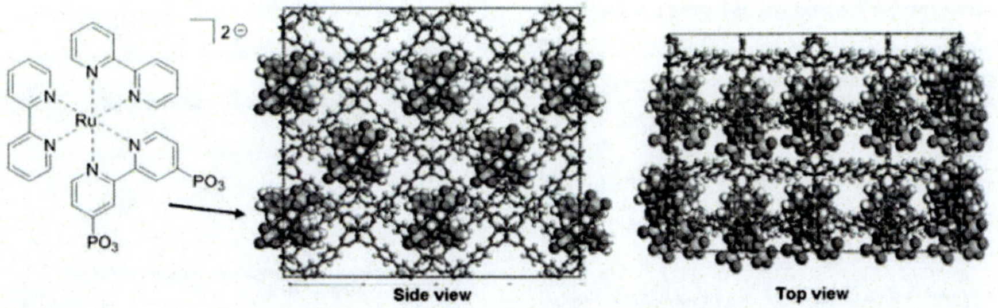

Fig. 1.3 Schematic representation of a MOF containing Ru^{II} complexes, as emitting guest molecules, within its channels [113]. Figure reprinted with kind permission; © 2014 American Chemical Society

complexes are of interest, e.g., for applications as fluorescent sensors, light-emitting materials, in non-linear optics or as heterogeneous catalysts for electro-/photo-catalytic water splitting [112]. One representative example is the host–guest system comprising of a MOF matrix, which resulted from the self-assembly of 1,4-benzenedicarboxylate (bdc) and Zn^{II} ions, and emissive Ru^{II}–polypyridyl complexes occupying the MOF's channels, as guest molecules (Fig. 1.3) [113]. Within the context of this book, these types of solid-state assemblies will be discussed separately from those that retain their polymeric nature also in solution (Chap. 6).

In polymer science, the definition of "metallopolymer" is strictly focused on the perception of a polymer chemist, that is the metal centers need to be integral parts of the polymeric architecture, either within the polymer backbone or side chains; alternatively, the metal centers might be attached to the macromolecular structure at certain binding sites. The binding strength, as expressed by the K_a value, quantifies the interaction between metal centers and their binding sites. The polymeric structure of metallopolymers is only retained in solution if the K_a value is high enough. In addition to this, metal-containing polymers should also possess physical properties that are characteristic for polymers (e.g., enhanced viscosity compared to that of the monomeric units, glassy solid state) [70].

Already in 1990s, Rehahn [71] and Ciardelli [114] suggested a classification of the broad field of metal-containing materials taking not only the polymer architecture but also the mode of metal binding into account (Fig. 1.4). Firstly, the metal species might be bound to the polymer architecture (i.e., the backbone, side chains or end groups) by electrostatic, coordinative or covalent interactions (type-1 materials). Secondly, type-2 materials feature metal centers as integral parts of their polymer backbones; thereby, the incorporation of the metal sites might arise from covalent binding (a) or metal-to-ligand coordination (b). As detailed by Ho and Wong [56], the closer the coupling of a metal center to the polymer backbone, the stronger the (photophysical) properties of the material are influenced—this prediction particularly holds true when dealing with π-conjugated backbones. Physical

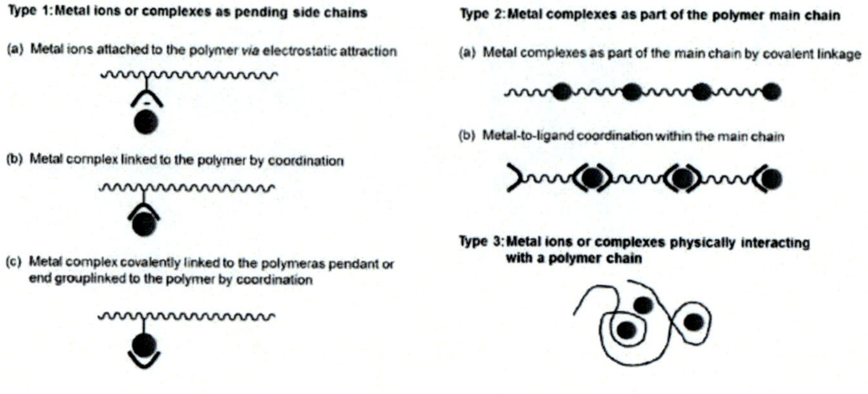

Fig. 1.4 Schematic representation of the potential architectures for metal-containing polymers [83]. Figure redrawn according to the literature; © 2011 The Royal Chemical Society

interactions between polymers and metal ions/complexes (or even metal nanoparticles) may generate hybrid materials, in which the so-called polymer chelate effect, as expressed by negative values of the Gibbs free energy (ΔG^0), represents the thermodynamic driving force for their formation and stabilization [108]. If metal-containing polymers are referred to as metallopolymers, it is obvious to use "ruthenopolymers" as a generic term for the range of Ru-containing polymers. These materials, which can be classified according to these considerations, are the subject of Chaps. 2 and 3; there, a further subdivision has been made taking into account the preparative method: Post-polymerization functionalization, conventional polymerization methods, and electropolymerization. Noteworthy, alternative classification schemes can be applied when distinguishing between the polymers' topologies (i.e., linear, grafted, hyperbranched or dendritic macromolecular structures) or their properties (e.g., conjugated *vs.* non-conjugated or kinetically stable vs. kinetically labile).

The main distinction between the latter two results from the different preparative methods typically used. In general, hyperbranched polymers are formed by one-pot (or single-pot) methodologies and, thus, such materials possess broad molar-mass distributions (referred to as dispersities, Đ). Whereas, dendrimers, are globular, highly branched, fractal-like polymers of well-defined 3D shape and size [57]. The iterative or stepwise synthesis thereby enables the precise control over the molar mass; however, the synthetic effort is significantly higher, in particular when aiming for dendrimers of higher generations. The cornerstone for research in this field was laid by Vögtle et al. in 1978 [115] and promoted *inter alia* by Newkome et al., who applied the robust and directional coordination chemistry of Ru^{II} to prepare metallodendrimers (for a representative example, see Fig. 1.5) [57]. Today, ruthenodendrimers are known where Ru^{II} complexes have been employed as cores, branching centers, building-block connectors, or terminal units (Fig. 1.6)

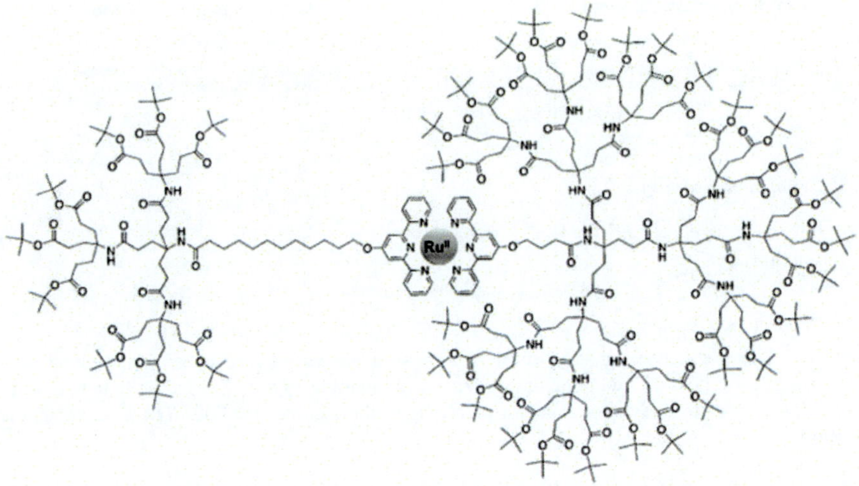

Fig. 1.5 Schematic representation of Newkome's "lock and key" dendrimer assembly [123]. Figure redrawn according to the literature; © 1995 Wiley-VCH

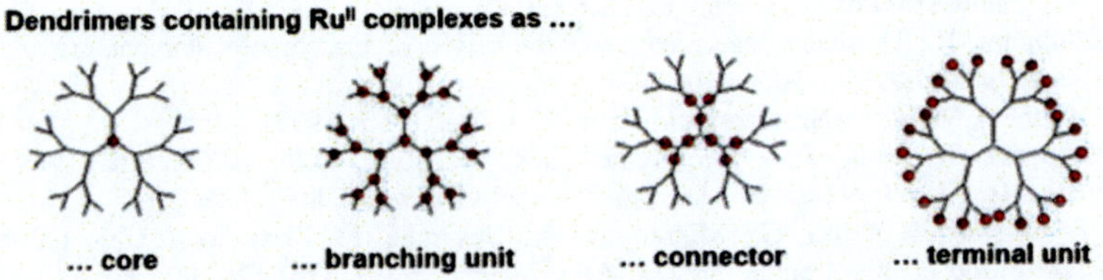

Fig. 1.6 Schematic representation of the various structural roles of Ru^{II} complexes within dendritic architectures [117]. Figure reproduced with kind permission; © 2007 The Royal Chemical Society

[116–121]. Moreover, the encapsulation of Ru species, as guests in dendrimer matrices, has been reported [122]. Research in these fields is summarized separately in Chap. 4.

For Ru-containing systems, the different macromolecular architectures can be obtained by utilizing the structural characteristics, which originate from the interactions between the metal centers and their binding sites, i.e. the ligands (Fig. 1.7). In this respect, coordination of Ru^{II} ions in a (pseudo-)octahedral fashion, e.g., by polypyridyl ligands, represents the most commonly used interaction, offering the advantages of a high directionality and chemical robustness.

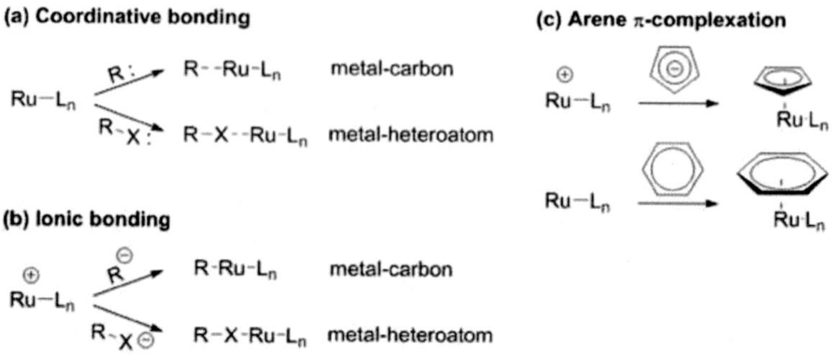

Fig. 1.7 Schematic representation of the main types of metal-to-ligand interactions, which are of relevance in the field of Ru-containing polymers (R: binding ligand, X: binding heteroatom, L_n: ancillary ligands) [61]. Scheme redrawn according to the literature; © 2007 The Royal Chemical Society

References

1. Staudinger, H.: Über Polymerisation. Ber. Dtsch. Chem. Ges. **53**(6), 1073 (1920)
2. Frey, H., Johann, T.: Celebrating 100 years of "polymer science": Hermann Staudinger's 1920 manifesto. Polym. Chem. **11**(1), 8 (2020)
3. Mülhaupt, R.: Hermann Staudinger and the origin of macromolecular chemistry. Angew. Chem. Int. Ed. **43**(9), 1054 (2004)
4. Carothers, W.H.: Polymerization. Chem. Rev. **8**(3), 353 (1931)
5. Schubert, S., Manners, I., Newkome, G.R., Schubert, U.S.: A Pandora's box of new materials: Metallopolymers. Macromol. Rapid Commun. **36**(7), 585 (2015)
6. Nyholm, L., Nyström, G., Mihranyan, A., Strømme, M.: Toward flexible polymer and paper-based energy storage devices. Adv. Mater. **23**(33), 3751 (2011)
7. Janoschka, T., Hager, M.D., Schubert, U.S.: Powering up the future: Radical polymers for battery applications. Adv. Mater. **24**(48), 6397 (2012)
8. Facchetti, A.: π-Conjugated polymers for organic electronics and photovoltaic cell applications. Chem. Mater. **23**(3), 733 (2010)
9. Li, G., Zhu, R., Yang, Y.: Polymer solar cells. Nat. Photon. **6**(3), 153 (2012)
10. Aamer, K.A., de Jeu, W.H., Tew, G.N.: Diblock copolymers containing metal complexes in the side chain of one block. Macromolecules **41**(6), 2022 (2008)
11. Kenawy, E.-R., Worley, S.D., Broughton, R.: The chemistry and applications of antimicrobial polymers: A state-of-the-art review. Biomacromol **8**(5), 1359 (2007)
12. Montemor, M.F.: Functional and smart coatings for corrosion protection: A review of recent advances. Surf. Coat. Technol. **258**, 17 (2014)
13. Hager, M.D., Greil, P., Leyens, C., van der Zwaag, S., Schubert, U.S.: Self-healing materials. Adv. Mater. **22**(47), 5424 (2010)
14. Hu, J., Zhu, Y., Huang, H., Lu, J.: Recent advances in shape–memory polymers: Structure, mechanism, functionality, modeling and applications. Prog. Polym. Sci. **37**(12), 1720 (2012)
15. Luo, S., Zhang, E., Su, Y., Cheng, T., Shi, C.: A review of NIR dyes in cancer targeting and imaging. Biomaterials **32**(29), 7127 (2011)
16. Rösler, A., Vandermeulen, G.W.M., Klok, H.-A.: Advanced drug delivery devices via self-assembly of amphiphilic block copolymers. Adv. Drug Delivery Rev. **53**(1), 95 (2001)

17. Brannon-Peppas, L., Blanchette, J.O.: Nanoparticle and targeted systems for cancer therapy. Adv. Drug Delivery Rev. **56**(11), 1649 (2012)
18. Luckachan, G., Pillai, C.K.S.: Biodegradable polymers: A review on recent trends and emerging perspectives. J. Polym. Environ. **19**(3), 637 (2011)
19. Liu, H.-K., Sadler, P.J.: Metal complexes as DNA intercalators. Acc. Chem. Res. **44**(5), 349 (2011)
20. Gill, M.R., Thomas, J.A.: Ruthenium(II) polypyridyl complexes and DNA-from structural probes to cellular imaging and therapeutics. Chem. Soc. Rev. **41**(8), 3179 (2012)
21. Eryazici, I., Moorefield, C.N., Newkome, G.R.: Square-planar Pd(II), Pt(II), and Au(III) terpyridine complexes: Their syntheses, physical properties, supramolecular constructs, and biomedical activities. Chem. Rev. **108**(6), 1834 (2008)
22. Gottschaldt, M., Schubert, U.S.: Prospects of metal complexes peripherally substituted with sugars in biomedicinal applications. Chem. Eur. J. **15**(7), 1548 (2009)
23. Cassoux, P., Valade, L., Kobayashi, H., Kobayashi, A., Clark, R.A., Underhill, A.E.: Molecular metals and superconductors derived from metal complexes of 1,3-dithiol-2-thione-4,5-dithiolate (dmit). Coord. Chem. Rev. **110**(1), 115 (1991)
24. Robertson, N., Cronin, L.: Metal bis-1,2-dithiolene complexes in conducting or magnetic crystalline assemblies. Coord. Chem. Rev. **227**(1), 93 (2002)
25. Chen, C.-T., Suslick, K.S.: One-dimensional coordination polymers: Applications to material science. Coord. Chem. Rev. **128**(1–2), 293 (1993)
26. Hang, T., Zhang, W., Ye, H.-Y., Xiong, R.-G.: Metal-organic complex ferroelectrics. Chem. Soc. Rev. **40**(7), 3577 (2011)
27. Sorace, L., Benelli, C., Gatteschi, D.: Lanthanides in molecular magnetism: Old tools in a new field. Chem. Soc. Rev. **40**(6), 3092 (2011)
28. Gatteschi, D., Bogani, L., Cornia, A., Mannini, M., Sorace, L., Sessoli, R.: Molecular magnetism, status and perspectives. Solid State Sci. **10**(12), 1701 (2008)
29. Bogani, L., Wernsdorfer, W.: Molecular spintronics using single-molecule magnets. Nat. Mater. **7**(3), 179 (2008)
30. Evans, O.R., Lin, W.: Crystal engineering of NLO materials based on metal−organic coordination networks. Acc. Chem. Res. **35**(7), 511 (2002)
31. Di Bella, S.: Second-order nonlinear optical properties of transition metal complexes. Chem. Soc. Rev. **30**(6), 355 (2001)
32. Grätzel, M.: Photoelectrochemical cells. Nature **414**(6861), 338 (2001)
33. Nazeeruddin, M.K., Klein, C., Liska, P., Grätzel, M.: Synthesis of novel ruthenium sensitizers and their application in dye-sensitized solar cells. Coord. Chem. Rev. **249**(13–14), 1460 (2005)
34. Bailey, D.C., Langer, S.H.: Immobilized transition-metal carbonyls and related catalysts. Chem. Rev. **81**(2), 109 (1981)
35. Leadbeater, N.E., Marco, M.: Preparation of polymer-supported ligands and metal complexes for use in catalysis. Chem. Rev. **102**(10), 3217 (2002)
36. Arimoto, F.S., Haven, A.C.: Derivatives of dicyclopentadienyliron. J. Am. Chem. Soc. **77**(23), 6295 (1955)
37. Baldwin, M.G., Johnson, K.E.: Free-radical polymerization of vinyl ferrocene. J. Polym. Sci., Part A-1: Polym. Chem. **5**(8), 2091 (1967)
38. Sasaki, Y., Walker, L.L., Hurst, E.L., Pittman, C.U., Jr.: Organometallic polymers. Part XXV. Preparation of polyvinylferrocene. J. Polym. Sci.: Polym. Chem. Ed. **11**(6), 1213 (1973)
39. Manners, I.: Synthetic Metal-Containing Polymers. Wiley-VCH, Weinheim (2004)
40. Whittell, G.R., Hager, M.D., Schubert, U.S., Manners, I.: Functional soft materials from metallopolymers and metallosupramolecular polymers. Nat. Mater. **10**(3), 176 (2011)
41. Winter, A., Schubert, U.S.: Synthesis and characterization of metallo-supramolecular polymers. Chem. Soc. Rev. **45**(19), 5311 (2016)

42. Liu, Y., Wang, Z., Zhang, X.: Characterization of supramolecular polymers. Chem. Soc. Rev. **41**(18), 5922 (2012)
43. Pastor, A., Martínez-Viviente, E.: NMR spectroscopy in coordination supramolecular chemistry: A unique and powerful methodology. Coord. Chem. Rev. **252**(21–22), 2314 (2008)
44. Baytekin, B., Baytekin, H.T., Schalley, C.A.: Mass spectrometric studies of non-covalent compounds: Why supramolecular chemistry in the gas phase? Org Biomol. Chem. **4**(15), 2825 (2006)
45. Raşa, M., Schubert, U.S.: Progress in the characterization of synthetic (supramolecular) polymers by analytical ultracentrifugation. Soft Matter **2**(7), 561 (2006)
46. Li, X., Chan, Y.-T., Casiano-Maldonado, M., Yu, J., Carri, G.A., Newkome, G.R., Wesdemiotis, C.: Separation and characterization of metallosupramolecular libraries by ion mobility mass spectrometry. Anal. Chem. **83**(17), 6667 (2011)
47. Eloi, J.-C., Chabanne, L., Whittell, G.R., Manners, I.: Metallopolymers with emerging applications. Mater. Today **11**(4), 28 (2008)
48. Abd-El-Aziz, A.S.: Organometallic polymers of the transition metals. Macromol. Rapid Commun. **23**(17), 995 (2002)
49. Abd-El-Aziz, A.S., Humphrey, M.G.: Towards oligoclusters—chemistry and properties. Macromol. Symp. **209**(1), 1 (2004)
50. Andres, P.R., Schubert, U.S.: New functional polymers and materials based on 2,2′:6′,2″-terpyridine metal complexes. Adv. Mater. **16**(13), 1043 (2004)
51. Hofmeier, H., Schubert, U.S.: Combination of orthogonal supramolecular interactions in polymeric architectures. Chem. Commun. (19), 2423 (2005)
52. Hofmeier, H., Schubert, U.S.: Recent developments in the supramolecular chemistry of terpyridine-metal complexes. Chem. Soc. Rev. **33**(6), 373 (2004)
53. Holliday, B.J., Swager, T.M.: Conducting metallopolymers: The roles of molecular architecture and redox matching. Chem. Commun. (1), 23 (2005)
54. Kingsborough, R.P., Swager, T.M.: Transition metals in polymeric π-conjugated organic frameworks. Prog. Inorg. Chem. **48**, 123 (2007)
55. Manners, I.: Putting metals into polymers. Science **294**(5547), 1664 (2001)
56. Ho, C.-L., Wong, W.-Y.: Metal-containing polymers: Facile tuning of photophysical traits and emerging applications in organic electronics and photonics. Coord. Chem. Rev. **255**(21–22), 2469 (2011)
57. Newkome, G.R., He, E., Moorefield, C.N.: Suprasupermolecules with novel properties: Metallodendrimers. Chem. Rev. **99**(7), 1689 (1999)
58. Nguyen, P., Gómez-Elipe, P., Manners, I.: Organometallic polymers with transition metals in the main chain. Chem. Rev. **99**(6), 1515 (1999)
59. Schubert, U.S, Newkome, G.R., Manners, I. (eds.): Metal-Containing and Metallo-supramolecular Polymers and Materials, vol. 928, ACS Washington, D. C. (2006)
60. Abd-El-Aziz, A.S., Manners, I. (eds.): Frontiers in Transition Metal-Containing Polymers, John Wiley & Sons, Hoboken (2007)
61. Williams, K.A., Boydston, A.J., Bielawski, C.W.: Main-chain organometallic polymers: Synthetic strategies, applications, and perspectives. Chem. Soc. Rev. **36**(5), 729 (2007)
62. Wong, W.-Y.: Luminescent organometallic poly(aryleneethynylene)s: Functional properties towards implications in molecular optoelectronics. Dalton Trans. (40), 4495 (2007)
63. Whittell, G.R., Manners, I.: Metallopolymers: New multifunctional materials. Adv. Mater. **19**(21), 3439 (2007)
64. Stanley, J.M., Holliday, B.J.: Luminescent lanthanide-containing metallopolymers. Coord. Chem. Rev. **256**(15–16), 1520 (2012)
65. Hardy, C.G., Zhang, J., Yan, Y., Ren, L., Tang, C.: Metallopolymers with transition metals in the side-chain by living and controlled polymerization techniques. Prog. Polym. Sci. **39** (10), 1742 (2014)

66. Ramanathan, M., Darling, S.B.: Nanofabrication with metallopolymers - recent developments and future perspectives. Polym. Int. **62**(8), 1123 (2013)
67. Chiper, M., Hoogenboom, R., Schubert, U.S.: Toward main chain metallo-terpyridyl supramolecular polymers: "The metal does the trick." Macromol. Rapid Commun. **30**(8), 565 (2009)
68. Schubert, U.S., Eschbaumer, C.: Macromolecules containing bipyridine and terpyridine metal complexes: Towards metallosupramolecular polymers. Angew. Chem. Int. Ed. **41**(16), 2892 (2002)
69. Hoogenboom, R., Fournier, D., Schubert, U.S.: Asymmetrical supramolecular interactions as basis for complex responsive macromolecular architectures. Chem. Commun. (2), 155 (2008)
70. Dobrawa, R.; Würthner, F.: Metallosupramolecular approach toward functional coordination polymers. J. Polym. Sci., Part A: Polym. Chem. **43**(21), 4981 (2005)
71. Rehahn, M.: Organic/inorganic hybrid polymers. Acta. Polym. **49**(5), 201 (1998)
72. Friese, V.A., Kurth, D.G.: Soluble dynamic coordination polymers as a paradigm for materials science. Coord. Chem. Rev. **252**(1–2), 199 (2008)
73. Friese, V.A., Kurth, D.G.: From coordination complexes to coordination polymers through self-assembly. Curr. Opin. Colloid Interface Sci. **14**(2), 81 (2009)
74. Fustin, C.A., Guillet, P., Schubert, U.S., Gohy, J.F.: Metallo-supramolecular block copolymers. Adv. Mater. **19**(13), 1665 (2007)
75. Gohy, J.-F.: Metallo-supramolecular block copolymer micelles. Coord. Chem. Rev. **253**(17–18), 2214 (2009)
76. Ulbricht, C., Beyer, B., Friebe, C., Winter, A., Schubert, U.S.: Recent developments in the application of phosphorescent iridium(III) complex systems. Adv. Mater. **21**(44), 4418 (2009)
77. Shunmugam, R., Gabriel, G.J., Aamer, K.A., Tew, G.N.: Metal-ligand-containing polymers: Terpyridine as the supramolecular unit. Macromol. Rapid Commun. **31**(9–10), 784 (2010)
78. Elemans, J.A.A.W., van Hameren, R., Nolte, R.J.M., Rowan, A.E.: Molecular materials by self-assembly of porphyrins, phthalocyanines, and perylenes. Adv. Mater. **18**(10), 1251 (2006)
79. Zhang, X., Wang, C.: Supramolecular amphiphiles. Chem. Soc. Rev. **40**(1), 94 (2011)
80. Li, S.-L., Xiao, T., Lin, C., Wang, L.: Advanced supramolecular polymers constructed by orthogonal self-assembly. Chem. Soc. Rev. **41**(18), 5950 (2012)
81. Yan, X., Wang, F., Zheng, B., Huang, F.: Stimuli-responsive supramolecular polymeric materials. Chem. Soc. Rev. **41**(18), 6042 (2012)
82. Ciferri, A.: Ionic mixed interactions in macromolecules. Chem. Eur. J. **16**(36), 10930 (2010)
83. Wild, A., Winter, A., Schlutter, F., Schubert, U.S.: Advances in the field of 2,2′:6′,2″-terpyridines. Chem. Soc. Rev. **40**(3), 1459 (2011)
84. Happ, B., Winter, A., Hager, M.D., Schubert, U.S.: Photogenerated avenues in macromolecules containing Re(I), Ru(II), Os(II), and Ir(III) metal complexes of pyridine-based ligands. Chem. Soc. Rev. **41**(6), 2222 (2012)
85. Marin, V., Holder, E., Hoogenboom, R., Schubert, U.S.: Functional ruthenium(II)- and iridium(III)-containing polymers for potential electro-optical applications. Chem. Soc. Rev. **36**(4), 618 (2007)
86. Tang, J., Sheng, Y., Hu, H., Shen, Y.: Macromolecular MRI contrast agents: Structures, properties and applications. Prog. Polym. Sci. **38**(3–4), 462 (2013)
87. Abd-El-Aziz, A.S., Todd, E.K.: Organoiron polymers. Coord. Chem. Rev. **246**(1), 3 (2003)
88. Xiang, J., Ho, C.-L., Wong, W.-Y.: Metallopolymers for energy production, storage and conservation. Polym. Chem. **6**(39), 6905 (2015)
89. Emsley, J.: Ruthenium In Nature's Building Blocks: An A-Z Guide to the Elements, p. 368. Oxford University Press, Oxford (2001)
90. Lewis, D.G.: The early life of Smithson Tennant FRS (1761–1815). Platinum Metals Rev. **55**(3), 196 (2011)

91. Usselman, M.C.: The platinum notebooks of William Hyde Wollaston. Platinum Metals Rev. **22**(3), 100 (1978)
92. Śniadecki, J. *Rosprawa o nowym metallu w surowey platynie odkrytym*, PhD Thesis, Vilnius University, 1808.
93. Osann, G.: Fortsetzung der Untersuchung des Platins vom Ural. Ann. Phys. **14**, 329 (1828)
94. Osann, G.: Berichtigung, meine Analyse des uralschen Platins betreffend. Ann. Phys. **15**, 158 (1829)
95. Weeks, M.E.: The discovery of the elements. Part VIII. The platinum metals. J. Chem. Educ. **9**(6), 1017 (1932)
96. Pitchkov, V.N.: The discovery of ruthenium. Platinum Metals Rev. **40**(4), 181 (1996)
97. https://www.hfdr.de/sub/persoenlichkeiten.htm. Last accessed 1 Nov 2020
98. Loferski, P.J., Ghalayini, Z.T., Singerling, S.A.: The 2016 Minerals Yearbook. U.S. Geological Survey (www.usgs.gov), pp. 57.51 (2016)
99. Noyori, R., Ohkuma, T.: Asymmetric catalysis by architectural and functional molecular engineering: practical chemo- and stereoselective hydrogenation of ketones. Angew. Chem. Int. Ed. **40**(1), 40 (2001)
100. Scholl, M., Ding, S., Lee, C.W., Grubbs, R.H.: Synthesis and activity of a new generation of ruthenium-based olefin metathesis catalysts coordinated with 1,3-dimesityl-4,5-dihydroimidazol-2-ylidene ligands. Org. Lett. **1**(6), 953 (1999)
101. O'Regan, B., Grätzel, M.: A low-cost, high-efficiency solar cell based on dye-sensitized colloidal TiO_2 films. Nature **353**(6346), 737 (1991)
102. Ott, I., Gust, R.: Non platinum metal complexes as anti-cancer drugs. Arch. Pharm. **340**(3), 117 (2007)
103. Burstall, F.H.: Optical activity dependent on co-ordinated bivalent ruthenium. J. Chem. Soc. (Res.) 173 (1936)
104. Zeitler, K.: Photoredox catalysis with visible light. Angew. Chem. Int. Ed. **48**(52), 9785 (2009)
105. Seddon, K.R.: Ruthenium: "A dance to the music of time." Platinum Metals Rev. **40**(3), 128 (1996)
106. Constable, E.C., Housecroft, C.E.: The early years of 2,2′-bipyridine—a ligand in its own lifetime. Molecules **24**(21), 3951 (2019)
107. Lara, P., Philippot, K., Chaudret, B.: Organometallic ruthenium nanoparticles: a comparative study of the influence of the stabilizer on their characteristics and reactivity. ChemCatChem **5**(1), 28 (2013)
108. Winter, A., Hager, M.D., Schubert, U.S.: Supramolecular Polymers. In: Möller, K., Matyjaszewski, M. (eds.) Polymer Science: A Comprehensive Reference. Elsevier: Amsterdam, pp 269 (2012)
109. Sakamoto, R., Takada, K., Pal, T., Maeda, H., Kambe, T., Nishihara, H.: Coordination nanosheets (CONASHs): Strategies, structures and functions. Chem. Commun. **53**(43), 5781 (2017)
110. James, S.L.: Metal-organic frameworks. Chem. Soc. Rev. **32**(5), 276 (2003)
111. Zhou, H.-C., Long, J.R., Yaghi, O.M.: Introduction to metal–organic frameworks. Chem. Rev. **112**(2), 673 (2012)
112. Wang, W., Xu, X., Zhou, W., Shao, Z.: Recent progress in metal-organic frameworks for applications in electrocatalytic and photocatalytic water splitting. Adv. Sci. **4**(4), 1600371 (2017)
113. Tang, Y., He, W., Lu, Y., Fielden, J., Xiang, X., Yan, D.: Assembly of ruthenium-based complex into metal-organic framework with tunable area-selected luminescence and enhanced photon-to-electron conversion efficiency. J. Phys. Chem. C **118**(44), 25365 (2014)
114. Ciardelli, F., Tsuchida, E., Wöhrle, D. (eds.): Macromolecule-Metal Complexes. Springer Verlag, Berlin (1996)
115. Buhleier, E., Wehner, W., Vögtle, F.: "Cascade"- and "nonskid-chain-like" syntheses of molecular cavity topologies. Synthesis (2), 155 (1978)

116. Gorman, C.: Metallodendrimers: Structural diversity and functional behavior. Adv. Mater. **10**(4), 295 (1998)
117. Hwang, S.-H., Shreiner, C.D., Moorefield, C.N., Newkome, G.R.: Recent progress and applications for metallodendrimers. New J. Chem. **31**(7), 1192 (2007)
118. Newkome, G.R., Moorefield, C.N.: From 1→3 dendritic designs to fractal supramacromolecular constructs: understanding the pathway to the Sierpiński gasket. Chem. Soc. Rev. **44**(12), 3954 (2015)
119. Moorefield, C.N., Schultz, A., Newkome, G.R.: From dendrimers to fractal polymers and beyond. Braz. J. Pharm. Sci. **49**, 67 (2013)
120. Newkome, G.R., Moorefield, C.N., Vögtle, F.: Dendritic Molecules: Concepts, Syntheses Perspectives. VCH, Weinheim (1996)
121. Newkome, G.R., Moorefield, C.N., Vögtle, F.: Dendrimers and Dendrons: Concepts, Syntheses Applications. Wiley-VCH, Weinheim (2001)
122. Antonels, N.C., Benjamin Williams, M., Meijboom, R., Haumann, M.: Well-defined dendrimer encapsulated ruthenium SCILL catalysts for partial hydrogenation of toluene in liquid-phase. J. Mol. Catal. A **421**, 156 (2016)
123. Newkome, G.R., Güther, R., Moorefield, C.N., Cardullo, F., Echegoyen, L., Pérez-Cordero, E., Luftmann, H.: Routes to dendritic networks: bis-dendrimers by coupling of cascade macromolecules through metal centers. Angew. Chem. Int. Ed. **34**(18), 2023 (1995)

Polymers Incorporating Ru Complexes

2

Abstract

Ru complexes might represent integral parts of a polymer—either within the backbone, the side chains or at the chain ends. Such polymers can be prepared via different synthetic routes, which have been adapted from conventional polymer science. The different synthetic strategies and types of Ru-containing polymers are discussed, taking also their characterization, properties and (potential) applications into account. In this respect, opto-electronics, photovoltaics, catalysis, and medicine have to be named as prominent areas of research.

2.1 General Considerations

As pointed out in Chap. 1,[1] Ru centers can be incorporated into polymeric architectures in different ways (Fig. 1.4). According to this classification scheme, type-3 materials contain small-molecule species embedded into a polymer matrix and the interaction mainly occurs via physical forces—Ru^{III} salts, Ru^{II} complexes, and Ru nanoparticles (NPs) have been utilized in this context. Examples for such Ru-containing polymers are given in Chap. 5 (Ru^{II} complexes) and Chap. 7 (Ru NPs). Whereas, type-1 and type-2 materials contain Ru species in the polymers' side

[1]Parts of this Chapter have been reproduced with permission from a previous Review Article (*Chem. Soc. Rev.* **2016**, *45*, 5311–5357) and Book Chapter (*Block Copolymers with Element Blocks: The Metal-Bisterpyridine Linkage* in *New Polymeric Materials Based on Element-Blocks*, Y. Chujo (Ed.), Springer Nature, Singapore, pp 307). © 2016 The Royal Chemical Society and 2019 Springer Nature, respectively.

© The Author(s), under exclusive license to Springer Nature Switzerland AG 2021

U. S. Schubert et al., *Ruthenium-Containing Polymers*,
https://doi.org/10.1007/978-3-030-75598-0_2

chains or backbone, respectively [1–3]. The electrostatic, coordinative or even pseudo-covalent linkage of the Ru centers to a polymer has a profound impact on the overall materials. Ho and Wong noted that the closer such a coupling is, the stronger the properties of the resulting material will be influenced; this holds particularly true for the photophysical properties in the case of π-conjugated backbones [4].

In the last decades, a range of methodologies for the synthesis of defined metal-containing polymers has been established. In general, all of those techniques, which have been established in the field of "organic" polymer science are also adaptive to their metal-containing counterparts. This holds notably true when dealing with Ru-based systems, though one has to be aware of some remaining restrictions. Thus, various protocols are available to prepare tailor-made Ru-containing polymers [2]. Prominent literature examples include polycondensation [5–8], ring-opening polymerization (ROP) [9–11], and electropolymerization methods [12–14]; moreover, the living ionic or controlled radical polymerization (CRP) methods can be utilized for some types of Ru-containing monomers [15, 16]. In all of these cases, the metallopolymers are formed directly in the course of the polymerization. As an alternative approach, pre-formed "organic" polymers can be modified by attaching appropriate Ru precursors to designated binding sites along the backbone [17] or side chains [18–20]; finally, Ru-containing residues might be attached to appropriately functionalized polymer chains using the so-called "grafting onto" strategy [21]. The five common approaches to synthesize Ru-containing (co)polymers are summarized in Fig. 2.1.

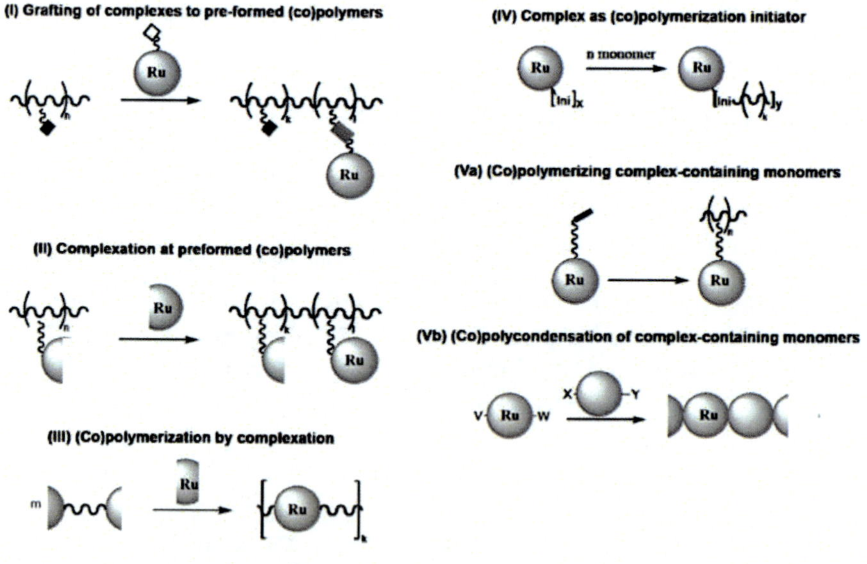

Fig. 2.1 Schematic representation of the general approaches available for the synthesis of metal-containing (co)polymers [22]. Scheme redrawn according to the literature; © 2009 The Royal Chemical Society

In this chapter, an in-depth analysis of these basic methods with respect to their scope and limitations will be given. Thereby, the different types of interactions, i.e., coordinative or ionic bonding as well as arene-π complexation will be considered (see Fig. 1.7). Not only the syntheses of respective ruthenopolymers will be evaluated but their properties and (proposed) applications will be summarized. Dendrimers and covalent organic frameworks (COFs) represent special polymer classes with respect to their unique structures and properties. For the sake of clarity, their combination with Ru species is thus presented independently (Chaps. 4 and 6, respectively). The electrochemical formation of ruthenopolymers has to be distinguished from the various solution-based methods: The former directly generates insoluble polymer films deposited onto electrode materials; whereas, the latter typically gives polymers, which are soluble and thus, can be handled and characterized in solution. Ruthenopolymers, derived from electropolymerization procedures will be considered in Chap. 3.

2.2 Synthesis, Characterization, and Properties of Ru-Containing Polymers

2.2.1 Ruthenopolymers, Derived from the Functionalization of Preformed (Co)polymers

When a (co)polymer has to be "decorated" with metal centers in the side chain, appropriate combinations of functionalities on the polymer and metal complex have to be employed. However, the defined side-chain post-polymerization modification of polymers is always challenging due to the low degrees-of-conversions, which might result from steric hindrance of the polymer backbone and/or the substrate to be grafted-onto [23]. Thus, efficient and high-yielding coupling reactions, running under mild conditions, avoiding the damage or degradation of the polymer structure, should ideally be applied [24]. The concept of "chemistry on the complex" was reviewed recently by Jäger, Schubert et al.; these authors focused on Ru^{II}-polypyridyl complexes, also taking the decoration of pre-formed polymers into account [25].

One crucial aspect to be emphasized in this context is that an excess of the complex-containing reagent should always be used when aiming for high degrees-of-conversion. This particular issue was already addressed in Meyer's pioneering work where Ru^{II}-containing polymers, such as **1** and **2**, were synthesized by nucleophilic substitution or amidation, respectively, on PS derivatives bearing adequate functional groups (PS: polystyrene, Fig. 2.2) [21, 26–36].

From copolymers **1** having different ratios of attached Ru^{II} and Os^{II} complexes, the one having a 4:1 ratio of these complexes was studied with respect to *intra*-chain electron- and energy-transfer processes (ElT and EnT, respectively). Photophysical measurements confirmed the Os^{II} centers to function as an energy trap; the ΔG^0 value for the EnT from the triple-excited $^*Ru^{II}$ sites to the Os^{II} centers was −0.36 eV.

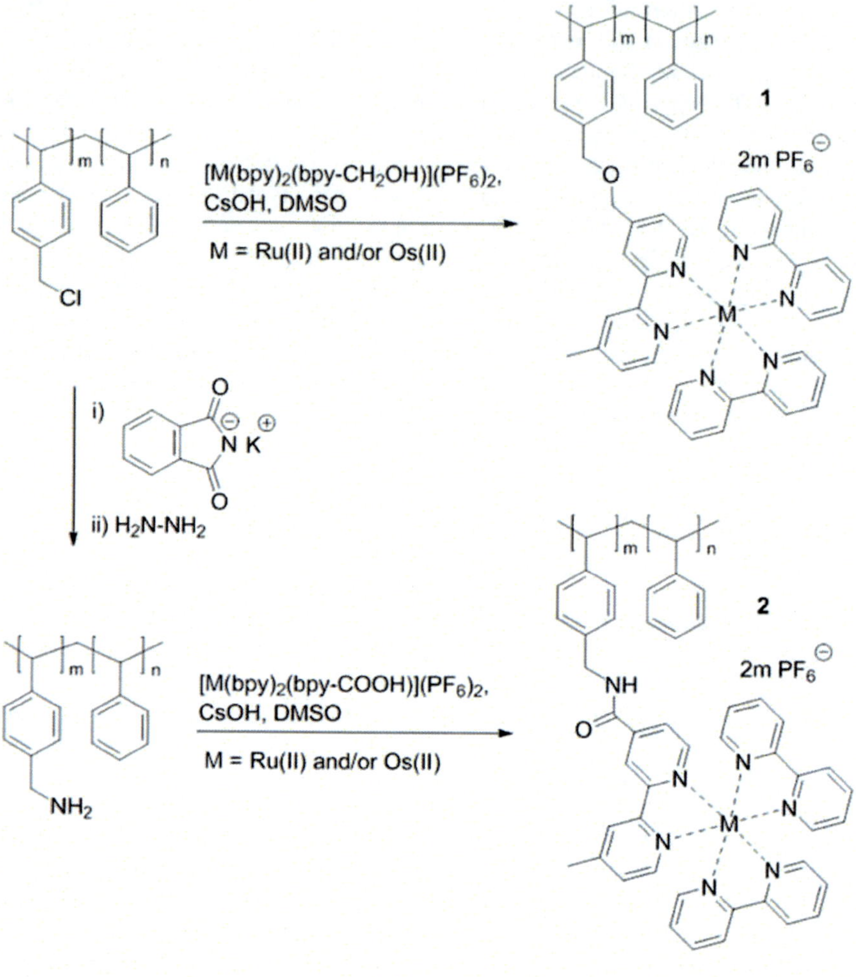

Fig. 2.2 Schematic representation of the synthesis of the side-chain functionalized polymers **1** and **2** via post-polymerization functionalization methods

Transient absorption/emission spectroscopic studies revealed a fast EnT process ($k > 2 \times 10^8$ s^{-1} in MeCN), in which only the Ru^{II} centers adjacent to Os^{II} were involved [21]. It was demonstrated later by applying laser-flash photolysis at high irradiation rates that multiphoton excitation of phenothiazine (PTZ), as an external reductive quencher in high concentrations, generated multiply reduced polymer strands (i.e., up to seven electrons were taken up; Fig. 2.3a) [28]. However, unproductive self-quenching was observed in polymers with a high degree of loading with Ru^{II} complexes [37, 38]: Upon excitation and electron-transfer quenching, a number of oxidative equivalents, such as Ru^{III} centers, were detected along the polymeric

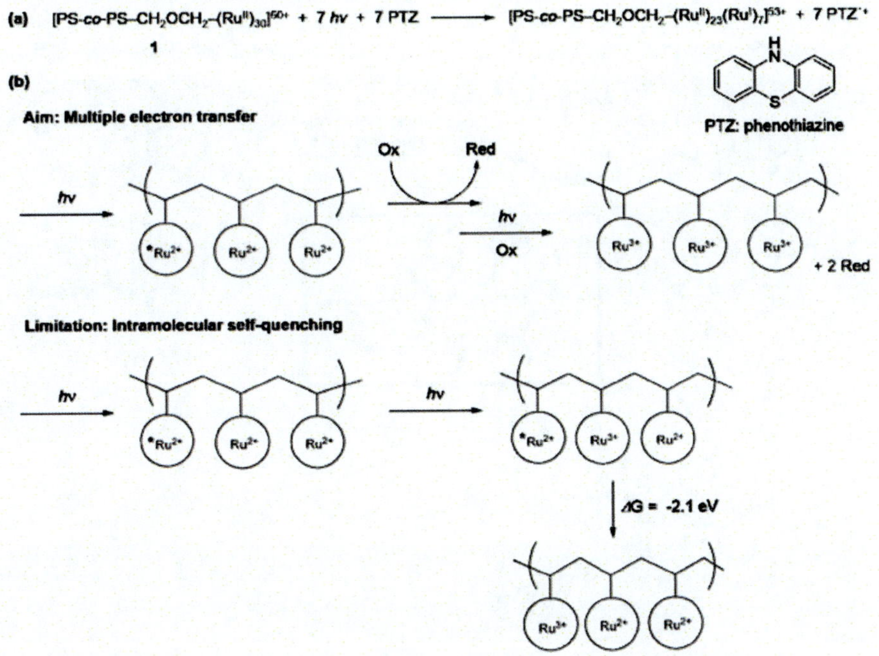

Fig. 2.3 a Reaction equation for the reductive quenching of the excited state in **1** by PTZ. **b** Schematic representation of the aimed and observed energy-/electron-transfer processes in **1** [41]. Figure redrawn according to the literature; © 2012 The Royal Chemical Society

strands. If the second excitation occurred at a Ru^{II} complex neighboring such a Ru^{III} site, the energetically favored electron transfer between these two entities proceeded readily ($\Delta G^0 = -2.1$ eV; Fig. 2.3b). Remarkably, significant differences in the photophysical behavior were observed in the related polymers **2** featuring amide linkages: Due to different orientations of the excited-state dipoles, the energy-migration rates were increased by two orders of magnitude relative to **1** (considering polymers of similar chemical composition) [35, 39, 40]. A comprehensive view on the excited-state dynamics of the Ru^{II}-containing polymers **2** was allowed by time-correlated single-photon counting with ps time resolution. In these experiments, the $^*Ru^{II}$ emission decay (at $\lambda_{max} = 640$ nm) as well as the growth and decay of the $^*Os^{II}$ emission (at $\lambda_{max} = 780$ nm) were monitored. Average rate constants for the nearest neighbor $^*Ru^{II} \rightarrow Os^{II}$ quenching ($k = 2 \times 10^8$ s^{-1}) and for the $^*Ru^{II} \rightarrow Ru^{II}$ energy migration ($k = 5 \times 10^7$ s^{-1}) were obtained, which explained the efficient quenching of the $^*Ru^{II}$ excited state [30].

As shown by Tew et al., the amidation of methacrylate-containing (co)polymers, in this case with Ru^{II}-tpy *bis*-complexes (tpy: 2,2′:6′,2″-terpyridine), can be achieved in high efficiency and under particularly mild conditions utilizing reactive ester moieties, e.g., *N*-hydroxysuccinimidyl esters (polymers **3**, Fig. 2.4a) [42, 43].

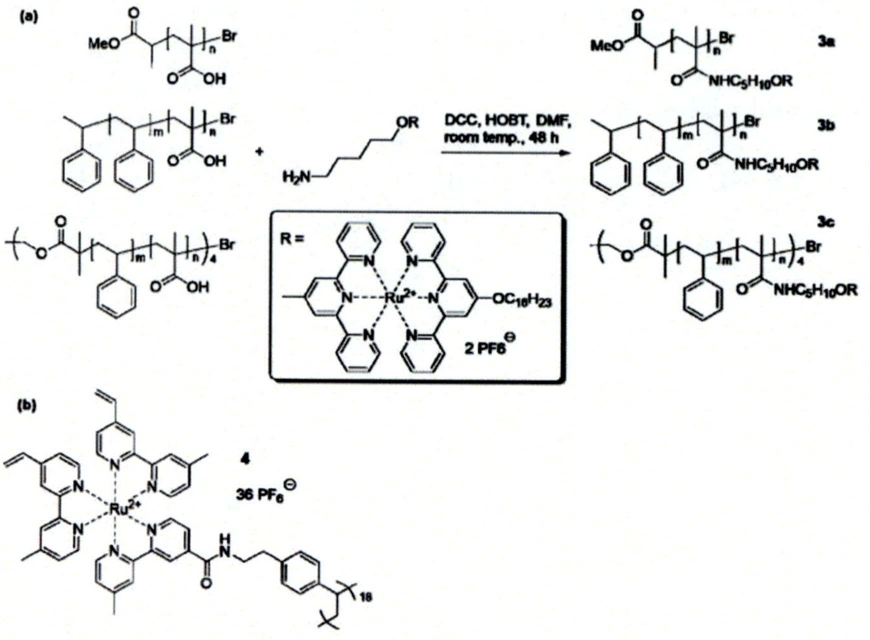

Fig. 2.4 a Schematic representation of the synthesis of Ru^{II}-containing polymers **3** involving the so-called "reactive ester chemistry". **b** Schematic representation of the vinyl-functionalized polymer **4** used by Meyer et al., as macromonomer, for the fabrication of electrode coatings

A similar approach was also followed by Meyer et al., who utilized aminoethyl-functionalized PS and a CO_2H-functionalized *tris*-bidentate Ru^{II} complex; the application of this metallopolymer (**4**, Fig. 2.4b) for electrode-coating purpose is discussed in Chap. 3 [44].

The homopolymers **3a** containing the long C_{16} alkyl tail exhibited a lyotropic liquid crystalline (LC) behavior in $CHCl_3$ in the temperature regime from 20 to 60 °C and, in the bulk, self-assembled into a hexagonal array of cylinders (Fig. 2.5a, b). Moreover, crystallization of the hydrophobic side chains in the solid state was concluded from wide-angle X-ray diffraction (WAXD) measurements (Fig. 2.5c). Neither such crystallization nor a lyotropic LC behavior was, however, observed for similar homopolymers missing the C_{16} alkyl chains. The diblock copolymer **3b** showed microphase separation in the bulk to give a structure-within-structure morphology, i.e., a so-called "cylinders in a sea of rods" motive (spacing of 38 and 5.7 nm, respectively); also in this case, crystallization of the C_{16} tails occurred [43]. In contrast, the star-shaped four-arm copolymer **3c** had crystalline C_{16} tails, but no evidence for a structure-within-structure motif was found.

In recent years, mainly the Cu^{I}-catalyzed azide-alkyne cycloaddition (CuCAAC, the so-called "click" reaction) proved its versatility in polymer science [23, 45, 46]. One example for this was reported by Schanze et al., who functionalized

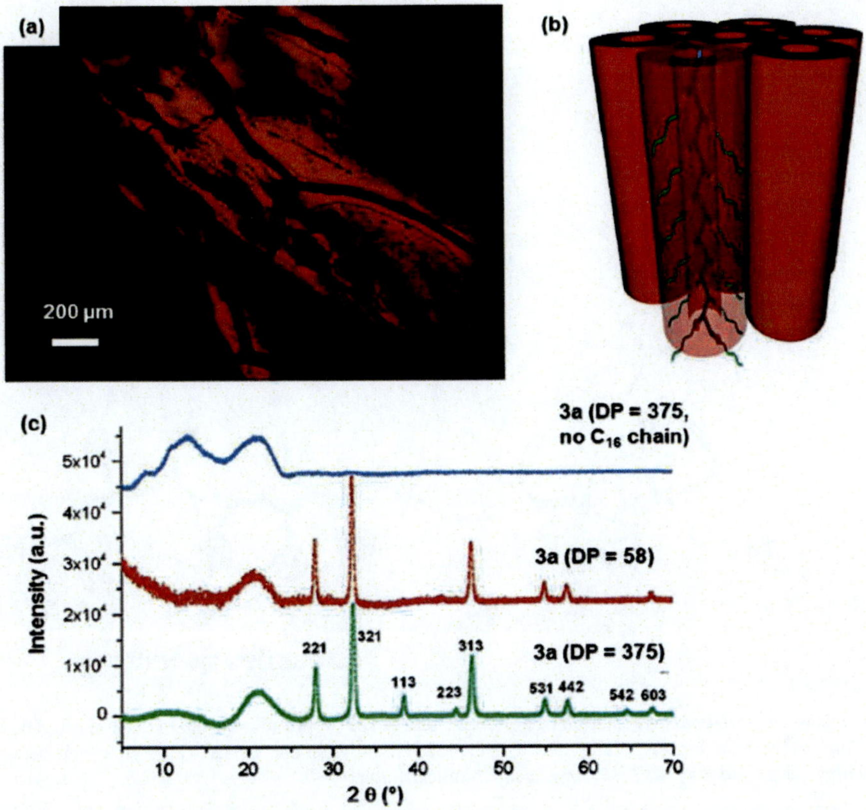

Fig. 2.5 **a** Polarized optical microscopy image of **3a** sheared by sliding the coverslip (DP of 375, 15 wt% in $CHCl_3$, room temp.). **b** Schematic representation of the hexagonal array of cylinders; **3a** is shown superimposed on one of the cylinders (the cylinders are not hollow but represent the amorphous nature of the backbone and the Ru^{II} complexes; whereas, the solid red represents the crystalline alkyl chains). **c** WAXD patterns of various homopolymers **3a** [42]. Figure reproduced with kind permission; © 2007 American Chemical Society

4-azidomethyl-styrene homopolymers with ethynyl-equipped *tris*-bidentate Ru^{II} complexes [the functionalized PS was obtained from a reversible addition-fragmentation chain-transfer (RAFT) polymerization] [47]. Almost quantitative conversion of the azido-functionalities was concluded from infrared (IR) spectroscopic studies. The chromophore-loaded polymers **5** acted as efficient light-harvesting arrays, with a single PS chain loaded with ca. 20 Ru^{II} units exhibiting molar absorptivity up to 350,000 M^{-1} cm^{-1}. Emission quantum yield and lifetime measurements on **5** revealed that the metal-to-ligand charge-transfer (MLCT) excited state was quenched relative to an appropriate Ru^{II} model system. Apparently, the thiol end-group on the polymers—arising from the utilized RAFT

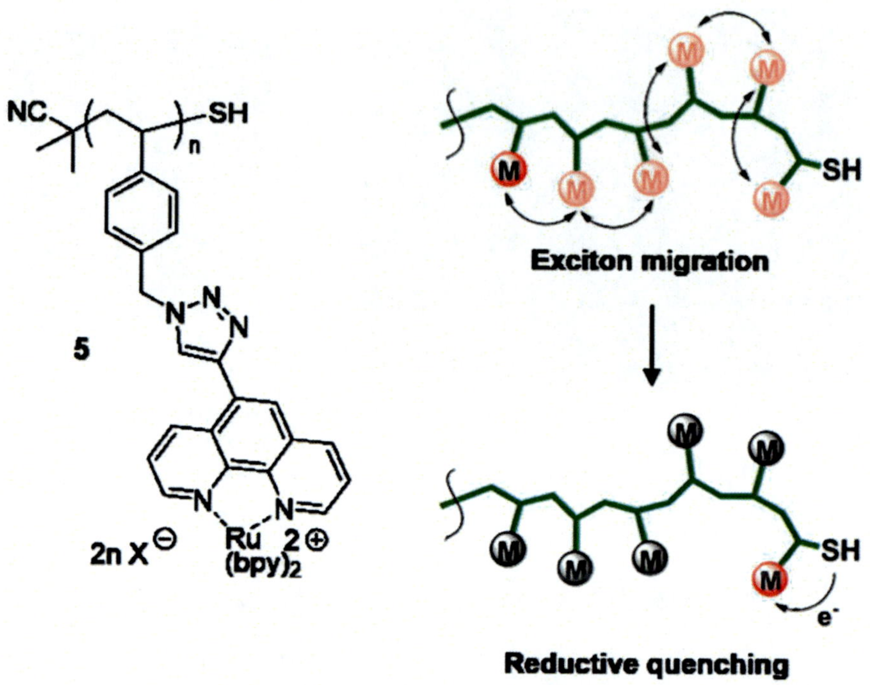

Fig. 2.6 Schematic representation of polymer **5** and of the proposed *intra*-chain exciton migration followed by reductive quenching involving the reactive polymer end group [47]. Figure reproduced with kind permission; © 2012 American Chemical Society

chain-transfer agent—opened this deactivation pathway via a charge-transfer mechanism (Fig. 2.6). Stern–Volmer quenching studies suggested that the emission of the polymers was also quenched efficiently by certain anions compared to models, therefore amplified quenching was assumed by the authors.

Polyphosphazenes represent one of the largest class of hybrid polymers, comprising of a backbone with alternating *P*- and *N*-atoms; however, discussing the long history of these interesting materials as well as the synthetic methodologies towards and the (potential) applications related to them is already a topic on its own —the reader is referred to more specialized review articles [1, 48, 49]. With respect to the scope of this chapter, only the post-polymerization modification of polyphosphazene-type materials is of relevance. Already in the 1960s, Allock et al. developed a versatile route to prepare the linear polyphosphazene that can be regarded as the "key polymer" in this field [50–52]. The treatment of such perchlorinated polyphosphazenes with organic nucleophiles allows one to replace the highly reactive P–Cl bonds by hydrolytically more stable ones in a straightforward fashion (Fig. 2.7). In order to achieve a maximum substitution of the Cl-atoms, the steric demand of the nucleophile has to be decreased, e.g., by using flexible alkyl

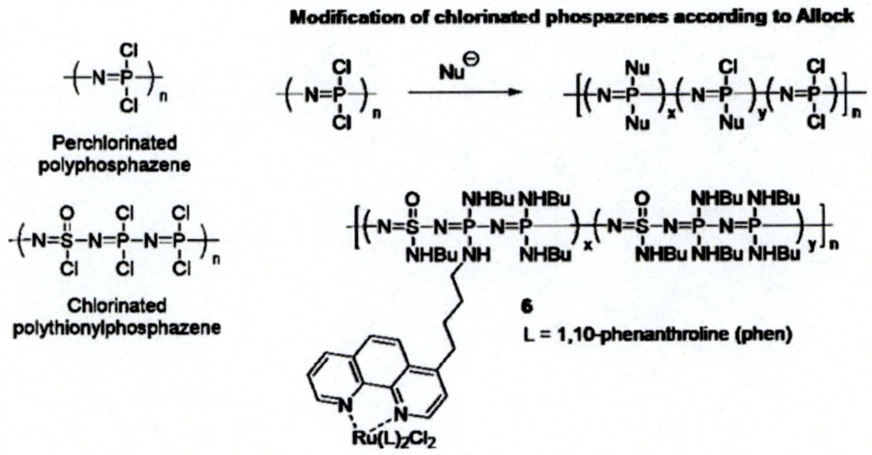

Fig. 2.7 Schematic representation of the post-polymerization modification of chlorinated polyphosphazenes

spacers. In line with these considerations, Winnik, Manners et al. grafted an aminoalkyl-functionalized Ru^{II}-phen *tris*-complex onto chlorinated polythionlyphosphazene (phen: 1,10-phenanthroline) [53]; the subsequent substitution of the remaining Cl-substituents by *n*-butylamine residues gave polymer **6** which was tested as an oxygen sensor, based on phosphorescence quenching (Fig. 2.7). In comparison to the analogous two-component system, i.e., the Ru^{II} dye imbedded in a polythionlyphosphazene matrix, a significantly higher sensitivity for oxygen detection was found.

Similar to the side-chain functionalization, the end-group modification of (block co)polymers can be conducted. The end group of a polymer chain, directly originating from the applied polymerization technique or obtained from subsequent post-polymerization functionalization reactions, can be reacted with suited metal complexes. This strategy was, for example, followed by Jäger et al. [54]: The terminal chloromethyl-groups of poly(triarylamine)s (PTARAs), originating from a nitroxide-mediated polymerization (NMP) using Hawker's universal initiator, were reacted with a hydroxyl-functionalized Ru^{II} complex to give the polymers **7** possessing a Ru^{II}-dqp *bis*-complex at one chain end [dqp: 2,6-di(quinoline-8-yl)-pyridine, Fig. 2.8a]. In these materials, the electron-rich polymer strands acted as efficient electron donors to reduce the photochemically generated Ru^{III} complexes with a driving force in the range of 200 mV. In subsequent work, these same authors attached a Ru^{II}-dqp dye to the terminus of a poly(naphthalene diimide) (PNDI) segment via various covalent linkages (**8**, Fig. 2.8b) [55]. Photophysical studies revealed an emission quenching efficiency of >86%, independent from the nature of the linkage, and indicated an EIT process from the Ru^{II} sensitizer to the acceptor-type polymer. Using hydrophilic substituents on the NDI chromophores,

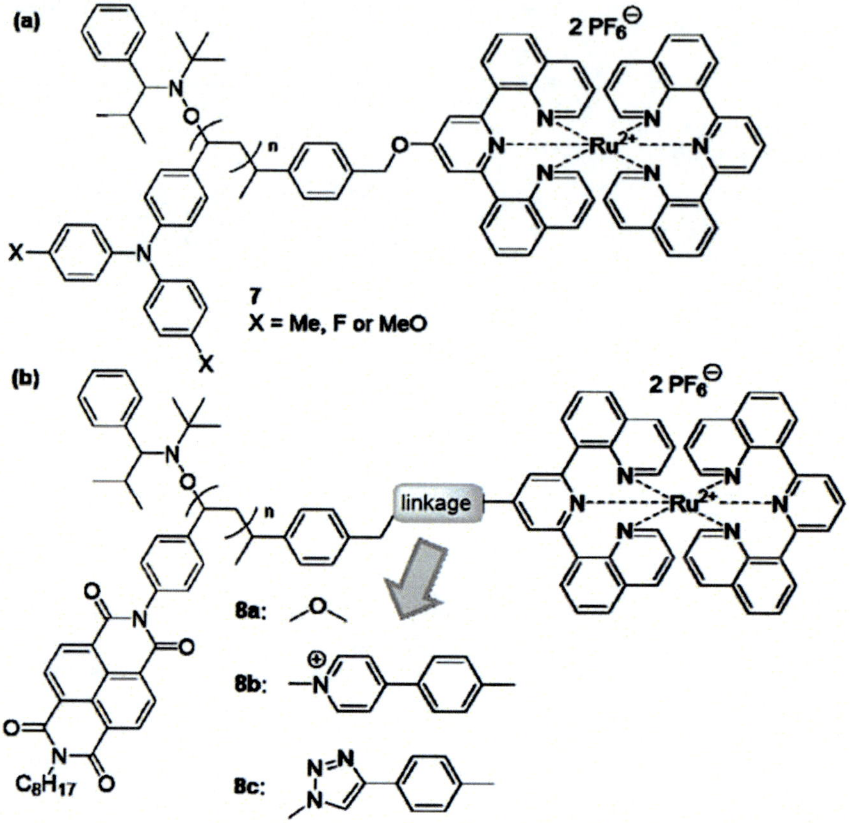

Fig. 2.8 Schematic representation of (a) PTARAs **7** and (b) PNDIs **8** bearing a redox-active Ru^{II} complex as one terminal group

such as short poly(ethylene glycol) (PEG) residues, improved the solubility of the materials in polar-protic environments and even water, thus enabling the handling of the photoactive materials in eco-friendly solvents [56].

The modularity of this approach was employed further to prepare a photoredox-active multidonor–photosensitizer–multiacceptor array (D_n–P–A_m, Fig. 2.9) [57]. Making use of orthogonal functionalities on the central Ru^{II}-dqp *bis*-complex, the PTARA and PNDI blocks were attached in a stepwise fashion. Upon photo-excitation, a very long-lived charge-separated state (CSS), presumably involving multiple contributing pathways, was populated (lifetime of 2,400 ns). Replacing the PTARA block by a π-conjugated poly(3,6-carbazole) chain gave a D_n–P–A_m triad, which featured, upon excitation of the central Ru^{II} fragment, an even longer lived CSS (ca. 7 μs) [58]. Apparently, the CSS was stabilized by the

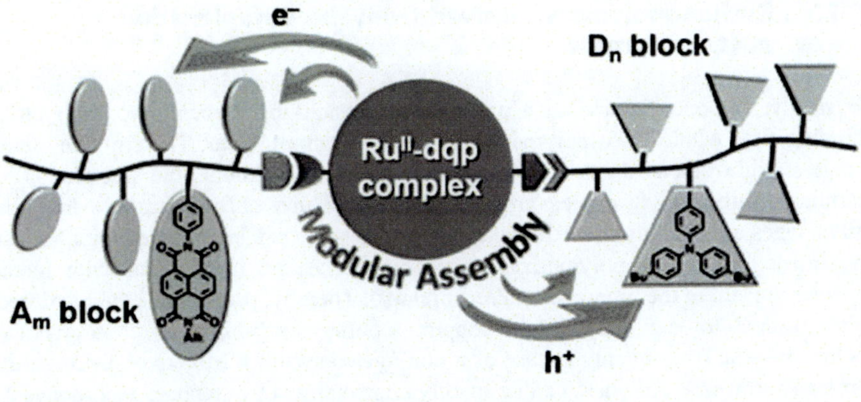

Fig. 2.9 Schematic representation of the photoredox-active D_n–P–A_m triad exhibiting a long-lived charge-separated state [57] Figure reproduced with kind permission; © 2017 The Royal Chemical Society

π-conjugated polymer, which offered a much higher charge-carrier mobility when compared, e.g., to the non-conjugated PTARA blocks.

Besides the previous examples in which various Ru^{II} complexes were attached covalently to the polymer backbone, a notable non-covalent system has also been reported. Carboxylated nanocellulose was utilized as a polyanionic support for the electrostatic binding of Ru^{II}-bpy *tris*-complexes, which possessed cationic amine groups [59]. The thusly obtained gel featured a bright photoluminescence and a highly sensitive response towards Ag NPs (detection limit of 1.11×10^{-5} mol L^{-1}). The applicability of this sensor was shown in the efficient detection of Ag NPs in deodorants and anti-microbial textiles. Noteworthy, nanocellulose represents an emerging platform in the design of sensors due to its natural abundance and bio-compatibility (many examples can be found in a recent review by Chen, Liu, Ni et al. [60]).

The last class of materials to be named in this section is the family of polymer-bound Ru^{II}(CO)-porphyrins, which are utilized, as catalysts, in the alkene epoxidation with 2,6-dichloropyridine-*N*-oxide, as oxidant, or the alkene cyclopropanation with diazo compounds [61, 62]. These materials were prepared by grafting ruthenoporphyrins onto a functionalized and cross-linked PS via nucleophilic substitution reactions. The immobilized catalysts efficiently supported the epoxidation of a broad range of substrates, electron-rich as well as electron-deficient alkenes: 1,5-Cyclooctadiene, *trans*-stilbene, α,*β*-unsaturated ketones, conjugated enynes, glycal and protected α-amino alkenes as well as common non-functionalized alkene (e.g., styrenes and norbornene).

2.2.2 Ruthenopolymers, Derived from the Complexation at (Co)polymers

Basically, the coordination of a proper metal-containing precursor to designated binding sites within a (co)polymer generates a metallopolymer. Thereby, the metal centers might be coordinated directly to the polymer backbone, the side chains or termini. Numerous examples are known making use of this strategy and the advantages are evident [2]: A broad range of polymers can be prepared with almost no restrictions to the polymerization technique (i.e., no interference with metal species present in the monomer and/or initiator). Thereby, properties can be tailored with respect to, e.g., chain/block length, topology, solubility or photophysical behavior—the latter mainly in case of π-conjugated systems. Moreover, the initially formed metal-free polymers can be readily characterized by standard techniques in polymer science: Size-exclusion chromatography (SEC), nuclear magnetic resonance (NMR) spectroscopy, mass spectrometry (MS), etc. The subsequently incorporated metal centers may be identified by comparing the chromatographic, spectroscopic and/or spectrometric properties of the final to those of the metal-free precursor polymer. Thus, this approach can be regarded as a highly modular strategy since both types of building blocks—the polymer and the metal-containing precursor—can be prepared with tailor-made properties and combined into one material in a straightforward fashion simply via coordination chemistry. Thus, it might be prospected that the incorporation into parallel synthesis and/or a combinatorial workflow should easily be possible. However, the ease of coordination always needs to be considered as a limiting factor. As pointed out in the previous Section, steric demand (e.g., if a bulky complex needs to be coordinated to a rigid polymer backbone, a branched polymer or even a cross-linked system) represents an issue to deal with and may hamper the coordination (at least) to some extent. The addition/coordination of Ru-containing precursors to a broad range of binding sites incorporated in polymer architectures has already been investigated.

The affinity of various transition-metal ions of groups 6 to 8 to arenes is commonly employed for the synthesis of metal-containing polymers. Considering the simplest case, the phenyl residues of plain PS have already proven their ability to coordinate appropriate precursor complexes in a η^6-type fashion. In the first available study in this respect, Pertici, Vitulli et al. reacted the Ru^0 precursor Ru(cod)(cot) with PS under hydrogenative conditions (cod: η^4-cycloocta-1,5-dienyl, cot: η^6-cycloocta-1,3,5-trienyl). These authors proposed that the cycloalkene ligands were, after their hydrogenation, replaced by phenyl rings of the PS. However, this assumption remains questionable, since the applied reaction conditions represent one prominent strategy to prepare Ru NPs (see Chap. 7). Thus, the formation of PS-stabilized Ru NPs, which were subsequently used as hydrogenation catalysts, is plausible. In a more defined approach, Chaudret et al. utilized three different organo-ruthenium precursors for the coordination to PS [63]. The polymers presumably contained pendent $[Ru(C_8H_{11})]^+$ (**9a**), $[RuCp^*]^+$ (Cp*: pentamethylcyclopentadienyl, **9b**) or $[RuH(PCy_3)_2]^+$ moieties (**9c**) coordinated to the phenyl rings (Fig. 2.10a); the latter represented the first polymer-supported metal–

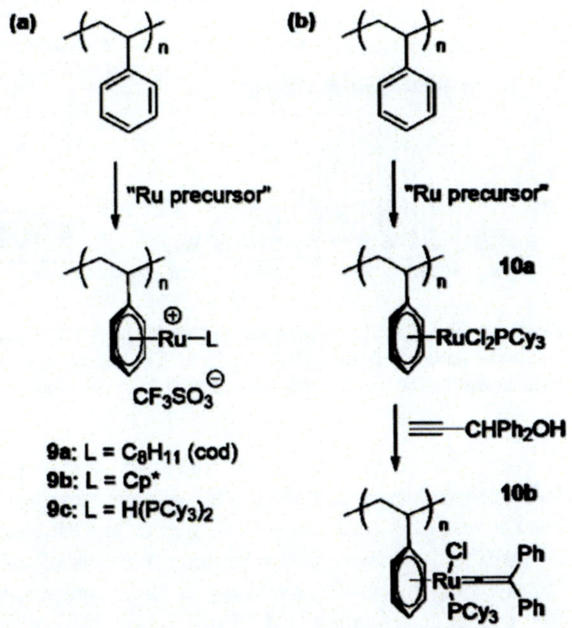

Fig. 2.10 Schematic representation of the synthesis of polymers **9** and **10** containing η^6-benzene-coordinated Ru^{II} species [63, 65].

hydrido complex reported in literature. The degree-of-functionalization correlated with the bulkiness of the attached complexes and decreased from 100 to 25% in the listed order. However, these materials were not studied further regarding, e.g., catalytic applications. The ability of the aforementioned organometallic precursors to coordinate to PS was utilized by Royo and Brintzinger in the mass spectrometric analysis of PS [64]. Commercial PS standards of various molar masses (M_n values in the range of 1,430 to 135,000 g mol^{-1}) were cationized by reaction with an equimolar amount of a $[Ru(Cp)]^+$ precursor (Cp: cyclopentadienyl). As a result, a significant increase in the signal intensity was observed in the matrix-assisted laser-desorption-ionization time-of-flight (MALDI-TOF) MS analysis of these samples when compared to using Ag^+, as ionization agent (Fig. 2.11). In a similar fashion, polypropene samples (M_n of 2,700 or 6,100 g mol^{-1}) were ionized by the treatment with the trinuclear $[Ru(Cp)]_3H_5$ precursor. This strategy enabled the first MALDI-TOF MS analysis of this particular apolar polymer.

A metal-containing polymer of the general structure poly(Ph-$RuCl_2PR_3$) (**10a**) was prepared by Akiyama and Kobayashi [65]; the polymer was converted further into the Ru-allenylidene system **10b**, which was employed, as a heterogeneous catalyst, in the ring-closing metathesis (RCM) reaction (Fig. 2.10b). Pertici et al. claimed the synthesis of a PS-based material in which each Ru^0 center was bound to two adjacent phenyl rings in a η^6-type fashion. The metallopolymer was prepared by reacting PS with the Ru(cod)(cot) complex under a H_2 atmosphere [66]. Later, the structure of this material was revised—indeed, colloidal Ru stabilized by a PS matrix was formed under those conditions.

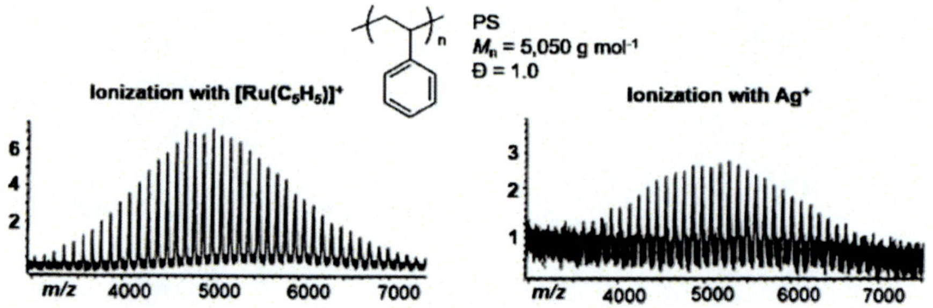

Fig. 2.11 MALDI-TOF mass spectra of PS (M_n = 5,050 g mol^{-1}, Đ = 1.0) after reaction with an equimolar amount of $[Ru(C_5H_5)]^+$ (left) or Ag^+ (right). In both cases, all-*trans*-retinoic acid was used as matrix [64]. Figure reproduced with kind permission; © 2002 Elsevier B.V.

Also, the coordination of transition metal precursors (i.e., carbonyl- or half-Sandwich-type complexes) to aromatic rings within a polymer main chain has been investigated to some extent. Regarding Ru-containing materials, the binding of $Ru_3(CO)_{12}$ clusters to the azulene moieties of π-conjugated thiophene-azulene copolymers, as reported by Wang et al., represents, to the best of our knowledge, the exclusive example (**11**, Fig. 2.12) [67]. According to X-ray diffraction (XRD) analysis of small-molecule reference complexes, different binding modes of the multinuclear clusters to the azulene entities were present. Scanning electron microscopy (SEM) imaging of the thin metal-free polymer films revealed smooth surfaces; whereas, sponge-like structures could be evidenced for the freshly prepared metallopolymers.

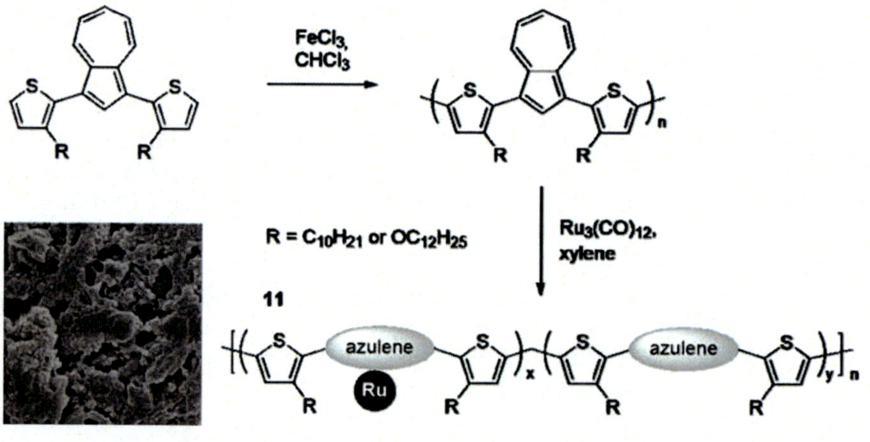

Fig. 2.12 Schematic representation of the synthesis of Ru^0-containing polymers by coordination of carbonyl clusters to azulene moieties. A representative SEM image of a spin-coated film of **11** is also shown (unknown scaling) [67]. Figure reprinted with kind permission; © 2003 American Chemical Society

The cyclopentadienyl-type anions represent, by far, the most common ligands to be found in (half-)Sandwich-type complexes; thereby, binding to transition metal ions typically occurs in a η^5-type mode. However, coordination of metal ions to Cp units within a polymer, either in the backbone or the side chain, has only been explored to a minor extent. This constraint can, most likely, be attributed to the difficulties in preparing such Cp-functionalized polymers. If available, a suitable metal precursor is reacted with the Cp moieties either in an appropriate solvent at reflux or under increased pressure [68]. However for Ru, this route has not yet been employed.

Phosphine ligands are known for their strong binding to transition metal ions and, consequently, this strategy has been adopted in order to attach metal complexes to polymers with pending phosphine moieties. A prominent example to be mentioned in this context is the immobilization of ruthenodrugs; such polymer-drug conjugates typically exhibit an improved drug-delivery performance compared to the analogous small-molecule drugs [69]. As summarized by Kostova [70], Ru-based compounds are highly prospective for medicinal application and, in particular, the Ru^{II}-(η^6-arene) complex RAPTA-C (**IV**, Fig. 1.2), featuring the phosphine ligand 1,3,5-triaza-7-phosphaadamantane (PTA), possess promising anti-cancer activity [71]. Blunden et al. prepared a poly(methyl methacrylate) (PMMA) derivative with PTA binding sites in the side chains, which were reacted with the [$RuCl_2$(*p*-cymene)] precursor complex to give the polymer-ruthenodrug conjugate **12** (Fig. 2.13) [20]. Due to steric crowding, a maximum loading of the pre-polymer of ca. 50% could be reached. In a second approach, the amphiphilic copolymer **13** was prepared and tested towards its anticancer activity [72]. Micellization and subsequent cytotoxicity and cell uptake of this conjugate were tested on the ovarian cancer cell lines A2780, A2780cis, and Ovcar-3. Confocal microscopy imaging confirmed that the micelles were taken up into the lysosome of the cells, being indicative for an endocytic pathway (Fig. 2.13). On average, a tenfold increase in toxicity was observed for **13** when compared to RAPTA-C itself.

One particular functionalized polymer has attracted significant attention for the binding of transition metal ions: PS equipped with diphenylphosphine moieties (PS-PPh_2), which is commercially available, can be regarded as the polymer-supported analogue to the ubiquitous tertiary triphenylphosphine ligand (PPh_3). Transition metal complexes of PPh_3 (and its derivatives) are widely used, as catalysts, for a broad range of organic reactions [73]. As depicted in Fig. 2.14, there are two important synthetic routes in order to attach metals ions to a phosphine-modified polymer (**14**): Phosphine substitution or direct formation of the phosphine complex by reaction with an appropriate precursor complex bearing (at least) one labile ligand, as the leaving group. The phosphine-equipped polymers can be obtained from lithiated or brominated pre-polymers [74].

Starting from pioneering work in the 1970s, many reports have been published, in particular, dealing with the binding of various catalytically active transition metal complexes to PS-PPh_2 (or analogous substrates, such as PS-CH_2PPh_2) [28, 30, 160, 161]. In early contributions to this field, the metal-containing polymers were

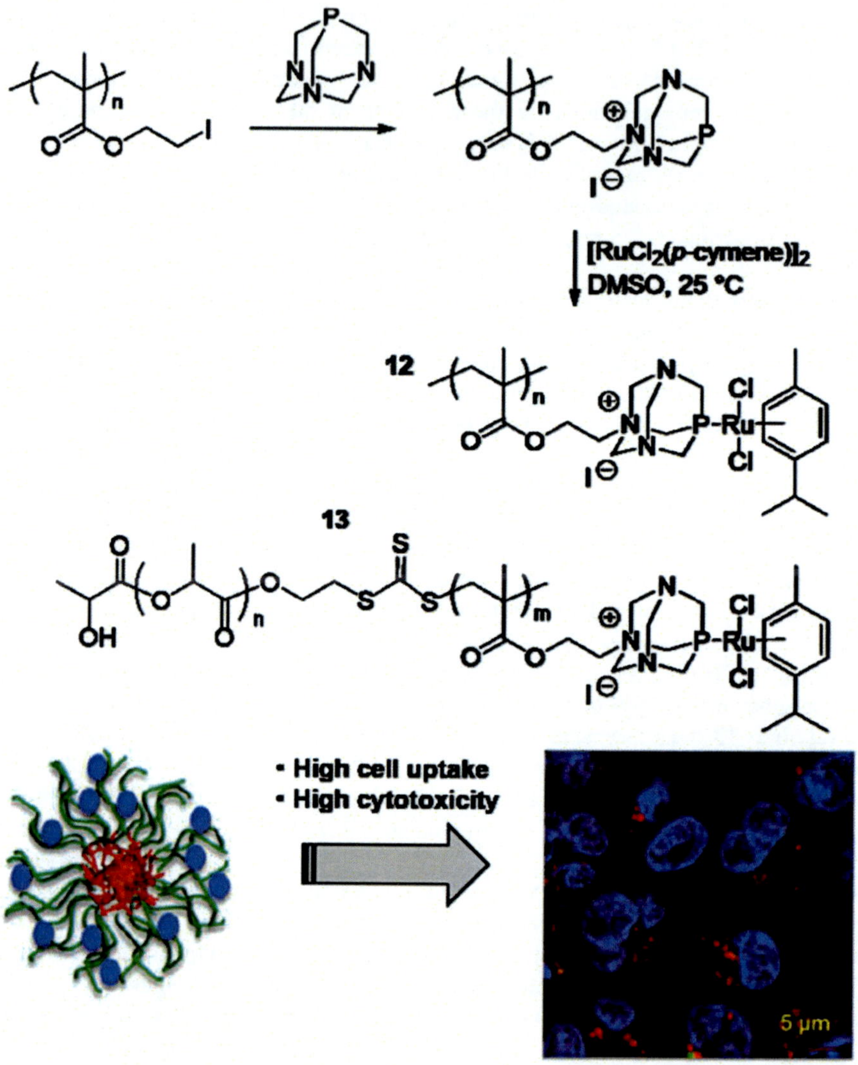

Fig. 2.13 Schematic representation of the synthesis of the (RAPTA-C)-polymer conjugate **12**. Micellization of the amphiphilic conjugate **13** and its cell uptake, as visualized by confocal microscopy imaging, is also shown [72]. Figure reproduced with kind permission; © 2013 American Chemical Society

typically generated in situ and were not (or only to a minor extent) structurally characterized. The exact composition of the heterogeneous catalysts remained, in most cases, unknown and, thus, their performance (e.g., selectivity and/or activity) often could not directly be correlated to that of their small-molecule analogues [68].

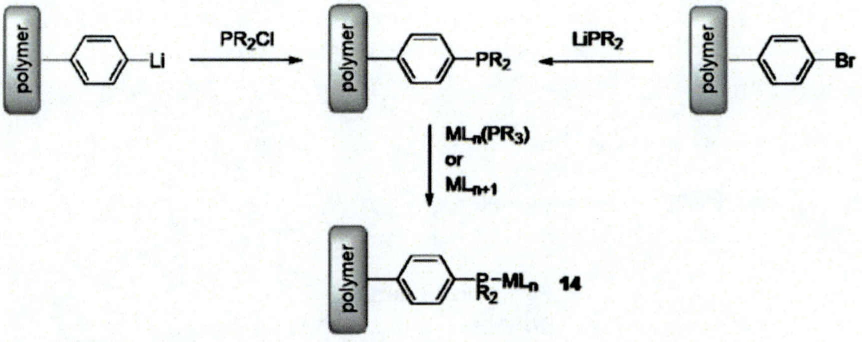

Fig. 2.14 Schematic representation of the synthesis of polymer-bound transition metal complexes via phosphine coordination

For example, $[(PPh_3)_2RuCl_2(CO)_2]$ was bound to PS-PPh_2 via a phosphine-ligand exchange, the degree-of-immobilization was estimated from elemental analysis [75]. Pitman et al. compared the catalytic performance of the material in the selective polyene-to-monoene hydrogenation to that of the small-molecule complex, which showed a significantly higher selectivity and efficiency. A similar observation was made by these same authors upon immobilization of the homogenous hydroformylation catalyst $[Ru(CO)_3(PPh_3)_2]$ onto PS-PPh_2 [76]. Gates et al. prepared a heterogeneous catalyst for the ethylene hydrogenation via a ligand-exchange reaction between $[H_4Ru_4(CO)_{12}]$ and phosphine-modified poly (styrene-*co*-divinylbenzene) membranes [77]. Displacement of up to four CO-ligands by the anchored phosphine units was assumed. The robustness of the obtained material under the typical hydrogenation conditions was monitored by IR spectroscopy. Moreover, olefin isomerization was catalyzed by a $[RuX(C_3H_5)(CO)_2]$ complex bound to PS-PPh_2 via its vacant coordination site [78]. Judkins et al. utilized ArgoGels™, as solid support, for attaching the $Ru_6C(CO)_{17}$ cluster, which is known to be a potent hydrogenation catalyst [79]. Contrary to earlier studies, a detailed characterization of the material was possible; e.g., ^{31}P magic-angle spinning (MAS) solid-state NMR corroborated the selective coordination of one phosphine ligand per metal cluster. The applied spectroscopic technique, offering a non-destructive investigation, represents an established tool to determine the degree-of-loading (or degradation) of PS-PPh_2 in a fast and quantitative manner [80].

More recently, a hypercross-linked polymer resin was prepared by copolymerizing PPh_3, benzene, and $CH_2(OMe)_2$ under Lewis-acidic conditions (Fig. 2.15) [81]. $RuCl_3$ was subsequently coordinated to the phosphine sites within the polymer network, thus generating a metallopolymer, which was used as a robust and reusable heterogeneous catalyst in various organic reactions (e.g., the synthesis of 2,4-diaryl-substituted pyridines from acetophenones, NH_4OAc and DMF). It was further shown that both the hydrogen-evolution reaction (HER) as well as the

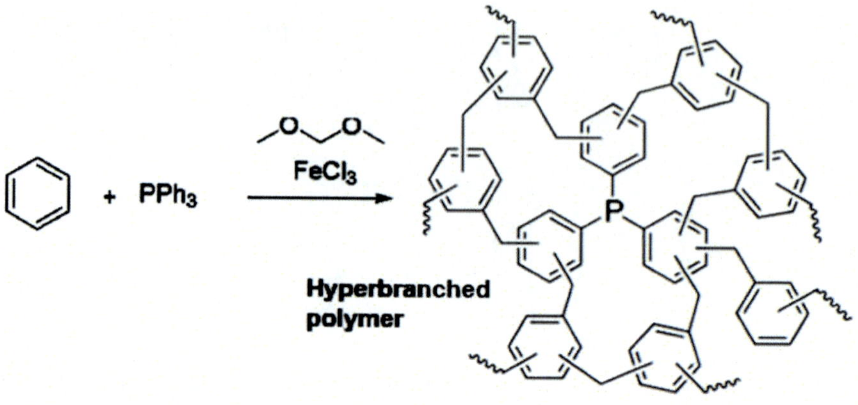

Fig. 2.15 Schematic synthesis of a hyperbranched polymer, as support for the immobilization of $RuCl_3$

ammonia borane ($NH_3{\cdot}BH_3$) reduction were catalyzed by this ruthenopolymer [82]. Thus, the electrochemical HER occurred with an overpotential of 61 mV at a current density of 10 mA cm^{-2}. The hydrolysis of $NH_3{\cdot}BH_3$ gave a maximum turnover frequency (TOF) of 402 min^{-1}. The optimum loading of the polymers with Ru^{III} ions in these experiments was 5 and 3.5 wt%, respectively [determined by X-ray photoelectron spectroscopy (XPS)]. The catalytic efficiency was rationalized by the high porosity of the material and stabilization of the in situ generated Ru NPs, as the presumably catalytically active species.

In a similar fashion, Ram et al. modified chloromethylated and cross-linked PS with 1,3-diaminopropane [83–87] or glycine [88] to introduce designated binding sites for, e.g., Ru^{III} ions. The materials were characterized with respect to their physicochemical properties, such as: surface area, moisture content, bulk density, swelling, crushing strength, morphology, and thermal stability. Moreover, UV/vis reflectance, IR, XPS, electron paramagnetic resonance (EPR) spectroscopy as well as XRD analysis were applied to confirm the coordination of the added $RuCl_3$. The polymers were tested successfully in the hydrogenation of olefins and nitroarenes. However, the stability of the heterogeneous catalysts was rather poor—presumably due to the ill-defined binding of the Ru^{III} centers to the pending amino groups. A formylated PS resin was employed, as substrate, for the formation of pending Schiff's base ligands (**15a/b**, Fig. 2.16) [89]. These resins, when loaded with Ru^{III} ions, were subsequently used as catalyst in the alkene epoxidation with *tert*-butyl hydroperoxide, as oxidant. Though the catalytic activity in terms of substrate conversion was found to be generally high, the scope concerning the product selectivity, was however low [90]. For example, the **15b**-based catalyst transferred *cis*-cyclooctene into the corresponding epoxide in 51% yield; in case of styrene both styrene oxide and benzaldehyde, as a secondary oxidation product, were obtained in a combined 95% yield; with cyclohexene as substrate, the major

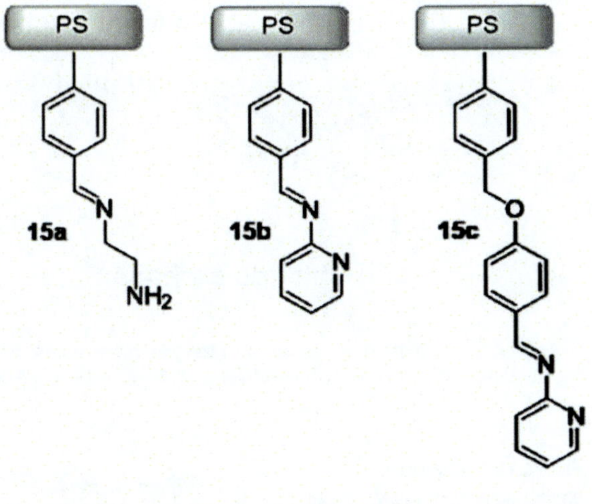

Fig. 2.16 Schematic representation of the PS resins **15** with pending Schiff's base ligands

product was 2-cyclohexene-1-one (64%); whereas, the epoxide was formed only to a minor content (2%). The Ru^{III}-loaded resin **15c** offered a good performance in oxidizing secondary alcohols to ketones, but appeared to be less selective when using primary alcohols or styrenes, as substrates [91].

What can be concluded from the early attempts to immobilize catalytically active ruthenium clusters on polymers? The performance of any polymer-bound catalysts will always rise and fall with the structurally defined nature of the active site and its stabilization by the macromolecular environment [92].

One catalytic application of immobilized transition metal complexes to be particularly highlighted in this context is the family of metathesis reactions: Ring-opening metathesis polymerization (ROMP), 1-alkyne polymerization, cyclopolymerization, acyclic diene metathesis (ADMET) polymerization, cross-metathesis (CM), tandem metathesis, enyne metathesis, and ring-opening cross metathesis. For this purpose, Ru^{II}-based complexes (the so-called Grubbs' catalysts) are commonly used and the immobilization by binding them to a polymeric support, thereby retaining the remarkable performance of the catalyst, is a highly active field of research on its own. As outlined by Buchmeiserx [93], Grubbs-type catalysts can basically be attached to a polymer via different positions: Modification/exchange of one of the neutral, two-electron donor ligands [i.e., the phosphine, the (substituted) pyridine or the *N*-heterocyclic carbene (NHC)], modification/exchange of the alkylidene ligand, exchange of the chloride ligands and, finally, (iv) utilization of non-covalent interactions (Fig. 2.17). One example for using this strategy has already been mentioned above—polymer **10b** contains the catalytically active species bound via arene-π-coordination (Fig. 2.10b) [65].

The first example for binding a Grubbs-type catalyst to cross-linked PS via phosphine units was reported by Grubbs et al. (Fig. 2.18) [94]; the decreased catalytic activity of **16** in ROMPs, relative to that found for the non-immobilized complex, was attributed to an incomplete phosphine-substitution, the phosphine

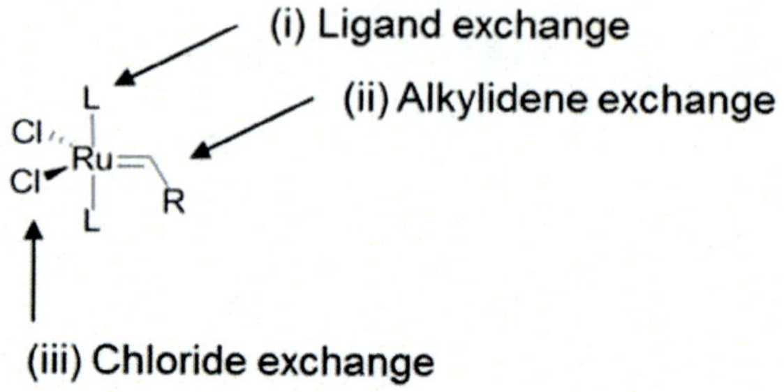

Fig. 2.17 Schematic representation of the general structure of Grubbs' catalyst; the potential sites for attaching the catalyst to a solid (organic/inorganic) support are highlighted

Fig. 2.18 Schematic representation of the synthesis of the polymer-supported Grubbs' catalyst **16** [94]

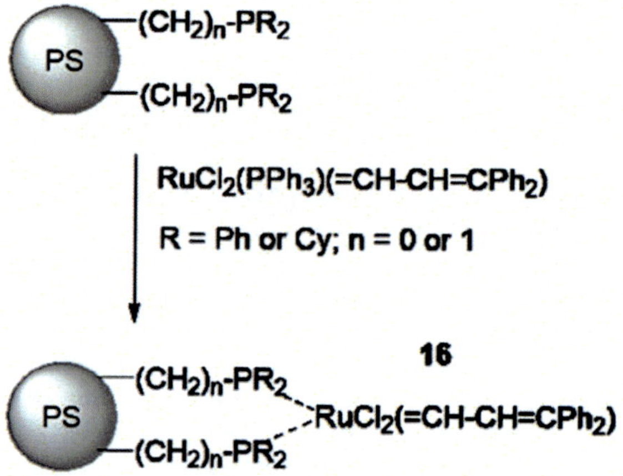

chelation effect, and limited diffusion. A similar strategy was followed by Melis et al. but using modified mesoporous silica, as the support [95].

The substitution of chloride ligands on Grubbs' catalysts by carboxylates was investigated by Mol et al. [96]; however, the opposing alkylidene ligand was only stabilized when strongly electron-withdrawing carboxylates, containing fluorinated or chlorinated chains, were utilized. These potent catalysts for the metathesis of linear alkenes as well as for the RCM were subsequently grafted onto cross-linked PS; thereby, activation with Ag^{I} ions was found to efficiently promote the coordination of the metal complex to the support [97]. In the following, Buchmeiser et al. extended this protocol by making also immobilized variants of catalysts of newer generations available; mainly ROMP-derived monolithic supports or Merrifield resins were employed in these studies [98–108]. Exemplary, the synthesis of the immobilized Grubbs-Hoveyda catalyst **17** is depicted in Fig. 2.19 [109].

Much earlier, the carbonyl complexes $Ru_3(CO)_{12}$ and $RuH_2(CO)(PPh_3)_3$ were attached to poly(acrylic acid), poly(methacrylic acid) or an ethylene-(maleic acid)-

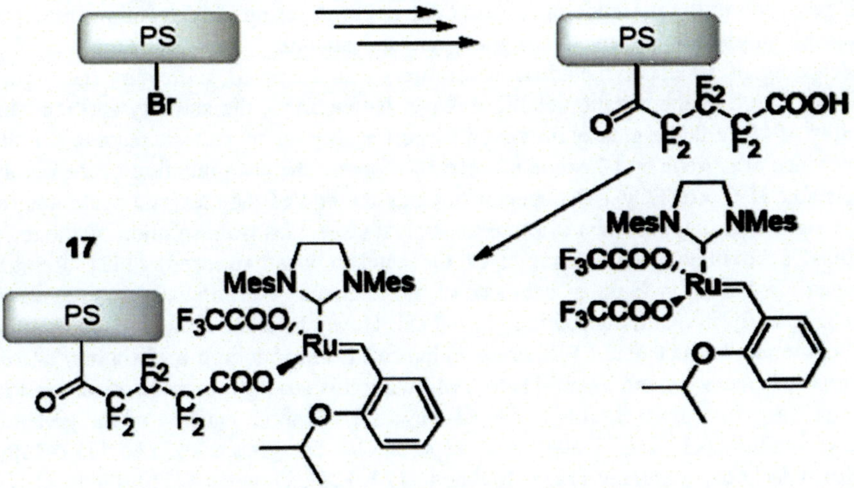

Fig. 2.19 Schematic representation of the synthesis of the immobilized Grubbs-Hoveyda catalyst **17** [109]

copolymer in order to establish polymer-supported catalysts for the olefin isomerization and hydrogenation, respectively; however, these authors neither addressed the synthesis and characterization of the materials nor evaluated their actual catalytic activity [110].

The alkylidene-exchange on the Grubbs' catalyst was first reported by Barrett et al. [111]. For this purpose, vinyl-functionalized PS was treated with Grubbs' 1st generation catalyst; thereby, styrene was released in a CM reaction leaving the complex attached to the polymer via a methylidene unit (Fig. 2.20a). Thus, the prepared heterogeneous catalyst **18** was stable as a solid sample but exhibited rapid

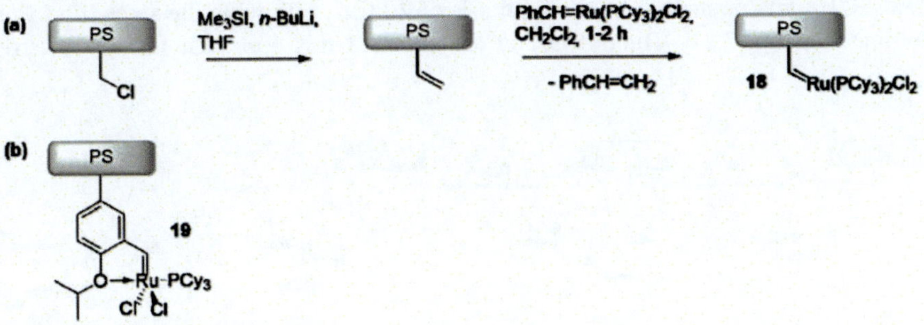

Fig. 2.20 **a** Schematic representation of immobilization of a Grubbs' catalyst via an alkylidene exchange (**18**). **b** Schematic representation of the immobilized catalyst **19**, which showed boomerang-like properties

degradation in the presence of solvents due to "back-biting" CM reactions between vacant vinyl moieties on the support and the anchored catalyst. Moreover, fast bleaching of **18** under metathesis conditions was observed, indicating the highly labile nature of the organometallic linkage. Noteworthy, the stability and recyclability of **18** could be greatly improved by performing the metathesis reactions in the presence of styrene or 1-hexene in order to suppress the accumulation of the highly unstable $H_2C{=}Ru(PCy_3)_2Cl_2$ species towards the end of the catalytic cycle and, at the same time, support the recapture of catalyst molecules from solution by the resin (thus, a boomerang-like behavior of the catalyst was proposed) [112]. Related materials, based on Grubbs' catalysts of more recent generations, were assembled subsequently in a similar fashion [113–115]. Hoveyda et al. demonstrated that the "boomerang"-effect could further be enhanced by introducing a chelating 2-(*iso*-propyloxy)benzylidene ligand [116]. Following this strategy, a range of (at least in some cases) partially soluble and well-recyclable Grubbs' catalyst of the generalized structure **19** (Fig. 2.20b) were reported by the groups of Yao [117–119], Hoveyda [120], Lamatay [121], Blechert [122–125], Dowden [126], Grela [127], Weberskirch [128], Bannwarth [129–132], Lee [133], etc.

The binding of Grubbs' catalysts to polymer chains via a NCH ligand represents an attractive alternative to the previous examples, since in this case the complex remains covalently attached to the polymer during metathesis reactions (thus, a boomerang-type behavior as for the alkylidene-bound systems is not required) [134]. The first polymer-supported Grubbs' catalysts, based on this type of binding (**20**), were reported by Blechert et al.; these authors reacted a NHC-functionalized Merrifield-type PS resin with $RuCl_2(PCy_3)_2({=}CHPh)$ (Fig. 2.21) [124, 135]. The coupled ROM-RCM and yne-ene cross-metathesis were used as test reactions in this work.

In order to reduce the reaction times and to overcome the diffusion-controlled reaction kinetics, monolithic SiO_2 rods were modified. For this purpose, tentacle-like polymer chains were grown from the catalyst-loaded rods via ROMP using a norbornene-type monomer, which contained cationic NHC precursor entities [98, 136, 137]. Subsequently, the catalyst was bound to the in situ generated free NCH units in analogy to the synthesis of **20** (i.e., following the steps *iii* to *v* as outlined in Fig. 2.21). Buchmeiser et al. extended this approach further towards

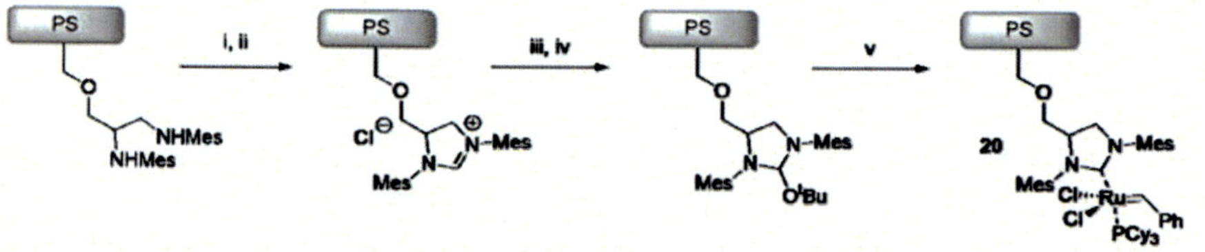

i) HC(OMe)3, HCOOH, toluene, 100 °C, 100 mbar, 15 h; ii) 0.1 M HCl, THF, room temp., 5 min; iii) TMSOTf, 2,6-lutidine,CH2Cl2, room temp., 30 min; iv) KOtBu, THF, room temp., 60 min; v) RuCl2(PCy3)2(=CHPh), toluene, 70-80 °C, 1 h

Fig. 2.21 Schematic representation of the synthesis of the polymer-supported Grubbs' catalyst **20**

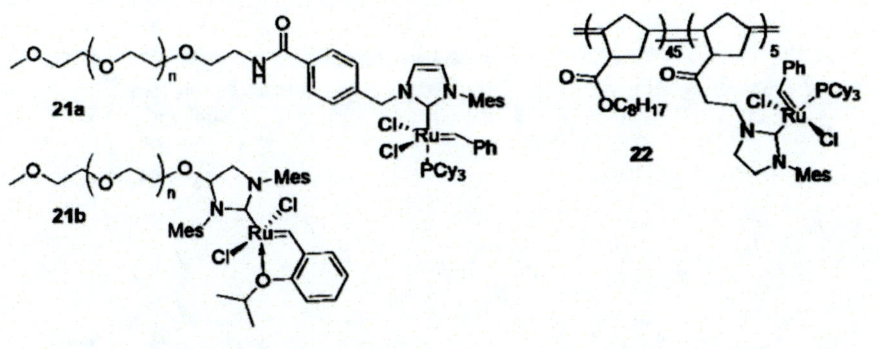

Fig. 2.22 Schematic representation of Grubbs' water-soluble metathesis complexes **21** and the poly(norborn-2-ene)-based material **22**

polymer-silica hybrid materials [138–140]. The water-soluble systems **21** were reported by Grubbs et al. (Fig. 2.22) [141, 142]. In this work, hetero-telechelic PEG was modified in a two-step procedure to give a terminal cationic NCH precursor, which was then utilized to bind a Grubbs' or Grubbs-Hoveyda-type catalyst, respectively. Catalyst **21b** featured a remarkable activity in aqueous olefin metathesis reactions, such as ROMP, RCM, and CM; whereas, its counterpart **21a** was unable to mediate the RCM under the same experimental conditions [142]. Finally, the poly(norborn-2-ene)-supported version of a Grubbs' catalyst (**22**) was reported by Weck et al.; these authors copolymerized *N*-mesityl-*N'*-(norborn-5-ene-2-ylcarbonyloxyundecyl) imidazolium bromide with octyl (norborn-5-ene-2-yl)carboxylate, followed by deprotonation and reaction with $RuCl_2(PCy_3)_2(=CHPh)$ (Fig. 2.22) [143].

The last type of ligand exchange that might be applied to coordinate Grubbs' catalysts to a preformed polymer is, according to Fig. 2.17, the substitution of the ancillary pyridine ligands. This rather unfrequented strategy was utilized by Mennecke et al. in order to immobilize a 3rd generation catalyst onto a cross-linked poly(4-vinylpyridine) (P4VP) support (**23**, Fig. 2.23a) [144]. In this study, commercial as well as synthetic resins were employed. The latter were prepared by the free-radical polymerization (FRP) of 4-vinylpyridine in the presence of a low content of 1,4-divinylbenzene (DVB). In a related study, Ru^{II}-salen complexes were coordinated to commercial P4VP, thereby forming the immobilized catalysts **24**, which were used in the olefination of aldehydes with diazo compounds [salen: *bis* (salicyliden)-ethylendiamine, Fig. 2.23b] [145]. As for the related metathesis catalysts, a significant bleaching and reduced activity compared to the homogeneous small-molecule catalysts in course of the respective application was observed. In the same context, the mesoporous and non-porous [146] as well as gelous polymer resins [147–151], which were loaded with the chiral Noyori-Ikariya catalyst, have to be mentioned (Fig. 2.23c). This particular catalyst is well known for its ability to catalyze the asymmetric hydrogenation of carbonyl compounds. Its heterogeneous

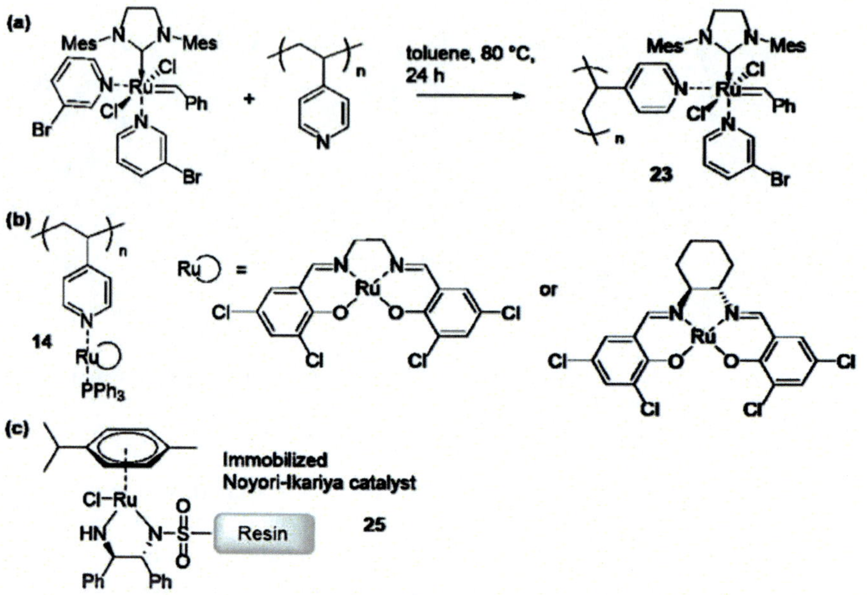

Fig. 2.23 Schematic representation of various immobilized Ru^{II} catalysts: **a** metathesis catalyst **23**, **b** Ru^{II}-salen complexes **24**, and **c** Noyori-Ikariya catalyst **25**

variants **25** were found to be highly efficient in this respect. Remarkably, if charged side chains were present, the catalysts could even be used in aqueous environments.

Recently, a comparative study was published by Xia et al. in which Hoveyda's *Z*-selective metathesis catalyst [152] was immobilized onto a PEGylated Merryfield resin [153] in three different ways (Fig. 2.24) [154]. In order to reduce steric crowding around the catalytic sites, the PEG chains were introduced as flexible spacer units between the PS resin core and the pending metal complexes. In

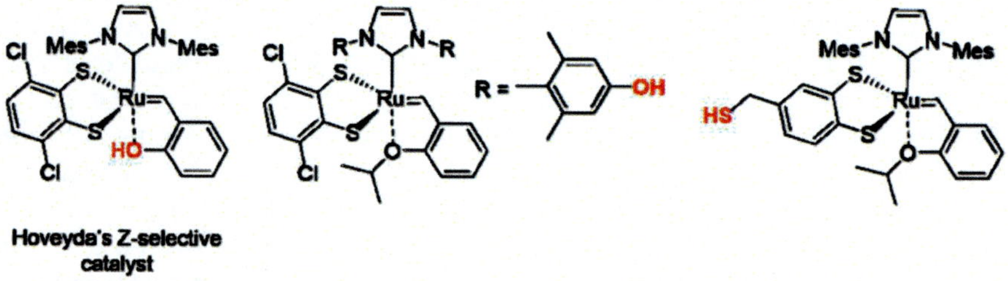

Fig. 2.24 Schematic representation of Hoveyda's *Z*-selective metathesis catalyst and two structural variants thereof (the position used for the immobilization onto a Merryfield resin are marked in red)

particular, the material in which the catalyst was attached to the resin via the NHC ligand offered a good performance in the metathesis of leaf alcohols in terms of substrate conversion and product selectivity; moreover, the heterogeneous catalyst could be reused up to five times without significant loss of activity.

In general, the coordination of transition metal ions to pyridyl moieties, as part of the polymer backbone or the side chains, represents a topic of research since the 1970s. Different structural motifs have been employed as binding site—varying in their denticity, rigidity as well as their binding strength. Most commonly, polypyridyl ligands, such as pyridine (py), bpy, and tpy, are used; recently, also some structural analogues have been introduced where (at least) one of the pyridine rings is replaced by a *N*-heteroaromatic five-membered ring (e.g., diazoles and triazoles).

2-Vinylpyridine and 4-vinylpyridine are versatile monomers, which can be polymerized in a highly controlled fashion into homopolymers as well as block copolymer architectures. As already shown, the pending py units can then be used for the coordination of transition metal ions (Fig. 2.23b). In very early contributions [68], the binding of transition-metal precursor complexes to poly(2-vinylpyridine) (P2VP) and P4VP was investigated. In this context, some Ru^0–carbonyl clusters, as designated catalysts for hydroformylation [i.e., $Ru_3(CO)_{12}$, $Ru_3(CO)_{10}(NO)_2$, Ru $(CO)_3(PPh_3)_2$] [155] or hydrogenation reactions {i.e., $[RuCl_2(CO)_3]_3$} [156] were utilized. Moreover, the preparation of P4VP-supported ruthenate, as oxidation catalyst, is also worth mentioning [157]; this material was obtained from Na_2RuO_4 and P4VP in aqueous 0.1 M NaOH. The strong binding of the metal center to the py sites was reflected by the low degree-of-bleaching. This finding is in sharp contrast to commercial polymer-supported ruthenates where anion-exchange resins, such as Amberlyst, are utilized (e.g., Johnson Mattey's FibreCat™) [158].

Different types of $Ru^{II/III}$ species have also been coordinated to polymer-bound py units. Besides the most straightforward system, i.e., binding of $RuCl_3$ to P4VP or P2VP [159, 160], in particular *cis*-$Ru(bpy)_2Cl_2$ was utilized for this purpose. Among others, Vos et al. and Meyer et al. combined this versatile precursor, as well as its heavier Os^{II} analog, with P4VP or poly(*N*-vinylimidazole) (PVI) (**26** and **27**, Fig. 2.25) [161–172]. Thereby, a wide range of redox-active metallopolymers, which only varied in their metal content, was obtained. Moreover, also partially

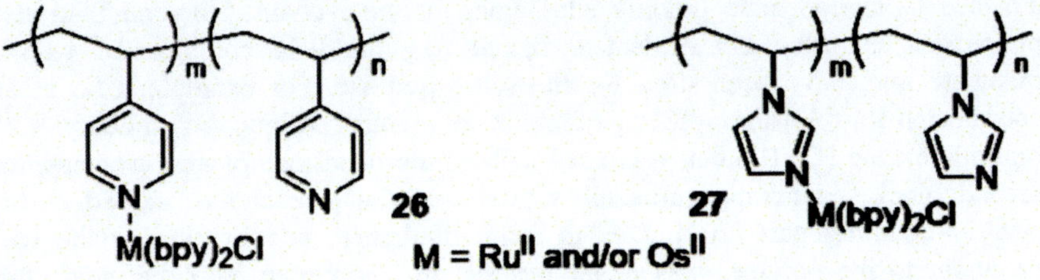

Fig. 2.25 Schematic representation of the P4VP- (**26**) and PVI-based metallopolymers (**27**)

quarternized P4VP was employed, as a macroligand, in order to improve the control over the electron diffusion within the systems [173]. Using a two-step approach both redox-active metal centers were incorporated into a single macromolecular material [168]. Here, the order of metal coordination was found to be crucial: The Os^{II} centers, requiring much harsher conditions for coordination, needed to be introduced first. Such bimetallic polymers exhibited different redox behavior of their Ru^{II} and Os^{II} centers towards desoxyribonucleic acid (DNA); also the two homopolymers were applied for the analyzing DNA, e.g., their hybridization or damage [174–180]. In-depth electrochemical studies detailed the influence of various electrolytes and solvents on the electrode processes [181–185]. Further stabilization of the metal-loaded polymers, deposited onto electrode materials, was achieved by solid-state reactions [186]. Binding of glucose oxidase, tyrosinase, laccase or peroxidase to electrode-immobilized metallopolymers, thereby generating redox-active enzyme sensors, was also reported [187–197]. The PS2-enzyme (PS2: photosystem-2) was wired around the electrodes and served as an electron donor for the metallopolymer matrix [198]. Finally, the Os^{II}-containing polymers were employed as electron relays in biofuel cells [199]. Due to a good wetting behavior, polymers **26** were, moreover, found to be excellent dispersants for carbon nanotubes (CNTs) [200]. The P4VP-derived polymer bearing $[Ru(bpy)_2]^{2+}$ units was further combined with a Dawson-type polyoxometallate (POM); in this hybrid material, the Ru^{II} chromophores served as sensitizer for the photoelectrocatalytic oxidation of alcohols by the POM [201].

NAMI-A is a ruthenodrug which was tested in phase-II clinical trial (though finally unsuccessfully) [202]. In order to improve the cytotoxicity and the cellular uptake of this potential anticancer agent, Stenzel et al. prepared a polymer conjugate thereof [203]. A copolymer containing imidazole units was synthesized and, subsequently, the Ru^{III} precursor was coordinated to the binding sites. Thusly, the obtained amphiphilic metallopolymer **28** assembled into spherical micelles, which were analyzed by transmission electron microscopy (TEM) imaging and dynamic light-scattering (DLS) measurements. The cytotoxic activity of **28** against various ovarian and pancreatic cancer cell lines was tested. Overall, **28** showed a ca. 1.5 times higher toxicity compared to NAMI-A itself. Moreover, the inhibition of metastasis by micellar **28** was also more efficient (as concluded from inhibitory tests on three different cell lines) (Fig. 2.26).

Most commonly, Ru^{II} precursors bearing two bidentate ligands (e.g., bpy) are attached to monodentate binding sites along polymer chains. Beyond that, also precursors, featuring one tridentate ligand (e.g., tpy) in combination with a bidentate one, have been used for the same purpose. For example, Cho et al. coordinated Ru(tpy)(dmbpy)Cl to a diblock copolymer (dmbpy: 4,4′-dimethyl-2,2′-bipyridine); the P2VP block was used to bind the metal complexes; whereas, the second block, containing carbazole moieties in the side chains, served as the electron-donating part [204]. Due to steric hindrance, coordination of the Ru^{II} precursor to the polymer was rather insufficient. As shown by Zeng et al., this confinement could be leveled out by using the less sterically crowded P4VP [in this

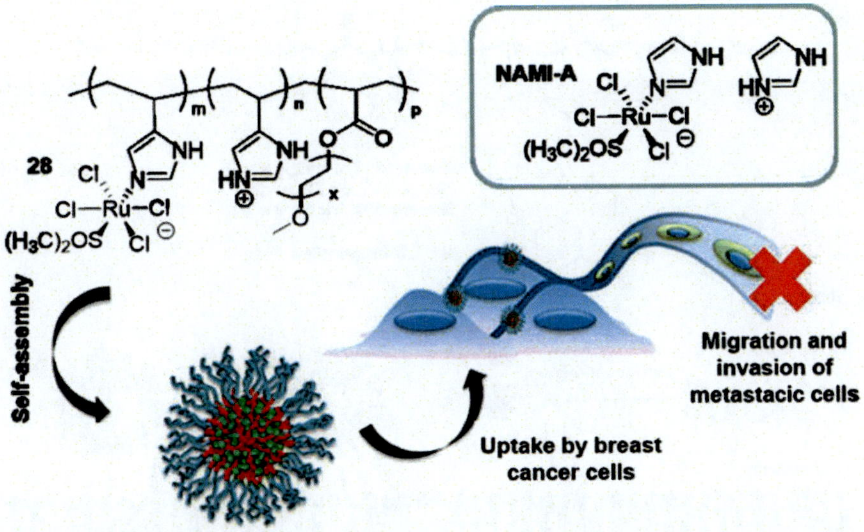

Fig. 2.26 Schematic representation of NAMI-A and it's immobilized variant **28**, which was used for the metastasis inhibition studies [203]. Figure reproduced with kind permission; © 2014 American Chemical Society

study, 2-(2-hydroxyphenyl)benzoxazole (box) was used as bidentate ligand instead of bpy] [205].

More recently, Wu et al. reported ruthenopolymers for the phototherapeutic treatment of cancer. In these studies, the photoactive Ru^{II} complexes were bound to nitrile moieties of the polymers (Fig. 2.27a) [206, 207]. Due to the amphiphilic nature of **29**, micellization into diverse nanostructures occurred (i.e., micelles, vesicles and larger aggregates—depending on the molecular weight of the polymers). The intrinsically labile Ru^{II}–nitrile bonds were stabilized against hydrolysis in these nanosized environments. Photophysical studies in diverse media demonstrated the stability even under physiological conditions [208]. The metallopolymer nanostructures, in particular the micelles, entered cancer cells where irradiation with red light ($\lambda = 656$ nm) was applied to induce both the photodissociation of the metallopolymer and sensitization of oxygen ($^1O_2$, Fig. 2.27b) [206]. This strategy enabled a dual anticancer activity: The released Ru^{II} complex offered a photoactivated chemotherapy (PACT); whereas, the $^1O_2$ was employed in the photodynamic therapy (PDT). For the micelles, vesicles, and larger aggregates, a decrease in the cell viability to 24%, 75%, and 80%, respectively, was observed. In a related study, these same authors attached two polymer chains via their terminal nitrile groups to a $[Ru(biq)_2]^{2+}$ dye (biq: 2,2′-*bis*quinoline). The resultant Ru^{II}-centered ABA-type triblock copolymer was, similar to the previous example, tested with respect to the dual PACT–PDT in vitro cancer treatment. Moreover, in vivo experiments in mice were conducted and the accumulation of the metallopolymer in

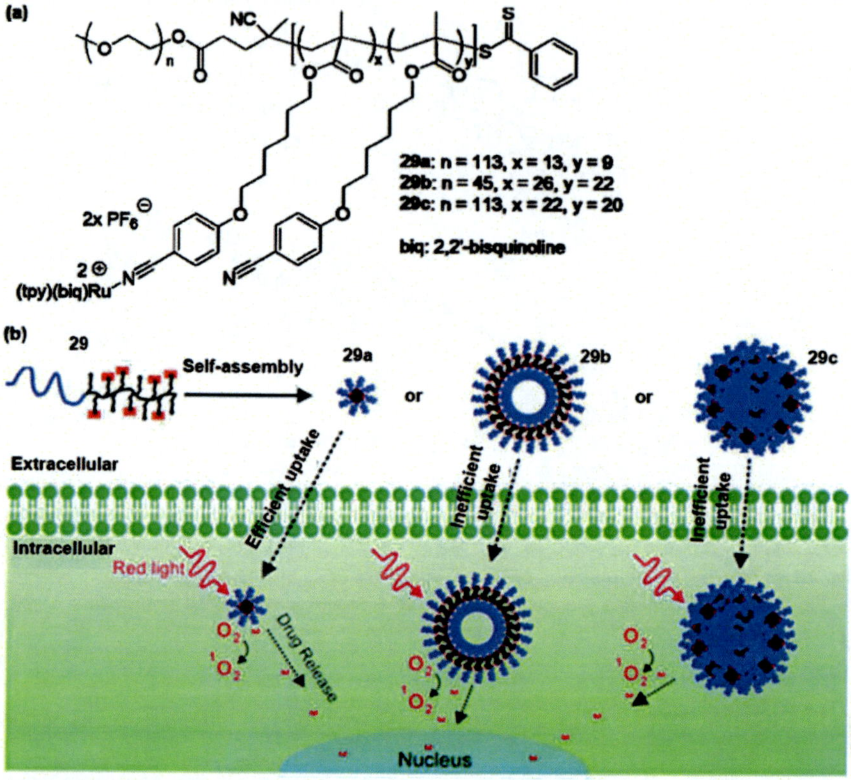

Fig. 2.27 a Schematic representation of Wu's photoresponsive ruthenopolymer **29**. **b** Schematic representation of the self-assembly of **29** into nanostructures, which served, after cellular uptake, as dual anticancer agents, i.e., release of a cytotoxic Ru^{II} drug and generation of $^1O_2$ [206]. Figure reproduced with kind permission; © 2018 The Royal Chemical Society

the tumor cells was corroborated by fluorescence imaging. Upon irradiation with red light, the tumor growth was successfully retarded whilst the healthy tissue and organs were hardly affected [209].

The coordination of transition-metal complexes or, simpler, transition-metal ions to bidentate N^N-type ligands as part of a polymer structure—as an integral part of the polymer backbone, within the side chain or as end group—represents a highly active field of research. The (presumably) first polymer having bpy moieties in the side chain was reported in the late 1970s by Neckers et al. [210–213]. In this initial study, various transition-metal salts, including $RuCl_3$, were added to the polymer to afford the corresponding metallopolymers. In most cases insoluble, probably densely cross-linked materials were obtained (Fig. 2.28). In related studies, vinyl-substituted bpy derivatives (or bpy-analogous substrates) were (co)polymerized by FRP or controlled radical polymerization (CRP) to give more defined pre-polymers for the

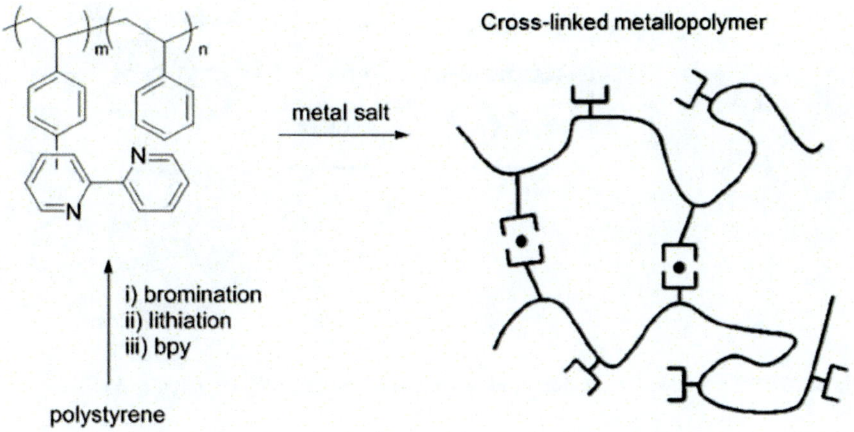

Fig. 2.28 Schematic representation of the synthesis of cross-linked metallopolymers by addition of a transition metal salt to a bpy-containing polymer

binding of metal ions [214–223]; alternatively, also the post-polymerization modification of polymers with bpy units was reported [224, 225].

Chujo et al. used a poly(*N*-acetylethylenimine) with pendant bpy moieties for gelation studies with various divalent transition metal ions; from these, the hydrogel obtained from the cross-linking with Ru^{II} ions was found to be highly robust due to the thermodynamic stability and kinetic inertness of the formed Ru^{II} *tris*-bipyridine complexes (the swelling degree of 14% remained constant when the hydrogel was kept in an aqueous environment for over one month) [226].

The concept of coordinating the *cis*-$Ru(bpy)_2Cl_2$ precursor to polymers with pendant bpy units was introduced by Kaneko et al. [227–233] and applied by various other groups who utilized different types of polymers as well as Ru^{II} precursor complexes [41, 214, 222, 234–237]. In particular, the more reactive precursors of the general formula $[Ru(N^{\wedge}N)_2(acetone)_2]^{2+}$ enabled the coordination of the Ru^{II} fragments to the polymers with higher efficiencies and under much milder conditions (N^N = bpy, dmbpy, phen, etc.) [237]. The partially cross-linked polymer **30** was obtained from poly[*N*-vinylcarbazole-*co*-(vinyl alcohol)] via a post-polymerization modification strategy (Fig. 2.29) [238]. The bpy entities within the cross-linking units were subsequently complexed by adding *cis*-$Ru(bpy)_2Cl_2$ to afford a thermally stable and luminescent material. Basically the same properties were also reported for a related non-cross-linked ruthenopolymer [239].

The bpy-functionalized polydiacetylenes, as reported by Camacho et al., are highly interesting due to their molecular structure [240]. The monomer synthesis involved an amidation reaction between carboxy-functionalized 1,3-diynes and 5-aminobipyridine. The subsequent polymerization was realized by UV- or γ-irradiation at room temperature. The insoluble polymers **31a**, possessing a polydiacetylene-type backbone, became soluble upon formation of Ru^{II} *tris*-

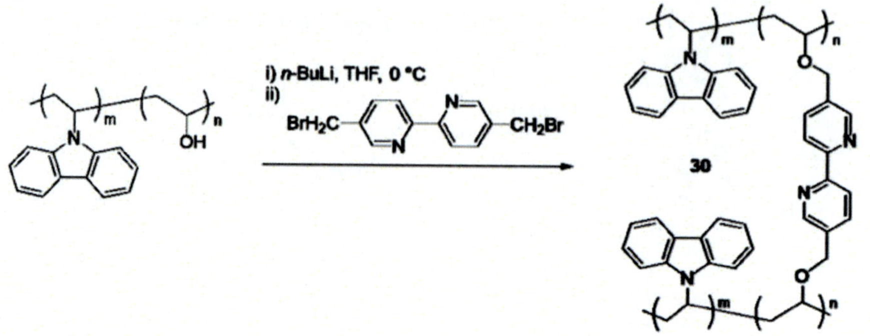

Fig. 2.29 Schematic representation of the synthesis of the cross-linked polymer **30**

bipyridine complexes within the side chains (**31b**, Fig. 2.30). These authors investigated the non-linear optical (NLO) properties of **31b** in order to establish an optical switch on the basis of the material's refractive nonlinearity. However, the intense linear absorption of **31b** at low energies prevented any optical switching (instead, complexation of **31a** with other divalent metal ions was, due to the resultant higher-energy absorption properties, far more efficient in this respect).

Energy- and electron-transfer processes within polymers containing pending, redox-active Ru^{II} *tris*-bpy complexes in combination with organic/organometallic donor and/or acceptor sites has emerged into a highly relevant topic of research within the last decades [41, 241]. Again, this field of research was pioneered by Kaneko et al., who introduced methylviologen (MV^{2+}) units or used metal colloids, as electron-acceptor moieties and external quenchers, respectively [230, 242–244].

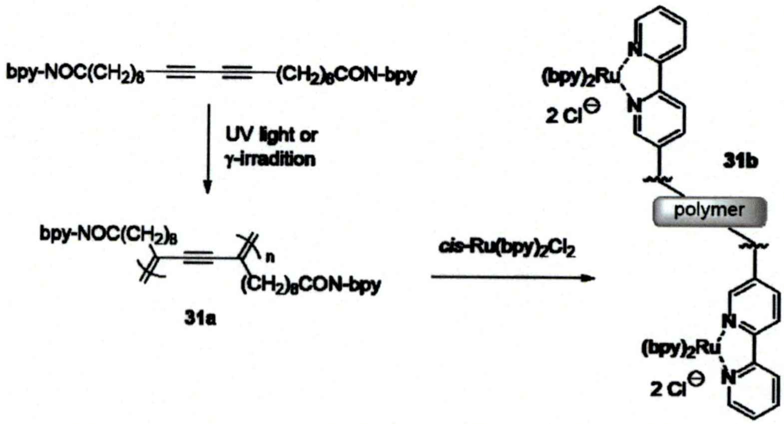

Fig. 2.30 Schematic representation of the synthesis of the Ru^{II}-containing polydiacetylene **31b**

Frechét et al. reported the synthesis of copolymers with pending coumarin and bpy moieties (Fig. 2.31) [18, 19]. The subsequent coordination of $[Ru(dmbpy)_2]^{2+}$ fragments to the bpy sites gave the corresponding ruthenopolymers **32**, which exhibited *intra*-chain energy transfer (dmbpy: 4,4-dimethyl-2,2-bipyridine). However, the synthetic strategy was found to be less efficient when compared to the alternative approach, i.e., the direct copolymerization using complex-containing monomers (see Sect. 2.5.1).

The (co)polymerization of ligand-containing monomers represents the most commonly used strategy to prepare (co)polymers with pending metal-binding sites. However, some notable examples were also reported where bpy derivatives were

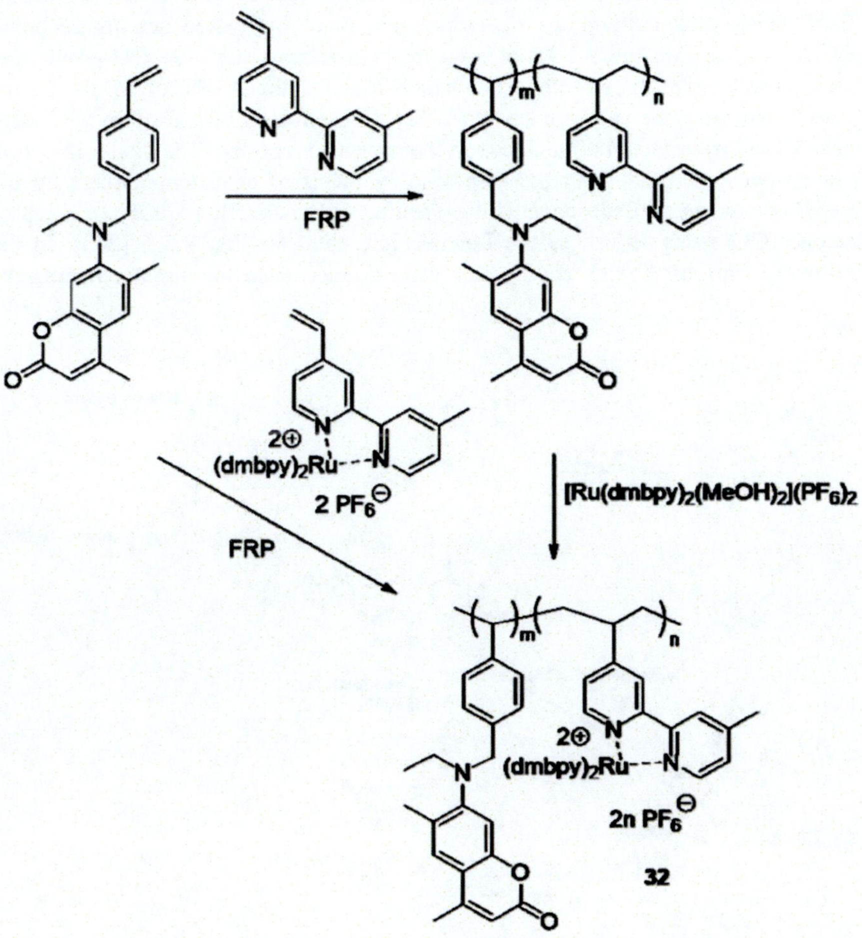

Fig. 2.31 Schematic representation of the two alternative synthetic routes towards copolymers **32**, containing pending coumarin dyes and Ru^{II} *tris*-bipyridine complexes [18, 19]

employed, as the initiator. In this respect, Fraser et al. reported the atom-transfer radical polymerization (ATRP) of styrene or methyl methacrylate (MMA), initiated by a chloromethylated bpy, and the subsequent coordination of Ru^{II} ions to obtain, for example, linear or star-shaped PSs containing a central $[Ru(bpy)_3]^{2+}$ complex (**33** and **34**, Fig. 2.32) [245–247]. In the following, a range of polymers with central or terminal bpy-type binding sites were prepared by using CRP techniques and the coordination of Ru^{II} precursor complexes was shown [248–251]. However, this convergent route to ruthenopolymers bears a considerable limitation, i.e., quantitative coordination of the precursor to the sterically demanding macroligands. In order to overcome this restriction, Fraser et al. explored an alternative strategy by utilizing an appropriately functionalized complex, as initiators, for the synthesis of metallo-supramolecular polymers of various architectures (see also Sect. 2.2.4) [245]. These same authors also developed the "dual" bpy-based initiator **35** being able to first initiate the ATRP of styrene; in a second step, the OH-group was derivatized to give an α-bromoester, which initiated the ATRP of MMA, in the presence of the same catalytic system [252]. Coordination of a $[Ru(bpy)_2]^{2+}$ fragment to the bpy-centered diblock copolymer was also reported. The dual initiator **35** was furthermore applied for the step-wise synthesis of block copolymers by the ATRP of styrene and subsequent ring-opening polymerization (ROP) of ε-caprolactone (CL) using Et_3Al, as the Lewis-acidic catalyst (Fig. 2.32) [253]. In the following, various diblock copolymers were derived from the initiator having the

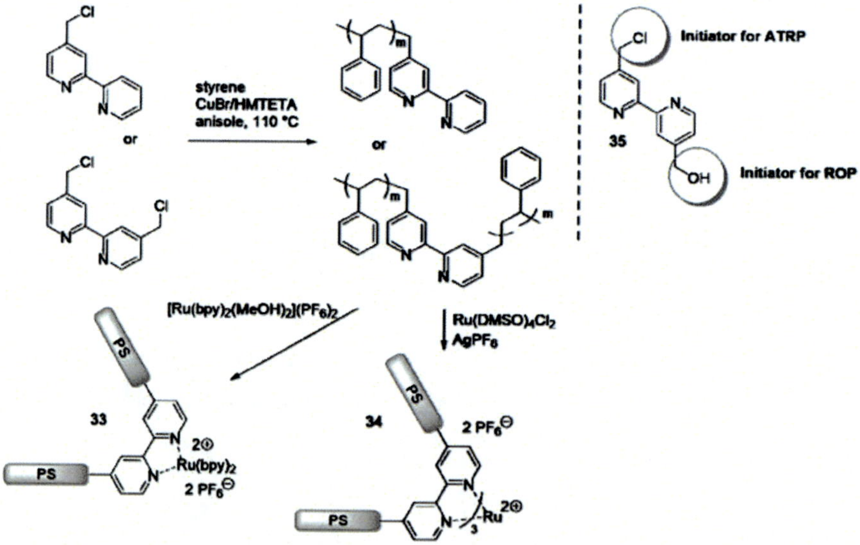

Fig. 2.32 Left: Schematic representation of the synthesis of $[Ru(bpy)_3]^{2+}$-centered linear (**33**) and star-shaped polymers (**34**). Right: Schematic representation of Fraser's bpy-based dual-functional initiator **35**

bpy binding site at the block junction [254]. Again, complexation with appropriate Ru^{II} precursor complexes afforded linear as well as star-shaped metallo-supramolecular polymers [255].

Remarkable antitumor activity, as revealed by very low IC_{50} values for various cancer cell lines, has been reported for cationic Ru^{II} complexes of the general formula $[RuCp(P)(N^{\wedge}N)]^{+}$ (P: monodentate phosphine ligand, N^N: *N,N'*-bidentate ligand) [256]. Valente et al. synthesized the macromolecular ruthenodrug **36**, which contained PLA chains, end-capped with D-glucose moieties (Fig. 2.33a) [257]. In-vitro tests against various human cancer cell lines gave IC_{50} values in the µmol range. Cellular distribution studies were performed and indicated a preferential accumulation of the drug in the cells' nuclei and membranes (Fig. 2.33b). The related ruthenopolymer **36**, missing the terminal D-glucose sites, changed the morphology of the F-actin cytoskeleton morphology and decreased the cells' clonogenic ability [258]. Apoptosis occurred at very low concentrations (in the same regime as known for cisplatin) and, even more important, the resistance of A2780cis cancer cell lines against platinum drugs could be overcome. The authors concluded that for **37** an action mode, different from the one of cisplatin, was operable.

Beyond bpy, a range of similar, structurally related bidentate chelators for metal ions has also been employed. A hydroxy-functionalized 2-(1*H*-1,2,3-triazol-4-yl) pyridine (trpy) ligand was synthesized by Schubert et al. via the CuCAAC reaction and, subsequently, applied to initiate the ROP of CL [259]. Such trpy ligands can be considered as bpy analogs where one of the pyridine rings is substituted by a 1*H*-1,2,3-triazol-4-yl moiety; the general coordination behavior to metal ions is thereby retained. The main advantage of these systems lies in the ease of synthesis, in particular in comparison to unsymmetrically functionalized bpy derivatives, which are more difficult to synthesize [237, 260]. Poly(ε-caprolactone) (PCL) macroligands with either trpy or bpy binding sites were applied for the syntheses of polymers containing a Ru^{II} complex, as terminal unit [259, 261].

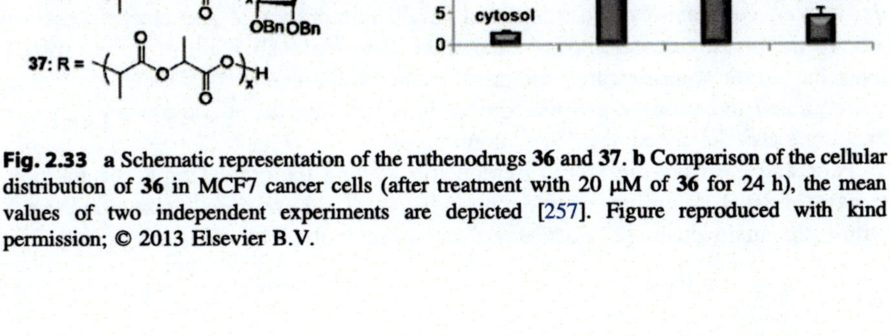

Fig. 2.33 a Schematic representation of the ruthenodrugs **36** and **37**. **b** Comparison of the cellular distribution of **36** in MCF7 cancer cells (after treatment with 20 µM of **36** for 24 h), the mean values of two independent experiments are depicted [257]. Figure reproduced with kind permission; © 2013 Elsevier B.V.

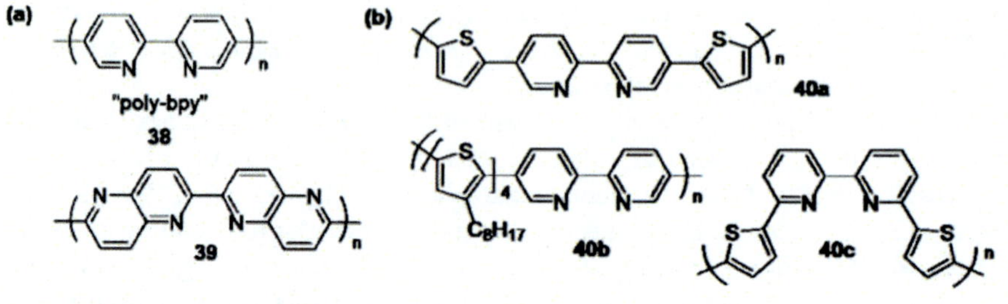

Fig. 2.34 Schematic representation of the structurally simple bpy-containing π-conjugated polymers **38–40**

The structurally most simple polymer with bpy units within the main chain is the plain poly(2,2′-bipyrin-5,5′-diyl), referred to as "poly-bpy" (**38a**, Fig. 2.34a) [262]. Due to solubility limitations as well as steric constraints, only a low to moderate loading of the polymer with Ru^{II} complexes was achieved [263]; the solubility of the polymer was increased significantly when short diethynyldiarylsilane spacers were placed between the binding sites [264, 265]. Yamamoto et al. also tested the ability of **38a** to bind $RuCl_3$ [266], an analogous approach was reported previously by Yanagida et al., who utilized a poly(pyridine-2,5-diyl) macroligand instead [267]. These two systems exhibited a similar catalytic activity in the photocatalytic HER from water; due to a significantly stronger binding of the Ru^{III} centers, the former material could be recycled and retained its catalytic activity over several consecutive HER experiments. A remarkable HER performance was achieved when **38a** was loaded with $[Ru(bpy)_2]^{2+}$ units and RuO_2 was added, as co-catalyst: Evolution of H_2 occurred readily from methanol—even in the absence of any sacrificial amine donor [263]. In the following, these same authors reported a library of π-conjugated polymers, which were prepared by organometallic dehalogenative polycondensation reactions (in total 26 examples) [268, 269]. From these, the poly (pyrimidin-5,5′-diyl) (**38b**) and the highly rigid poly(1,5-naphthyridin-2,6-diyl) (**39**) were used, as ligands, for the complexation of Ru^{II} precursor complexes (Fig. 2.34a). The polythiophene-type polymers with incorporated Ru^{II} *tris*-bipyridine complexes were prepared by Swager, Guillerez et al. (Fig. 2.34b) [270, 271]. The hybrid materials were obtained by complexation of the linear oligothiophene-*alt*-bipyridine copolymers **40**, which were obtained from electropolymerization (**40a**) or cross-coupling polymerization reactions (**40b**); related to this, also angular derivatives, such as **40c**, were reported [272]. Photophysical and electrochemical studies on the Ru^{II}-containing materials revealed a delocalization of the π-orbitals over the entire π-electronic structures; cyclic voltammetry (CV) measurements corroborated the strong electronic interactions between the π-conjugated polymer backbone and the coordinated Ru^{II} complexes.

Peng and Yu were the first to particularly address the critical aspects regarding the synthesis of π-conjugated polymers with bpy units (or their metal complexes) within the main chain [273]: The Pd^0-catalyzed polyaddition, using appropriate monomers, should give π-conjugated polymers with metal binding sites—however, catalyst deactivation, e.g., by interaction with bpy-containing monomers, might hamper or even fully retard the polymerization. Thus, utilization of pre-formed complex-containing monomers often represents a more versatile synthetic approach and only scattered examples are known, where this strategy to obtain polymers with uncomplexed ligand sites has been used (see also Sect. 2.5.1b) [274–276]. Following a different approach, Rasmussen et al. synthesized the so-called "polyaza-bpy" (**41**) by polycondensation of 5,5′-diamino-2,2′-bipyridine with glyoxal; this organo-soluble polymer was subsequently used for the coordination of $[Ru(bpy)_2]^{2+}$ fragments (Fig. 2.35) [277]. The particular interest in using a polyazine-type structure arouse from the special properties of the parent polymer $[–CH{=}N–N{=}CH–]_n$, which is both isostructural and isoelectronic to polyacetylene but stable under environmental conditions. Euler et al. reported high conductivities for iodine-doped polyazine samples (σ values up to 1.3 Ω^{-1} cm^{-1}) [278]. This study indicates that polycondensations are well-suited for the synthesis of π-conjugated polymers bearing ligand sites within the backbone (or side chains), since any undesired interaction of the metal catalyst with the reagents is avoided. Most commonly, Wittig- or Horner-Wadsworth-Emmons-type (HWE) polycondensations with dialdehyde and *bis*(phosphonium salt) or *bis*phosphonate reactants,

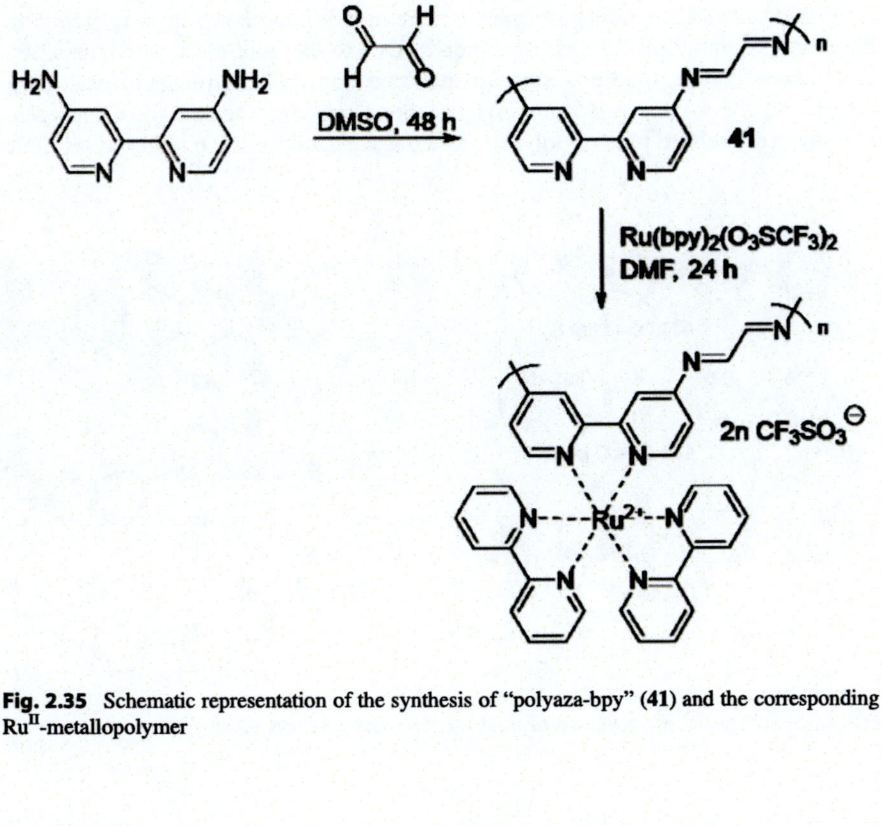

Fig. 2.35 Schematic representation of the synthesis of "polyaza-bpy" (**41**) and the corresponding Ru^{II}-metallopolymer

respectively, are used for this purpose. Some years ago, a specialized review devoted to the synthetic routes towards π-conjugated polymers was published by Grimsdale et al. [279].

Biswas et al. reacted bpy with pyromellitic dianhydride under highly forcing conditions (i.e., $ZnCl_2$, 240 °C); the proposed polycondensate **42** was utilized for the coordination of various transition metal ions (Fig. 2.36) [280, 281]. Structurally more defined polymers with bpy units within the main chain—a poly(benzobisoxazole) (**43a**), poly(benzobisthiazole) (**43b**) and poly(pyridobisimidazole) (**44**) —were investigated by Chan et al. [17] as well as Tan et al. (Fig. 2.36) [282]. It was demonstrated that the solubility of the **43a/b** was improved significantly when *cis*-$Ru(bpy)_2Cl_2$ was coordinated to the binding sites along the rigid-linear polymer chain; along with this, the electron-transport efficiency increased by two orders of magnitude up to 10^{-5} cm^2 V^{-1} s^{-1} [17]. A related type of material was obtained with bidentate 5,5′-bisimidazole units within the backbone (**45**, Fig. 2.36) [283]. The Ru^{II}-complexed polymer had a M_n value of > 50,000 g mol^{-1}, according to SEC analysis; elemental analysis indicated that ca. 60% of the binding sites were occupied by Ru^{II} centers. Metallopolymer films were deposited onto Pt disc electrodes and an increase in the electron-transport rate, compared to films prepared by electropolymerization of vinyl-functionalized Ru^{II} complexes (e.g., **1** in Fig. 3.2), was observed in the impedance measurements. It was suggested that the electronic communication between the conjugated backbone and coordinated Ru^{II} centers facilitated the electron transport (thus, a mere hopping mechanism, as operative in the electropolymerized film, was excluded). Due to the poor electronic interaction in the excited state, dual emission arising from both the coordinated metal complexes and the π-conjugated polymer backbone was observed [284]. Chan et al. investigated a related type of polymer, as a macroligand for the binding of multiple

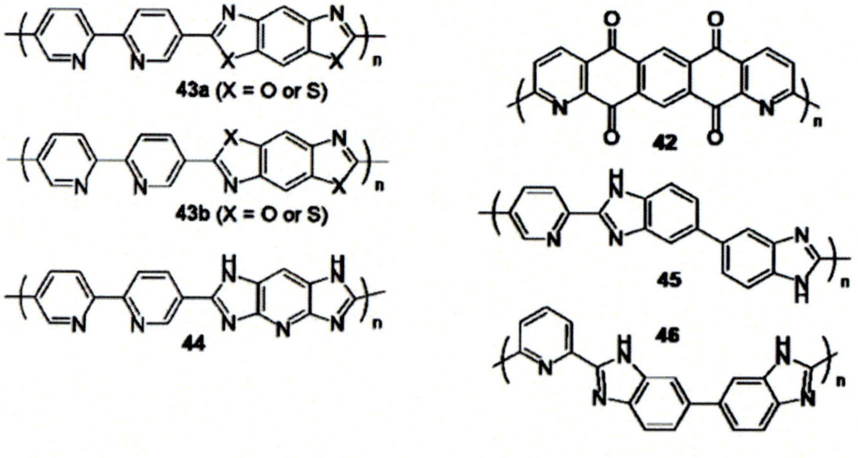

Fig. 2.36 Schematic representation of rigid π-conjugated polymers **42–46**

Ru^{II} centers. This polymer, comprising tridentate 2,6-*bis*(benzimidazol-2-yl)pyridine sites along the polymer backbone, was prepared via a polycondensation route (**43**, Fig. 2.36) [285]. The complexation with Ru^{II} complexes (elemental analysis indicated a loading of ca. 90%) gave a metallopolymer with broad absorption and photosensitivity beyond 500 nm; thus, a large photocurrent response was recorded upon irradiation of this ruthenopolymer with visible light.

The polyamides **44** were prepared by a polycondensation between aromatic diamines and 2,2′-bipyridine-5,5′-dicarboxylic acid (Fig. 2.37a) [286, 287]. Due to the pronounced rigidity of the structures, sufficient degrees-of-polymerization (DPs) were only reached in a mixture of hexamethylphosphoramide (HMPA) and LiCl; LC behavior of **44** in conc. H_2SO_4 or in HMPA/LiCl was observed (Fig. 2.37a). A similar library of metallopolymers featuring LC behavior was obtained from the polyesters **45** and, *inter alia*, Ru^{II} ions (Fig. 2.37b) [287, 288]. The Ru^{II}-containing polymers exhibited red-light emission (λ_{em} = 700 nm) and were processed into light-emitting devices (in those cases where homogeneous thin films could be fabricated via spin-coating).

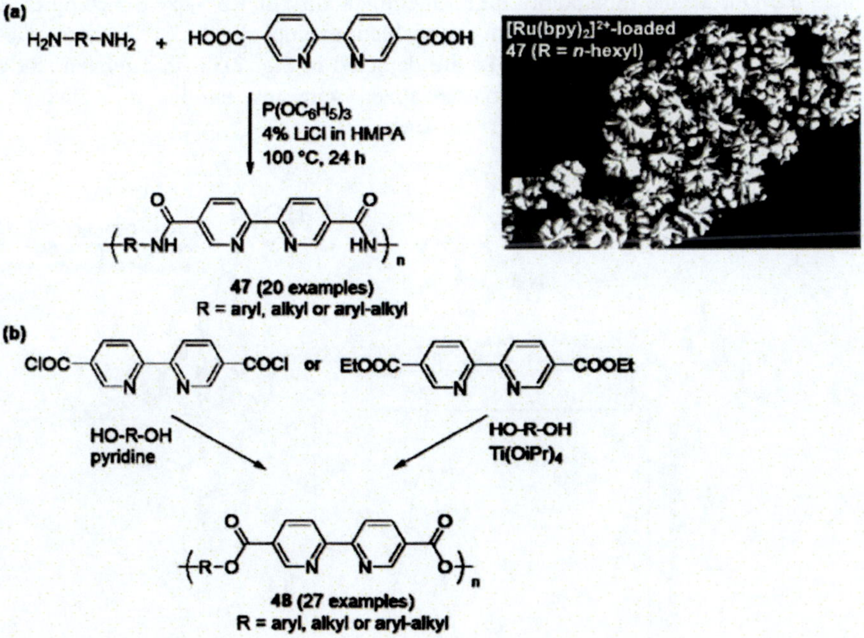

Fig. 2.37 Schematic representation of the synthesis of polyamides (**47**) and polyesters (**48**) via polycondensation. The polarized optical micrograph (POM) image of a Ru^{II}-containing metallopolymer (**47** with R = *n*-hexyl) is also depicted (4.4 wt% in HMPA containing 4 wt% LiCl, 25 °C; 400× magnification) [287]. Figure reproduced with kind permission; © 2000 American Chemical Society

Undisputed, tpy is the most prominent tridentate ligand that has been utilized in the field of metallopolymer science [289]. In analogy to its bidentate counterparts, the (co)polymerization of appropriate monomers, using radical polymerization techniques, represents an established strategy and many synthetic protocols can be found in the literature. The first report in this respect dates back to 1988 when Potts and Usifer (co)polymerized vinyl-substituted terpyridine ligands via FRP and studied the coordination of some divalent transition-metal ions to these polymers [290]. Hanabusa et al. utilized tpy-containing monomers for the free-radical copolymerization with styrene, MMA, vinyl acetate or acrylic acid [291, 292]; at low contents of the tpy monomer, soluble random copolymers were obtained. In the following, Schubert et al. also prepared tpy-containing polymers by FRP [293–296]. Besides the coordination of small-molecule Ru^{III} *mono*-tpy precursor complexes (**50a/b**) to the PMMA-based copolymer **47**, the grafting of PEG (**50c**) or PLA (**50d**), each bearing a terminal tpy unit, onto the side-chain functionalized polymer was carried out making use of the directional coordination chemistry of the $Ru^{II/III}$ ions (Fig. 2.38) [294]. The stepwise assembly allowed the selective and directed formation of heteroleptic Ru^{II} *bis*-terpyridine complexes—the key requirement for preparing the well-defined metallo-supramolecular graft copolymers **51**. Due to the remarkably high kinetic stability of Ru^{II}–tpy *bis*-complexes, their analysis by SEC measurements is enabled; exemplary, the SEC traces of **46** and the derived graft copolymer **51d** are depicted in Fig. 2.38. In a related study, poly(vinyl chloride) PVC was functionalized with tpy entities in a two-step

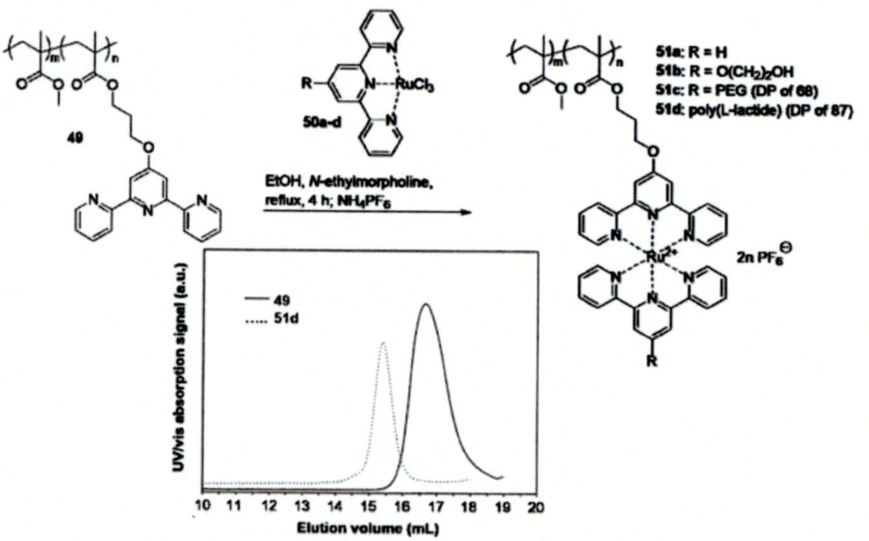

Fig. 2.38 Schematic representation of the synthesis of the Ru^{II}-containing graft copolymers **51**. The SEC traces of the metal-free pre-polymer **49** and ruthenopolymer **51d** are also shown (THF as eluent) [294]. Figure reproduced with kind permission; © 2002 Wiley-VCH

procedure; besides the metallo-supramolecular crosslinking also the grafting of Ru^{II} complexes onto the polymer was investigated [297]. The methodology was adopted later to modify cellulose or dextran, as two examples for biopolymers, with Ru^{II}-tpy *bis*-complexes [298–300]. The highly branched homopolysaccharide dextran was first functionalized with tpy entities by esterification (reaching degrees-of-substitution of ca. 0.3); subsequently, various Ru^{II} precursor complexes (e.g., **50a**) were grafted onto the biopolymers [299]. Compared to Rehahn's study on "ruthenocellulose" [300] significantly milder reaction conditions were required to obtain the analogous "ruthenodextran" materials. In the same context, Hassan et al. reported the surface modification of cellulose nanocrystals with Ru^{II}-tpy *bis*-complexes [298]. Nanocellulose represents an emerging material for potential applications in materials science or biomedical engineering [301]. For example, highly fluorescent nanocrystals were obtained when a perylene dye was grafted to the tpy-functionalized cellulose via directional Ru^{III}/Ru^{II} coordination chemistry (Fig. 2.39) [302].

The implementation of, for example, copolymers with tpy units in the side chain in sophisticated applications, however, often requires a precise polymer architecture, which is not accessible via FRP. Thus, the various CRP techniques have moved in the focus of interest (i.e., ATRP, RAFT or NMP). Tew et al. showed, that ATRP with tpy-functionalized monomers is, in contrast to RAFT and NMP, often difficult to perform, due to considerable deactivation of the Cu^{I} catalyst by complexation [303, 304]. However, Tzanetos et al. demonstrated that, when using CuBr/bpy, as the catalytic system, the ATRP of such monomers can be conducted in a controlled fashion [305]. Moreover, the organo-soluble, CNT-polymer conjugate **52** was obtained by the polymerization of a tpy-equipped styrene monomer using an ATRP initiator immobilized onto CNTs; subsequently, Ru^{II} precursor complexes were grafted onto the side chains of **52** (Fig. 2.40) [306, 307]. The functionalization of the CNTs was monitored by UV/vis absorption and Raman spectroscopy. The amphiphilic copolymer **53**, prepared by O'Reilly et al. via NMP,

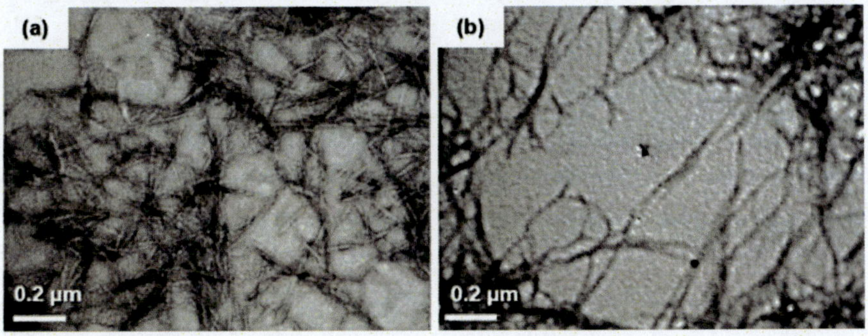

Fig. 2.39 TEM images of (a) parent cellulose nanocrystals and (b) after modification with tpy moieties [302]. Figure reproduced from ref. [302], copyright 2012 Elsevier B.V

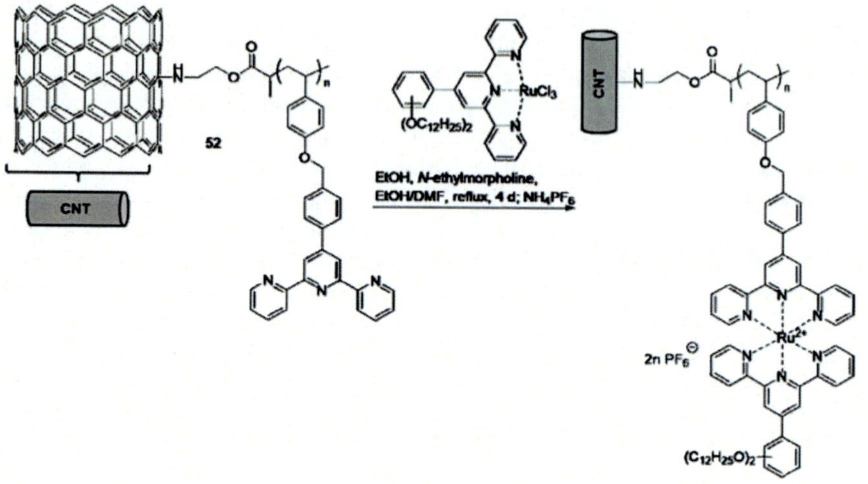

Fig. 2.40 Schematic representation of the synthesis of the CNT-ruthenopolymer conjugates according to Stefopoulos et al. [306]

assembled into spherical micelles in THF-water mixtures [308]. The hydrophilic corona of these assemblies was subsequently cross-linked using amidation chemistry (Fig. 2.41). The thusly obtained nanoparticles (having a diameter of ca. 60 nm according to DLS and TEM measurements) could be loaded with metal ions, such as Ru^{II}, without any alteration in size and shape of the particles. In related work, Ge and Liu assembled spherical micelles from a tpy-functionalized double-hydrophilic block copolymer [309]. These micelles were then stabilized by cross-linking the tpy moieties in the core via Ru^{II} ion complexation. The metallo-supramolecular micelles exhibited rich stimuli-responsive properties, such as pH-responsive cores and thermo-responsive shells. Moreover, decomplexation could be realized in the presence of a strong competitive ligand [i.e., hydroxyethylethylene diaminetriacetic acid sodium salt (HEEDTA)]. It was demonstrated that the responsiveness of the micelles could be used for the encapsulation and controlled release of hydrophobic guest molecules (i.e., Nile red, as model substrate).

As for the aforementioned bpy-containing systems (see Figs. 2.34, 2.35, and 2.36), also a variety of π-conjugated polymers with tpy units in the pending side chains has been reported. In this respect, two types of conjugated polymers with grafted tpy moieties can be distinguished: The π-conjugation is extended over the entire molecule, including side chains, or the π-conjugation into the side chains is interrupted by heteroatoms and/or alkyl spacers. For the synthesis of the fully π-conjugated polymers, the various polyaddition and polycondensation strategies are applicable, i.e., Pd^0-catalyzed cross-coupling and Wittig-type reactions [5, 289]. However, no defined metallopolymers were derived from these highly conjugated materials. In general, the coordination of transition- or rare-earth-metal ions to the

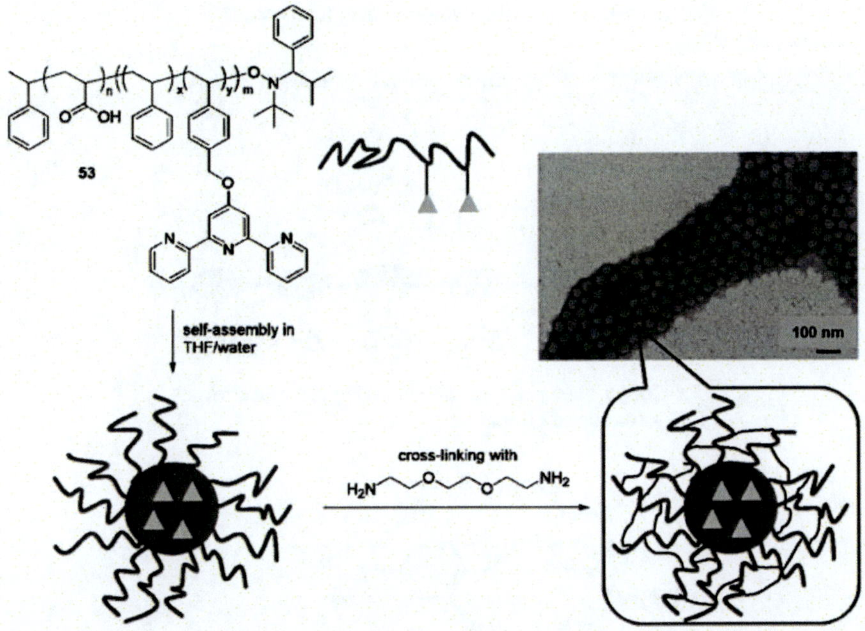

Fig. 2.41 Schematic representation of the synthesis of tpy-functionalized nanoparticles from amphiphilic the block copolymer **53** [308] Figure reproduced with kind permission; © 2008 American Chemical Society

pending binding sites was investigated; in many cases, fluorescence quenching or ionochromism was observed, even at low added ion concentrations. Thus, these materials might be considered as sensitive macromolecular sensors for metal ions, making use of the "turn off" fluorescence behavior (Fig. 2.42a) [310]. One representative example is the conjugated polymer **54** as prepared by Kimura et al. by a Wittig-type polycondensation (M_w = 4,000 g mol^{-1}; Fig. 2.42b) [311]. A variety of divalent main-group and transition metal ions, such as Ru^{II}, induced an efficient quenching of the fluorescence upon addition to **54** (presumably due to inter- and intrachain EnT and ElT pathways).

Also for the second case—the tpy moieties are attached to a π-conjugated backbone via non-conjugated, flexible spacers—examples can be found in the literature [312, 313]. Chan et al. utilized a Suzuki cross-coupling approach to synthesize a poly(dialkoxyphenylene-*alt*-thiophene) and poly(dialkylfluorene-*alt*-thiophene) with pending tpy moieties (**55** and **56**, respectively; M_n of *ca.* 1,400 g mol^{-1}) [313]. Targeting photovoltaic applications, the incorporated tpy moieties were subsequently used for the preparation of $[Ru(tpy)(NCS)_3]^-$ complexes (Fig. 2.43). Thus, the final metallopolymers can be considered as hybrid-type materials combining the active species of dye-sensitized solar cells (DSSCs, see dye

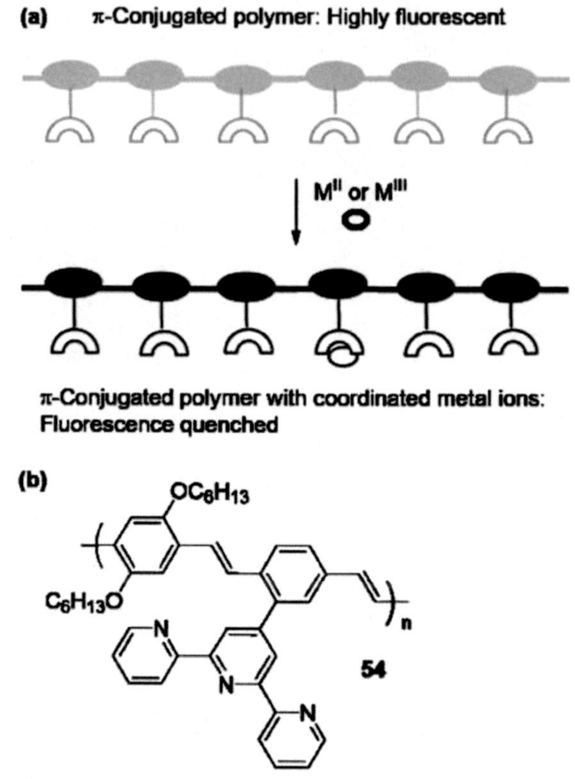

Fig. 2.42 Schematic representation of (**a**) the metal-induced fluorescence quenching in π-conjugated polymers and (**b**) of a representative polymer (**54**) used in this respect

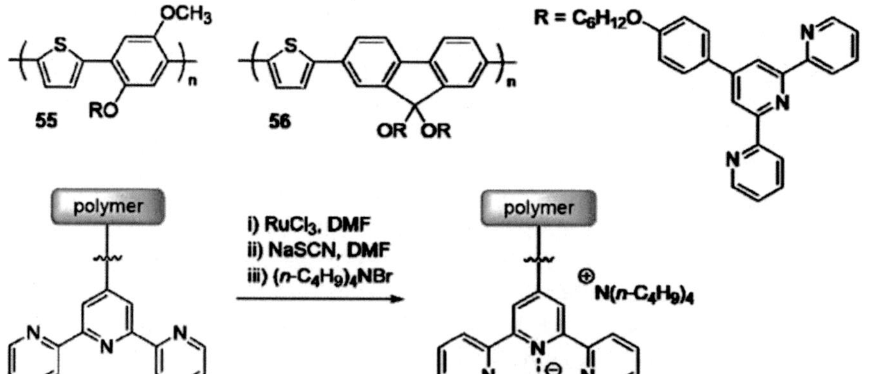

Fig. 2.43 Schematic representation of the synthesis of π-conjugated polymers with pending [Ru(tpy)(NCS)$_3$]$^-$ complexes

III in Fig. 1.2) with those of bulk heterojunction solar cells (BHJ; a conjugated polymer that is typically blended with a fullerene derivative, as acceptor). Upon illumination with simulated AM1.5 solar light, the fabricated photovoltaic cells gave short circuit currents, open circuit voltages, and power conversion efficiencies of 1.53 to 2.58 mA cm^{-2}, 0.12 to 0.24 V, and 0.084 to 0.12%, respectively. In all these experiments, the device configuration was ITO|polymer/C_{60}|Al (ITO: indium tin oxide).

One important application of metal-to-ligand coordination in the field of polymer science is the ability to assemble thermodynamically and kinetically stable metallo-supramolecular block copolymer architectures [314, 315]. Basically, this concept aims to combine the characteristic features of block copolymers (e.g., phase separation due to the immiscibility of the constituent blocks) with the intrinsic properties of metal complexes (i.e., reversibility of bonding as well as photophysical, electrochemical or magnetic properties). As recently pointed out, [316] the charged character of the incorporated metal complexes might have considerable impact on the physical properties which might, thus, significantly differ when comparing metallo-supramolecular block copolymers with their covalent analogues. As a consequence, the metal complexes in such materials not only represent simple metallo-supramolecular linkages, but might even be regarded as very short block/segment themselves. Such block copolymer architectures stand in one line with the range of metal-containing or hybrid materials for which Chujo coined the term "polymeric materials containing element blocks" [317]. At closer inspection, some prerequisites can be identified which exclude many types of metal–ligand combinations from being suited for the formation of block copolymer architectures of the general formula A-[M]-B [318]. Firstly, the constituent blocks need to be arranged linearly and coordination isomers cannot be formed. Secondly, the coordination of macromolecular components should give stable heteropolymeric assemblies; whereas, the two possible homopolymeric species, which are A–[M]–A and B–[M]–B, should not be obtained under these reaction conditions (Fig. 2.44). This approach of forming metallo-supramolecular block copolymers by complexation of macromolecular subunits represents a versatile alternative to the preparation of such materials by attaching the constituent blocks to the orthogonal

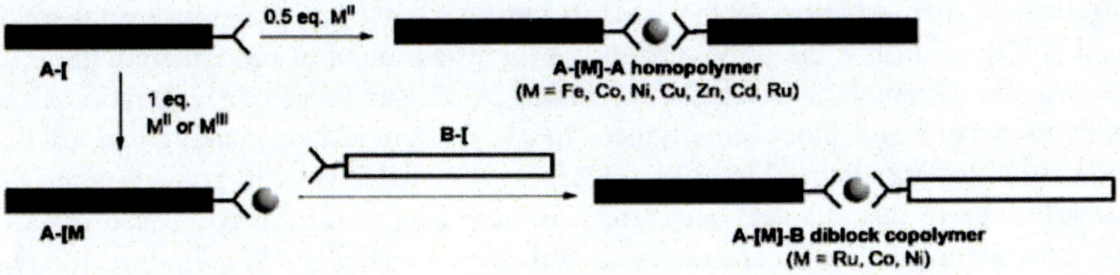

Fig. 2.44 Schematic representation of the synthesis of metallo-supramolecular homopolymers and diblock copolymers. Figure redrawn according to the literature; © 2016 The Royal Chemical Society

functionalities present in a small-molecule complex {e.g., the $[Ru(dqp)_2]^{2+}$-centered diblock copolymers, as depicted in Fig. 2.9, was prepared this way}.

The synthesis of metallo-supramolecular block copolymers typically proceeds via selective, stepwise complexation methods, which have been adapted from the established coordination chemistry of small-molecules counterparts. However, this general concept applies only to a very limited number of metal–ligand combinations and has, to the best of our knowledge, only been realized for *bis*-tridentate polypyridyl complexes (e.g., with tpy ligands). Concerning the formation of stable heteroleptic [(A-tpy)M(tpy-B)] complexes, only a handful of transition metal ions are applicable: Ru^{II}, Os^{II}, Ir^{III}, Co^{III}, Ni^{II} and, in very special cases, Zn^{II} (A-tpy and B-tpy denote differently substituted tpy ligands) [289, 316]. From these, the Ru^{II} ions are, by far, the most commonly utilized centers when aiming for the assembly of kinetically stable metallo-supramolecular copolymers [319–347].

The strategy outlined in Fig. 2.44 can only be employed when suitable macroligands, equipped with a tpy moiety at one chain end, are provided. Such macroligands can be synthesized via two routes: Usage of a ligand-containing initiator for the polymerization or post-polymerization functionalization of a pre-formed polymer. For these strategies many examples can be found in literature and have extensively been summarized elsewhere [289]. At this point, the relevant ligand-containing initiators will only be introduced briefly. Besides the various OH-functionalized ligands that have been utilized to initiate the Sn^{II}-catalyzed ROP of cyclic esters (i.e., ε-caprolactone, L-lactide), [348–350] the initiators for the family of CRP reactions are worth mentioning—the derivatives used for the polymerization of diverse types of monomers under ATRP (**57**) [351] or RAFT polymerization conditions (**58** and **59**) are summarized in Fig. 2.45 [352, 353]. As one example, O'Reilly et al. employed a *bis*-functionalized btp initiator for the RAFT polymerization of styrene, methyl methacrylate, and *tert*-butyl acrylate, allowing the simultaneous growth of two polymer chains; the coordination of the resultant macroligand **62** to Ru^{II} ions gave four-armed, star-shaped, metallopolymers [btp: 2,6-*bis*(1*H*-1,2,3-trizol-4-yl)pyridine; Fig. 2.46] [354]. In 1999, Hawker et al. introduced TIPNO as an universal initiator for the NMP of diverse monomers (TIPNO: 2,2,5-trimethyl-4-phenyl-3-azahexane nitroxide) [355]. Schubert et al. modified this initiator by grafting a tpy ligand onto chloromethyl-functionalized TIPNO; [356] this particular derivative (**60**, Fig. 2.45) represented the first efficient ligand-equipped initiator for the NMP featuring a high functional-group tolerance and ability to initiate the polymerization of a broad range of monomers with good control over the molecular weight. Accordingly, a broad range of tpy-functionalized homopolymers and block copolymers has been synthesized making use of this powerful NMP initiator [336, 338, 342, 344, 357–361]. Though going beyond the actual scope of this chapter, one further development of the TIPNO-based initiator **60** is noteworthy: The heterotelechelic derivative **61** (Fig. 2.45) combined the two most common orthogonal supramolecular entities, i.e., a tpy ligand for metal-ion complexation and a self-complementary quadruple *H*-bonding array; Schubert et al. employed this initiator for the NMP of styrenes [362, 363].

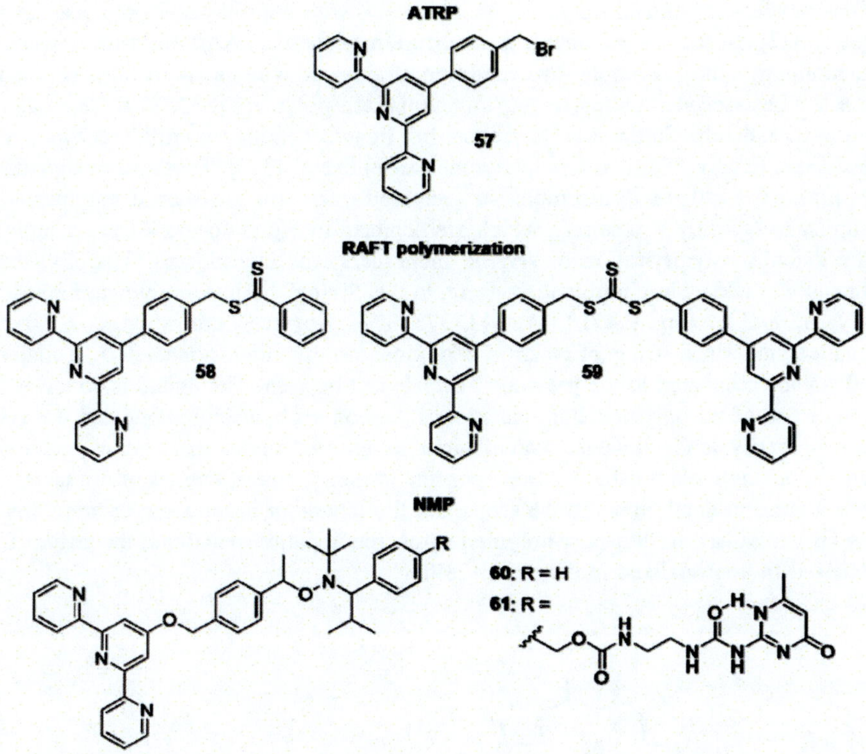

Fig. 2.45 Schematic representation of the various terpyridine-containing initiators for CRPs

As an alternative to using ligand-equipped initiators, metal-binding sites can also be tethered to the termini of pre-formed polymers. In this respect, a highly versatile method was elaborated by Schubert et al., who reacted 4′-chloro-tpy with OH-functionalized polymers, i.e., basically performing an aromatic nucleophilic substitution under basic conditions [289]. A literature survey shows that those polymers having a robust, non-reactive backbone can thusly be functionalized. Owing to the special properties of the respective polymers (e.g., solubility, reactivity, etc.), the reaction conditions still need to be adjusted in each case. Accordingly, the following types of polymers could be functionalized: PEG [322, 325, 364–367], PS [322, 368], PVPs [323], poly(ethylene-*co*-butylene) (PEB) [322], poly(ex-tetrahydrofuran) [369], poly(2-oxazoline)s [369], poly (dimethylsiloxane) (PDMS) [346] and, finally, Pluronics™ [370]. In the case of more sensitive polymers, milder reaction conditions need to be employed in order to avoid degradation of the polymer backbone. Such OH-terminated polymers can, for instance, be functionalized further by reaction with an isocyanate-substituted ligand [331]. Other mild coupling methods rely on the esterification or amidation of

OH- or NH_2-terminated polymers, respectively, with appropriate ligand reactants [371, 372]. In the same context, the utilization of the CuCAAC reaction is worth mentioning—the 1,3-dipolar cycloaddition of an azide with an alkyne also enabled the synthesis of various tpy- or btp-functionalized polymers [373, 374]. The latter allowed Kakuchi, Satoh et al. to prepare btp ligands bearing two different types of polymers [e.g., a PCL and a poly(styrene oxide) chain] [373]. These macroligands, in turn, were utilized to assemble Ru^{II}-centered miktoarm star-shaped polymers—similar to O'Reilly's materials, which are depicted in Fig. 2.46. As a last example, the anionic polymerization of styrene monomers was investigated. Thereby, the ligand was added, as a terminating agent, to the "living" lithiated polymer chains to give **63** and **64** (Fig. 2.47) [334, 343, 375–377]. However, this strategy is rather challenging, since stringent reaction conditions are required; nonetheless, a major advantage compared to the previous methods can be seen. The ligand is connected to one end of the polymer chain via a robust carbon–carbon single bond and not via a supposedly labile linkage, which might be cleaved under the forcing reaction conditions required for the Ru^{II} ion coordination (e.g., ethers, amines or urethanes). For a more in-depth view on the scope and limitations of introducing terminal tpy, as an exemplar, via the post-polymerization modification reactions, the reader is referred to a specialized monograph [289].

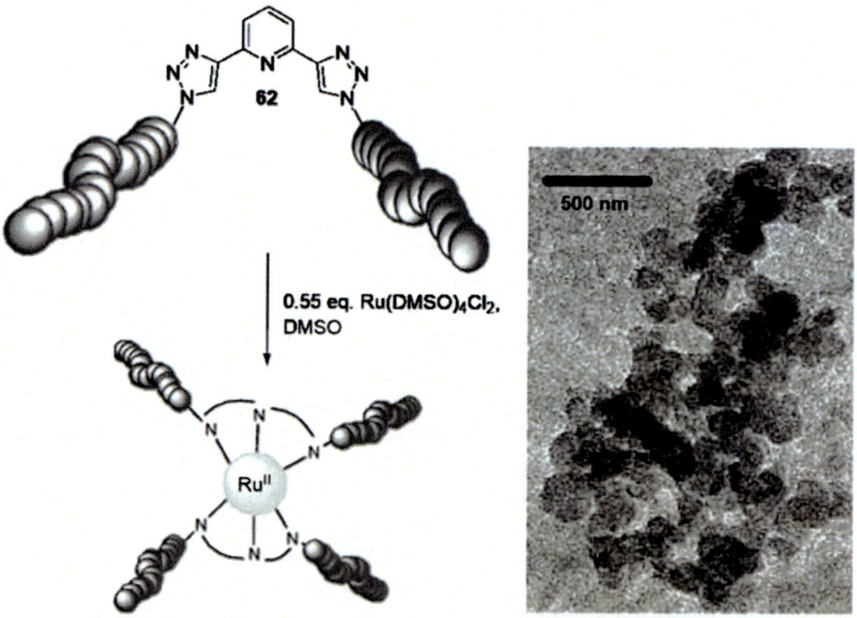

Fig. 2.46 Schematic representation of the formation of four-armed metallostars according to O'Reilly et al.; a representative TEM image of the metallostars is also shown [354]. Figure reprinted with kind permission, © 2010 The Royal Chemical Society

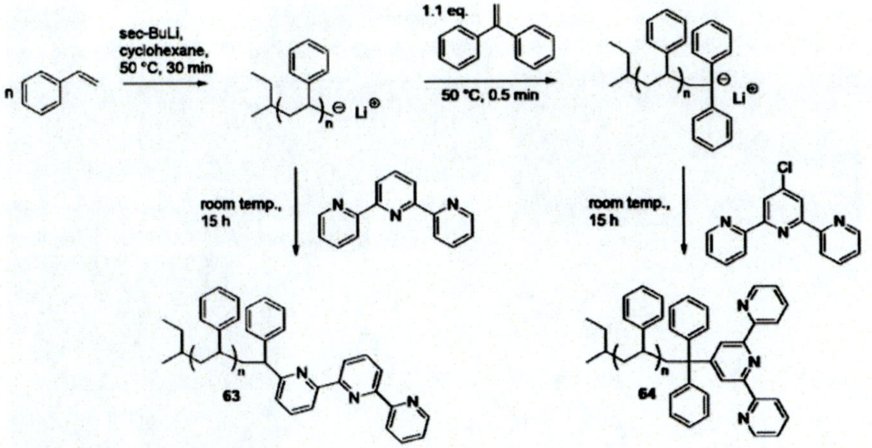

Fig. 2.47 Schematic representation of the synthesis of the tpy-functionalized PS-macroligands **63** and **64** by anionic polymerization

From a broad range of polymers, end-functionalized with a tpy ligand, the PEG monomethyl ethers (MPEGs) represent the most common system used for the complexation of transition-metal ions. This remarkably interest results from the special characteristics of the PEG chain (i.e., hydrophilicity, biocompatibility, softness, etc.). The formation of A–[Ru^{II}]–A homopolymers is well-documented in the literature; however, no potential application of these metallopolymers has yet been suggested. Still, the A–[Ru^{II}]–A homopolymers serve as multi-purpose models and reference materials to which more advanced metallo-supramolecular architectures are often compared [345, 347, 352, 378]. Studies devoted to their pH-sensitivity [325] as well as their characterization by mass spectrometry [379], SEC [328], capillary zone electrophoresis [380] or analytical ultracentrifugation [340] are worth mentioning at this stage.

In accordance with the general procedure (Fig. 2.44), Schubert, Gohy et al. prepared a plethora of block copolymer architectures by linking different polymers, end-functionalized with a tpy ligand, by directed Ru^{II} complexation. As reviewed by Gohy and Fustin, the phase separation of the incompatible polymer segments either in solution (i.e., micellization in a selective solvent) or in the solid state (i.e., microphase separation) represented the center of this research [314, 315]. As one example, MPEG-[Ru]-PS diblock copolymers (**65**) were intensively studied in this respect (Fig. 2.48). In an aqueous environment, **65** assembled into spherical micelles in which the hydrophobic PS chains formed the core, surrounded by a corona of the water-soluble PEG chains (Fig. 2.48a); in such micelles, the cationic Ru^{II} complexes were located on the core-corona interface [321, 330, 333, 339, 381]. In the solid or molten state, the aggregation/segregation behavior of **65** depended not only on the dimensions of the constituent blocks (Fig. 2.48b), [329, 332] but

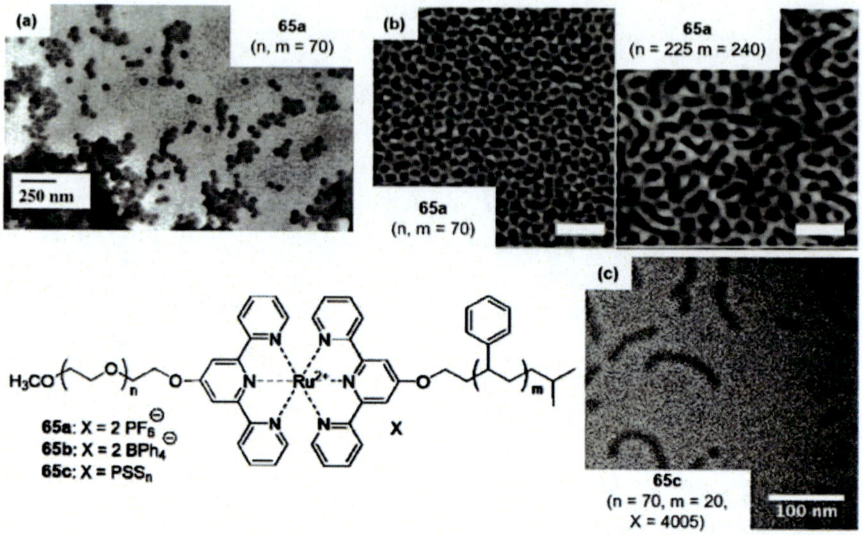

Fig. 2.48 Schematic representation of the metallo-supramolecular diblock copolymer **65**. **a** TEM image of the micelles formed by **65a** in aqueous medium [321]; **b** AFM phase images of thin films obtained from spin-casting **65a** from toluene (scale bar: 100 nm) [332]; **c** Cryo-TEM image of **65c** in aqueous medium [452]. Figures reprinted with kind permission; © 2002 American Chemical Society, 2004 The Royal Chemical Society and 2009 The Royal Chemical Society, respectively

also on the nature of the counterion: For small counterions, spherical micelles were obtained; whereas, worm-like assemblies were formed in the presence of polyanions, such as poly(styrene sulfonate) (PSS, Fig. 2.48c) [335, 382]. Gohy, Schubert et al. demonstrated that the virtually stable $[Ru(tpy)_2]^{2+}$ linkage can be cleaved oxidatively with Ce^{IV} at low pH values [337, 383]. Applying a two-step protocol, nanoporous films were fabricated: The MPEG-[Ru]-PS diblock copolymer was first processed into the solid state where phase separation gave well-ordered cylindrical domains; subsequently, the water-soluble MPEG-tpy block were removed oxidatively.

In the same context, also a variety of other Ru^{II}-containing diblock copolymers were studied: MPEG-[Ru]-PDMS, [346] MPEG-[Ru]-PEB, [319, 326] MPEG-[Ru]-PFDMS, [331] PNIPAM-[Ru]-PS [352] or PS-[Ru]-PAA [357] [PFDMS: poly(ferrocenyldimethylsilane), PAA: poly(acrylic acid)]. So-called multicompartment micelles might be obtained when amphiphilic ABC triblock copolymers assemble in an appropriate selective solvent [384, 385]. Gohy, Schubert et al. applied this concept for Ru^{II}-containing block copolymer architectures, such as the triblock copolymers PS-*b*-P2VP-[Ru]-MPEG and PS-*b*-PTFMS-[Ru]-MPEG [PTFMS: poly(4-trifluoromethylstyrene)]; [323, 344] further, these same authors also studied the self-assembly behavior of the tetrablock copolymer **66** (Fig. 2.49) [342].

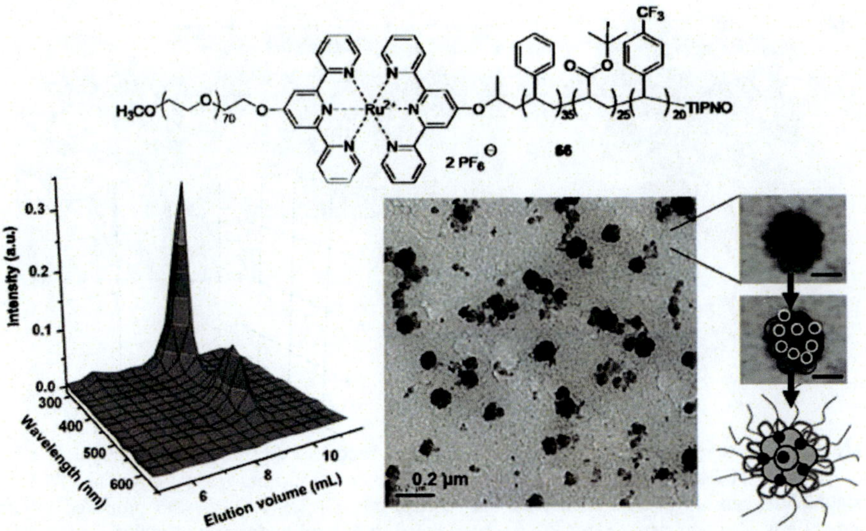

Fig. 2.49 Schematic representation of the metallo-supramolecular tetrablock copolymer **66**. The 3D SEC trace, using a photodiode array (PDA) detector, is also depicted. The TEM image of **66** in *iso*-propanol confirmed the formation of micelles; from the magnified image (scale bar: 50 nm) the structure of the micelles was corroborated (schematic representation of the micelle: PS cores, as grey spheres; PEG chains, as thin black lines; P^{t}BA chains, as thick grey lines; and PTFMS chains, as thick black lines) [342]. Figure reprinted with kind permission, © 2009 The Royal Chemical Society

A special type of biohybrid material, which has to be mentioned here, are proteins (or enzymes) to which a hydrophobic synthetic polymer has been linked; such conjugates are commonly referred to as "giant amphiphiles" or "giant surfactants" (Fig. 2.50) [386–388]. In addition to all the covalent systems, the approach as reported by Velonia et al. is of relevance in the context of Ru^{II}-containing polymers [389, 390]. Basically, the same synthetic protocol as established for the aforementioned metal-supramolecular block copolymers was applied to conjugate a PS block with either bovine serum albumin (BSA) or lipase-A from *Candida antarctica* (CALB). For this purpose, the macromolecular building blocks were equipped with a tpy ligand and assembled in a stepwise fashion (according to Fig. 2.44). The aggregation of metallo-supramolecular amphiphiles was investigated by TEM imaging and significant differences were observed when comparing the obtained superstructures to those formed by the corresponding "covalent" amphiphiles (Fig. 2.50b). It was later shown by Thordarson et al., that also cytochrome-c could be modified with tpy [391] or $[Ru(tpy)_2]^{2+}$ complexes [392, 393]. These bioconjugates were obtained in high purity and could potentially be utilized in bioelectronic devices, where redox-active protein-based bioconjugates are known to play a key role.

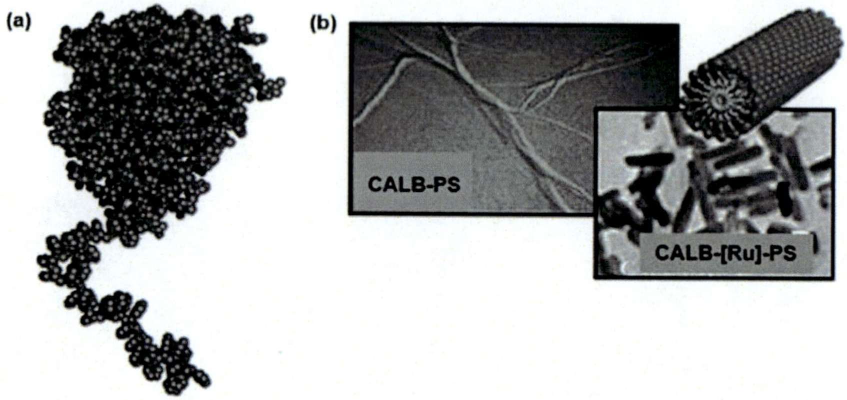

Fig. 2.50 **a** Representation of a computer-generated model of a giant lipase-PS amphiphile (PS block: M_n = 40,000 g mol^{-1}, molecular volume: 25 nm^3); **b** TEM images of a covalent CALB-PS amphiphile and its metallo-supramolecular counterpart CALB-[Ru]-PS (the proposed tubular structure is also depicted) [386]. Figure reproduced with kind permission; © 2006 Springer Int. Publ.

Oligonuclear assemblies represent one further type of block copolymer architectures, which have been formed by selective Ru^{II} complexation of appropriate macromolecular building blocks. While the aforementioned A-[Ru]-A homopolymers or A-[Ru]-B diblock copolymers as well as the multinuclear chain-extended polymers, as detailed in Sects. 2.2.3, have been the subject of intense research, their oligonuclear relatives, such as the dinuclear triblock copolymers of the general formula A-[Ru]-B-[Ru]-A, have up-to-now attracted significantly less attention. Schubert et al. synthesized a small library of such assemblies utilizing the rigid-linear ligands **67**, as the central segment. Such ligands represent typical monomers for the synthesis of coordination polymers via the "polymerization by complexation" approach (see Sect. 2.2.3) [3, 5]. For the preparation of the dinuclear assemblies basically the same method was applied as for the related diblock species, using MPEG-tpy, as the flexible, outer block. Accordingly, the triblock copolymers comprising of a coil-rod-coil topology were formed in a two-step synthesis using $Ru(DMSO)_4Cl_2$, as the reactive precursor complex (Fig. 2.51) [345, 347]. Similarly, the synthesis of star-block copolymer architectures was realized starting from analogous three- or four-armed rigid ligands [378]. In an aqueous environment, all these materials gave nanosized assemblies, which were analyzed by TEM imaging and DLS measurements. However, these authors could not yet clarify how the molecular structure of the central segments governs the size and shape of the resultant nanostructures.

To the end of this Section, some ligands attached to polymers, other than the common polypyridyl-type ones, have to be mentioned. In order to avoid the dissociation of Ru^{II}-salen complexes from the polymer, as observed when the complex

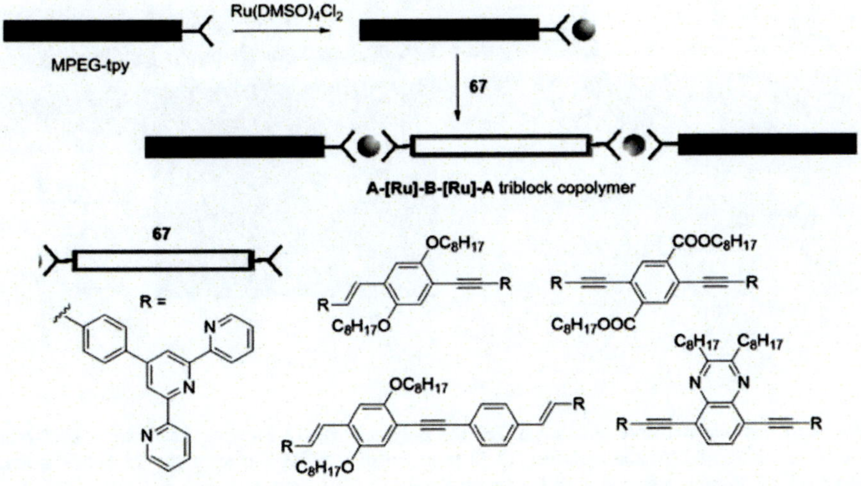

Fig. 2.51 Schematic representation of the synthesis of ABA-type triblock copolymers

is bound to P4VP via Ru^{II}-pyridine coordination, [145] Jones et al. utilized styrene-functionalized salen ligands for the copolymerization with DVB under FRP conditions [394]. The obtained cross-linked polymers were subsequently loaded with Ru^{II} ions to give immobilized $[Ru(salen)(py)_2]$ complexes (Fig. 2.52a). The highly active, selective and recyclable heterogeneous catalysts **68** were employed in the asymmetric cyclopropanation of terminal olefins; the catalytic activity of the material could be restored by treatment with pyridine prior to a further usage. A similar synthetic strategy was followed by Iwasa et al., who investigated a

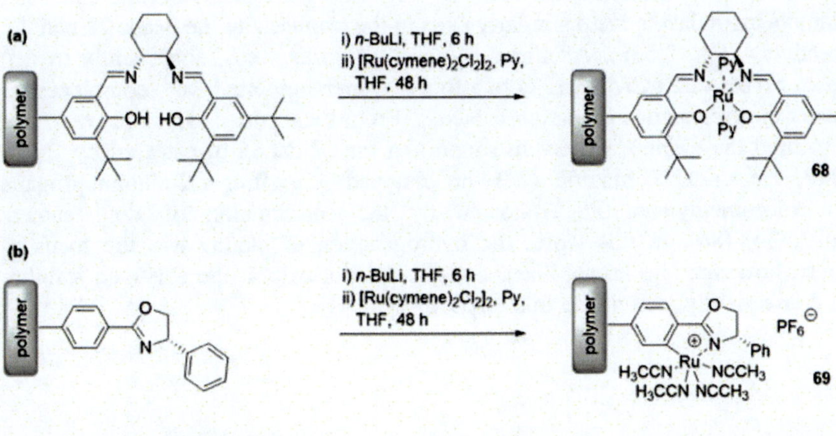

Fig. 2.52 Schematic representation of the synthesis of polymer-bound Ru^{II}-salen (**68**) and Ru^{II}-phenyloxazoline complexes (**69**)

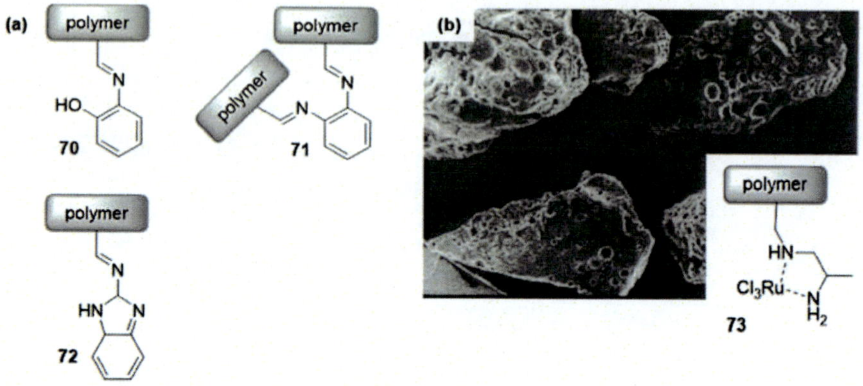

Fig. 2.53 Schematic representation of (**a**) the polymers **70–72** functionalized with Schiff-type bases and of (**b**) the ruthenopolymer **73** (a representative TEM image of the **73** beads is also shown) [86]. Figure reprinted with kind permission; © 1997 Taylor & Francis Group

cross-linked polymer network with pending chiral phenyloxazoline moieties [395]. These sites acted as C^N-type ligands to bind Ru^{II} centers (Fig. 2.52b). These macroporous metallopolymers **69** were also utilized, as heterogeneous catalysts, in cyclopropanation reactions. Noteworthy, the immobilization of the former two well-know Ru^{II} catalysts onto polymers represents a less common strategy—typically, such catalysts are anchored onto silica; however, the formation, characterization and application of silica-supported materials goes beyond the scope of this book.

The structurally more simple Schiff-type bases were derived from formyl-functionalized cross-linked PS and 2-aminophenol [396]. Besides the preparation of functionalized resins **70**, also 1,2-phenylenediamine and 2-aminobenzimidazole were similarly used in the synthesis of the beads **71** and **72**, respectively (Fig. 2.53a). All these polymer networks were subsequently treated with an excess of $RuCl_3 \times xH_2O$ to afford polymer-supported Ru^{III} complexes, as potential heterogeneous oxidation catalysts. From these, the catalyst derived from **70** featured the highest activity in supporting the oxidation of catechol by H_2O_2. Earlier, a less defined material (**73**) was obtained by grafting 1,2-diaminopropane onto chloromethylated PS, followed by the coordination of Ru^{III} centers (Fig. 2.53b) [86]. In this work, the hydrogenation of olefins was the focus of interest; however, the immobilized catalyst did not exhibit the expected stability and, consequently, activity in this respect.

2.2.3 Main-Chain Ruthenopolymers, Classified as Metallo-Supramolecular Materials

The formation of ruthenopolymers via the (co)polymerization by the coordination of Ru^{II} ions represents an important approach and, accordingly, numerous examples can be found in literature. When aiming for a selective preparation of linear metallo-supramolecular assemblies via a bidirectional chain-growth mechanism, often referred to as "supramolecular polymerization", [397] the utilized building blocks have to meet some structural requirements: Firstly, the metal center has to offer two vacancies in its coordination sphere, ideally on opposing sides (e.g., the *trans*-positions in square-planar or octahedral complexes). The organic building blocks, in turn, must possess two appropriate binding sites for metal ions (such ligands are referred to as ditopic or telechelic ligands). Last but not least, the binding strength, i.e., the affinity of the ligand to coordinate to a metal ion, needs to be considered. This parameter is of utmost importance, since a sufficiently high thermodynamic driving force, as expressed by a high binding strength, is required to drive the polymerization towards high DP values. A classification of a broad range of designated organic ligands can be conducted according to Fig. 1.7. Thus, the supramolecular polymerization might proceed via coordinative bonding, ionic/covalent bonding or arene-π complexation [398].

As elaborated in diverse reviews [318, 397, 399, 400], three fundamental mechanistic pathways have been proposed in order to describe supramolecular polymerizations: Isodesmic supramolecular polymerization (IDP), ring-chain-mediated supramolecular polymerization, and (anti-)cooperative supramolecular polymerization. Independent from the particular pathway the process is following, some common and general aspects need to be addressed here. Though some mechanistic parallels can be deduced when analyzing the formation of "covalent" polymers via polyaddition- or polycondensation-type reactions and supramolecular polymers, one striking difference has to be named is the reversibility-of-the-polymerization. Thus, depolymerization has generally been considered when dealing with "non-covalent" polymers. In other words, supramolecular polymers in solution might be sensitive towards changes of the environment (e.g., changes in temperature, concentration, pH value, etc.), possibly resulting in a decrease of the DP due to dissociation of the supramolecular bonds. For all types of polymers that have been prepared by polyaddition/polycondensation reactions, the DP, i.e., the molecular weight, can be calculated from the monomer conversion (p). This relationship is the well-known Carothers' equation (Eq. 2.1) [401]

$$\mathrm{DP} = 1/(1-\mathrm{p}) \tag{2.1}$$

The DP of any supramolecular polymer depends, at least to some extent, on the strength of the non-covalent interactions which, in turn, can be quantified by the association constant (K_a). As shown in Fig. 2.54, supramolecular polymers of high molecular weight can only be obtained if (a) high K_a values are given and (b) high

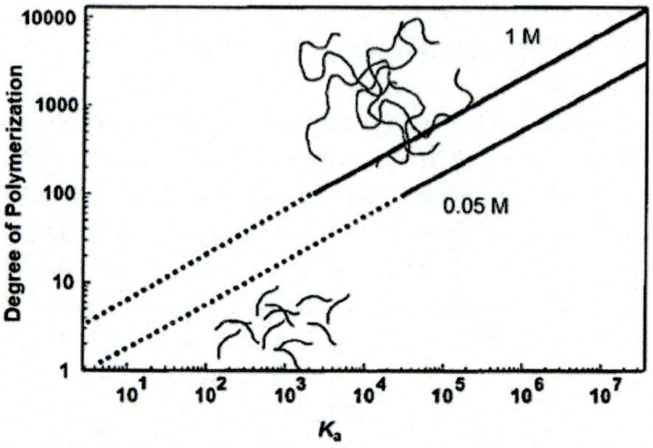

Fig. 2.54 Theoretical plot of the DP *versus* the association constant (K_A in M^{-1}) for an isodesmic supramolecular polymerization at two different concentrations [397]. Figure reproduced with kind permission; © 2001 American Chemical Society

monomer concentrations are provided. However, the influence of other parameters on the polymerization, such as temperature and applied reaction conditions, still have to be considered. From the wide range of metal ions that are basically applicable in metallo-supramolecular polymerizations, the Ru^{II} ions appear to be particularly suited for this purpose, since the Ru^{II} ion typically shows a high binding affinity with diverse organic ligands. Admittedly, the robustness of this interaction is accompanied with rather harsh reaction conditions being required to establish these environments. Therefore, the formation of Ru^{II}-containing metallo-supramolecular polymers via the complexation of monomer ligands is, in practice, much more challenging compared to most other metal ions, since the limitations arising from available synthetic methods need to be overcome. This holds particularly true when aiming for polymers with high molecular weights.

The reliable determination of the molecular weight represents a critical issue, which always needs to be addressed when dealing with metallo-supramolecular polymers. As detailed elsewhere [402], the analytical methods that are commonly employed for characterizing "classical" macromolecules (e.g., SEC and mass spectrometry) are demanding to be applied due to the kinetically labile nature of the non-covalent interactions. However, several spectroscopic techniques (e.g., NMR or UV/vis absorption) as well as calorimetry, viscometry or analytical ultracentrifugation (AUC) can well be employed to investigated the supra-macromolecular structures [403]. In the context of supramolecular ruthenopolymers, SEC still represents an appropriate tool to estimate the molecular weight—dissociation of the polymer chains under the measurement conditions is, due to the high kinetic stability, typically not observed [364]. NMR spectroscopy on the active ^{99}Ru nucleus ($I = 5/2$, abundancy of 12.72%, chemical shift range of 9,000 ppm) represents a useful—though scarcely employed—technique for the study of small variations around the metal center and has, for example, been used to identify geometrical isomerism in mono- and oligonuclear Ru^{II} complexes [404, 405]. However, high sample concentrations are required to overcome the low resonance frequencies and to level out the acoustic waves generated in the spectrometer probe. With the high-performance spectrometers available today, ^{99}Ru NMR spectroscopy is worth being revisited and tested for the characterization, in particular of Ru^{II}-containing supramolecular polymers.

2.2.3.1 Main-Chain Ruthenopolymers, Formed Via Coordinative Bonding

Polymers of this type can, in principle, be assembled from the interactions between electron-poor metal centers and telechelic organic ligands possessing neutral σ-donor moieties: Amines, carbenes, ethers, imines, ketones, nitriles, phosphines, pyridines, etc. [398]. Since these groups differ in their donor strength, the stability of the resulting metal-to-ligand coordinative bonds will vary over a broad spectrum. As one extreme, very labile metallopolymers will already depolymerize in the presence of polar solvents (e.g., DMSO, MeCN, or acetone). In contrast, very stable metallopolymers will not exhibit any considerable change in their macromolecular structure—even under the influence of an external stimulus (e.g., temperature, light, ultrasound, pH value, etc.). In general, the robustness of a metal-to-ligand interaction can be increased in two ways: Utilizing donor ligands with high affinities towards metal centers and chelation (multidentate ligands enable a much stronger binding than monodentate ones).

Since a rather weak binding to neutral organic monodentates has to be expected, even for $Ru^{II/III}$ ions, the stability of metallopolymers, based on such interactions, will intrinsically be low—at least in solution. However, a variety of 1D/2D coordination homo-/heterometallic polymers, in which the metal centers are bridged via a CN^- ligand, has been reported. Such materials are of particular interest due to their electronically and/or magnetically coupled metal centers. The example depicted in Fig. 2.55a, containing Ru^{III} and Ni^{II} ions in a strictly alternating order, showed a low-temperature ferromagnetic coupling between the metal centers (however, a long-range magnetic order was absent). [406] Representing solid-state materials only, the synthesis and application of such coordination polymers will not be detailed further; a specialized review in this field was published by Ohba and Okawa some time ago [407]. A different type of coordination mode was reported by Cowie, Singleton et al. who prepared the linear coordination polymers **74** from the interaction of the dimeric *μ*-acetate-bridged Ru^I precursor complex $[Ru(CO)_2(O_2CCH_3)(NCCH_3)]_2$ with various diphosphine or disulfide ligands (Fig. 2.55b) [408]. The polymeric nature of the barely soluble assemblies was concluded from IR measurements as well as X-ray structure analyses [409]. More recently, Clark and Burzlaff reported the formation of 1D coordination polymers comprising dinuclear $[M_2(OAc)_4]$ fragments and bridging *bis*-imidazole ligands (M = Cu, Rh, Ru, Cr, or Mo) [410]. Density-functional theory (DFT) calculations were conducted to estimate the strength of the M–M bonds. The K_a value for the interaction with the bridging ligand decreased with increasing metal-to-metal interaction; the Ru–Ru bond had a

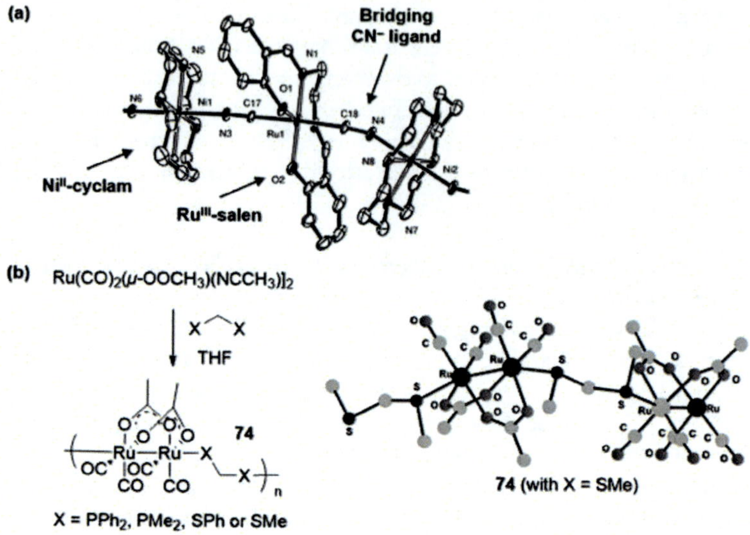

Fig. 2.55 **a** Solid-state structure of a supramolecular polymer comprising alternating Ru^{III} and Ni^{II} centers, bridged by CN^- ligands (counterions are omitted for clarity) [413]. **b** Schematic representation of the synthesis of the Ru^I-containing coordination polymers **74** (the solid-state assembly with X = SMe is also shown) [409]. Figure reproduced with kind permission; © 2006 Elsevier B.V. and 1990 American Chemical Society

strong double-bond character (effective bond order of 1.7) and, thus, the coordination polymer was rather labile. Later, coordination polymers similar to **74** were assembled in which a range of ditopic *N*-heterocycles was employed. For these, effective bond orders in range of 0.65 to 0.7 were found. [411] For a more in-depth overview on coordination polymers incorporating metal–carbonyl clusters, the reader is referred to Shieh's recent review [412].

One further polymer, possessing a carbido-carbonyl cluster in the backbone, was assembled from $[Ru_6C(CO)_{17}]$ and the ditopic $Ph_2PC{\equiv}CPPh_2$ ligand in refluxing THF [414–416]. The resulting metallopolymer $\{[Ru_6C(CO)_{15}]Ph_2PC{\equiv}CPPh_2\}_n$ was soluble in CH_2Cl_2 and was characterized by ^{31}P NMR and IR spectroscopy, elemental analysis and TEM imaging; the latter suggested a DP of up to 1000. Upon treatment with an electron beam, the metallopolymer was converted into a nanoparticle chain and, finally, Ru nanowires with a width between 100 and 300 nm were obtained (Fig. 2.56) [414].

The formation of metallo-supramolecular polymers incorporating planar macrocyclic building blocks (MACs), such as metalloporphyrins or metallophthalocyanines, represents a relatively common approach in metallo-supramolecular chemistry. Additional ligands are coordinated to the transition-metal ion in the center of the organic framework; thereby, these ligands occupy the vacant axial positions within the octahedral coordination sphere of the metal ions. On the basis

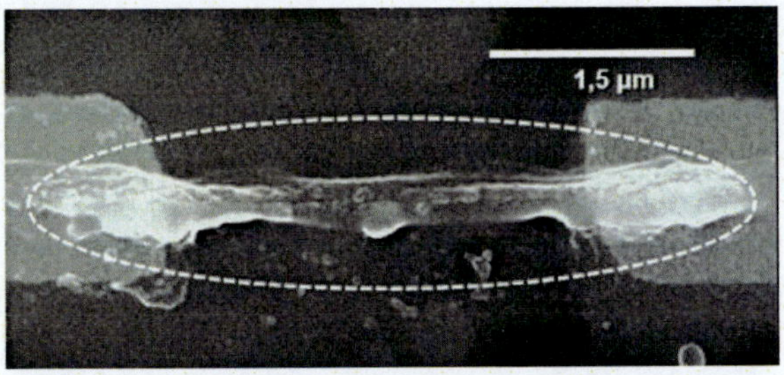

Fig. 2.56 SEM image of a Ru nanowire deposited onto two Au contacts [414]. Figure reproduced with kind permission; © 2000 The Royal Chemical Society

of this particular interaction, three different metallopolymer architectures can—according to Abd-El-Aziz et al.—be distinguished (Fig. 2.56) [417].

However, for the type-B and type-C assemblies (Fig. 2.57), no examples incorporating Ru^{II} centers have yet been reported. In contrast to this, the linear type-A assembly of structurally rather simple metalloporphyrins in the so-called Shish-Kebab fashion (Fig. 2.57) via telechelic monodentate anionic (e.g., cyanides, thiocyanates or acetylides) or neutral ligands (e.g., pyrazine or 4,4′-bpy) attracted considerable interest already in late 1970s. In order to assemble supramolecular conductive wires, those MACs were polymerized with various short bridging ligands. Thereby, the electronic interaction between the stacked MACs could be modulated via the nature of the bridging ligand: Not only the metal-to-metal distance but also the chemical structure of the ligand was found to be of significance;

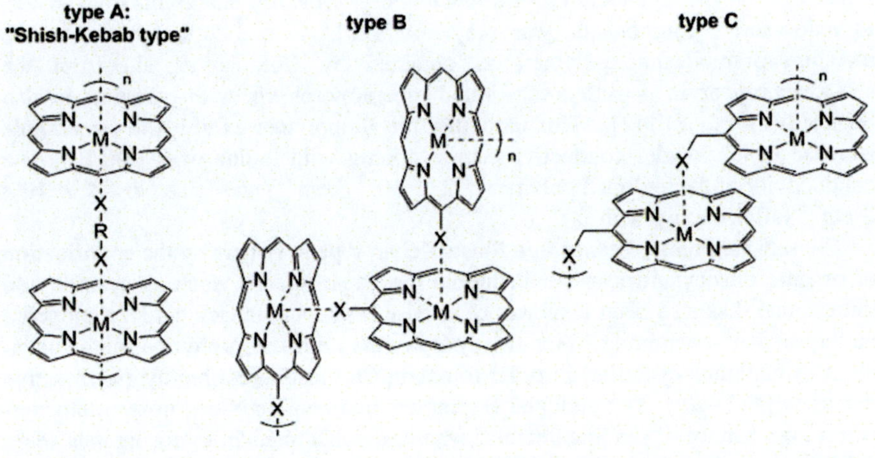

Fig. 2.57 Schematic representation of the three general types of metallopolymers incorporating centers of metalloporphyrins

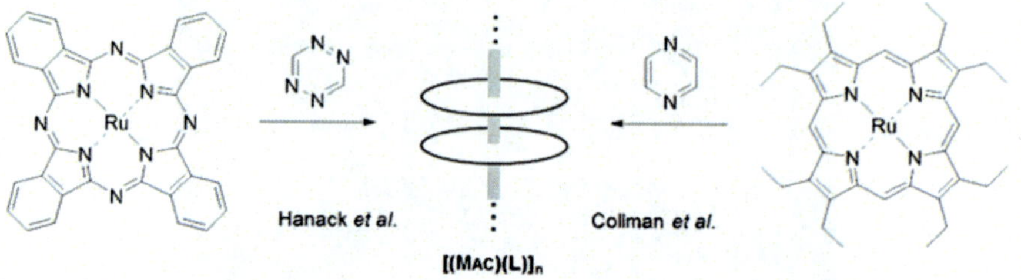

Fig. 2.58 Schematic representation of the formation of Shish-Kebab-type metallo-supramolecular polymers from Ru^{II}-centered MACs

for example, neutral *N*-heteroaromatic ligands interact with the metal centers in a coordinative fashion; whereas, the acetylide dianion $[C{\equiv}C]^{2-}$ establishes two σ-bonds and, as a consequence, a much stronger electronic coupling of the metal centers can be achieved. In particular; Hanack, a pioneer in the field of porphyrin and phthalocyanine chemistry, contributed to this field [418]. For example, Hanack et al. employed phthalocyanine or naphthalocyanine, complexed with a divalent group-8 metal ion, to obtain Shish-Kebab-type assemblies with various ditopic *N*-heterocyclic ligands, as the bridge (e.g., tetrazine or pyrazine; Fig. 2.58). These metallo-supramolecular polymers exhibited an increased conductivity relative to the corresponding MACs (e.g., utilizing tetrazine, as the bridge, led to an increase of the conductivity by a factor of 10^7) [419]. The intrinsic conductivity, in the range of 10^{-2} S cm^{-1}, was mainly attributed to the low oxidation potentials of the bridging ligand; the low-lying lowest unoccupied molecular orbital (LUMO) level of the ligand facilitated a charge-transfer process which, in turn, induced the intrinsic conductivity. Following the same strategy for biaxial coordination, a library of Shish-Kebab-type architectures with metalloporphyrins or metallophthalocyanines, as redox-active components, was prepared [420]. As one further example, the metallo-supramolecular polymer, as obtained by Collman et al., from the self-assembly of an octaethyl-substituted ruthenoporphyrin with pyrazine, is also depicted (Fig. 2.58) [421]. This particular metallopolymer exhibited a remarkable increase in its powder conductivity upon doping with iodine or (NO)(PF_6): The conductivity of the undoped polymer (1.2×10^{-8} S cm^{-1}) was raised to 1.4×10^{-2} S cm^{-1} (after doping with I_2).

The self-assembly of MACs into Shish-Kebab-type polymers by the coordination of organic diisocyanides or small ditopic inorganic ligands, such as cyanide and thiocyanate, has also been a subject of intense research. Hanack et al. employed a variety of Ru^{II}-containing MACs (i.e., porphyrins, tetraazaporphyrins, phthalocyanines or naphthalocyanines) in order to obtain (in most cases) highly electroactive materials [422–429]. The undoped assemblies had conductivities in the characteristic range for weak semiconductors; whereas, substituted bridging ligands (e.g., 2,3,5,6-tetramethyl-1,4-tetrazine) decreased the conductivity compared to unsubstituted ones (e.g., 1,4-tetrazine)—this finding was attributed to a reduced stacking

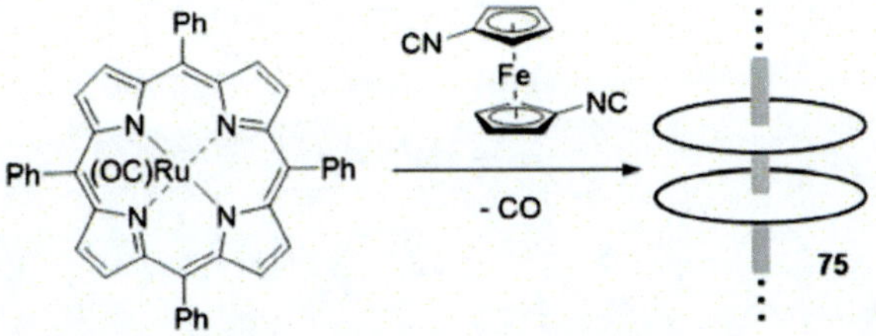

Fig. 2.59 Schematic representation of the synthesis of the Shish-Kebab-type assembly **75**

ability of the former due to steric repulsion [429]. More recently, Nemykin et al. intended to increase the electronic communication between Ru^{II}–MACs in Shish-Kebab-type assemblies by introducing redox-active moieties within the bridging ligand [430]. However, the supramolecular polymer **75** was obtained as an insoluble material, which could be characterized only by elemental analysis and IR spectroscopy (Fig. 2.59). Photophysical and electrochemical studies on bi- and trinuclear model complexes suggested a weak long-range Fe–Fe coupling within these materials.

The formation of Ru^{II}-containing polymers by the reaction of appropriate Ru^{II} or Ru^{III} precursor complexes with neutral *bis*-bidentate ligands is basically possible, but has, up to now, only been applied sporadically. More conveniently, such metallopolymers can be derived from polyaddition or polycondensation reactions utilizing pre-formed *bis*-functionalized Ru^{II} complexes and, if desired, appropriately modified organic building blocks (denoted as method Vb, see Fig. 1); this strategy is discussed later in Sect. 2.5.2b. In contrast to this, a broad range of hyper-branched or dendritic metallo-supramacromolecular architectures has been realized making use of the directed assembly of bidentate polypyridyl ligands with Ru^{II} ions and the kinetically inert character of the resulting complexes [431]. The field of ruthenodendrimers is discussed separately in Chap. 4. As one example in this respect, the metallodendrimer **76**, comprising of ten Ru^{II} *tris*-bidentate polypyridyl complexes, is depicted in Fig. 2.60 (more complex architectures with >20 Ru^{II} centers can also be found in literature) [431–439].

Rehahn et al. were the first to tackle the synthesis of Ru^{II}-containing polymers via the "polymerization-by-complexation" route. [440–442] In this initial study, the oligomeric Ru^{III} precursor complex $[(R_2bpy)RuCl_3]_x$ (R=H or C_6H_5) was reacted with tetrapyridophenazine, as the telechelic ligand (Fig. 2.61). These authors pointed out that a careful precursor synthesis was of utmost importance, i.e., preclusion of the presence of other potentially coordinating $Ru^{II/III}$ species, to finally obtain the metallopolymers **77** in high purity. NMR spectroscopic end-group analysis and small-angle X-ray scattering (SAXS) measurements on the soluble materials were conducted in order to estimate the DP; for example, M_w values of up to 50,000 g mol^{-1} were determined for **77b**. Due to the high kinetic stability of the Ru^{II} complexes within the polymer backbone, disassembly in solution did not occur —not even in the presence of polar, potentially coordinating solvents. Basically,

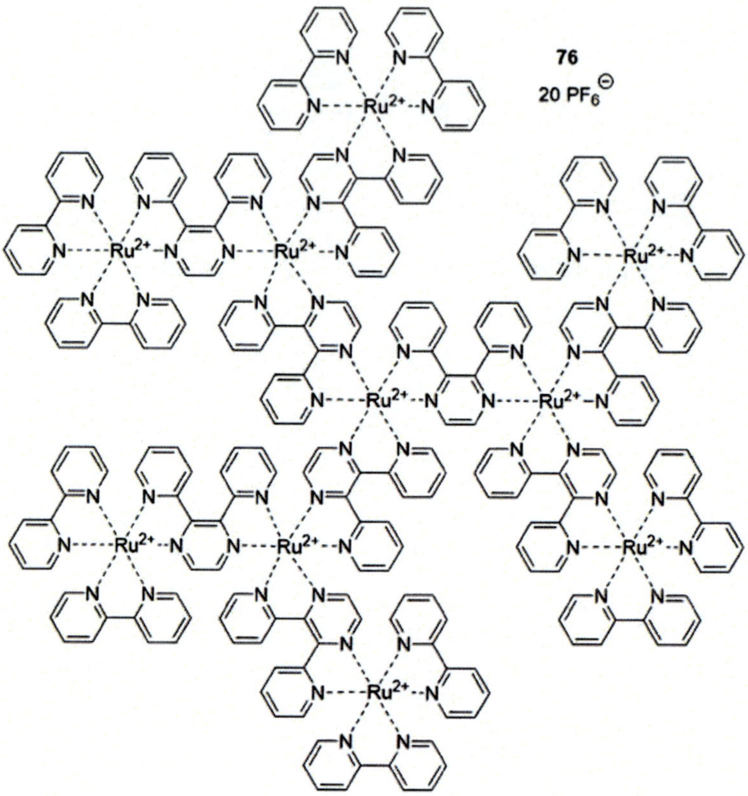

76

20 $PF_6^{\ominus}$

Fig. 2.60 Schematic representation of the ruthenodendrimer **76** featuring ten Ru^{II} centers

heteroleptic *tris*-bidentate complexes of octahedral geometry feature a metal-centered chirality with two possible enantiomers—the Δ- and the Λ-form (an issue which can be bypassed by using *bis*-tridentate complexes instead, vide infra). The metallo-supramolecular polymerization, as described by Rehahn et al., offered no stereoselectivity and, thus, the molecular structure of **77** was less defined in this respect (random distribution of Δ and Λ enantiomers along the chain). Accordingly, no helical arrangement could be evidenced and the metallopolymers adopted randomly coiled chain conformations only.

In contrast to Rehahn's work, Fendler et al. employed a different strategy to obtain a polymer with $[Ru(bpy)_2]^{2+}$ complexes within the backbone [443]. First, a pre-polymer was prepared by the reaction of an alkyl-bridged ditopic bpy ligand with $RuCl_3 \times xH_2O$ in an equimolar ratio; the final polymer **70** was then obtained by adding an excess of bpy to the pre-polymer in order to fully saturate the vacant coordination sites of the Ru^{II} centers (Fig. 2.62). Basically, when dealing with flexible telechelic ligands, as in the cited work, one has to be conscious that not

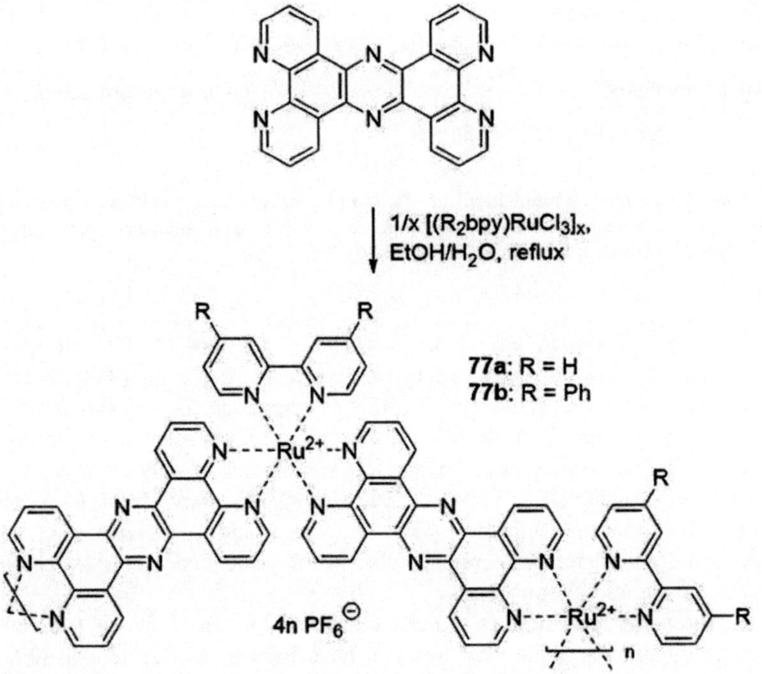

Fig. 2.61 Schematic representation of the synthesis of metallopolymer **77** based on Ru^{II} *tris*-bidentate complexes

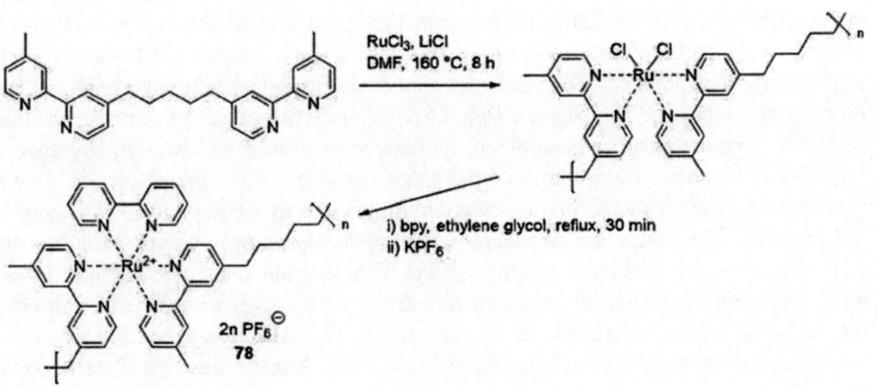

Fig. 2.62 Schematic representation of the synthesis of the coordination polymer **78** containing alkyl-ridged Ru^{II}–bpy tris-complexes within the backbone

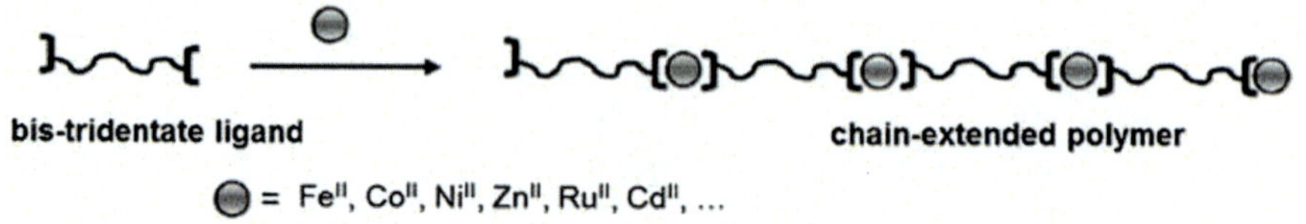

Fig. 2.63 Schematic representation of the formation of linear metallopolymers via the self-assembly of telechelic *bis*-tridentate ligands with divalent transition-metal ions [445]. Figure reproduced with kind permission; © 2009 Wiley-VCH

only linear assemblies might be formed—the entropy-driven metallo-macrocyclization might arise as an issue to deal with [444]. Neither did the authors address this issue nor was that of relevance for the anticipated application, i.e. formation of molecular rectifiers based on bilayers of redox-active polymers. The multilayer rectifier, which functioned at low voltages, had the configuration ITO|(PDDA/TiO_2)$_{10}$|(PDDA)|(GO/**78**)$_{30}$|Au [PDDA: poly(diallyldimethylammonium chloride), GO: graphite oxide]. Remarkably, electron transfer occurred efficiently despite the absence of a direct contact between the TiO_2 particles and the metallopolymer.

The aforementioned reports represent the few examples where telechelic *bis*-bipyridine ligands (or analogous systems) have been employed to assemble linear ruthenopolymers by complexation. Telechelic *bis*-tridentate polypyridyl ligands are, however, more in the focus of interest. From these, the family of *bis*-tpy ligands has to be foregrounded due to their ability to readily give metallo-supramolecular polymers by coordination of low-valent transition-metal ions [3, 5, 445]. A broad range of substrates, flexible as well as rigid-rod, has been utilized and their applicability in various fields of research has been tested, e.g. opto-electronics, sensing, self-healing materials, etc. The generalized synthesis of such metallopolymers is outlined in Fig. 2.63. As already indicated, the linkage between the terminal tpy sites and the organic linker via the 4-positions of the central pyridine rings, in concert with the preference of many transition metal ions to organize a (pseudo-)octahedral coordination geometry, enables the formation of linear assemblies; thereby, the formation of enantiomers due to metal-centered chirality is not possible. The formation of macrocyclic assemblies can, however, often not be fully excluded, in particular when dealing with flexible telechelic ligands [446–453]. The ring-chain equilibrium can be influenced by adjusting the reaction conditions; thus macrocyclization, as an entropically driven process, becomes favored when high-dilution conditions are employed [454]. A more detailed discussion of the underlying mechanism, i.e., the ring-chain mediated supramolecular polymerization, can be found elsewhere [318, 397].

One of the first libraries of ditopic tpy ligands and their application, as building blocks, for the formation of the linear metallopolymers **79** was published by Colbran et al. in the late 1990 [455, 456]. These authors tested two orthogonal approaches to reach this goal (Fig. 2.64): The first variant relied on a

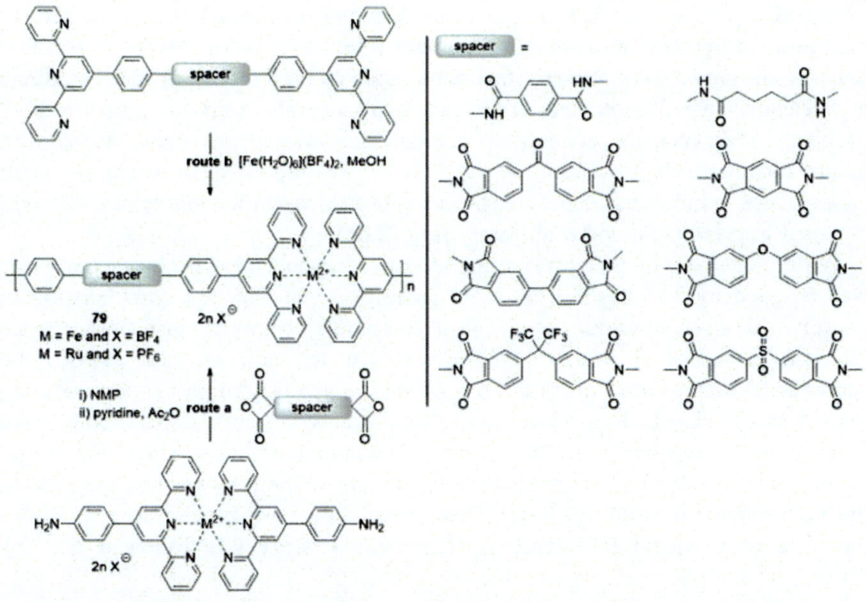

Fig. 2.64 Schematic representation of the synthesis of the linear Fe^{II}- and Ru^{II}-containing metallopolymers **79** with amido- or imido-linkages between the complexes

polycondensation between pre-formed homoleptic complexes [corresponding to the "complexation first" approach, denoted as "method Vb" in Fig. 1; see also Sect. 2.5.1]. As an alternative, the metallo-supramolecular polymerization of ditopic amide- and imide-linked *bis*-tpy ligands with Fe^{II} or Ru^{II} ions was also tested. When using Fe^{II} ions, both approaches gave metallo-supramolecular polymers of comparable molecular weight (according to ^{1}H NMR spectroscopic end-group analysis). However, when using Ru^{II} ions, the second variant appeared to be less appropriate due to the relatively harsh reaction conditions required to assemble $[Ru(tpy)_2]^{2+}$ complexes. Based on these observations, Chan et al. also applied the "complexation first" strategy to obtain a variety of Ru^{II}-containing polymers of similar structure, as the materials prepared by Colbran et al. (Fig. 2.64) [457].

The metallo-supramolecular self-assembly of various OEG-bridged telechelic *bis*-tpy ligands in the presence of Fe^{II} ions was studied independently by the Constable and Schubert groups [OEG: oligo(ethylene glycol)] [447, 448, 452]. At room temperature, this process gave a mixture of defined metallo-macrocyclic species along with linear assemblies. At more forcing conditions, the entropy-driven assembly into metallo-macrocycles, containing three or four metal centers, became predominant. However, the equilibration between cyclic and linear assemblies could be frozen if Ru^{II} ions, forming kinetically stable *bis*-complexes, were used. Schubert et al. reported on the optimization of the reaction conditions to enable the selective metallo-supramolecular polymerization of the flexible, ditopic ligand **80** by using $RuCl_3 \times xH_2O$ under reducing conditions, i.e., in a solvent mixture of *n*-butanol and *N*-ethylmorpholine (Fig. 2.65) [446]. It was shown that linear chain growth was strongly favored as long as high monomer concentrations were ensured (metallopolymers of high molecular weights were identified by SEC analysis). Moreover, the concurrent macrocyclization pathway could be blocked almost quantitatively by adding 10 mol% of MPEG-tpy, as chain-stopping agent; under these conditions, the $A(B_n)A$-type pseudo-triblock copolymer **81** was obtained in good yield and high purity (Fig. 2.65).

A different type of metallo-supramolecular chain-extended triblock copolymer was reported by McLaughlin et al. by making use of Ru^{II} ion complexation in concert with DNA hybridization [458]. Two complementary 20-mer DNA strands (denoted as A and B) were functionalized with tpy units at their 5′-ends, the subsequent complexation with Ru^{II} ions gave the corresponding homopolymers (A-[Ru]-A and B-[Ru]-B, Fig. 2.66). Mixing of these two macromolecular complexes induced the self-assembly via DNA hybridization into long linear arrays. The length of these chain-extended materials could be adjusted simply by varying the ratio of the homopolymeric building blocks. Thus, the self-assembly process strictly obeyed the rules of a typical IDP reaction. The same strategy was followed later by

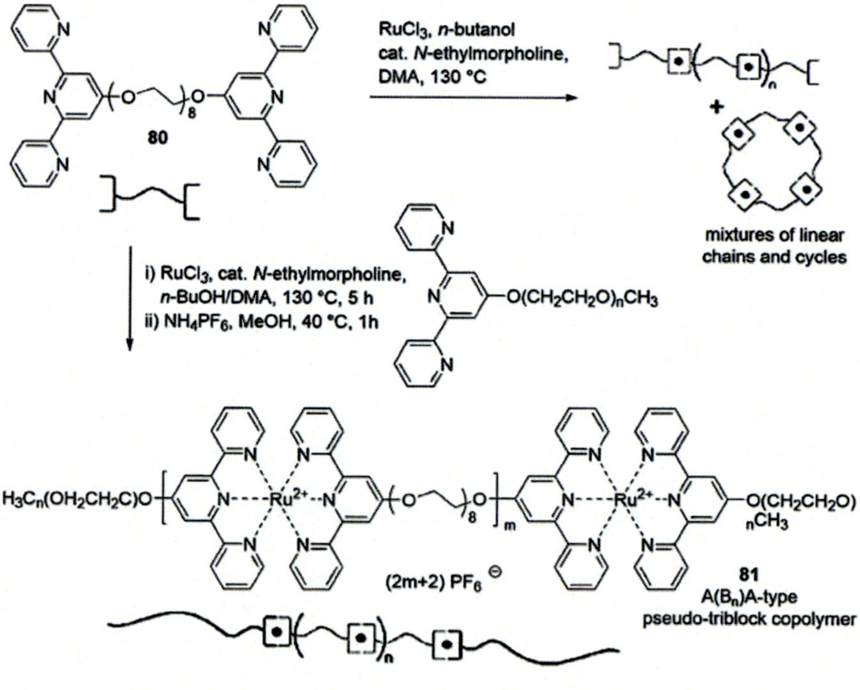

Fig. 2.65 Schematic representation of the synthesis of linear and/or cyclic metallo-supramolecular materials via self-assembly of the telechelic ligand **80** with Ru^{II} ions

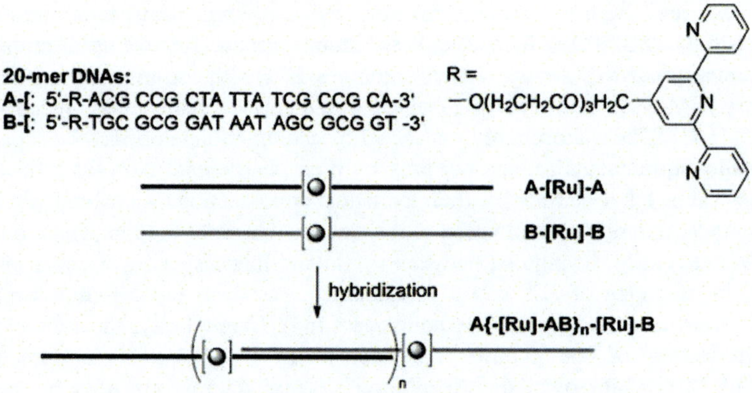

Fig. 2.66 Schematic representation of two 20-mer DNA fragments (A and B), each end-functionalized with a tpy, and of their self-assembly with Ru^{II} into the A-[Ru]-A and B-[Ru]-B homopolymers as well as the chain-extended A{-[Ru]-AB}$_n$-[Ru]-B triblock copolymer [458]. Figure redrawn according to the literature; © 2003 The Royal Chemical Society

Defrancq et al., who synthesized similar oligonucleotide-based homopolymers with Fe^{II} and Ru^{II} ions [459]. From these, the self-assembly of A-[Ru]-A with B-[M]-B (M = Ru or Fe), at a 1:1 ratio, was investigated by non-denaturizing gel electrophoreses as well as TEM imaging; the latter technique revealed rod-like filaments with an average length of ca. 200 nm (Fig. 2.67). Besides these examples for strictly linear Ru^{II}-containing polymers, recently DNA assemblies with pending Ru^{II} complexes have also been reported [460, 461]. The concept of using metal complexes for the construction of 2D/3D DNA arrays has been detailed elsewhere by Ghosh and Defrancq [462].

It has to be pointed out, that substantial progress has also been made in the last decade regarding the structural characterization of sophisticated metallo-supramolecular architectures (i.e., discrete and stoichiometric assemblies)

Fig. 2.67 TEM image of a oligonucleotide-based A{-[Ru]-AB}$_n$-[Ru]-B triblock copolymer (scale bar: 200 nm) [459]. Figure reproduced with kind permission; © 2009 The Royal Chemical Society

in a fast and highly accurate fashion by applying electrospray ionization time-of-flight (ESI-TOF) MS. The Wesdemiotis group, for example, employs a multidimensional MS set-up, which combines ESI with ion-mobility mass spectrometry (IM-MS) and energy-resolved (gradient) tandem mass spectrometry (gMS^2) [463]. Under thoroughly optimized measurement conditions, the separation of metallo-supramolecular ions not only by their charge and mass (as reflected by the *m*/*z* values) but further by their collision cross-section is enabled; the latter corresponds in good approximation to the size and shape of the ion. Thus, one can, for instance, easily distinguish between cyclic and linear species of equal *m*/*z* values. A broad range of 2D- and 3D-assemblies, based on tpy *bis*-complexes, has already been analyzed by this methodology [464]. Exemplarily, the synthesis and characterization of the symmetrical nanosphere $[Ru_6(\mathbf{82})_4]^{12+}$ is depicted in Fig. 2.68 [465]. Moreover, high-resolution electron microscopy imaging, such as TEM and ultrahigh vacuum scanning-tunneling electron microscopy (UHV STEM), have proven their versatility to determine the size and shape of such materials [464]. Newkome's fractional polymers, e.g., the Sierpiński's hexagonal gasket, [466] represent exampled in this respect.

Not only telechelic ligands with short, monodisperse spacers but also those having a macromolecular bridge between the terminal binding sites were employed

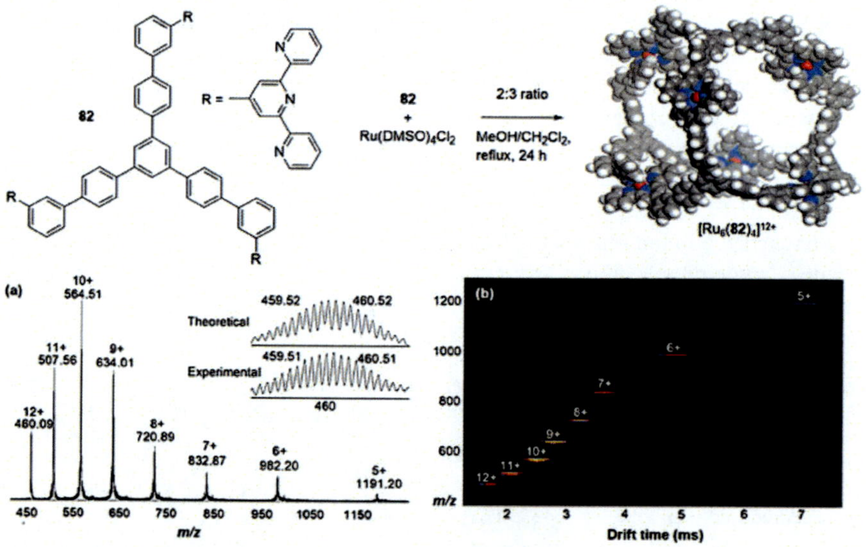

Fig. 2.68 Top: Schematic representation of the tpy-based building block **82** and its self-assembly with Ru^{II} ions into the metallo-supramolecular nanosphere $[Ru_6(\mathbf{82})_4]^{12+}$. Bottom: **a** ESI–MS spectrum of the nanosphere (the inset shows the isotope pattern of the +12 species); **b** ESI-TWIM-MS plot of the nanosphere (*m*/*z* *vs.* drift time) with the charge states of intact assemblies marked [465]. Figure reproduced with kind permission; © 2014 American Chemical Society

83

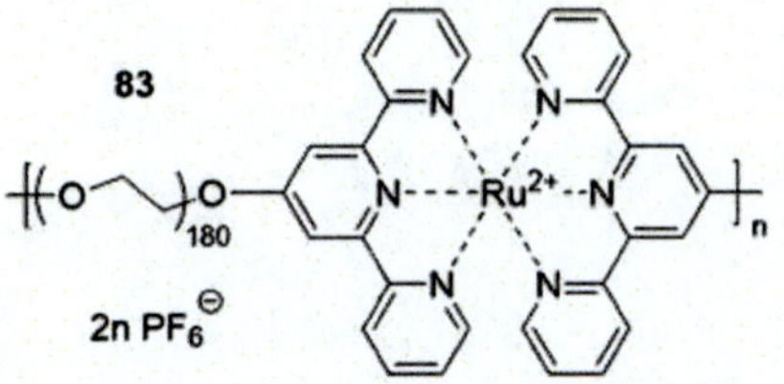

Method	Molar mass (g mol^{-1})	DP
SEC	M_n = 136,700	15
AUC	M_{AUC} = 143,000	16
Viscosimetry	M_V = 123,000	14

Fig. 2.69 Schematic representation the Ru^{II}-containing chain-extended polymer **83** along with some characterization data

to assemble metallo-supramolecular polymers [289, 445]. A tpy-functionalized PEG macroligand, for example, was polymerized with Ru^{II} ions to give the chain-extended polymer **83**, which was characterized via a set of independent analytical techniques (Fig. 2.69) [364–367]; these measurements revealed a molecular weight in the range of ca. 130,000 g mol^{-1} (DP of ca. 15). These results demonstrate the reliability of the various techniques and, thus, one is not limited to a single method in order to determine the molecular weight of metal-containing polymers.

A broad range of π-conjugated chromophores flanked by two tpy units has been prepared by cross-coupling and/or condensation chemistry [5]. These telechelic ligands featuring different degrees-of-flexibility and π-conjugation represent versatile starting materials for the assembly of rigid-linear coordination polymers by complexation with transition-metal ions. This field of research, which is often referred to as "molecular wires" was pioneered by Ziessel et al. [467, 468], followed by many other groups [63, 462, 471–518, 518]. The metallo-supramolecular polymerization of such ligands proceeds smoothly, at least when employing the divalent first-row transition-metal ions (e.g., Fe^{II}, Co^{II}, Ni^{II}, Zn^{II}) [5]. On the contrary, the analogous Ru^{II}-containing materials are much more difficult to prepare, since more forcing reaction conditions are needed—this finding is in line with the observations made in preparing metallopolymers, based on the aforementioned *bis*-bidentate ligands (e.g., **77** in Fig. 2.61). Representatively, the preparation of the ruthenopolymers **84** and **85**, according to Rehahn et al., is depicted in Fig. 2.70 [470]. The related copolymer **86**, featuring donor- and acceptor-type and chromophores in a strictly alternating fashion, was prepared by Schubert et al. making use of Rehahn's original protocol. [471] These authors conducted viscometry measurements ([η] = 90 cm [3] g^{-1}), SEC (M_n = 38,000 g mol^{-1}), and AUC analysis (M_{AUC} = 20,000 g mol^{-1}) to prove the formation of a high-molecular-weight material.

In order to gain further insight into the electrostatic interactions between the polycationic assemblies and their accompanying anions, detailed studies were performed on various polycationic ruthenopolymers (e.g., **87** in Fig. 2.71). By utilizing Fremy's nitrosodisulfonate anion, Rehahn, Jeschke et al. evidenced by

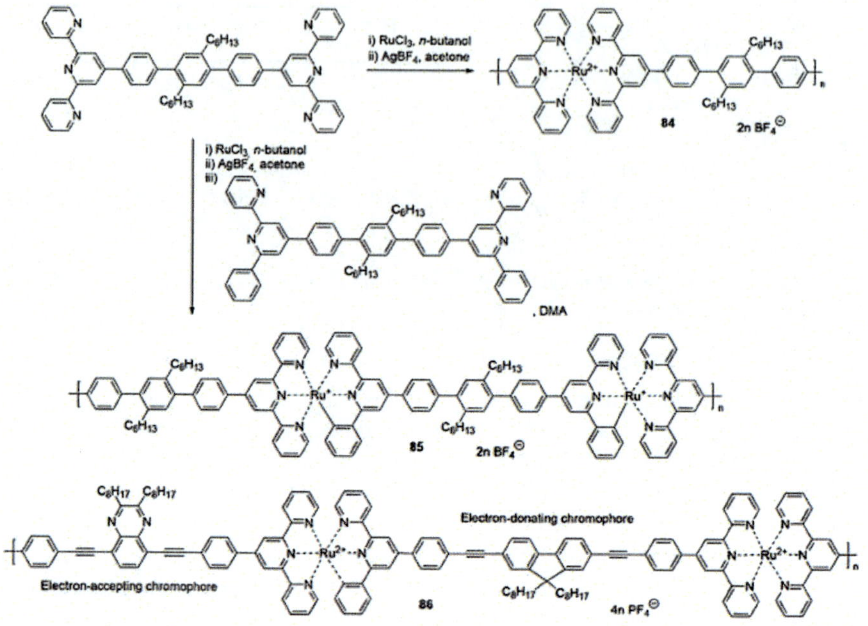

Fig. 2.70 Schematic representation of the synthesis of the Ru^{II}-containing polymers **84–86**

continuous-wave as well as double-electron EPR measurements that the spatial distribution of the spin-carrying dianion accurately mirrored the charge distribution along the polymer chain [472]. Furthermore, the counterion cloud around the [Ru(tpy)$_2$]$^{2+}$ sites featured an almost Gaussian-type radial distribution with a width of ca. 0.5 nm (Fig. 2.71).

Various Ru^{II}-containing polymers, based on tpy *bis*-complexes, have been processed from solution into the solid state and studied regarding potential applications, such as photovoltaics [471] or electrochromic materials [473, 474]. Basically, the phenomenon of electrochromism arises from reversible changes of the UV/vis/NIR absorption bands upon switching between (at least) two redox states by applying an external electric field (NIR: near infrared) [475–477]. These so-called electrochromic materials are of interest for diverse applications: data storage, optical memory, smart windows, color displays, etc. Among others (e.g., π-conjugated polymers or metal oxides), metallopolymers have attracted considerable interest [478]. Depending on the oxidation state of the incorporated transition-metal ions, often well-defined redox processes and characteristic absorption spectra can be obtained; thus, these materials are highly suited for this particular application. Thin-solid films featuring electrochromic behavior might be obtained via different techniques: Layer-by-layer assembly [479, 480], processing from solution (e.g., spin-coating, drop-casting or inkjet-printing) [473, 481] or reductive/oxidative electropolymerization (see Chap. 4) [482–487].

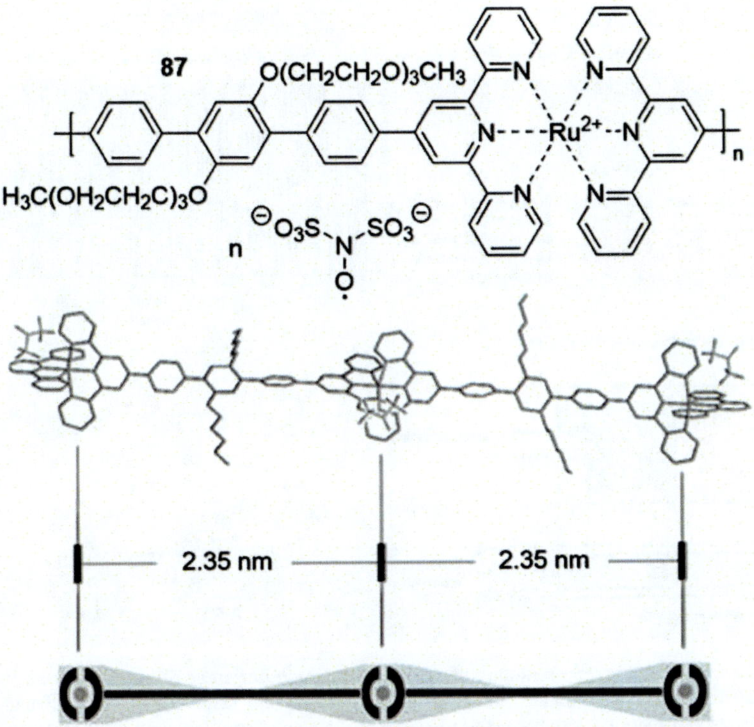

Fig. 2.71 Top: Schematic representation of the Ru^{II}-containing polymer **87**. Bottom: Schematic representation of the possible localized binding of the radical anions to the Ru^{II} centers of **87**. The distances between the anions, extracted from the EPR data, agreed with the calculated distances between two Ru^{II} centers. The schematic plot shows the most probable attachment sites for the radical anions (gray area); the relatively broad distribution of spin-probe distances can be rationalized by the degrees-of-freedom of the electrostatic attachment [472]. Figure reproduced with kind permission; © 2004 Wiley-VCH

Higuchi et al. addressed the issue of color, which can be achieved by variation of the metal center, structural modification of the ligand or both. As depicted in Fig. 2.72, the ruthenopolymers **88**, assembled from the respective ditopic ligands and Ru(DMSO)$_4$Cl$_2$, were reddish in color and exhibited a good switchability under redox conditions, i.e., the optical intensity of the MLCT absorption of the polymers was restored with a loss of up to 40% over 4000 redox cycles [488, 489]. In the same context, also other ruthenopolymers with structural variations on the ligands were studied [490–493]. These authors also studied the influence of the counterion on the electrochromic performance of the material [494]. Thereby, bulky anions (classified as "soft" according to Pearson's concept) gave a reversible elctrochromic behavior [e.g., $(CF_3SO_2)_2N$, CF_3SO_3 , ClO_4 , PF_6 , and BF_4]; whereas, the hard anions (e.g., Cl or CO_3^{2}) formed stable contact-ion pairs with the electrolyte

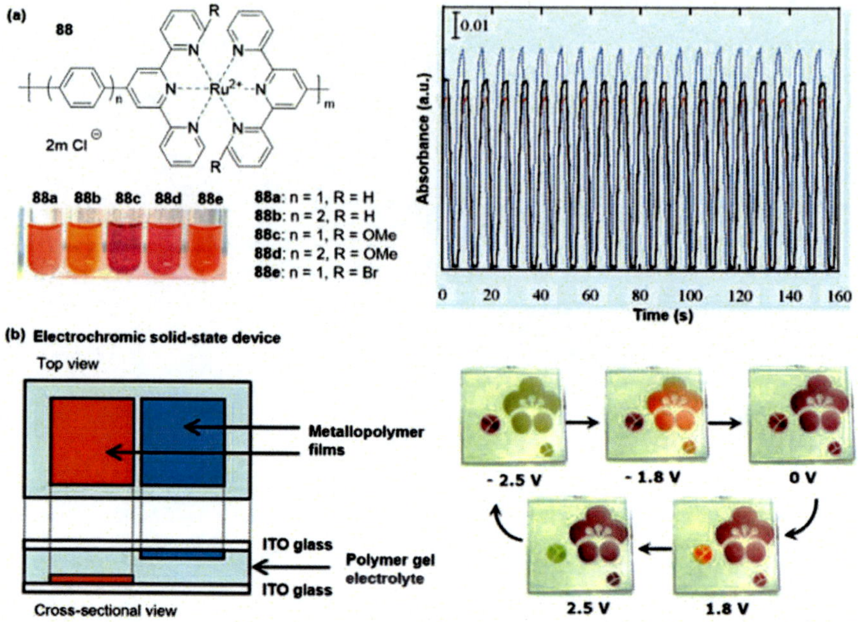

Fig. 2.72 **a** Schematic representation of the Ru^{II}-containing polymers **88** and their color in dilute solution; the MLCT absorbance change of a thin film of **88d** at potentials of + 1.5 and 0.0 V versus Ag/AgCl with a 4 s delay between switching is also shown [the first 40 (blue), middle 40 (black) and last 40 (red) of 4000 redox cycles are depicted] [488]. **b** Schematic representation of the solid-state device structure and the multi-color electrochromic change of a random Ru^{II}/Fe^{II}-copolymer film in the device (ligand as in **88a**) [473]. Figures reproduced with kind permission; © 2008 American Chemical Society and 2009 Nature Publishing Group, respectively

and, thus, hampered the diffusion-controlled redox processes. In particular, polymer **88a** with $(CF_3SO_2)_2N^-$ or ClO_4^- anions featured the highest durability and fastest response times (3.69 and 0.98 s, respectively), the former also gave the highest color-efficiency value ($\eta = 270.3$ cm^2 C^{-1}) Furthermore, the incorporation of two different metal centers, such as Ru^{II} and Fe^{II}, into a statistical metallo-supramolecular copolymer enabled the fabrication of an electrochromic solid-state device, which showed five different displays that could be activated by changing the applied potentials between −2.5 and + 2.5 V (Fig. 2.72b) [473]. Moreover, Higuchi et al. synthesized some metallopolymers with Fe^{II} and Ru^{II} centers in a strictly alternating order by applying a two-step assembly strategy (again, the same ligand as in **88a** was used) [495]. When fabricated into electrochromic devices, these polymers exhibited high optical contrast ratios (e.g., 68% at 508 nm) and very fast response times of only seconds; the η values for the polymers were 242.1 and 188.2 cm^2 C^{-1} at different wavelengths.

Based on the same materials—metallopolymers containing Fe^{II} and/or Ru^{II} ions—electrochromic multi-patterned (MPD) and a tube-shaped displays (TSD) were

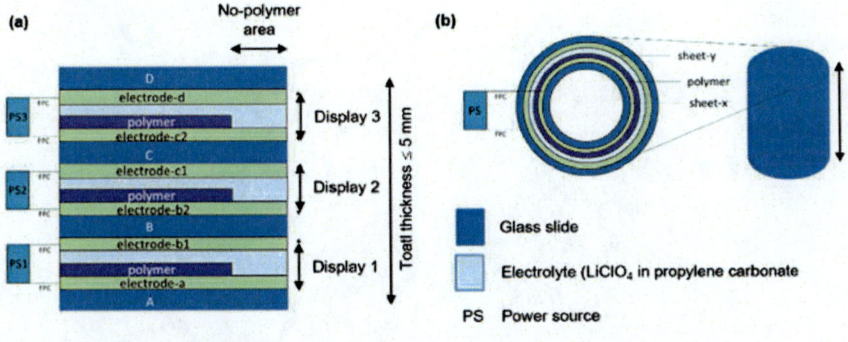

Fig. 2.73 Schematic representation of (**a**) an electrochromic multi-patterned and (**b**) a tube-shaped display (MPD and TSD, respectively) [496]. Figure reproduced with kind permission; © 2018 J-Stage

prepared [496]. The MPD, capable of producing eight different images, was assembled from three single electrochromic layers on glass slides, coated with ITO on both sides (total thickness $\leq$ 5 mm). The TSD, however, was fabricated from two flexible ITO-covered poly(ethylene terephthalate) (PET) foils and two tubular glass slides. The composition of the devices is depicted in Fig. 2.73. Both displays showed a very good electrochromic behavior, as revealed by high optical contrast ratios as well as fast response and coloration times. For example, the bleaching and coloring times (T_b and T_c, respectively) of the MPD were 1.7 and 2.0 s at 500 nm (i.e., in the range of the MLCT absorption band of the Ru^{II} complexes).

Recently, Higuchi et al. combined the Fe^{II}/Ru^{II} mixed-metal polymer, as working electrode, with Prussian blue, as counter electrode, into an electrochromic device with panchromatic properties [497]. A three-color scheme was accessible in potential range between −1.3 and 2.2 V with transmittance changes (ΔT) of 52.7, 46.9, and 28.0% at 503, 580, and 690 nm, respectively. For the coloring and bleaching processes (at 503 and 580 nm), very fast response times of <0.5 s were found. Noteworthy, the employed PMMA-based electrolyte reduced the leakage; an issue related to the commonly used liquid electrolytes. Addition of multi-walled carbon nanotubes (MWCNTs) to the metallopolymer significantly enhanced the memory effect of the device, which was attributed to a transient adsorption of the ClO_4^- anions to the MWCNTs. The authors identified the latter two findings to be crucial to improve the long-term stability of electrochromic devices.

In a different type of application, the electrostatic interaction of various high-molecular-weight polymers **88** with DNA was investigated [498, 499]. UV/vis absorption spectroscopy was used to study the binding of calf thymus (ct) DNA to polycationic **88a/b** [498]. Circular dichroism (CD) measurements suggested an unwinding of the DNA helix upon coordination, as revealed by a decrease of the CD signal intensity. Due to a collaborative effect of the neighboring Ru^{II} centers, binding constants in the range of 1.7 to 7.3×10^6 M^{-1} were observed; these

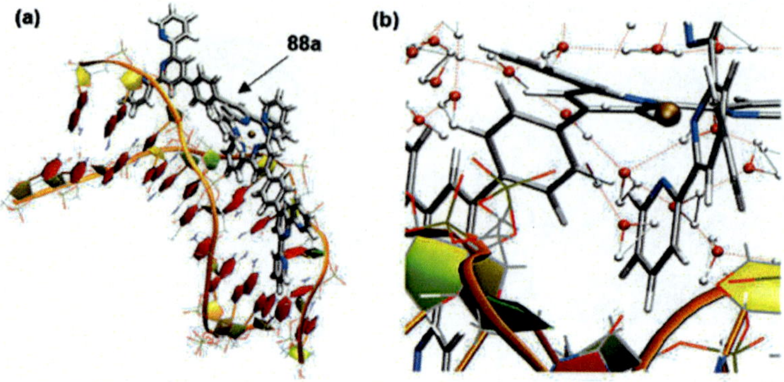

Fig. 2.74 Representation of computationally optimized structures of DNA and repeating unit of **88a**, the polymer chain is placed into the minor groove of the double-helix: (**a**) The whole computational model and (**b**) a detail of the interaction of the Ru^{II} center with one phosphate anion is depicted [498]. Figure reproduced with kind permission; © 2011 The Royal Chemical Society

values were similar or even higher than those reported for related small-molecule metal complexes. According to AFM imaging, supported by QM/MM computational modeling, the stable structure of the conjugate based on a groove binding with the Ru^{II} centers strongly interacting with the phosphate anions of the DNA (QM/MM: quantum mechanics/molecular mechanics, see Fig. 2.74). However, the cytotoxic activity of the Ru^{II}-containing conjugates was significantly lower when compared to the results obtained for the analogous Fe^{II}-based systems [499]; thus, the utilization only of the latter in this respect was followed further on [500].

A few examples are known where chiral *bis*-tpy ligands served, as building blocks, in metallo-supramolecular polymerizations. With respect to Ru^{II}-containing systems, the assemblies **89** and **90**, as reported by Abruña et al., are of relevance (Fig. 2.75) [501]. However, in contrast to the corresponding Fe^{II}-containing systems [502] no analysis on the structure of the presumably helical polymer chains by, e.g., CD spectroscopy or microscopy imaging was conducted. Finally, the structural modification of the basic tridentate tpy framework needs also to be mentioned at this stage (Fig. 2.76). This currently highly popular strategy was, for example, followed by Chandrasekar et al. who investigated the rigid ditopic bpp-type ligand **91** where the outer pyridine rings of the parent tpy moiety were substituted by (1*H*-pyrazol-1-yl) ones [503–505]. In a related study, Schubert et al. employed ditopic ligands bearing tridentate btp-type units (**92**) as binding sites for the self-assembly with Ru^{II} ions [506].

Very recently, Sahu, Bandyopadhyay et al. reported a ruthenopolymer, based on the hexadentate Schiff-base ligand **93** [507]. As concluded from FT-IR measurements, only the enol form of **93** was incorporated into the polymer and XPS analysis suggested that the oxidation state of the Ru centers was +II. The photophysical and electrochemical properties of the assembly were studied and the *I*–

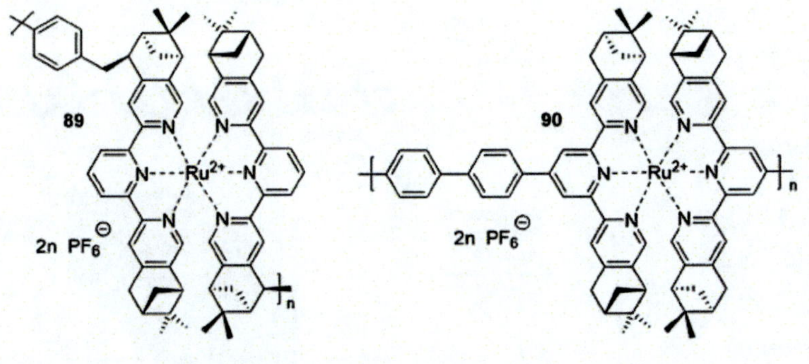

Fig. 2.75 Schematic representation of the chiral metallopolymers **89** and **90**

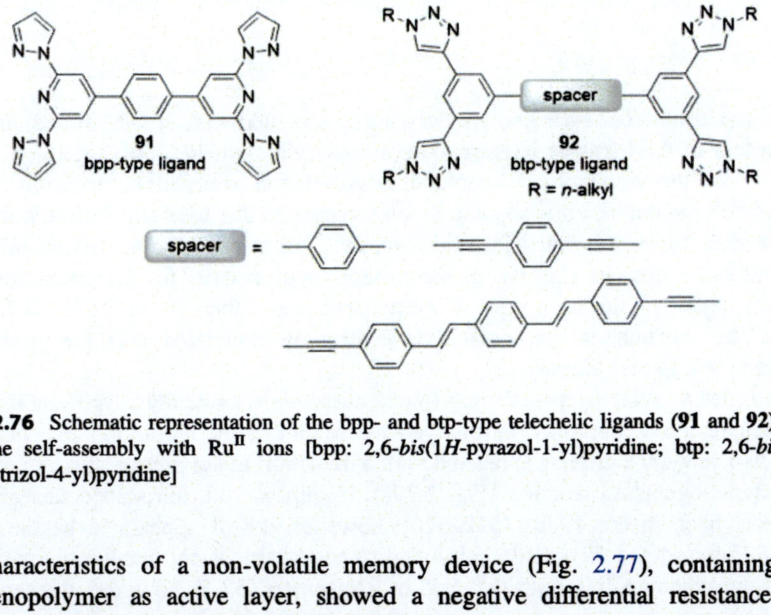

Fig. 2.76 Schematic representation of the bpp- and btp-type telechelic ligands (**91** and **92**) used for the self-assembly with Ru^{II} ions [bpp: 2,6-*bis*(1*H*-pyrazol-1-yl)pyridine; btp: 2,6-*bis*(1*H*-1,2,3-trizol-4-yl)pyridine]

V characteristics of a non-volatile memory device (Fig. 2.77), containing the ruthenopolymer as active layer, showed a negative differential resistance and resistive switching—the memoristive behavior was much more pronounced and sustained as for the analogous Fe^{II}-containing polymer.

2.2.3.2 Main-Chain Ruthenopolymers, Formed via Ionic or Covalent Bonding

Metallopolymers can not only be obtained from the coordinative interaction between neutral ditopic ligands and metal centers, but also the formation of metal-to-carbon bonds might be utilized for basically the same purpose. In particular, two types of metallopolymers possessing such pseudo-covalent

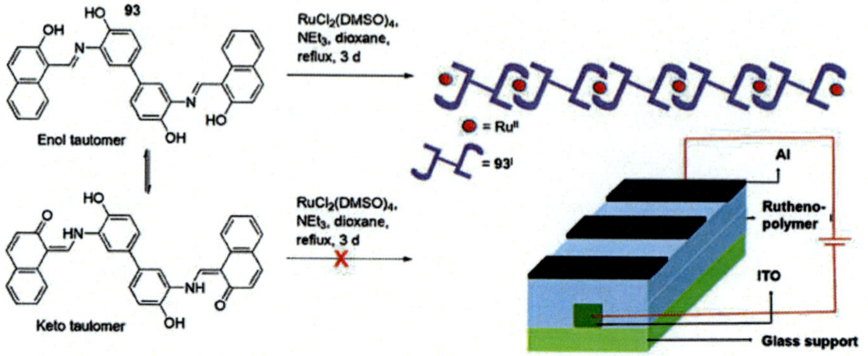

Fig. 2.77 Schematic representation of the synthesis of a Ru^{II}-containing metallo-supramolecular polymer, based on ligand **93**. The configuration of a non-volatile memory device, incorporating the ruthenium polymer, is also shown [507]. Figure reproduced with kind permission; © 2020 Wiley–VCH

metal-to-carbon bonds are known: poly(metal acetylide)s [4, 508–518] and, though being less in the focus of interest, poly(metal arylide)s (Fig. 2.78) [519].

By far, the rigid-rod Pt^{II}-containing poly(metal acetylide)s, in short polyplatinynes, are the best investigated representatives in this field and their application in various types of opto-electronic set-ups, such as polymer solar cells or light-emitting devices (LEDs), or their usage as precursors for the preparation of metallic nanoparticles by pyrolysis is extensively described [4, 511, 516, 517, 520, 521]. The research on these materials is strongly associated with the groups of Raithby, Wong and Harvey.

In order to synthesize such poly(metal acetylide)s, typically a "polymerization by complexation" approach is followed where an organic chromophore, equipped with two acetylene groups, is reacted with a transition-metal precursor complex in a dehydrohalogenation reaction (Fig. 2.79a). Hagihara et al. introduced this general protocol back in the 1970s [522–526]; however, it still appears to be the most straightforward and frequently employed route towards such metallopolymers—in particular when dealing with Pt^{II}, Pd^{II} or Rh^{I} centers [527]. To widen the scope of

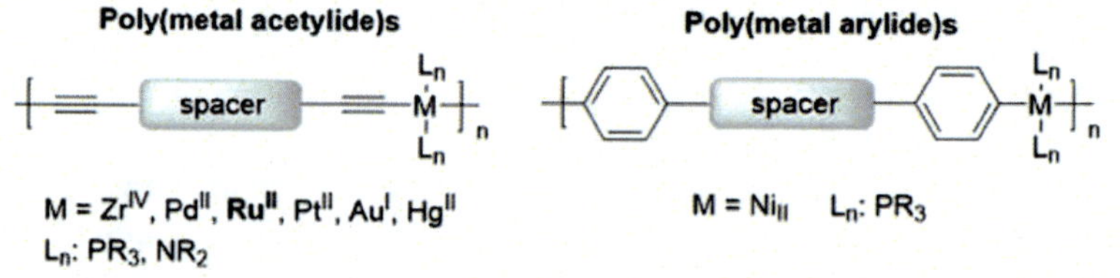

Fig. 2.78 Schematic representation of the generalized structure of poly(metal acetylide)s and poly (metal arylide)s

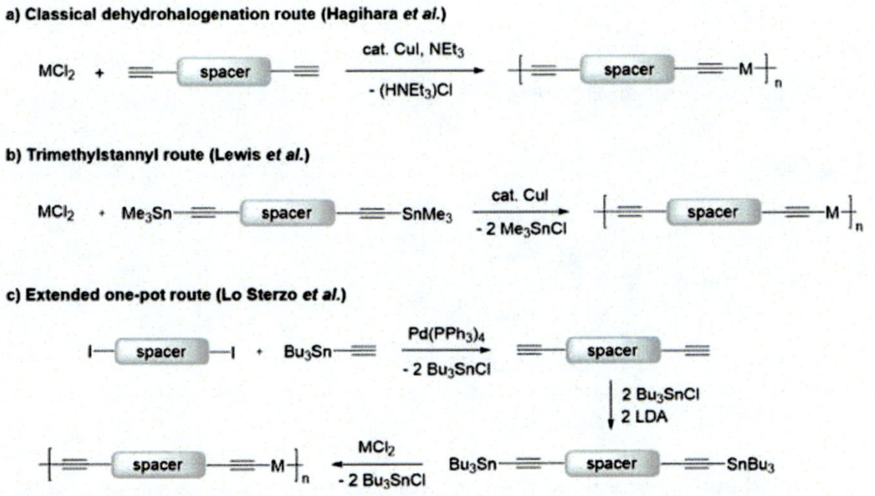

Fig. 2.79 Schematic representation of the general synthetic routes towards poly(metal acetylide) s: **a** Hagihara's original dehydrohalogenation route, **b** Lewis' trimethylstannyl route, and **c** Lo Sterzo's extended one-pot route [527] Figure redrawn according to the literature; © 2010 Wiley–VCH

the initial protocol to other transition-metal ions, i.e., of groups 8 to 10, two metathesis-type methods were developed by Lewis et al. [528] as well as Lo Sterzo et al. (Fig. 2.79b, c) [529–531]. Both strategies rely on the activation of the aromatic diacetylides as trialkyltin-derivatives, which subsequently undergo a metal-exchange with the precursor complexes [527]. The Ru^{II}-containing polymers **94** and **95**, as representative examples, are depicted in Fig. 2.80 [532–535]. As concluded from UV/vis and FT-IR measurements, supported by molecular orbital calculations, both the nature of the ancillary ligands (i.e., CO or phosphines) as well as of the bridging dialkyne unit influenced the degree of π-electron delocalization within the rod-like polymers [535]. Very recently, Komber, Blacque et al. reported the solution-processable ruthenopolymer **94d**, which was prepared via a Cu-free dehydrohalogenation route [536]. On the basis of SEC and NMR end-group analysis, a DP of ca. 32 was estimated. According to AFM imaging, the spin-coated film exhibited a smooth surface. Electrochemical studies on these films corroborated a metal-centered two-electron oxidation process. In order to further analyze the electronic properties, an organic field-effect transistor (OFET) device was fabricated. The transfer and output characteristics of this device suggested semiconductor-type properties of **94d**.

In extension to the previous examples, Dixneuf et al. used a step-wise protocol to synthesize the heterobimetallic polymetallyne **96** in the high yield of 84% [537]. For this purpose, a *bis*(alkynyl) Ru^{II}-complex, functionalized with two ethynyl moieties, was prepared and reacted subsequently with $Pd(Bu_3P)_2Cl_2$ in a classical

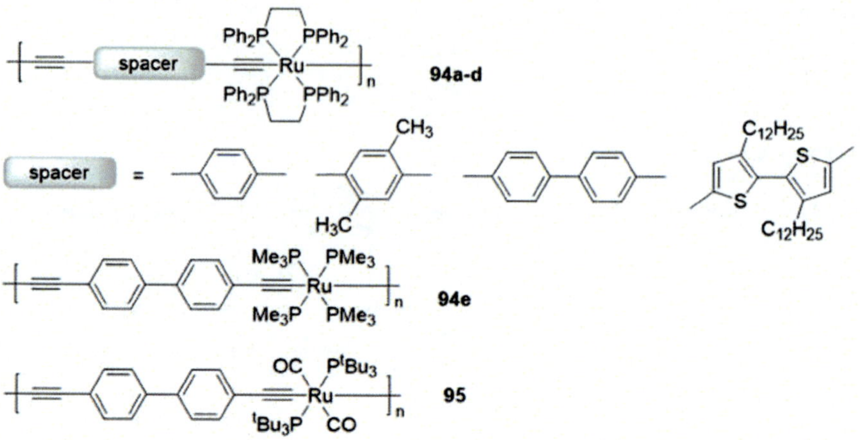

Fig. 2.80 Schematic representation of the Ru^{II}-containing poly(metal acetylidene)s **94** and **95**

dehydrohalogenation according to Hagihara's original protocol. The resulting heterobimetallic polymer was characterized by SEC ($M_n = 7{,}800$ g mol^{-1}, Đ = 1.89) and the ^{31}P NMR spectrum revealed the presence of two chemically different *P*-centers with the expected integral ratio of 2:1. Besides these Ru^{II}-containing polymetallynes, related materials were prepared by hydroboration polymerization [538] and electropolymerization [539]—details on ruthenopolymers prepared via the latter method are given in Chap. 3 (Fig. 2.81).

The carbonyl-hydrido complex [RuHCl(CO)(PPh_3)$_3$] was reacted with aromatic diynes to give the dinuclear five-coordinate Ru^{II}–vinyl complexes **97** [540]. In a second step, the coordinatively unsaturated metal centers of **97** were linked with monodentate telechelic ligands, i.e., aromatic isonitriles or 4,4′-bpy, to afford polymers **98** and **99** (Fig. 2.82). Due to the poor solubility of these materials, characterization via SEC or NMR spectroscopy was obsolete and only solid-state IR spectroscopy as well as elemental analysis could be applied to conclude on the identity of the proposed structures.

Metallacyclopentadienes of the group-4 metals, i.e., incorporating Ti^{II} or Zr^{IV} centers, represent important building blocks for the synthesis of a wide range of hetero- and carbocyclic compounds [541], including polymers [542–544]. The latter are also known as poly(arylene metallacyclopentadiene)s; as a common structural feature, the metallacyclopentadiene moieties, comprising a *d*-block element, are part of the polymer main chain. These materials can be considered as organometallic analogs to π-conjugated polymers containing heterocycles with *p*-block heteroatoms (e.g., pyrroles, thiophenes or selenophenes). The synthesis of such polymers typically follows a metallacycling polymerization (MCP), which is based on the successive metallacyclization of conjugated diacetylene derivatives with an appropriate transition-metal precursor species [545, 546]. In particular,

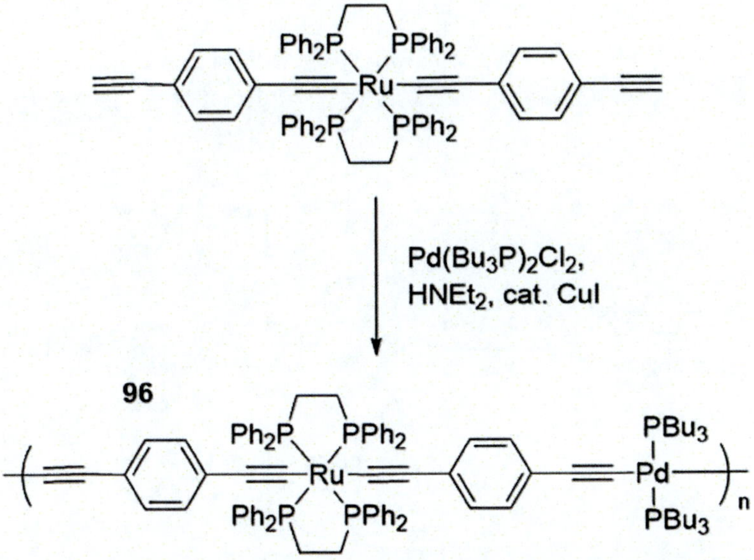

96

Fig. 2.81 Schematic representation of the synthesis of the heterobimetallic poly(metal acetylide) **96**

polymers with cobaltacyclopentadiene or, with relevance in the context of this chapter, ruthenacyclopentadiene entities within the polymer backbone have been reported [547, 548]. The latter material was obtained in a MCP reaction from 4,4′-diethynylbiphenyl and [Ru(*n*-hexyl-Cp)(cod)Br] at 0 °C (**100**, Fig. 2.83) [549]. Only the 2,5-disubstituted isomer was obtained revealing a remarkable regioselectivity of the metallacyclization step—this behavior was already shown before for various small-molecule ruthenacyclopentadienes [550–552]. However, polymer **100** exhibited a pronounced air-sensitivity and a low thermal stability; nonetheless, SEC analysis, under inert conditions, could be applied to estimate the molar mass (M_n = 3,400 g/mol, Đ = 1.70); the equilibrium constant for the MCP reaction was calculated to be 15. In the reduced state, intramolecular ferromagnetic interactions between the Ru centers were concluded from electrochemical studies and EPR spectroscopy in frozen THF.

Besides the aforementioned metallopolymers, which are exclusively based on covalent metal-to-carbon single bonds, i.e., relying on monodentate binding, some examples have been reported where this interaction is supported by additional metal-to-ligand coordination. Typical examples in this respect are the bi- or tridentate cyclometallating ligands that are widely employed for the coordination of Ru^{II} or Ir^{III} centers and allow, due to their strong electronic coupling with the metal centers, an efficient manipulation of the opto-electronic properties of the resulting complexes. Structurally, such ligands represent analogues of the classical polypyridyl ligands (e.g., bpy, tpy) in which a pyridine ring is replaced by a benzene

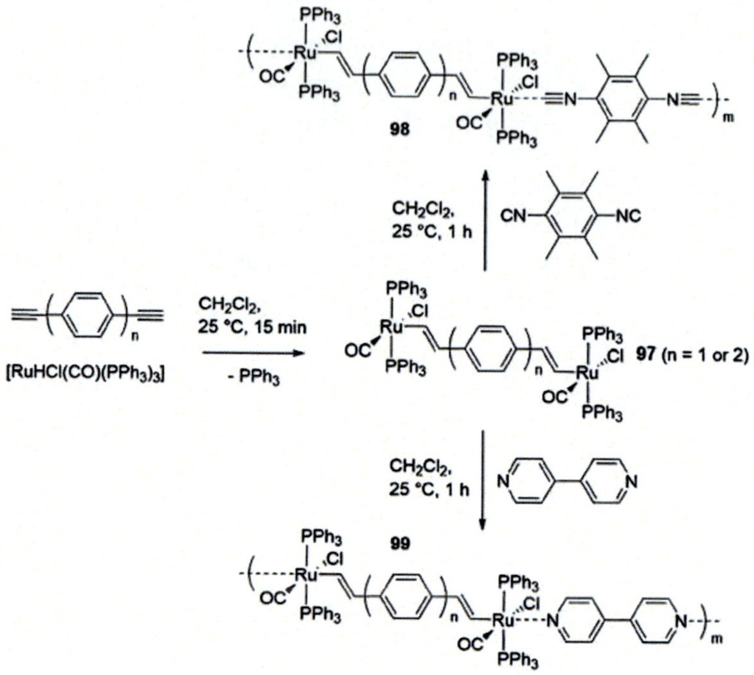

Fig. 2.82 Schematic representation of the synthesis of the coordinatively unsaturated dinuclear complex **97** and its utilization in the preparation of ruthenopolymers **98** and **99**

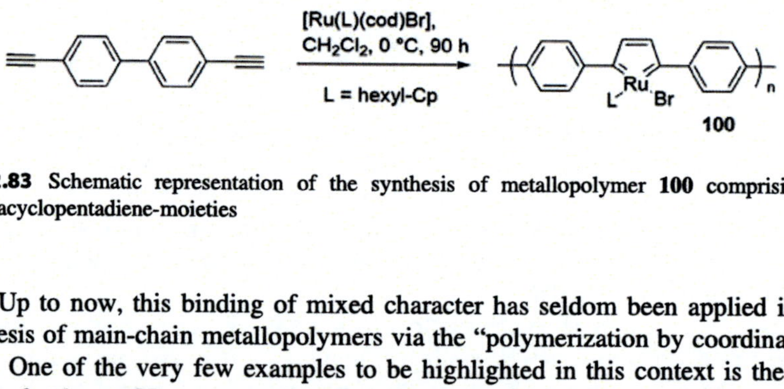

Fig. 2.83 Schematic representation of the synthesis of metallopolymer **100** comprising of ruthenacyclopentadiene-moieties

ring. Up to now, this binding of mixed character has seldom been applied in the synthesis of main-chain metallopolymers via the "polymerization by coordination" route. One of the very few examples to be highlighted in this context is the synthesis of polymer **85**, as reported by Rehahn et al. (Fig. 2.70) [470]. These authors noted that the main reason for this constraint arose from the harsh reaction conditions that were required for Ru^{II} *bis*-tridentate complexation (i.e., high temperature, long reaction time); accordingly, the formation of polymers was hampered. It was demonstrated that an alternative route, i.e., the (co)polymerization of complex monomers (denoted as method Vb, Fig. 1), was more applicable when metallopolymers featuring such complexes within the main chain had to be prepared [4, 22].

Van Koten and co-workers proposed the formation of various linear metallopolymers based on telechelic pincer-type ligand systems [553]. However, in the case of Pd^{II} and Pt^{II} ions, the formation of such assemblies was only concluded from titration experiments—attempts to isolate these materials failed. In contrast, for analogous systems incorporating Ru^{II} center only some monodisperse oligonuclear pincer complexes have been reported [554]. To the best of our knowledge, linear polymers bearing Ru^{II}-pincer complexes within the backbone have not yet been reported.

2.2.3.3 Main-Chain Ruthenopolymers, Formed via Arene-Π Complexation

The last class of metallopolymers, which might be accessible via a "(co)polymerization by coordination" approach, are those that are formed by arene-π complexation. A plethora of polymers comprising π-coordinated metal centers is reported, either having these entities within the main chain or in the side chains [417, 518, 555–559]. In most cases, however, the synthesis is carried out by (co) polymerizing (half-)sandwich-type complexes, which are equipped with appropriate functional groups, i.e., the polymerization methods Va/b are applied (Fig. 1). In contrast to this, only scattered examples are known where polymers with metallocene entities within the main chain have been synthesized by a direct metallo-supramolecular polymerization. As outlined by Manners, this family of polymers, possessing arene-π complexes within their backbone, can be sub-classified further into the so-called polydecker sandwich assemblies (types A and B) and linear polymetallocenes (types C and D) [556]. The basic structures of these four systems are depicted in Fig. 2.84.

For all four types, some examples have been reported, though mainly incorporating Fe^{II} centers (other transition-metal centers have only been introduced sporadically) [318]. To the best of our knowledge, Ru^{II}-containing polydecker sandwich complexes or polymetallocenylenes have not yet been obtained via a direct metallo-supramolecular polymerization approach. How can this be rationalized? For the group-8 elements, the synthetic ease to prepare the parent metallocenes decreases in the order Fe >> Ru > Os; in other words, ferrocenes can be prepared on a large scale in a rather straightforward fashion; whereas, much harsher reaction conditions are required for Ru and Os, as iron's heavier homologous [560]. Apparently, such protocols are rather inadequate for the formation of defined polymers relying on arene-π complexation. As shown by Neuse in 1981 [561], type-D polyruthenocenylenes (Fig. 2.84) can be prepared, at least in special cases by applying homo- or cross-coupling reactions of appropriately 1,1′-disubstituted ruthenocene derivatives (see Sect. 2.5.2a).

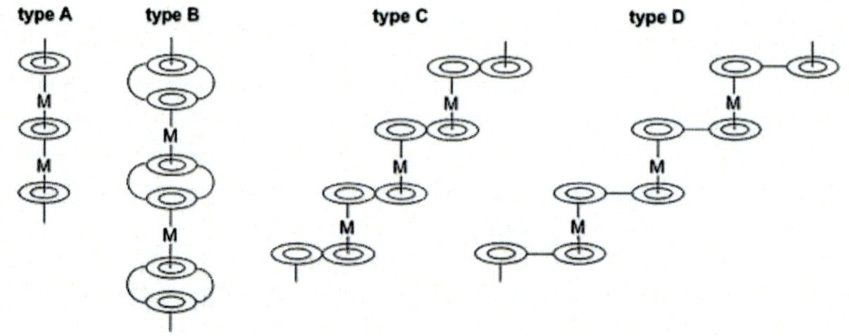

Fig. 2.84 Schematic representation of Manners' classification of polymer containing arene-π-complexes within the backbone: Polydecker sandwich assemblies (types A and B) and linear polymetallocenes (types C and D) [556]

2.2.4 Ruthenopolymers, Derived from Ru^{II}-Containing Initiators

Molecules bearing designated metal binding sites can be utilized, as initiators, in various living/controlled polymerizations, such as the ROP of cyclic esters or the family of CRP techniques (i.e., ATRP, NMP, and RAFT). As explained previously in Sect. 2.2.2, this strategy enables the synthesis of well-defined polymers, functionalized at the chain end(s) with a potential ligand site. However, some established polymerization techniques are not compatible with this approach. For example, a cationic ring-opening polymerization (CROP) of 2-oxazolines with an initiator, bearing a polypyridyl-type ligand, will proceed only with poor control due to a possible electrophilic attack of the propagating cationic species on the *N*-atoms of the pyridyl rings. Therefore, a divergent route—using preformed metal complexes, as initiators, for polymerizations—was developed (Fig. 2.85). Thereby, the intrinsic limitations of the alternative convergent strategy with respect to undesired side-reactions can, at least to some extent, be overcome [562, 563].

With respect to the CROP of 2-oxazolines, the reports by Fraser et al. are of particular importance. A variety of *bis*-, tetra*kis*- and hexa*kis*(halomethyl)-functionalized Ru^{II}–bpy *tris*-complexes was employed, as initiators, for the formation of well-defined metal-centered star-shaped poly(2-oxazoline)s (**101**, Fig. 2.86a, b) [564–567]. Efficient initiation of the polymerization from all functionalities was concluded from the first-order kinetic plots (Fig. 2.86c). In all cases, the rate of monomer consumption was fairly proportional to the number of present initiator sites. The calculated propagation rates were in the typical range for a 2-oxazoline polymerization. Different 2-R-2-oxazoline monomers (i.e., R = ethyl, methyl, phenyl, and 2-undecyl) were used in these studies and gave homopolymers as well as block copolymer chains attached to the central $[Ru(bpy)_2]^{2+}$ sites.

Poly(ethylenimine) (PEI), as a polycationic polymer, is considered as the "gold standard" in gene-delivery applications due to its ability to electrostatically interact

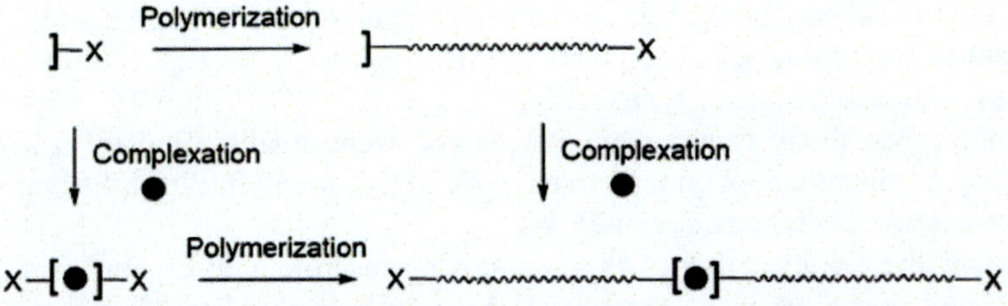

Fig. 2.85 Schematic representation of the synthesis of metal-containing polymers via convergent or divergent route using a metal-free or metal-containing initiator, respectively (X denotes the initiating functionality) [562]. Figure redrawn according to the literature; © 2001 Wiley–VCH

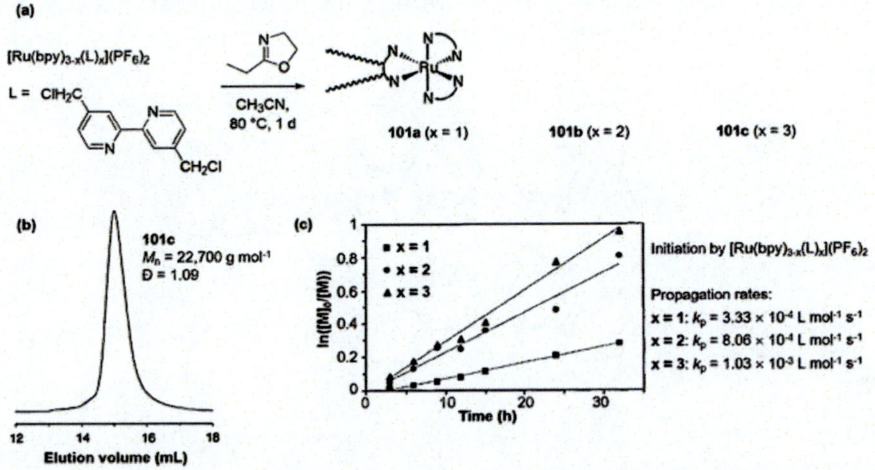

Fig. 2.86 **a** Schematic representation of the synthesis of star-shaped poly(2-ethyl-2-oxazoline)s **101** using Ru^{II} *tris*-bipyridine initiators. **b** SEC trace of the six-arm star-shaped polymer **101c** (eluent: DMF w/0.4% Et_3N, 50 °C). **c** Comparison of the first-order kinetics plots for polymerization of 2-ethyl-2-oxazoline (EtOx) under identical reaction conditions, using the various $[Ru(bpy)_{3-x}(L)_x](PF_6)_2$ initiators ($[initiator]_0$ = 8.3 mM, $[EtOx]_0$ = 5 M) [565]. Figure reproduced with kind permission; © 1999 American Chemical Society

with the polyanionic DNA strands and, thereby, protect them [568]. In addition to this, PEI features a remarkable buffering capacity, which arises from the secondary amino groups along the polymer chain. Thus, multiple protonation is enabled at low pH values and even under physiological conditions (for this reason, PEI is considered as a "proton sponge") [569]. Two different topologies are commonly utilized in biomedical applications: Branched and linear PEI. The latter can readily be prepared by acidic hydrolysis of the pending amide groups in poly(2-oxazolines) [570]. Against this background, Fraser et al. converted the star-shaped polymer

101c into the corresponding PEI derivative by hydrolysis with aq. HCl [571]. The resultant polymer was tested, as a vector, with respect to its applicability in gene delivery. For this purpose, transfection studies on LNCaP prostate cancer cells, using the green fluorescent protein, as marker, were conducted. The transfection efficiency of the ruthenopolymer was ca. 50% of that found for linear PEI under the otherwise same experimental conditions.

Besides the complexation of PCL-containing macroligands by Ru^{II} ions [261], the ROP of cyclic esters by appropriate hydroxy-functionalized Ru^{II} polypyridyl complexes was also investigated. In this respect, Schubert et al. utilized a heteroleptic Ru^{II} *bis*-tridentate complex bearing a hydroxymethyl-moiety in the Sn^{II}-catalyzed ROP of ε-caprolactone (Fig. 2.87) [572]. These authors prepared the same Ru^{II}-PCL conjugate **102** via both the convergent and divergent routes and proposed that the latter would be preferable in those cases where the complexation step requires harsh reaction conditions which, in turn, might result in a (partial) decomposition of the polymer chain.

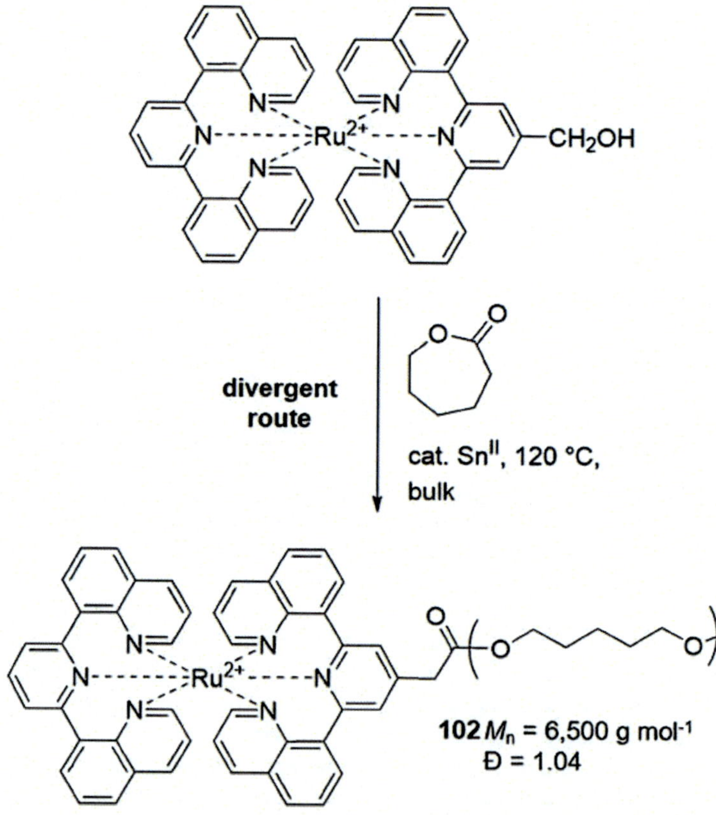

Fig. 2.87 Schematic representation of the Ru^{II}-PCL conjugate **102** via the divergent route

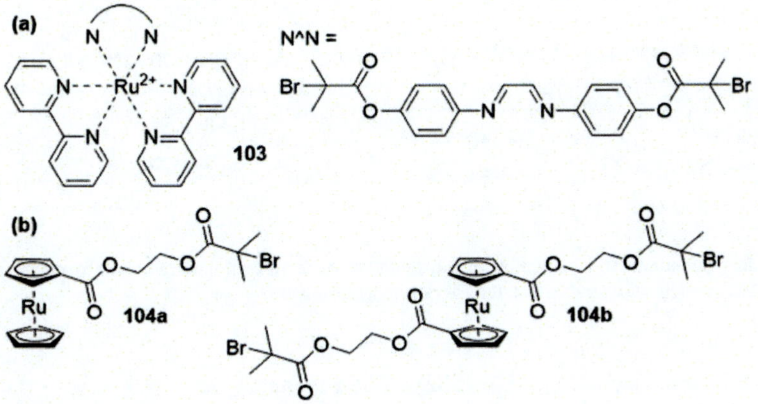

Fig. 2.88 Schematic representation of the Ru^{II}-containing ATRP initiators **103** and **104**

The ATRP of styrene, initiated by various chloromethylated Ru^{II}–bpy *tris*-complexes was reported by Fraser et al. in 1998 [573]. Well-defined star-shaped Ru^{II}-centered polymers were obtained demonstrating the compatibility of the ATRP process with cationic initiators. Furthermore, the ATRP of MMA-type monomers was conducted utilizing a $[Ru(bpy)_3]^{2+}$ complex with pending α-bromoester groups, as initiating moieties [574–577]. Chan et al. prepared, *inter alia*, the Ru^{II} complex **103**, which contained a bis(α-bromoester)-functionalized diamine ligand (Fig. 2.88a); [578] this complex successfully initiated the ATRP of MMA or styrene, following first-order kinetics with a polymerization rate of ca. 10^{-5} s^{-1}. Moreover, the thusly obtained metallopolymers served as efficient photosensitizers in the visible region and photoconductivity in the order of 10^{-10} Ω^{-1} cm^{-1} with a quantum yield (Φ) of ca. 10^{-6} were observed for the tin polymer films after doping with triphenylamine (TPA).

Recently, the mono- and bifunctional ruthenocene-containing ATRP initiators **104** were reported by Craig, Tang et al., thus widening the field of scope of Ru^{II}-containing initiators (Fig. 2.88b) [579]. The ATRP of MMA afforded the corresponding PMMAs with a terminal or central ruthenocene $[Ru(Cp)_2]$ moiety, respectively. The latter was investigated with respect to its mechanochemistry, i.e., to induce chain scission at the basically robust metallocene sites by sonication. After one hour of sonication, the M_n value was decreased from 76,400 to 52,600 g mol^{-1}. ^1H NMR analysis indicated that ca. 60% of the $Ru(Cp)_2$ sites were cleaved—despite the fact that $Ru(Cp)_2$ is significantly more stable against homo- or heterolytic dissociation compared to ferrocene (Fc).

The Ru^{II} *tris*-bipyridine complex **105**, equipped with two pending chain-transfer moieties, was reported by Ghiggino et al. (Fig. 2.89) [43, 580]; This complex was employed in the AIBN-initiated RAFT polymerization of a styrene-functionalized coumarin dye. Moreover, star-shaped polymers with pending 9,10-diphenylantharcene (DPA) moieties were synthesized from the same

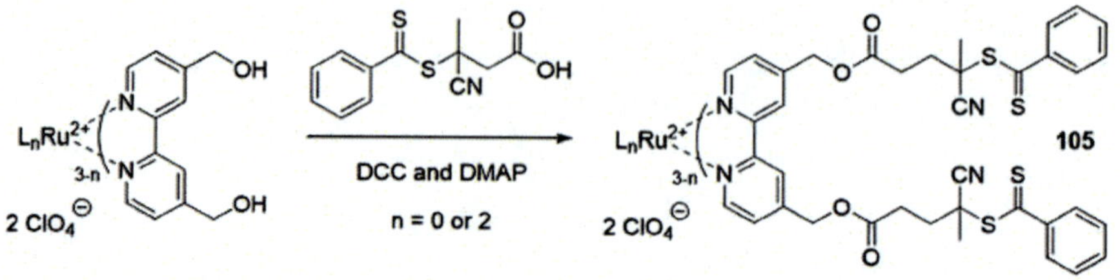

Fig. 2.89 Schematic representation of the synthesis of the metallo-supramolecular chain-transfer agent **105**, which was used in the RAFT polymerizations

chain-transfer agent [581, 582]. Steady-state as well as transient emission/absorption experiments were conducted in order to investigate the intramolecular EnT processes following the photo-excitation of the $[Ru(bpy)_3]^{2+}$ core and DPA units. Thereby, *intra*-chain triplet energy transfer from the Ru^{II} core to the DPA chromophores along the polymer chain was found to be the more favored pathway compared to the reverse singlet energy-transfer process. These authors attributed this finding to the long triplet-state lifetime of both the Ru^{II} complex and the polymer. Furthermore, triplet–triplet annihilation (TTA) and delayed fluorescence from the DPA moieties was evidenced (Fig. 2.90). TTA represents a versatile strategy towards energy upconversion, where generally a high-energy photon is produced following the absorption of two low-energy photons [583].

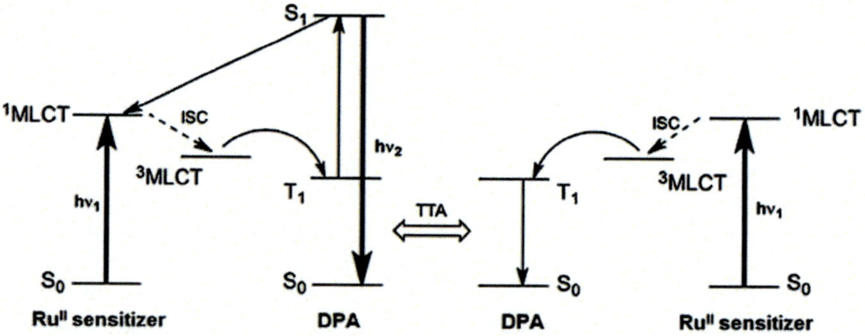

Fig. 2.90 Schematic representation of the proposed mechanism of the energy-upconversion process using a Ru^{II} complex and DPA, as sensitizer and emitter, respectively [581]. Figure reproduced with kind permission; © 2013 Elsevier B.V

2.2.5 Ruthenopolymers, Derived from the (Co) polymerization of Ru^{II}-Containing Monomers

The (co)polymerization of Ru^{II}-containing monomers represents a widely used strategy when aiming for the incorporation of such complexes into polymeric architectures. In order to discuss the broad range of materials reported in the literature up to now, a further classification is required. Besides the structural discrimination, i.e., distinguishing between polymers having the metal centers within the backbone or somehow bound to the polymer chain (i.e., either within the pending side chains or coordinated directly to the backbone), the employed polymerization method also needs to be considered. Independent from the architecture of the metallopolymers, either conventional polymerization techniques or an electrochemical polymerization might be applied. The latter typically generates insoluble metal-containing polymer films directly onto conductive substrates. Herein, ruthenopolymers derived from conventional "organic" polymerizations are discussed (denoted as methods Va/b, Fig. 1). Those Ru^{II}-containing polymers, which have been generated by an electropolymerization, are dealt with separately in Chap. 3.

2.2.5.1 Polymers Bearing Ru^{II} Centers in the Side Chains or Attached to the Backbone

(a) Ruthenocene-type and related monomers

In contrast to the broad range of polymers with ferrocene moieties in the side chain, their ruthenocene-containing counterparts have been investigated only to a minor extend. Besides the greater costs when just comparing the involved metals (i.e., iron vs. ruthenium) also the increased synthetic effort required to prepare ruthenocene, $Ru(Cp)_2$, as well as its reduced reactivity have to be considered. In 1966, the latter issue was addressed by Neuse, who by the same time, also reported on the first ruthenocene-containing polymer [584]. In contrast to ferrocene, the polycondensation of ruthenocene and different aldehydes proceeded only at elevated temperatures; moreover, higher Zn^{II} catalyst loading and longer reaction times were required. The ruthenopolymer **106** was obtained as a random mixture of 1,2-; 1,3-; and 1,1′-linked $Ru(Cp)_2$ species (Fig. 2.91). Due to the off-white to light grey color of the polymer (the corresponding ferrocene-based systems were yellow to dark brown), such materials could be useful in applications requiring stability towards high-energy UV radiation. Remarkably, the thermal stability was increased significantly when going from Fe^{II}- to Ru^{II}-containing polymer as shown by thermogravimetric analysis (TGA; i.e., 10% vs. 15% weight loss at 450 °C).

The utilization of vinylferrocene, as monomer, in the polymerizations was already investigated in the late 1950s, shortly after the discovery of parent ferrocene [585–587]. Using AIBN, as the initiator for the FRP, gave poly(vinylferrocene) (PVFc) with low molecular weight and high dispersity. Though Wilkinson isolated ruthenocne almost at the same time as ferrocene was first described [588], its vinyl derivative was employed, as monomer, in the FRP much later [589–591].

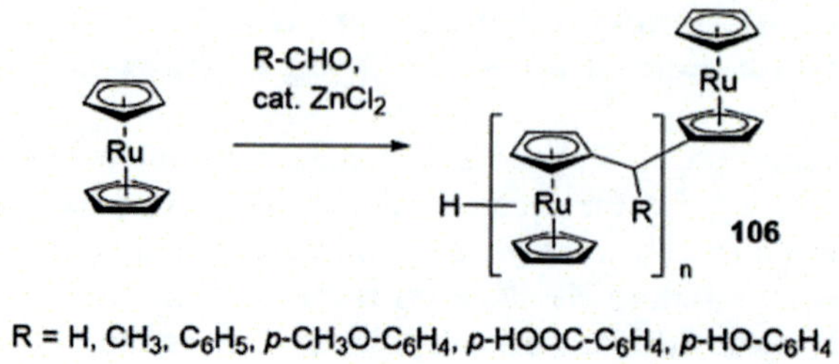

Fig. 2.91 Schematic representation of Neuse's synthesis of the first ruthenocene-containing polymer (**106**)

Homopolymerization as well as the copolymerization with acrylates, styrene or *N*-vinylpyrrolidinone was reported by Sheats et al. [representatively, the synthesis of poly(vinylruthenocene) **107** is depicted in Fig. 2.92a]. Pittman proposed that the redox-active metallocene center adjacent to the reactive double bond might quench the radical in the polymerization via an intramolecular ElT process between the metal center and active chain end (Fig. 2.92b) [592]. Due to the higher oxidation potential of ruthenocene, relative to ferrocene, this process became retarded; however, chain-transfer reactions occurred during the polymerization, which led to a significant branching of the polymer as revealed by the high dispersity (Đ $\leq$ 6). A library of well-defined copolymers, based on vinylruthenocene and styrene monomers, with molar masses (M_n) in the range of 0.8 to 2 $\times$ 10^5 g mol^{-1}, was reported by Pittman et al. [592, 593]. Subsequent oxidation with DDQ gave

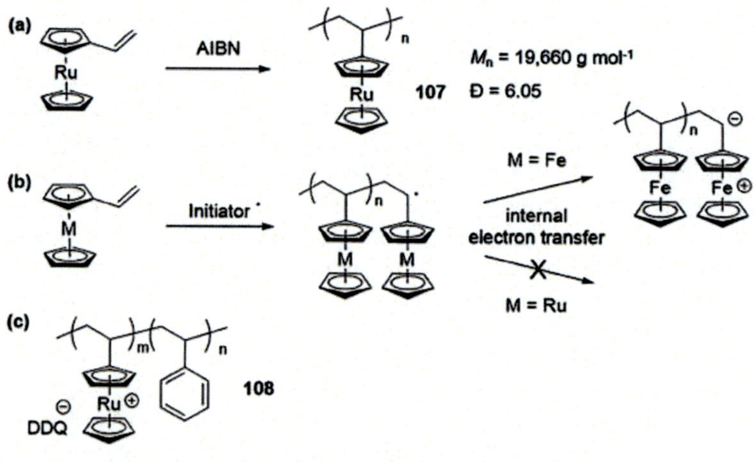

Fig. 2.92 Schematic representation of (**a**) the synthesis of poly(vinylruthenocene) **107** via FRP, (**b**) the FRP quenching due to an internal electron-transfer process and (**c**) the ruthenocenium-containing copolymer **108**

ruthenocenium-containing polymers (**108**; DDQ: 2,3-dichloro-5,6-dicyano-1,4-benzoquinone; Fig. 2.92c); however, the conductivity of these materials was, in contrast to their ferrocene-containing counterparts, not investigated. It is worth mentioning, that vinylosmocene was also utilized, as a comonomer, in this study—this represents one of the very few examples where this particular monomer was employed to prepare osmocene-containing polymers.

Allock et al. employed the $Ru(Cp)_2$-functionalized fluorinated cyclophosphazenes **109** and **110**, as monomers, in the thermal ROP (Fig. 2.93) [10, 594]. In all cases, the labile fluorine atoms were subsequently replaced by trifluoroethoxy moieties, thus affording hydrolytically stable and organo-soluble materials. In contrast to the corresponding Fc-substituted cyclophosphazene, monomer **109** readily polymerized at 300 °C in absence of any activation agent, such as $(NPCl_2)_2$. The transannular monomer **110** gave a homopolymer of high molecular weight ($M_n > 2 \times 10^6$ g mol^{-1} according to SEC analysis and viscometry measurements). Moreover, copolymerization of **110** with the analogous Fc derivative gave the heterobimetallic random copolymer **111** (Fig. 2.93).

The first controlled/living polymerization of a ruthenocene-containing monomer was reported by Buchmeiser and Schrock in 1995 [595]. These authors utilized a Schrock-type Mo^{VI} imido-alkylidene complex, as catalyst, for the polymerization of ethynlyruthenocene under mild conditions; pyridine-4-carboxaldehyde served as the terminating agent (Fig. 2.94a). This method afforded the polyacetylene-type polymers **112** with molecular weights (M_n) ranging from 6,700 to 16,400 g mol^{-1} (Đ values ca. 1.19). The chain growth exclusively occurred via an α-addition and kinetic studies confirmed the good control over the polymerization process (Fig. 2.94b). Due to steric crowding and the resulting distortion of the linear polyacetylene backbone, **112** could not be considered as a highly π-conjugated material. In general, the degree-of-conjugation and, thereby, the conductivity of

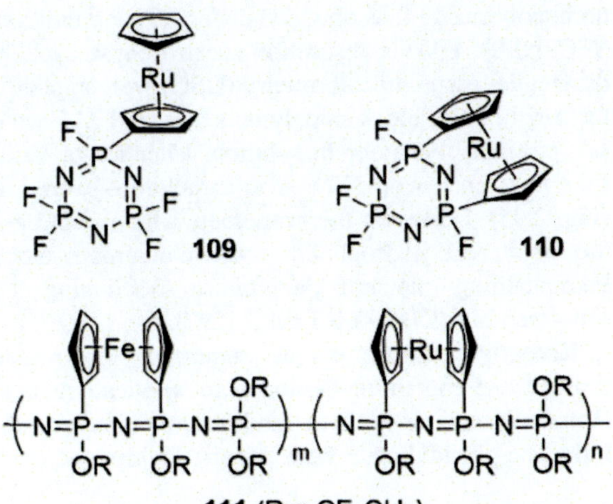

111 ($R = CF_3CH_2$)

Fig. 2.93 Schematic representation of the cyclophosphazene monomers **109** and **110** as well as of the polyphosphazene **111** possessing two different metallocenyl substituents

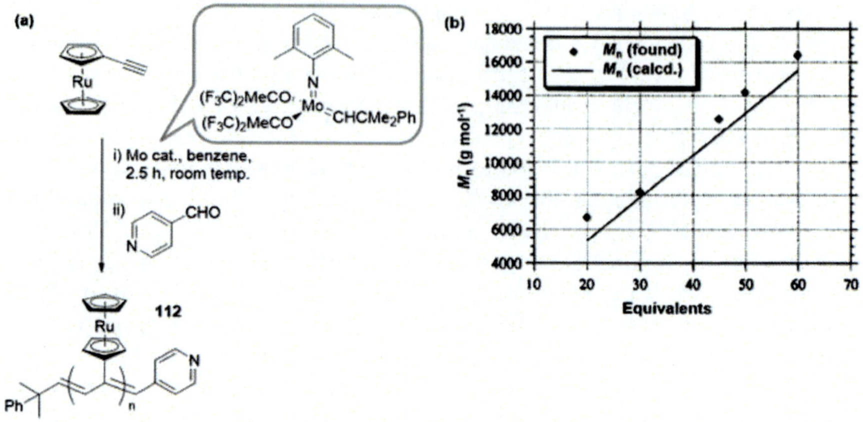

Fig. 2.94 a Schematic representation of the synthesis of poly(ethynylruthenocene) **112**. **b** Linear plot of the obtained molecular weight (M_n) of **112** against the calculated one [595]. Figure reproduced with kind permission; © 1995 American Chemical Society

polyacetylenes can be increased by oxidative or reductive doping; however, studies on how this applies to **112** were not reported by these authors.

Today, polymetallocenes are typically prepared by utilizing living and controlled polymerization methods, anionic as well as radical. For this purpose (meth)acrylic and styrenic metallocene derivatives are commonly employed—with the vast majority being ferrocene or cobaltocenium derivatives [596]. The first (and presumably so far only) example, where well-defined side-chain ruthenocene-containing polymers and block copolymers were prepared by ATRP and RAFT, was reported by Tang et al. (Fig. 2.95) [16]. Both techniques allowed the polymerization of **113** to proceed in a highly controlled fashion. The homopolymer **114**, as obtained under ATRP conditions, was characterized in detail: SEC, NMR, UV/vis absorption spectroscopy, cyclic voltammetry, TGA as well as differential scanning calorimetry (DSC) were applied as analytical tools. Moreover, the amphiphilic block copolymers **115** and **116** were investigated regarding their self-assembly behavior in solution. Micelles of various morphologies—including spherical and worm-like nanostructures—were observed by TEM imaging (Fig. 2.95). However, the properties, which might arise from the crystallization of the hard and hydrophobic metal-containing block, were not studied [some Fc-containing materials exhibit a fascinating "living" crystallization-driven self-assembly (CDSA) behavior [597].

Recently, Wurm et al. reported on ruthenocenyl glycidyl ether, a ruthenocene-containing monomer for applications in anionic polymerizations [598]. Homopolymers as well as random copolymers with ethylene oxide (EO) and/or ferrocene glycidyl ether were prepared with good control over the molecular weight

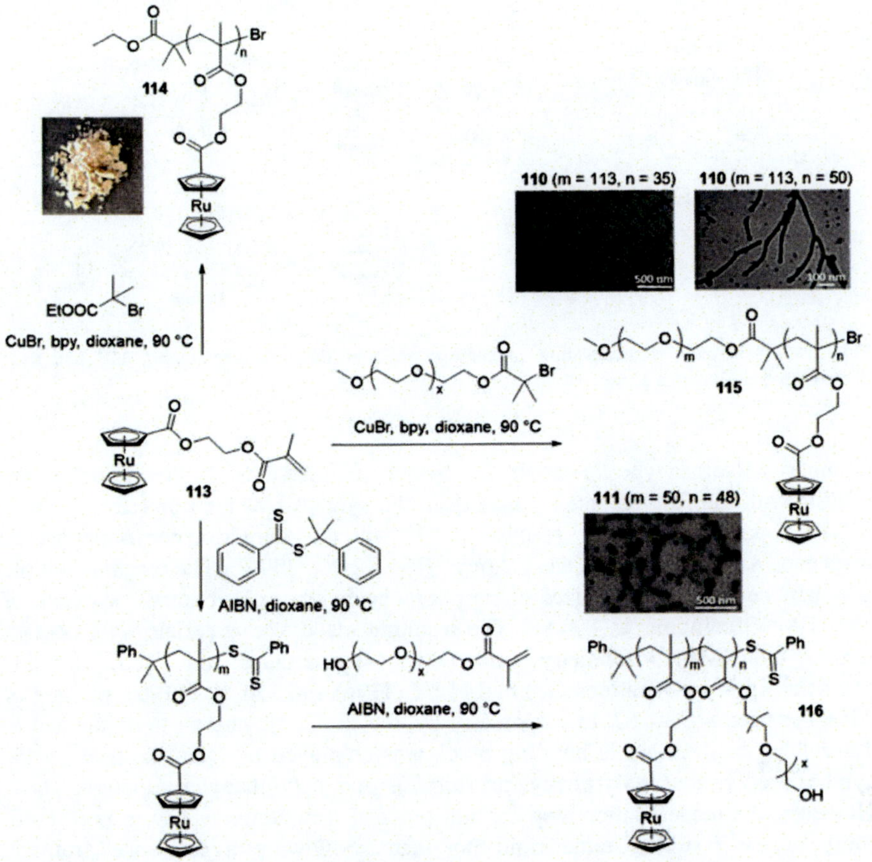

Fig. 2.95 Schematic representation of the synthesis of ruthenocene-containing homo- (**114**) and diblock copolymers (**115** and **116**) via ATRP or RAFT polymerization. The TEM images of the polymer micelles formed in aqueous medium are also depicted [16]. Figure reproduced with kind permission; © 2013 American Chemical Society

distribution (Đ ≤ 1.15). The copolymers with EO were water soluble and showed the characteristic redox-activity associated to ruthenocene.

Segal investigated the applicability of the cationic, η^6-arene-containing $Ru(Cp)^+$ complex **117** in the synthesis of polymeric aromatic ethers [599]. Basically, the nucleophilic aromatic substitution reaction on chloride-functionalized benzenes can be accelerated significantly by coordinating metal centers in a η^6-fashion to the aromatic ring. In particular, the $Fe(Cp)^+$ unit has widely been utilized for this purpose and Abd-El-Aziz et al. have reported a wide range of metal-containing polymers, based on this concept [417, 600]. It has been demonstrated, that the demetallation of the metallopolymers represents a versatile method for preparing

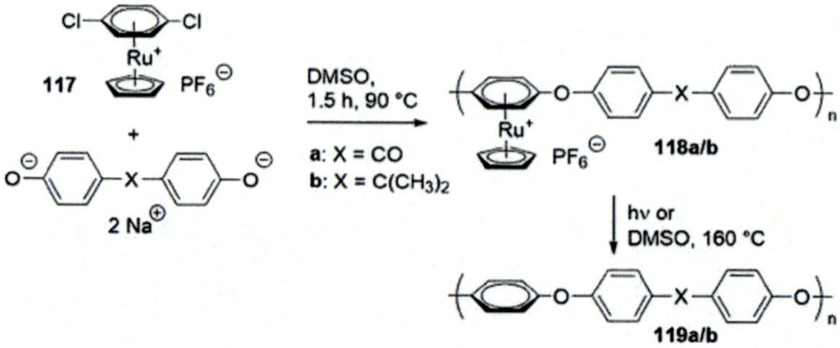

Fig. 2.96 Schematic representation of the synthesis of the ruthenopolymers **118** and their demetallized derivatives **119**

aromatic polyether-type materials of diverse architectures. In the initial study dealing with the Ru^{II}-containing materials, the synthesis of the poly(ether-ketone) **118a** and polyether **118b** by reaction of **117** with 4,4′-dihydroxybenzophenone or bisphenol-A, respectively, was shown (Fig. 2.96) [599]. The organo-soluble polymers could be demetallized either photochemically or by thermal treatment of their DMSO solutions at 160 °C. The resulting metal-free materials were characterized by NMR, viscometry, and DSC measurements; the latter revealed glass-transition temperatures (T_g) of 144 °C (**119a**) and 115 °C (**119b**), melting of **119a** occurred at 335 °C. In a related study, the Ru(Cp*)$^+$ unit was coordinated to 1,4-dichlorobenzene—this building block was employed by Dembek et al. in the synthesis of various poly(phenylene sulfide)s and poly(phenylene oxide)s. [601] However, the demetallation step did not proceed quantitatively due to poor solubility of the resulting rather-rigid polymers. Following a step-wise protocol, Abd-El-Aziz et al. introduced both Fe(Cp)$^+$ as well as the Ru(Cp*)$^+$ units along polymer backbones; two selected examples for such heterobimetallic polymers (**120**) are depicted in Fig. 2.97 [602].

(a) ***Monomers containing Ru^{II}-polypyridyl complexes***

In contrast to the scattered reports on the synthesis of polymers with pending ruthenocene moieties, a broad range of polymers with Ru^{II}–polypyridyl complex within the side chains has been reported—many of them were prepared by polymerizing complex-containing monomers. Besides the early studies, where mainly the FRP was utilized for this purpose, today particularly the controlled/living methodologies are in the focus of research.

Meyer et al. demonstrated that intramolecular energy and/or electron transfer may occur in appropriately functionalized polymers, such as **121**, which contained pending $[Ru(bpy)_3]^{2+}$ and/or $[Os(bpy)_3]^{2+}$ complexes as well as anthracene moieties (Fig. 2.98) [27, 34, 35]. Upon excitation of the Ru^{II} centers, a rapid long-range EnT towards the Os^{II} sites occurred ($k > 1 \times 10^8\ s^{-1}$); for this process, a

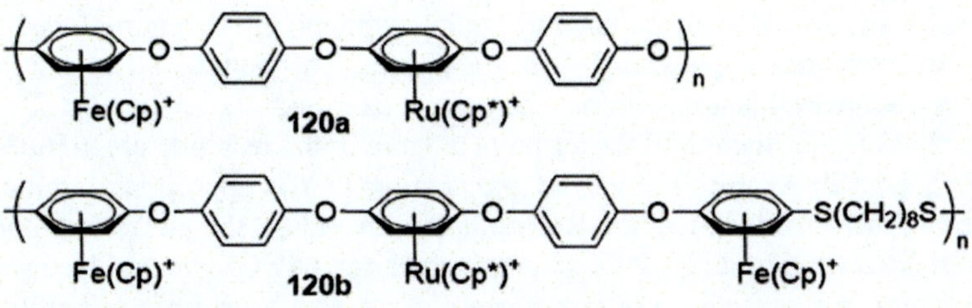

Fig. 2.97 Schematic representation of the heterobimetallic polymers **120**

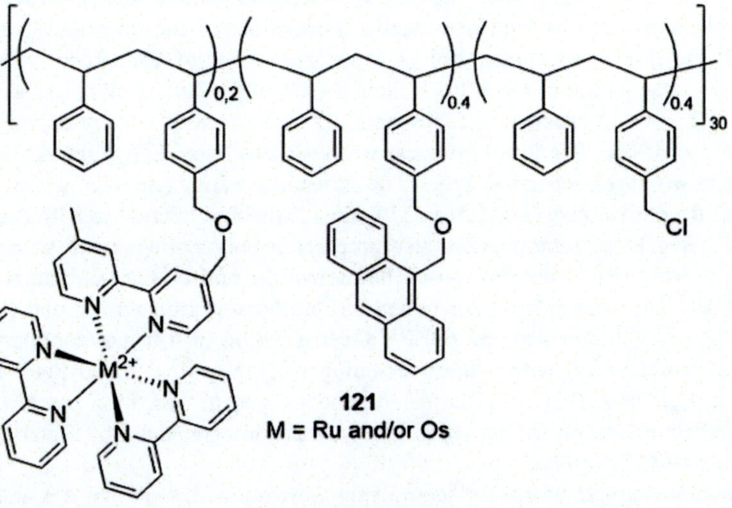

Fig. 2.98 Schematic representation of the Ru^{II}/Os^{II}-containing copolymer **121**

cascade-like mechanism was identified in which the anthracene units acted as energy relays. Laser-flash photolysis and emission quantum yield measurements revealed an almost quantitative quenching of the photo-excited Ru^{II} centers; whereby, the EnT process proceeded on the timescale of the laser pulse ($\sim$5 ns). However, for the copolymers missing the anthracene moieties, the metal-to-metal EnT process was absent [27, 34, 35]. Thus, substituted polystyrene represents an appropriate framework for constructing multifunctional polymer architectures via the stepwise (co)polymerization of the respective monomers. An alternative synthetic route, which is the post-polymerization functionalization (i.e., application of methods I or II, Fig. 1), basically suffers from poor control over the relative distribution of functional groups along the polymer chain.

These same authors addressed this issue by making use of a solid-phase peptide synthesis. This strategy has hardly been employed in the field of metal-containing materials—presumably due to the substantial synthetic effort required when compared to conventional polymerization techniques. The peptide strands were prepared in a stepwise fashion with precise control over both the spatial distribution of the functional side chains and the tertiary structure of the resulting chain [603–605]. The oligopeptide synthesis involved preparation of the appropriate amino acids containing the sensitizer, i.e., the Ru^{II} complex, as well as the electron-donor and -acceptor units (in this study, PTZ and anthraquinone (ANQ) were utilized as donor and acceptor, respectively). For this purpose, the *L*-enantiomer of azidoproline was reduced to the amine; subsequently, the amine group was *tert*-butoxycarbonyl-protected and the proline derivative was coupled to the carboxylic derivatives of the three photoactive units. Solid-state peptide synthesis gave the 13-residue oligoproline **122** (Fig. 2.99), which adopted a helical structure with threefold symmetry in aqueous media. Formation of the so-called polyproline type-II motif was concluded from CD measurements of the amide region [604, 605]. The quenching of the MLCT excited state of the Ru^{II} centers was dominated by a PTZ → $^{*}Ru^{II}$ reductive ElT within 10 to 20 ns with solvent-dependent efficiencies of >90%. The former process was followed by an ElT to the ANQ units to populate a charge-separated state. The efficiency of its population was strongly solvent dependent and varied from 33% (in 1,2-dichloroethane) to 86% (in MeCN) [605]. Later, these same authors also prepared a polypeptide strand with two different pending Ru^{II} complexes—a photosensitizer and a water-oxidation catalyst site [606]. The material was integrated into a photoelectrochemical cell (PEC) for the oxygen-evolution reaction (OER). Over a period of 8 h, the continuous electrolysis produced O_2 with a turnover number (TON) of ca. 8,000 (64% Faradaic efficiency, TOF of 0.28 s^{-1}), the heterogeneous system was, by a factor of 10, the more efficient catalyst for the OER than typical homogeneous Ru^{II} water-oxidation catalysts (WOCs) alone.

The first example where Ru^{II} complexes were utilized, as monomers, in a radical polymerization was reported by Fréchét et al. [18, 19]. Utilizing FRP gave the copolymers **32** containing a coumarin moieties and Ru^{II}–bpy *tris*-complexes, as pending side chains. These authors compared these polymers to those prepared via an alternative route, i.e., the complexation of (co)polymers (method II; the synthesis of **32** via both routes has already been shown above, see Fig. 2.31). The reduced solubility of the latter arose from the partial cross-linking of the polymer chains during the complexation step. The bichromophoric materials featured enhanced absorption and luminescence properties when compared to analogous small-molecule dyads due to a highly efficient EnT between the electron-donating coumarin dyes and the Ru^{II} centers (>95%). Moreover, the terpolymer **123**, comprising of two different donor-type chromophores in combination with the Ru^{II} complexes, was synthesized via FRP (Fig. 2.100); also, this polymer offered an efficient EnT from the organic donor sites to the metal centers [18].

Schubert et al. investigated the FRP of methacrylate-based monomers, including a $[Ru(bpy)_3]^{2+}$ complex; different monomer-to-initiator ratios were used and, thus, the Ru^{II} content in the resultant polymers was varied over a wide range from 1.3 up to 22% [607]. These copolymers (**124**) revealed the photophysical properties of

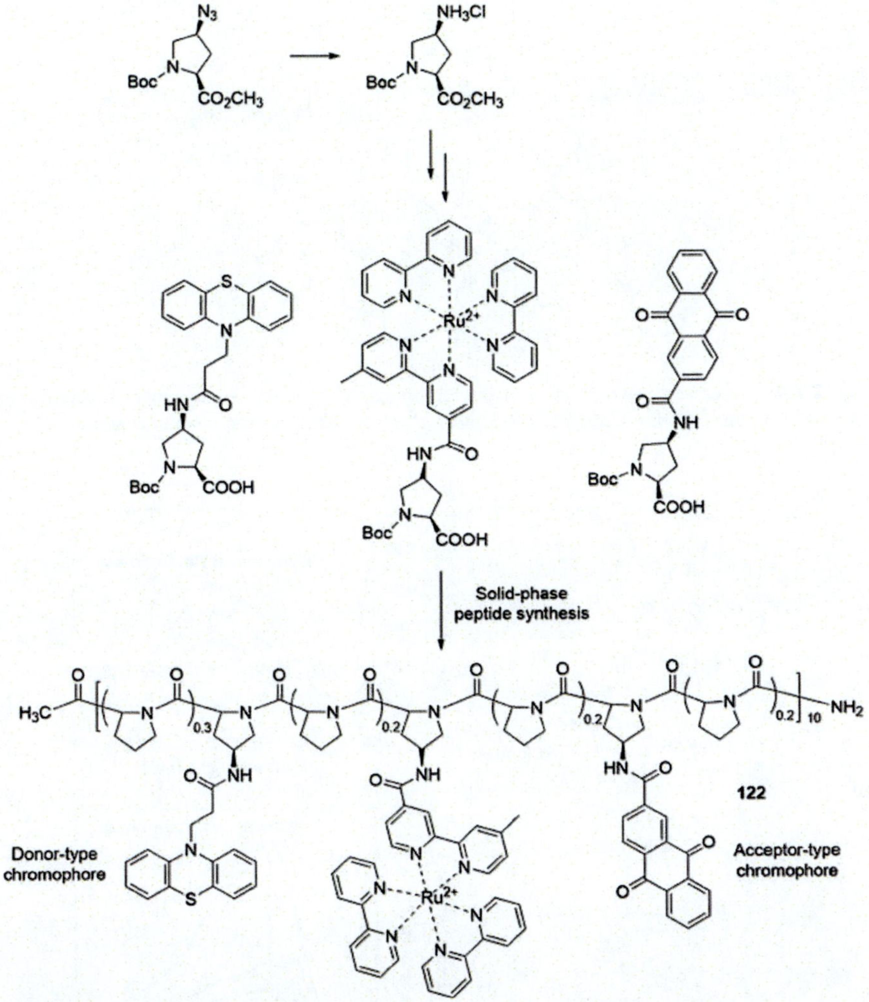

Fig. 2.99 Schematic representation of the synthesis of the polypeptide **122** with donor, sensitizer and acceptor moieties, as side-chain residues

parent $[Ru(bpy)_3]^{2+}$ without any interference by the PMMA backbone. Moreover, **124** was processed into thin-solid films, either by spin-coating or inkjet-printing; thereby, aggregation of the complexes was efficiently suppressed even for the polymers with high Ru^{II} contents. With increasing film thickness, the emission and absorption intensity of **124** decreased almost linearly (Fig. 2.101) [608].

As summarized by Zhou, Ding et al., self-oscillating polymers (SOPs) represent important materials to mimic natural self-oscillating phenomena [609]. Against this background, Yoshida et al. applied polymer-bound $[Ru(bpy)_3]^{2+}$ complexes, as

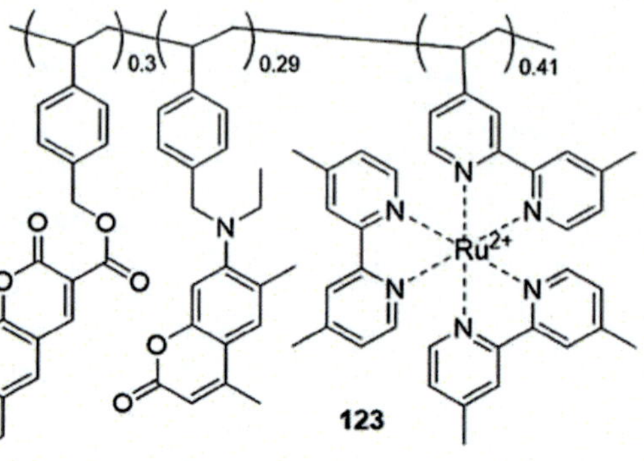

Fig. 2.100 Schematic representation of the terpolymer **123**, which contains two different laser dyes (as electron-donating chromophores) and Ru^{II} complexes, as pending side chains

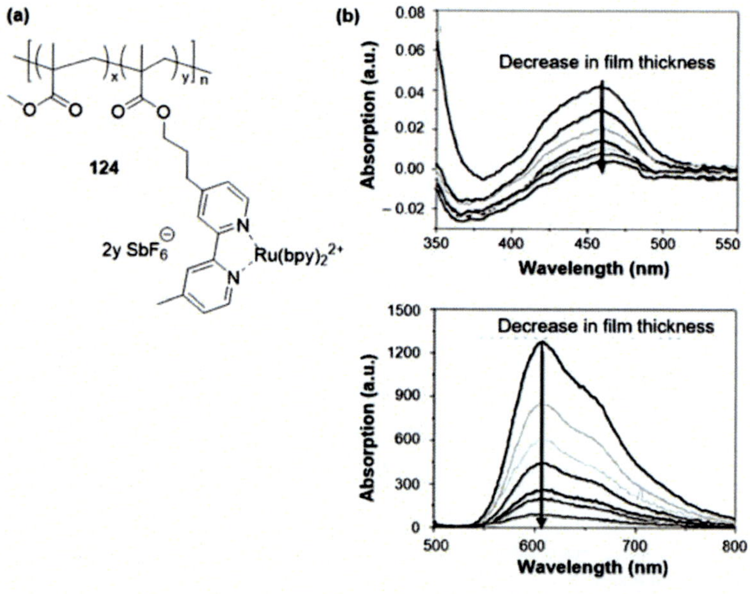

Fig. 2.101 **a** Schematic representation of the PMMA-based copolymers **124**. **b** UV/vis absorption and emission spectra of the ink-jet printed copolymer films with variable film thickness ranging from 120 to 900 nm [608]. Figure reproduced with kind permission; . © 2005 Wiley-VCH

catalyst, for the self-oscillating Belousov-Zhabotinsky (BZ) reaction [610–612]. For this purpose, these authors copolymerized the complex-containing monomer with *N*-isopropylacrylamide in the presence of a cross-linking agent to obtain the thermo-responsive Ru^{II}-containing polymer gel **125**. Immersing this SOP gel into a Ce^{IV}/Ce^{III} solution induced the BZ reaction; periodical swelling and deswelling of the gel, at a constant temperature, was observed without any external stimulus. Depending on the composition of the environment, the chemical energy was transduced into mechanical energy and vice versa to drive the oscillation over a period of ca. 5 min. Later, SOPs containing additional sites, such as acrylamidomethylpropane sulfonate (AMPS), as a proton reservoir, and methacrylamidopropyltrimethylammonium chloride (MAPTAC), as an oxidizing species, were prepared [613, 614]. This particular polymer design allowed self-oscillation only to occur in the presence of maleic anhydride (alternatively, bio-related acids, such as citric acid, could be utilized for the same purpose). By introducing additional photochromic spirobenzopyran units, as side chains, the SOP could be switched between a soluble and an insoluble state (due to the reversible photo-induced isomerism of the spirobenzopyran sites) [615] (Fig. 2.102).

In subsequent work, Yoshida et al. developed autonomous mass-transport systems in which the peristaltic motion of the surfaces of SOP gel sheets [616–619] or

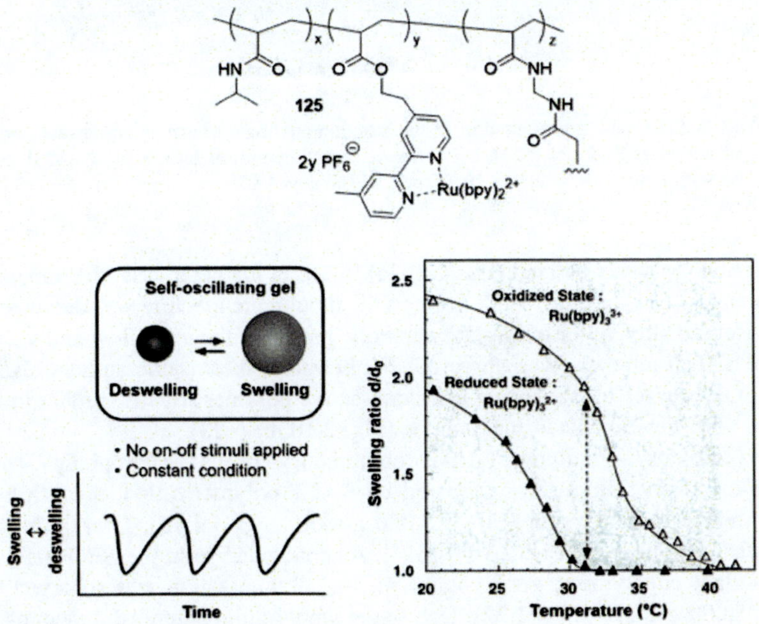

Fig. 2.102 Schematic representation of the thermo-responsive Ru^{II}-containing SOP gel **125**. Bottom left: Representation of the reversible swelling/deswelling behavior of an ideal self-oscillating gel. Bottom right: Equilibrium swelling ratio d/d_0 (d: characteristic length of the gel, d_0: length in the shrunken state) of **125** as a function of temperature under different conditions of a reduced Ru^{II} state (in a Ce^{III} solution) and an oxidized Ru^{III} state (in a Ce^{IV} solution) [611]. Figure reproduced with kind permission; © 1999 AIP Publishing LLC

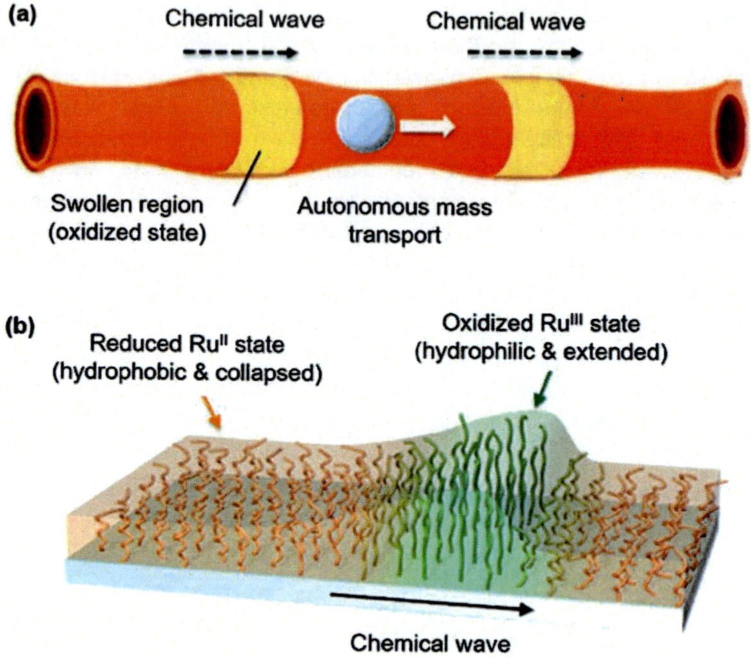

Fig. 2.103 **a** Schematic representation of the autonomous mass transport due to the peristaltic motions of a tubular SOP gel [621]. **b** Schematic representation of a SOP brush [623]. Figures reproduced with kind permission; © 2012 and 2013 Wiley–VCH

tubular SOP gels was utilized (Fig. 2.103a) [620]. In the latter case, the transport of CO_2 bubbles [621] or latex beads through the tubular architecture was demonstrated [622]. Remarkably, a Poiseuille-like parabolic profile of the mass flow and very low Reynolds (RE) numbers was observed. Furthermore, these same authors also prepared SOP brushes by anchoring the redox-active polymer strands, similar to those of **125**, onto functionalized glass slides (Fig. 2.103b) [623].

The diblock copolymer **126**, comprising of a hydrophilic and a thermo-responsive block, was prepared by RAFT polymerization, using a hydrophilic macro-RAFT agent (M_n = 5,000 g mol^{-1}, Đ = 1.15), and subsequent chain-end modification (Fig. 2.104) [624]. As concluded from the SEC analysis of the resultant copolymers, good control of the polymerization was achieved (**126**: M_n = 25,000 g mol^{-1}, Đ = 1.22). Due to the amphiphilic nature of the copolymer, micellization occurred in aqueous media—in the reduced as well as oxidized state. The critical aggregation temperatures were determined and the oxidized copolymer was found to disassemble into unimers at a slightly higher temperature (27.5 vs. 25 °C). This subtle difference was applied in the reversible switching between micelles and unimers under the conditions of the BZ reaction at room temperature (i.e., HNO_3, $NaBrO_3$, and MA). The temporal evolution of the structure was

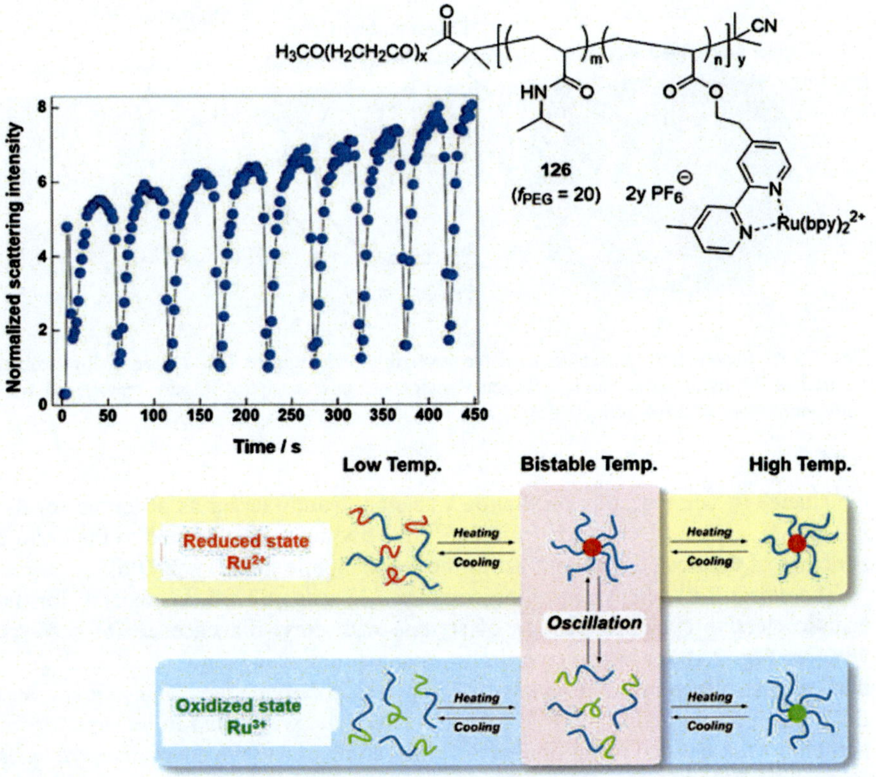

Fig. 2.104 Schematic representation of the amphiphilic diblock copolymer **126**. The time-resolved DLS data reveal the reversible switching between micelles (high scattering intensity) and unimers (low scattering intensity) under the conditions of the BZ reaction at room temperature, i.e., in a bistable regime [624]. Figure reproduced with kind permission; © 2013 The Royal Chemical Society

analyzed by time-resolved DLS measurements and exhibited a remarkable periodicity of the process (Fig. 2.104). Decreasing the fraction of hydrophilic segment gave a diblock copolymer, which afforded large vesicles in water (average diameter of ca. 1 μm) [625]. Also, these assemblies displayed a reversible switching between the vesicular and unimer states, driven by the BZ reaction (Fig. 2.105).

Finally, Yoshida et al. also reported biomimetic cross-linked polymersomes, which were formed from the multi-functional diblock copolymer **127** [626]. As in the previous example, the polymer assembled into vesicles in an aqueous medium. The vesicles, in turn, were stabilized by photochemical cross-linking of the vinyl groups within the shell. The resultant architectures showed, under the conditions of the BZ reaction, an autonomous oscillation with respect to their volume and shape (e.g., buckling deformations). According to time-lapse optical microscopy imaging,

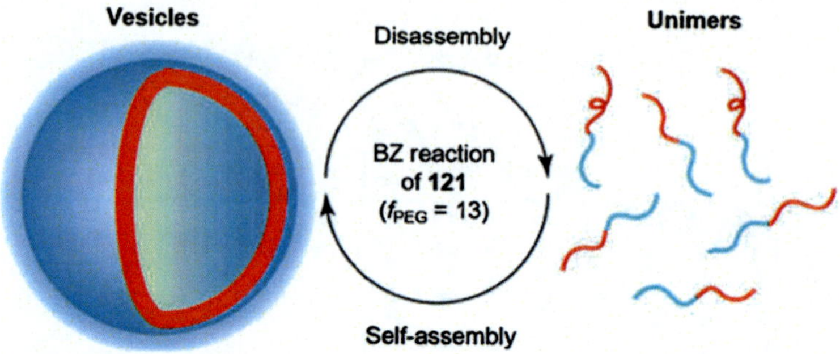

Fig. 2.105 Schematic representation of the reversible switching of **126**, having a low weight fraction of the hydrophilic block, between vesicles and unimers [625]. Figure reproduced with kind permission; © 2014 Wiley-VCH

the change in volume upon oscillation was significantly larger as observed for the initial SOP gels, e.g., obtained from **125**. This finding was attributed to the hollow structure of the vesicles in contrast to the more mobile gel (Fig. 2.106).

The chiral Ru^{II}–phenyloxazoline complex **128** was utilized, as catalyst, for the enantioselective cyclopropanation of styrene with ethyl diazoacetate (66% yield, 98% ee; Fig. 2.107a) [395]. In order to establish polymer-immobilized analogs of this catalyst, the acrylate derivative of **123** was copolymerized with styrene and DVB under FRP conditions (Fig. 2.107b). This route was found to be superior for preparing well-defined metallopolymers when compared to the alternative synthesis via method II, i.e., the loading of a ligand-equipped resin with Ru^{II} centers. Both the macroporous (**124a**) as well as macroreticular material (**124b**) were tested, as heterogeneous catalysts, in cyclopropanation reactions. In particular, **124a** offered

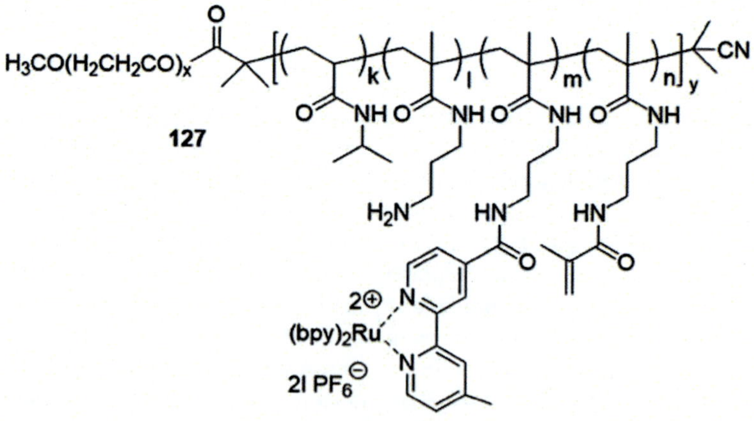

127

Fig. 2.106 Schematic representation of the multi-functional deblock copolymer **127**

excellent reactivity and enantioselectivity in inter- as well as intramolecular cyclopropanations. Presumably, **124b** suffered from its lower internal surface area due to decreased porosity. Moreover, **124a** could be recycled and retained its catalytic activity and selectivity even after ten applications or after several months of storage under ambient conditions. Under performance, no considerable leaching of the catalytically active Ru^{II} units from the polymer was detected.

The issue of preparing heterogeneous Ru^{II}-containing catalysts was also addressed by Mayoral et al. [627]. These authors synthesized the chiral immobilized catalyst **125** via two different methods (Fig. 2.107c): Copolymerization of an in situ generated complex-containing monomer and complexation of the ligand-containing resin. However, the material obtained from the first route possessed considerable Ru^{II} impurities; whereas, the second method allowed only a complexation of 50 to 60% of the binding sites—the low porosity of the polymer network prevented higher degrees-of-loading. As in the previous example, the catalysts showed a reasonable enantioselectivity in the cyclopropanation of styrene (*ee* values up to 85%). It was demonstrated later that more defined and, even more important, more potent catalysts can be obtained when flexible spacer units were introduced between the resin and ligand sites (these heterogeneous catalysts gave yields and *ee* values up to 70 and 99%, respectively) [628].

The FRP of monomers containing Ru^{II}–polypyridyl complexes is favorable, in particular when aiming for cross-linked materials, as in the previous examples, or when the complex sites need to be incorporated into a polymer regardless of the resulting microstructure or dispersity. On the other hand, the various CRP

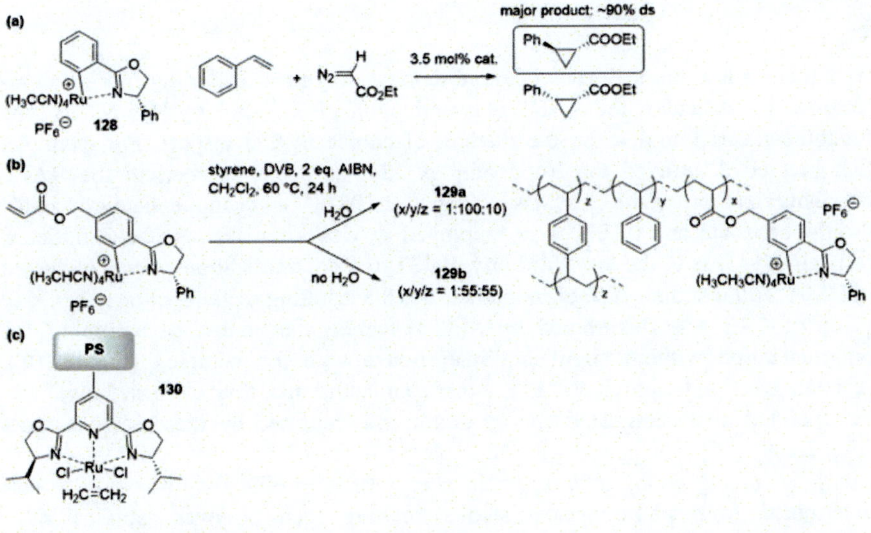

Fig. 2.107 **a** Schematic representation of the chiral Ru^{II} catalyst **128** for styrene epoxidation. **b** Schematic representation of the synthesis of the immobilized catalyst **129** via FRP. **c** Schematic representation of Mayoral's heterogeneous catalyst **130**

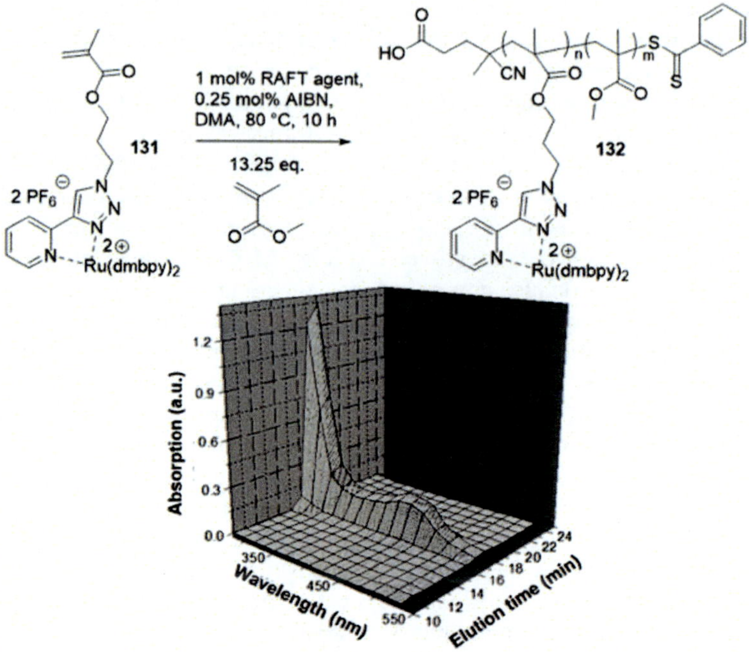

Fig. 2.108 Schematic representation of the RAFT copolymerization of **131**. The 3D-SEC plot of the resulting ruthenopolymer **132** is also shown [629]. Figure reproduced with kind permission; © 2009 Elsevier B.V.

techniques allow the synthesis of more defined polymers, including blocky architectures. In particular, the RAFT polymerization of Ru^{II}–polypyridyl MMA-type monomers was found to be the method of choice in this respect. For example, Schubert et al. utilized the Ru^{II} complex **131**, as a comonomer, in the RAFT copolymerization with MMA (Fig. 2.108); 4-(cyanopentanoic acid)-4-dithiobenzoate and AIBN were employed as chain-transfer agent and initiator, respectively. Due to the low solubility of **131**, the polymerizations were conducted in DMA with less than 10% of monomer. For the resulting metallopolymer **132**, a Đ value of 1.13 was determined by SEC revealing the controlled nature of the polymerization without significant interference with the complex (Fig. 2.108). A related set of copolymers **133**, where an additional thiazole-based dye was incorporated as a designated energy donor, was reported by these same authors (Fig. 2.109) [15]. The pending Ru^{II} complexes were structure-optimized to serve as energy-accepting units; an *intra*-chain energy-transfer efficiency of ca. 70% was determined from photophysical studies making such polymers applicable as light-harvesting antennas, e.g., in artificial photosynthetic systems. However, the content of the utilized Ru^{II} monomers had to be kept below 5 mol% in order to

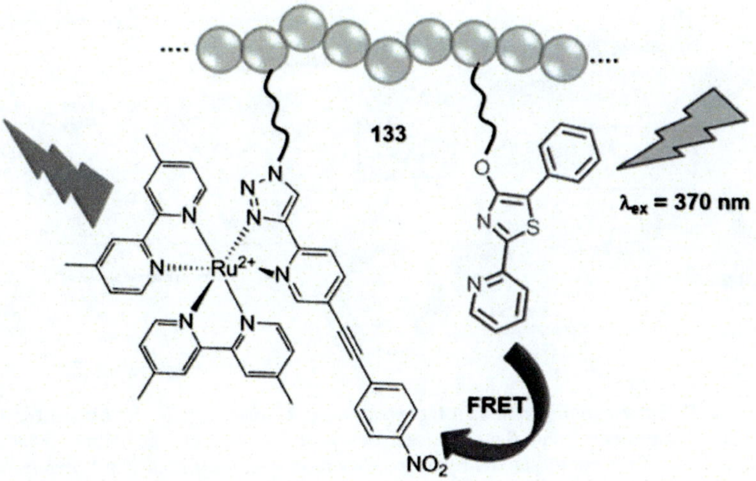

Fig. 2.109 Schematic representation of the donor–acceptor copolymer **133** exhibiting FRET-type *intra*-chain energy transfer [15]. Figure reproduced with kind permission; © 2011 American Chemical Society

guarantee the initiation; at higher ratios, the polymerization was retarded presumably due to side reactions involving the NO_2-substituents.

In order to employ Ru^{II}-equipped copolymers in photocatalytic applications, Schubert, Dietzek, Rau et al. prepared the water-soluble polymers **134** containing photoactive Ru^{II} or Ir^{III} complexes, as sensitizers, a MV^{2+}-type electron mediator as well as triethylene glycol methyl ether chains, as solubilizer (Fig. 2.110) [630]. As a proof-of-concept, hydrogen-evolution experiments were performed in an aqueous medium with Et_3N and colloidal platinum being added, as sacrificial donor and hydrogen-evolving catalyst, respectively. However, only the Ir^{III}-containing systems were found to be catalytically active due to a better spectral overlap of the blue LED irradiation light with the UV/vis absorption bands of the complexes.

A systematic study concerning the synthesis of Ru^{II}-containing copolymers via a RAFT polymerization was performed by Schubert et al. [631] Thereby, the applicability of various methods, according to Fig. 1, to obtain a copolymer with pending Ru^{II} complexes (**135**) was studied. In detail, either ethynyl-substituted Ru^{II} complexes were attached to the side chains of a preformed polymer using the Sonogashira cross-coupling reaction (i.e., method I), tpy-containing monomers were copolymerized and subsequently complexed (i.e., method II) or, thirdly, a polymerizable Ru^{II} complex was directly copolymerized (i.e., method Va). In general, all three methods gave the highly substituted and well-defined PMMA-based copolymers **135** (Fig. 2.111). From these experimental results, these authors concluded that method II was most applicable, since this approach not only enabled the straightforward removal of unreacted educts and/or side products

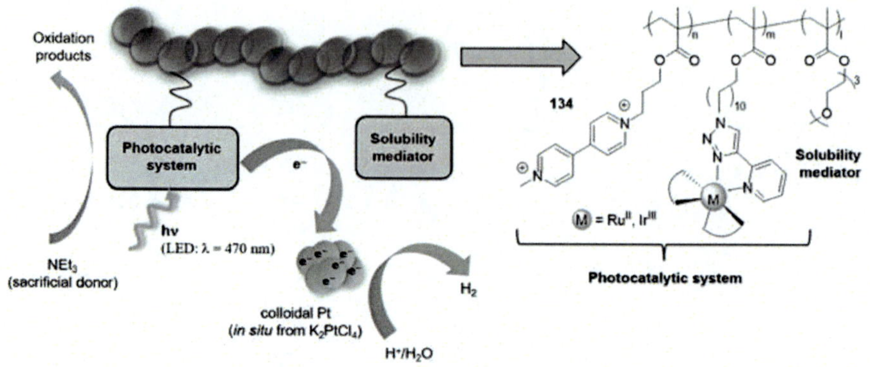

Fig. 2.110 Schematic representation of copolymer **134** containing a photosensitizer and an electron relay (together forming the photocatalytic system) as well as a solubility mediator. The working-scheme of the photocatalytic hydrogen evolution is also depicted [630]. Figure redrawn according to the literature; © 2015 Wiley-VCH

but also offered the highest yields. However, when aiming for materials with a high Ru^{II} contents method Va still appeared to be favorable, since steric crowding hampered a full conversion when the grafting-onto or complexation routes were followed.

Besides the utilization of CRPs to obtain Ru^{II}-containing (co)polymers directly from polymerizable polypyridyl complexes, also the ROMP of analogous norbornene-type monomers was investigated. In 2004, Weck et al. [11] as well as Sleiman et al. [632] independently reported the Ru^{II}-containing (co)polymers **136** and **137** (Fig. 2.112). These studies showed that different types of Grubbs' catalysts might be utilized to prepare well-defined polymers with Ru^{II}-polypyridyl complexes in the side chains—even the characteristics for a living polymerization were observed without interference by the metal centers. Thus, sequential addition of monomers allows the formation of diblock copolymers **137** with different relative block ratios and total chain lengths. The self-assembly behavior of amphiphilic **137** in solution was studied; thereby, both the polymer composition as well as self-assembly conditions were varied. Nanostructures including interconnecting bilayers, micelles, nanobowls, nanotubes or vesicles were observed by microscopy imaging [633]. For example, **137** (with m,n = 20) gave nanotubes, spherical bilayers or vesicles depending on the solvent composition (i.e., the applied MeCN/toluene ratio, Fig. 2.113). Using appropriate quenching agents enabled the end-group modification. For example, a protein-binding biotin entity was attached to one chain end (**138**) [634]. At a DP of 45 for both blocks, the solution self-assembly in a MeCN/H_2O mixture gave spherical core–shell micelles; the Ru^{II}-containing block represented the shell and the terminal biotin residues were located on the surface. In the presence of Streptavidin, possessing four potent biotin binding sites, cross-linking of the individual micelles occurred (Fig. 2.114).

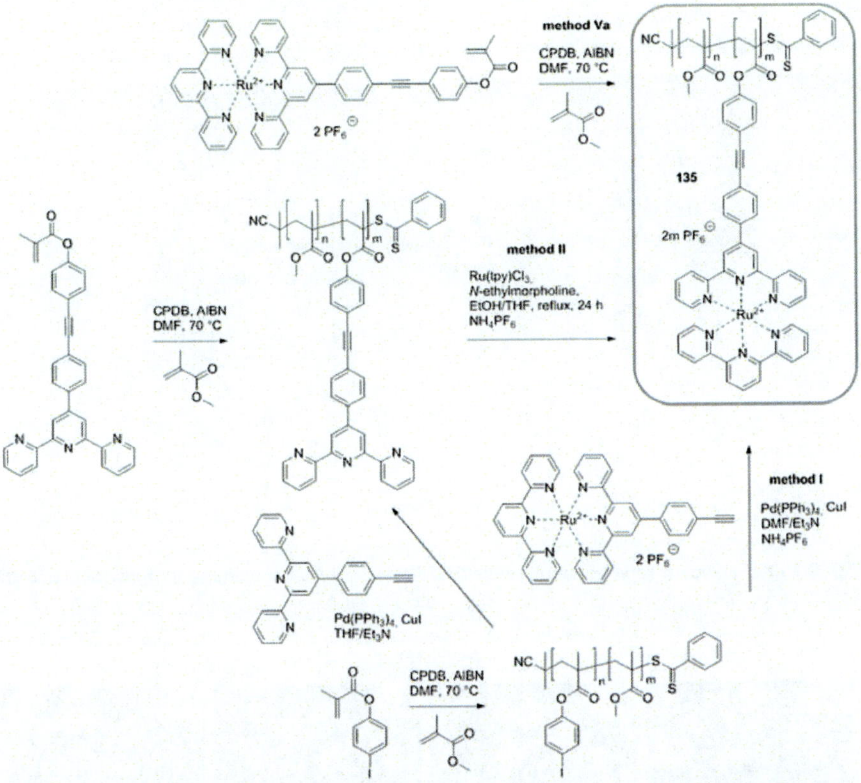

Fig. 2.111 Schematic representation of the synthesis of the Ru^{II}-containing copolymer **135** via three different routes (denoted as methods I, II and Va, see Fig. 1)

In a related study, Sleiman et al. prepared the triblock copolymer **139** comprising of a Ru^{II}-containing block, a hydrophilic side-chain PEG block, and a biotin-containing block [635]. The self-assembly in aqueous MeCN gave star-shaped micelles in which the metal-rich core was surrounded by the biocompatible PEG corona and the biotin markers on the surface (Fig. 2.115). In contrast to the aforementioned copolymer **138**, the number of biotin moieties along the polymer could be simply tuned by adjusting the monomers' feed ratio.

Finally, anion-exchange membranes (AEMs) comprised of cationic Ru^{II}–polypyridyl complexes were reported by Tew et al. (Fig. 2.116) [9, 636]: For this purpose, a water-soluble norbornene-type Ru^{II}-containing monomer was copolymerized with dicyclopentadiene via ROMP, using a 2nd generation Grubbs' catalyst. Cross-linking of the resulting copolymer **140** was achieved by a subsequent ROMP reaction involving the incorporated dicyclopentadiene units. In order to analyze how an increase of the hydrophobicity or cross-linking density impacts the AEM properties, the monomer ratio (m:n) was varied over a wide range. In general,

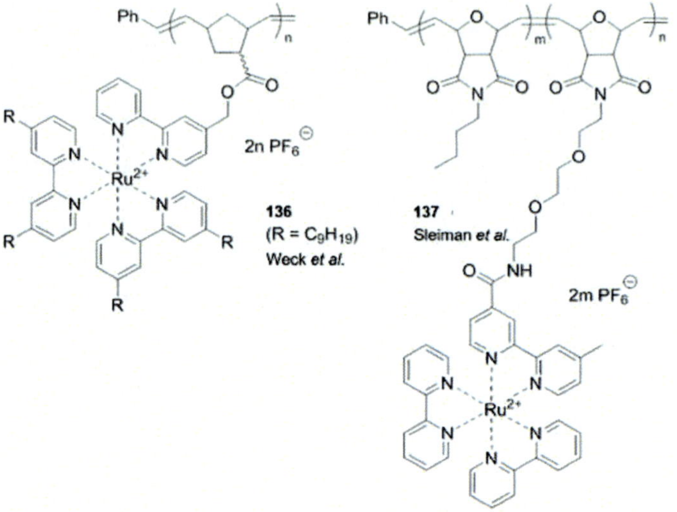

Fig. 2.112 Schematic representation of the ROMP-derived Ru^{II}-containing (co)polymers **136** and **137**

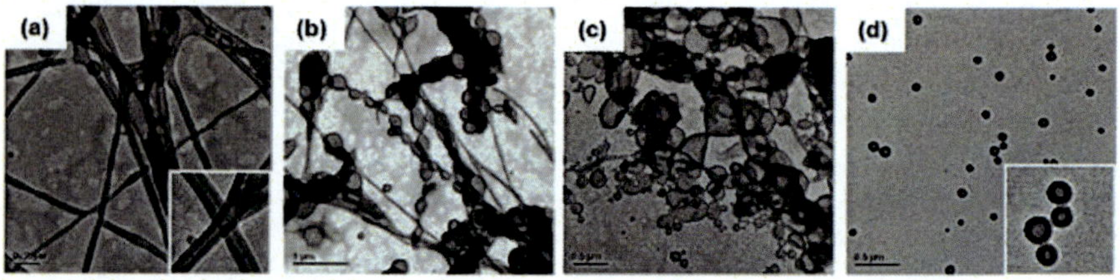

Fig. 2.113 TEM images of **137** (m,n = 20) in MeCN with increasing toluene content: **a** 20% toluene, **b** 35% toluene, **c** 40% toluene and (d) 70% toluene; the insets show magnified images [633]. Figure reproduced with kind permission; © 2007 American Chemical Society

all AEMs **140** displayed properties that were similar to those of traditional AEMs, based on quaternary ammonium salts [e.g., conductivity (σ) and mechanical stability; see Fig. 2.116]. The membranes also showed remarkable tolerance towards MeOH—expressed by a low swelling behavior—and bases, thus making this system attractive for applications in fuel cells.

As one example for (formally) π-conjugated polymers, synthesis of a poly (phenylacetylene) with pending $[Ru(tpy)_2]^{2+}$ complexes has to be mentioned. This polymer **142** was reported by Schubert et al., who utilized the ethynyl-functionalized Ru^{II} complex **141**, as comonomer, in the polymerization of ethynylbenzene (Fig. 2.117) [637]. Hereby, DMA was employed as solvent

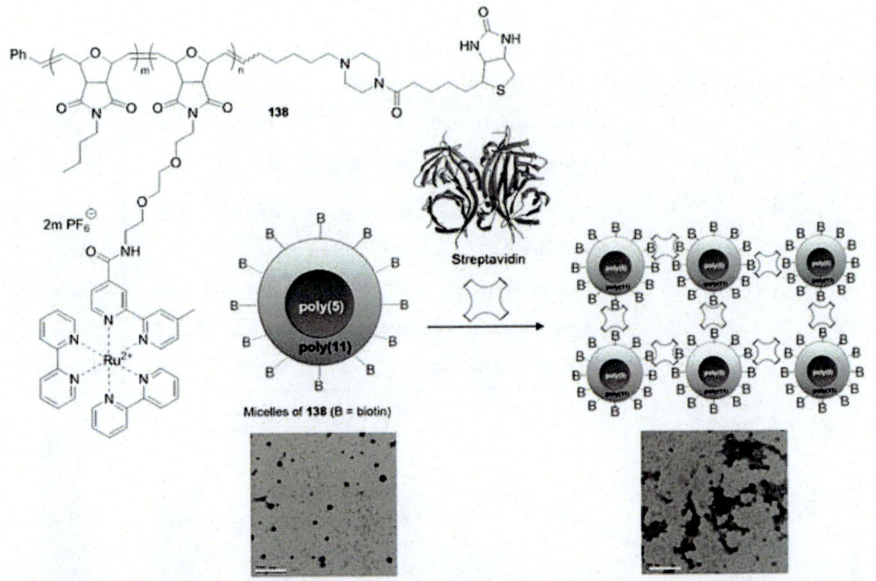

Fig. 2.114 Schematic representation of the biotin-functionalized copolymer **138**. The supramolecular cross-linking, based on protein binding, is also depicted. The TEM images with a scaling of 0.2 µm show the individual (left) and cross-linked micelles (right) [634]. Figure reproduced with kind permission; © 2005 American Chemical Society

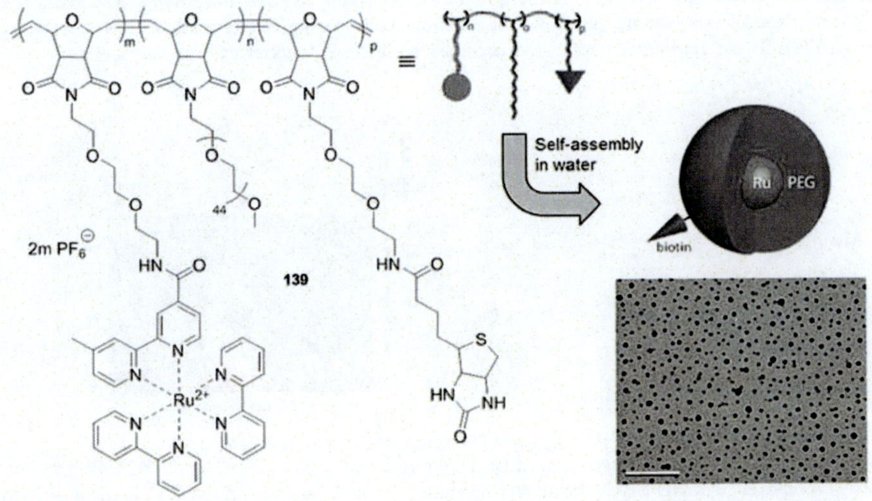

Fig. 2.115 Schematic representation of the ROMP-derived triblock copolymer **139** and it's self-assembly into star-shaped micelles in an aqueous medium. [635] Figure reproduced with kind permission; © 2005 American Chemical Society

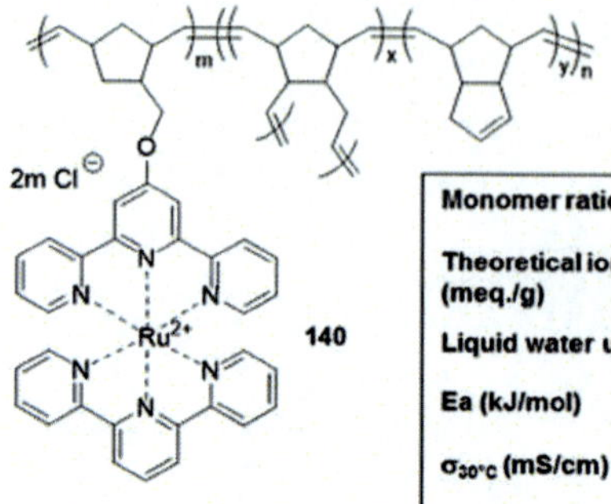

Monomer ratio (m:n)	1:2	1:4	1:10
Theoretical ion exchange capacity (meq./g)	2.0	1.4	1.0
Liquid water uptake (wt%)	432	126	30
Ea (kJ/mol)	14	16	22
$\sigma_{30°C}$ (mS/cm)	19.6	28.6	14.1
Tensile stress at break (MPa)	0.6 ± 0.1	6.2 ± 0.4	27.0 ± 0.5
Tensile strain at break (%)	53 ± 2	80 ± 11	87 ± 16

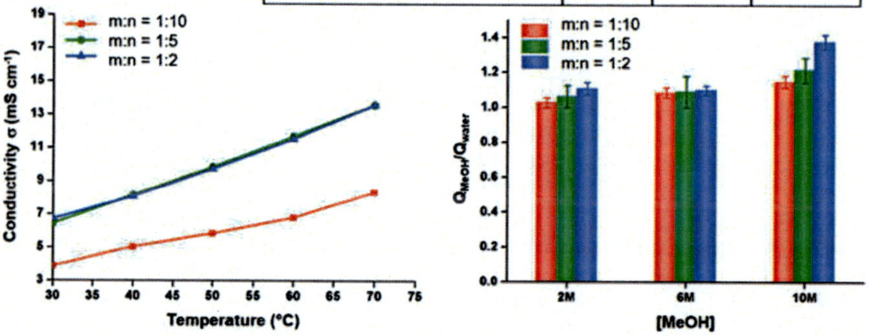

Fig. 2.116 Schematic representation of the Ru^{II}-containing AEM material **140**. Some characteristics at various m:n ratios are also summarized. The two plots show the anion conductivity (σ) as a function of temperature (left) and the swelling behavior in MeOH relative to that in water (right) [9]. Figure reproduced with kind permission; © 2012 American Chemical Society

Fig. 2.117 Schematic representation of the synthesis of poly(phenylacetylene) **142** with pending Ru^{II} complexes. The 3D-SEC plot of **142** revealing the incorporation of the Ru^{II} complexes into the polymer structure is also shown (eluent: DMA with 0.08 mol% NH_4PF_6). [637] Figure reproduced with kind permission; © 2014 Wiley-VCH

offering both a good solubility of the monomers and compatibility with the polymerization conditions; a Rh^{I} catalyst was chosen, since such catalysts are superior to the alternative tungsten-based ones when dealing with aromatic monomers. After purification, copolymer **142** was obtained with a narrow molecular-weight distribution (M_n = 21,500 g mol^{-1}, Đ = 1.97); the rather low content of incorporated Ru^{II} complexes (exp.: 4%, calcd.: 10%) was attributed to steric effects and/or interactions with the catalyst. For **142**, an increase in the excited-state lifetime, compared to $[Ru(tpy)_2]^{2+}$ reference complexes, was observed (i.e., 4.4 ns for **142** vs. 120 ps for parent $[Ru(tpy)_2]^{2+}$, indicating an electronic stabilization of the ^{3}MLCT state by the (partially) conjugated polymer backbone. Noteworthy, a Ru-incorporated polyacetylene was reported by Sponsler et al., who performed the acetylene metathesis polymerization using high concentrations of a 2nd generation Grubb's catalyst [638]. The resultant metallated polymers were significantly more stable under ambient conditions and, after doping with iodine, as conductive as parent polyacetylene. Moreover, Kunitake, Yashima et al. prepared poly(phenylacetylene)s with pending Ru^{II} *tris*-bidentate complexes [639]. When using enantiomerically pure complexes, as monomers, the polymers adopted helical structures in solution and in the solid state; CD spectroscopy and AFM imaging provided evidence for this behavior (Fig. 2.118).

Porous cross-linked polymers (PCPs) incorporating Ru^{II} (**143a**) or Ir^{III} complexes (**143b**) were prepared by the Co^0-catalyzed trimerization of monomers according to Fig. 2.119 [640]. The amorphous materials were investigated by microscopy imaging and XRD analysis; the porosity was investigated by N_2 sorption measurements at 77 K and Brunauer–Emmett–Teller (BET) surface areas

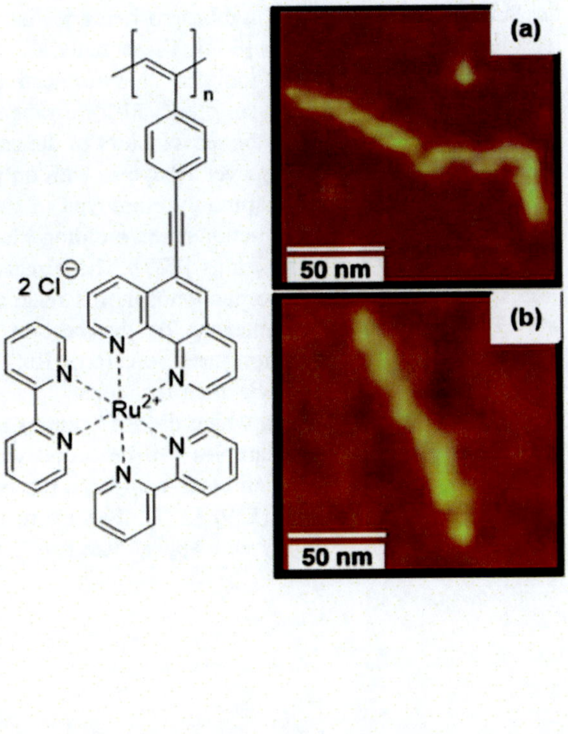

Fig. 2.118 Schematic representation of a poly (phenylacetylene) with pending chiral Ru^{II} complexes. Representative AFM images of the polymers are also shown; the incorporated complexes had *Δ*- or *Λ*-configuration (**a** and **b**, respectively) [639]. Figure reproduced with kind permission; © 2004 Wiley Periodicals Inc

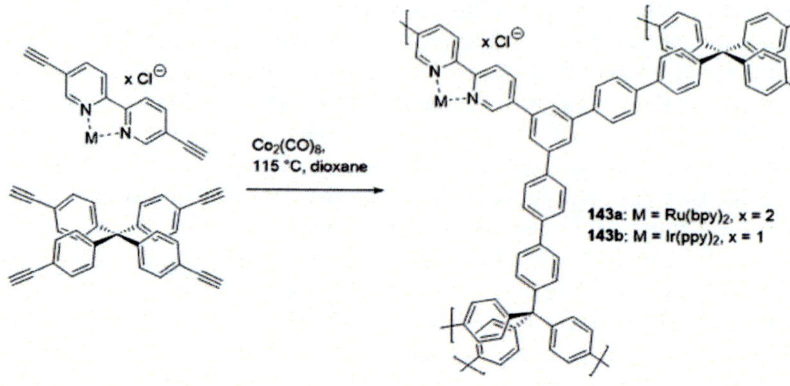

Fig. 2.119 Schematic representation of the synthesis of the PCPs **143**

of 1348 $m^2\ g^{-1}$ (**143a**) and 1547 $m^2\ g^{-1}$ (**143b**) were determined. The PCPs were found to be thermally stable up to 350 °C in air, almost insoluble in organic solvents, and hydrolytically robust against strong acids. The photocatalytic activity in various organic reactions was explored (e.g., in the aza-Henry reaction, α-arylation of bromomalonate or the oxyamination of aldehydes), the heterogeneous catalysts showed a similar performance when compared to the analogous homogenous small-molecule complexes, but could be recycled and reused without loss-of-activity over several runs. In subsequent studies, these same authors prepared photocatalytically active Ru^{II}-loaded PCPs via Sonogashira- or Glaser-type cross-coupling procedures [641, 642]. Very recently, PCPs were also prepared by the Suzuki cross-coupling of a dibromo-functionalized $[Ru(bpy)_3]^{2+}$ complex and a tetra-boronate building block [643]. These materials were applied, as heterogeneous photocatalysts, in the light-induced enantioselective α-alkylation of aldehydes. High yields and *ee* values were observed for a range of substrates (72–95% yield and 90–95% *ee*). Moreover, the recyclability of the catalyst was explored: The PCP retained its enantioselectivity over 10 cycles with only a slight decrease of activity.

Polycondensations are commonly employed in the field of π-conjugated polymers where appropriately functionalized building blocks are reacted, for example, in Wittig- or HWE-type reactions [279]. These reactions, which require a strong base for deprotonating the organophosphorus reagent, are typically incompatible with Ru^{II}–polypyridyl complexes. To the best of our knowledge, pre-formed complexes of the late transition-metal ions (e.g., Ru^{II}, Ir^{III}, Os^{II}, etc.) have not been incorporated into polymers via such reactions.

However, for polymers in which the Ru^{II} centers are directly coordinated to the polymer backbone some examples can be found where other polycondensations have been applied. Rasmussen et al. compared the two different routes leading to Ru^{II}-loaded "polyaza-bpy" (**139**) [277]. The alternative synthesis relying on the complexation of polyaza-bpy with appropriate Ru^{II} precursor complexes, according

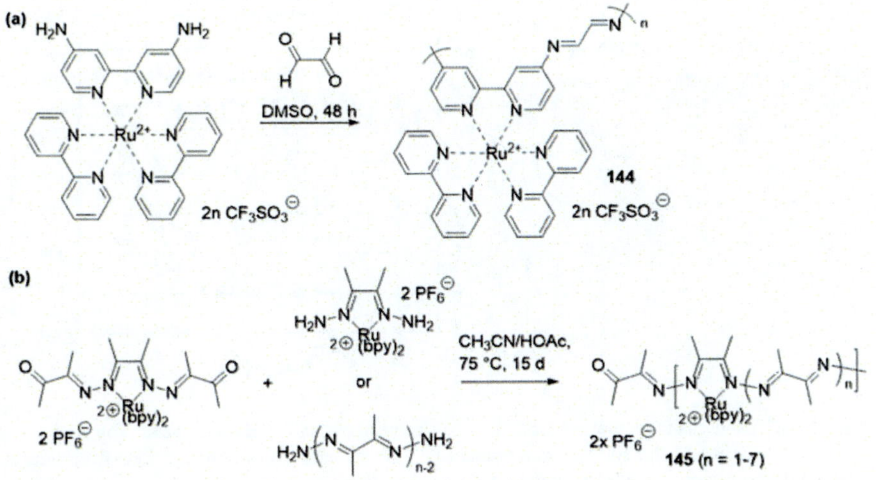

Fig. 2.120 Schematic representation of the synthesis of **144** and **145** via a polycondensation

to Method II, has been shown earlier (**41**, Fig. 2.35). In this context, condensation of a diamino-functionalized $[Ru(bpy)_3]^{2+}$ complex with glyoxal, in stoichiometric amounts, gave ruthenopolymer **144** (Fig. 2.120a); however, no characterization data regarding the molecular structure of **144** were given. These authors claimed that this approach should be highly versatile for constructing advanced metal-containing polymers making use of a step-wise growth of the polymer chain, thereby being able to incorporate different types of organic chromophores or metal complexes in a defined order. A related series of polyazine-type polymers with incorporated Ru^{II} complexes was reported by Cheng and Euler (Fig. 2.120b) [644]. After a reaction time of 15 days, the deep-red and organo-soluble polymers **145** were obtained in moderate (20–40%) yields; a low average DP of 4.5 was estimated from ^{1}H NMR spectroscopic end-group analysis. These authors investigated the influence of the distance between the Ru^{II} centers on the photophysical and electrochemical properties. According to these studies, the Ru^{II} sites had a long-range influence onto the polymer backbone upto at least 15 Å. As revealed by UV/vis-spectroelectrochemical measurements, the first oxidized state of the polymers, a ligand-based process, showed a spectroscopic signature, which was assigned to an intervalence charge-transfer (IVCT) transition. Since this spectral feature was absent in small-molecule reference complexes, it was concluded that the IVCT involved ligand-ligand transition spanning over several repeat units of the polymers.

The polyester-type ruthenopolymer **146** was prepared from a *bis*(hydroxymethyl)-functionalized Ru^{II} complex and dodecanoyl dichloride (Fig. 2.121) [645]. According to ^{1}H NMR end-group analysis, a M_n of ca. 5,500 g mol^{-1} was estimated. Nonetheless, homogeneous thin films could be casted from solution. The

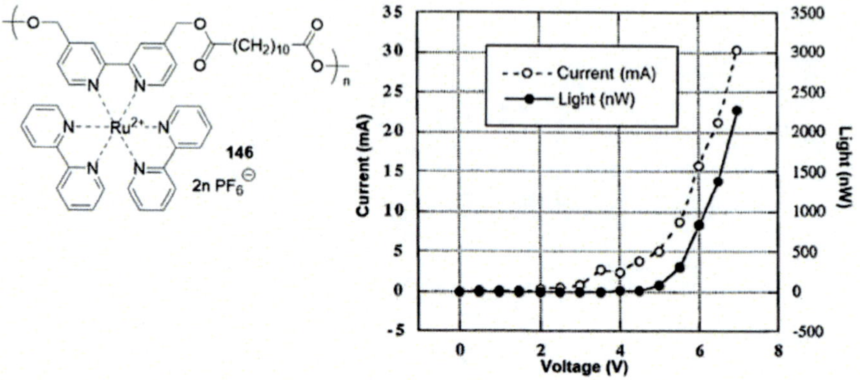

Fig. 2.121 Schematic representation of the polyester-based polymer **146**. The I-V- and L-V-characteristics of the ELD device containing **146** are also shown [645]. Figure reproduced with kind permission; © 1997 American Chemical Society

polymer was fabricated into a single-layer electroluminescent device (ELD) comprised of an ITO-coated glass anode and an Al cathode, the thickness of the active polymer layer was 650 Å. The device featured an orange-red emission under a forward bias with a maximum luminance of ca. 80 cd m^{-2} (2400 nW) at a current density of 750 mA cm^{-2}; the external quantum efficiency of this device was estimated to be ca. 0.01% (Fig. 2.121). Moreover, a multi-layer assembly was prepared utilizing polycationic **146** in combination with PAA, as a polyanion—for this set-up, external quantum efficiencies close to 1% (i.e., having a ratio of nW:mA in the 6000:1 range) were reached.

In 1996, Peng and Yu utilized a Pd^0-catalyzed cross-coupling methodology for the synthesis of π-conjugated polymers—with or without coordinated Ru^{II} centers. In this initial study, a Heck-type cross-coupling between diiodo-functionalized building blocks and DVB was conducted to afford polymers **147** with intrinsic viscosities ([η]) in the range of 0.48 to 1.60 dL g^{-1} (Fig. 2.122) [273]. The highest photoconductivity was found for copolymer **147b** in which the Ru^{II} complexes apparently enhanced the photosensitivity at longer wavelengths and, thereby, supported the photo-induced charge-separation process (Fig. 2.123). Aiming for metallopolymers featuring a large photorefractive effect, copolymers containing additional NLO-chromophores were prepared via the same methodology (**148** and **149**, Fig. 2.124); the polymer samples showed a large net optical gain (>200 cm^{-1}) at a zero electric field [646]. A similar concept was utilized by Chan et al. in the synthesis of conjugated polymers incorporating dipyridophenazine units, as designated coordination site for the Ru^{II} centers [647] or of a series of poly(*p*-phenylenevinylene)s (PPVs) with pending Ru^{II} complexes {either of the $[Ru(bpy)_3]^{2+}$- or the $[Ru(tpy)_2]^{2+}$-type} [648]. Those materials exhibited yellow photo- and electroluminescence at ambient temperatures. Spin-coated samples on ITO-coated substrates gave, upon illumination at 490 nm, conductivity values in the order of

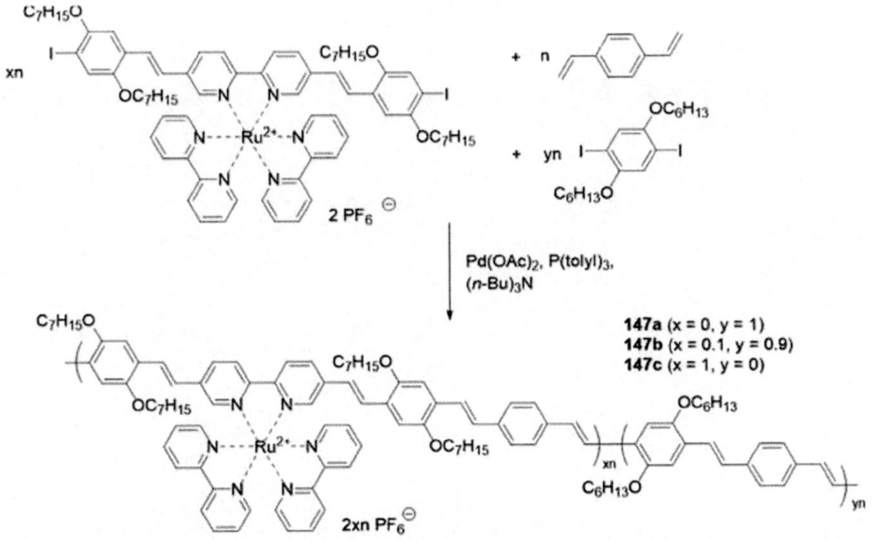

Fig. 2.122 Schematic representation of the synthesis of polymer **147** via a Heck-type cross-coupling polymerization

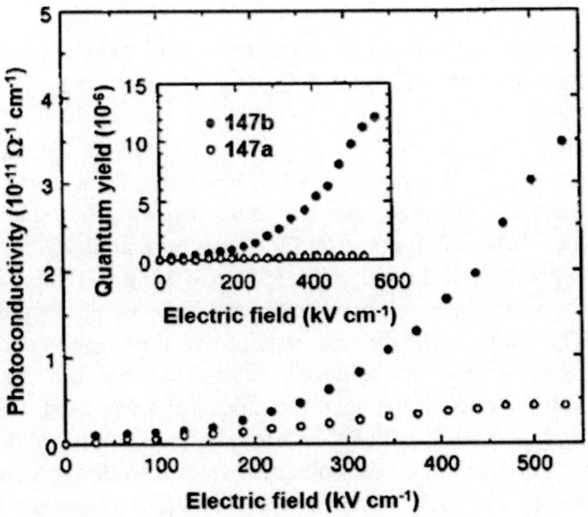

Fig. 2.123 Electric-filed dependency of the photoconductivity and quantum yield (inset) for copolymers **142a/b** [273]. Figure reproduced with kind permission; © 1996 American Chemical Society

10^{-12} Ω^{-1} cm^{-1}. Single-layer LEDs were produced from the polymers (device configuration: ITO|Ru^{II}-PPV|Al) with a turn-on voltage of 5 V and a maximum luminance of 360 cd m^{-1}.

Besides the synthesis of Ru^{II}-containing polymers via a Heck-type polymerization, three other important types of Pd^0-catalyzed cross-coupling procedure were

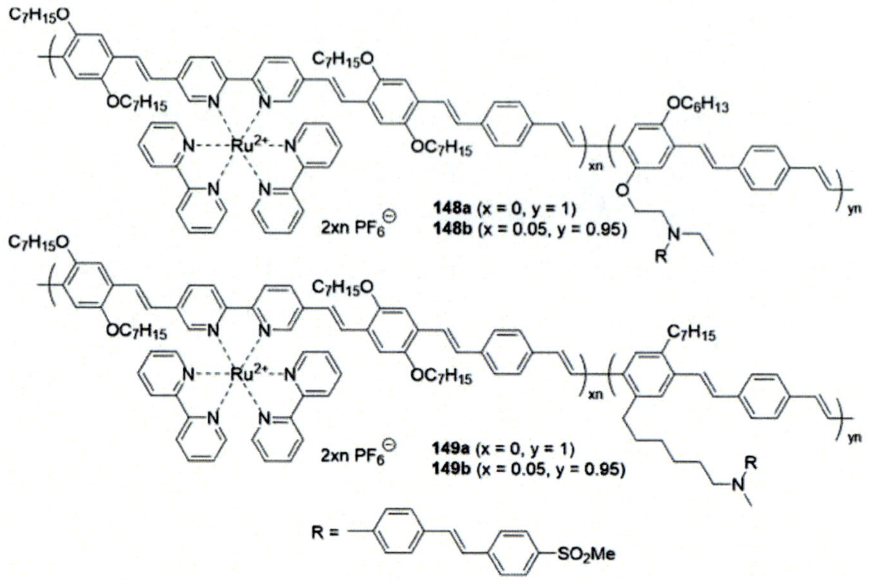

Fig. 2.124 Schematic representation of the ruthenopolymers **148** and **149**, which contain additional NLO-chromophores

applied for the same purpose: Suzuki (**150** and **151**, Fig. 2.125), [649–651]. Sonogashira (**152** and **153**, Fig. 2.126) [652, 653] and Stille (**154** and **155**, Fig. 2.127) [271, 654–657]. For polymers **150–153**, the DP was estimated from elemental analysis, i.e., from the bromine content, as well as SEC measurements (eluent: DMSO containing 0.1 M $LiPF_6$). Concerning the latter method, pullulan or dextran standards gave the more realistic M_n values (e.g., **150b**: M_n = 23,400 g mol^{-1}, Đ = 2.4) compared to those estimated when relying on linear PS or MMA calibrations. The polymers **152g** and **153g** were also investigated further regarding their non-linear optical properties and potential applications in solar cells [653]. The NLO behavior was studied by time-resolved degenerate four-wave mixing measurements in solutions as well as on thin films. A time response of the non-linearity, faster than the pulse durations used, was observed and χ2-values up to 4.9×10^{-13} cm^2 W^{-1} were measured in the solid state.

Polymers **154** were obtained from a Stille-type cross-coupling reaction according to Fig. 2.127. For **154a**, detailed electrochemical studies showed that such metallopolymers could be doped either in a *p*- or *n*-type fashion due to the alternating order of electron-rich (i.e., the thiophene rings) and electron-poor units (i.e., the Ru^{II} complexes). These measurements further suggested that conductivity arouse from the redox properties rather than electronic structure of the π-conjugated backbone. In other words, the hopping of localized charges via a self-exchange process appeared to be more plausible than an extended delocalization of charges.

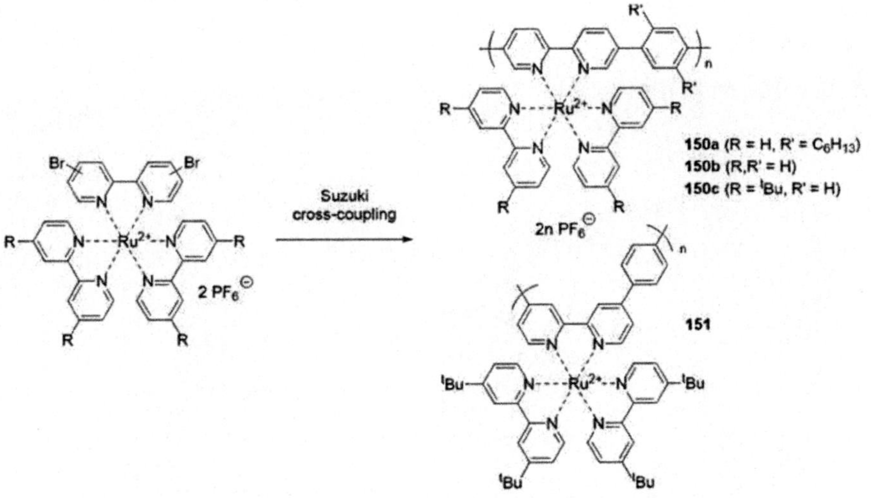

Fig. 2.125 Schematic representation of the ruthenopolymers **150** and **151**, which were prepared by a Suzuki cross-coupling polymerization

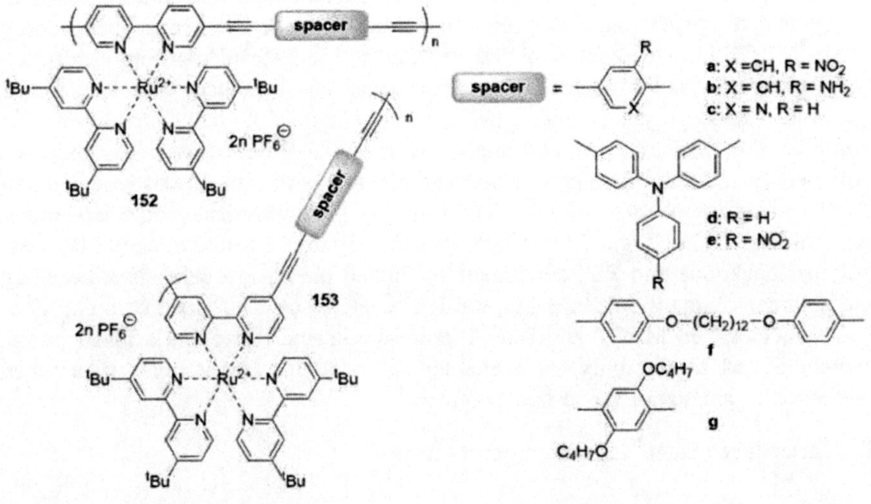

Fig. 2.126 Schematic representation of the ruthenopolymers **152** and **153**, which were prepared by a Sonogashira cross-coupling polymerization

For **154b/c**, containing an oliothiophene linkage between the Ru^{II} centers, detailed electrochemical and photophysical studies were conducted to elaborate the degree-of-electronic interactions between the metal sites and π-conjugated

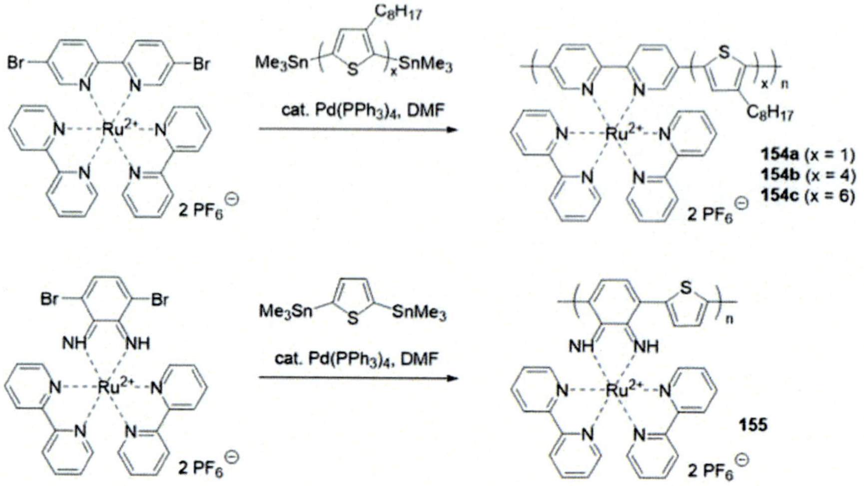

Fig. 2.127 Schematic representation of the ruthenopolymers **154** and **155**, which were prepared by a Stille cross-coupling polymerization

segments [656]. It was shown that the ^{3}MLCT excited state was slightly higher in energy than the $^3\pi,\pi^*$-state of the oligothiophene backbone. As a consequence, only weak ^{3}MLCT luminescence was detected and the long-lived transient absorption (TA) spectrum was dominated by the $^3\pi,\pi^*$-state. Photo-induced bimolecular ElT reactions of the polymers with MV^{2+}, as an external oxidative quencher, was observed. TA spectroscopy indicated that the ElT process gave the oxidized polymers [i.e., the transient spectra were dominated by the characteristic features of a poly(3-octylthiophene) polaron]. The proposed photophysical model is summarized in Fig. 2.128. It was shown later that the electronic communication between polymer backbone and Ru^{II} complexes facilitated the charge delocalization along the polymer chains [658]. In detail, the Ru^{II} centers acted as "electronic gates" or "hole injectors" in MLCT and LMCT processes, respectively. As a result, redox switching and conductivity were enabled at much lower potentials than when compared to analogous metal-free polymers.

(b) Miscellaneous Ru^{II}-containing monomers

Ruthenium-carbonyl complexes or clusters are known for their catalytic activity in a wide range of organic transformations. The immobilization of such species generates stable and (potentially) recyclable macrocatalysts—a highly active field of research already since the 1970s [68]. Typically, these materials are prepared by coordinating an appropriate precursor complex/cluster to the designated binding sites present in a pre-formed polymer (examples for this are shown in Sect. 2.2.2). On the other hand, Pittman et al. investigated a wide range of structurally simple vinyl-functionalized η^5-cyclopentadienylmetal carbonyl complexes, as potential

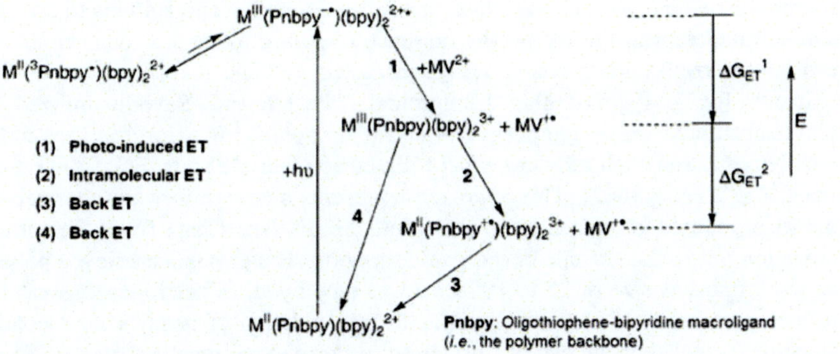

Fig. 2.128 Schematic representation of the photophysical model for the photo-induced ElT between **154b/c** and MV^{2+} [656]. Figure reproduced with kind permission; © 2000 American Chemical Society

monomers in the FRP [592]; however, monomers containing ruthenium-carbonyl units were not included in these studies.

In 1981, Nakamura et al. reported (vinyl ether)-functionalized tricarbonyl([1–4]-η^4-1,3-pentadiene)-Ru^0 and its utilization, as monomer, in the cationic polymerization [659]. In the presence of a Lewis acid, such as $Et_2O \times BF_3$ or $TiCl_4$, the ruthenopolymer **156** was obtained as yellow, air-stable elastomer (M_n = 15,00 g mol^{-1}, Đ = 2.4; Fig. 2.129). Further treatment with HCl at −70 °C afforded the corresponding polymer-bound chloro(tricarbonyl)-η^3-allylruthenium species **157**, as evidenced by ^{1}H NMR spectroscopy. The thin-solid films of **156**, obtained by drop-casting from a CH_2Cl_2 solution, exhibited an insulator-like behavior [i.e., direct current conductivity (σ_{dc}) < 10^{-10} S cm^{-1}] [660]. Doping with iodine increased the film conductivity significantly by some orders of magnitude (σ_{dc} = 1.4×10^{-3} S cm^{-1} in the doped state). However, the maximum conductivity, as observed for the analogous Fe^0-containing polymer ($\sigma_{dc} = 3.6 \times 10^{-3}$ S cm^{-1}) was not reached, not even at high degrees-of-doping.

A variety of metal-containing porphyrins—as mimics of the natural metalloenzymes—has been employed to catalyze oxidation reactions. For example, Ru^{II}(CO)

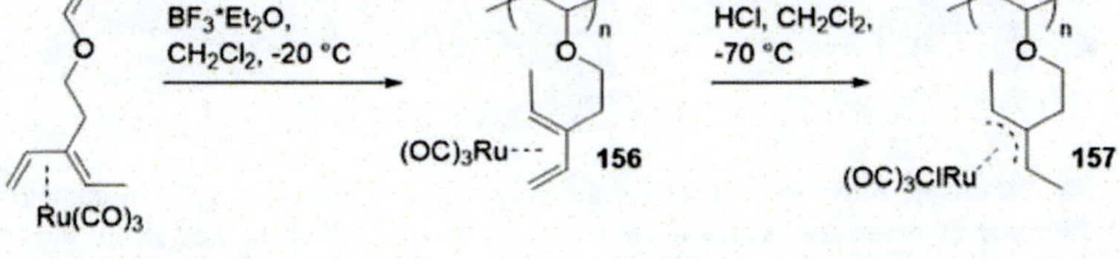

Fig. 2.129 Schematic representation of the synthesis of Ru^0-containing polymers **156** and **157**

(porphyrin) efficiently catalyzes the conversion of olefins and sulfides to epoxides and sulfoxides, respectively (in the presence of strong mineral acids even alkanes can be oxidized); in this study, 1,3-disubstituted pyridine *N*-oxides were used, as oxidants [661]. Against this background, Nestler and Severin utilized the vinyl-substituted *meso*-tetraphenylporphyrin complex **158a** for the free-radical copolymerization with ethylene glycol dimethacrylate (**158a** to EGDMA ratio of 1:99, Fig. 2.130a) [662]. This approach represents a straightforward alternative to the incorporation of ruthenoporphyrins into MOFs (see Chap. 6) or their immobilization onto solid organic or inorganic supports. The ground and sieved polymer network (particle size of 25 to 100 µm) was employed, as heterogeneous catalyst, in the epoxidation of olefins as well as in the oxidation of benzylic or secondary alcohols (1 mol% of the catalyst was used in these experiments). Later, **158b** was

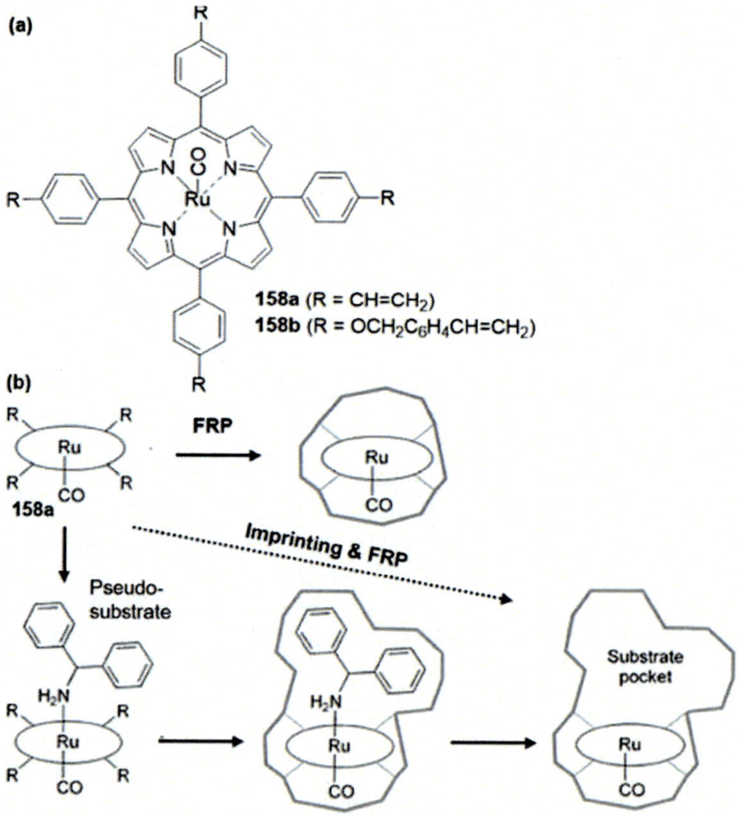

Fig. 2.130 Schematic representation of (**a**) the polymerizable ruthenoporphyrin complexes **158** and (**b**) their copolymerization with EGDMA via conventional FRP as well as by the so-called molecular imprinting technology [663]. Figure reproduced with kind permission; © 2005 Wiley-VCH

also copolymerized with EGDMA (monomer ratio of 1:400) via FRP; the resultant polymer network had a BET surface area of 409 $m^2\ g^{-1}$ and an average pore size of 54 Å [663]. Also this polymer was highly active in catalyzing the oxidation of secondary alcohols (>90% conversion with 0.5 mol% catalyst after 6 h). Moreover, the activity and selectivity of this heterogeneous catalyst was improved by applying the so-called molecular imprinting technique [664]. Here, a pseudo-substrate, which is coordinatively bound to the active site of the catalyst during the polymerization, is utilized. After polymerization, the pseudosubstrate is selectively cleaved, thereby generating an imprint (i.e., a substrate pocket) in the direct proximity to the active site (Fig. 2.130b). The catalytic activity of the molecularly imprinted polymer (MIP) was increased significantly when compared to the one obtained by traditional FRP (e.g., the oxidation of anthracene to anthraquinone was enhanced by a factor of 15.7). Making use of the remarkable, enzyme-like selectivity and specificity of MIPs, a diverse range of applications has been established, e.g., chemical sensing, analyte separation/storage, drug delivery, and heterogeneous catalysis [665–667].

Ferrand et al. employed a structural variant of **158**, as macrocatalyst, in the carbene-transfer reaction. In order to allow this reaction to proceed enantioselectively, a chiral ruthenoporphyrin, possessing styryl-functionalized *C*2-symmetric substituents in the *meso*-positions, was copolymerized with styrene and EGDMA (or DVB) and FRP conditions [668]. The resultant polymeric networks were characterized by IR spectroscopy as well as TEM imaging. The asymmetric addition of ethyl diazoacetate or diazoacetonitrile to styrene derivatives was conducted using these polymers as heterogeneous catalysts—the reaction proceeded under mild conditions and gave *trans*-substituted cyclopropanes with high *ee* values (up to 90%). Later, the scope of this immobilized ruthenoporphyrin catalyst was extended by these same authors to the epoxidation of olefins and oxidation of organic sulfides; however, for the latter application, the analogous Fe^{II}-containing polymers were found to be superior with respect to both yield and enantioselectivity [669].

Beyond ruthenoporphyrins, the concept of molecular imprinting was also applied to other types of Ru^{II}-containing monomers. In 1999, Polburn and Severin reported on a MIP Noyori–Ikariya-type catalyst for the hydrogenation of benzophenone [670]. For this purpose, the precursor complex **159** was copolymerized with EGDMA, in an 1:99 ratio, in the presence of $CHCl_3$, as the porogen, to give a highly cross-linked, though porous, polymer (Fig. 2.131). After grinding and wet-sieving, the supporting phosphinato ligand was cleaved using a solution of $BnNEt_3Cl$ in MeOH to obtain the final MIP catalyst **160**; this material exhibited a significant rate-enhancement (compared to the non-catalyzed reaction) and high selectivity for reducing benzophenone in comparison to other ketone substrates. In subsequent work, a MIP was reported in which the active catalyst was locked in the matrix via two ligands to avoid conformational flipping during and after polymerization; this more rigid MIP catalyst displayed a high hydrogenation selectivity for benzophenone compared to other ketones and even a certain degree-of-regioselectivity was found [671].

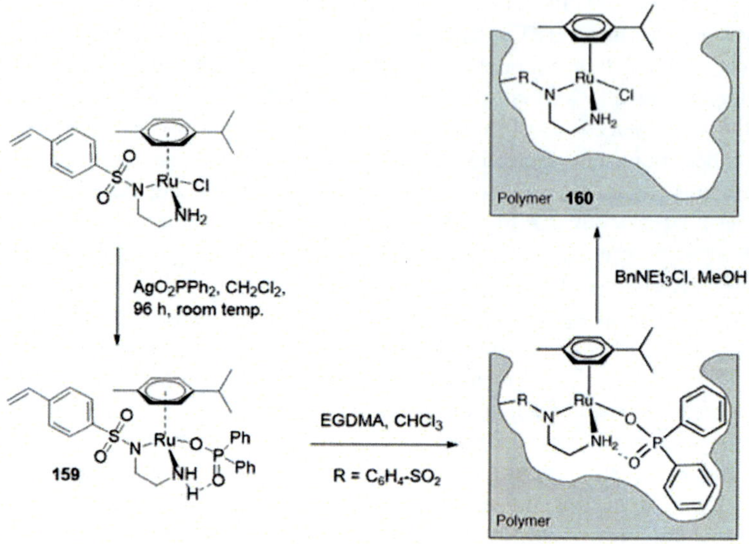

Fig. 2.131 Schematic representation of the synthesis of the MIP Noyori–Ikariya catalyst **160**

In the same context, the ion-imprinted polymer (IIP) **156**, designed for recognition/pre-concentration of Ru^{III} ions was prepared by Godlewska-Żyłkiewiczvia et al. via an AIBN-initiated bulk polymerization of methacrylic acid and EGDMA with Ru^{III}-tetra*kis*(allyl acetoacetate), as the templating agent (Fig. 2.132) [672]. Ion-imprinted polymers (IIPs) have emerged recently as promising materials for the selective solid-phase extraction of (transition) metal ions [673]. The quantitative removal of the Ru^{III} ions from the cross-linked polymer was achieved upon treatment with 0.3 M solution of thiourea in 0.1 M HCl [672]. The BET surface area and average pore diameter for the Ru^{III}-containing and the final metal-free material (**161**) were 1.03 and 2.39 $m^2\ g^{-1}$ as well as 10.53 and 11.18 Å, respectively. Subsequently, the IIP was utilized, as a solid support, in the extraction of Ru^{III} ions from different types of environmental samples prior to electrothermal atomic absorption spectrometric (AAS) determination. Thereby, a detection limit of 0.32 ng mL^{-1} was reached (relative standard deviation of 2.5%). Moreover, **161** exhibited a high selectivity for Ru^{III} ions against other transition metal ions (i.e., Pd^{II}, Pt^{IV}, Rh^{III}, Co^{II}, Ni^{II}, and Fe^{III}); in particular, very high selectivity values of >70 were obtained for Ru^{III} versus Ni^{II} as well as for Ru^{III} vs. Co^{II}. In related work, these same authors also copolymerized styrene, DVB, and 4-vinylpyridine in the presence of a Ru^{III}–thiosemicarbazone precursor complex [674]. The resultant IIP offered an even higher selectivity for the binding of Ru^{III} ions and a lower detection limit of 0.26 ng L^{-1}. The bulk copolymerization of MMA and EGDMA, using Ru^{III}–thiobarbiturate, as a templating agent, under the otherwise same reaction conditions gave an IIP for which a Ru^{III} detection limit of 0.21 ng L^{-1},

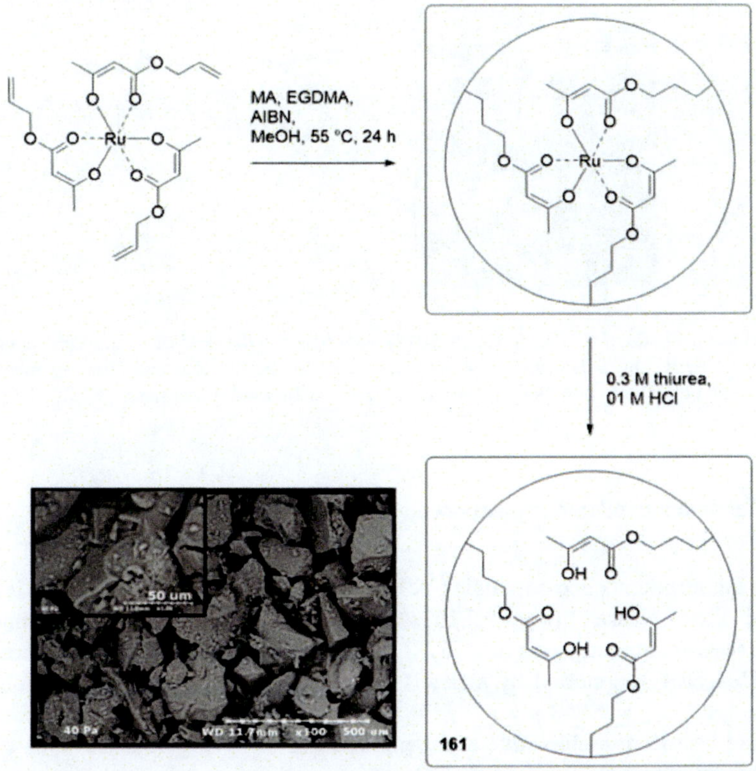

Fig. 2.132 Schematic representation of the synthesis of the IIP **161** using Ru^{III}-tetrakis(allyl acetoacetate), as templating agent. A representative SEM image of **161** is also shown (100- and 1000-fold magnification) [672]. Figure reproduced with kind permission; © 2012 Elsevier B.V

according to electrothermal atomic absorption spectrometry, was reported [675]. A chitosan-based IIP with 2-pyridylthiourea binding sites was published by Monier et al. [676]. As in the previous examples, this bio-based material exhibited a high selectivity for the binding of Ru^{III} ions and robustness: The IIP retained its performance over (at least) eight adsorption/desorption cycles. For an overview on the recent progress in using IIPs in the adsorption and/or separation of ions, the reader is referred to a specialized review by Dai, Yan et al. [677].

2.2.5.2 Polymers Bearing Ru^{II} Centers Within the Backbone

In order to synthesize polymers with ruthenium centers within the backbone via method Vb (Fig. 1), two fundamentally different approaches can be followed: Either a pre-formed complex is (co)polymerized in a polycondensation/-addition-type reaction, thereby making use of appropriate functional groups, or the ROP of strained monomers is employed.

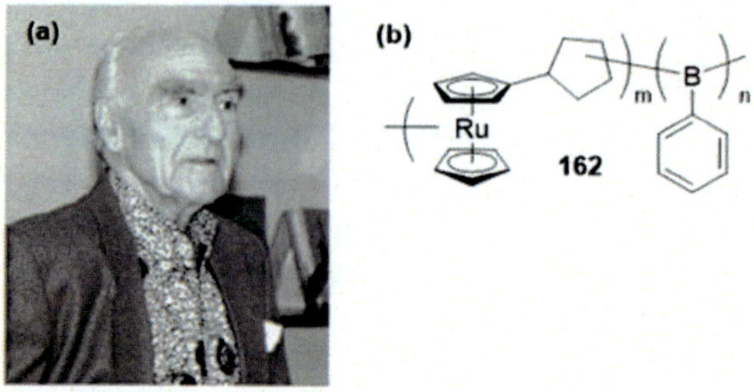

Fig. 2.133 a Eberhard W. Neuse (1925–2014)—a pioneer in the field of metallocene-containing polymers [679]. **b** Schematic representation of an early example for a main-chain ruthenocene-containing polymer. Figure reproduced with kind permission; © 2015 Springer Nature

(a) Monomers containing ruthenocene or related moieties

As for other ruthenocene-containing polymers, pioneering work in this field is again related to Neuse (Fig. 2.133a), who reported on a main-chain ruthenocene-containing polymer [678]. The reaction of a ruthenocene derivative with $C_6H_5BCl_2$, supported by a Lewis acid, afforded a material of low molecular weight (**162**, M_n = 2,850 g mol^{-1}, Fig. 2.133b). However, the proposed chemical structure has been doubted due to the unlikeliness of having both Ru–C as well as B–B bonds within the main chain [2].

According to Neuse, the first attempts to oligomerize or even polymerize ruthenocene were made in the early 1970s when the Ullmann (cross-)coupling reaction of iodo- and 1,1′-diiodoruthenocene gave *bis*-ruthenocene besides some higher oligomers (n = 2–4) [561]. The different strategies are summarized in Fig. 2.134. It was shown, that in the melt a slightly higher DP of ca. 5 could be achieved. However, solidification of the melt and dehalogenation, thereby forming pristine ruthenocene, represented the crucial limitations of this strategy (to the best, polyruthenocene **163** with a M_n of 1,180 g mol^{-1} was obtained in only 1% yield). Though the reaction of the analogous ferrocene derivatives was found to be highly suited in preparing polyferrocene, the cross-coupling of TMEDA-stabilized 1,1′-dilithioruthenocene with 1,1′-diiodoruthenocene mainly afforded oligomers; neither catalysis by first-row transition-metal ions nor cuprate activation of the dilithio-species gave any significant improvement. The combined oxidative and thermal coupling, however, increased the yield up to 70%; nonetheless, the DP of the isolated polyruthenocene remained low (i.e., M_n values ranging from 1,240 to 2,500 g mol^{-1}). Consistently with the electronic properties of the group-8 metallocenes, the nucleophilicity of the metallated species is strongly decreasing from ferrocene via ruthenocene to osmocene. Thus, these coupling procedures appear to be hardly suited for preparing polyruthenocene of high molecular weight.

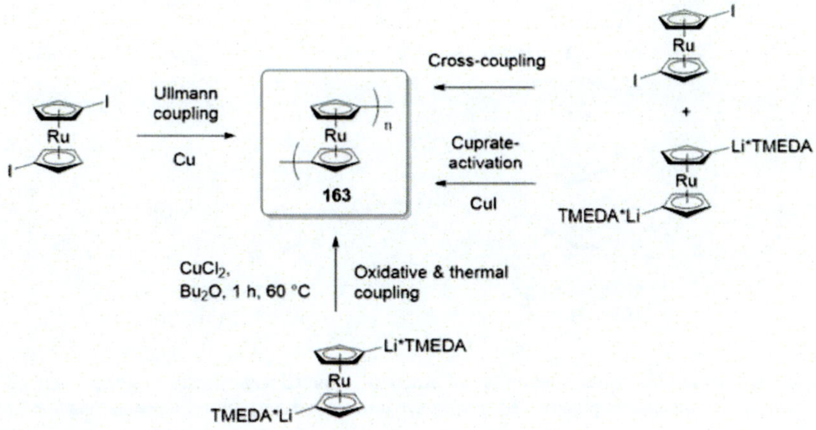

Fig. 2.134 Schematic representation of the synthesis of polyruthenocene **163** via different homo- and cross-coupling procedures

In contrast to the direct polymerization of metallocenes, the utilization of strained [*n*]metallocenophanes is much more favorable: Many examples have been published, where either the thermal or, in particular, the highly-controlled anionic ROP allows the synthesis of metallocene-containing polymers—even when materials of high molecular weights are targeted [518, 556, 680]. Basically, the tendency of a [*n*]metallocenophanes to undergo a ROP is directly related to inherent ring strain. The release of ring strain and relaxation of the tilted cyclopentadienyl rings back to the coplanar arrangement represents the main driving force for the ROP. An indication of the extent of the strain present in a [*n*]metallocenophane can be derived from X-ray structure analysis with the angle α between the two planes of the tilted ligands being the measure (Fig. 2.135a). For a fixed type of transition-metal sandwich complex, the degree-of-tilting mainly depends on the size/length of the bridging element (i.e., small bridging elements implicate larger tilting angles). The first strained Ru^{II}-containing system to be employed in the ROP was the [2]ruthenocenophane **164**, as reported by Manners et al. [681]. In contrast to the hydrocarbon-bridged [2]ferrocenophane, which underwent ROP only at 300 °C, **164a** polymerized already at 220 °C in the melt (Fig. 2.135b). This increased reactivity can be attributed to the size of the sandwiched metal center in the [2]metallocenophane. The separation between the cyclopentadienyl rings in ruthenocene is larger than in ferrocene (3.68 Å vs. 3.32 Å); consequently, the strain in [*n*]ruthenocenophanes is intrinsically higher [682]. The resulting polymer **165a** was almost insoluble; whereas, its methyl-substituted analog **165b** was readily soluble and, thus, could be characterized by NMR and SEC analysis (bimodal distribution: $M_{n,1} = 12{,}000$ g mol^{-1} and $Đ_1 = 3.59$, $M_{n,2} = 4{,}400$ g mol^{-1} and $Đ_2 = 2.89$). With increasing length of the bridge, the ring-strain is decreased and, consequently, the reactivity with respect to a ROP vanishes. Thus, trisilyl-bridged

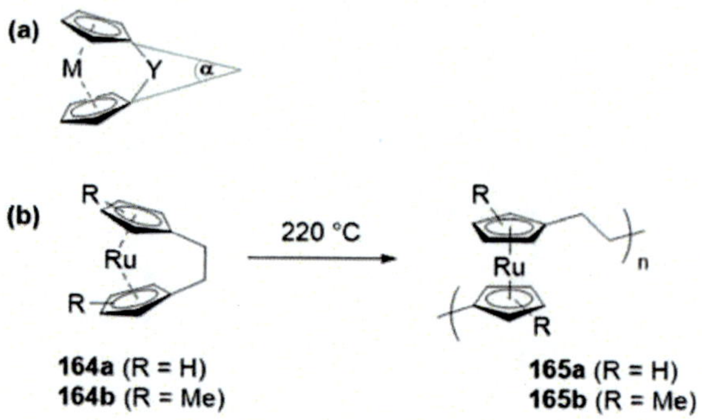

Fig. 2.135 Schematic representation (**a**) of strained [*n*]metallocenophases in general and (**b**) of the ROP of [2]ruthenocenophanes **164** giving the corresponding poly(ruthenocenophane)s **165**

[3]ruthenocenophanes did not polymerize and even thermally stable [3]osmocenophanes, as the heavier analogs, became available [683].

Due their very high ring strain, silicon-bridged [1]ruthenocenophanes **166** readily decompose and, consequently such materials do not represent appropriate monomers for preparing defined polymers. However, with increasing size of the bridging atom, [1]ruthenocenophanes become stable (i.e., **167** and **168**) and could, at least in case of the tin-bridged derivative **168**, be polymerized (sealed tube, 200 °C, 4.5 h; Fig. 2.136a) [684]. The resulting ruthenopolymer **169** was stable to air and moderately soluble in organic solvents (e.g., toluene, CH_2Cl_2 or THF). DSC analysis revealed a T_g of 221 °C; according to TGA measurements, decomposition started at 270 °C and the ceramic yield at 900 °C was 32%.

Müller et al. introduced various group-13 elements into the bridge of [1]ruthenocenophanes [685]. Utilizing the Me_2Ntsi ligand—providing an intramolecular coordination via the *N*-donor and steric shielding through the bulky trimethylsilyl groups—allowed the synthesis of the Al- and Ga-bridged derivatives **170** (Fig. 2.136b). In order to test the applicability of these monomers in the ROP, DSC experiments were performed; from these, the melting point (as endothermic peak), the onset temperature for the thermal ROP and the enthalpy of ring-strain release can generally be estimated. However, the ROP could not be observed, presumably due to the efficient shielding of the molecule preventing any sufficient initiation of the ROP. Thus, the structures were revised to reduce the steric shielding to a minimum, while retaining sufficient stabilization of the [1]ruthenocenophane moiety [686]. The new monomers, for which a computer-aided structure design was applied, could then be used in the ROP, where **171a** already underwent ROP at ambient temperature under the conditions of the monomer synthesis; whereas, **171b** could be isolated and employed in the ROP thereafter (Fig. 2.136c). The ruthenopolymers **172** were analyzed by NMR spectroscopy and DLS. From the latter, the hydrodynamic radii (R_h), molecular weights (M_w), and DPs could be

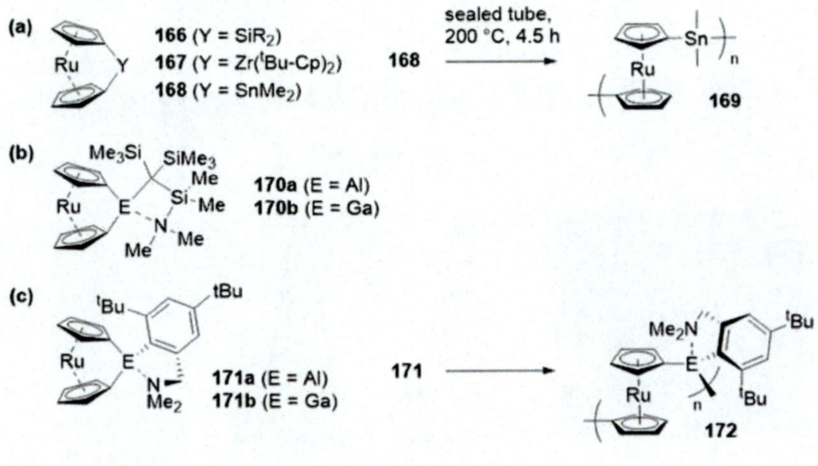

Fig. 2.136 Schematic representation of the strained [1]ruthenocenophanes **166–168** (a), **170** (b) and **171** (c) and their applicability in thermal ROPs

estimated (**172a**: 1.33 ± 0.25 nm, 8.07 ± 3.0 kDa, 16 ± 6; **172b**: 1.50 ± 0.37 nm, 10.1 ± 4.8 kDa, 19 ± 9).

As pointed out before, the CDSA as well as the polymerization-induced self-assembly (PISA) represent the most common strategies in the field of modern block copolymer self-assembly studies. In particular, the CDSA of block copolymers comprising of a hard, crystallizable block have attracted significant attention [687]. Recently, Tang et al. reported a highly versatile approach to combine PISA and CDSA [688]: The so-called ROMPI-CDSA method, a one-pot reaction, offered the advantage to induce the CDSA process under the conditions of a proceeding ROMP (Fig. 2.137). Ferrocenenophane and ruthenocenophane monomers (**173a/b**) were employed to afford the PNR-*b*-p**173** diblock copolymers (**174**, PNR: polynorbornene); the dehydroabietic side chains were introduced to improve the solubility of the soft PNR block and to increase the T_g beyond 100 °C (thus, suppressing temperature effects on the self-assembly process). In the case of **173b**, the ROMPI-CDSA gave well defined nanostructures, which were studied by TEM imaging (Fig. 2.137). Similar to the PMMAs comprising of a central $Ru(Cp)_2$ unit (see Sect. 2.2.4), the diblock copolymers were also studied regarding their mechanochemistry. The sonication-induced chain scission occurred selectively at the metallocene sites, as revealed by a gradual decrease of the molecular weights [579].

Polymers, containing metal–carbonyl half-sandwich complexes within the backbone, represent a different class of metallopolymers to be named at this point. According to Wang et al., three sub-classes can be distinguished for these materials: (a) metallopolymers containing metal–metal bonds (e.g., Tyler's photodegradable molybopolymers [689]), (b) metallopolymers containing main-group elements (e.g.,

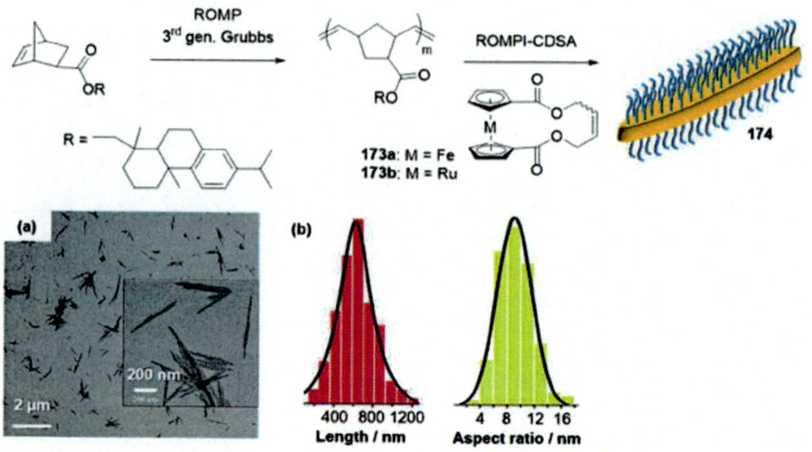

Fig. 2.137 Schematic representation of the synthesis of diblock copolymers **174** via an one-pot ROMPI-CDSA process. A representative TEM image **(a)** and the corresponding size distributions **(b)** of PNR_{31}-*b*-p**173b**$_{49}$ are also shown [688]. Figure reproduced with kind permission; © 2019 The Royal Chemical Society

Si, Ge or Sn), and (c) metallopolymers with metal–acyl bonds [typically prepared via a migration-insertion polymerization (MIP)] [690]. However, only the latter exhibit an adequate stability to enable applications in material's science.

Regarding ruthenopolymers, the ROP of disila-metallacyclobutanes has been applied. Sharma and Pannell demonstrated that the iron-containing polymer **175a** could readily be synthesized from an appropriate half-sandwich precursor complex (Fig. 2.138) [691]. In a similar fashion, the analogous ruthenopolymer **175b** was obtained; however, a different type of precursor was required. The precursor **176** was prepared from $[(\eta^5\text{-}C_5H_5)Ru(CO)_2]_2$ and the thermal ROP to afford **175b** occurred at a temperature of 150 °C (Fig. 2.138) [692]. The ruthenopolymer, which was sparsely soluble in hot THF, was analyzed by SEC—using a PS calibration, a DP of ca. 220 was estimated for **175b** (M_n = 7430 g mol^{-1}, Đ = 1.6). Moreover, ^{29}Si NMR spectroscopy gave further evidence for the proposed molecular structure—the resonances at 0.37 and −18.39 ppm were assigned to the Si atoms neighboring the Ru center and the Cp ring, respectively.

(b) ***Monomers containing Ru^{II}-polypyridyl complexes***

For the synthesis of polymers containing Ru^{II}–polypyridyl complexes within the backbone basically two types of reactions according to method Vb can be applied (Fig. 1): Polycondensations and metal-catalyzed (cross-)coupling procedures. The common advantage of utilizing method Vb (as an alternative to the orthogonal method III, see Sect. 2.3.1) is that pre-formed Ru^{II} complexes bearing appropriate

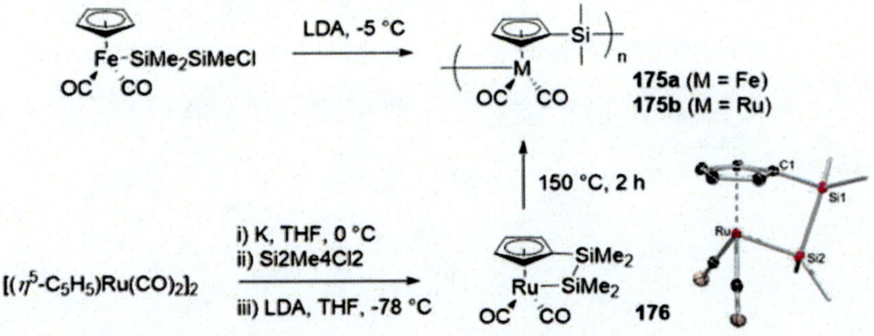

Fig. 2.138 Schematic representation of the synthesis of the metallopolymers **175**. The solid-state structure of monomer **176** is also shown [692]. Figure reproduced with kind permission; © 2014 American Chemical Society

functional groups can serve as monomers in the polymer synthesis; thereby, the limitations arising from the rather drastic reaction conditions that are necessary to assemble multiple Ru^{II}-polypyridyl complexes can, at least to some extent, be circumvent.

When aiming for main-chain metallopolymers possessing Ru^{II} *tris*-bidentate complexes one further issue has to be addressed which is chirality. The formation of heteroleptic complexes of the general structure $[Ru(L^1)_2(L^2)]^{2+}$ (L^1 and L^2 represent bidentate ligands) will always yield a mixture of coordination enantiomers, referred to as the *Δ*- and the *Λ*-form (in such complexes the Ru^{II} center is chiral). Therefore, the synthesis of **77**, as shown in Fig. 2.61, produced a structurally less-defined material due to a random distribution of *Δ*- and the *Λ*-configured complexes along the chain. As a consequence, those ruthenopolymers adopted a randomly coiled globular structure [440–442]. Lincoln et al. established the first protocol for the chiral resolution of such complexes as the example of *rac*-**177** utilizing sodium arsenyl L-(+)-tartrate (Fig. 2.139) [693]; today, chiral resolution can conveniently be achieved by high-pressure liquid chromatography (HPLC) using chiral stationary phases, such as cyclofructan [694]. In the following, chiral Ru^{II} *tris*-bidentate complexes were utilized for the synthesis of enantio- and diastereomerically pure oligonuclear Ru^{II}-containing assemblies by Lehn et al. [695], Tor et al. [696], as well as MacDonnell et al. [697, 698]. The latter authors employed the chiral synthons *Δ*-**178** and *Δ*-**179** for the synthesis of ruthenopolymer *Δ*-**180** (DP > 13, Fig. 2.140); in a similar fashion also *Λ*-**180** as well as the alternating *meso*-copolymer *ΛΔ*-**180** were prepared from the corresponding chiral Ru^{II} building blocks **178** and **179** [699]. CD spectroscopy corroborated that the stereochemistry of the individual complexes was retained under the relatively harsh polycondensation conditions. Topochemical analyses led to the conclusion that *meso*-**180** had a *zig-zag* structure in which the Ru^{II} centers adopted an almost coplanar arrangement. The two homochiral polymers *Λ*-**180** and *Δ*-**180** were proposed to exist in random-coil structures, due to the pre-defined complex geometries [700].

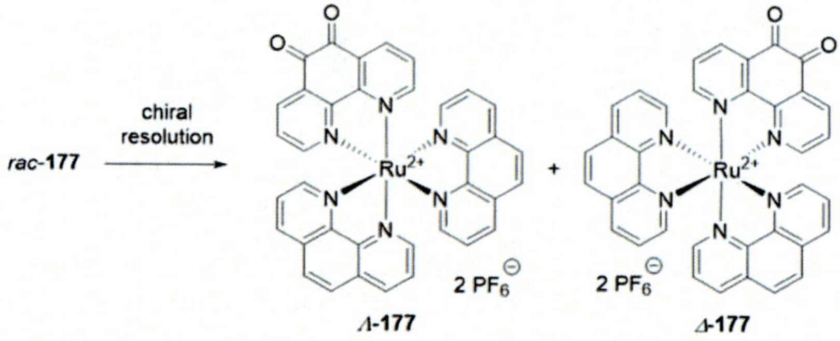

Fig. 2.139 Schematic representation of chiral resolution of *rac*-**177**

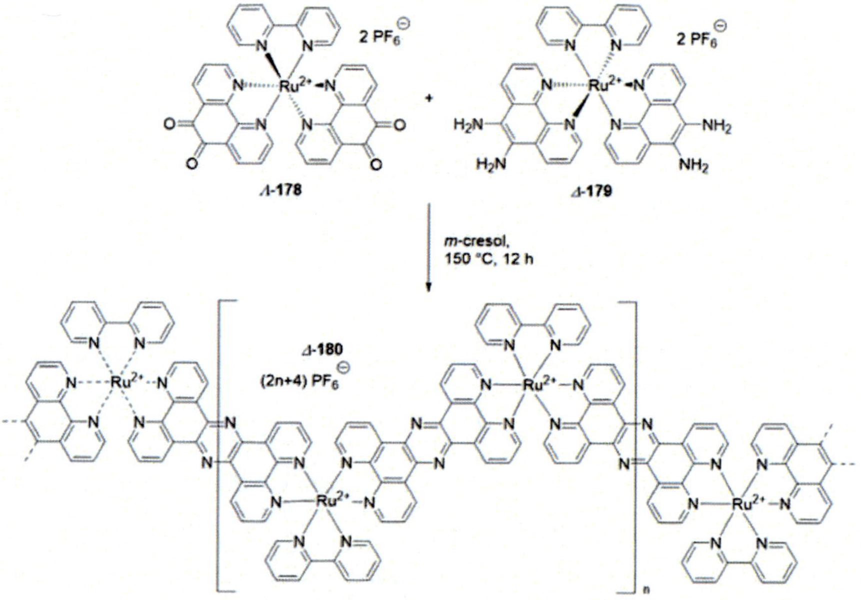

Fig. 2.140 Schematic representation of the synthesis of the homochiral ruthenopolymer *Δ*-**180** starting from two chiral *Δ*-configured Ru^{II} *tris*-bidentate complexes

In contrast to their *tris*-bidentate counterparts, the octahedral *bis*-tridentate complexes do not show chirality and are, therefore, suited for incorporation into linear metallopolymers. As already shown in Fig. 2.64, Chan et al. employed a homoleptic amino-functionalized Ru^{II}–terpyridine *bis*-complex for the synthesis of polymers **76**. These polyimides were prepared in two-step polycondensations using different dianhydrides, as monomers: First, a ring-opening addition of the amino group to the anhydride in NMP gave a polyamic acid, as an intermediate (due to the low reactivity of the complex's amino group, longer reaction times and higher temperatures were required when compared to typical reactions between dianhydrides and primary amines). The subsequent imidization step was conducted by heating the polyamic acid solution with a mixture of pyridine and Ac_2O. The resultant polymers possessed electron- and hole-carrier mobilities in the range of 10^{-4} cm^2 V^{-1} s^{-1}; apparently, the electron-withdrawing imide moieties contributed to the charge-transport process. When fabricated into single-layered LEDs, ruthenopolymers **76** exhibited electroluminescence – the external quantum efficiency and maximum luminance of these devices were 0.1% and 120 cd m^{-2}, respectively.

Besides these polycondensation approaches also various Pd^0-catalyzed cross-coupling procedures were utilized to obtain pseudo-π-conjugated polymers with incorporated $[Ru(tpy)_2]^{2+}$ complexes within the backbone. For example, Chan et al. utilized the Heck cross-coupling polymerization for the synthesis of a series of Ru^{II}-containing copolymers **181** (Fig. 2.141) [701, 702]. Both the charge mobility and photoconductivity of these materials could be correlated to the content of Ru^{II} complexes within the polymer chain: At the same external electric field, the photoconductivity increased with increasing Ru^{II} content, thereby enhancing the photosensitivity of the materials.

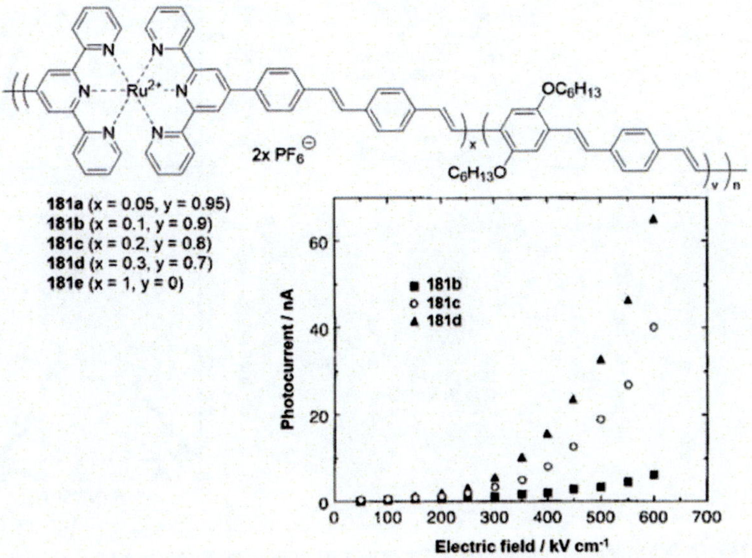

Fig. 2.141 Schematic representation of the Ru^{II}-containing polymers **181** prepared by Heck cross-coupling polymerization. The plot shows the photocurrent response of selected polymers as a function of the external electric field [701]. Figure reproduced with kind permission; © 1997 American Chemical Society

The Ru^{II}-containing polymers **183** were synthesized by a Suzuki cross-coupling polymerization using the bromo-functionalized complexes **182**, as comonomer (Fig. 2.142a) [703]. DSSCs using **182** or **183**, as sensitizers, were fabricated and revealed that the CO_2H-groups (as present in **182b** and **183b**) were beneficial to enable an efficient binding to the TiO_2 semiconductor. The DSSC devices containing complex **182b** or polymer **183b** gave power conversion efficiencies up to 0.13% (the performance of the device based on **183a** was significantly lower, Fig. 2.142b). Similar results in terms of device performance were obtained when TiO_2 nanotubes were employed, as semiconductor; however, nitrogen doping of the nanotubes did not significantly improve the device performance. The incident photo-to-electron conversion efficiency (IPCE) plots, as depicted exemplary for the system **183b** on TiO_2 nanotubes (Fig. 2.142c), resembled the UV/vis absorption spectrum of the ruthenopolymer; thus, sensitization was attributed mainly to the incorporated Ru^{II} complexes.

The homoleptic perfluorophenyl-substituted Ru^{II} complex **184** was employed by Schubert et al. in a thiol-*para*-fluorine aromatic substitution reaction for the synthesis of homometallic (**185**) as well as heterobimetallic polymers (**187**, Fig. 2.143)

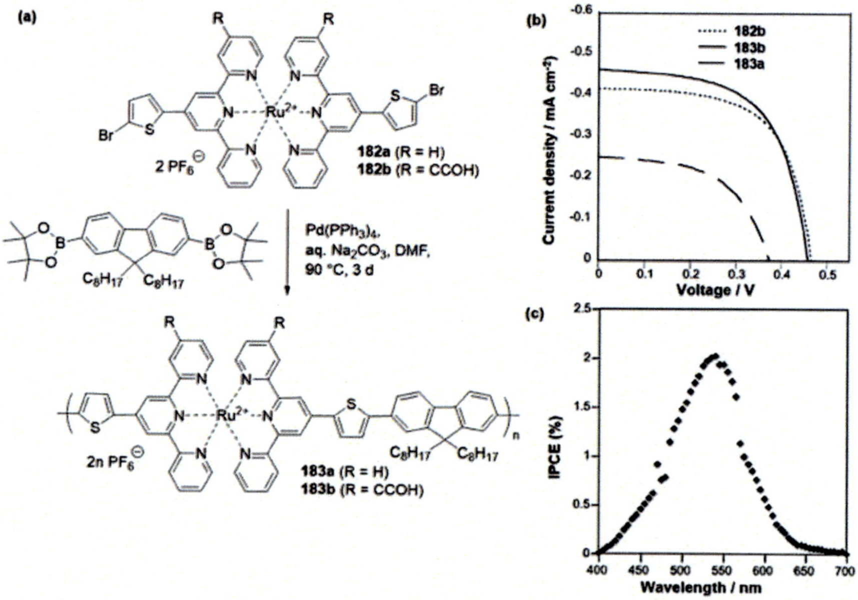

Fig. 2.142 **a** Schematic representation of the synthesis of polymers **183** by Suzuki cross-coupling polymerization. **b** Current–voltage characteristics of DSSCs fabricated from TiO_2 nanoparticles (the devices were illuminated with simulated AM1.5 solar light with an intensity of 100 mW cm^{-2}). **c** Plot of the IPCE as the function of incident light wavelength for the DSSC fabricated from TiO_2 nanotubes coated with **183b** [703]. Figure reproduced with kind permission; © 2011 Wiley-VCH

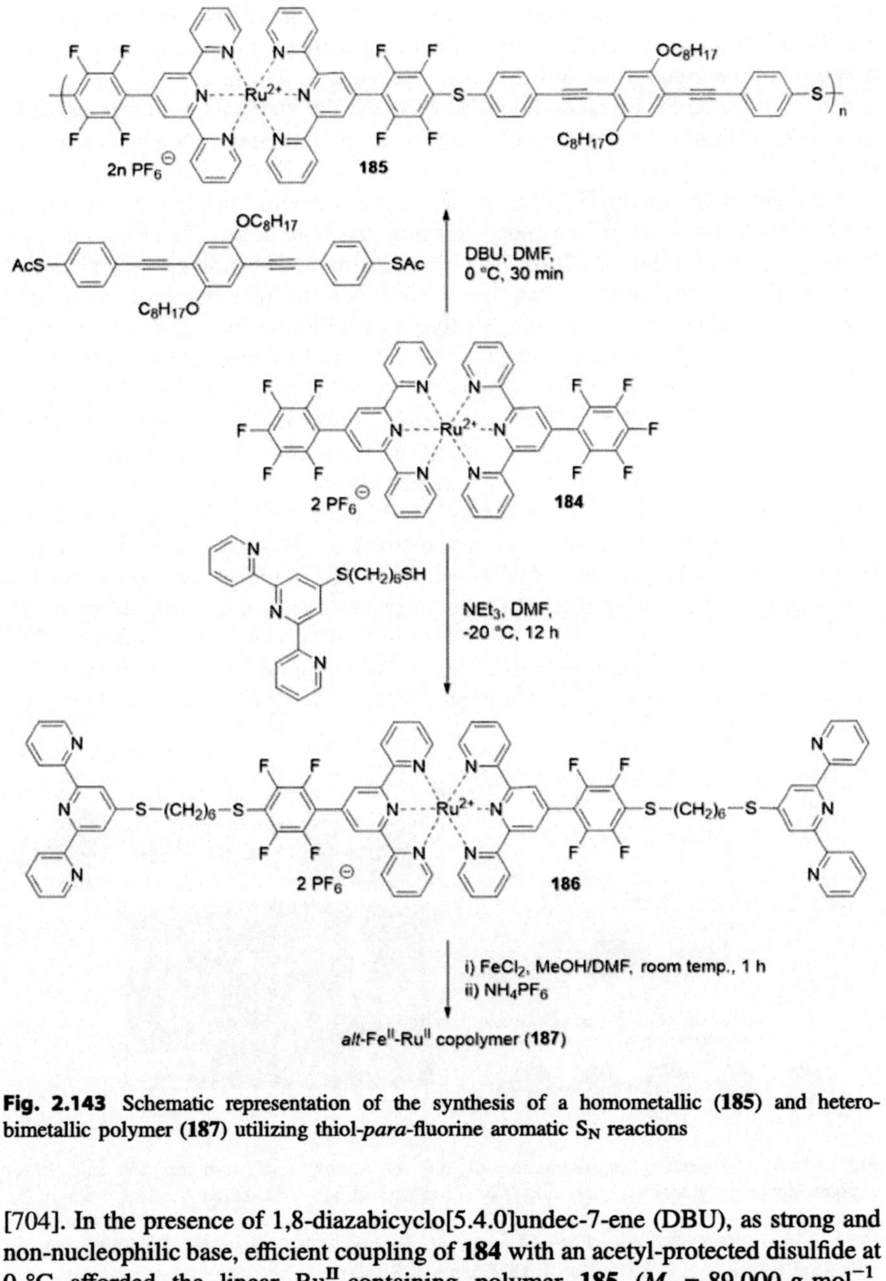

Fig. 2.143 Schematic representation of the synthesis of a homometallic (**185**) and hetero-bimetallic polymer (**187**) utilizing thiol-*para*-fluorine aromatic S_N reactions

[704]. In the presence of 1,8-diazabicyclo[5.4.0]undec-7-ene (DBU), as strong and non-nucleophilic base, efficient coupling of **184** with an acetyl-protected disulfide at 0 °C afforded the linear Ru^{II}-containing polymer **185** ($M_n = 89{,}000\ g\ mol^{-1}$, $Đ = 2.41$). When using phosphacene-type bases, shorter polymer chains were obtained and side-reactions, as concluded from the multimodal SEC traces, were observed. Furthermore, reaction of **184** with 6-(2,2′:6′,2″-terpyridin-4′-ylthio)

hexane-1-thiol gave the Ru^{II} complex **186**, which possessed two binding sites for further metal-ion complexation. The subsequent metallo-supramolecular polymerization with Fe^{II} ions generated the metallopolymer **187** with a strictly alternating order of Ru^{II} and Fe^{II} centers. Such heterobimetallic materials represent versatile precursor materials for the formation of, e.g., iron-ruthenium alloy-nanoparticles [705].

Ureidopyrimidinone (UPy), as a self-complementary quadruple *H*-bonding system, represents one of the most versatile arrays to assemble supramolecular multi-component structures [706]. The orthogonality of *H*-bonding interactions and metal-to-ligand coordination has been employed to prepare multi-responsive materials. Schubert et al. pioneered this field by combining the complexation of tpy ligands with Upy dimerization; small-molecule as well as macromolecular building blocks were employed in these studies [349, 707]. More recently, Haga et al. synthesized the homoleptic Ru^{II} complex **188**, which assembled into linear supramolecular polymers in non-coordinating solvents [708]. According to DLS measurements, these assemblies had a size of several μm. The self-assembly into linear polymers was also investigated by AFM imaging; for this purpose, **188** was spin-coated onto highly-ordered pyrolytic graphite (HOPG). Nanowires with a height and length in the range of 3 nm and 5–10 μm, respectively, were observed. Remarkably, the supramolecular polymer enabled the solubilization of single-walled carbon nanotubes (SWCNTs) in $CHCl_3$, simply by sonication. The resultant hybrid materials were characterized by photophysical and electrochemical measurements as well as microscopy imaging (Fig. 2.144).

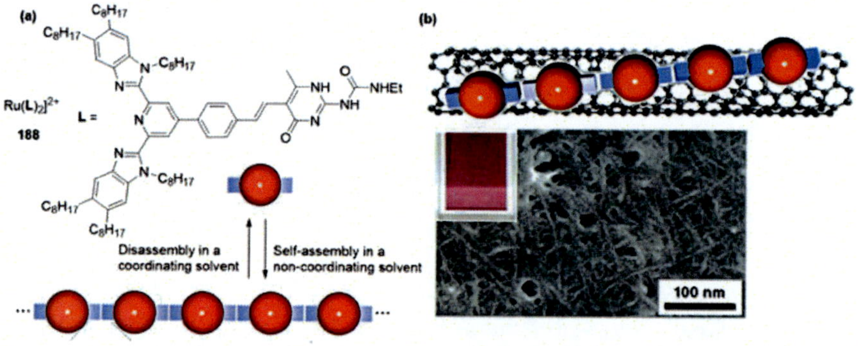

Fig. 2.144 **a** Schematic representation of the self-assembly of complex **188** into linear supramolecular polymers via quadruple *H*-bonding interactions. **b** Schematic representation of the interaction between poly-**188** and SWCNTs; a representative photograph of the solution and a SEM image of SWCNT@poly-**188** on a silica substrate are also shown [708]. Figure reproduced with kind permission; © 2018 The Royal Chemical Society

(c) ***Miscellaneous monomers***

The so-called knitting polymerization represents a relatively new method to prepare hyper-cross-linked polymers in a straightforward fashion, i.e., simply utilizing a Lewis-acidic catalyst (e.g., $AlCl_3$ or $FeCl_3$) and a cross-linking agent [e.g., CH_2Cl_2 or $CH_2(OMe)_2$] [709]. In order to incorporate catalytically active metal complexes into such materials, Lai, Huang et al. [710] as well as Maya, Iglesias et al. [711] employed the Ru^{II} complexes **189** and **190**, respectively (Fig. 2.145). The former, introduced by these same authors as single-site catalyst for the CO_2 reduction [712], was polymerized in benzene to afford the complex-containing porous polymer network [710]. This heterogeneous catalyst was found to be highly active in the dehydrogenation of formic acid in an aqueous or organic environment. The selective and sustainable formation of H_2 and CO_2 proceeded with a high TON of 145,300 in a continuous experiment (the small-molecule complex **189** gave a TON of 5,600 under the same conditions). In a similar fashion, **190** was converted into a knitted polymer using $AlCl_3$, as Lewis acid (also a polymer with biphenyl, as cross-linking agent, was prepared) [711]. As concluded from solid-state ^{13}C NMR, IR and XPS spectroscopy, decomposition of the Ru^{II} complex did not occur; a sponge-like morphology was observed by microscopy imaging (Fig. 2.146). The surface area of both polymers, in the range of 250–300 $m^2\ g^{-1}$, was calculated from the N_2 adsorption/desorption isotherms. The porous polymers were tested, as heterogeneous catalysts, in the imidation of alcohols under basic conditions. Both high conversion and very good chemoselectivity was found for a range of substrates. The catalyst could be recovered and used in up to six runs without significant decrease-of-activity; leaching of Ru ions was no issue.

One further porous polymer was prepared McGrier et al., who reacted Ru^{II}-tetra*kis*(4-formylphenyl)porphyrin with 2,5-diamino-1,4-benzenedithiol [713]. The obtained polymer network **191** (Fig. 2.147) was characterized thoroughly regarding its composition, morphology as well as photophysical and thermal properties. TGA measurements revealed a thermal stability up to 380 °C, XRD analysis and SEM imaging indicated a cloud-like morphology of the amorphous material. Moreover, the surface area was determined (655 $m^2\ g^{-1}$) and high CO_2 uptake capacities of 105 and 71 mg g^{-1} were measured at 273 and 298 K, respectively. As a

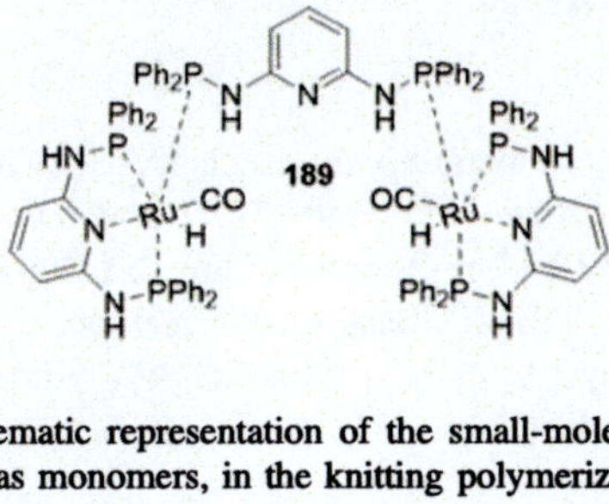

[Ru(PPh3)3(CO)ClH]

190

Fig. 2.145 Schematic representation of the small-molecule Ru^{II} catalysts **189** and **190**, which were employed, as monomers, in the knitting polymerization

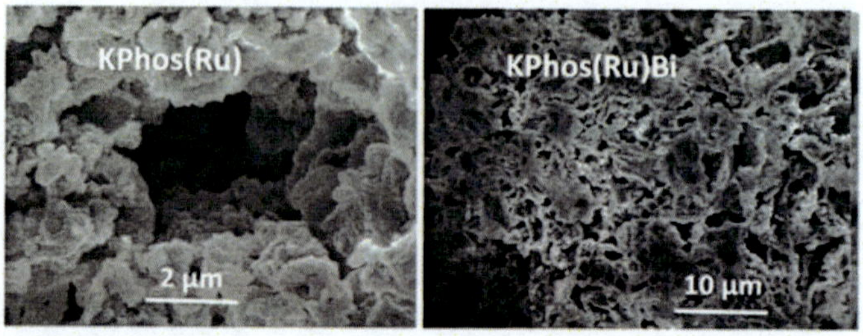

Fig. 2.146 Representative SEM images of the knitted polymers, derived from **190** alone (left) and **190**/biphenyl (right) [711]. Figure reproduced with kind permission; © 2019 The Royal Chemical Society

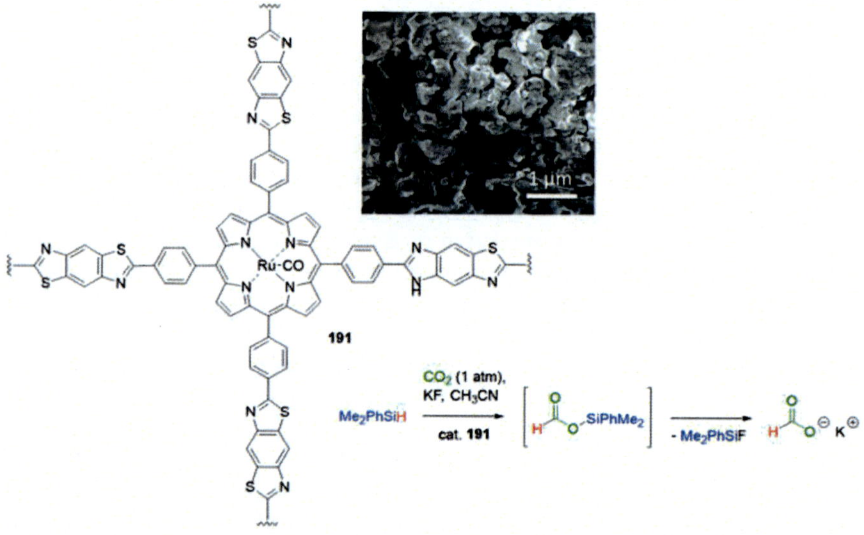

Fig. 2.147 Schematic representation of the porous ruthenopolymer **191**, which was employed, as heterogeneous catalyst, in the CO_2 reduction. A representative SEM image of **191** showing the amorphous, cloud-like morphology is also depicted [713]. Figure reproduced with kind permission; © 2019 The Royal Chemical Society

proof-of-concept, **191** was applied, as heterogeneous catalyst, to reduce CO_2 to formate via a silyl formate intermediate, moderate TON and TOF values were found for this test reaction (67 and 17 h^{-1}, respectively; Fig. 2.147). Related to this, a microporous polymer, comprising Tröger's base-type moieties to with Ru^{III} ions were coordinated, was employed by Liu et al. in the CO_2/H_2 capture and hydrogenation of CO_2 to afford formic acid (TON of 2,254 at 40 °C) [714].

2.3 Concluding Remarks

Discovered in 1844 by Carl Ernst Claus, ruthenium represents one of the least naturally abundant elements on earth. Due to its special properties, most of the ruthenium produced today (>50%) is used in wear-resistant electrical contacts and thick-film resistors as well as in platinum alloys [715]. Beyond these, also its application in the fields of catalysis, solar energy conversion and data storage is of significant relevance. In order to establish more robust, soluble and/or processable Ru-containing materials for these purposes, their combination with polymers has been realized—either by covalent linkage or blending into hybrid materials. Has this concept been found fruitful? From a synthetic point-of-view, the implementation of Ru into polymers has experienced a dramatic improvement. Tailor-made "ruthenopolymers" of whatever architecture are now available by making use of a rich toolbox of synthetic inorganic, organic, organometallic, and macromolecular chemistry. The advanced analytical techniques, which have emerged in the last years, enable the in-depth characterization of the polymers in solution as well as the solid state; the high stability of the Ru-containing materials in comparison to other types of metallopolymers facilitates their analysis further. What remains is the critical question regarding the potential applications of the ruthenopolymers. Here, the price and abundancy of the metal need to be addressed again. This being the limiting factors, a usage in large-scale applications becomes quite unlikely (e.g., bulk polymer applications, large-area coatings). In a few selected fields however, Ru-containing polymers have shown a very promising performance: Heterogeneous catalysis and redox-active electrode coatings. Since rather little amounts of ruthenium are required for these purposes, these applications can be expected to emerge further. From these, the ruthenopolymers addressing the needs of the twenty-first century will certainly be in the foreground: Water splitting, CO_2 reduction, and light-to-energy conversion have to be named in this context. Basically, the same holds true for biomedical applications, where RAPTA-C represents one of the few established ruthenodrugs [716]. With more potent small-molecule candidates, e.g., for cancer therapy, being identified their conjugates with polymers should enter the clinical trials as soon as possible. Though not having been covered in this review, Ru^{II}–polypyridyl complexes are of importance due to their ability to bind structure- and site-specifically to DNA. The reversible binding modes include electrostatic interactions, intercalation, and groove binding [717]; the different ways of irreversible, i.e., covalent, binding of such complexes to DNA have already been subject of various reviews [718–721]. In particular, the intercalation of Ru^{II} dyes into DNA might be adaptable also to synthetic polymers giving rise to new types of functional/switchable materials. Ruthenoproteins represent one further type of natural polymers that have been modified with Ru^{II} complexes, in particular focusing on electron-transfer process within these assemblies [722].

To conclude, the combination of ruthenium with polymers is particularly fruitful for those photophysical, electrochemical or biomedical cutting-edge applications, where the costs and poor abundancy of the metal are (at least) leveled out by the performance of the material.

References

1. Rehahn, M.: Organic/inorganic hybrid polymers. Acta Polym. **49**(5), 201 (1998)
2. Hardy, C.G., Zhang, J., Yan, Y., Ren, L., Tang, C.: Metallopolymers with transition metals in the side-chain by living and controlled polymerization techniques. Prog. Polym. Sci. **39** (10), 1742 (2014)
3. Winter, A., Schubert, U.S.: Synthesis and characterization of metallo-supramolecular polymers. Chem. Soc. Rev. **45**(19), 5311 (2016)
4. Ho, C.-L., Wong, W.-Y.: Metal-containing polymers: Facile tuning of photophysical traits and emerging applications in organic electronics and photonics. Coord. Chem. Rev. **255**(21–22), 2469 (2011)
5. Wild, A., Winter, A., Schlutter, F., Schubert, U.S.: Advances in the field of 2,2′:6′,2″-terpyridines. Chem. Soc. Rev. **40**(3), 1459 (2011)
6. Chan, W.K.: Metal containing polymers with heterocyclic rigid main chains. Coord. Chem. Rev. **251**(17–20), 2104 (2007)
7. Fukumoto, H., Yamane, K., Kase, Y., Yamamoto, T.: π-Conjugated poly(aryleneethynylene)s consisting of salophen and Ni–salophen units in the π-conjugated main chain: Preparation and chemical properties. Macromolecules **43**(24), 10366 (2010)
8. Wong, W.-Y., Harvey, P.D.: Recent progress on the photonic properties of conjugated organometallic polymers built upon the *trans*-bis(*para*-ethynylbenzene)bis(phosphine) platinum(II) chromophore and related derivatives. Macromol. Rapid Commun. **31**(8), 671 (2010)
9. Zha, Y., Disabb-Miller, M.L., Johnson, Z.D., Hickner, M.A., Tew, G.N.: Metal-cation-based anion exchange membranes. J. Am. Chem. Soc. **134**(10), 4493 (2012)
10. Allcock, H.R., Lavin, K.D., Riding, G.H.: Ring-opening polymerization of metallocene cyclophosphazene derivatives. Macromolecules **18**(6), 1340 (1985)
11. Carlise, J. R., Weck, M.: Side-chain functionalized polymers containing bipyridine coordination sites: Polymerization and metal-coordination studies. J. Polym. Sci., Part A: Polym. Chem. **42**(12), 2973 (2004)
12. Friebe, C., Hager, M.D., Winter, A., Schubert, U.S.: Metal-containing polymers *via* electropolymerization. Adv. Mater. **24**(3), 332 (2012)
13. Yang, J., Sykora, M., Meyer, T.J.: Electropolymerization of vinylbipyridine complexes of ruthenium(II) and osmium(II) in SiO_2 sol-gel films. Inorg. Chem. **44**(10), 3396 (2005)
14. Friebe, C., Schulze, B., Görls, H., Jäger, M., Schubert, U.S.: Designing cyclometalated ruthenium(II) complexes for anodic electropolymerization. Chem. Eur. J. **20**(8), 2357 (2014)
15. Happ, B., Schäfer, J., Menzel, R., Hager, M.D., Winter, A., Popp, J.R., Beckert, R., Dietzek, B., Schubert, U.S.: Synthesis and resonance energy transfer study on a random terpolymer containing a 2-(pyridine-2-yl)thiazole donor-type ligand and a luminescent $[Ru(bpy)_2(2\text{-(triazol-4-yl)pyridine})]^{2+}$ chromophore. Macromolecules **44**(16), 6277 (2011)
16. Yan, Y., Zhang, J., Qiao, Y., Ganewatta, M., Tang, C.: Ruthenocene-containing homopolymers and block copolymers *via* ATRP and RAFT polymerization. Macromolecules **46**(22), 8816 (2013)
17. Yu, S.C., Gong, X., Chan, W.K.: Synthesis and characterization of poly(benzobisoxazole)s and poly(benzobisthiazole)s with 2,2′-bipyridyl units in the backbone. Macromolecules **31** (17), 5639 (1998)

18. Serin, J., Schultze, X., Adronov, A., Fréchet, J.M.J.: Synthesis and study of the absorption and luminescence properties of polymers containing $Ru(bpyMe_2)_3^{2+}$ chromophores and coumarin laser dyes. Macromolecules **35**(14), 5396 (2002)
19. Schultze, X., Serin, J., Adronov, A., Frechet, J.M.J.: Light harvesting and energy transfer in a ruthenium-coumarin-2 copolymer. Chem. Commun. (13), 1160 (2001)
20. Blunden, B.M., Thomas, D.S., Stenzel, M.H.: Macromolecular ruthenium complexes as anti-cancer agents. Polym. Chem. **3**(10), 2964 (2012)
21. Jones, W.E., Baxter, S.M., Strouse, G.F., Meyer, T.J.: Intrastrand electron and energy transfer between polypyridyl complexes on a soluble polymer. J. Am. Chem. Soc. **115**(16), 7363 (1993)
22. Ulbricht, C., Beyer, B., Friebe, C., Winter, A., Schubert, U.S.: Recent developments in the application of phosphorescent iridium(III) complex systems. Adv. Mater. **21**(44), 4418 (2009)
23. Fournier, D., Hoogenboom, R., Schubert, U.S.: Clicking polymers: A straightforward approach to novel macromolecular architectures. Chem. Soc. Rev. **36**(8), 1369 (2007)
24. Feng, C., Li, Y., Yang, D., Hu, J., Zhang, X., Huang, X.: Well-defined graft copolymers: from controlled synthesis to multipurpose applications. Chem. Soc. Rev. **40**(3), 1282 (2011)
25. Mede, T., Jäger, M., Schubert, U.S.: "Chemistry-on-the-complex": Functional Ru(II) polypyridyl-type sensitizers as divergent building blocks. Chem. Soc. Rev. **47**(20), 7577 (2018)
26. Younathan, J.N., McClanahan, S.F., Meyer, T.J.: Synthesis and characterization of soluble polymers containing electron- and energy-transfer reagents. Macromolecules **22**(3), 1048 (1989)
27. Strouse, G.F., Worl, L.A., Younathan, J.N., Meyer, T.J.: Long-range energy transfer in a soluble polymer by an energy-transfer cascade. J. Am. Chem. Soc. **111**(25), 9101 (1989)
28. Worl, L.A., Jones, W.E., Strouse, G.F., Younathan, J.N., Danielson, E., Maxwell, K.A., Sykora, M., Meyer, T.J.: Multiphoton, multielectron transfer photochemistry in a soluble polymer. Inorg. Chem. **38**(11), 2705 (1999)
29. Dupray, L.M., Meyer, T.J.: Synthesis and characterization of amide-derivatized, polypyridyl-based metallopolymers. Inorg. Chem. **35**(21), 6299 (1996)
30. Fleming, C.N., Dupray, L.M., Papanikolas, J.M., Meyer, T.J.: Energy transfer between Ru(II) and Os(II) polypyridyl complexes linked to polystyrene. J. Phys. Chem. A **106**(10), 2328 (2002)
31. Fleming, C.N., Maxwell, K.A., DeSimone, J.M., Meyer, T.J., Papanikolas, J.M.: Ultrafast excited-state energy migration dynamics in an efficient light-harvesting antenna polymer based on Ru(II) and Os(II) polypyridyl complexes. J. Am. Chem. Soc. **123**(42), 10336 (2001)
32. Friesen, D.A., Kajita, T., Danielson, E., Meyer, T.J.: Preparation and photophysical properties of amide-linked, polypyridylruthenium-derivatized polystyrene. Inorg. Chem. **37**(11), 2756 (1998)
33. Olmsted, J., III; McClanahan, S. F., Danielson, E., Younathan, J. N., Meyer, T. J. : Electron and energy shuttling between redox sites on soluble polymers. J. Am. Chem. Soc. **109**(11), 3297 (1987)
34. Baxter, S.M., Jones, W.E., Jr., Danielson, E., Worl, L., Strouse, G., Younathan, J., Meyer, T. J.: Photoinduced electron and energy transfer in soluble polymers. Coord. Chem. Rev. **111**, 47 (1991)
35. Huynh, M.H.V., Dattelbaum, D.M., Meyer, T.J.: Exited state electron and energy transfer in molecular assemblies. Coord. Chem. Rev. **249**(3–4), 457 (2005)
36. Walsh, J.J., Zeng, Q., Forster, R.J., Keyes, T.E.: Highly luminescent Ru(II) metallopolymers: Photonic and redox properties in solution and as thin films. Photochem. Photobiol. Sci. **11**(10), 1547 (2012)

37. Worl, L.A., Strouse, G.F., Younathan, J.N., Baxter, S.M., Meyer, T.J.: Production and storage of multiple, photochemical redox equivalents on a soluble polymer. J. Am. Chem. Soc. **112**(21), 7571 (1990)
38. Shaw, G.B., Papanikolas, J.M.: Triplet-triplet annihilation of excited states of polypyridyl Ru(II) complexes bound to polystyrene. J. Phys. Chem. B **106**(24), 6156 (2002)
39. Dupray, L.M., Devenney, M., Striplin, D.R., Meyer, T.J.: An antenna polymer for visible energy transfer. J. Am. Chem. Soc. **119**(42), 10243 (1997)
40. Fleming, C.N., Brennaman, M.K., Papanikolas, J.M., Meyer, T.J.: Efficient, long-range energy migration in RuII polypyridyl derivatized polystyrenes in rigid media Antennae for artificial photosynthesis. Dalton Trans. (20), 3903 (2009)
41. Happ, B., Winter, A., Hager, M.D., Schubert, U.S.: Photogenerated avenues in macromolecules containing Re(I), Ru(II), Os(II), and Ir(III) metal complexes of pyridine-based ligands. Chem. Soc. Rev. **41**(6), 2222 (2012)
42. Aamer, K.A., Tew, G.N.: Supramolecular polymers containing terpyridine–metal complexes in the side chain. Macromolecules **40**(8), 2737 (2007)
43. Aamer, K.A., de Jeu, W.H., Tew, G.N.: Diblock copolymers containing metal complexes in the side chain of one block. Macromolecules **41**(6), 2022 (2008)
44. Kajita, T., Leasure, R.M., Devenney, M., Friesen, D., Meyer, T.J.: Electropolymerized films of macromeric assemblies. Inorg. Chem. **37**(19), 4782 (1998)
45. Golas, P.L., Matyjaszewski, K.: Marrying click chemistry with polymerization: Expanding the scope of polymeric materials. Chem. Soc. Rev. **39**(4), 1338 (2010)
46. Binder, W.H., Sachsenhofer, R.: 'Click' chemistry in polymer and material science: An update. Macromol. Rapid Commun. **29**(12–13), 952 (2008)
47. Sun, Y., Chen, Z., Puodziukynaite, E., Jenkins, D.M., Reynolds, J.R., Schanze, K.S.: Light harvesting arrays of polypyridine ruthenium(II) chromophores prepared by reversible addition-fragmentation chain-transfer polymerization. Macromolecules **45**(6), 2632 (2012)
48. Gleria, M., De Jaeger, R.: Polyphosphazenes: A review. Top. Curr. Chem. **250**, 165 (2005)
49. Teasdale, I., Brüggemann, O.: Polyphosphazenes: Multifunctional, biodegradable vehicles for drug and gene delivery. Polymers **5**(1), 161 (2013)
50. Allcock, H.R., Kugel, R.L.: Synthesis of high polymeric alkoxy- and aryloxyphosphonitriles. J. Am. Chem. Soc. **87**(18), 4216 (1965)
51. Allcock, H.R., Kugel, R.L.: Phosphonitrilic compounds. Part VII. High molecular weight poly(diaminophosphazenes). Inorg. Chem. **5**(10), 1716 (1966)
52. Allcock, H.R., Kugel, R.L., Valan, K.J.: Phosphonitrilic compounds. Part VI. High molecular weight poly(alkoxy- and aryloxyphosphazenes). Inorg. Chem. **5**(10), 1709 (1966)
53. Wang, Z., McWilliams, A.R., Evans, C.E.B., Lu, X., Chung, S., Winnik, M.A., Manners, I.: Covalent attachment of Ru(II) phenanthroline complexes to polythionylphosphazenes: the development and evaluation of single-component polymeric oxygen sensors. Adv. Funct. Mater. **12**(6–7), 415 (2002)
54. Schroot, R., Friebe, C., Altuntas, E., Crotty, S., Jäger, M., Schubert, U.S.: Nitroxide-mediated polymerization of styrenic triarylamines and chain-end functionalization with a ruthenium complex: toward tailored photoredox-active architectures. Macromolecules **46**(6), 2039 (2013)
55. Schroot, R., Schlotthauer, T., Schubert, U.S., Jäger, M.: Modular assembly of poly (naphthalene diimide) and Ru(II) dyes for an efficient light-induced charge separation in hierarchically controlled polymer architectures. Macromolecules **49**(6), 2112 (2016)
56. Schroot, R., Schlotthauer, T., Jäger, M., Schubert, U.S.: Hydrophilic poly(naphthalene diimide)-based acceptor–photosensitizer dyads: Toward water-processible modular photoredox-active architectures. Macromol. Chem. Phys. **218**(6), 1600534 (2017)
57. Schlotthauer, T., Schroot, R., Glover, S., Hammarström, L., Jäger, M., Schubert, U.S.: A multidonor–photosensitizer–multiacceptor triad for long-lived directional charge separation. Phys. Chem. Chem. Phys. **19**(42), 28572 (2017)

58. Schroot, R., Schlotthauer, T., Dietzek, B., Jäger, M., Schubert, U.S.: Extending long-lived charge separation between donor and acceptor blocks in novel copolymer architectures featuring a sensitizer core. Chem. Eur. J. **23**(65), 16484 (2017)
59. Ruiz-Palomero, C., Soriano, M.L., Valcárcel, M.: Gels based on nanocellulose with photosensitive ruthenium bipyridine moieties as sensors for silver nanoparticles in real samples. Sen. Actuators B **229**, 31 (2016)
60. Dai, L., Wang, Y., Zou, X., Chen, Z., Liu, H., Ni, Y.: Ultrasensitive physical, bio, and chemical sensors derived from 1-, 2-, and 3-D nanocellulosic materials. Small **16**(13), 1906567 (2020)
61. Yu, X.-Q., Huang, J.-S., Yu, W.-Y., Che, C.-M.: Polymer-supported ruthenium porphyrins: versatile and robust epoxidation catalysts with unusual selectivity. J. Am. Chem. Soc. **122** (22), 5337 (2000)
62. Zhang, J.-L., Che, C.-M.: Soluble polymer-supported ruthenium porphyrin catalysts for epoxidation, cyclopropanation and aziridination of alkenes. Org. Lett. **4**(11), 1911 (2002)
63. Chaudret, B., Chung, G., Huang, Y.-S.: Synthesis and characterization of soluble polycationic ruthenium polystyrene derivatives [PSTRuL]CF_3SO_3. J. Chem. Soc., Chem. Commun. (10), 749 (1990)
64. Royo, E., Brintzinger, H.-H.: Mass spectrometry of polystyrene and polypropene ruthenium complexes. A new tool for polymer characterization. J. Organomet. Chem. **663**(1), 213 (2002)
65. Akiyama, R., Kobayashi, S.: A novel polymer-supported arene–ruthenium complex for ring-closing olefin metathesis. Angew. Chem. Int. Ed. **41**(14), 2602 (2002)
66. Pertici, P., Vitulli, G., Carlini, C., Ciardelli, F.: Versatile polystyrene-ruthenium hydrogenation catalysts. J. Mol. Catal. **11**(2), 353 (1981)
67. Wang, F., Lai, Y.-H., Han, M.Y.: Post-coordination of multinuclear transitional metal clusters to azulene-based polymers: A novel strategy for tuning properties in π-conjugated polymers. Org. Lett. **5**(25), 4791 (2003)
68. Bailey, D.C., Langer, S.H.: Immobilized transition-metal carbonyls and related catalysts. Chem. Rev. **81**(2), 109 (1981)
69. Blunden, B.M., Stenzel, M.H.: Incorporating ruthenium into advanced drug delivery carriers – an innovative generation of chemotherapeutics. J. Chem. Technol. Biotechnol. **90**(7), 1177 (2015)
70. Kostova, I.: Ruthenium complexes as anticancer agents. Curr. Med. Chem. **13**(9), 1085 (2006)
71. Gasser, G., Ott, I., Metzler-Nolte, N.: Organometallic anticancer compounds. J. Med. Chem. **54**(1), 3 (2010)
72. Blunden, B.M., Lu, H., Stenzel, M.H.: Enhanced delivery of the RAPTA-C macromolecular chemotherapeutic by conjugation to degradable polymeric micelles. Biomacromol **14**(12), 4177 (2013)
73. Leadbeater, N.E., Marco, M.: Preparation of polymer-supported ligands and metal complexes for use in catalysis. Chem. Rev. **102**(10), 3217 (2002)
74. Farrall, M.J., Frechet, J.M.J.: Bromination and lithiation: Two important steps in the functionalization of polystyrene resins. J. Org. Chem. **41**(24), 3877 (1976)
75. Pittman, C.U., Jr., Smith, L.R.: Hanes, R. M. Catalytic reactions using polymer-bound *vs.* homogeneous complexes of nickel, rhodium, and ruthenium. J. Am. Chem. Soc. **97**(7), 1742 (1975)
76. Pittman, C.U., Jr., Wilemon, G.M.: 1-Pentene hydroformylation catalyzed by polymer-bound ruthenium complexes. J. Org. Chem. **46**(9), 1901 (1981)
77. Otero-Schipper, Z., Lieto, J., Gates, B.C.: Ethylene hydrogenation catalyzed by polymer-bound tetraruthenium clusters. J. Catal. **63**(1), 175 (1980)
78. Carlini, C., Braca, G., Ciardelli, F., Sbrana, G.: Complexes of ruthenium(II) and rhodium(I) with N-and P-donor macromolecular ligands: Preparation and catalytic isomerisation of olefins. J. Mol. Catal. **2**(6), 379 (1977)

79. Judkins, C.M.G., Knights, K.A., Johnson, B.F.G., Miguel, Y.R.d., Raja, R., Thomas, J.M.: Immobilisation of ruthenium cluster catalysts *via* novel derivatisations of ArgoGel resins. Chem. Commun. (24), 2624 (2001)
80. Rentsch, D., Hany, R., Barthélémy, S., Steinauer, R.: Quantitative determination of loadings and oxidation products of polystyrene-bound phosphines using ^{31}P MAS NMR. J. Comb. Chem. **5**(5), 610 (2003)
81. Jia, Z., Wang, K., Tan, B., Gu, Y.: Ruthenium complexes immobilized on functionalized knitted hypercrosslinked polymers as efficient and recyclable catalysts for organic transformations. Adv. Synth. Catal. **359**(1), 78 (2017)
82. Xu, C., Wang, H., Wang, Q., Wang, Y., Zhang, Y., Fan, G.: Ruthenium coordinated with triphenylphosphine-hyper-crosslinked polymer: An efficient catalyst for hydrogen evolution reaction and hydrolysis of ammonia borane. Appl. Surface Sci. **466**, 193 (2019)
83. Shah, J.N., Gokak, D.T., Ram, R.N.: Preparation, characterization and catalytic activity of polymer-supported trimethylenediamine-ruthenium complex. J. Mol. Catal. **60**(2), 141 (1990)
84. Shah, J.N., Ram, R.N.: Hydrogenation of nitrobenzene using polymer-supported ruthenium (III) complexes as catalysts. J. Mol. Catal. **77**(3), 235 (1992)
85. Shah, J.N., Ram, R.N.: Hydrogenation of cyclohexene using polymer-supported ruthenium (III) complexes as catalysts. J. Mol. Catal. **83**(1), 67 (1993)
86. John, J., Dalal, M.K., Patel, D.R., Ram, R.N.: Preparation, properties and catalytic application of polymer-bound Ru(III) complexes. J. Macromol. Sci., Part A **34**(3), 489 (1997)
87. John, J., Dalal, M.K., Ram, R.N.: Oxidation of cyclohexane using polymer bound Ru(III) complexes as catalysts. Stud. Surf. Sci. Catal. **110**, 1165 (1997)
88. Patel, D.R., Dalal, M.K., Ram, R.N.: Preparation characterization and catalytic activity of polymer supported Ru(III) complexes. J. Mol. Catal. A: Chem. **109**(2), 141 (1996)
89. Antony, R., Tembe, G.L., Ravindranathan, M., Ram, R.N.: Polymer supported Ru(III) complexes, synthesis and catalytic activity. Polymer **39**(18), 4327 (1998)
90. Antony, R., Tembe, G.L., Ravindranathan, M., Ram, R.N.: Synthesis and catalytic property of poly(styrene-*co*-divinylbenzene) supported ruthenium(III)-2-aminopyridyl complexes. Eur. Polym. J. **36**(8), 1579 (2000)
91. Trakarnpruk, W., Kanjina, W.: Preparation, characterization and oxidation catalysis of polymer-supported ruthenium and cobalt complexes. Ind. Eng. Chem. Res. **47**(3), 964 (2008)
92. Jones, C.W.: On the stability and recyclability of supported metal-ligand complex catalysts: myths, misconceptions and critical research needs. Top. Catal. **53**(13), 942 (2010)
93. Buchmeiser, M.R.: Polymer-supported well-defined metathesis catalysts. Chem. Rev. **109** (2), 303 (2008)
94. Nguyen, S.T., Grubbs, R.H.: The syntheses and activities of polystyrene-supported olefin metathesis catalysts based on $Cl_2(PR_3)_2Ru{=}CH{-}CH{=}CPh_2$. J. Organomet. Chem. **497**(1–2), 195 (1995)
95. Melis, K., De Vos, D., Jacobs, P., Verpoort, F.: ROMP and RCM catalysed by $(R_3P)_2Cl_2Ru{=}CHPh$ immobilised on a mesoporous support. J. Mol. Catal. A: Chem. **169** (1–2), 47 (2001)
96. Buchowicz, W., Ingold, F., Mol, J.C., Lutz, M., Spek, A.L.: Novel ruthenium(II) carboxylates as catalysts for alkene metathesis. Chem. Eur. J. **7**(13), 2842 (2001)
97. Nieczypor, P., Buchowicz, W., Meester, W.J.N., Rutjes, F.P.J.T., Mol, J.C.: Synthesis and application of a new polystyrene-supported ruthenium carbene catalyst for alkene metathesis. Tetrahedron Lett. **42**(40), 7103 (2001)
98. Lubbad, S., Mayr, B., Mayr, M., Buchmeiser, M.R.: Monolithic systems: from separation science to heterogeneous catalysis. Macromol. Symp. **210**(1), 1 (2004)

99. Krause, J.O., Lubbad, S., Nuyken, O., Buchmeiser, M.R.: Monolith- and silica-supported carboxylate-based Grubbs–Herrmann-type metathesis catalysts. Adv. Synth. Catal. **345**(8), 996 (2003)
100. Krause, J.O., Nuyken, O., Wurst, K., Buchmeiser, M.R.: Synthesis and reactivity of homogeneous and heterogeneous ruthenium-based metathesis catalysts containing electron-withdrawing ligands. Chem. Eur. J. **10**(3), 777 (2004)
101. Krause, J.O., Lubbad, S.H., Nuyken, O., Buchmeiser, M.R.: Heterogenization of a modified Grubbs-Hoveyda catalyst on a ROMP-derived monolithic support. Macromol. Rapid Commun. **24**(15), 875 (2003)
102. Yang, L., Mayr, M., Wurst, K., Buchmeiser, M.R.: Novel metathesis catalysts based on ruthenium 1,3-dimesityl-3,4,5,6-tetrahydropyrimidin-2-ylidenes: Synthesis, structure, immobilization, and catalytic activity. Chem. Eur. J. **10**(22), 5761 (2004)
103. Krause, J.O., Zarka, M.T., Anders, U., Weberskirch, R., Nuyken, O., Buchmeiser, M.R.: Simple synthesis of poly(acetylene) latex particles in aqueous media. Angew. Chem. Int. Ed. **42**(48), 5965 (2003)
104. Krause, J.O., Nuyken, O., Buchmeiser, M.R.: Factors relevant for the ruthenium–benzylidene-catalyzed cyclopolymerization of 1,6-heptadyines. Chem. Eur. J. **10**(8), 2029 (2004)
105. Krause, J.O., Wang, D., Anders, U., Weberskirch, R., Zarka, M.T., Nuyken, O., Jäger, C., Haarer, D., Buchmeiser, M.R.: Stereoselective cyclopolymerization of diynes: Smart materials for electronics and sensors. Macromol. Symp. **217**(1), 179 (2004)
106. Halbach, T.S., Krause, J.O., Nuyken, O., Buchmeiser, M.R.: Stereoselective cyclopolymerization of polar 1,6-heptadiynes by novel, tailor-made ruthenium-based metathesis catalysts. Macromol. Rapid Commun. **26**(10), 784 (2005)
107. Kotre, T., Zarka, M.T., Krause, J.O., Buchmeiser, M.R., Weberskirch, R., Nuyken, O.: Design and application of amphiphilic polymeric supports for micellar catalysis. Macromol. Symp. **217**(1), 203 (2004)
108. Halbach, T.S., Mix, S., Fischer, D., Maechling, S., Krause, J.O., Sievers, C., Blechert, S., Nuyken, O., Buchmeiser, M.R.: Novel ruthenium-based metathesis catalysts containing electron-withdrawing ligands: Synthesis, immobilization, and reactivity. J. Org. Chem. **70** (12), 4687 (2005)
109. Braddock, D.C., Tanaka, K., Chadwick, D., Böhm, V.P.W., Roeper, M.: Vacuum-driven anionic ligand exchange in Buchmeiser–Hoveyda–Grubbs ruthenium(II) benzylidenes. Tetrahedron Lett. **48**(30), 5301 (2007)
110. Braca, G., Carlini, C., Ciardelli, F., Sbrana, G.: at *6th International Conference on Catalysis*, p. A43. London (1976)
111. Barrett, A.G.M., Cramp, S.M., Roberts, R.S.: ROMP-spheres: A novel high-loading polymer support using cross metathesis between vinyl polystyrene and norbornene derivatives. Org. Lett. **1**(7), 1083 (1999)
112. Ahmed, M., Barrett, A.G.M., Braddock, D.C., Cramp, S.M., Procopiou, P.A.: A recyclable 'boomerang' polymer-supported ruthenium catalyst for olefin metathesis. Tetrahedron Lett. **40**(49), 8657 (1999)
113. Ahmed, M., Arnauld, T., Barrett, A.G.M., Braddock, D.C., Procopiou, P.A.: Second generation recyclable `boomerang' polymer supported catalysts for olefin metathesis: Application of Arduengo carbene complexes. Synlett **2000**(7), 1007 (2000)
114. Jafarpour, L., Nolan, S.P.: Simply assembled and recyclable polymer-supported olefin metathesis catalysts. Org. Lett. **2**(25), 4075 (2000)
115. Jafarpour, L., Heck, M.-P., Baylon, C., Lee, H.M., Mioskowski, C., Nolan, S.P.: Preparation and activity of recyclable polymer-supported ruthenium olefin metathesis catalysts. Organometallics **21**(4), 671 (2002)
116. Kingsbury, J.S., Hoveyda, A.H.: Regarding the mechanism of olefin metathesis with sol-gel-supported Ru-based complexes bearing a bidentate carbene ligand. Spectroscopic evidence for return of the propagating Ru carbene. J. Am. Chem. Soc. **127**(12), 4510 (2005)

117. Yao, Q.: A soluble polymer-bound ruthenium carbene complex: a robust and reusable catalyst for ring-closing olefin metathesis. Angew. Chem. Int. Ed. **39**(21), 3896 (2000)
118. Yao, Q., Rodriguez Motta, A.: Immobilization of the Grubbs second-generation ruthenium-carbene complex on poly(ethylene glycol): a highly reactive and recyclable catalyst for ring-closing and cross-metathesis. Tetrahedron Lett. **45**(11), 2447 (2004)
119. Yao, Q., Zhang, Y.: Poly(fluoroalkyl acrylate)-bound ruthenium carbene complex: a fluorous and recyclable catalyst for ring-closing olefin metathesis. J. Am. Chem. Soc. **126**(1), 74 (2004)
120. Garber, S.B., Kingsbury, J.S., Gray, B.L., Hoveyda, A.H.: Efficient and recyclable monomeric and dendritic Ru-based metathesis catalysts. J. Am. Chem. Soc. **122**(34), 8168 (2000)
121. Varray, S., Lazaro, R., Martinez, J., Lamaty, F.: New soluble-polymer bound ruthenium carbene catalysts: synthesis, characterization and application to ring-closing metathesis. Organometallics **22**(12), 2426 (2003)
122. Connon, S.J., Dunne, A.M., Blechert, S.: A self-generating, highly active and recyclable olefin-metathesis catalyst. Angew. Chem. Int. Ed. **41**(20), 3835 (2002)
123. Blechert, S., Connon, S.J.: Olefin metathesis catalysts based on ring-opening-prepared polymer-supported transition metal complexes. Chem. Abstr. **140**, 236227 (2004)
124. Randl, S., Buschmann, N., Connon, S.J., Blechert, S.: Highly efficient and recyclable polymer-bound catalyst for olefin metathesis reactions. Synlett (10), 1547 (2001)
125. Connon, S.J., Blechert, S.: A solid-supported phosphine-free ruthenium alkylidene for olefin metathesis in methanol and water. Bioorg. Med. Chem. Lett. **12**(14), 1873 (2002)
126. Dowden, J., Savovic, J.: Olefin metathesis in non-degassed solvent using a recyclable, polymer supported alkylideneruthenium. Chem. Commun. (1), 37 (2001)
127. Grela, K., Tryznowski, M., Bieniek, M.: A PS-DES immobilized ruthenium carbene: a robust and easily recyclable catalyst for olefin metathesis. Tetrahedron Lett. **43**(50), 9055 (2002)
128. Zarka, M.T., Nuyken, O., Weberskirch, R.: Polymer-bound, amphiphilic Hoveyda-Grubbs-type catalyst for ring-closing metathesis in water. Macromol. Rapid Commun. **25**(8), 858 (2004)
129. Koç, F., Michalek, F., Rumi, L., Bannwarth, W., Haag, R.: Catalysts on functionalized polymer chips (PC) as recyclable entities. Synthesis (19), 3362 (2005)
130. Michalek, F., Mädge, D., Rühe, J., Bannwarth, W.: The activity of covalently immobilized Grubbs-Hoveyda type catalyst is highly dependent on the nature of the support material. J. Organomet. Chem. **691**(24–25), 5172 (2006)
131. Bannwarth, W., Michalek, F., Rühe, J.: Polymer-supported transition metal complexes and use thereof in catalysis of olefin metathesis. Chem. Abstr. **147**, 53271 (2007)
132. Michalek, F., Mädge, D., Rühe, J., Bannwarth, W.: Ring-closure metathesis in supercritical carbon dioxide as sole solvent with use of covalently immobilized ruthenium catalysts. Eur. J. Org. Chem. (3), 577 (2006)
133. Chen, S.-W., Kim, J.H., Song, C.E., Lee, S.-G.: Self-supported oligomeric Grubbs/Hoveyda-type Ru-carbene complexes for ring-closing metathesis. Org. Lett. **9**(19), 3845 (2007)
134. Sommer, W.J., Weck, M.: Supported N-heterocyclic carbene complexes in catalysis. Coord. Chem. Rev. **251**(5–6), 860 (2007)
135. Schürer, S.C., Gessler, S., Buschmann, N., Blechert, S.: Synthesis and application of a permanently immobilized olefin-metathesis catalyst. Angew. Chem. Int. Ed. **39**(21), 3898 (2000)
136. Mayr, M., Mayr, B., Buchmeiser, M.R.: Monolithic materials: new high-performance supports for permanently immobilized metathesis catalysts. Angew. Chem. Int. Ed. **40**(20), 3839 (2001)
137. Buchmeiser, M.R.: Applications of metathesis in heterogeneous catalysis and separation sciences. J. Mol. Catal. A: Chem. **190**(1–2), 145 (2002)

138. Mayr, M., Wang, D., Kröll, R., Schuler, N., Prühs, S., Fürstner, A., Buchmeiser, M.R.: Monolithic disk-supported metathesis catalysts for use in combinatorial chemistry. Adv. Synth. Catal. **347**(2–3), 484 (2005)
139. Buchmeiser, M.R., Seeber, G., Mupa, M., Bonn, G.K.: ROMP-Based, highly hydrophilic poly(7-oxanorborn-2-ene-5,6-dicarboxylic acid)-coated silica for analytical and preparative scale high-performance ion chromatography. Chem. Mater. **11**(6), 1533 (1999)
140. Mayr, M., Buchmeiser, M.R., Wurst, K.: Synthesis of a silica-based heterogeneous second generation Grubbs catalyst. Adv. Synth. Catal. **344**(6–7), 712 (2002)
141. Gallivan, J.P., Jordan, J.P., Grubbs, R.H.: A neutral, water-soluble olefin metathesis catalyst based on an N-heterocyclic carbene ligand. Tetrahedron Lett. **46**(15), 2577 (2005)
142. Hong, S.H., Grubbs, R.H.: Highly active water-soluble olefin metathesis catalyst. J. Am. Chem. Soc. **128**(11), 3508 (2006)
143. Sommer, W.J., Weck, M.: Poly(norbornene)-supported N-heterocyclic carbenes as ligands in catalysis. Adv. Synth. Catal. **348**(15), 2101 (2006)
144. Mennecke, K., Grela, K., Kunz, U., Kirschning, A.: Immobilisation of the Grubbs-III olefin metathesis catalyst with polyvinyl pyridine (PVP). Synlett (19), 2948 (2005)
145. Syukri, S., Sun, W., Kühn, F.E.: Immobilization of ruthenium(II) salen complexes on poly (4-vinylpyridine) and their application in catalytic aldehyde olefination. Tetrahedron Lett. **48** (9), 1613 (2007)
146. Sun, Q., Jin, Y., Zhu, L., Wang, L., Meng, X., Xiao, F.-S.: Superhydrophobic, chiral, and mesoporous TsDPEN copolymer coordinated to ruthenium species as an efficient catalyst for asymmetric transfer hydrogenation. Nano Today **8**(4), 342 (2013)
147. Itsuno, S.: Immobilization method of chiral catalysts onto amphiphilic polymer supports. J. Synth. Org. Chem., Jpn. **67**(10), 1025 (2009)
148. Sugie, H., Hashimoto, Y., Haraguchi, N., Itsuno, S.: Synthesis of polymer-immobilized TsDPEN ligand and its application in asymmetric transfer hydrogenation of cyclic sulfonimine. J. Organomet. Chem. **751**, 711 (2014)
149. Haraguchi, N., Tsuru, K., Arakawa, Y., Itsuno, S.: Asymmetric transfer hydrogenation of imines catalyzed by a polymer-immobilized chiral catalyst. Org. Biomol. Chem. **7**(1), 69 (2009)
150. Arakawa, Y., Haraguchi, N., Itsuno, S.: Design of novel polymer-supported chiral catalyst for asymmetric transfer hydrogenation in water. Tetrahedron Lett. **47**(19), 3239 (2006)
151. Arakawa, Y., Chiba, A., Haraguchi, N., Itsuno, S.: Asymmetric transfer hydrogenation of aromatic ketones in water using a polymer-supported chiral catalyst containing a hydrophilic pendant group. Adv. Synth. Catal. **350**(14–15), 2295 (2008)
152. Koh, M.J., Khan, R.K.M., Torker, S., Yu, M., Mikus, M.S., Hoveyda, A.H.: High-value alcohols and higher-oxidation-state compounds by catalytic *Z*-selective cross-metathesis. Nature **517**(7533), 181 (2015)
153. Wang, Z., Yang, R., Zhu, J., Zhu, X.: PEG-related polymer resins as synthetic supports. Sci. Chin. Chem. **53**(9), 1844 (2010)
154. Xia, L., Peng, T., Wang, G., Wen, X., Zhang, S., Wang, L.: Grubbs catalysts immobilized on Merrifield resin for metathesis of leaf alcohols by using a convenient recycling approach. ChemistryOpen **8**(1), 45 (2019)
155. Gray, H.B., Rembaum, A., Gupta, A.: Photochemical preparation of olefin addition catalysts. Chem. Abstr. **90**, 173430q (1978)
156. Braca, G., Sbrana, G., Carlini, C., Ciardelli, F.: Catalytic activity of ruthenium(II) and rhodium(I) complexes with linear macromolecular ligands. Catal., Proc. Int. Symp. 307 (1975)
157. Friedrich, H.B., Singh, N.: The very efficient oxidation of alcohols by poly(4-vinylpyridine)-supported sodium ruthenate. Tetrahedron Lett. **41**(20), 3971 (2000)
158. Hinzen, B.V., Ley, S.: Polymer supported perruthenate (PSP): a new oxidant for clean organic synthesis. J. Chem. Soc., Perkin Trans. **1**(13), 1907 (1997)

159. Clear, J.M., Kelly, J.M., Vos, J.G.: Polyvinylpyridine complexes of ruthenium(III) chloride. Makromol. Chem. **184**(3), 613 (1983)
160. Roda, J., Hrabák, F.: Darstellung und Eigenschaften der Koordinationsverbindungen von Poly(vinylpyridin)en mit den Chloriden von Ru, Rh und Pd. Makromol. Chem. **177**(1), 1 (1976)
161. Westmoreland, T.D., Calvert, J.M., Murray, R.W., Meyer, T.J.: A photoelectrode based on visible photolysis of a metallopolymer film. J. Chem. Soc., Chem. Commun. (2), 65 (1983)
162. Calvert, J.M., Caspar, J.V., Binstead, R.A., Westmoreland, T.D., Meyer, T.J.: Metallopolymer photochemistry. Photophysical, photochemical and photoelectrochemical properties of $(bpy)^2Ru(II)$ sites bound to poly(4-vinylpyridine). J. Am. Chem. Soc. **104**(24), 6620 (1982)
163. Clear, J.M., Kelly, J.M., O'Connell, C.M., Vos, J.G.: Preparation, characterization, and photochemical properties of bis(2,2′-bipyridyl)ruthenium(II) complexes of 4-vinylpyridine, poly-4-vinylpyridine, and copolymers. J. Chem. Res. **9**, 260 (1981)
164. Haas, O., Kriens, M., Vos, J.G.: An electrochemical investigation of photosubstitution processes at glassy-carbon electrodes coated with polymer-bound ruthenium complexes. J. Am. Chem. Soc. **103**(6), 1318 (1981)
165. Haas, O., Vos, J.G.: Preparation, characterisation and electrochemistry of electrodes coated with poly(4-vinylpyridine) bound ruthenium complexes. J. Electroanal. Chem. Interfacial Electrochem. **113**(1), 139 (1980)
166. Geraty, S.M., Vos, J.G.: Synthesis, characterisation and photochemical properties of a series of ruthenium containing metallopolymers based on poly-N-vinylimidazole. J. Chem. Soc., Dalton Trans. (12), 3073 (1987)
167. Haas, O., Zumbrunnen, H.R., Vos, J.G.: Electrochemical and photochemical behaviour of poly(4-vinylpyridine) bound ruthenium complexes attached to electrode surfaces Part IV. Electrochim. Acta **30**(11), 1551 (1985)
168. Forster, R.J., Vos, J.G.: Synthesis, characterization, and properties of a series of osmium- and ruthenium-containing metallopolymers. Macromolecules **23**(20), 4372 (1990)
169. Forster, R.J., Vos, J.G.: Redox site loading, electrolyte concentration and temperature effects on charge transport and electrode kinetics of electrodes modified with osmium con. J. Electroanal. Chem. Interfacial Electrochem. **314**(1–2), 135 (1991)
170. Doherty, A.P., Stanley, M.A., Arana, G., Koning, C.E., Brinkhuis, R.H.G., Vos, J.G.: The effect of the nature of the polymer backbone on the stability and the analytical response of polymer-modified electrodes. Electroanalysis **7**(4), 333 (1995)
171. Calvert, J.M., Meyer, T.J.: Poly(pyridyl)ruthenium(II) complexes of poly(4-vinylpyridine). Synthesis, characterization and thin-film electrochemistry of a series of bis(bipyridine) metallopolymers. Inorg. Chem. **21**(11), 3978 (1982)
172. Clear, J.M., Kelly, J.M., Pepper, D.C., Vos, J.G.: Preparation, reactions and photoreaction of Bis(bipyridyl)poly-4-vinylpyridine ruthenium(II) complexes. Inorg. Chim. Acta **33**, L139 (1979)
173. Oh, S.M., Faulkner, L.R.: Luminescent and redox probes of structure and dynamics in quaternized poly(4-vinylpyridine) films on electrodes. J. Am. Chem. Soc. **111**(15), 5613 (1989)
174. Dennany, L., Forster, R.J., White, B., Smyth, M., Rusling, J.F.: Direct electrochemiluminescence detection of oxidized DNA in ultrathin films containing $[Os(bpy)_2(PVP)_{10}]^{2+}$. J. Am. Chem. Soc. **126**(28), 8835 (2004)
175. Mugweru, A., Wang, B., Rusling, J.: Voltammetric sensor for oxidized DNA using ultrathin films of osmium and ruthenium metallopolymers. Anal. Chem. **76**(18), 5557 (2004)
176. Mugweru, A., Rusling, J.: Studies of DNA damage inhibition by dietary antioxidants using metallopolyion/DNA sensors. Electroanalysis **18**(4), 327 (2006)
177. Kavanagh, P., Leech, D.: Redox polymer and probe DNA tethered to gold electrodes for enzyme-amplified amperometric detection of DNA hybridization. Anal. Chem. **78**(8), 2710 (2006)

178. Hvastkovs, E.G., So, M., Krishnan, S., Bajrami, B., Tarun, M., Jansson, I., Schenkman, J.B., Rusling, J.F.: Electrochemiluminescent arrays for cytochrome P450-activated genotoxicity screening. DNA damage from benzo[*a*]pyrene metabolites. Anal. Chem. **79**(5), 1897 (2007)
179. Dennany, L., Wallace, G.G., Forster, R.J.: Luminescent metal complexes within polyelectrolyte layers: Tuning electron and energy Transfer. Langmuir **25**(24), 14053 (2009)
180. Song, B., Pan, S., Tang, C., Li, D., Rusling, J.F.: Voltammetric microwell array for oxidized guanosine in intact ds-DNA. Anal. Chem. **85**(22), 11061 (2013)
181. Ju, H., Gong, Y., Zhu, H.: Electrolyte effects on electrochemical properties of osmium complex polymer modified electrodes. Anal. Sci. **17**(1), 59 (2001)
182. O'Mullane, A.P., Macpherson, J.V., Unwin, P.R., Cervera-Montesinos, J., Manzanares, J.A., Frehill, F., Vos, J.G.: Measurement of lateral charge propagation in $[Os(bpy)_2(PVP)_nCl]Cl$ thin films: A scanning electrochemical microscopy approach. J. Phys. Chem. B **108**(22), 7219 (2004)
183. Wang, B., Brown, S., Rusling, J.F.: Electrolyte effects on charge-transport behavior of $[Os(bpy)_2(PVP)_{10}Cl]Cl$ and $[Ru(bpy)_2(PVP)_{10}Cl]Cl$ redox polymers in ultra-thin films of polyions. Electroanalysis **17**(18), 1601 (2005)
184. Fei, J.J., Hu, S.S.: Osmium redox polymer film electrode and its electrocatalytic properties towards the oxidation of epinephrine: Electrochemical quartz-crystal microbalance and voltametric characterization. Russ. J. Electrochem. **41**(12), 1296 (2005)
185. Gallaway, J.W., Calabrese Barton, S.A.: Kinetics of redox polymer-mediated enzyme electrodes. J. Am. Chem. Soc. **130**(26), 8527 (2008)
186. Doherty, A.P., Buckley, T., Kelly, D.M., Vos, J.G.: Stabilization of the redox polymer $[Os(bipy)_2(PVP)_{10}Cl]Cl$ by *in-situ* chemical cross-linking. Electroanalysis **6**(7), 553 (1994)
187. Iwuoha, E.I., Smyth, M.R., Vos, J.G.: Amperometric glucose sensor containing nondiffusional osmium redox centers: Analysis of organic-phase responses. Electroanalysis **6**(11–12), 982 (1994)
188. Pravda, M., Adeyoju, O., Iwuoha, E.I., Vos, J.G., Smyth, M.R., Vytras, K.: Amperometric glucose biosensors based on an osmium(II/III) redox polymer-mediated electron transfer at carbon paste electrodes. Electroanalysis **7**(7), 619 (1995)
189. Robinson, G., Leech, D., Smyth, M.R.: Electrically 'wired' tyrosinase enzyme inhibition electrode for the detection of respiratory poisons. Electroanalysis **7**(10), 952 (1995)
190. Trudeau, F., Daigle, F., Leech, D.: Reagentless mediated laccase electrode for the detection of enzyme modulators. Anal. Chem. **69**(5), 882 (1997)
191. Iwuoha, E.I., Leister, I., Miland, E., Smyth, M.R., Ó Fágáin, C.: Reactivities of organic-phase biosensors. Part 1. Enhancement of the sensitivity and stability of amperometric peroxidase biosensors using chemically modified enzymes. Anal. Chem. **69** (8), 1674 (1997)
192. Jezkova, J., Iwuoha, E.I., Smyth, M.R., Vytras, K.: Stabilization of an osmium bis-bipyridyl polymer-modified carbon paste amperometric glucose biosensor using polyethyleneimine. Electroanalysis **9**(13), 978 (1997)
193. Park, T.-M., Iwuoha, E.I., Smyth, M.R.: Development of a sol-gel enzyme inhibition-based amperometric biosensor for cyanide. Electroanalysis **9**(14), 1120 (1997)
194. Daigle, F., Leech, D.: Reagentless tyrosinase enzyme electrodes: Effects of enzyme loading, electrolyte pH, ionic strength, and temperature. Anal. Chem. **69**(20), 4108 (1997)
195. Kim, H.-H., Mano, N., Zhang, Y., Heller, A.: A miniature membrane-less biofuel cell operating under physiological conditions at 0.5 V. J. Electrochem. Soc. **150**(2), A209 (2003)
196. Ferry, Y., Leech, D.: Amperometric detection of catecholamine neurotransmitters using electrocatalytic substrate recycling at a laccase electrode. Electroanalysis **17**(2), 113 (2005)
197. Tsujimura, S., Murata, K., Akatsuka, W.: Exceptionally high glucose current on a hierarchically structured porous carbon electrode with "wired" flavin adenine dinucleotide-dependent glucose dehydrogenase. J. Am. Chem. Soc. **136**(41), 14432 (2014)

198. Badura, A., Guschin, D., Esper, B., Kothe, T., Neugebauer, S., Schuhmann, W., Rögner, M.: Photo-induced electron transfer between photosystem-2 *via* cross-linked redox hydrogels. Electroanalysis **20**(10), 1043 (2008)
199. Rasmussen, M., Ritzmann, R.E., Lee, I., Pollack, A.J., Scherson, D.: An implantable biofuel cell for a live insect. J. Am. Chem. Soc. **134**(3), 1458 (2012)
200. Frehill, F., In het Panhuis, M., Young, N.A., Henry, W., Hjelm, J., Vos, J.G.: Microscopy and spectroscopy of interactions between metallopolymers and carbon nanotubes. J. Phys. Chem. B **109**(27), 13205 (2005)
201. Zhu, J., Walsh, J.J., Bond, A.M., Keyes, T.E., Forster, R.J.: Ruthenium metallopolymer: Dawson polyoxomolybdate α-$[Mo_{18}O_{54}(SO_4)_2]^{4-}$ adduct films: Sensitization for visible photoelectrocatalysis. Langmuir **28**(37), 13536 (2012)
202. Alessio, E.: Thirty years of the drug candidate NAMI-A and the myths in the field of ruthenium anticancer compounds: A personal perspective. Eur. J. Inorg. Chem. **2017**(12), 1549 (2017)
203. Blunden, B.M., Rawal, A., Lu, H., Stenzel, M.H.: Superior chemotherapeutic benefits from the ruthenium-based anti-metastatic drug NAMI-A through conjugation to polymeric micelles. Macromolecules **47**(5), 1646 (2014)
204. Cho, Y.-S., Lee, H.-K., Lee, J.-S.: Well-defined ruthenium-coordinated block copolymers of 2-vinylpyridine with 2-(*N*-carbazolyl)ethyl methacrylate and their optical properties. Macromol. Chem. Phys. **203**(17), 2495 (2002)
205. Zeng, Q., McNally, A., Keyes, T.E., Forster, R.J.: Three colour electrochromic metallopolymer based on a ruthenium phenolate complex bound to poly(4-vinyl)pyridine. Electrochem. Commun. **10**(3), 466 (2008)
206. Sun, W., Zeng, X., Wu, S.: Photoresponsive ruthenium-containing polymers: potential polymeric metallodrugs for anticancer phototherapy. Dalton Trans. **47**(2), 283 (2018)
207. Sun, W., Parowatkin, M., Steffen, W., Butt, H.-J., Mailänder, V., Wu, S.: Ruthenium-containing block copolymer assemblies: Red-light-responsive metallopolymers with tunable nanostructures for enhanced cellular uptake and anticancer phototherapy. Adv. Healthcare Mater. **5**(4), 467 (2016)
208. Chen, M., Sun, W., Kretzschmann, A., Butt, H.-J., Wu, S.: Nanostructured polymer assemblies stabilize photoactivatable anticancer ruthenium complexes under physiological conditions. J. Inorg. Biochem. **207**, 111052 (2020)
209. Sun, W., Li, S., Häupler, B., Liu, J., Jin, S., Steffen, W., Schubert, U.S., Butt, H.-J., Liang, X.-J., Wu, S.: An amphiphilic ruthenium polymetallodrug for combined photodynamic therapy and photochemotherapy *in vivo*. Adv. Mater. **29**(6), 1603702 (2017)
210. Card, R.J., Neckers, D.C.: Preparation of polymer-bound bipyridine and some of its transition metal complexes. J. Am. Chem. Soc. **99**(23), 7733 (1977)
211. Card, R.J., Neckers, D.C.: Poly(styryl)bipyridine: synthesis and formation of transition-metal complexes and some of their physical, chemical, and catalytic properties. Inorg. Chem. **17**(9), 2345 (1978)
212. Tamagaki, S., Card, R.J., Neckers, D.C.: (Polystyrylbipyridine)(tetracarbonyl)tungsten. An active, reusable heterogeneous catalyst for metathesis of internal olefins. J. Am. Chem. Soc. **100**(21), 6635 (1978)
213. Card, R.J., Liesner, C.E., Neckers, D.C.: Poly(styryl)bipyridinepalladium complexes as heterogeneous catalysts for hydrogenation of alkenes and alkynes. J. Org. Chem. **44**(7), 1095 (1979)
214. Furue, M., Sumi, K., Nozakura, S.-I.: Synthesis and luminescence properties of tris (bipyridineruthenium(II)-containing vinyl polymers. Chem. Lett. **10**(10), 1349 (1981)
215. Pitt, C.G., Bao, Y., Seltzman, H.H.: The synthesis of polymers containing the 2,2′-bipyridine ligand. J. Polym. Sci., Part C: Polym. Lett. **24**(1), 13 (1986)
216. Bao, Y.T., Pitt, C.G.: Polymeric amines and their copper(II) complexes: Catalysts for the hydrolysis of organophosphate esters. J. Polym. Sci., Part A: Polym. Chem. **28**(4), 741 (1990)

217. Newkome, G.R., Yoneda, A.: Synthesis, characterization, and complexation of poly (5-vinyl-6,6′-dimethyl-2,2′-bipyridine). Makromol. Chem., Rapid Commun. **4**(8), 575 (1983)
218. Newkome, G.R., Yoneda, A.: Methyl (*E*)-3-[5′-(6′, 6″-dimethyl-2′,2″-bipyridinyl)]acrylate: Synthesis and copolymerization with styrene. Makromol. Chem., Rapid Commun. 6(6), 451 (1985)
219. Lei, Z.-Q., Wang, Y.-P.: Oxidation of alkylbenzenes catalyzed by polymer-bound 2,2′-bipyridine-iron complexes. React. Polym. **16**(2), 223 (1992)
220. Antonietti, M., Lohmann, S., Eisenbach, C.D., Schubert, U.S.: Synthesis of metal-complexing latices *via* polymerization in microemulsion. Macromol. Rapid Commun. **16**(4), 283 (1995)
221. Happ, B., Pavlov, G.M., Perevyazko, I., Hager, M.D., Winter, A., Schubert, U.S.: Induced charge effect by Co(II) complexation on the conformation of a copolymer containing a bidentate 2-(1,2,3-triazol-4-yl)pyridine chelating unit. Macromol. Chem. Phys. **213**(13), 1339 (2012)
222. Marin, V., Holder, E., Schubert, U.S.: Polymeric ruthenium bipyridine complexes: New potential materials for polymer solar cells. J. Polym. Sci., Part A: Polym. Chem. **42**(2), 374 (2004)
223. Sandmann, B., Happ, B., Kupfer, S., Schacher, F.H., Hager, M.D., Schubert, U.S.: The self-healing potential of triazole-pyridine-based metallopolymers. Macromol. Rapid Commun. **36**(7), 604 (2014)
224. Trumbo, D.L., Marvel, C.S.: Synthesis of acetylene-terminated polymers using supported palladium(II) catalysts. J. Polym. Sci., Part A: Polym. Chem. **25**(3), 847 (1987)
225. Fischer, C.-H.: Synthesis of 4,4′-bis(tridecyl)- and 4-methyl-4′-tridecyl-2,2′-bipyridine and a 2,2′-bipyridine-substituted polystyrene: an easier method for the preparation of 2,2′-bipyridines with long-chain substituents. J. Photochem. **24**(4), 415 (1984)
226. Chujo, Y., Sada, K., Saegusa, T.: Synthesis of bipyridyl-branched polyoxazoline and its gelation by means of metal coordination. Polym. J. **25**(6), 599 (1993)
227. Kaneko, M., Nemoto, S., Yamada, A., Kurimura, Y.: Preparation and photochemical behavior of polystyrene pendant tris(bipyridyl)ruthenium(II) complex. Inorg. Chim. Acta **44**, L289 (1980)
228. Kurimura, Y., Shinozaki, N., Ito, F., Uratani, Y., Shigehara, K., Tsuchida, E., Kaneko, M., Yamada, A.: The synthesis and luminescence quenching of the water-soluble polymer-supported tris(2,2′-bipyridine)ruthenium(II) analogue. Bull. Chem. Soc. Jpn. **55** (2), 380 (1982)
229. Kaneko, M., Yamada, A.: Photoinduced charge separation at solid–liquid interface utilizing polymer pendant $Ru(bpy)_3Cl_2$ complex. Inorg. Chim. Acta **45**, L73 (1980)
230. Kaneko, M., Ochiai, M., Kinosita, K., Yamada, A.: Photoinduced electron transfer from polystyrene pendant tris(2,2′-bipyridyl)ruthenium(II) complex to methylviologen. J. Polym. Sci.: Polym. Chem. Ed. **20**(4), 1011 (1982)
231. Kaneko, M., Yamada, A., Tsuchida, E., Kurimura, Y.: Preparation and photochemical reactivity of poly(styrene-co-vinylbipyridine) pendant $Ru(bpy)_3^{2+}$ complex. J. Polym. Sci.: Polym. Lett. Ed. **20**(11), 593 (1982)
232. Kaneko, M.: Photosensing device composed of bilayer membranes of polymer-pendant Ru $(bpy)_3^{2+}$ and Prussian blue. J. Macromol. Sci., Part A **24**(3–4), 357 (1987)
233. Kaneko, M., Nakamura, H.: Effect of polymer on the quenching of the excited state of copolymer-pendant $Ru(bpy)_3^{2+}$ containing carboxylate groups. Macromolecules **20**(9), 2265 (1987)
234. Taniguchi, T., Kuroki, M., Miyashita, T.: Electron transfer quenching of tris(2,2′-bipyridine) ruthenium(II) complex derivatives covalently linked to poly(*N*-isopropylacrylamide) in aqueous solutions. Colloid Polym. Sci. **274**(8), 717 (1996)
235. Taniguchi, T., Miyashita, T.: Photoelectrochemical responses of polymer Langmuir-Blodgett films containing tris(2,2′-bipyridine)ruthenium(II) complex. Chem. Lett. **26**(4), 295 (1997)

236. Marin, V., Holder, E., Meier, M.A.R., Hoogenboom, R., Schubert, U.S.: A mixed ruthenium polypyridyl complex containing a PEG-bipyridine macroligand. Macromol. Rapid Commun. **25**(7), 793 (2004)
237. Marin, V., Holder, E., Hoogenboom, R., Schubert, U.S.: Functional ruthenium(II)- and iridium(III)-containing polymers for potential electro-optical applications. Chem. Soc. Rev. **36**(4), 618 (2007)
238. Farah, A.A., Pietro, W.J.: Synthesis and characterization of partially crosslinked poly(*N*-vinylcarbazole-vinylalcohol) copolymers with polypyridyl Ru(II) luminophores: Potential materials for electroluminescence. Polym. Bull. **43**(2–3), 135 (1999)
239. Farah, A.A., Veinot, J.G.C., Najman, M., Pietro, W.J.: Redox active, multi-chromophore Ru (II) polypyridyl-carbazole copolymers: Synthesis and characterization. J. Macromol. Sci., Part A **37**(11), 1507 (2000)
240. Camacho, M.A., Kar, A.K., Lindsell, W.E., Murray, C., Preston, P.N., Wherrett, B.S.: Synthesis and nonlinear optical properties of functionalised polydiacetylenes and their complexes with transition metals. J. Mater. Chem. **9**(6), 1251 (1999)
241. Kasem, K.K.: Photo-electrochemistry at polymer/semiconductor interface characterization of surface modified CdS based photovoltaic cells. J. Mater. Sci. **34**(21), 5237 (1999)
242. Hou, X.-H., Kaneko, M., Yamada, A.: Quenching of photo-excited state of copolymer pendant $Ru(bpy)_3^{2+}$ complexes with methylviologen. J. Polym. Sci., Part A: Polym. Chem. **24**(11), 2749 (1986)
243. Zhao, F., Zhang, J., Hou, X., Abe, T., Kaneko, M.: Quenching of photoluminescence from copolymer pendant $Ru(bpy)_3^{2+}$complexes by colloidal Prussian Blue. J. Chem. Soc., Faraday Trans. **94**(2), 277 (1998)
244. Kaneko, M., Yamada, A., Tsuchida, E., Kurimura, Y.: Novel effect of neighboring groups on the quenching of the excited state of a polymer-pendant $Ru(bpy)_3^{2+}$ by methyl viologen. J. Phys. Chem. **88**(6), 1061 (1984)
245. Wu, X., Fraser, C.L.: Architectural diversity *via* metal template-assisted polymer synthesis: A macroligand chelation approach to linear and star-shaped polymeric ruthenium tris (bipyridine) complexes. Macromolecules **33**(11), 4053 (2000)
246. Wu, X., Fraser, C.L.: The importance of macroligand molecular weight and solvent polarity in modulating metal core reactivity in heteroleptic polymeric ruthenium tris(bipyridine) complex synthesis. Macromolecules **33**(21), 7776 (2000)
247. Fraser, C.L., Smith, A.P., Wu, X.: Metal template-assisted block copolymer synthesis: Use of solvent polarity to control chain conformation and reactivity at the metal core. J. Am. Chem. Soc. **122**(37), 9026 (2000)
248. Peter, K., Thelakkat, M.: Synthesis and characterization of bifunctional polymers carrying tris(bipyridyl)ruthenium(II) and triphenylamine units. Macromolecules **36**(6), 1779 (2003)
249. Zhou, G., Harruna, I.I.: Synthesis and characterization of tris(2,2′-bipyridine)ruthenium(II)-centered polystyrenes *via* reversible addition−fragmentation chain transfer (RAFT) polymerization. Macromolecules **37**(19), 7132 (2004)
250. Haque, S.A., Handa, S., Peter, K., Palomares, E., Thelakkat, M., Durrant, J.R.: Supermolecular control of charge transfer in dye-sensitized nanocrystalline TiO_2 films: Towards a quantitative structure–function relationship. Angew. Chem. Int. Ed. **44**(35), 5740 (2005)
251. Zhou, G., He, J., Harruna, I.I.: Synthesis and characterization of tris(2,2′-bipyridine) ruthenium-cored star-shaped polymers *via* RAFT polymerization. J. Polym. Sci., Part A: Polym. Chem. **45**(18), 4225 (2007)
252. Smith, A.P., Fraser, C.L.: Luminescent polymeric ruthenium complexes with polystyrene-*b*-poly(methyl methacrylate) macroligands: The sequential activation of initiator sites for blocks generated by parallel polymerization mechanisms. J. Polym. Sci., Part A: Polym. Chem. **40**(23), 4250 (2002)

253. Smith, A.P., Fraser, C.L.: Metal-centered heteroarm polymers: Chelation of polystyrene-*b*-polycaprolactone with a bipyridine donor at the block junction. Macromolecules **35**(3), 594 (2001)
254. Smith, A.P., Fraser, C.L.: Synthesis of bipyridine-centered diblock copolymers. Macromolecules **36**(8), 2654 (2003)
255. Smith, A.P., Fraser, C.L.: Ruthenium-centered heteroarm stars by a modular coordination approach: Effect of polymer composition on rates of chelation. Macromolecules **36**(15), 5520 (2003)
256. Moreno, V., Font-Bardia, M., Calvet, T., Lorenzo, J., Avilés, F.X., Garcia, M.H., Morais, T. S., Valente, A., Robalo, M.P.: DNA interaction and cytotoxicity studies of new ruthenium (II) cyclopentadienyl derivative complexes containing heteroaromatic ligands. J. Inorg. Biochem. **105**(2), 241 (2011)
257. Valente, A., Garcia, M.H., Marques, F., Miao, Y., Rousseau, C., Zinck, P.: First polymer "ruthenium-cyclopentadienyl" complex as potential anticancer agent. J. Inorg. Biochem. **127**, 79 (2013)
258. Moreira, T., Francisco, R., Comsa, E., Duban-Deweer, S., Labas, V., Teixeira-Gomes, A.-P., Combes-Soia, L., Marques, F., Matos, A., Favrelle, A., et al.: Polymer "ruthenium-cyclopentadienyl" conjugates - new emerging anticancer drugs. Eur. J. Med. Chem. **168**, 373 (2019)
259. Happ, B., Friebe, C., Winter, A., Hager, M.D., Hoogenboom, R., Schubert, U.S.: 2-(1*H*–1,2,3-Triazol-4-yl)-pyridine ligands as alternatives to 2,2′-bipyridines in ruthenium(II) complexes. Chem. Asian J. **4**(1), 154 (2009)
260. Schulze, B., Schubert, U.S.: Beyond click chemistry - supramolecular interactions of 1,2,3-triazoles. Chem. Soc. Rev. **43**(8), 2522 (2014)
261. Marin, V., Holder, E., Hoogenboom, R., Schubert, U.S.: Mixed iridium(III) and ruthenium (II) polypyridyl complexes containing poly(ε-caprolactone)-bipyridine macroligands. J. Polym. Sci., Part A: Polym. Chem. **42**(17), 4153 (2004)
262. Yamamoto, T., Zhou, Z.-H., Kanbara, T., Maruyama, T.: Preparation and properties of poly (2,2′-bipyridine-5,5′-diyl). Chem. Lett. **19**(2), 223 (1990)
263. Yamamoto, T., Yoneda, Y., Maruyama, T.: Ruthenium and nickel complexes of a π-conjugated electrically conducting polymer chelate ligand, poly(2,2′-bipyridine-5,5″-diyl), and their chemical and catalytic reactivity. J. Chem. Soc., Chem. Commun. (22), 1652 (1992)
264. Tokura, S., Yashuda, T., Segawa, Y., Kira, M.: Novel σ-π alternating polymers having 2,2′-bipyridyl in the polymer backbone and their ruthenium complexes. Chem. Lett. **26**(11), 1163 (1997)
265. Corriu, R.J.P., Douglas, W.E., Yang, Z.-X., Karakus, Y., Cross, G.H., Bloor, D.: Preparation of diphenylsilylene polymers containing main-chain acetylene and (hetero)aromatic groups: $\chi(2)$ Non-linear optical and other properties. J. Organomet. Chem. **455**(1–2), 69 (1993)
266. Yamamoto, T., Maruyama, T., Zhou, Z.-H., Ito, T., Fukuda, T., Yoneda, Y., Begum, F., Ikeda, T., Sasaki, S.: π-Conjugated poly(pyridine-2,5-diyl), poly(2,2′-bipyridine-5,5′-diyl), and their alkyl derivatives. Preparation, linear structure, function as a ligand to form their transition metal complexes, catalytic reactions, n-type electrically conducting properties, optical properties, and alignment on substrates. J. Am. Chem. Soc. **116**(11), 4832 (1994)
267. Matsuoka, S., Kohzuki, T., Nakamura, A., Pac, C., Yanagida, S.: Efficient visible-light-driven photocatalysis. Poly(pyridine-2,5-diyl)-catalysed hydrogen photoevolution and photoreduction of carbonyl compounds. J. Chem. Soc., Chem. Commun. (8), 580 (1991)
268. Yamamoto, T., Nakajima, H., Hayashida, N., Shiraishi, K., Kokubo, H.: π-Conjugated polymers prepared by organometallic polycondensation and metal complexes of the polymers. Polym. Adv. Technol. **11**(8–12), 658 (2000)

269. Hayashida, N., Yamamoto, T.: Synthesis of Ru- and Os-complexes of π-conjugated oligomers of 2,2′-bipyridine and 5,5′-bipyrimidine. Optical properties and catalytic activity for photoevolution of H_2 from aqueous media. Bull. Chem. Soc. Jpn. **72**(5), 1153 (1999)
270. Zhu, S.S., Swager, T.M.: Design of conducting redox polymers: A polythiophene-Ru $(bipy)_3{}^{2n+}$ hybrid material. Adv. Mater. **8**(6), 497 (1996)
271. Trouillet, L., De Nicola, A., Guillerez, S.: Synthesis and characterization of a new soluble, structurally well-defined conjugated polymer alternating regioregularly alkylated thiophene oligomer and 2,2′-bipyridine units: Metal-free form and Ru(II) complex. Chem. Mater. **12** (6), 1611 (2000)
272. Jenkins, I.H., Rees, N.G., Pickup, P.G.: Conducting bipyridine-bithiophene copolymers. Chem. Mater. **9**(5), 1213 (1997)
273. Peng, Z., Yu, L.: Synthesis of conjugated polymers containing ionic transition metal complexes. J. Am. Chem. Soc. **118**(15), 3777 (1996)
274. Vinogradova, S.V., Vasnev, V.A., Keshtov, M.L.: Condensation metallosupramolecular polymers containing main-chain 2,2′-bipyridine and 2,2′:6′,2″-terpyridine moieties. Polym. Sci. Ser. B **49**(11–12), 267 (2007)
275. Egbe, D.A.M., Klemm, E.: Preparation of new rigid backbone conjugated organic polymers with large fluorescence quantum yields. Macromol. Chem. Phys. **199**(12), 2683 (1998)
276. Al-Higari, M., Birckner, E., Heise, B., Klemm, E.: Synthesis of some novel heteroarylene ethynylene polymers containing 1,10-phenanthroline in the polymer backbone. J. Polym. Sci., Part A: Polym. Chem. **37**(23), 4442 (1999)
277. Rasmussen, S.C., Thompson, D.W., Singh, V., Petersen, J.D.: Controlled synthesis of a new, soluble, conjugated metallopolymer containing ruthenium chromophoric units. Inorg. Chem. **35**(12), 3449 (1996)
278. Hauer, C.R., King, G.S., McCool, E.L., Euler, W.B., Ferrara, J.D., Youngs, W.J.: Structure of 2,3-butanedione dihydrazone and IR study of higher polyazines: a new class of polymeric conductors. J. Am. Chem. Soc. **109**(19), 5760 (1987)
279. Grimsdale, A.C., Leok Chan, K., Martin, R.E., Jokisz, P.G., Holmes, A.B.: Synthesis of light-emitting conjugated polymers for applications in electroluminescent devices. Chem. Rev. **109**(3), 897 (2009)
280. Majumdar, A., Biswas, M.: Polymer anchored metal complexes. Polym. Bull. **26**(2), 145 (1991)
281. Majumdar, A., Biswas, M.: Polymer anchored metal complexes. Polym. Bull. **26**(2), 151 (1991)
282. Tan, L.-S., Burkett, J.L., Simko, S.R., Alexander, J.M.D.: New aromatic benzazole polymers. Part 3. Synthesis of rigid-rod benzobisazole polymers with main-chain 2,2′-bipyridine-5,5′-diyl units. Macromol. Rapid Commun. **20**(1), 16 (1999)
283. Cameron, C.G., Pickup, P.G.: A conjugated polymer/redox polymer hybrid with electronic communication between metal centres. Chem. Commun. (3), 303 (1997)
284. O'Reilly, E.J., Dennany, L., Griffith, D., Moser, F., Keyes, T.E., Forster, R.J.: Ground and excited state communication within a ruthenium containing benzimidazole metallopolymer. Phys. Chem. Chem. Phys. **13**(15), 7095 (2011)
285. Yu, S.C., Hou, S., Chan, W.K.: Synthesis, metal complex formation and electronic properties of a novel conjugate polymer with a tridentate 2,6-bis(benzimidazol-2-yl)pyridine ligand. Macromolecules **32**(16), 5251 (1999)
286. Yu, S.C., Chan, W.K.: Synthesis and liquid crystal properties of bipyridine-containing aromatic polyamides. Macromol. Rapid Commun. **18**(3), 213 (1997)
287. Yu, S.C., Hou, S., Chan, W.K.: Synthesis and properties of polyamides and polyesters on the basis of 2,2′-bipyridine-5,5′-dicarboxylic acid and the corresponding polymer–ruthenium complexes. Macromolecules **33**(9), 3259 (2000)
288. Hanabusa, K., Higashi, J.-I., Koyama, T., Shirai, H., Hojo, N., Kurose, A.: Synthesis and properties of thermotropic liquid-crystalline polymers containing transition metals. Makromol. Chem. **190**(1), 1 (1989)

289. Winter, A., Newkome, G.R., Schubert, U.S.: Terpyridine-based Materials. Wiley-VCH, Weinheim (2011)
290. Potts, K.T., Usifer, D.A.: Polymers and polymer-metal complexes containing pendent 2,2′:6′,2″-terpyridinyl ligands. Macromolecules **21**(7), 1985 (1988)
291. Hanabusa, K., Nakamura, A., Koyama, T., Shirai, H.: Synthesis, polymerization, copolymerization, and transition-metal coordination of 4-(2,2′:6′,2″-terpyridin-4′-yl)styrene and its polymers and copolymers. Makromol. Chem. **193**(6), 1309 (1992)
292. Hanabusa, K., Nakano, K., Koyama, T., Shirai, H., Hojo, N., Kurose, A.: Synthesis and properties of polymeric iron(II)-complexes containing terpyridinyl groups. Makromol. Chem. **191**(2), 391 (1990)
293. Heller, M., Schubert, U.S.: Polystyrene with pendant mixed functional ruthenium(II)-terpyridine complexes. Macromol. Rapid Commun. **23**(7), 411 (2002)
294. Schubert, U.S., Hofmeier, H.: Metallo-supramolecular graft copolymers: A novel approach toward polymer-analogous reactions. Macromol. Rapid Commun. **23**(9), 561 (2002)
295. Hofmeier, H., El-ghayoury, A., Schubert, U.S.: Combination of supramolecular cross-linking with covalent cross-linking through epoxide ring-opening including gel studies. e-Polymers **3**(1), 676 (2003)
296. Hofmeier, H., Schubert, U.S.: Supramolecular branching and crosslinking of terpyridine-modified copolymers: Complexation and decomplexation studies in diluted solution. Macromol. Chem. Phys. **204**(11), 1391 (2003)
297. Meier, M.A.R., Schubert, U.S.: Terpyridine-modified poly(vinyl chloride): Possibilities for supramolecular grafting and crosslinking. J. Polym. Sci., Part A: Polym. Chem. **41**(19), 2964 (2003)
298. Hassan, M.L., Moorefield, C.M., Elbatal, H.S., Newkome, G.R.: New metallo-supramolecular terpyridine-modified cellulose functional nanomaterials. J. Macromol. Sci., Part A **49**(4), 298 (2012)
299. Wild, A., Hornig, S., Schlütter, F., Vitz, J., Friebe, C., Hager, M.D., Winter, A., Schubert, U. S.: Complexation of terpyridine-containing dextrans: Toward water-soluble supramolecular structures. Macromol. Rapid Commun. **31**(9–10), 921 (2010)
300. Laube, B., Schmelz, O., Rehahn, M., Wunderlich, W.: Metallo-celluloses. Part 2. Synthesis of cellulose derivatives bearing lateral bis-terpyridine ruthenium(II) complexes. e-Polymers **3**(1), 067 (2003)
301. Abitbol, T., Rivkin, A., Cao, Y., Nevo, Y., Abraham, E., Ben-Shalom, T., Lapidot, S., Shoseyov, O.: Nanocellulose, a tiny fiber with huge applications. Curr. Opin. Biotechnol. **39**, 76 (2016)
302. Hassan, M.L., Moorefield, C.M., Elbatal, H.S., Newkome, G.R., Modarelli, D.A., Romano, N.C.: Fluorescent cellulose nanocrystals via supramolecular assembly of terpyridine-modified cellulose nanocrystals and terpyridine-modified perylene. Mater. Sci. Eng., B 177(4), 350 (2012)
303. Aamer, K.A., Tew, G.N.: Synthesis of terpyridine-containing polymers with blocky architectures. Macromolecules **37**(5), 1990 (2004)
304. Tew, G.N., Aamer, K.A., Shunmugam, R.: Incorporation of terpyridine into the side chain of copolymers to create multi-functional materials. Polymer **46**(19), 8440 (2005)
305. Tzanetos, N.P., Andreopoulou, A.K., Kallitsis, J.K.: Side-chain terpyridine polymers through atom transfer radical polymerization and their ruthenium complexes. J. Polym. Sci., Part A: Polym. Chem. **43** 20), 4838 (2005)
306. Stefopoulos, A.A., Pefkianakis, E.K., Papagelis, K., Andreopoulou, A.K., Kallitsis, J.K.: Carbon nanotubes decorated with terpyridine-ruthenium complexes. J. Polym. Sci., Part A: Polym. Chem. **47**(10), 2551 (2009)
307. Müller, M., Papagelis, K., Maultzsch, J., Stefopoulos, A.A., Pefkianakis, E.K., Andreopoulou, A.K., Kallitsis, J.K., Thomsen, C.: Raman spectroscopy of single wall carbon nanotubes functionalized with terpyridine-ruthenium complexes. Phys. Status Solidi B **246**(11–12), 2721 (2009)

308. Ievins, A.D., Wang, X., Moughton, A.O., Skey, J., O'Reilly, R.K.: Synthesis of core functionalized polymer micelles and shell cross-linked nanoparticles. Macromolecules **41**(9), 2998 (2008)
309. Ge, Z., Liu, S.: Facile fabrication of multistimuli-responsive metallo-supramolecular core cross-linked block copolymer micelles. Macromol. Rapid Commun. **34**(11), 922 (2013)
310. Fan, L.-J., Zhang, Y., Murphy, C.B., Angell, S.E., Parker, M.F.L., Flynn, B.R., Jones, W.E., Jr.: Fluorescent conjugated polymer molecular wire chemosensors for transition metal ion recognition and signaling. Coord. Chem. Rev. **253**(3–4), 410 (2009)
311. Kimura, M., Horai, T., Hanabusa, K., Shirai, H.: Fluorescence chemosensor for metal ions using conjugated polymers. Adv. Mater. **10**(6), 459 (1998)
312. Zhang, Y., Murphy, C.B., Jones, W.E.: Poly[*p*-(phenyleneethynylene)-*alt*-(thienyleneethynylene)] polymers with oligopyridine pendant groups: Highly sensitive chemosensors for transition metal ions. Macromolecules **35**(3), 630 (2001)
313. Cheng, K.W., Mak, C.S.C., Chan, W.K., Ng, A.M.C., Djurišić, A.B.: Synthesis of conjugated polymers with pendant ruthenium terpyridine trithiocyanato complexes and their applications in heterojunction photovoltaic cells. J. Polym. Sci., Part A: Polym. Chem. **46**(4), 1305 (2008)
314. Gohy, J.-F.: Metallo-supramolecular block copolymer micelles. Coord. Chem. Rev. **253**(17–18), 2214 (2009)
315. Fustin, C.A., Guillet, P., Schubert, U.S., Gohy, J.F.: Metallo-supramolecular block copolymers. Adv. Mater. **19**(13), 1665 (2007)
316. Winter, A., Schubert, U.S.: Block Copolymers with Element Blocks: The Metal-Bisterpyridine Linkage. In: Chujo, Y. (ed.) New Polymeric Materials Based on Element-Blocks, pp. 307. Springer, Singapore (2019)
317. Chujo, Y., Tanaka, K.: New polymeric materials based on element-blocks. Bull. Chem. Soc. Jpn. **88**(5), 633 (2015)
318. Winter, A., Hager, M.D., Schubert, U.S.: Supramolecular polymers. In: Möller, K., Matyjaszewski, M. (eds.) Polymer Science: A Comprehensive Reference, pp. 269. Elsevier, Amsterdam (2012)
319. Gohy, J.-F., Lohmeijer, B.G.G., Schubert, U.S.: Reversible metallo-supramolecular block copolymer micelles containing a soft core. Macromol. Rapid Commun. **23**(9), 555 (2002)
320. Gohy, J.-F., Lohmeijer, B.G.G., Varshney, S.K., Schubert, U.S.: Covalent *vs.* metallo-supramolecular block copolymer micelles. Macromolecules **35**(19), 7427 (2002)
321. Gohy, J.-F., Lohmeijer, B.G.G., Schubert, U.S.: Metallo-supramolecular block copolymer micelles. Macromolecules **35**(12), 4560 (2002)
322. Lohmeijer, B.G.G., Schubert, U.S.: Supramolecular engineering with macromolecules: An alternative concept for block copolymers. Angew. Chem. Int. Ed. **41**(20), 3825 (2002)
323. Gohy, J.-F., Lohmeijer, B.G.G., Varshney, S.K., Décamps, B., Leroy, E., Boileau, S., Schubert, U.S.: Stimuli-responsive aqueous micelles from an ABC metallo-supramolecular triblock copolymer. Macromolecules **35**(26), 9748 (2002)
324. Lohmeijer, B.G.G., Schubert, U.S.: Playing LEGO with macromolecules: Design, synthesis, and self-organization with metal complexes. J. Polym. Sci., Part A: Polym. Chem. **41**(10), 1413 (2003)
325. Lohmeijer, B.G.G., Schubert, U.S.: Water-soluble building blocks for terpyridine-containing supramolecular polymers: Synthesis, complexation, and pH stability studies of poly(ethylene oxide) moieties. Macromol. Chem. Phys. **204**(8), 1072 (2003)
326. Gohy, J.-F., Lohmeijer, B.G.G., Schubert, U.S.: From supramolecular block copolymers to advanced nano-objects. Chem. Eur. J. **9**(15), 3472 (2003)
327. Lohmeijer, B.G.G., Schlaad, H., Schubert, U.S.: Synthesis and thermal properties of diblock copolymers utilizing non-covalent interactions. Macromol. Symp. **196**(1), 125 (2003)
328. Meier, M.A.R., Lohmeijer, B.G.G., Schubert, U.S.: Characterization of defined metal-containing supramolecular block copolymers. Macromol. Rapid Commun. **24**(14), 852 (2003)

329. Al-Hussein, M., Lohmeijer, B.G.G., Schubert, U.S., de Jeu, W.H.: Melt morphology of polystyrene−poly(ethylene oxide) metallo-supramolecular diblock copolymer. Macromolecules **36**(25), 9281 (2003)
330. Regev, O., Gohy, J.-F., Lohmeijer, B.G., Varshney, S., Hubert, D.W., Frederik, P., Schubert, U.: Dynamic light scattering and cryogenic transmission electron microscopy investigations on metallo-supramolecular aqueous micelles: evidence of secondary aggregation. Colloid Polym. Sci. **282**(4), 407 (2004)
331. Gohy, J.-F., Lohmeijer, B.G.G., Alexeev, A., Wang, X.-S., Manners, I., Winnik, M.A., Schubert, U.S.: Cylindrical micelles from the aqueous self-assembly of an amphiphilic poly (ethylene oxide)-*b*-poly(ferrocenylsilane) (PEO-*b*-PFS) block copolymer with a metallo-supramolecular linker at the block junction. Chem. Eur. J. **10**(17), 4315 (2004)
332. Lohmeijer, B.G.G., Wouters, D., Yin, Z., Schubert, U.S.: Block copolymer libraries: modular versatility of the macromolecular Lego™ system. Chem. Commun. (24), 2886 (2004)
333. Mayer, G., Vogel, V., Lohmeijer, B.G.G., Gohy, J.-F., Van Den Broek, J.A., Haase, W., Schubert, U.S., Schubert, D.: Metallo-supramolecular block copolymer micelles: Improved preparation and characterization. J. Polym. Sci., Part A: Polym. Chem. **42**(17), 4458 (2004)
334. Guerrero-Sanchez, C., Wouters, D., Fustin, C.-A., Gohy, J.-F., Lohmeijer, B.G.G., Schubert, U.S.: Structure−property study of diblock copolymer micelles: Core and corona radius with varying composition and degree of polymerization. Macromolecules **38**(24), 10185 (2005)
335. Al-Hussein, M., de Jeu, W.H., Lohmeijer, B.G.G., Schubert, U.S.: Phase behavior of the melt of oolystyrene−poly(ethylene oxide) metallo-supramolecular diblock copolymer with bulky counterions. Macromolecules **38**(7), 2832 (2005)
336. Lohmeijer, B.G.G., Schubert, U.S.: The LEGO toolbox: Supramolecular building blocks by nitroxide-mediated controlled radical polymerization. J. Polym. Sci., Part A: Polym. Chem. **43**(24), 6331 (2005)
337. Fustin, C.A., Lohmeijer, B.G.G., Duwez, A.S., Jonas, A.M., Schubert, U.S., Gohy, J.F.: Nanoporous thin films from self-assembled metallo-supramolecular block copolymers. Adv. Mater. **17**(9), 1162 (2005)
338. Ott, C., Lohmeijer, B.G.G., Wouters, D., Schubert, U.S.: Terpyridine-terminated homo and biblock copolymer LEGO units by nitroxide-mediated radical polymerization. Macromol. Chem. Phys. **207**(16), 1439 (2006)
339. Guillet, P., Fustin, C.-A., Lohmeijer, B.G.G., Schubert, U.S., Gohy, J.-F.: Study of the influence of the metal−ligand complex on the size of aqueous metallo-supramolecular micelles. Macromolecules **39**(16), 5484 (2006)
340. Raşa, M., Tziatzios, C., Lohmeijer, B.G., Schubert, D., Schubert, U.: Analytical ultracentrifugation studies on terpyridine-end-functionalized poly(ethylene oxide) and polystyrene systems complexed *via* Ru(II) ions. Prog. Colloid Polym. Sci. **131**, 165 (2006)
341. Guillet, P., Fustin, C.-A., Mugemana, C., Ott, C., Schubert, U.S., Gohy, J.-F.: Tuning block copolymer micelles by metal-ligand interactions. Soft Matter **4**(11), 2278 (2008)
342. Gohy, J.-F., Ott, C., Hoeppener, S., Schubert, U.S.: Multicompartment micelles from a metallo-supramolecular tetrablock quatercopolymer. Chem. Commun. (40), 6038 (2009)
343. Ott, C., Kranenburg, J.M., Guerrero-Sanchez, C., Hoeppener, S., Wouters, D., Schubert, U. S.: Supramolecular assembly *via* noncovalent metal coordination chemistry: Synthesis, characterization, and elastic properties. Macromolecules **42**(6), 2177 (2009)
344. Ott, C., Hoogenboom, R., Hoeppener, S., Wouters, D., Gohy, J.-F., Schubert, U.S.: Tuning the morphologies of amphiphilic metallo-supramolecular triblock terpolymers: from spherical micelles to switchable vesicles. Soft Matter **5**(1), 84 (2009)
345. Chiper, M., Winter, A., Hoogenboom, R., Egbe, D.A.M., Wouters, D., Hoeppener, S., Fustin, C.-A., Gohy, J.-F., Schubert, U.S.: Synthesis and micellization of coil−rod−coil ruthenium(II) terpyridine assemblies. Macromolecules **41**(22), 8823 (2008)
346. Landsmann, S., Winter, A., Chiper, M., Fustin, C.-A., Hoeppener, S., Wouters, D., Gohy, J.-F., Schubert, U.S.: Poly(dimethylsiloxane)-substituted 2,2′:6,2″-terpyridines: Synthesis and

characterization of new amphiphilic supramolecular diblock copolymers. Macromol. Chem. Phys. **209**(16), 1666 (2008)
347. Schlütter, F., Pavlov, G.M., Gohy, J.-F., Winter, A., Wild, A., Hager, M.D., Hoeppener, S., Schubert, U.S.: Synthesis, characterization, and micellization studies of coil-rod-coil and ABA ruthenium(II) terpyridine assemblies with π-conjugated electron acceptor systems. J. Polym. Sci., Part A: Polym. Chem. **49**(6), 1396 (2011)
348. Heller, M., Schubert, U.S.: Optically active supramolecular poly(L-lactide)s end-capped with terpyridine. Macromol. Rapid Commun. **22**(16), 1358 (2001)
349. Hofmeier, H., Hoogenboom, R., Wouters, M.E.L., Schubert, U.S.: High molecular weight supramolecular polymers containing both terpyridine metal complexes and ureidopyrimidinone quadruple hydrogen-bonding units in the main chain. J. Am. Chem. Soc. **127**(9), 2913 (2005)
350. Winter, A., Schubert, U.S.: New polyester-based trerpyridine macroligands and their blue iron(II) complexes. Macromol. Chem. Phys. **208**(18), 1956 (2007)
351. Pefkianakis, E.K., Tzanetos, N.P., Chochos, C.L., Andreopoulou, A.K., Kallitsis, J.K.: End-functionalization of semiconducting species with dendronized terpyridine–Ru(II)–terpyridine complexes. J. Polym. Sci., Part A: Polym. Chem. **47**(7), 1939 (2009)
352. Zhou, G., Harruna, I.I.: Synthesis and characterization of bis(2,2′:6′,2″-terpyridine) ruthenium(II)-connected diblock polymers *via* RAFT polymerization. Macromolecules **38** (10), 4114 (2005)
353. Zhang, L., Zhang, Y., Chen, Y.: Synthesis of bis(2,2′:6′,2″-terpyridine)-terminated telechelic polymers by RAFT polymerization and ruthenium–polymer complexation thereof. Eur. Polym. J. **42**(10), 2398 (2006)
354. Munuera, L., O'Reilly, R.K.: Using metal-ligand interactions for the synthesis of metallostar polymers. Dalton Trans. **39**(2), 388 (2010)
355. Benoit, D., Chaplinski, V., Braslau, R., Hawker, C.J.: Development of a universal alkoxyamine for "living" free radical polymerizations. J. Am. Chem. Soc. **121**(16), 3904 (1999)
356. Lohmeijer, B.G.G., Schubert, U.S.: Expanding the supramolecular polymer LEGO system: Nitroxide-mediated living free-radical polymerization as a tool for mono- and telechelic polystyrenes. J. Polym. Sci., Part A: Polym. Chem. **42**(16), 4016 (2004)
357. Ievins, A.D., Moughton, A.O., O'Reilly, R.K.: Synthesis of hollow responsive functional nanocages using a metal–ligand complexation strategy. Macromolecules **41**(10), 3571 (2008)
358. Guillet, P., Mugemana, C., Stadler, F.J., Schubert, U.S., Fustin, C.-A., Bailly, C., Gohy, J.-F.: Connecting micelles by metallo-supramolecular interactions: Towards stimuli responsive hierarchical materials. Soft Matter **5**(18), 3409 (2009)
359. Ott, C., Hoogenboom, R., Schubert, U.S.: Post-modification of poly(pentafluorostyrene): A versatile "click" method to create well-defined multifunctional graft copolymers. Chem. Commun. (30), 3516 (2008)
360. Ott, C., Ulbricht, C., Hoogenboom, R., Schubert, U.S.: Metallo-supramolecular materials based on amine-grafting onto polypentafluorostyrene. Macromol. Rapid Commun. **33**(6–7), 556 (2012)
361. Mugemana, C., Guillet, P., Hoeppener, S., Schubert, U.S., Fustin, C.-A., Gohy, J.-F.: Metallo-supramolecular diblock copolymers based on heteroleptic cobalt(III) and nickel(II) bis-terpyridine complexes. Chem. Commun. **46**(8), 1296 (2010)
362. Mansfeld, U., Hager, M.D., Hoogenboom, R., Ott, C., Winter, A., Schubert, U.S.: Advanced supramolecular initiator for nitroxide-mediated polymerizations containing both metal-ion coordination and hydrogen-bonding sites. Chem. Commun. (23), 3386 (2009)
363. Mansfeld, U., Winter, A., Hager, M.D., Hoogenboom, R., Günther, W., Schubert, U.S.: Orthogonal self-assembly of stimuli-responsive supramolecular polymers using one-step prepared heterotelechelic building blocks. Polym. Chem. **4**(1), 113 (2013)

364. Meier, M.A.R., Hofmeier, H., Abeln, C.H., Tziatzios, C., Rasa, M., Schubert, D., Schubert, U.S.: First GPC results of terpyridine based chain extended supramolecular polymers: comparison with viscosity and analytical ultracentrifugation. e-Polymers **6**(1), 209 (2006)
365. Schubert, U.S., Hien, O., Eschbaumer, C.: Functionalized polymers with metal complexing segments: A simple and high-yield entry towards 2,2′:6′,2″-terpyridine-based oligomers. Macromol. Rapid Commun. **21**(16), 1156 (2000)
366. Hofmeier, H., Schmatloch, S., Wouters, D., Schubert, U.S.: Linear terpyridine-ruthenium(II) poly(ethylene glycol) coordination polymers. Macromol. Chem. Phys. **204**(18), 2197 (2003)
367. Chiper, M., Hoogenboom, R., Schubert, U.S.: Ruthenium(II) ions triggered direct supramolecular polymerization of bis-terpyridine poly(ethylene glycol): New Insights on synthesis and optimization. e-Polymers **8**(1), 1881 (2008)
368. Lohmeijer, B.G.G., Schubert, U.S.: Synthesis of metallo-supramolecular block-copolymers. PMSE Prepr. **85**, 460 (2001)
369. Chiper, M., Hoogenboom, R., Schubert, U.S.: New terpyridine macroligands as potential synthons for supramolecular assemblies. Eur. Polym. J. **46**(2), 260 (2010)
370. Chiper, M., Hoeppener, S., Schubert, U.S., Fustin, C.-A., Gohy, J.-F.: Self-assembly behavior of bis(terpyridine) and metallo-bis(terpyridine) pluronics in dilute aqueous solutions. Macromol. Chem. Phys. **211**(21), 2323 (2010)
371. Beck, J.B., Ineman, J.M., Rowan, S.J.: Metal/ligand-induced formation of metallo-supramolecular polymers. Macromolecules **38**(12), 5060 (2005)
372. Chiper, M., Fournier, D., Hoogenboom, R., Schubert, U.S.: Thermosensitive and switchable terpyridine-functionalized metallo-supramolecular poly(*N*-isopropylacrylamide). Macromol. Rapid Commun. **29**(20), 1640 (2008)
373. Xiao, N., Chen, Y., Shen, X., Zhang, C., Yano, S., Gottschaldt, M., Schubert, U.S., Kakuchi, T., Satoh, T.: Synthesis of miktoarm star copolymer Ru(II) complexes by click-to-chelate approach. Polym. J. **45**(2), 216 (2013)
374. Winter, A., Wild, A., Hoogenboom, R., Fijten, M.W.M., Hager, M.D., Fallahpour, R.-A., Schubert, U.S.: Azido- and ethynyl-substituted 2,2′:6′,2″-terpyridines as suitable substrates for click reactions. Synthesis (09), 1506 (2009)
375. Henderson, I.M., Hayward, R.C.: Synthesis of end-functionalized polystyrene by direct nucleophilic addition of polystyryllithium to bipyridine or terpyridine. Macromolecules **43** (7), 3249 (2010)
376. Ott, C., Pavlov, G.M., Guerrero-Sanchez, C., Schubert, U.S.: Alternating terpyridine-endfunctionalized copolymers of styrene and diphenylethylene *via* anionic polymerization techniques: a detailed characterization study. J. Polym. Sci., Part A: Polym. Chem. **47**(14), 3691 (2009)
377. Pavlov, G.M., Amorós, D., Ott, C., Zaitseva, I.I., Garcia de la Torre, J., Schubert, U.S.: Hydrodynamic analysis of well-defined flexible linear macromolecules of low molar mass. Macromolecules **42**(19), 7447 (2009)
378. Gohy, J.-F., Chiper, M., Guillet, P., Fustin, C.-A., Hoeppener, S., Winter, A., Hoogenboom, R., Schubert, U.S.: Self-organization of rod-coil tri- and tetra-arm star metallo-supramolecular block copolymers in selective solvents. Soft Matter **5**(15), 2954 (2009)
379. Meier, M.A.R., Lohmeijer, B.G.G., Schubert, U.S.: Relative binding strength of terpyridine model complexes under matrix-assisted laser desorption/ionization mass spectrometry conditions. J. Mass Spectrom. **38**(5), 510 (2003)
380. Oudhoff, K.A., Schoenmakers, P.J., Kok, W.T.: Characterization of metallo bis(terpyridine) diblock polymers by nonaqueous capillary zone electrophoresis. Chromatographia **60**(7–8), 475 (2004)
381. Vogel, V., Gohy, J.-F., Lohmeijer, B.G.G., Van Den Broek, J.A., Haase, W., Schubert, U.S., Schubert, D.: Metallo-supramolecular micelles: Studies by analytical ultracentrifugation and electron microscopy. J. Polym. Sci., Part A: Polym. Chem. **41**(20), 3159 (2003)

382. Guillet, P., Fustin, C.-A., Wouters, D., Hoeppener, S., Schubert, U.S., Gohy, J.-F.: Amphiphilic brushes from metallo-supramolecular block copolymers. Soft Matter **5**(7), 1460 (2009)
383. Fustin, C.-A., Guillet, P., Misner, M.J., Russell, T.P., Schubert, U.S., Gohy, J.-F.: Self-assembly of metallo-supramolecular block copolymers in thin films. J. Polym. Sci., Part A: Polym. Chem. **46**(14), 4719 (2008)
384. Lutz, J.-F., Laschewsky, A.: Multicompartment micelles: has the long-standing dream become a reality? Macromol Chem. Phys. **206**(8), 813 (2005)
385. Moughton, A.O., Hillmyer, M.A., Lodge, T.P.: Multicompartment block polymer micelles. Macromolecules **45**(1), 2 (2012)
386. Thordarson, P., Le Droumaguet, B., Velonia, K.: Well-defined protein–polymer conjugates—synthesis and potential applications. Appl. Microbiol. Biotechnol. **73**(2), 243 (2006)
387. Velonia, K.: Protein-polymer amphiphilic chimeras: recent advances and future challenges. Polym. Chem. **1**(7), 944 (2010)
388. Reynhout, I.C., Cornelissen, J.J.L.M., Nolte, R.J.M.: Synthesis of Polymer−Biohybrids: From Small to Giant Surfactants. Acc. Chem. Res. **42**(6), 681 (2009)
389. Velonia, K., Thordarson, P., Andres, P.R., Schubert, U.S., Rowan, A.E., Nolte, R.J.M.: Polymer-protein giant amphiphiles by metal-to-ligand coordination. Polym. Prepr. (Am. Chem. Soc., Div. Polym. Chem.) **44**, 648 (2003)
390. Velonia, K., Thordarson, P., Schubert, U.S., Rowan, A.E., Nolte, R.M.J.: Biohybrid giant amphiphiles by metal-to-ligand coordination. Polym. Prepr. (Am. Chem. Soc., Div. Polym. Chem.) **45**, 401 (2004)
391. Peterson, J.R., Thordarson, P.: Optimising the purification of terpyridine-cytochrome c bioconjugates. Chiang Mai J. Sci. **36**(2), 236 (2009)
392. Peterson, J.R., Smith, T.A., Thordarson, P.: Photoinduced reduction of catalytically and biologically active Ru(II)bisterpyridine-cytochrome c bioconjugates. Chem. Commun. **19**, 1899 (2007)
393. Peterson, J.R., Smith, T.A., Thordarson, P.: Synthesis and room temperature photo-induced electron transfer in biologically active bis(terpyridine)ruthenium(II)-cytochrome c bioconjugates and the effect of solvents on the bioconjugation of cytochrome c. Org. Biomol. Chem. **8**(1), 151 (2010)
394. Gill, C.S., Venkatasubbaiah, K., Jones, C.W.: Recyclable polymer- and silica-supported ruthenium(II)-salen bis-pyridine catalysts for the asymmetric cyclopropanation of olefins. Adv. Synth. Catal. **351**(9), 1344 (2009)
395. Abu-Elfotoh, A.-M., Phomkeona, K., Shibatomi, K., Iwasa, S.: Asymmetric inter- and intramolecular cyclopropanation reactions catalyzed by a reusable macroporous-polymer-supported chiral ruthenium(II)/phenyloxazoline complex. Angew. Chem. Int. Ed. **49**(45), 8439 (2010)
396. Pearly, S.C., Sridevi, N., Mohammed Yusuff, K.K.: Characterization and catalytic activity of polymer supported ruthenium Schiff base complexes towards catechol oxidation. J. Appl. Polym. Sci. **105**(3), 997 (2007)
397. De Greef, T.F.A., Smulders, M.M.J., Wolffs, M., Schenning, A.P.H.J., Sijbesma, R.P., Meijer, E.W.: Supramolecular polymerization. Chem. Rev. **109**(11), 5687 (2009)
398. Williams, K.A., Boydston, A.J., Bielawski, C.W.: Main-chain organometallic polymers: Synthetic strategies, applications, and perspectives. Chem. Soc. Rev. **36**(5), 729 (2007)
399. Brunsveld, L., Folmer, B.J.B., Meijer, E.W., Sijbesma, R.P.: Supramolecular polymers. Chem. Rev. **101**(12), 4071 (2001)
400. Ciferri, A.: Supramolecular polymerizations. Macromol. Rapid Commun. **23**(9), 511 (2002)
401. Carothers, W.H.: Polymers and polyfunctionality. Trans. Faraday Soc. **32**, 39 (1936)
402. Yang, L., Tan, X., Wang, Z., Zhang, X.: Supramolecular polymers: historical development, preparation, characterization and functions. Chem. Rev. **115**(15), 7196 (2015)
403. Liu, Y., Wang, Z., Zhang, X.: Characterization of supramolecular polymers. Chem. Soc. Rev. **41**(18), 5922 (2012)

404. Brevard, C., Granger, P.: Ruthenium NMR spectroscopy: a promising structural and analytical tool. General trends and applicability to organometallic and inorganic chemistry. Inorg. Chem. **22**(3), 532 (1983)
405. Orellana, G., Kirsch-De Mesmaeker, A., Turro, N.J.: Ruthenium-99 NMR spectroscopy of ruthenium(II) polypyridyl complexes. Inorg. Chem. **29**(4), 882 (1990)
406. Duimstra, J.A., Stern, C.L., Meade, T.J.: Magnetic and structural properties of a cyano-bridged 1D ruthenium(III)-nickel(II) coordination polymer. Polyhedron **25**(14), 2705 (2006)
407. Ohba, M., Ōkawa, H.: Synthesis and magnetism of multi-dimensional cyanide-bridged bimetallic assemblies. Coord. Chem. Rev. **198**(1), 313 (2000)
408. Sherlock, S.J., Cowie, M., Singleton, E., Steyn, M.M.d.V.: A convenient route to an uncommon class of polymeric and binuclear ruthenium(I) complexes containing diphosphine and dithioether ligands. Organometallics **7**(7), 1663 (1988)
409. Hilts, R.W., Sherlock, S.J., Cowie, M., Singleton, E., Steyn, M.M.d.V.: Polymeric ruthenium(I) and binuclear ruthenium(I) and osmium(I) carboxylate-bridged complexes containing diphoshine, diarsine, arsinophosphine, diphosphine chalcogenide, and dithioether ligands: The structure of $[Ru_2(CO)_4(.\mu\text{-}O_2CMe)_2(MeSCH_2SMe)]_2xTHF$. Inorg. Chem. **29** (17), 3161 (1990)
410. Fritsch, N., Wick, C.R., Waidmann, T., Dral, P.O., Tucher, J., Heinemann, F.W., Shubina, T.E., Clark, T., Burzlaff, N.: Multiply bonded metal(II) acetate (rhodium, ruthenium, and molybdenum) complexes with the *trans*-1,2-*bis*(*N*-methylimidazol-2-yl)ethylene ligand. Inorg. Chem. **53**(23), 12305 (2014)
411. Fritsch, N., Wick, C.R., Waidmann, T., Pflock, S., Dral, P.O., Tucher, J., Steiner, C., Shubina, T.E., Maier, S., Clark, T., et al.: 1D Chains of diruthenium tetracarbonyl sawhorse complexes. Eur. J. Inorg. Chem. **2018**(1), 54 (2018)
412. Shieh, M., Liu, Y.-H., Li, Y.-H., Lin, R.Y.: Metal carbonyl cluster-based coordination polymers: diverse syntheses, versatile network structures, and special properties. CrystEngComm **21**(48), 7341 (2019)
413. Duimstra, J.A., Stern, C.L., Meade, T.J.: Magnetic and structural properties of a cyano-bridged 1D ruthenium(III)–nickel(II) coordination polymer. Polyhedron **25**(14), 2705 (2006)
414. Johnson, B.F.G., Sanderson, K.M., Shephard, D.S., Ozkaya, D., Zhou, W., Ahmed, H., Thomas, M.D.R., Gladden, L., Mantle, M.: Electron-beam induced formation of nanoparticle chains and wires from a ruthenium cluster polymer. Chem. Commun. **14**, 1317 (2000)
415. Thomas, M.D.R., Ahmed, H., Sanderson, K.M., Shephard, D.S., Johnson, B.F.G., Zhou, W.: Electron-beam-induced conduction in a ruthenium carbonyl nanoparticle polymer. Appl. Phys. Lett. **76**(13), 1773 (2000)
416. Thomas, M.D.R., Ahmed, H., Sanderson, K.M., Shephard, D.S., Johnson, B.F.G., Ozkaya, D., Sharma, N., Humphreys, C.: Effects of electron-beam exposure on a ruthenium nanocluster polymer. J. Appl. Phys. **90**(2), 947 (2001)
417. Abd-El-Aziz, A.S., Shipman, P.O., Boden, B.N., McNeil, W.S.: Synthetic methodologies and properties of organometallic and coordination macromolecules. Prog. Polym. Sci. **35**(6), 714 (2010)
418. Subramanian, L.R.: Tribute to Professor Dr Michael Hanack. J. Porphyrins Phthalocyanines **4**(3), 300 (2000)
419. Hanack, M., Dürr, K., Lange, A., Osío Barcina, J., Pohmer, J., Witke, E.: Synthesis of low band gap polymers - a challenge for organic chemists. Synth. Metals **71**(1–3), 2275 (1995)
420. Hanack, M., Deger, S., Lange, A.: Bisaxially coordinated macrocyclic transition metal complexes. Coord. Chem. Rev. **83**, 115 (1988)
421. Collman, J.P., McDevitt, J.T., Yee, G.T., Leidner, C.R., McCullough, L.G., Little, W.A., Torrance, J.B.: Conductive polymers derived from iron, ruthenium, and osmium metalloporphyrins: The shish-kebab approach. Proc. Nat. Acad. Sci. **83**(13), 4581 (1986)

422. Hanack, M., Hirsch, A., Kamenzin, S., Thies, R., Vermehren, P.: Soluble bridged phthalocyaninato transition metal complexes. Synth. Metals **42**(3), 2633 (1991)
423. Hanack, M., Knecht, S., Witke, E., Haisch, P.: Synthesis and properties of soluble metallophthalocyanines. Synth. Metals **55**(2–3), 873 (1993)
424. Hanack, M., Knecht, S., Pollcy, R., Subramanian, L.R.: Axially 1,4-diisocyanobenzene bridged substituted iron(II) phthalocyanines and 2,3-naphthalocyanines. Synth. Metals **80** (2), 183 (1996)
425. Stuzhin, P.A., Vagin, S.I., Hanack, M.: Synthesis and spectral properties of bisaxially coordinated (octaphenyltetraazaporphyrinato)ruthenium(II) complexes. Inorg. Chem. **37** (11), 2655 (1998)
426. Knecht, S., Polley, R., Hanack, M.: Synthesis of oligomeric axially bridged ruthenium phthalocyanines and 2,3-naphthalocyanines. Appl. Organomet. Chem. **10**(8), 649 (1996)
427. Fernández-Rodríguez, O., Fernández-Lázaro, F., Cabezón, B., Hanack, M., Torres, T.: Iron-and rutheniumtriazolehemiporphyrazines as molecular subunits for the preparation of semiconducting polymers. Synth. Metals **84**(1–3), 369 (1997)
428. Hanack, M., Hees, M., Witke, E.: Synthesis of bridged soluble phthalocyaninatoruthenium complexes. New J. Chem. **22**(2), 169 (1998)
429. Hanack, M., Kamenzin, S., Kamenzin, C., Subramanian, L.R.: Synthesis and properties of axially disubstituted monomeric and oligomeric phthalocyaninato ruthenium(II) compounds. Synth. Metals **110**(2), 93 (2000)
430. Nemykin, V.N., Dudkin, S.V., Fathi-Rasekh, M., Spaeth, A.D., Rhoda, H.M., Belosludov, R.V., Barybin, M.V.: Probing electronic communications in heterotrinuclear Fe–Ru–Fe molecular wires formed by ruthenium(II) tetraphenylporphyrin and isocyanoferrocene or 1,1′-diisocyanoferrocene ligands. Inorg. Chem. **54**(22), 10711 (2015)
431. Newkome, G.R., He, E., Moorefield, C.N.: Suprasupermolecules with novel properties: Metallodendrimers. Chem. Rev. **99**(7), 1689 (1999)
432. Serroni, S., Denti, G., Campagna, S., Juris, A., Ciano, M., Balzani, V.: Arborols based on luminescent and redox-active transition metal complexes. Angew. Chem. Int. Ed. Engl. **31** (11), 1493 (1992)
433. Denti, G., Campagna, S., Serroni, S., Ciano, M., Balzani, V.: Decanuclear homo- and heterometallic polypyridine complexes: Syntheses, absorption spectra, luminescence, electrochemical oxidation, and intercomponent energy transfer. J. Am. Chem. Soc. **114** (8), 2944 (1992)
434. Campagna, S., Denti, G., Serroni, S., Juris, A., Venturi, M., Ricevuto, V., Balzani, V.: Dendrimers of nanometer size based on metal complexes: Luminescent and redox-active polynuclear metal complexes containing up to twenty-two metal centers. Chem. Eur. J. **1**(4), 211 (1995)
435. Serroni, S., Campagna, S., Denti, G., Keyes, T.E., Vos, J.G.: A tetranuclear ruthenium(II) complex containing both electron-rich and electron-poor bridging ligands. Absorption spectrum, luminescence, redox behavior, and intercomponent energy rransfer. Inorg. Chem. **35**(15), 4513 (1996)
436. Serroni, S., Juris, A., Venturi, M., Campagna, S., Resino, I.R., Denti, G., Credi, A., Balzani, V.: Polynuclear metal complexes of nanometre size. A versatile synthetic strategy leading to luminescent and redox-active dendrimers made of an osmium(II)-based core and ruthenium (II)-based units in the branches. J. Mater. Chem. **7**(7), 1227 (1997)
437. Arakawa, R., Matsuo, T., Nozaki, K., Ohno, T., Haga, M.-A.: Analysis of multiply charged Ions of ruthenium(II) tetranuclear complexes by electrospray ionization mass spectrometry. Inorg. Chem. **34**(9), 2464 (1995)
438. Haga, M.-A., Ali, M., Arakawa, R.: Proton-induced switching of electron transfer pathways in dendrimer-type tetranuclear $RuOs_3$ complexes. Angew. Chem. Int. Ed. Engl. **35**(1), 76 (1996)

439. Moucheron, C., Kirsch-De Mesmaeker, A., Dupont-Gervais, A., Leize, E., Van Dorsselaer, A.: Synthesis and characterization by electrospray mass spectrometry of a novel dendritic heptanuclear complex of ruthenium(II). J. Am. Chem. Soc. **118**(50), 12834 (1996)
440. Knapp, R., Schott, A., Rehahn, M.: A novel synthetic strategy toward soluble, well-defined ruthenium(II) coordination polymers. Macromolecules **29**(1), 478 (1996)
441. Kelch, S., Rehahn, M.: High-molecular-weight ruthenium(II) coordination polymers: Synthesis and solution properties. Macromolecules **30**(20), 6185 (1997)
442. Kelch, S., Rehahn, M.: Soluble ruthenium(II) coordination polymers bearing bulky side groups. Macromolecules **31**(13), 4102 (1998)
443. Cassagneau, T., Fendler, J.H., Johnson, S.A., Mallouk, T.E.: Self-assembled diode junctions prepared from a ruthenium tris(bipyridyl) polymer, n-type TiO_2 nanoparticles, and graphite oxide sheets. Adv. Mater. **12**(18), 1363 (2000)
444. Constable, E.C., Harris, K., Housecroft, C.E., Neuburger, M., Schaffner, S.: A metallopolymer case-history: Polymer, ring or ligand reaction? Chem. Commun. (42), 5360 (2008)
445. Chiper, M., Hoogenboom, R., Schubert, U.S.: Toward main chain metallo-terpyridyl supramolecular polymers: "The metal does the trick." Macromol. Rapid Commun. **30**(8), 565 (2009)
446. Meier, M.A.R., Wouters, D., Ott, C., Guillet, P., Fustin, C.-A., Gohy, J.-F., Schubert, U.S.: Supramolecular ABA triblock copolymers *via* a polycondensation approach: Synthesis, characterization, and micelle formation. Macromolecules **39**(4), 1569 (2006)
447. Andres, P.R., Schubert, U.S.: Formation of metallo-polymers and -macrocycles by complexation of alkyl-linked di-terpyridines with iron(II) Ions. Synthesis (08), 1229 (2004)
448. Andres, P.R., Schubert, U.S.: Metallo-polymerization/-cyclization of a C16-bridged di-terpyridine ligand and iron(II) Ions. Macromol. Rapid Commun. **25**(15), 1371 (2004)
449. Constable, E.C., Hermann, B.A., Housecroft, C.E., Neuburger, M., Schaffner, S., Scherer, L. J.: 2,2':6',2"-Terpyridine-4'(1'*H*)-thione: A missing link in metallosupramolecular chemistry. New J. Chem. **29**(11), 1475 (2005)
450. Constable, E.C., Housecroft, C.E., Neuburger, M., Schaffner, S., Shardlow, E.J.: Formation of a [2+2]-heterotetranuclear macrocycle from reaction of a platina-homoditopic ligand with iron(II). Dalton Trans. (2), 234 (2005)
451. Constable, E.C., Housecroft, C.E., Neuburger, M., Schaffner, S., Smith, C.B.: Formation of [2+2] diruthenium(II) metallomacrocycles from ligands containing 2,2':6',2"-terpyridine domains linked through flexible polyethyleneoxy spacers. Dalton Trans. (13), 2259 (2005)
452. Constable, E.C., Housecroft, C.E., Smith, C.B.: Self-assembly of two discrete polynuclear iron(II) metallomacrocycles from a ligand containing two 2,2′:6′,2″-terpyridine binding domains. Inorg. Chem. Commun. **6**(8), 1011 (2003)
453. Schwarz, G., Bodenthin, Y., Geue, T., Koetz, J., Kurth, D.G.: Structure and properties of dynamic rigid-rod-like metallo-Ssupramolecular polyelectrolytes in solution. Macromolecules **43**(1), 494 (2010)
454. Knops, P., Sendhoff, N., Mekelburger, H.-B., Vögtle, F.: High dilution reactions - new synthetic applications. Top. Curr. Chem. **161**, 1 (1992)
455. Storrier, G.D., Colbran, S.B.: Transition-metal complexes of 4'(4-anilino)-2,2′:6′,2″-terpyridine (and derivatives): Versatile building blocks for construction of metallooligomers and macromolecules. J. Chem. Soc., Dalton Trans. (10), 2185 (1996)
456. Storrier, G.D., Colbran, S.B., Craig, D.C. Bis[4′-(4-anilino)-2,2′:6′,2″-terpyridine] transition-metal complexes: electrochemically active monomers with a range of magnetic and optical properties for assembly of metallo oligomers and macromolecules. J. Chem. Soc., Dalton Trans. (17), 3011 (1997)
457. Ng, W.Y., Gong, X., Chan, W.K.: Electronic and light-emitting poroperties of some polyimides based on bis(2,2′:6′,2″-terpyridine) ruthenium(II) complex. Chem. Mater. **11**(4), 1165 (1999)
458. Stewart, K.M., McLaughlin, L.W.: Design and synthesis of DNA-tethered ruthenium complexes that self-assemble into linear arrays. Chem. Commun. (23), 2934 (2003)

459. Ghosh, S., Pignot-Paintrand, I., Dumy, P., Defrancq, E.: Design and synthesis of novel hybrid metal complex-DNA conjugates: key building blocks for multimetallic linear DNA nanoarrays. Org. Biomol. Chem. **7**(13), 2729 (2009)
460. Irvoas, J., Noirot, A., Chouini-Lalanne, N., Reynes, O., Garrigues, J.-C., Sartor, V.: Programmable multimetallic linear nanoassemblies of ruthenium-DNA conjugates. RSC Adv. **2**(25), 9538 (2012)
461. Irvoas, J., Noirot, A., Chouini-Lalanne, N., Reynes, O., Sartor, V.: DNA three-way junction-ruthenium complex assemblies. New J. Chem. **37**(8), 2324 (2013)
462. Ghosh, S., Defrancq, E.: Metal-Complex/DNA Conjugates: A Versatile Building Block for DNA Nanoarrays. Chem. Eur. J. **16**(43), 12780 (2010)
463. Guo, K., Guo, Z., Ludlow, J.M., Xie, T., Liao, S., Newkome, G.R., Wesdemiotis, C.: Characterization of metallosupramolecular polymers by top-down multidimensional mass spectrometry methods. Macromol. Rapid Commun. **36**(17), 1539 (2015)
464. Chakraborty, S., Newkome, G.R.: Terpyridine-based metallosupramolecular constructs: tailored monomers to precise 2D-motifs and 3D-metallocages. Chem. Soc. Rev. **47**(11), 3991 (2018)
465. Xie, T.-Z., Liao, S.-Y., Guo, K., Lu, X., Dong, X., Huang, M., Moorefield, C.N., Cheng, S. Z.D., Liu, X., Wesdemiotis, C., et al.: Construction of a highly symmetric nanosphere *via* a one-pot reaction of a tristerpyridine ligand with Ru(II). J. Am. Chem. Soc. **136**(23), 8165 (2014)
466. Newkome, G.R., Wang, P., Moorefield, C.N., Cho, T.J., Mohapatra, P.P., Li, S., Hwang, S.-H., Lukoyanova, O., Echegoyen, L., Palagallo, J.A., et al.: Nanoassembly of a fractal polymer: A molecular "Sierpinski hexagonal gasket" . Science **312**(5781), 1782 (2006)
467. Harriman, A., Ziessel, R.: Making photoactive molecular-scale wires. Chem. Commun. (15), 1707 (1996)
468. Romero, F.M., Ziessel, R., Dupont-Gervais, A., Van Dorsselaer, A.: Monitoring the iron(II)-induced self-assembly of preorganized tritopic ligands by electrospray mass spectrometry: unique formation of metallomacrocycles. Chem. Commun. (4), 551 (1996)
469. Shokouhi Mehr, H., Romano, N.C., Altamimi, R., Modarelli, J.M., Modarelli, D.A.: Core substituted naphthalene diimide - metallo bisterpyridine supramolecular polymers: synthesis, photophysics and morphology. Dalton Trans. **44**(7), 3176 (2015)
470. Schmelz, O., Rehahn, M.: Synthesis of rodlike ruthenium(II) coordination polymers having metal-nitrogen and metal-carbon bonds in their main chains. e-Polymers **2**(1), 659 (2002)
471. Wild, A., Schlütter, F., Pavlov, G.M., Friebe, C., Festag, G., Winter, A., Hager, M.D., Cimrová, V., Schubert, U.S.: π-Conjugated donor and donor–acceptor metallopolymers. Macromol. Rapid Commun. **31**(9–10), 868 (2010)
472. Hinderberger, D., Schmelz, O., Rehahn, M., Jeschke, G.: Electrostatic site attachment of divalent counterions to rodlike ruthenium(II) coordination polymers characterized by EPR spectroscopy. Angew. Chem. Int. Ed. **43**(35), 4616 (2004)
473. Higuchi, M.: Electrochromic organic-metallic hybrid polymers: Fundamentals and device applications. Polym. J. **41**(7), 511 (2009)
474. Higuchi, M.: Stimuli-responsive metallo-supramolecular polymer films: Design, synthesis and device fabrication. J. Mater. Chem. C **2**(44), 9331 (2014)
475. Mortimer, R.J.: Electrochromic materials. Chem. Soc. Rev. **26**(3), 147 (1997)
476. Sonmez, G.: Polymeric electrochromics. Chem. Commun. (42), 5251 (2005)
477. Beaujuge, P.M., Reynolds, J.R.: Color control in π-conjugated organic polymers for use in electrochromic devices. Chem. Rev. **110**(1), 268 (2010)
478. Zhong, Y.-W.: Electrochromism within transition-metal coordination complexes and polymers. In: Mortimer, R.J., Rosseinsky, D.R., Monk, P.M.S. (eds.) Electrochromic Materials and Devices. Wiley-VCH, Weinheim, pp 185 (2013)
479. Maier, A., Cheng, K., Savych, J., Tieke, B.: Double-electrochromic coordination polymer network films. ACS Appl. Mater. Interfaces **3**(7), 2710 (2011)

480. Motiei, L., Lahav, M., Freeman, D., van der Boom, M.E.: Electrochromic behavior of a self-propagating molecular-based assembly. J. Am. Chem. Soc. **131**(10), 3468 (2009)
481. Chen, B.-H., Kao, S.-Y., Hu, C.-W., Higuchi, M., Ho, K.-C., Liao, Y.-C.: Printed multicolor high-contrast electrochromic devices. ACS Appl. Mater. Interfaces **7**(45), 25069 (2015)
482. Chen, X.-Y., Yang, X., Holliday, B.J.: Photoluminescent europium-containing inner sphere conducting metallopolymer. J. Am. Chem. Soc. **130**(5), 1546 (2008)
483. Hjelm, J., Handel, R.W., Hagfeldt, A., Constable, E.C., Housecroft, C.E., Forster, R.J.: Conducting polymers containing in-chain metal centers: Electropolymerization of oligothienyl-substituted M(tpy)2 complexes and in situ conductivity studies, M = Os(II), Ru(II). Inorg. Chem. **44**(4), 1073 (2005)
484. Yao, C.-J., Yao, J., Zhong, Y.-W.: Metallopolymeric films based on a biscyclometalated ruthenium complex bridged by 1,3,6,8-tetra(2-pyridyl)pyrene: Applications in near-infrared electrochromic windows. Inorg. Chem. **51**(11), 6259 (2012)
485. Yao, C.-J., Zhong, Y.-W., Nie, H.-J., Abruña, H.D., Yao, J.: Near-IR electrochromism in electropolymerized films of a biscyclometalated ruthenium complex bridged by 1,2,4,5-tetra (2-pyridyl)benzene. J. Am. Chem. Soc. **133**(51), 20720 (2011)
486. Milum, K.M., Kim, Y.N., Holliday, B.J.: Pt-[NCN] pincer conducting metallopolymers that display redox-attenuated metal-ligand interactions. Chem. Mater. **22**(8), 2414 (2010)
487. Powell, A.B., Bielawski, C.W., Cowley, A.H.: Design, synthesis and study of main chain poly(N-heterocyclic carbene) complexes: Applications in electrochromic devices. J. Am. Chem. Soc. **132**(29), 10184 (2010)
488. Han, F.S., Higuchi, M., Kurth, D.G.: Metallosupramolecular polyelectrolytes self-assembled from various pyridine ring-substituted bisterpyridines and metal ions: Photophysical, electrochemical, and electrochromic properties. J. Am. Chem. Soc. **130**(6), 2073 (2008)
489. Han, F.S., Higuchi, M., Ikeda, T., Negishi, Y., Tsukuda, T., Kurth, D.G.: Luminescence properties of metallo-supramolecular coordination polymers assembled from pyridine ring functionalized ditopic bis-terpyridines and Ru(II) ions. J. Mater. Chem. **18**(38), 4555 (2008)
490. Higuchi, M., Akasaka, Y., Ikeda, T., Hayashi, A., Kurth, D.: Electrochromic solid-state devices using organic-metallic hybrid polymers. J. Inorg. Organomet. Polym. Mater. **19**(1), 74 (2009)
491. Li, J.-H., Higuchi, M.: Substituent effects on metallo-supramolecular coordination polymers. J. Inorg. Organomet. Polym. Mater. **20**(1), 10 (2010)
492. Muronoi, Y., Zhang, J., Higuchi, M., Maki, H.: Electrochemical switching of luminescence in Ru(II)-based metallosupramolecular polymer device. Chem. Lett. **42**(7), 761 (2013)
493. Kanao, M., Higuchi, M.: Synthesis of metallo-supramolecular polymers with bis-NNN-tridentate ligand for electrochromic devices. J. Photopolym. Sci. Technol. **28**(3), 363 (2015)
494. Zhang, J., Hsu, C.-Y., Higuchi, M.: Anion effects to electrochromic properties of Ru-based metallo-supramolecular polymers. J. Photopolym. Sci. Technol. **27**(3), 297 (2014)
495. Hu, C.-W., Sato, T., Zhang, J., Moriyama, S., Higuchi, M.: Multi-colour electrochromic properties of Fe/Ru-based bimetallo-supramolecular polymers. J. Mater. Chem. C **1**(21), 3408 (2013)
496. Yoshida, T., Higuchi, M.: Diversity in design of electrochromic devices with metallo-supramolecular polymer: Multi-patterned and tube-shaped displays. J. Photopolym. Sci. Technol. **31**(3), 343 (2018)
497. Hsiao, L.-Y., Chang, T.-H., Lu, H.-C., Wang, Y.-C., Lu, Y.-A., Ho, K.-C., Higuchi, M.: A panchromatic electrochromic device composed of Ru(II)/Fe(II)-based heterometallo-supramolecular polymer. J. Mater. Chem. C **7**(25), 7554 (2019)
498. Li, J., Futera, Z., Li, H., Tateyama, Y., Higuchi, M.: Conjugation of organic-metallic hybrid polymers and calf-thymus DNA. Phys. Chem. Chem. Phys. **13**(11), 4839 (2011)
499. Li, J., Murakami, T., Higuchi, M.: Metallo-supramolecular polymers: versatile DNA binding and their cytotoxicity. J. Inorg. Organomet. Polym. Mater. **23**(1), 119 (2013)

500. Rana, U., Chakraborty, C., Pandey, R.K., Hossain, M.D., Nagano, R., Morita, H., Hattori, S., Minowa, T., Higuchi, M.: Selective DNA recognition and cytotoxicity of water-soluble helical metallosupramolecular polymers. Bioconjug. Chem. **27**(10), 2307 (2016)
501. Barron, J.A., Glazier, S., Bernhard, S., Takada, K., Houston, P.L., Abruña, H.D.: Photophysics and redox behavior of chiral transition metal polymers. Inorg. Chem. **42**(5), 1448 (2003)
502. Bernhard, S., Takada, K., Díaz, D.J., Abruña, H.D., Mürner, H.: Enantiomerically pure chiral coordination polymers: Synthesis, spectroscopy, and electrochemistry in solution and on surfaces. J. Am. Chem. Soc. **123**(42), 10265 (2001)
503. Rajadurai, C., Fuhr, O., Kruk, R., Ghafari, M., Hahn, H., Ruben, M.: Above room temperature spin transition in a metallo-supramolecular coordination oligomer/polymer. Chem. Commun. (25), 2636 (2007)
504. Basak, S., Hui, P., Boodida, S., Chandrasekar, R.: Micropatterning of metallopolymers: Syntheses of back-to-back coupled octylated 2,6-bis(pyrazolyl)pyridine ligands and their solution-processable coordination polymers. J. Org. Chem. **77**(7), 3620 (2012)
505. Narayana, Y.S.L.V., Basak, S., Baumgarten, M., Müllen, K., Chandrasekar, R.: White-emitting conjugated polymer/inorganic hybrid spheres: Phenylethynyl and 2,6-bis (pyrazolyl)pyridine copolymer coordinated to Eu(tta)$_3$. Adv. Funct. Mater. **23**(47), 5875 (2013)
506. Schulze, B., Friebe, C., Hoeppener, S., Pavlov, G.M., Winter, A., Hager, M.D., Schubert, U. S.: Ruthenium(II) metallo-supramolecular polymers of click-derived tridentate ditopic ligands. Macromol. Rapid Commun. **33**(6–7), 597 (2012)
507. Oberoi, D., Dagar, P., Shankar, U., Sahu, S., Bandyopadhyay, A.: Stabilization of pristine low resistive state in a ruthenium(II)-polymer of a hexadentate Schiff base ligand: Synthesis, characterization, and application as non-volatile memory device. Eur. J. Inorg. Chem. **2020** (32), 3048 (2020)
508. Wong, W.-Y.: Metallopolyyne polymers as new functional materials for photovoltaic and solar cell applications. Macromol. Chem. Phys. **209**(1), 14 (2008)
509. Wong, W.-Y.: Luminescent organometallic poly(aryleneethynylene)s: Functional properties towards implications in molecular optoelectronics. Dalton Trans. (40), 4495 (2007)
510. Wong, W.-Y., Ho, C.-L.: Di-, oligo- and polymetallaynes: Syntheses, photophysics, structures and applications. Coord. Chem. Rev. **250**(19–20), 2627 (2006)
511. Wong, W.-Y., Ho, C.-L.: Organometallic photovoltaics: A new and versatile approach for harvesting solar energy using conjugated polymetallaynes. Acc. Chem. Res. **43**(9), 1246 (2010)
512. Puddephatt, R.J.: Macrocycles, catenanes, oligomers and polymers in gold chemistry. Chem. Soc. Rev. **37**(9), 2012 (2008)
513. Wolf, M.: Recent advances in conjugated transition metal-containing polymers and materials. J. Inorg. Organomet. Polym. Mater. **16**(3), 189 (2006)
514. Wong, W.-Y.: Mercury alkynyls as versatile templates for new organometallic materials and polymers. Coord. Chem. Rev. **251**(17–20), 2400 (2007)
515. Wong, W.-Y.: Optical properties and photophysics of platinum-containing poly(aryleneethynylene)s. In: Macromolecules Containing Metal and Metal-Like Elements, pp 289. Wiley (2010)
516. Zhou, G.-J., Wong, W.-Y.: Organometallic acetylides of Pt(II), Au(I) and Hg(II) as new generation optical power limiting materials. Chem. Soc. Rev. **40**(5), 2541 (2011)
517. Wong, W.-Y., Ho, C.-L.: Rigid-rod polymetallaynes. In: Frontiers in Transition Metal-Containing Polymers, pp. 247. Wiley (2006)
518. Nguyen, P., Gómez-Elipe, P., Manners, I.: Organometallic polymers with transition metals in the main chain. Chem. Rev. **99**(6), 1515 (1999)
519. Guo, X.A., Sturge, K.C., Hunter, A.D., Williams, M.C.: Molecular weight determination and establishment of a rodlike structure for organonickel polymers - [Ni(PR$_3$)$_2$ArF]$_n$. Macromolecules **27**(26), 7825 (1994)

520. Ho, C.-L., Wong, W.-Y.: Charge and energy transfers in functional metallophosphors and metallopolyynes. Coord. Chem. Rev. **257**(9–10), 1614 (2013)
521. Ho, C.-L., Yu, Z.-Q., Wong, W.-Y.: Multifunctional polymetallaynes: Properties, functions and applications. Chem. Soc. Rev. **45**(19), 5264 (2016)
522. Fujikura, Y., Sonogashira, K., Hagihara, N.: Preparation and UV spectra of some oligomer-complexes composed of platinum group metals and conjugated poly-yne systems. Chem. Lett. **4**(10), 1067 (1975)
523. Sonogashira, K., Takahashi, S., Hagihara, N.: A new extended chain polymer, poly[*trans*-bis (tri-*n*-butylphosphine)platinum 1,4-butadiynediyl]. Macromolecules **10**(4), 879 (1977)
524. Sonogashira, K., Kataoka, S., Takahashi, S., Hagihara, N.: Studies of poly-yne polymers containing transition metals in the main chain. Part III. Synthesis and characterization of a poly-yne polymer containing mixed metals in the main chain. J. Organomet. Chem. **160**(1), 319 (1978)
525. Sonogashira, K., Fujikura, Y., Yatake, T., Toyoshima, N., Takahashi, S., Hagihara, N.: Syntheses and properties of *cis*- and *trans*-dialkynyl complexes of platinum(II). J. Organomet. Chem. **145**(1), 101 (1978)
526. Takahashi, S., Kariya, M., Yatake, T., Sonogashira, K., Hagihara, N.: Studies of poly-yne polymers containing transition metals in the main chain. Part 2. Synthesis of poly[*trans*-bis (tri-*n*-butylphosphine)platinum 1,4-butadiynediyl] and evidence of a rodlike structure. Macromolecules **11**(6), 1063 (1978)
527. Wild, A., Winter, A., Hager, M.D., Schubert, U.S.: Synthetic strategies to conjugated main-chain metallopolymers. In: Chujo, Y. (ed.) Conjugated Polymer Synthesis, pp. 251. Wiley-VCH (2010)
528. Khan, M.S., Davies, S.J., Kakkar, A.K., Schwartz, D., Lin, B., Johnson, B.F.G., Lewis, J.: Synthesis of monomeric, oligomeric and polymeric σ-acetylide complexes of platinum, palladium, nickel and rhodium. J. Organomet. Chem. **424**(1), 87 (1992)
529. Lo Sterzo, C.: The wonder of palladium catalysis: From carbon-carbon to metal-carbon bond formation. An opportunity of getting astonishment from reality. Synlett (11), 1704 (1999)
530. Antonelli, E., Rosi, P., Lo Sterzo, C., Viola, E.: A convenient short cut from aromatic iodides to alkynylstannanes and their use for the straightforward preparation of polyacetylene and polymetallaacetylene polymers. J. Organomet. Chem. **578**(1–2), 210 (1999)
531. Pizzoferrato, R., Berliocchi, M., Di Carlo, A., Lugli, P., Venanzi, M., Micozzi, A., Ricci, A., Lo Sterzo, C.: Improvement of the extended one-pot (EOP) procedure to form poly (aryleneethynylene)s and investigation of their electrical and optical properties. Macromolecules **36**(7), 2215 (2003)
532. Atherton, Z., Faulkner, C.W., Ingham, S.L., Kakkar, A.K., Khan, M.S., Lewis, J., Long, N. J., Raithby, P.R.: Rigid rod σ-acetylide complexes of iron, ruthenium and osmium. J. Organomet. Chem. **462**(1–2), 265 (1993)
533. Davies, S.J., Johnson, B.F.G., Lewis, J., Raithby, P.R.: Synthesis of mononuclear and oligomeric ruthenium(II) acetylides. J. Organomet. Chem. **414**(2), C51 (1991)
534. Faulkner, C.W., Ingham, S.L., Khan, M.S., Lewis, J., Long, N.J., Raithby, P.R.: Ruthenium (II) σ-acetylide complexes: Monomers, dimers and polymers. J. Organomet. Chem. **482**(1–2), 139 (1994)
535. Khan, M.S., Kakkar, A.K., Ingham, S.L., Raithby, P.R., Lewis, J., Spencer, B., Wittmann, F., Friend, R.H.: Synthesis and electronic structure of rigid rod octahedral Ru-σ-acetylide complexes. J. Organomet. Chem. **472**(1–2), 247 (1994)
536. Ho, P.-Y., Komber, H., Horatz, K., Tsuda, T., Mannsfeld, S.C.B., Dmitrieva, E., Blacque, O., Kraft, U., Sirringhaus, H., Lissel, F.: Synthesis and characterization of a semiconducting and solution-processable ruthenium-based polymetallayne. Polym. Chem. **11**(2), 472 (2020)
537. Lavastre, O., Even, M., Dixneuf, P.H., Pacreau, A., Vairon, J.-P.: Novel ruthenium- or iron-containing tetraynes as precursors of mixed-metal oligomers. Organometallics **15**(6), 1530 (1996)

538. Matsumi, N., Chujo, Y., Lavastre, O., Dixneuf, P.H.: Facilitated dπ–pπ* transition in a novel organoboron π-conjugated polymer including a ruthenium–phosphine complex. Organometallics **20**(12), 2425 (2001)
539. Zhu, Y., Millet, D.B., Wolf, M.O., Rettig, S.J.: Models for conjugated metal acetylide polymers: Ruthenium oligothienylacetylide complexes. Organometallics **18**(10), 1930 (1999)
540. Jia, G., Wan Fung, W., Yeung, R.C.Y., Xia, H.P.: Dimeric and polymeric ruthenium complexes with Ru-vinyl linkages. J. Organomet. Chem. **539**(1), 53 (1997)
541. Yan, X., Xi, C.: Conversion of zirconacyclopentadienes into metalloles: Fagan-Nugent reaction and beyond. Acc. Chem. Res. **48**(4), 935 (2015)
542. Sahoo, P.K., Swain, S.K.: Synthesis of zirconocene-acetylene and zirconocene-diacetylene polymer. J. Polym. Sci., Part A: Polym. Chem. **37**(21), 3899 (1999)
543. Zhou, W.-M., Tomita, I.: Synthesis and properties of novel germole-containing polymers by reaction of bis(cyclopentadienyl)titanacyclopentadiene-containing polymer with germanium tetrachloride. Polym. Bull. **61**(5), 603 (2008)
544. Zhou, W.-M., Tomita, I.: Synthesis and properties of stannole-containing polymers by reaction of bis(cyclopentadienyl)titanacyclopentadiene-containing polymers and tin(IV) chloride. J. Inorg. Organomet. Polym. Mater. **19**(1), 113 (2009)
545. Nitschke, J.R., Zürcher, S., Tilley, T.D.: New zirconocene-coupling route to large, functionalized macrocycles. J. Am. Chem. Soc. **122**(42), 10345 (2000)
546. Tomita, I., Lee, J.-C., Endo, T.: Reaction of organocobalt polymers with isocyanide. Synthesis of novel poly[(η^5-cyclopentadienyl)(η^5-amino-cyclopentadienyl) cobalticinium-diyl-alt-biphenyl-4,4′-diyl)]s. J. Organomet. Chem. **611**(1–2), 570 (2000)
547. Nishihara, H., Kurashina, M., Murata, M.: Organometallic conducting polymers synthesized by metallacycling polymerization. Macromol. Symp. **196**(1), 27 (2003)
548. Kurashina, M., Murata, M., Nishihara, H.: Synthesis and physical properties of π-conjugated metallacycle polymers of cobalt and ruthenium. Macromol. Symp. **209**(1), 141 (2004)
549. Kurashina, M., Murata, M., Watanabe, T., Nishihara, H.: Synthesis of poly(biphenylene ruthenacyclopentatrienylene), a new organometallic conducting polymer with ferromagnetic interaction in its reduced state. J. Am. Chem. Soc. **125**(41), 12420 (2003)
550. Albers, M.O., De Waal, D.J.A., Liles, D.C., Robinson, D.J., Singleton, E., Wiege, M.B.: The novel cyclodimerization of phenylacetylene at a ruthenium(II) centre. The synthesis and X-ray structural characterization of the first metallacyclopentatriene, [η^5-(C_5H_5)Ru ($C_4Ph_2H_2$)Br], and its facile conversion into metallacyclopentadienes. J. Chem. Soc., Chem. Commun. (22), 1680 (1986)
551. Ernst, C., Walter, O., Dinjus, E., Arzberger, S., Görls, H.: Structural characterization of Cp*Ru-intermediates of phenylacetylene cyclotrimerization. J. Prakt. Chem. **341**(8), 801 (1999)
552. Yamada, Y., Mizutani, J., Kurihara, M., Nishihara, H.: Synthesis of a new bis(ferrocenyl) ruthenacyclopentatriene compound with a significant inter-metal electronic communication. J. Organomet. Chem. **637–639**, 80 (2001)
553. Steenwinkel, P., Kooijman, H., Smeets, W.J.J., Spek, A.L., Grove, D.M., van Koten, G.: Intramolecularly stabilized 1,4-phenylene-bridged homo- and heterodinuclear palladium and platinum organometallic complexes containing N, C, N-coordination motifs; η^1-SO_2 coordination and formation of an organometallic arenium ion complex with two Pt–C σ-bonds. Organometallics **17**(24), 5411 (1998)
554. Gagliardo, M., Amijs, C.H.M., Lutz, M., Spek, A.L., Havenith, R.W.A., Hartl, F., van Klink, G.P.M., van Koten, G.: Diorganoruthenium complexes incorporating noninnocent [$C_6H_2(CH_2ER_2)_2$–3,5]$_2^{2-}$ (E = N, P) bis-pincer bridging ligands: Synthesis, spectroelectrochemistry, and DFT studies. Inorg. Chem. **46**(26), 11133 (2007)
555. Abd-El-Aziz, A.S., Manners, I. (eds.): Frontiers in Transition Metal-Containing Polymers. Wiley, Hoboken (2007)
556. Manners, I.: Synthetic Metal-Containing Polymers. Wiley-VCH, Weinheim (2004)

557. Whittell, G.R., Hager, M.D., Schubert, U.S., Manners, I.: Functional soft materials from metallopolymers and metallosupramolecular polymers. Nat. Mater. **10**(3), 176 (2011)
558. Saleem, M., Yu, H., Wang, L., Zain ul, A., Khalid, H., Akram, M., Abbasi, N.M., Huang, J.: Review on synthesis of ferrocene-based redox polymers and derivatives and their application in glucose sensing. Anal. Chim. Acta **876**, 9 (2015)
559. Pietschnig, R.: Polymers with pendant ferrocenes. Chem. Soc. Rev. **45**(19), 5216 (2016)
560. Ingram, G., Jaitner, P., Schwarzhans Karl, E.: Synthesis and characterization of hetero-oligometallocenes containing ruthenocene and osmocene. Z. Naturforschung B **45** (6), 781 (1990)
561. Neuse, E.W.: Polymetallocenylenes - recent developments. J. Macromol. Sci., Part A **16**(1), 3 (1981)
562. Schubert, U.S., Heller, M.: Metallo-supramolecular initiators for the preparation of novel functional architectures. Chem. Eur. J. **7**(24), 5252 (2001)
563. Hoogenboom, R., Schubert, U.S.: The use of (metallo-)supramolecular initiators for living/controlled polymerization techniques. Chem. Soc. Rev. **35**(7), 622 (2006)
564. Lamba, J.J.S., Fraser, C.L.: Synthesis of metal-centered star-shaped polyoxazolines using Fe (II) and Ru(II) tris-bipyridine derivatives as multifunctional Initiators. J. Am. Chem. Soc. **119**(7), 1801 (1997)
565. McAlvin, J.E., Fraser, C.L.: Polymerization of 2-ethyl-2-oxazoline using di-, tetra-, and hexafunctional ruthenium tris(bipyridine) metalloinitiators. Macromolecules **32**(21), 6925 (1999)
566. McAlvin, J.E., Scott, S.B., Fraser, C.L.: Synthesis of thermochromic iron(II) tris(bipyridine)-centered star-shaped polyoxazolines and their bipyridine-centered macroligand Counterparts. Macromolecules **33**(19), 6953 (2000)
567. Park, C., McAlvin, J.E., Fraser, C.L., Thomas, E.L.: Iron cluster and microstructure formation in metal-centered star block copolymers: Amphiphilic iron tris(bipyridine)-centered polyoxazolines. Chem. Mater. **14**(3), 1225 (2002)
568. Godbey, W.T., Wu, K.K., Mikos, A.G.: Poly(ethylenimine) and its role in gene delivery. J. Control. Release **60**(2), 149 (1999)
569. Benjaminsen, R.V., Mattebjerg, M.A., Henriksen, J.R., Moghimi, S.M., Andresen, T.L.: The possible "proton sponge" effect of polyethylenimine (PEI) does not include change in lysosomal pH. Mol. Ther. **21**(1), 149 (2013)
570. Brissault, B., Kichler, A., Guis, C., Leborgne, C., Danos, O., Cheradame, H.: Synthesis of linear polyethylenimine derivatives for DNA transfection. Bioconjug. Chem. **14**(3), 581 (2003)
571. Fiore, G.L., Edwards, J.M., Payne, S.J., Klinkenberg, J.L., Gioeli, D.G., Demas, J.N., Fraser, C.L.: Ruthenium(II) tris(bipyridine)-centered poly(ethylenimine) for gene delivery. Biomacromol **8**(9), 2829 (2007)
572. Schulze, M., Jäger, M., Schubert, U.S.: Poly(ε-caprolactone) decorated with one room-temperature red-emitting ruthenium(II) complex: Synthesis, characterization, thermal and optical properties. Macromol. Rapid Commun. **33**(6–7), 579 (2012)
573. Collins, J.E., Fraser, C.L.: Transition metals as templates for multifunctional initiators: Bulk atom-transfer radical polymerization of styrene using di-, tetra- and hexafunctional ruthenium tris(bipyridine) reagents. Macromolecules **31**(19), 6715 (1998)
574. Johnson, R.M., Corbin, P.S., Ng, C., Fraser, C.L.: Poly(methyl methacrylates) with ruthenium tris(bipyridine) cores via $NiBr_2(PR_3)_2$-catalyzed atom transfer radical polymerization (ATRP). Macromolecules **33**(20), 7404 (2000)
575. Viau, L., Even, M., Maury, O., Haddleton, D.M., Le Bozec, H.: New star-shaped metallo-polymeric chromophores. Macromol. Rapid Commun. **24**(10), 630 (2003)
576. Viau, L., Even, M., Maury, O., Haddleton, D.M., Le Bozec, H.: Synthesis of star-shaped metallo-polymeric chromophores by atom transfer radical polymerization. Compt. Rend. Chim. **8**(8), 1298 (2005)

577. Farah, A.A., Pietro, W.J.: Synthesis and characterization of multifunctional polymers *via* atom-transfer radical polymerization of *N*-(ω′-alkylcarbazolyl) methacrylates initiated by Ru (II) polypyridyl chromophores. J. Polym. Sci., Part A: Polym. Chem. **43**(23), 6057 (2005)
578. Chan, S.H., Lam, L.S.M., Tse, C.W., Man, K.Y.K., Wong, W.T., Djurišić, A.B., Chan, W. K.: Photosensitivity of functional polystyrene and poly(methyl methacrylate) synthesized by controlled radical polymerization. Macromolecules **36**(15), 5482 (2003)
579. Sha, Y., Zhang, Y., Xu, E., McAlister, C.W., Zhu, T., Craig, Stephen L., Tang, C.: Generalizing metallocene mechanochemistry to ruthenocene mechanophores. Chem. Sci. **10** (19), 4959 (2019)
580. Chen, M., Ghiggino, K.P., Launikonis, A., Mau, A.W.H., Rizzardo, E., Sasse, W.H.F., Thang, S.H., Wilson, G.J.: RAFT synthesis of linear and star-shaped light harvesting polymers using di- and hexafunctional ruthenium polypyridine reagents. J. Mater. Chem. **13** (11), 2696 (2003)
581. Tilley, A.J., Kim, M.J., Chen, M., Ghiggino, K.P.: Photo-induced energy transfer in ruthenium-centred polymers prepared by a RAFT approach. Polymer **54**(12), 2865 (2013)
582. Boutin, P.C., Ghiggino, K.P., Kelly, T.L., Steer, R.P.: Photon upconversion by triplet-triplet annihilation in Ru(bpy)$_3$- and DPA-functionalized polymers. J. Phys. Chem. Lett. **4**(23), 4113 (2013)
583. Zhao, J., Ji, S., Guo, H.: Triplet-triplet annihilation based upconversion: From triplet sensitizers and triplet acceptors to upconversion quantum yields. RSC Adv. **1**(6), 937 (2011)
584. Neuse, E.W.: Metallocene polymers. Parrt XV. Polymers containing the ruthenocene system. J. Organomet. Chem. **6** (1), 92 (1966)
585. Arimoto, F.S., Haven, A.C.: Derivatives of dicyclopentadienyliron. J. Am. Chem. Soc. **77** (23), 6295 (1955)
586. Baldwin, M.G., Johnson, K.E.: Free-radical polymerization of vinyl ferrocene. J. Polym. Sci., Part A-1: Polym. Chem. **5**(8), 2091 (1967)
587. Sasaki, Y., Walker, L.L., Hurst, E.L., Pittman, C.U., Jr.: Organometallic polymers. Part XXV. Preparation of polyvinylferrocene. J. Polym. Sci.: Polym. Chem. Ed. **11**(6), 1213 (1973)
588. Wilkinson, G.: The preparation and some properties of ruthenocene and ruthenicinium salts. J. Am. Chem. Soc. **74**(23), 6146 (1952)
589. Sheats, J.E., Hessel, F., Tsarouhas, L., Podejko, K.G., Porter, T., Kool, L.B., Nolen, R.L., Jr.: Use of organometallic polymers as pre-heat shields for targets in inertial-confinement nuclear fusion. PMSE Preprints **49**, 363 (1983)
590. Sheats, J.E., Willis, T.C.: Synthesis and characterization of polymers and copolymers of vinylruthenocene. Org Coatings Plastics Prepr **41**, 33 (1979)
591. Willis, T.C., Sheats, J.E.: Synthesis and characterization of polymers and copolymers of vinylruthenocene. J. Polym. Sci.: Polym. Chem. Ed. **22**(5), 1077 (1984)
592. Pittman, C.U., Jr.: The discovery of metallocene- and metallocene-like addition polymers. J. Inorg. Organomet. Polym. Mater. **15**(1), 33 (2005)
593. Pittman, C.U., Jr., Jayaraman, T.V., Priester, R.D., Jr., Rausch, M.D., Macomber, D.W., Hart, W.P.: New organometallic monomers. Sythesis and polymerization behavior. Polym. Prepr. (Am. Chem. Soc., Div. Polym. Chem.) **23**(2), 73 (1982)
594. Allcock, H.R., Riding, G.H., Lavin, K.D.: Polymerization of new metallocenylphosphazenes. Macromolecules **20**(1), 6 (1987)
595. Buchmeiser, M., Schrock, R.R.: Synthesis of polyenes that contain metallocenes via the living polymerization of ethynylferrocene and ethynylruthenocene. Macromolecules **28**(19), 6642 (1995)
596. Hardy, C.G., Ren, L., Zhang, J., Tang, C.: Side-chain metallocene-containing polymers by living and controlled polymerizations. Isr. J. Chem. **52**(3–4), 230 (2012)
597. Hailes, R.L.N., Oliver, A.M., Gwyther, J., Whittell, G.R., Manners, I.: Polyferrocenylsilanes: Synthesis, properties and applications. Chem. Soc. Rev. (2016)

598. Alkan, A., Gleede, T., Wurm, F.R.: Ruthenocenyl glycidyl ether: A ruthenium-containing epoxide for anionic polymerization. Organometallics **36**(16), 3023 (2017)
599. Segal, J.A.: Aromatic polyether synthesis via activated in a ruthenium(II) complex of p-dichlorobenzene. J. Chem. Soc., Chem. Commun. (19), 1338 (1985)
600. Abd-El-Aziz, A.S., Strohm, E.A.: Transition metal-containing macromolecules: En route to new functional materials. Polymer **53**(22), 4879 (2012)
601. Dembek, A.A., Fagan, P.J., Marsi, M.: Synthesis of soluble, organometallic poly(phenylene sulfide) and poly(phenylene oxide) derivatives via transition-metal-activated nucleophilic substitution. Macromolecules **26**(11), 2992 (1993)
602. de Denus, C.R., Hoffa, L.M., Todd, E.K., Abd-El-Aziz, A.S.: Synthesis of oligomeric and polymeric ethers containing the Cp^*Ru^+ and $CpFe^+$ metal moieties. J. Inorg. Organomet. Polym. Mater. **10**(4), 189 (2000)
603. McCafferty, D.G., Bishop, B.M., Wall, C.G., Hughes, S.G., Mecklenberg, S.L., Meyer, T.J., Erickson, B.W.: Synthesis of redox derivatives of lysine and their use in solid-phase synthesis of a light-harvesting peptide. Tetrahedron **51**(4), 1093 (1995)
604. Slate, C.A., Striplin, D.R., Moss, J.A., Chen, P., Erickson, B.W., Meyer, T.J.: Photochemical energy transduction in helical proline arrays. J. Am. Chem. Soc. **120**(19), 4885 (1998)
605. Striplin, D.R., Reece, S.Y., McCafferty, D.G., Wall, C.G., Friesen, D.A., Erickson, B.W., Meyer, T.J.: Solvent dependence of intramolecular electron transfer in a helical oligoproline assembly. J. Am. Chem. Soc. **126**(16), 5282 (2004)
606. Ryan, D.M., Coggins, M.K., Concepcion, J.J., Ashford, D.L., Fang, Z., Alibabaei, L., Ma, D., Meyer, T.J., Waters, M.L.: Synthesis and electrocatalytic water oxidation by electrode-bound helical peptide chromophore–catalyst sssemblies. Inorg. Chem. **53**(15), 8120 (2014)
607. Holder, E., Meier, M.A.R., Marin, V., Schubert, U.S.: New soluble functional polymers by free-radical copolymerization of methacrylates and bipyridine ruthenium complexes. J. Polym. Sci., Part A: Polym. Chem. **41**(24), 3954 (2003)
608. Tekin, E., Holder, E., Marin, V., de Gans, B.-J., Schubert, U.S.: Ink-jet printing of luminescent ruthenium- and iridium-containing polymers for applications in light-emitting devices. Macromol. Rapid Commun. **26**(4), 293 (2005)
609. Zhou, H., Zheng, Z., Wang, Q., Xu, G., Li, J., Ding, X.: A modular approach to self-oscillating polymer systems driven by the Belousov-Zhabotinsky reaction. RSC Adv. **5** (18), 13555 (2015)
610. Yoshida, R., Takahashi, T., Yamaguchi, T., Ichijo, H.: Self-oscillating gels. Adv. Mater. **9** (2), 175 (1997)
611. Yoshida, R., Kokufuta, E., Yamaguchi, T.: Beating polymer gels coupled with a nonlinear chemical reaction. Chaos **9**(2), 260 (1999)
612. Yoshida, R., Yamaguchi, T., Kokufuta, E.: New intelligent polymer gels: a self-oscillating gel with pacemaking and actuating functions. J. Artif. Organs **2**(2), 135 (1999)
613. Hara, Y., Yoshida, R.: Self-oscillating polymer fueled by organic acid. J. Phys. Chem. B **112** (29), 8427 (2008)
614. Hara, Y., Yoshida, R.: Damping behavior of the aggregation-disaggregation self-oscillation of a polymer chain. Macromol. Rapid Commun. **30**(19), 1656 (2009)
615. Yamamoto, T., Yoshida, R.: Self-oscillation of polymer and photo-regulation by introducing photochromic site to induce LCST changes. React. Funct. Polym. **73**(7), 945 (2013)
616. Murase, Y., Maeda, S., Hashimoto, S., Yoshida, R.: Design of a Mass Transport Surface Utilizing Peristaltic Motion of a Self-Oscillating Gel. Langmuir **25**(1), 483 (2009)
617. Yoshida, R., Murase, Y.: Self-oscillating surface of gel for autonomous mass transport. Colloids Surf. B: Biointerfaces **99**, 60 (2012)
618. Murase, Y., Hidaka, M., Yoshida, R.: Self-driven gel conveyer: Autonomous transportation by peristaltic motion of self-oscillating gel. Sens. Actuators B: Chem. **149**(1), 272 (2010)

619. Murase, Y., Takeshima, R., Yoshida, R.: Self-driven gel conveyer: Effect of interactions between loaded cargo and self-oscillating gel surface. Macromol. Biosci. **11**(12), 1713 (2011)
620. Tamate, R., Mizutani Akimoto, A., Yoshida, R.: Recent advances in self-oscillating polymer material systems. Chem. Rec. **16**(4), 1852 (2016)
621. Shiraki, Y., Yoshida, R.: Autonomous intestine-like motion of tubular self-oscillating gel. Angew. Chem. Int. Ed. **51**(25), 6112 (2012)
622. Shiraki, Y., Akimoto, A.M., Miyata, T., Yoshida, R.: Autonomous pulsatile flow by peristaltic motion of tubular self-oscillating gels. Chem. Mater. **26**(19), 5441 (2014)
623. Masuda, T., Hidaka, M., Murase, Y., Akimoto, A.M., Nagase, K., Okano, T., Yoshida, R.: Self-oscillating polymer brushes. Angew. Chem. Int. Ed. **52**(29), 7468 (2013)
624. Ueki, T., Shibayama, M., Yoshida, R.: Self-oscillating micelles. Chem. Commun. **49**(62), 6947 (2013)
625. Tamate, R., Ueki, T., Shibayama, M., Yoshida, R.: Self-oscillating vesicles: Spontaneous cyclic structural changes of synthetic diblock copolymers. Angew. Chem. Int. Ed. **53**(42), 11248 (2014)
626. Tamate, R., Ueki, T., Yoshida, R.: Self-beating artificial cells: Design of cross-linked polymersomes showing self-oscillating motion. Adv. Mater. **27**(5), 837 (2015)
627. Cornejo, A., Fraile, J.M., García, J.I., García-Verdugo, E., Gil, M.J., Legarreta, G., Luis, S. V., Martínez-Merino, V., Mayoral, J.A.: The first immobilization of pyridine-bis(oxazoline) chiral ligands. Org. Lett. **4**(22), 3927 (2002)
628. Cornejo, A., Fraile, J.M., García, J.I., Gil, M.J., Luis, S.V., Martínez-Merino, V., Mayoral, J. A.: A flexible and versatile strategy for the covalent immobilization of chiral catalysts based on pyridinebis(oxazoline) ligands. J. Org. Chem. **70**(14), 5536 (2005)
629. Happ, B., Friebe, C., Winter, A., Hager, M.D., Schubert, U.S.: Click chemistry meets polymerization: Controlled incorporation of an easily accessible ruthenium(II) complex into a PMMA backbone *via* RAFT copolymerization. Eur. Polym. J. **45**(12), 3433 (2009)
630. Happ, B., Kübel, J., Pfeffer, M.G., Winter, A., Hager, M.D., Dietzek, B., Rau, S., Schubert, U.S.: Towards hydrogen evolution Initiated by LED light: 2-(1*H*–1,2,3-triazol-4-yl) pyridine-containing polymers as photocatalyst. Macromol. Rapid Commun. **36**(7), 671 (2015)
631. Breul, A.M., Schäfer, J., Friebe, C., Schlütter, F., Paulus, R.M., Festag, G., Hager, M.D., Winter, A., Dietzek, B., Popp, J., et al.: Synthesis and characterization of poly(methyl methacrylate) backbone polymers containing side-chain pendant ruthenium(II) bis-terpyridine complexes with an elongated conjugated system. Macromol. Chem. Phys. **213**(8), 808 (2012)
632. Chen, B., Sleiman, H.F.: Ruthenium bipyridine-containing polymers and block copolymers *via* ring-opening metathesis polymerization. Macromolecules **37**(16), 5866 (2004)
633. Metera, K.L., Sleiman, H.: Luminescent vesicles, tubules, bowls and star micelles from ruthenium-bipyridine block copolymers. Macromolecules **40**(10), 3733 (2007)
634. Chen, B., Metera, K., Sleiman, H.F.: Biotin-terminated ruthenium bipyridine ring-opening metathesis polymerization copolymers: Synthesis and self-assembly with Streptavidin. Macromolecules **38**(4), 1084 (2005)
635. Sankaran, N.B., Rys, A.Z., Nassif, R., Nayak, M.K., Metera, K., Chen, B., Bazzi, H.S., Sleiman, H.F.: Ring-opening metathesis polymers for biodetection and signal amplification: synthesis and self-assembly. Macromolecules **43**(13), 5530 (2010)
636. Pawar, M.S., Zha, Y., Disabb-Miller, M.L., Johnson, Z.D., Hickner, M.A., Tew, G.N.: Metal-ligand based anion exchange membranes. In: Polymer Composites for Energy Harvesting, Conversion and Storage, pp. 127. American Chemical Society (2014)
637. Breul, A.M., Kübel, J., Häupler, B., Friebe, C., Hager, M.D., Winter, A., Dietzek, B., Schubert, U.S.: Synthesis and characterization of poly(phenylacetylene)s with Ru(II) bis-terpyridine complexes in the side chain. Macromol. Rapid Commun. **35**(7), 747 (2014)

638. Schuehler Sherwood, D.E., Williams, J.E., Sponsler, M.B.: Ruthenium-incorporated, metathesis-active polyacetylene. J. Polym. Sci., Part A: Polym. Chem. **47**(4), 1061 (2009)
639. Sakurai, S.-i., Ohira, A., Suzuki, Y., Fujito, R., Nishimura, T., Kunitake, M., Yashima, E.: Synthesis and property of helical poly(phenylacetylene)s bearing chiral ruthenium complexes and real space imaging of meso- and nanoscopic structures by atomic force microscopy. J. Polym. Sci., Part A: Polym. Chem. **42**(18), 4621 (2004)
640. Xie, Z., Wang, C., deKrafft, K.E., Lin, W.: Highly stable and üorous cross-linked polymers for efficient photocatalysis. J. Am. Chem. Soc. **133**(7), 2056 (2011)
641. Wang, C., Xie, Z., deKrafft, K.E., Lin, W.: Light-harvesting cross-linked polymers for efficient heterogeneous photocatalysis. ACS Appl. Mater. Interfaces **4**(4), 2288 (2012)
642. Wang, J.-L., Wang, C., deKrafft, K.E., Lin, W.: Cross-linked polymers with exceptionally high $Ru(bipy)_3^{2+}$ loadings for efficient heterogeneous photocatalysis. ACS Catal. **2**(3), 417 (2012)
643. Luo, Y., Xu, Z.-Y., Wang, H., Sun, X.-W., Li, Z.-T., Zhang, D.-W.: Porous $Ru(bpy)_3^{2+}$-linked polymers for recyclable photocatalysis of enantioselective alkylation of aldehydes. ACS Macro Lett. **9**(1), 90 (2020)
644. Cheng, M., Euler, W.B.: Synthesis and solution characterization of $[Ru(bpy)_2]^{2+}$ modified polyazines. Inorg. Chem. **42**(17), 5384 (2003)
645. Lee, J.-K., Yoo, D., Rubner, M.F.: Synthesis and characterization of an electroluminescent polyester containing the Ru(II) complex. Chem. Mater. **9**(8), 1710 (1997)
646. Peng, Z., Gharavi, A.R., Yu, L.: Synthesis and characterization of photorefractive polymers containing transition metal complexes as photosensitizer. J. Am. Chem. Soc. **119**(20), 4622 (1997)
647. Chan, W.K., Ng, P.K., Gong, X., Hou, S.: Synthesis and electronic properties of conjugated polymers based on rhenium or ruthenium dipyridophenazine complexes. J. Mater. Chem. **9** (9), 2103 (1999)
648. Wong, C.T., Chan, W.K.: Yellow light-emitting poly(phenylenevinylene) incorporated with pendant ruthenium bipyridine and terpyridine complexes. Adv. Mater. **11**(6), 455 (1999)
649. Frank, W., Pautzsch, T., Klemm, E.: Bipyridinylene-based conjugated polymer containing a ruthenium(II) bipyridine metal complex synthesized by Suzuki coupling. Macromol. Chem. Phys. **202**(12), 2535 (2001)
650. Pautzsch, T., Klemm, E.: Synthesis and characterization of alt-bipyridinylene-phenylene copolymers containing ruthenium(II) complexes. J. Polym. Sci., Part A: Polym. Chem. **42** (12), 2911 (2004)
651. Ng, P.K., Gong, X., Wong, W.T., Chan, W.K.: Quinoxaline-based conjugated polymers containing ruthenium(II) bipyridine metal complex. Macromol. Rapid Commun. **18**(12), 1009 (1997)
652. Pautzsch, T., Blankenburg, L., Klemm, E.: Design and synthesis of Ru(II)-bipyridyl-containing conjugated polymers. J. Polym. Sci., Part A: Polym. Chem. **42**(3), 722 (2004)
653. Hotzel, M., Urban, S., Egbe, D.A.M., Pautzsch, T., Klemm, E.: Degenerate four-wave mixing measurements of the third-order nonlinearity of poly(aryleneethynylenes) and their model substances. J. Opt. Soc. Am. B **19**(11), 2645 (2002)
654. Trouillet, L., Lapkowski, M., Stephan, O., Guillerez, S.: Synthesis, electrochemical and electroluminescent properties of oligothiophene-based conjugated polymers. Synth. Metals **109**(1–3), 277 (2000)
655. Koizumi, T.-A., Hayashi, H., Shiozaki, S., Yamamoto, T.: $[Ru(bpy)_2(bqdiBr_2)](PF_6)_2$ bearing a 3,6-dibromo-1,2-benzoquinone diimine ligand ($bqdiBr_2$ = $Br\text{-}C_6H_2(NH)_2\text{-}Br$): Synthesis and its cross coupling reactions with organostannanes and organoboronic acids. Inorg. Chim. Acta **421**, 427 (2014)
656. Walters, K.A., Trouillet, L., Guillerez, S., Schanze, K.S.: Photophysics and electron transfer in poly(3-octylthiophene) alternating with Ru(II)- and Os(II)-bipyridine complexes. Inorg. Chem. **39**(24), 5496 (2000)

657. Cheung, W.K., Mak, C.S.K., Chan, W.K.: Heteroleptic ruthenium(II) complex-containing polymers and their bandgap tuning and photosensitizing properties. Macromol. Rapid Commun. **33**(6–7), 585 (2012)
658. Tito, D.N., Kluson, P.: Amperometric analysis of Ru π-conjugated polymers: Prospective co-supports for enantioselective Ru catalysts. J. Phys. Chem. C **120**(38), 21228 (2016)
659. Yasuda, H., Morita, Y., Noda, I., Sugi, K., Nakamura, A.: Organometallic polymers containing tricarbonyl(η4-diene)-ruthenium(0) and -iron(0) as pendant groups. J. Organomet. Chem. **205**(2), C9 (1981)
660. Yasuda, H., Noda, I., Miyanaga, S., Nakamura, A.: Novel semiconducting organometallic polymers containing $M(CO)_3$(1,3-diene) groups (M = Fe, Ru). Macromolecules **17**(11), 2453 (1984)
661. Naota, T., Takaya, H., Murahashi, S.-I.: Ruthenium-catalyzed reactions for organic synthesis. Chem. Rev. **98**(7), 2599 (1998)
662. Nestler, O., Severin, K.: A ruthenium porphyrin catalyst immobilized in a highly cross-linked polymer. Org. Lett. **3**(24), 3907 (2001)
663. Burri, E., Öhm, M., Daguenet, C., Severin, K.: Site-isolated porphyrin catalysts in imprinted polymers. Chem. Eur. J. **11**(17), 5055 (2005)
664. Alexander, C., Davidson, L., Hayes, W.: Imprinted polymers: artificial molecular recognition materials with applications in synthesis and catalysis. Tetrahedron **59**(12), 2025 (2003)
665. Vasapollo, G., Sole, R.D., Mergola, L., Lazzoi, M.R., Scardino, A., Scorrano, S., Mele, G.: Molecularly imprinted polymers: Present and future prospective. Int. J. Mol. Sci. **12**(9), 5908 (2011)
666. Haupt, K., Mosbach, K.: Molecularly Imprinted Polymers and Their Use in Biomimetic Sensors. Chem. Rev. **100**(7), 2495 (2000)
667. Wulff, G.: Enzyme-like catalysis by molecularly imprinted polymers. Chem. Rev. **102**(1), 1 (2002)
668. Ferrand, Y., Le Maux, P., Simonneaux, G.: Macroporous chiral ruthenium porphyrin polymers: a new solid-phase material used as a device for catalytic asymmetric carbene transfer. Tetrahedron Asymm. **16**(23), 3829 (2005)
669. Ferrand, Y., Daviaud, R., Le Maux, P., Simonneaux, G.: Catalytic asymmetric oxidation of sulfide and styrene derivatives using macroporous resins containing chiral metalloporphyrins (Fe, Ru). Tetrahedron Asymm. **17**(6), 952 (2006)
670. Polborn, K., Severin, K.: Molecular imprinting with an organometallic transition state analogue. Chem. Commun. (24), 2481 (1999)
671. Polborn, K., Severin, K.: Biomimetic catalysis with immobilised organometallic ruthenium complexes: substrate- and regioselective transfer hydrogenation of ketones. Chem. Eur. J. **6** (24), 4604 (2000)
672. Godlewska-Żyłkiewicz, B., Zambrzycka, E., Leśniewska, B., Wilczewska, A.Z.: Separation of ruthenium from environmental samples on polymeric sorbent based on imprinted Ru(III)-allyl acetoacetate complex. Talanta **89**, 352 (2012)
673. Shakerian, F., Kim, K.-H., Kwon, E., Szulejko, J.E., Kumar, P., Dadfarnia, S., Haji Shabani, A.M.: Advanced polymeric materials: Synthesis and analytical application of ion imprinted polymers as selective sorbents for solid phase extraction of metal ions. TrAC **83**, 55 (2016)
674. Zambrzycka, E., Kiedysz, U., Wilczewska, A.Z., Leśniewska, B., Godlewska-Żyłkiewicz, B.: A novel ion imprinted polymer as a highly selective sorbent for separation of ruthenium ions from environmental samples. Anal. Meth. **5**(12), 3096 (2013)
675. Zambrzycka, E., Godlewska-Żyłkiewicz, B.: A new ion imprinted polymer based on Ru(III)-thiobarbituric acid complex for solid phase extraction of ruthenium(III) prior to its determination by ETAAS. Microchim. Acta **181**(9), 1019 (2014)
676. Monier, M., Abdel-Latif, D.A., Youssef, I.: Preparation of ruthenium(III) ion-imprinted beads based on 2-pyridylthiourea modified chitosan. J. Colloid Interface Sci. **513**, 266 (2018)

677. Chen, L., Dai, J., Hu, B., Wang, J., Wu, Y., Dai, J., Meng, M., Li, C., Yan, Y.: Recent progresses on the adsorption and separation of ions by imprinting routes. Sep. Purif. Rev. **49** (4), 265 (2020)
678. Neuse, E.W.: Polymeric ferrocene and ruthenocene compounds with boron in the main chain. J. Macromol. Sci. Chem. **2**(4), 751 (1968)
679. Bariyanga, J.; Carraher, C. E. In memory of Dr. Eberhard W. Neuse (1925–2014). J. Inorg. Organomet. Polym. Mater. **25**(3), 327 (2015)
680. Bhattacharjee, H., Müller, J.: Metallocenophanes bridged by group 13 elements. Coord. Chem. Rev. **314**, 114 (2016)
681. Nelson, J.M., Lough, A.J., Manners, I.: Synthesis and ring-opening polymerization of highly strained, ring-tilted [2]ruthenocenophanes. Angew. Chem. Int. Ed. **33**(9), 989 (1994)
682. Herbert, D.E., Mayer, U.F.J., Manners, I.: Strained metallocenophanes and related organometallic rings containing π-hydrocarbon ligands and transition-metal centers. Angew. Chem. Int. Ed. **46**(27), 5060 (2007)
683. Braunschweig, H., Damme, A., Hammond, K., Mager, J.: Synthesis and structure of new [3] silametallocenophanes of group 8 metals. Organometallics **31**(17), 6317 (2012)
684. Vogel, U., Lough, A.J., Manners, I.: Isolation of [1]ruthenocenophanes: synthesis of polyruthenocenylstannanes by ring-opening polymerization. Angew. Chem. Int. Ed. **43**(25), 3321 (2004)
685. Schachner, J.A., Tockner, S., Lund, C.L., Quail, J.W., Rehahn, M., Müller, J.: [1] Metallocenophanes (M = Fe, Ru) of heavier group 13 elements (E = Al, Ga): synthesis, characterization and ring-opening polymerization. Organometallics **26**(18), 4658 (2007)
686. Bagh, B., Schatte, G., Green, J.C., Müller, J.: Understanding the reactivity of strained sandwich compounds with aluminum or gallium in bridging positions: experiments and DFT calculations. J. Am. Chem. Soc. **134**(18), 7924 (2012)
687. Tritschler, U., Pearce, S., Gwyther, J., Whittell, G.R., Manners, I.: 50th Anniversary Perspective: Functional Nanoparticles from the Solution Self-Assembly of Block Copolymers. Macromolecules **50**(9), 3439 (2017)
688. Sha, Y., Rahman, M.A., Zhu, T., Cha, Y., McAlister, C.W., Tang, C.: ROMPI-CDSA: Ring-opening metathesis polymerization-induced crystallization-driven self-assembly of metallo-block copolymers. Chem. Sci. **10**(42), 9782 (2019)
689. Shultz, G.V., Tyler, D.R.: Preparation of functionalized organometallic metal–metal bonded dimers used in the synthesis of photodegradable polymers. J. Inorg. Organomet. Polym. Mater. **19**(4), 423 (2009)
690. Cao, K., Murshid, N., Wang, X.: Synthesis of main-chain metal carbonyl organometallic macromolecules (MCMCOMs). Macromol. Rapid Commun. **36**(7), 586 (2015)
691. Sharma, H.K., Pannell, K.H.: The substituent-dependent base-treatment chemistry of (η^5-C_5H_5)$Fe(CO)_2SiR_2SiR_2Cl$: Formation of 1,2-disila-3-metallacyclobutanes, their ring-opened polymers, migrations and substitutions. Chem. Commun. **22**, 2556 (2004)
692. Braunschweig, H., Dellermann, T., Dewhurst, R.D., Mies, J., Radacki, K., Stellwag-Konertz, S., Vargas, A.: Strained ansa half-sandwich complexes of ruthenium and osmium and a non-iron metallopolymer by ring-opening polymerization. Organometallics **33**(7), 1536 (2014)
693. Hiort, C., Lincoln, P., Norden, B.: DNA binding of Δ- and Λ-$[Ru(phen)_2DPPZ]^{2+}$. J. Am. Chem. Soc. **115**(9), 3448 (1993)
694. Shu, Y., Breitbach, Z.S., Dissanayake, M.K., Perera, S., Aslan, J.M., Alatrash, N., MacDonnell, F.M., Armstrong, D.W.: Enantiomeric separations of ruthenium(II) polypyridyl complexes using HPLC with cyclofructan chiral stationary phases. Chirality **27**(1), 64 (2015)
695. Wärnmark, K., Thomas, J.A., Heyke, O., Lehn, J.-M.: Stereoisomerically controlled inorganic architectures: Synthesis of enantio- and diastereo-merically pure ruthenium-palladium molecular rods from enantiopure building blocks. Chem. Commun. (6), 701 (1996)

696. Tzalis, D., Tor, Y.: Stereochemically-defined supramolecular architectures: diastereomerically-pure multi-Ru(II) complexes. J. Am. Chem. Soc. **119**(4), 852 (1997)
697. MacDonnell, F.M., Kim, M.-J., Bodige, S.: Substitutionally inert complexes as chiral synthons for stereospecific supramolecular syntheses. Coord. Chem. Rev. **185–186**, 535 (1999)
698. Bodige, S., Torres, A.S., Maloney, D.J., Tate, D., Kinsel, G.R., Walker, A.K., MacDonnell, F.M.: First-generation chiral metallodendrimers: Stereoselective synthesis of rigid D_3-symmetric tetranuclear ruthenium complexes. J. Am. Chem. Soc. **119**(43), 10364 (1997)
699. Chen, J., MacDonnell, F.M.: Stereoselective synthesis of chiral coordination polymers with partial control of the rigid main chain topology. Chem. Commun. (24), 2529 (1999)
700. MacDonnell, F. M.; Meser Ali, M. D.; Kim, M.-J. Robust Chiral Nanostructures: Global Chirality in Supramolecules Constructed from Enantiopure Ruthenium(II) Trisdiimine Building Blocks. Comm. Inorg. Chem. **22**(3–4), 203 (2000)
701. Ng, W.Y., Chan, W.K.: Synthesis and photoconducting properties of poly (p-phenylenevinylene)containing a bis(terpyridine)ruthenium(II) complex. Adv. Mater. **9** (9), 716 (1997)
702. Chan, W.K., Gong, X., Ng, W.Y.: Photoconductivity and charge transporting properties of metal-containing poly(p-phenylenevinylene)s. Appl. Phys. Lett. **71**(20), 2919 (1997)
703. Chan, H.T., Mak, C.S.K., Djurišić, A.B., Chan, W.K.: Synthesis of ruthenium complex containing conjugated polymers and their applications in dye-sensitized solar cells. Macromol. Chem. Phys. **212**(8), 774 (2011)
704. Wild, A., Winter, A., Hager, M.D., Görls, H., Schubert, U.S.: Perfluorophenyl-terpyridine ruthenium complex as monomer for fast, efficient and mild metallopolymerizations. Macromol. Rapid Commun. **33**(6–7), 517 (2012)
705. Liu, W.-J., Qian, T.-T., Jiang, H.: Bimetallic Fe nanoparticles: Recent advances in synthesis and application in catalytic elimination of environmental pollutants. Chem. Eng. J. **236**, 448 (2014)
706. Goor, O.J.G.M., Hendrikse, S.I.S., Dankers, P.Y.W., Meijer, E.W.: From supramolecular polymers to multi-component biomaterials. Chem. Soc. Rev. **46**(21), 6621 (2017)
707. Hofmeier, H., El-ghayoury, A., Schenning, A.P.H.J., Schubert, U.S.: New supramolecular polymers containing both terpyridine metal complexes and quadruple hydrogen bonding units. Chem. Commun. (3), 318 (2004)
708. Oka, S., Ozawa, H., Yoshikawa, K., Ikeda, T., Haga, M.-A.: Hydrogen-bonded metallo-supramolecular polymers based on ruthenium or iron complexes for the selective extraction of single-walled carbon nanotubes. Dalton Trans. **47**(40), 14195 (2018)
709. Li, B., Gong, R., Wang, W., Huang, X., Zhang, W., Li, H., Hu, C., Tan, B.: A new strategy to microporous polymers: Knitting rigid aromatic building blocks by external cross-linker. Macromolecules **44**(8), 2410 (2011)
710. Wang, X., Ling, E.A.P., Guan, C., Zhang, Q., Wu, W., Liu, P., Zheng, N., Zhang, D., Lopatin, S., Lai, Z., et al.: Single-site ruthenium pincer complex knitted into porous organic polymers for dehydrogenation of formic acid. ChemSusChem **11**(20), 3591 (2018)
711. Valverde-González, A., Marchal, G., Maya, E.M., Iglesias, M.: A step forward in solvent knitting strategies: ruthenium and gold phosphine complex polymerization results in effective heterogenized catalysts. Catal. Sci. Technol. **9**(17), 4552 (2019)
712. Guan, C., Pan, Y., Ang, E.P.L., Hu, J., Yao, C., Huang, M.-H., Li, H., Lai, Z., Huang, K.-W.: Conversion of CO_2 from air into formate using amines and phosphorus-nitrogen PN^3P-Ru(II) pincer complexes. Green Chem. **20**(18), 4201 (2018)
713. Eder, G.M., Pyles, D.A., Wolfson, E.R., McGrier, P.L.: A ruthenium porphyrin-based porous organic polymer for the hydrosilylative reduction of CO_2 to formate. Chem. Commun. **55**(50), 7195 (2019)
714. Yang, Z.-Z., Zhang, H., Yu, B., Zhao, Y., Ji, G., Liu, Z.: A Tröger's base-derived microporous organic polymer: design and applications in CO_2/H_2 capture and hydrogenation of CO_2 to formic acid. Chem. Commun. **51**(7), 1271 (2015)

715. Emsley, J.: *Ruthenium* In *Nature's Building Blocks: An A-Z Guide to the Elements*, p. 368. Oxford University Press, Oxford (2001)
716. Page, S.: Ruthenium compounds as anticancer agents. Educ. Chem. **49**(1) 26 (2012)
717. Gill, M.R., Thomas, J.A.: Ruthenium(II) polypyridyl complexes and DNA-from structural probes to cellular imaging and therapeutics. Chem. Soc. Rev. **41**(8), 3179 (2012)
718. Clarke, M.J.: Ruthenium metallopharmaceuticals. Coord. Chem. Rev. **236**(1–2), 209 (2003)
719. Clarke, M.J., Zhu, F., Frasca, D.R.: Non-platinum chemotherapeutic metallopharmaceuticals. Chem. Rev. **99**(9), 2511 (1999)
720. Zhang, C.X., Lippard, S.J.: New metal complexes as potential therapeutics. Curr. Opin. Chem. Biol. **7**(4), 481 (2003)
721. Bruijnincx, P.C.A., Sadler, P.J.: New trends for metal complexes with anticancer activity. Curr. Opin. Chem. Biol. **12**(2), 197 (2008)
722. Winkler, J.R., Gray, H.B.: Electron transfer in ruthenium-modified proteins. Chem. Rev. **92** (3), 369 (1992)

3 Electropolymerization—An Item-Centered View on Ruthenopolymers

Abstract

The electrochemical polymerization of Ru-containing monomers generates polymers on semiconductor or electrode surfaces. The different techniques to prepare such thin metallopolymer films are summarized in this chapter. As before, the proposed applications, such as opto-electronics, catalysis or sensing, are evaluated.

3.1 Introduction

In Chap. 2, a range of potential applications related to the diverse types of Ru-containing polymers has been addressed. In many cases, the materials have to be transferred from solution into the solid state in order to fabricate thin films, e.g., as coatings on electrode or nanoparticle surfaces. In addition to the classical solution-based deposition techniques (e.g., spin-coating, inkjet-printing, doctor-blading or drop-casting), the electropolymerization represents a versatile approach for the direct coating of a substrate by the in-situ formation of the polymer. Thus, this method offers some important advantages compared to the common solution-based polymerization procedures [1, 2]. In particular, solubility problems associated with the polymer's structure are known to hamper its processing into thin-solid films; this issue can be bypassed, since electropolymerization enables the formation of the polymers directly on the desired surfaces. Conveniently, only the reactive monomers need to be dissolved and treated electrochemically (depending on the structure of the monomer, anodic or cathodic processes are applied). The time frame required for such a polymerization is generally rather short and the basic technical equipment can be used for electropolymerizing a wide range of monomers. A further beneficial aspect to be mentioned is

© The Author(s), under exclusive license to Springer Nature Switzerland AG 2021
U. S. Schubert et al., *Ruthenium-Containing Polymers*,
https://doi.org/10.1007/978-3-030-75598-0_3

scalability, i.e., the thickness of the electro-deposited films can simply be adjusted by the polymerization time (at least in a good approximation).

A historical perspective on this topic was published by Abruña in 1988 [3]. This author briefly summarized the early attempts to chemically modify electrodes. The term "chemically modified electrode" was initially used by Murray et al. in 1975 [4] and refers to electrodes on which molecules were immobilized by adsorption, chemisorption or electronic interactions. Oyama and Anson demonstrated that charged transition-metal complexes were able to bind to electrode surfaces, which were coated with highly charged polymer films. For example, $[Fe(CN)_6]^{3-}$ was efficiently bound to a protonated poly(4-vinylpyridine) (P4VP) film, as electrode coating; the same concept was used to immobilize $[Ru(NH_3)_6]^{3+}$ via electrostatic interactions with a deprotonated poly(acrylic acid) (PAA) film [5, 6]. These same authors later introduced poly(styrene sulfonate) (PSS), as polyanionic electrode coating; this polyelectrolyte-coated electrode was capable of binding transition-metal complexes, including $[Ru(NH_3)_6]^{3+}$ and $[Ru(bpy)_3]^{2+}$ [7]. At the same time, Nafion™, as a perfluorinated ion-exchange polymer, emerged as an electrode-coating material and, in initial studies, the electrostatic binding of $[Ru(bpy)_3]^{2+}$ was reported (for more details on this topic, see Chap. 5) [8]. According to Murray, these studies marked the transition from simple chemically modified electrodes [bearing a (few) monolayer(s) of molecules] to the so-called polymer-modified electrodes [9]. The pioneering report on the electropolymerization of some polypyridyl complexes of either Ru^{II} or Fe^{II} ions via their vinyl-substituents has to be seen in the same context. The formed metallopolymer films were highly adherent and showed the photophysical and electrochemical properties of their parent mnomers [10]. In the following, the applicability of this approach to a broad range of transition-metal complexes was shown and a range of opto-electronic or electrochemical applications was established [1].

When aiming for the electropolymerization of transition-metal complexes in general, the functionalization with an electropolymerizable group (e.g., vinyl, pyrrole, aniline or thiophene groups) represents a requirement. As for the different types of "conventional" metallopolymers (Chap. 1), also different scenarios have to be considered when dealing with electropolymerized ones. Firstly, the metal centers can be bound to the formed polymer film via a flexible, non-conjugated spacer. Alternatively, the complexes might directly be functionalized with the electropolymerizable group—thereby, the metal centers will be closely bound to the electro-generated polymer chain or even become an integral part of the polymer backbone. For these two cases a considerable electronic interaction between the metal centers and the (conjugated) polymer chain as well as between adjacent metal sites can be observed (thereby, one has to expect the largest influences for structures of the latter type). According to Wolf, these three systems are classified as metallopolymers of the type I, II, and III, respectively (Fig. 3.1) [2].

Thereby, the interaction of the metal complexes with the polymer backbone or with each other will be rather low (type I) or considerably strong (type III > type II). The forth class of systems features polymer backbones exclusively comprised of metal centers (e.g., Ru^0); as expected, the degree of metal-to-metal interaction will

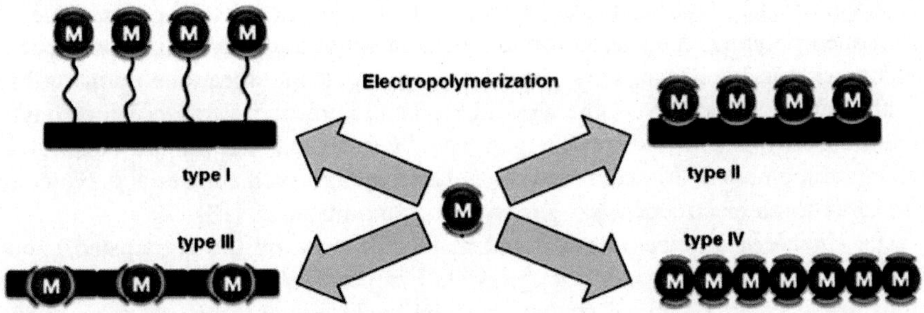

Fig. 3.1 Schematic representation of the four different types of materials accessible via electropolymerization of metal-containing monomers [1]. The classification is in line with the one suggest by Wolf [2]. Figure reproduced with kind permission; © 2012 Wiley–VCH

be the highest in such type-IV materials. Noteworthy, not only linear polymer chains are accessible via electropolymerization methods—also cross-linked polymer networks can be formed when suitable multiply functionalized monomers are employed.

In here, ruthenopolymers, which were prepared by electropolymerizations, will be discussed. Their classification has been made on the basis of the applied electropolymerization method (i.e., anodic vs. cathodic) and the reactive functional group(s) of the monomer. From the range of cathodic electropolymerization reactions, the reductive polymerization of vinyl groups represents, by far, the most common example. On the other hand, the oxidative coupling of electron-rich functionalities is the characteristic of anodic electropolymerizations [here, 5-membered heterocycles (e.g., pyrroles, thiophenes, furans) or electron-rich aromatic compounds are involved]. Before detailing this, some general considerations regarding electropolymerization itself will be made [1].

3.2 Experimental Remarks

When dealing with electropolymerization, one first has to consider the electrochemical technique, which is applied for this purpose. In this respect, one has to distinguish between potentio- and galvano-static methods on the one hand and potentiodynamic methods on the other. López-Palacios et al. pointed out that the polymerization mechanism is not affected by the chosen method—holding either the voltage or current constant and cycling the other basically gives polymers of identical chemical structures; whereas, the solid-state morphology of the prepared material strongly depends on the applied electrochemical method [11]. However, one can hardly predict which methods preferentially generates the more homogeneous and surface-adhesive films—these properties will always depend, at least to a certain extent, on the chemical and physical nature of the material under investigation [12]. One further difference is related to the electrochemical state of the formed polymer. A material formed via a galvanostatic polymerization process remains charged and can, thus, experience structural modifications upon further electrochemical treatment. Otherwise, the potentiodynamic electropolymerization generates polymers in its non-charged ground state and, due to the continuous cycling of the potential in course of the polymerization, such a material is typically insensible to an electrochemical post-synthesis modification [13].

The choice of an electrode material mainly depends on the anticipated application of the electropolymerized material. For proof-of-concept experiments or fundamental electrochemical studies, e.g., by cyclic voltammetry (CV) or differential pulse voltammetry (DVP), standard working electrodes are conveniently used: Glassy carbon (GC), platinum, gold, etc. However, when a spectroscopic characterization is required or an opto-electronic application is anticipated, one has to change to transparent electrode materials—typically glass or polymer substrates, which have been coated with an indium tin oxide (ITO) or a fluorine-doped tin oxide (FTO) layer [14]. Polymer-based light-emitting diodes (PLEDS) as well as electrochromic and photovoltaic devices represent typical applications to be mentioned in this context. Beyond this, also other electrode materials, for energy storage or photoelectrochemical purposes, are utilized: Carbon materials [e.g., carbon nanotubes (CNTs)] [15] or semiconductor photoanodes and photocathodes [16].

Due to their intrinsic insolubility, electrochemically prepared polymers cannot be characterized by the common solution-based analytical methods [e.g., size-exclusion chromatography (SEC), NMR spectroscopy, and mass spectrometry]. However, the electrochemical polymerization process itself offers the advantage of an in-situ characterization. First information regarding the growth of the polymer film and changes of the electrochemical response can be already derived from the current–voltage characteristics or the CV curves in the case of potentio-/galvano-static procedures or potentiodynamic ones, respectively. In addition to this, some non-electrochemical methods are available allowing one to monitor the electropolymerization process online. UV/vis and IR spectroelectrochemistry gives insight into the electropolymerization process itself and may aid to identify potential intermediates; whereas, the growth of the electropolymerized film can be followed directly by electrochemical quartz crystal microbalance (EQCM) measurements. With respect to the post-synthesis film characterization, a range of analytical methods can be applied. These include the various electrochemical methods (i.e., CV, DPV, differential pulse polarography (DPP), and in-situ conductivity measurements). From the broad range of spectroscopic techniques, in particular UV/vis, NIR, IR, Raman, and electron paramagnetic resonance (EPR) spectroscopy can be used for the polymer film characterization; these methods can also be implemented into a spectroelectrochemical set-up. Electron microscopy imaging, such as transmission and scanning electron microscopy (TEM and SEM, respectively), allows one to study the morphology the polymer films; whereas, versatile information regarding the molecular structure and elemental composition can be obtained from NMR and X-ray photoelectron spectroscopy (XPS) measurements, respectively.

3.3 Reductive Electropolymerization of Vinyl-Containing Monomers

Zhong et al. suggested a categorization of Ru^{II} complexes according to their polymerization efficiency and the stability of the derived film [17]: Complexes with bidentate or tridentate ligands that can easily be electropolymerized into stable films (Figs. 3.2 and 3.3), complexes with low polymerization efficiency as single monomers (though, their copolymerization with other vinyl-functionalized monomers will still readily occur; Fig. 3.4) and those complexes that can be electropolymerized only into unstable films. [$Ru(4\text{-vpy})_4Cl_2$] is a typical example in this respect (4-vpy: 4-vinylpyridine) [18, 19]; further Ru^{II} complexes, which show a similar behavior, are depicted in Fig. 3.5. Since the exact polymerization efficiency for a specific monomer is difficult to define, a rule-of-thumb shall be given to conclude if a polymerization will occur efficiently or not: Though representing a crude approximation only, the surface coverage (Γ) with the electro-generated film should ideally be $>10^{-8}$ mol cm^{-2}; this value can be calculated according to Eq. 3.1 [thereby, Q denotes the charge under the metal-based redox wave, n is the number of electrons required for reducing a molecule, F represents Faraday's constant (96,485 C mol^{-1}) and A describes the area of the electrode in cm^2].

$$\Gamma = Q/n \cdot F \cdot A \tag{3.1}$$

In general, reductive electropolymerizations are considered to be initiated anodically with a subsequent radical–radical chain propagation—mechanistically this pathway, which is depicted in Fig. 3.6, is reminiscent of the one of classical free-radical polymerization (FRP) [20]. In order to achieve a highly efficient polymerization and to guarantee the formation of a rather dense cross-linked polymer network, at least two vinyl groups per complex monomer are advisable. Accordingly, monomers with only a single vinyl-group, such as Ru^{II} complexes **15** or **16**, represent improper monomers for an electrochemical homopolymerization, but they may well be utilized, as co-monomers, in electropolymerizations [10, 21]. One further aspect to be considered is the ligand's denticity—with increasing denticity, the kinetic stability of the complexes is increased (owing to the so-called chelate effect). In other words, monodentate ligands, such as pyridine (py) are only bound loosely; whereas, 2,2'-bipyridine (bpy) or 2,2':6',2"-terpyridine (tpy) as bidentate and tridentate ligand, respectively, give kinetically stable complexes with Ru^{II} ions. Thus, the stability of an electropolymerized material also depends on the kinetic stability of the complexes, which are incorporated. Naturally, these general considerations are not limited to vinyl-functionalized complexes but hold true for all electropolymerizable groups and complexes to which they are attached.

In the earliest available studies dealing with electrochemically generated electrode coatings, Ru^{II} species, in-situ generated from the reduction of a Ru^{III}–EDTA precursor complex, were coordinated to a graphite electrode, which was coated with P4VP or polyacrylnitrile of high molar masses (EDTA:

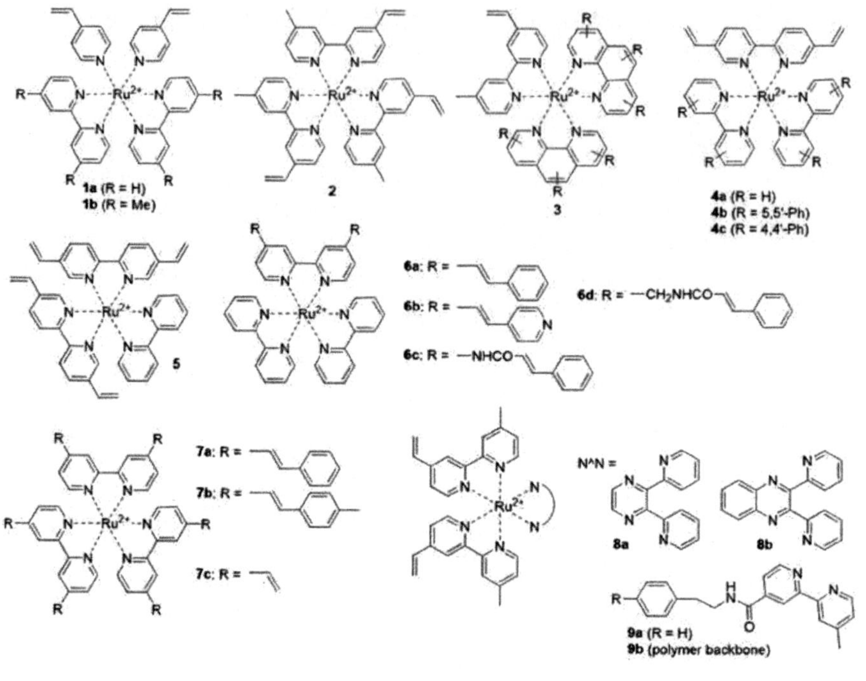

Fig. 3.2 Schematic representation of the Ru^{II} complexes **1–9**; these complexes can easily be electropolymerized via their vinyl-groups (the counterions are omitted for clarity)

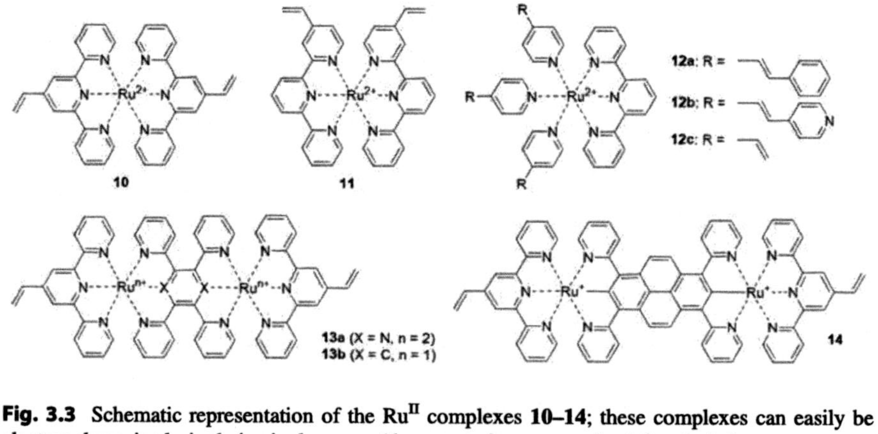

Fig. 3.3 Schematic representation of the Ru^{II} complexes **10–14**; these complexes can easily be electropolymerized via their vinyl-groups (the counterions are omitted for clarity)

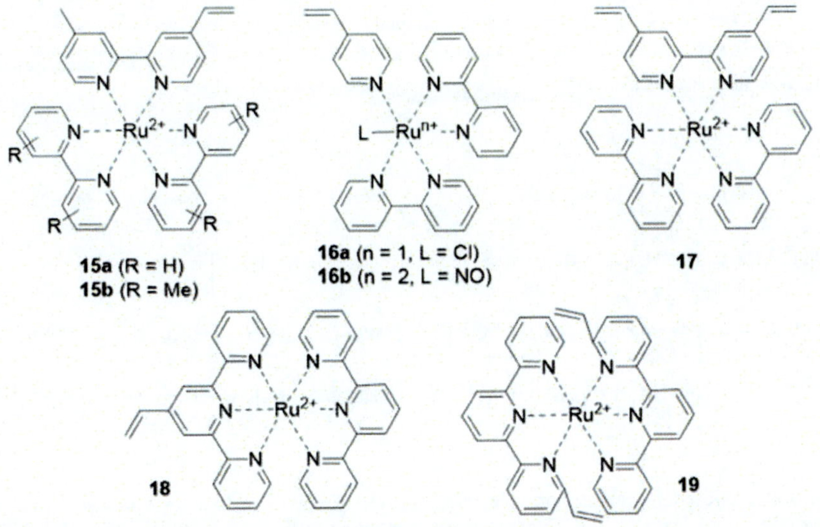

Fig. 3.4 Schematic representation of the Ru^{II} complexes **15–19**; these complexes are improper monomers for the electrochemical homopolymerization via their vinyl-groups (the counterions are omitted for clarity)

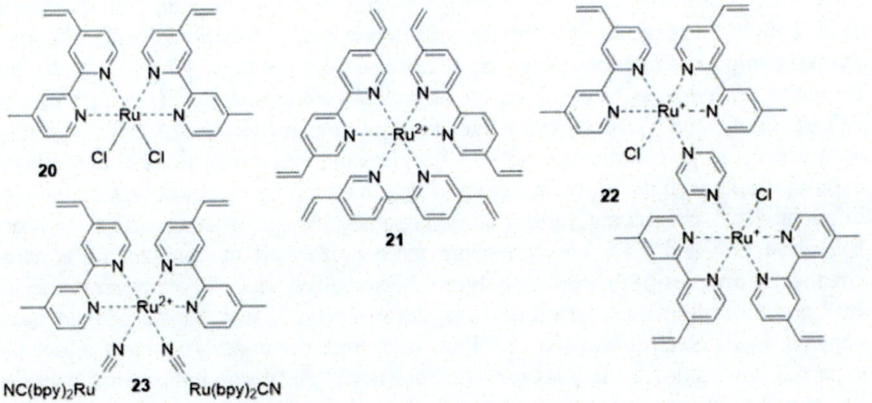

Fig. 3.5 Schematic representation of the Ru^{II} complexes **20–23**; these complexes can easily be electropolymerized via their vinyl-groups, but the resultant films are unstable (the counterions are omitted for clarity)

ethylenediaminetetraacetate) [22–24]. According to electrochemical measurements, a surface coverage of *ca.* 4.5×10^{-9} mol cm^{-2} was obtained, thus being about 50 times higher than expected for a simple monolayer of Ru^{II} complexes. Though a

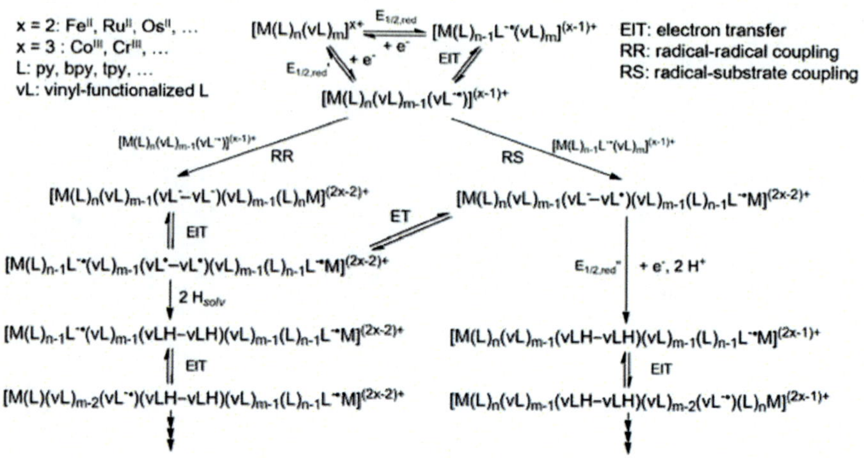

Fig. 3.6 Proposed mechanism for the reductive electropolymerization of vinyl-functionalized polypyridyl complexes [1, 20]. Figure redrawn according to the literature; © 2012 Wiley–VCH

high loading of polymer-modified electrodes with various types of transition metal ion species was achieved, Anson's approach mainly suffered from the rather ill-defined nature of the complexes formed by the in-situ reductive electrochemistry.

In their pioneering work on electropolymerized Ru^{II}-containing films, Abruña et al. utilizing a range of structurally well-defined vinyl-functionalized Ru^{II}–polypyridyl complexes, as monomers (**1a, 2, 15a** and **16a**; see Figs. 3.2 and 3.4), for the formation of stable, adherent films on various electrode materials (i.e., Pt, vitreous carbon, SnO_2, and TiO_2) by reductive cross-linking in MeCN [0.1 M $NEt_4(ClO_4)$ was added, as a conducting salt] [10]. Thereby, randomly mixed single-layer copolymers as well as bilayer schemes were generated by electropolymerizing two different Ru^{II} monomers either simultaneously or in a consecutive fashion. Extensive CV and XPS measurements were performed to analyze the electrochemical properties of the metallopolymer-coated electrodes. Two different types of Ru^{II} centers within the copolymer film, derived from **1** and **16a**, exhibited independent, facile electron-transfer (EIT) reactions with the underlying electrode at the expected potentials. In the bilayer-coated electrode, electronic interactions between the two layers were observed at certain potentials [10]. Complex **2** was electropolymerized onto a Pt flag electrode (dimension of 3 × 3 mm) and the electro-generated chemiluminescence (ECL) was investigated [25]. The phenomenon of ECL was already described earlier for $[Ru(bpy)_3]^{2+}$ in solution (though appearing only in close vicinity of the electrode) [26] as well as for the same complex electrostatically bound onto a Nafion™-coated electrode (**24**, Fig. 3.7a) [8]. Upon pulsing, orange luminescence—arising from an EIT process between electro-generated Ru^{I} and Ru^{III} states—was observed; the emission maximum was blue-shifted by *ca.* 15 nm relative to the luminescence spectrum of the parent monomer (Fig. 3.7b). However, the ECL was rather short-lived (*ca.* 20 min of continuous pulsing at ±1.5 V *vs.* Ag/AgCl with pulse durations of 1 s). As concluded from analogous ECL investigations on polymer-bound $[Ru(bpy)_3]^{2+}$ in solution, the low stability of the device was attributed to an irreversible reduction of the Ru^{II} centers to the Ru^{I} state [27]. Significant differences in electronic conductivity and ECL were observed comparing solvent-swollen and dry films [28]. The dried polymer films revealed concentration gradients of Ru^{III}, Ru^{II}, and Ru^{I} centers between the electrodes; the different states were formed by applying a constant potential for the oxidation/reduction of the Ru^{II} sites in the solvent-swollen films. Subsequently, the gradients were "frozen" by evaporation of the MeCN and cooling the films under a permanent inter-electrode bias of 2.6 V. The films were sensitive towards changes in the voltage bias and featured a diode-type behavior, *i.e.*, they were conductive and emissive at biases above 2.6 V. The highest observed quantum efficiency for solid-state ECL was comparable to that detected from the solvent-swollen films at 0 °C (Φ_{ECL} of 0.0003). However, the solvent-swollen films were less affected by changes in the applied voltage bias and did not reveal the diode-like characteristics. Moreover, potential step chronoamperometry was utilized to determine the charge transport through electropolymerized films of **2** giving an effective diffusion coefficient (D_{ct}) of 2×10^{-2} cm^2 s^{-1} [21]. Kaufman et al. suggested a migration of electrons via a successive self-exchange mechanism [29], i.e., a hopping between neighboring oxidized and reduced metal centers; on the basis of this assumption, an electron self-exchange rate of 1.7×10^5 M^{-1} s^{-1} was estimated.

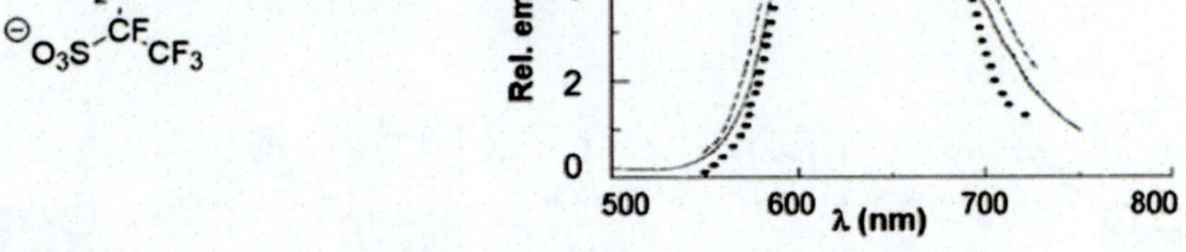

Fig. 3.7 **a** Schematic representation of the Nafion™ membrane loaded with $[Ru(bpy)_3]^{2+}$ complexes (**24**). **b** Normalized emission spectrum of **24** in MeCN (solid line), ECL for 1 mM **24** in 0.1 M $(n\text{-}Bu_4N)ClO_4$/MeCN (dotted line) and surface ECL for a Pt electrode coated with a film of electropolymerized **2** (dashed line) [25]. Figure reproduced with kind permission; © 1982 American Chemical Society

Meyer et al. applied photolysis on thin films of electropolymerized **1a/b** on a Pt disk; thereby, the weak Ru^{II}–py coordinative bonds were cleaved exposing the underlying substrate (Fig. 3.8) [30]. When utilizing a masking technique, a patterned surface could be obtained. Subsequently, $[Os(vbpy)_3]^{2+}$ (vbpy: 4-methyl-4'-vinyl-bpy) was polymerized selectively into the exposed regions of the electrode giving a laterally resolved, two-component, metallopolymer film (i.e., having discrete, spatially segregated regions of Ru^{II}- and Os^{II}-containing polymer). Finally, the remaining Ru^{II}-containing film was removed by photolysis (selectivity was guaranteed due to the inert character of the Os^{II} complexes towards photolysis). This protocol produced an image of the original mask in which the Os^{II}-containing polymer remained (Fig. 3.8). In a related study, the polymer film prepared from $[Ru(vbpy)_2(py)_2]^{2+}$ was photolyzed in the presence of a chloride salt in order to abstract the weakly bound py ligands [31]. The thusly treated surface was investigated by XPS enabling one to distinguish between photolyzed and non-photolyzed regions of a masked film. Both the UV/vis absorption as well as the redox behavior changed significantly upon ligand abstraction. This strategy, as suggested by Meyer et al., provided a basis for the fabrication of bicomponent, electrochromic polymer films on optically transparent electrodes of ITO-coated glass slides.

Meyer, Murray et al. pointed out that all the Ru^{II}-containing films are—due to their polycationic and highly cross-linked nature—hardly permeable for cations and bulky, neutral molecules. However, small solvent molecules or anions can migrate through the pores and react with the complex sites. In this context, the electropolymerization of monomer *cis*-**20** is worth mentioning [21]: The polymer film exhibited a number of anodic waves in CV measurements that were attributed to a (partial) displacement of the chloride ligands by solvent molecules, thereby generating polymer-bound $[Ru(bpy)_2(MeCN)Cl]^+$ and $[Ru(bpy)_2(MeCN)_2]^{2+}$ species.

Murray et al. demonstrated that the reductive electropolymerization of Ru^{II}–polypyridyl complexes is not restricted to monomers having terminal $-HC=CH_2$ groups [20]. In special cases, also internal $C=C$ bonds might serve as electropolymerizable groups (e.g., complexes **6** or **7** which bear stilbenic substituents, Fig. 3.2). Remarkably, steric crowding around the olefinic moieties did not necessarily inhibit the polymerization (though slower rates compared to analogous Ru^{II} complexes with vbpy or vpy ligands were observed). However, effective

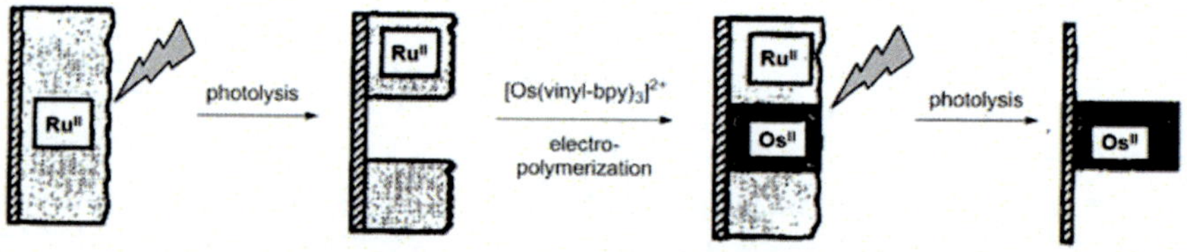

Fig. 3.8 Schematic representation of the generation of an electrode patterned with an Os^{II}-containing polymer via an electropolymerization-photolysis approach [30]. Figure reproduced with kind permission; © 1990 American Chemical Society

conjugation between the olefinic bond and the coordinating ligand was required. If the conjugation was interrupted, e.g., by an insulating CH_2-group, electropolymerization no longer occurred: **6c** could be readily electropolymerized; whereas, **6d** was inert in this respect (Fig. 3.2).

Besides these examples, the mixed-ligand complexes **3** having phen ligands could serve as monomers in reductive electropolymerization processes (phen: 1,10-phenanthroline; Fig. 3.2). These complexes were efficiently polymerized following an initial two-electron reduction, even when the phen ligand was highly substituted. Under the same conditions, the parent $[Ru(phen)_3]^{2+}$ complex did not electropolymerize, thus electrochemical homo-coupling between phen ligands could be excluded. Therefore, the reaction was suggested to proceed via radical–radical cross-coupling between the vinyl and phen moieties (the 4,7-positions were found to be the most reactive in this respect).

Zhong et al. introduced 5,5'-divinyl-2,2'-bipyridine, as ligand for Ru^{II} complexes, thereby reducing the synthetic effort as involved in the synthesis of the more common vbpy [32]. The resulting Ru^{II} complexes **4** and **5** were readily polymerized into stable films on the electrode surfaces, even in the case of the more bulky derivatives **4b/c**; instead, the homoleptic derivative (**7c**) gave a film that lost its electrochemical activity upon repetitive anodic cycling [33]. The electropolymerization of **4a** proceeded in a highly controlled fashion, thus this example is worth a more detailed discussion. Complex **4a** featured a reversible $Ru^{II/III}$ wave at + 1.37 V in concert with three ligand-based reduction couples (i.e., −1.09, −1.40, and −1.63 V vs. Ag/AgCl; Fig. 3.9a). Upon repeated scanning at a Pt disk electrode in the potential window of −0.60 to −1.55 V, a gradual and continuous increase of the current, indicative for the deposition of a film of poly-**4a** on the electrode surface, became apparent (Fig. 3.9b). The CV of poly-**4a**—recorded in a fresh supporting electrolyte solution at a scan rate of 100 mV s^{-1}—is depicted in Fig. 3.9c. All waves, i.e., the $Ru^{II/III}$ as well as the three ligand-based reduction couples, were retained with the potential for the $Ru^{II/III}$ couple not being shifted after polymerization (thus, the coordination sphere of the Ru^{II} centers remained intact). At slow scan rates, the potential split between the anodic and cathodic peaks of the $Ru^{II/III}$ wave of the polymer was small (ΔE_{pa-pc} = 59 mV); however, with increasing scan rate, the curves became spoiled due to limitations in the ElT kinetics (Fig. 3.9d). Both the anodic and cathodic currents were linearly dependent on the scan rate—a characteristic for redox processes occurring on electrode surface (Fig. 3.9e). The surface coverage of the films could simply be adjusted via the applied electropolymerization time (Fig. 3.9f highlights the CVs of the metallopolymer films obtained after 15, 30, and 45 potential cycles between −0.60 and −1.55 V of 0.5 mM **4a** in MeCN, corresponding to $\Gamma = 2.3 \times 10^{-9}$, 5.7×10^{-9}, and 14.3×10^{-9} mol cm^{-2}, respectively). Moreover, SEM imaging was applied to probe the surface morphology of the obtained metallopolymer films on ITO electrodes (Fig. 3.9g). In all cases, small domains of irregular shapes were observed on top of the electropolymerized films. Despite these artifacts, the surface of poly-**4a** was smooth compared to the rougher morphology imaged, e.g., for poly-**5** and poly-**7c** (also the contact of the latter two polymers to the ITO surface was much looser).

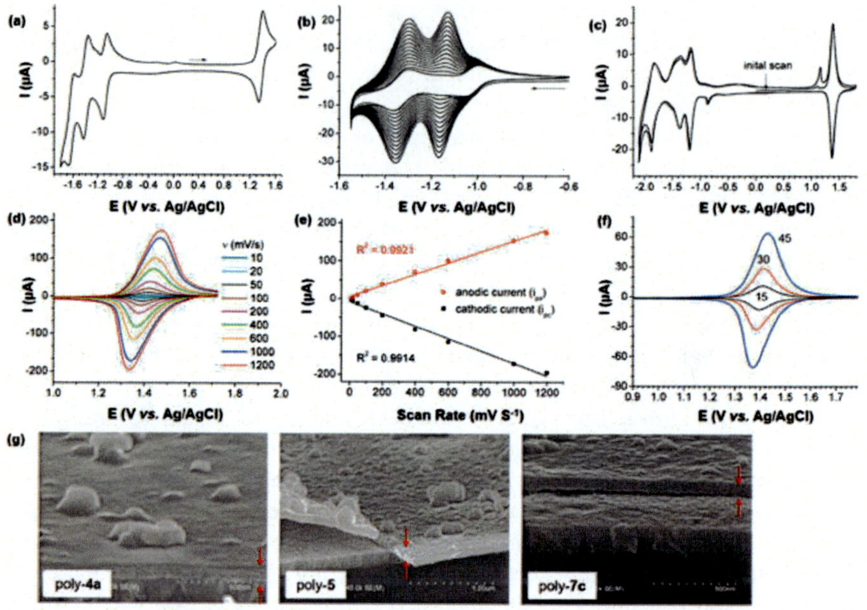

Fig. 3.9 **a** CV of **4a** in MeCN at a Pt electrode. **b** Reductive electropolymerization of **4a** (0.5 mM in MeCN) on a Pt disk electrode (Ø = 2 mm) by 15 repeated cyclic potential scans at 100 mV s^{-1} between −0.60 and −1.55 V versus Ag/AgCl in 0.1 M (n-Bu_4N)ClO_4/MeCN. **c** CV of the thusly obtained poly-**4a** at a scan rate of 100 mV s^{-1} in a clean supporting electrolyte solution. **d** Anodic CVs of poly-**4a** at different scan rates (10 to 1200 mV s^{-1}). **e** Linear dependence of the peak currents as a function of the scan rate. **f** CVs of poly-**4a** obtained after 15, 30, and 45 electropolymerization cycles (scan rate: 100 mV s^{-1}). **g** SEM images of poly-**4a**, poly-**5**, and poly-**7c** (from left to right) on ITO-coated glass electrodes with surface coverages of 1.0 × 10^{-8}, 1.3 × 10^{-8}, and 0.3 × 10^{-8} mol cm^{-2}, respectively (polymeric layers are indicated by arrows) [33]. Figure reproduced with kind permission; © 2012 American Chemical Society

Utilizing the analogous 4,4'-divinyl-substituted ligand in the coordination sphere of Ru^{II} (**17**), a very low polymerization efficiency was observed. This drastic decrease compared to the high efficiency of **4a** was attributed to steric rather than electronic effects: Due to the proximity of the two vinyl groups, chain propagation was inhibited; presumably, intramolecular radical–radical coupling represented a much faster reaction pathway.

The Ru^{II}-polypyridyl complexes **8** with vacant binding sites were electropolymerized by Nalos and Brewer onto Pt or GC disk electrodes [34]. The resultant materials were proposed to be of interest due to their ability to further modify the surface by the coordination of metal ions; therefore, such electrodes might be of importance in view of potential applications in the field of electro-analysis. This field of research has been reviewed by Ruiz, García et al.; these authors summarized the various polymer-coated electrodes which have, *inter alia*, been applied for the sensing of heavy-metal cations [35]. However, no such study based on poly-**8a**, poly-**8b** or related polymers is yet available.

Abruña et al. investigated the electropolymerization of homoleptic vinyl-functionalized tpy complexes; besides the Ru^{II}-containing complexes **10** and **11** (Fig. 3.3), the corresponding complexes based on a range of first-row transition-metal ions as well as Os^{II} were also utilized in this context [36–38]. Presumably for steric reasons, the electropolymerization of **11** proceeded *ca.* 150 times slower than for **10** [36]. The polymerization progress was studied by EQCM and admittance measurements of the quartz crystal resonator [38]. It was concluded from changes in the frequency that the polymerization was initiated by a ligand-reduction step (except for the Ni^{II}-containing complexes, where a metal-centered reduction appeared to be involved). The increase in the resistance parameter of the electrical equivalent circuit of the resonator with potential cycling indicated the growth of polymer films on the electrode surface. The redox processes of the electropolymerized films were accompanied with diffusion of charge-balancing anions as well as solvent molecules into the porous film.

In general, the direct reduction of CO_2 at electrode surfaces requires very large overpotentials, since the formation of high-energy intermediates, e.g., the radical-anion $CO_2^{\cdot -}$, is involved [39]. As summarized by Dubois, electro-catalysts might be employed in order to bypass this issue by decreasing the activation-energy barrier. Against this background, various electropolymerized metallopolymer films, including poly-**10**, poly-**11**, and poly-**19**, were studied with respect to CO_2 reduction [37]. It was shown that the catalytic activity was controlled by the nature of the metal ion: Complexes $[M(tpy)_2]^{n+}$ featuring metal-based redox processes at negative potentials (e.g., Cr^{III} or Co^{II}) were found to be far more active than those where the relevant redox processes were ligand-based only (e.g., Fe^{II}, Ru^{II}, Os^{II}); moreover, a strong decrease in catalytic activity was observed for the divalent group-8 metal ions in the order $Fe^{II} >> Ru^{II} > Os^{II}$. Therefore, the ruthenopolymers where not considered further as potential electrocatalyst for CO_2 reduction [40].

The oligo-nuclear complexes **22** and **23** (Fig. 3.5) were transferred into thin metallopolymer films via reductive electropolymerization [41]. Remarkably, their mixed-valence nature and photochemical properties were retained upon film formation. However, dissociation was observed when the coated electrodes were emerged and scanned in coordinating solvents. Nonetheless, these examples of electropolymerization indicate how discrete molecular assemblies of complex photophysical behavior might be incorporated into structured polymer films with highly controllable chemical and physical properties. This finding is of particular relevance regarding opto-electronic applications of such films, e.g., in electrochromic devices (vide infra).

The dinuclear complexes **13** and **14** were electropolymerized successfully by Abruña et al. as well as Zhong et al. (Fig. 3.3) [42–44]. From these, **13a** was electropolymerized by scanning the potential beyond the two-electron reduction wave at −1.5 V (corresponding to the reversible reduction of the bridging ligand); at less negative potentials, no polymerization could be observed [42]. A surface

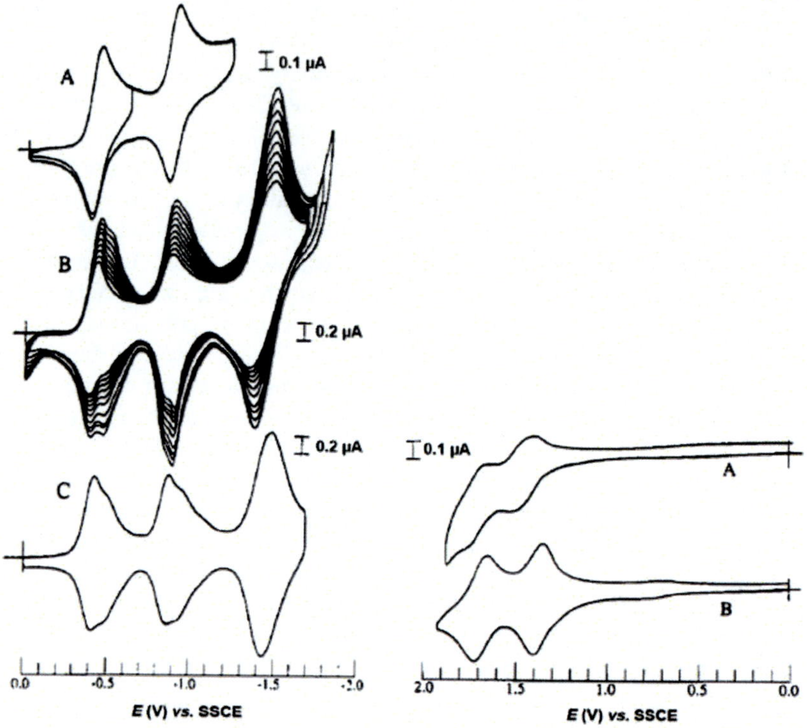

Fig. 3.10 Left: CVs at a Pt disk electrode in contact with a solution of **13a** scanned over two regions (0.0 to –0.6 V as well as 0.0 to −1.2 V, panel A), scanned from 0.0 to −1.7 V (panel B) and CV obtained at a Pt disk electrode modified poly-**13a** (panel C). Right: CVs at a Pt disk electrode of **13a** in solution (panel A) and of poly-**13a** (panel B). For all CVs: 0.1 M (n-Bu_4N)PF_6 in MeCN [42]. Figure reproduced with kind permission; © 1993 American Chemical Society

coverage of $\Gamma = 9.7 \times 10^{-10}$ mol cm^{-2} was calculated on the basis of CV measurements in the positive region (Fig. 3.10).

Various types of ruthenopolymers, mainly based on Ru^{II}–tpy *bis*-complexes, have been studied regarding their applicability, as active materials, in electrochromic devices [45]. As already outlined in Chap. 2, the phenomenon of electrochromism arises from reversible changes of the UV/vis/NIR absorption bands upon switching between at least two redox states by applying an external electric field (NIR: near infrared) [46–48]. These so-called electrochromic materials are of interest for a range of potential applications: Data storage, optical memory, smart windows, color displays, etc. (noteworthy, not only metallopolymers, but also π-conjugated polymers or metal oxides have attracted considerable interest in this respect) [49]. In the case of metallopolymers, the incorporated metal centers often exhibit, depending on the oxidation state, well-defined redox processes and characteristic absorption spectra; these properties make metallopolymers highly attractive materials for this particular application. Thin solid films featuring electrochromic behavior might be obtained via different techniques: Layer-by-layer assembly [50, 51], processing from solution (e.g., spin-coating, drop-casting or inkjet-printing) [52, 53] or electropolymerization [43, 44, 54–57].

In contrast to Higuchi's work, which is focused on the fabrication of electrochromically active films by processing pre-formed metallo-supramolecular polymers from solution onto electrode substrates (Chap. 2) [52], Zhong et al. followed an electropolymerization approach basically for the same purpose. The dinuclear cyclometallated complexes **13b** [43] and **14** [44] represent, to the best of our knowledge, the first examples in this respect. The stable polymer films exhibited similar mixed-valence characteristics as their small-molecule analogs. Multi-step electrochromism in the NIR-regime was observed for the thin poly-**13b** films on ITO-coated glass slides (dimension of 20 × 8 mm, $\Gamma = 5.5 \times 10^{-9}$ mol cm^{-2}). Such a behavior is of relevance when aiming, e.g., for applications in fiber-optic communication [58]. The CV curves of poly-**13b**, as depicted in Fig. 3.11a, featured two well-separated $Ru^{II/III}$ redox waves at +0.15 and +0.56 V versus Ag/AgCl; the stability of the electro-generated films was concluded from scanning over 1000 cycles, where the peak current decreased only to *ca.* 76% of its initial value. The electrochromic properties of poly-**13b** were studied by spectro-electrochemical measurements. When the applied potential was increased gradually to +0.4 V, a mixed-valence state was generated, which showed pronounced NIR intervalence charge-transfer (IVCT) bands at *ca.* 1165 nm (Fig. 3.11c), accompanied by a color change from deep blue to pink (Fig. 3.11b). Increasing the potential, further up to +1.0 V, led to a decrease of the IVCT band (Fig. 3.11d); thereby, the film experienced a second color change from pink to green (Fig. 3.11b). The entire tricolor-change process from blue via pink to green was highly reversible. Moreover, the electrochromic behavior of poly-**13b** was investigated by double-potential step chronoamperometry where for the first one-electron oxidation step a contrast ratio (ΔT%) of 40% was obtained at 1165 nm (Fig. 3.11e). The coloration efficiency (CE) at this wavelength was 250 cm^2 C^{-1}, as calculated according to Eq. 3.2 [OD denotes the optical density, Q_d is the injected/ejected charge density (C cm^{-2}), and T_b and T_c describe the transmittance in the bleached and colored states at the indicated wavelength].

$$CE(\lambda) = \Delta OD/O_d \text{ with } \Delta OD = \log[T_b/T_c] \quad (3.2)$$

A very short response time for the contrast ratio to reach over 90% of its maximum was observed for both the oxidation process (6 s) and the reverse reduction process (5 s, Fig. 3.11f). For the second one-electron oxidation step, a comparable contrast ratio was reached (ΔT% = 37%) and the response times for the oxidation and reverse reduction process were 4 and 8 s, respectively. Remarkably, a pronounced optical memory effect was found (i.e., the ability to retain the transmittance after the applied potential was turned off): The pink color of poly-**13b** films, after the first one-electron oxidation, remained for several days—thus, the

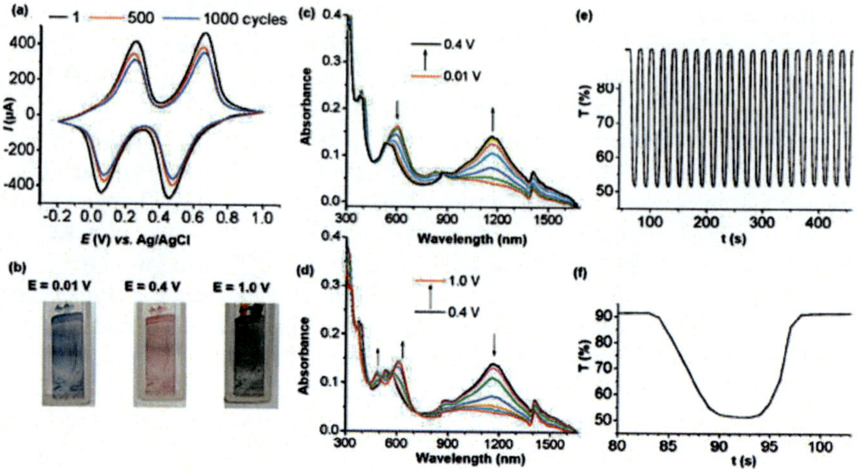

Fig. 3.11 **a** CV curves of poly-**13b** film on ITO-coated glass after 1, 500, and 1000 scans at 100 mV s^{-1}. **b** Representation of poly-**13b** films exhibiting different colors at various applied potentials versus Ag/AgCl. **c** and **d** Changes in the UV/vis/NIR absorption spectra of poly-**13b** upon stepwise applying potential from +0.01 to + 0.4 V (**c**) and from +0.4 to +1.0 V (**d**) versus Ag/AgCl, respectively. **e** and **f** Changes in the transmittance at 1165 nm as a function of time during the double-potential step chronoamperometry measurements between –0.2 V and +0.4 V [43]. Figure reproduced with kind permission; © 2011 American Chemical Society

material exhibited a bistable character (however, the memory time for the second step, only 50 min, was significantly shorter). Such values become even more meaningful when compared to other materials, such as the linear metallopolymers, as reported by Higuchi, Kurth et al., which possessed shorter memory times of up to 15 min [59].

In a similar fashion, poly-**14** also showed an electrochromic behavior; due to the increased π-conjugation in the bridging ligand, the $Ru^{II/III}$ redox potentials (+0.47 and +0.67 V *vs.* Ag/AgCl) as well as the absorption maximum of the IVCT band (λ = 1900 nm) were shifted compared to poly-**13b** [44]. As a result from the altered opto-electronic properties, also the reversible tricolor-change process gave different colors upon the two-step oxidation/reduction (i.e., blue ↔ brown ↔ orange, Fig. 3.12a); the best contrast ratio for the first step was 35%, the corresponding response times were remarkably fast with 6 and 2 s for the oxidation and reverse reduction process, respectively (similar values were also obtained for the second step). The CE at 2050 nm was calculated to be 220 $cm^2\ C^{-1}$. The observed memory times for the first and second step were *ca.* 100 and 10 min, respectively (Fig. 3.12b).

In the following, Zhong et al. reported on a variety of further vinyl-functionalized mono- as well as dinuclear Ru^{II} complexes, which were utilized in the reductive electropolymerization. In all cases, stable metallopolymer films, coated onto electrodes, were obtained and a rich electrochromic behavior was

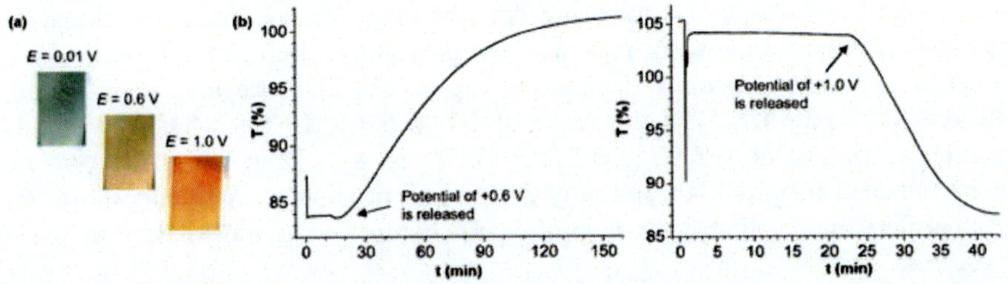

Fig. 3.12 **a** Representation of poly-**14** films exhibiting different colors at various applied potentials *vs.* Ag/AgCl. **b** Optical memory effect of poly-**14** on ITO-coated glass after the potential at +0.6 V (first step) or +1.0 V versus Ag/AgCl (second step) was released [44]. Figure reproduced with kind permission; © 2012 American Chemical Society

observed. For instance, the dinuclear 1,4-(pyridin-2-yl)benzene-bridged complex **23** was electropolymerized and investigated further (Fig. 3.13a) [60]. In the mixed-valent state, an IVCT transition at 1300 nm was dominating the absorption spectrum. At this wavelength, a high optical contrast ratio of 41% with a CE of 200 $cm^2\ C^{-1}$ was achieved. A dependency of the electrochromic behavior on the surface coverage was observed—the highest contrast ratio was obtained for a film with $\Gamma = 6.0 \times 10^{-9}$ mol cm^{-2} (Fig. 3.13b).

In the previous example, the electropolymerization gave (cross-linked) type-III materials, according to Wolf's classification (see Fig. 3.1); whereas, the Ru^{II} complexes **24** (Fig. 3.14a) were used to obtain linear type-II polymers [61]. All four homopolymers as well as a copolymer of **24b** and **24c** were prepared on GC as well as ITO-coated glass electrodes and showed electrochromism in the visible region

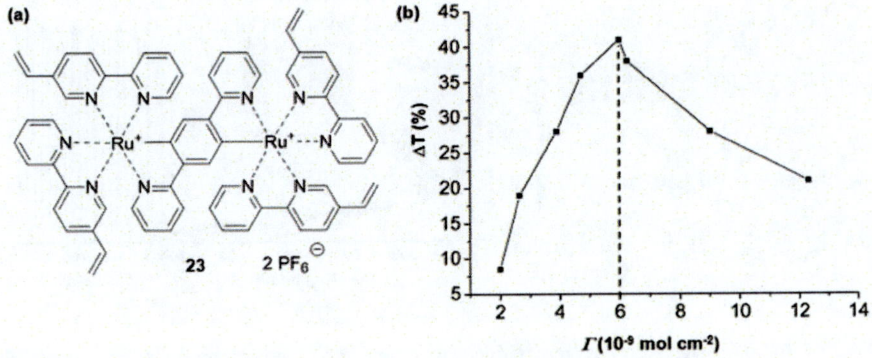

Fig. 3.13 **a** Schematic representation of the dinuclear Ru^{II} complex **23**. **b** Plot of ΔT% at 1300 nm versus Γ for poly-**23** during the electrochromic switching between −0.1 and +0.4 V versus Ag/AgCl [60]. Figure reproduced with kind permission; © 2014 American Chemical Society

with good reversibility and moderate contrast ratios. Representatively, the spectroelectrochemical analysis of poly-**24b** and poly-**24c** as well as the corresponding changes in transmittance are depicted in Fig. 3.14b. The measured apparent diffusion coefficients (D_{app}) in the range of 1.1 to 6.9 × 10^{-10} cm^2 s^{-1} were comparable to that of poly-**2** (D_{app} = 2.2 × 10^{-10} cm^2 s^{-1}) [21]; higher D_{app} values were obtained for poly-**24b** and poly-**24d** due to the higher charge density at the Ru^{II} centers that facilitated the flow of charge-compensating ions during the redox processes.

The electrochromic behavior of electropolymerized films might be utilized further to construct molecular devices, e.g., for data storage applications. In 2016, Zhong et al. published an Account on this special topic focusing on the preparation of Ru^{II}-containing electropolymerized films featuring multistate NIR electrochromism with good contrast ratios and long optical retention times [62].

In this context, the diarylamino-functionalized Ru^{II} complex **25** was reductively electropolymerized (Fig. 3.15) [63]. The obtained films featured two well-defined redox processes at +0.32 and +0.68 V versus Ag/AgCl, i.e., three well-separated redox states could be addressed. These singly- and doubly-oxidized states revealed

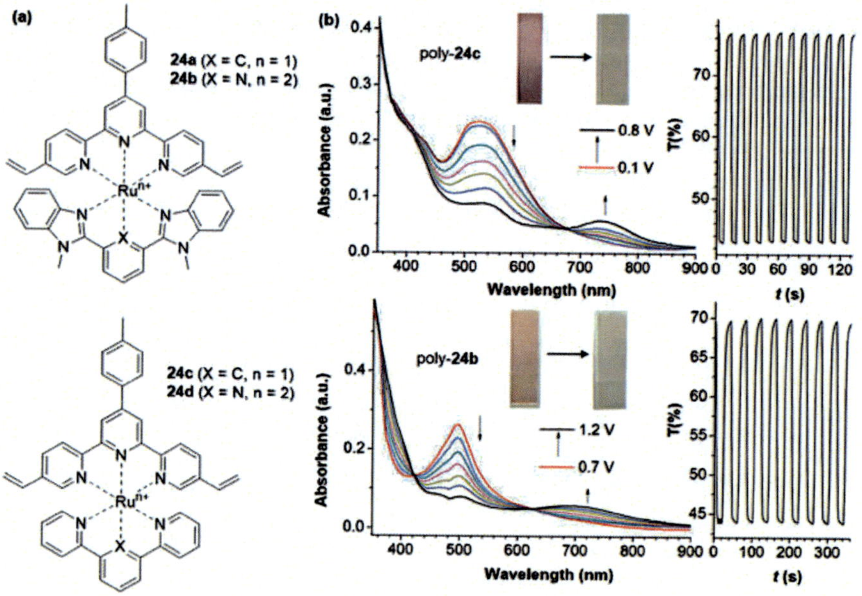

Fig. 3.14 a Schematic representation of the Ru^{II} complexes **24**. **b** vis/NIR Spectro-electrochemical analysis of poly-**24c** (top) and poly-**24b** (bottom); the insets show the color changes of the films. The transmittance changes were monitored at 520 nm between +0.1 V and +0.8 V versus Ag/AgCl for poly-**24c** and at 500 nm between +0.7 V and +1.2 V versus Ag/AgCl for poly-**24b** [61]. Figure reproduced with kind permission; © 2013 The Royal Chemical Society

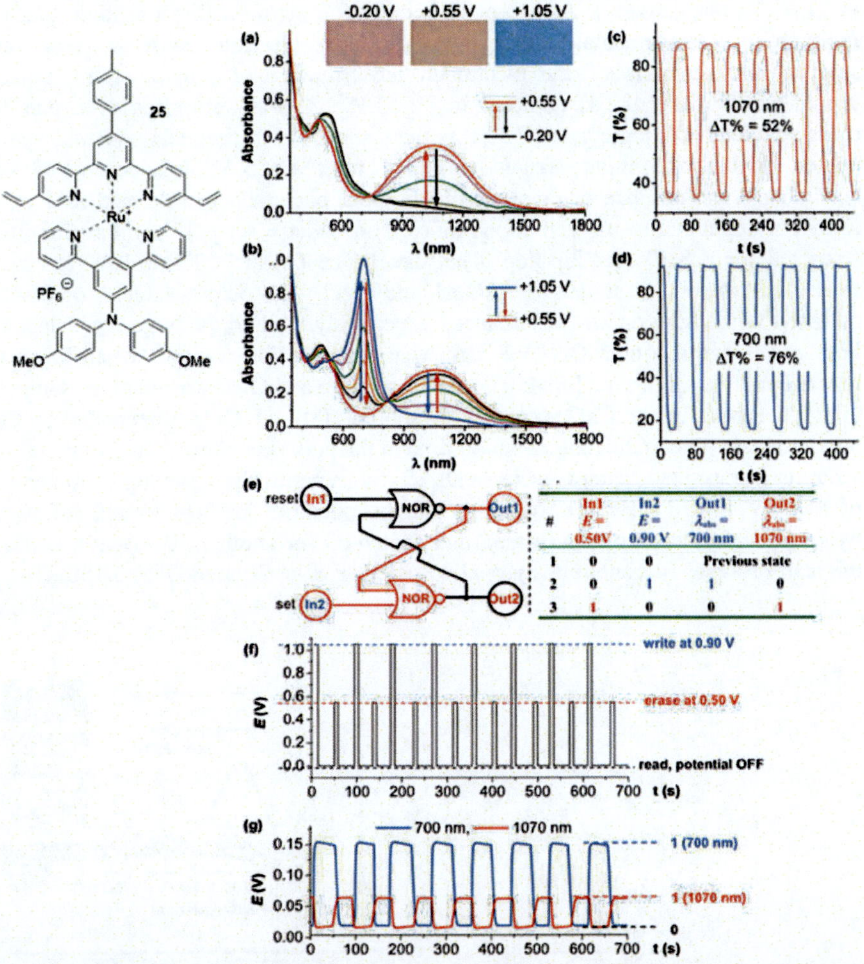

Fig. 3.15 Schematic representation of the Ru^{II} complex **25**. **a** and **b** Changes in the vis/NIR absorption behavior of poly-**25** upon scanning between −0.2 and +0.55 V as well as between +0.55 and 1.05 V; the inset shows the different film colors at the respective potentials. **c** and **d** Transmittance changes of poly-**25** during electrochromic switching between in the same potential windows as in (**a** and **b**). **e** and **f** Logic circuit and corresponding truth table of the set/reset flip-flop logic. **f** and **g** Switching of the input potentials (vs. Ag/AgCl) and output absorbance at 700 and 1020 nm of poly-**25** [63]. Figure reproduced with kind permission; © 2015 The Royal Chemical Society

intense absorption bands at 1070 and 700 nm, respectively (Fig. 3.15a, b). Investigation of the highly reversible NIR electrochromism gave the best ΔT values of 52% and 76% at those absorption bands (three-color scheme: purple ↔ brown ↔ sky blue; Fig. 3.15c, d). Remarkably long retention times were observed for all three stages (i.e., infinity at −0.20 V, 4 h at +0.55 V, and 30 min at +1.05 V, respectively). As a proof-of-concept, a set/reset flip-flop logic gate—representing a fundamental building block in digital electronic systems—was constructed on the basis of the electrochromic behavior of poly-**25**. This binary sequential logic circuit, as depicted in Fig. 3.15e, f, was comprised of a pair of cross-coupled NOR logic gates. Thereby, the applied electrochemical potentials were treated as the two input signals (In1 and reset: +0.55 V; In2, set: +1.05 V); accordingly, the absence or presence of the input potentials was defined as 1 or 0 (only one input can be applied at one time). The "reset = set = 1" combination was restricted for a Set/Reset flip-flop. The absorbance (*A*) at 700 and 1070 nm was treated as two output signals (i.e., Out1 and Out2); the output value 1 or 0 was assigned to the high or low absorbance, respectively. Due to the huge ΔA values at both wavelengths, a high ON/OFF ratio was ensured. For example, when In1 was low and In2 was high (i.e., Input: 01), Out1 was high and Out2 was low (i.e., Output: 10). When both In1 and In2 were low (i.e., Input: 00), the Output depended on the previous state (due to the long retention time of the polymeric film). Thus, continuous write-read-erase-read cycles were realized by repetitively applying a potential of +1.05 V for 10 s (write), switching off the potential for 30 s (read), followed by +0.55 V for another 10 s (erase), and switch-off for another 30 s (read). During the read process, no continuously applied potential was required (Fig. 3.15f, g).

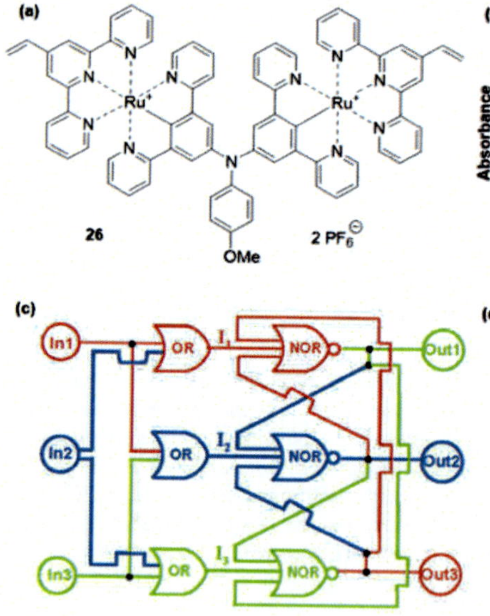

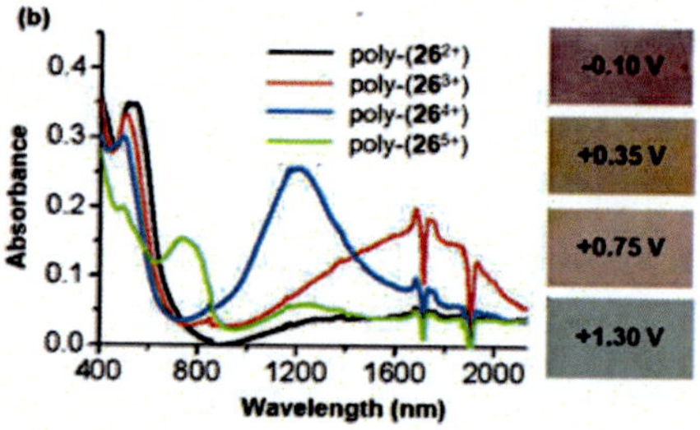

(d)

#	In1 E = 0.35 V	In2 E = 0.75 V	In3 E = 1.30 V	Out1 λ_{abs} = 750 nm	Out2 λ_{abs} = 1170 nm	Out3 λ_{abs} = 1680 nm
			flip-flap-flop			
1	0	0	0	Previous state		
2	0	0	1	1	0	0
3	0	1	0	0	1	0
4	1	0	0	0	0	1
			ternary memory			
1	0	0	0	Previous state		
2	0	0	1	1	0	-1
3	0	1	0	0	1	0
4	1	0	0	-1	-1	1

Fig. 3.16 **a** Schematic representation of the dinuclear Ru^{II} complex **26**. **b** UV/vis/NIR Absorption spectra and film pictures of poly-**26** at different redox states. **c** Schematic representation of the logic circuit of flip-flap-flop. **d** Truth table of the flip-flap-flop or ternary memory device, based on poly-**26** [64]. Figure reproduced with kind permission; © 2015 Wiley–VCH

In the extension of this research, the dinuclear Ru^{II} complex **26** was utilized as monomer for the reductive electropolymerization (Fig. 3.16a) [64]. The ruthenopolymer films displayed multistate NIR electrochromism with low operational potentials, good contrast ratios, and long retention times (Fig. 3.16b). Based on these properties, three flip-flop memory devices could be constructed by using any two of the in-total three redox processes of poly-**26** (in all three cases the ON/OFF ratio was *ca.* 6). A more advanced flip-flap-flop and ternary memory device was realized making use of all three redox processes (Fig. 3.16c, d): The three electrochemical inputs (In1: 0.35 V, In2: 0.75 V, In3: 1.30 V) were first converted into three new input sequences (I1, I2, and I3) with three OR gates. Subsequently, three cross-coupled NOR gates provided the memory and three optical outputs (Out1 at 750 nm, Out2 at 1170 nm, and Out3 at 1680 nm). However, the entire circuit represented an equivalent circuit only, since the three OR gates, converting the input states, remained formal constructs and were not real logic gates implemented in the molecular system.

The ruthenopolymer **9b** (Fig. 3.2), synthesized by Meyer et al. from poly[(4-aminomethyl)styrene] and a COOH-functionalized Ru^{II} complex via conventional amidation, was utilized further, as a coating material for various electrode materials [65]. As shown in Chap. 2, the grafting of Ru^{II} complexes to a preformed polymer via appropriate functional groups represents one versatile strategy to prepare well-defined ruthenopolymers. AFM and SEM imaging were used to analyze the morphology of the metallopolymer films deposited on a HOPG electrode (HOPG: highly-oriented pyrolytic graphite, Fig. 3.17). Compared to electropolymerized films obtained from analogous small-molecule Ru^{II} complexes (such as **1**, **2** or **9a**), poly-**9b** exhibited a significantly higher surface roughness (**9a**: R_q = 15 nm vs. **9b**: R_q = 200 nm) and a microporous structure.

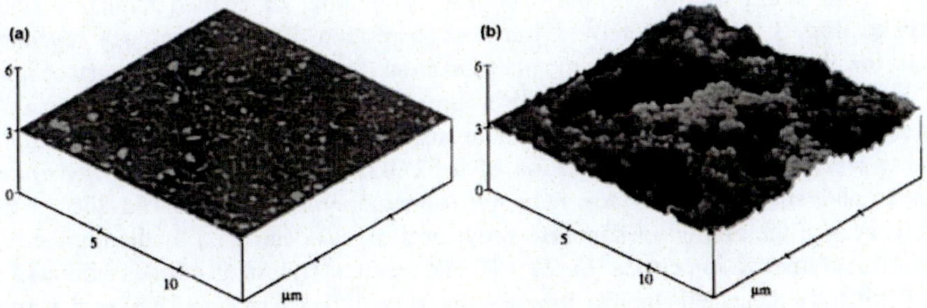

Fig. 3.17 3D-AFM images (12.5 × 12.5 µm, tapping mode) of poly-**9a** (left) and poly-**9b** (right). Both films were prepared by reductive electropolymerization onto HOPG with a similar surface coverage ($\Gamma \approx 5 \times 10^{-9}$ mol cm^{-2}) [65]. Figure reproduced with kind permission; © 1998 American Chemical Society

In general, the long-term performance of dye-sensitized solar cells or photo-electrochemical cells (DSSCs and DSPEC, respectively) is, at least to some extent, reflected by the stability of the binding of the chromophores (or the chromophore-catalyst assemblies) to the metal-oxide surface, such as TiO_2 nanoparticles, over a wide pH regime [66]. For example, the binding of carboxylate moieties to metal-oxide surfaces is robust in non-aqueous medium, but highly labile in the presence of water (basically, the same holds true for phosphonate-binding—hydrolysis from the surface will occur at pH values of <5). To bypass this, a protecting overlayer can be coated around the dye-loaded nanoparticles. This can simply be achieved by binding a hydrophobic polymer to the substrate, which prevents water molecules from reaching the nanoparticles (e.g., PMMA was coated onto TiO_2 nanoparticles loaded with a Ru^{II} dye simply by applying the dip-coating technique) [67, 68]. The atomic-layer deposition (ALD) technique represents one further method to protect the dye-loaded electrodes with a "physical barrier", i.e., the active layer is covered with a protective layer (e.g., of Al_2O_3 or TiO_2) [69–71]. Alternatively, reductive or oxidative electropolymerization, offering a chemical stabilization, might be employed for the same purpose [72–75]. Meyer et al. followed the latter approach by grafting an overlayer of **2** onto a monolayer of surface-anchored $[Ru(bpy)_3]^{2+}$ complexes (these were bound to nanocrystalline TiO_2 via their carboxylate moieties) [72]. A significant increase in thermal stability for the surface-bound Ru^{II} complex was found over a wide pH regime (from highly acidic to strongly basic) without bleaching of chromophores over a three week period (under the same conditions the non-protected surface-bound complexes showed quantitative desorption in less than one hour). The photostability of nanocrystalline TiO_2, loaded with an analogous phosphonate-equipped Ru^{II}–polypyridyl complex, could be raised by a factor of 30 when a protective polymer layer was formed, i.e., when the surface-bound dye molecules were copolymerized with $[Fe(vtpy)_2]^{2+}$ (vtpy: 4'-vinyl-tpy; Fig. 3.18a) [73]. Thereby, the outer-to-inner Fe^{II}/Ru^{II} ratio could simply be controlled via the number of applied reductive scan cycles (Fig. 3.18b, c). Moreover, transient-absorption (TA) spectroscopy revealed that the electron injection from the photo-excited Ru^{II} complexes on the surface into the semi-conductor was followed by a directional electron transfer from the Fe^{II} centers within the outer layer. Even more stable electrodes for DSPEC applications were fabricated by functionalization of a FTO|TiO_2 electrode with a vinylsilane layer and subsequent surface reductive electropolymerization of **4a** [76]. The thickness of the grafted mesoporous polymer film was controlled by the number of electropolymerization cycles (Fig. 3.19). The resultant hybrid electrodes exhibited a significantly increased stability over a wide pH regime (from 1 to 12.5) in the dark and, even more important, under illumination.

A remarkable electrochemically controlled radical polymerization was reported by Meyer et al. [74]. The polymerization, following a radical chain-growth mechanism, occurred within the cavities of mesoporous TiO_2 nanoparticles. First, the Ru^{II} complex **27** (Fig. 3.20a) was adsorbed onto the nanoparticles, giving a maximum surface coverage of *ca.* 4.8×10^{-8} mol cm^{-2}. Subsequently, the electropolymerization of **1a** was achieved by scanning from 0 to −2.0 V at a scan rate

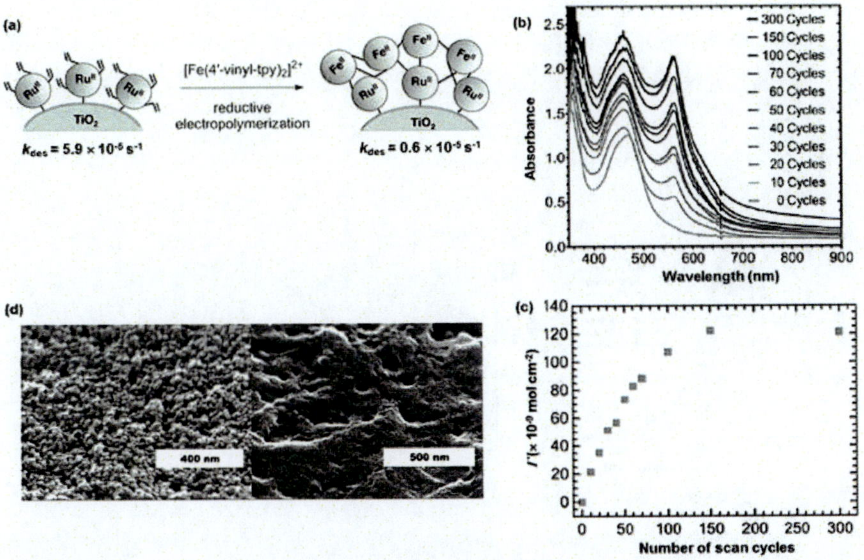

Fig. 3.18 Schematic representation of the stabilization of dye-loaded TiO_2 nanoparticles by an electropolymerized overlayer. **b** Change in the UV/vis absorption behavior with increasing number of electrochemical scan cycles. **c** Increase in surface coverage with increasing number of scan cycles. **d** SEM images of the overlayer-coated TiO_2 nanoparticles after 50 (left) and 300 (right) scan cycles [73]. Figure reproduced with kind permission; © 2013 American Chemical Society

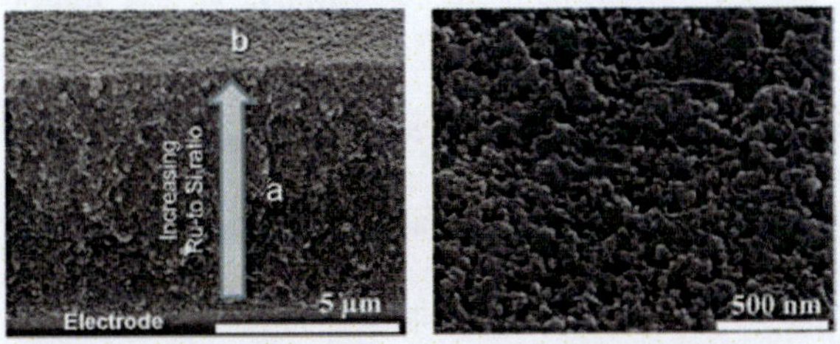

Fig. 3.19 SEM images of the FTO|TiO_2|poly-**4a** electrode (20 electropolymerization cycles were applied). Left: Profile image in which (a) and (b) denote the bulk and the surface of the mesoporous polymer film, respectively [76]. Right: Zoomed-in view of the surface of the mesoporous film. Figure reproduced with kind permission; © 2019 American Chemical Society

of 100 mV s^{-1}. On-line UV/vis absorption spectroscopy (Fig. 3.20b) as well as XPS measurements on dried samples corroborated the controlled growth of oligomeric strands from the surface-bound initiators according to the proposed mechanism (Fig. 3.20c). In related work, also the electropolymerization of **2** within

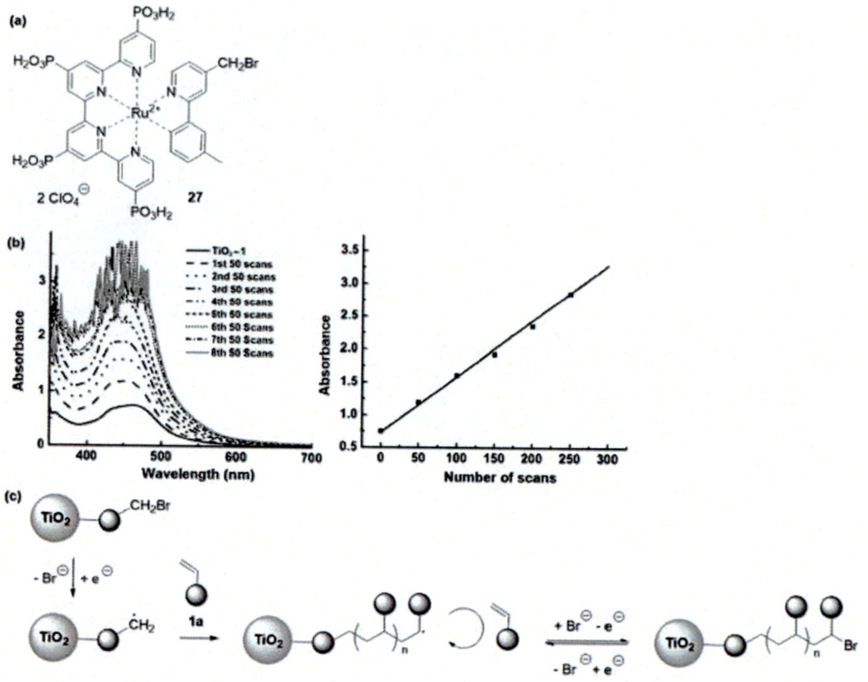

Fig. 3.20 a Schematic representation of the Ru^{II} initiator dye **27**. **b** UV/vis absorption spectra and scan-number dependence of the electropolymerization of **1a** initiated from TiO_2-adsorbed **27**. **c** Schematic representation of the proposed mechanism for the electrochemically controlled radical polymerization [74]. Figure reproduced with kind permission; © 2014 Wiley–VCH

the pores of SiO_2 sol–gel films, deposited on conductive Sn^{IV}-doped ITO-coated glass slides, was investigated [77].

Besides the protective effect, appropriate electropolymerized overlayers on Ru^{II}-loaded metal-oxide electrodes can be utilized to fabricate DSPECs for the light-driven water splitting [78]. One assembly, as published by Meyer et al., comprised of core–shell SnO_2/TiO_2 nanoparticles with phosphonate-bound Ru^{II} complexes which, in turn, were copolymerized with the Ru^{II}-bda complex **28** via their vinyl groups (bda: 2,2'-bipyridine-6,6'-dicarboxylate) [79]. Upon illumination with simulated sunlight (100 mW cm^{-2}) in a neutral phosphate buffer and an applied bias of 0.4 V versus SCE, oxidative water splitting occurred at the Pt cathode (Fig. 3.21): Over a photolysis period of 5 min, the photoanode generated O_2 with a faradaic efficiency of 22%. The instability of the device was attributed to the lability of the surface-bound chromophore in its oxidized state leading to a gradual decrease in photocurrent. High photocurrent densities of *ca.* 1.4 mA cm^{-2} and *ca.* 0.4 mA cm^{-2} were observed when utilizing nanoporous TiO_2 and Fe_2O_3 nanorods, respectively, as the anode materials [80].

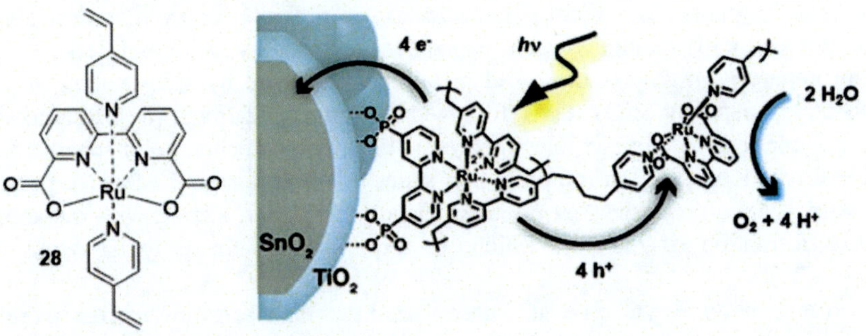

Fig. 3.21 Schematic representation of the electropolymerizable Ru^{II}-bda photocatalyst **28** and of the working scheme of the fabricated DSPEC [79]. Figure reproduced with kind permission; © 2014 American Chemical Society

Ruthenoporphyrins represent commonly used catalysts, *e.g.*, for the epoxidation of olefins or oxidation of organic sulfides by 1,3-disubstituted pyridine *N*-oxides [81]. As shown in Chap. 1, Ru^{II}(CO)(porphyrin) has been immobilized by using FRP, thus forming a heterogeneous catalyst [82]. The electropolymerization of the Fe^{IV}-porphyrin derivative **29**, equipped with four vinyl-functionalized Ru^{II}-complexes in the periphery, represents an alternative to the classical immobilization-by polymerization route (Chap. 2) [83]. The electropolymerization of **29** onto a GC disk electrode gave a robust electroactive polymer film. In the presence of aqueous acid, poly-**29** electrocatalytically epoxidized olefins at positive potentials from the formal Fe^{IV} state (however, under the acidic conditions the initially formed epoxides were directly transformed further into the corresponding *trans*-1,2-diols, Fig. 3.22).

The thermal polymerization of vinylic compounds onto electrode materials represents a versatile alternative to electropolymerization methods, in particular when dealing with acrylates [45]. If the acrylate group is decoupled from the

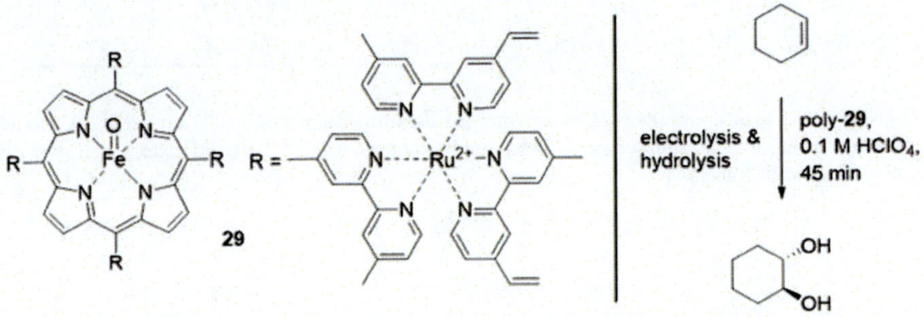

Fig. 3.22 Schematic representation of the electrocatalytic olefin hydroxylation

aromatic fragment, the electropolymerization is difficult (or even impossible) to achieve. One early example in this respect was reported by Elliot and Redepenning, who prepared a highly cross-linked metallopolymer film by spin-coating **30** onto optically transparent SnO_2 slides, followed by thermal treatment [84]. Remarkably, the polymer exhibited a rich electrochromic behavior—a seven-color scheme could be realized depending on the oxidation states of the robust and adhesive polymer network. Mayo proposed that the thermal polymerization started with the spontaneous formation of radicals, which, in turn, initiated the polymerization [85] (Fig. 3.23).

More recently, Reynolds et al. demonstrated that the electropolymerization might prevail if the acrylate moieties are separated from the Ru^{II} complex by very short spacers. Accordingly, complexes **31c/d** were electropolymerized reductively; whereas, derivatives **31a/b** required thermal treatment to obtain the corresponding metallopolymer films [86]. Remarkably, these films exhibited electrochromism (EC) as well as electroluminescence (EL), thus their incorporation into single-layer dual functional devices was enabled. Depending on the structure of the Ru^{II} complex, the electrochromic devices gave different color schemes, such as red ↔ yellow ↔ green in the case of poly-**31a** (Fig. 3.24a). When applied as emitters in light-emitting electrochemical cells (LECs), intense red–orange to deep red electroluminescence with luminance values up to 135 cd m^{-2} was detected (Fig. 3.24b). Additionally, a difunctional device was fabricated and exhibited both electrochromism and electroluminescence from a single metallopolymer layer, poly-**31a**, under the respective performance conditions.

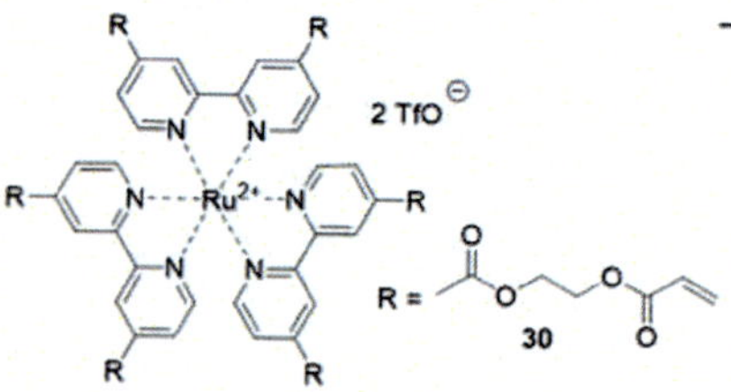

Oxidation state	Potential range (V vs. SEC electrode)	Solution color
+II	+1.53 to −0.66	Organge
+I	−0.66 to −0.75	Purple
0	−0.75 to −0.91	Blue
−I	−0.91 to −1.37	Green-blue
−II	−1.37 to −1.57	Brown
−III	−1.57 to −1.82	Rust
−IV	−1.82 to −2.50	Cherry red

Fig. 3.23 Schematic representation of the acrylate-functionalized Ru^{II} complex **30**, which exhibited a multicolor electrochromic behavior in solution and in the solid state, i.e., thermally polymerized thin film [84]

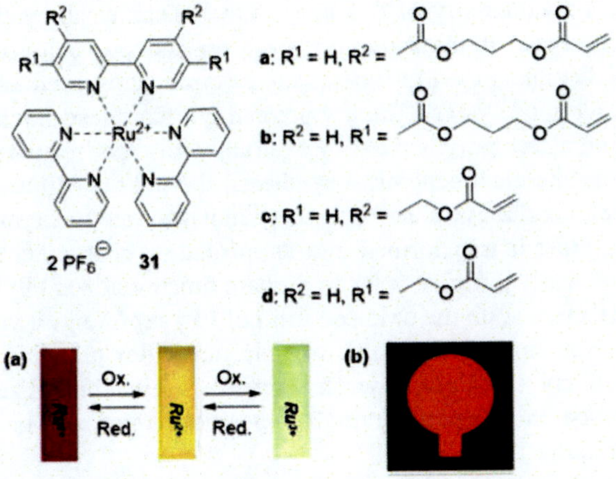

Fig. 3.24 Schematic representation of the acrylate-functionalized Ru^{II} complexes **31**, which showed electrochromism as well as electroluminescence. Representatively, the EC color scheme (**a**) and the EL (**b**) of the devices containing poly-**31a**, as active layer, are depicted [86]. Figure reproduced with kind permission; © 2012 American Chemical Society

3.4 Oxidative Electropolymerization of Pyrrole-Containing Monomers

Polypyrrole (PPy) represents one of the most studied conducting polymers, which can easily be prepared via an oxidative electropolymerization pathway (though the exact mechanism is still in a controversy) [87]. Compared to other conducting polymers some advantages can be listed: Environmental stability, good redox properties, and the ability to obtain high electrical conductivities [88]. However, the latter strongly depends on the way of preparation, i.e., on the particular electropolymerization conditions. Moreover, electrochemically or photophysically active substituents, such as transition metal complexes, can be introduced in the 1-position enabling a modification of the parent PPy [89].

The chemical oxidative polymerization of pyrrole (and its derivatives) can be performed in a wide range of solvents utilizing transition-metal salts (e.g., $FeCl_3$), halogens or organic molecules, as oxidant [88]. Though the layer-by-layer deposition of the thus formed PPy (and of some related conductive polymers) offers the advantage to obtain more regular polymer structures, which exhibit higher degrees-of-crystallinity [90], the electrochemical polymerization of pyrrole has emerged as an alternative strategy. The benefits of the latter are basically those that have been summarized in the introduction of this Chapter (vide supra). The first report on the electrochemical synthesis of PPy was already published in 1968 by Dall'Ollio et al.; these authors prepared a PPy film by electrolyzing a pyrrole solution in the presence of H_2SO_4 [91]. The black powder, deposited onto the electrode, had a conductivity of 8 S cm^{-1}. Later, Diaz et al. synthesized PPy by the direct electrolysis of the monomer from an aqueous solution [92–94]. The thusly formed flexible PPy film possessed a significantly increased conductivity (100 S cm^{-1}). Vernitskaya and Efimov pointed out that these studies represent the start for utilizing electropolymerized PPy (along with other related polymers) [88]. As a result from the electrochemical synthesis, the polymer films are obtained in their doped, i.e., conductive state; if further reductive electrochemical treatment is applied, the polymer is transformed into its insulating, charge-neutral form. There are two fundamentally different ways to prepare functionalized PPy films [89]. The anion-exchange process on the oxidized form of PPy represents a versatile strategy, which is, however, limited to anionic organic molecules or transition-metal complexes. Alternatively, substituted pyrrole derivatives might be utilized (the presence of substituents on the *N*-atom or in 3-/4-positions does not interfere with the electropolymerization).

The first anodic electropolymerization of a pyrrole-substituted Ru^{II} complex (**32a**) was reported by Moutet et al. in 1984 [95], followed by the utilization of **32b** and **33–34** for the same purpose soon thereafter (Fig. 3.25) [95–97]. Concerning the stability of the electropolymerized films, the same holds true as for those films obtained from the reductive electropolymerization of vinyl-derivatives: At least two pyrrole-units per complex as well as a *tris*-bidentate (or *bis*-tridentate) coordination sphere around the Ru^{II} center are required; very few examples are known where stable films were obtained from complexes only with monodentate ligands [98]. However, the first electrochemically generated PPy film did not exhibit the typical electroactivity and conductivity—a finding that was mainly attributed to steric constraints arising from the bulky Ru^{II} complexes, as pending groups. The observed redox conductivity was attributed to an electron-hopping processes between metal- or ligand-based redox sites. The photoconductivity could be enhanced when the films were irradiated in the presence of a diazonium salt, as an irreversible quencher [99]. Remarkably, such diazonium quenchers could also be utilized in the photopolymerization of pyrrole-functionalized Ru^{II} complexes in solution [100, 101]. Complex **33a**, which only gave unstable polymer films upon anodic electropolymerization, could be polymerized instead due to pyrrole oxidation by the photogenerated Ru^{III} species (Fig. 3.26); the efficiency of this process was estimated to be *ca.* 64%. Soluble metallopolymers were also obtained with O_2, as the quencher (here, the oxidizing Ru^{III} species was formed by an in-situ electron-transfer quenching of the photo-excited $[Ru(bpy)]^{2+}$ species by oxygen). Laguitton-Pasquier et al. investigated the photophysical properties of soluble polypyrroles with pending Ru^{II}–bpy *tris*-complexes—prepared photo- or electrochemically [102]. In general, the spectral properties and temperature dependency of their excited-state lifetimes were similar to those found for the parent $[Ru(bpy)_3]^{2+}$ complex; however, the emission quantum yields (Φ) and the time-resolved emission decays were strongly affected by the polymer structure: Increasing the cross-linking density of the polymers, for example, led to a pronounced decrease of Φ. For the non-cross-linked polymers, the excited states were easily accessible for external quenchers [e.g., methylviologen (MV^{2+}) or *N*-methyl-phenothiazine (PTZ), as

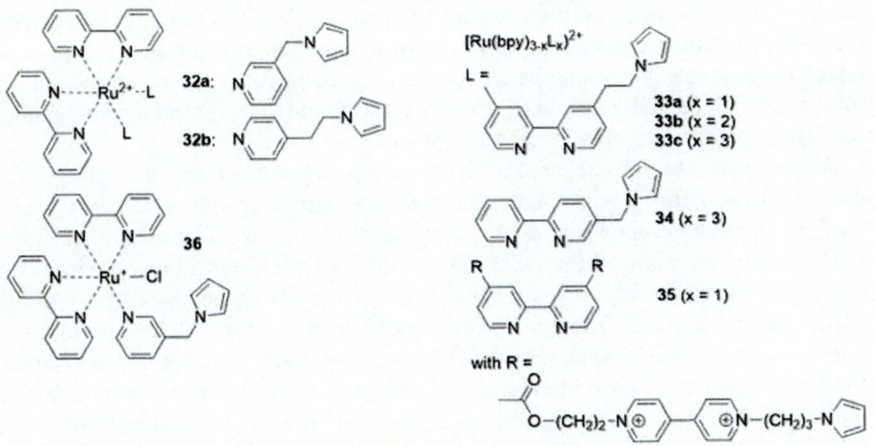

Fig. 3.25 Schematic representation of the pyrrole-functionalized Ru^{II} complexes **33–36** (the counterions are omitted for clarity)

electron acceptor and donor, respectively], thus supporting the model that polymer chains existed in a randomly-coiled structure. When the MV^{2+}-units were attached covalently to the Ru^{II} chromophores, photo-induced intramolecular ElT from the photo-excited Ru^{II} centers to the pending MV^{2+}-groups was evidenced [103].

When copolymerizing Ru^{II} complexes, such as **32a**, with 3-methylthiophene the diffusion coefficient for electron hopping was strongly increased; in particular, the thiophene-rich copolymers exhibited an increased conductivity based on the thiophene-pyrrole linkages within the polymer film; however these films were highly susceptible to oxidative degradation—a behavior that blocked the conjugated pathway required for ElT processes between Ru^{II} sites [104, 105].

Some of the key issues concerning the properties of Ru^{II}-containing PPy films were addressed, e.g., employing the homoleptic complex **33c**, as the monomer [101, 106]. This particular complex was electropolymerized in the presence of various supporting electrolytes having large anions. Utilizing bulky anions, such as 4-alkylbenzenesulfonates, strongly enhanced the permeability of the polymer films; in contrast, more bulky polycyclic aromatic anions had an opposing effect due to

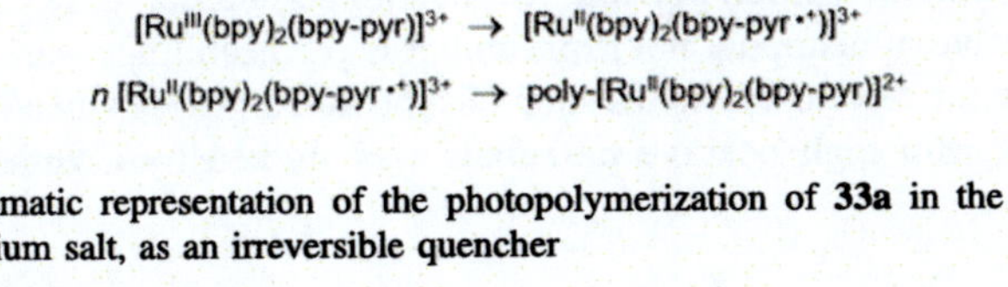

$$[Ru^{II}(bpy)_2(bpy\text{-}pyr)]^{2+*} + ArN_2^+ \rightarrow [Ru^{III}(bpy)_2(bpy\text{-}pyr)]^{3+} + ArN_2^{\bullet}$$

$$ArN_2^{\bullet} \rightarrow Ar^{\bullet} + N_2 \qquad Ar^{\bullet} + SH \rightarrow ArH \ (S = \text{solvent})$$

$$[Ru^{III}(bpy)_2(bpy\text{-}pyr)]^{3+} \rightarrow [Ru^{II}(bpy)_2(bpy\text{-}pyr^{\bullet+})]^{3+}$$

$$n\,[Ru^{II}(bpy)_2(bpy\text{-}pyr^{\bullet+})]^{3+} \rightarrow \text{poly-}[Ru^{II}(bpy)_2(bpy\text{-}pyr)]^{2+}$$

Fig. 3.26 Schematic representation of the photopolymerization of **33a** in the presence of an aromatic diazonium salt, as an irreversible quencher the poor structural order of the polymer film, its permeability was low. Moreover, the polymer-coated electrodes prepared by a potentiostatic process exhibited a better permeability when compared to those obtained from the potentiodynamic method. Finally, a high cross-linking density was found to be essential when aiming for highly porous and, thus, permeable films.

Meyer, Moutet et al. prepared a library of electropolymerized films using three different monomers—**33c**, as photosensitizer, and pyrrole-functionalized MV^{2+} and PTZ, as electron acceptor and donor, respectively [107]; the sequential deposition of the monomers allowed the fabrication of bi- and trilayered films; whereas, the copolymerization of any two (or all three) components gave the corresponding multi-component films. Though a good control over the film thickness and composition was achieved, the anticipated ElT processes could not clearly be identified. In related work, these same authors also investigated the photocurrent production of the copolymer films, *i.e.*, the electron injection into the electrode upon photolysis of the grafted polymers in the presence of $N(CH_2CH_2OH)_3$ (TEOA), as the sacrificial donor [108]. The sensitizer-acceptor assembly **35** was employed by Deronzier et al. in order to achieve a spatial separation of photo-generated charged-separated states by assembling polymer films on electrode surfaces [109]. This design aimed to hamper the electron back-transfer and to improve the long-term storage of redox energy, much more efficiently as in the aforementioned multi-layer approach. The electropolymerization of **35** onto ITO-coated glass slides gave a stable, adherent film missing the electroactivity of the PPy backbone. This electrode was utilized, as a photocathode, for the oxidation of TEOA. Under optimized conditions, a stable photocurrent with an intensity of 12 $\mu A\ cm^{-2}$ was reached (in the aforementioned copolymer design [108], the intensity was *ca.* 15 times lower).

Walder et al. fabricated two "isomeric" photoelectrodes by sequential electropolymerization of **33b**, **37**, and **38** onto a GC electrode (Fig. 3.27) [110]. When the trilayer assembly of GC/poly-**37**/poly-**38**/poly-**33b** was emerged into a TEOA-containing electrolyte solution and, at an applied potential of −0.2 V versus SCE, illuminated with white light, an anodic photocurrent of 3.5 $\mu A\ cm^{-2}$ was observed. The performance of the "isomeric" assembly, i.e., the device with an inverted layer order of poly-**37** and poly-**38**, was—due to a mismatch of the redox potentials—lower by a factor of ten.

The electrochemical copolymerization of the chloro-complex **36** with pyrrole was performed at various **36**-to-pyrrole ratios revealing that the ionic conductivity (i.e., the permeability of the films) was increased and the electronic conductivity was decreased compared to bare PPy films due to the attached Ru^{II} sites [111]. The pyrrole-rich polymers had conductivities in the range of $10^{-2}\ S\ cm^{-1}$. When aiming for a high degree-of-loading with Ru^{II} sites, both a low pyrrole concentration as well as a high polymerization potential were required—under these conditions, only barely conductive ruthenopolymer films with low permeabilities were obtained. The issue of losing the electroactivity of such *N*-substituted (PPy)s when potentiostating the electrodes with high positive potentials was studied by Cosnier et al. [112].

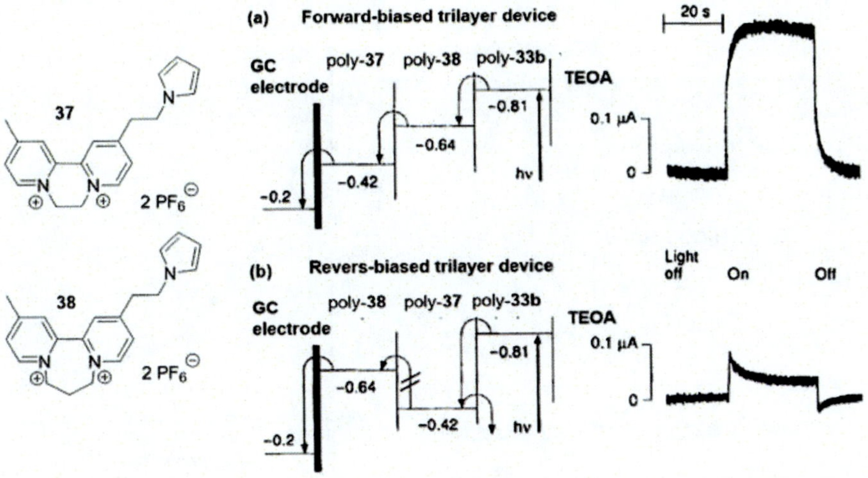

Fig. 3.27 Schematic representation of the diquat monomers **37** and **38**. The photocurrent experiments using the two different tri-layer assemblies in a forward- and reverse-biased intrinsic diode (**a** and **b**) are also shown [110]. Figure reproduced with kind permission; © 1994 The Royal Chemical Society

These authors showed that the characteristic PPy-based redox couple was generally lost as a consequence of the resistance of the films.

The electrochemical recognition of chloride ions by a Ru^{II}-containing PPy-coated Pt electrode was demonstrated by Moutet et al. [113]. The heteroleptic complex **39** (Fig. 3.28a) was investigated in solution and as electropolymerized film by UV/vis absorption spectroscopy with respect to the capability of binding halide ions. In the presence of Cl^-, a significant shift in potential of the first one-electron reduction, assigned to the reduction of the amide-containing ligand, was found (Fig. 3.28b, c); in contrast, no detectable influence of iodide or bromide was observed, while the electroactivity of the film was fully transformed by fluoride. In a similar fashion, pyrrole-functionalized Ru^{II} complexes with additional pending aza-crown-moieties were electropolymerized and investigated with respect to the electrochemical recognition of alkali-metal ions [114]. Depending on the size of the coordinating macrocycle, either Li^+ and Na^+ or K^+ were potentiometrically recognized. This work by Moutet et al. contrasts the earlier reports by Mortimer et al., who realized cation recognition only in solution but not in electropolymerized films [115–117].

Cosnier et al. fabricated a photoelectrochemical immunosensor by the oxidative electropolymerization of the biotin-labeled Ru^{II} complex **40** (Fig. 3.29a) [118]. This approach enabled the immobilization of avidin and the biotinylated cholera toxin as well as the photoelectrochemical detection of the corresponding antibody. In the same context, these same authors also realized the label-free detection and quantification of the anti-cholera toxin antibody in the presence of $[Co(NH_3)_5Cl]$

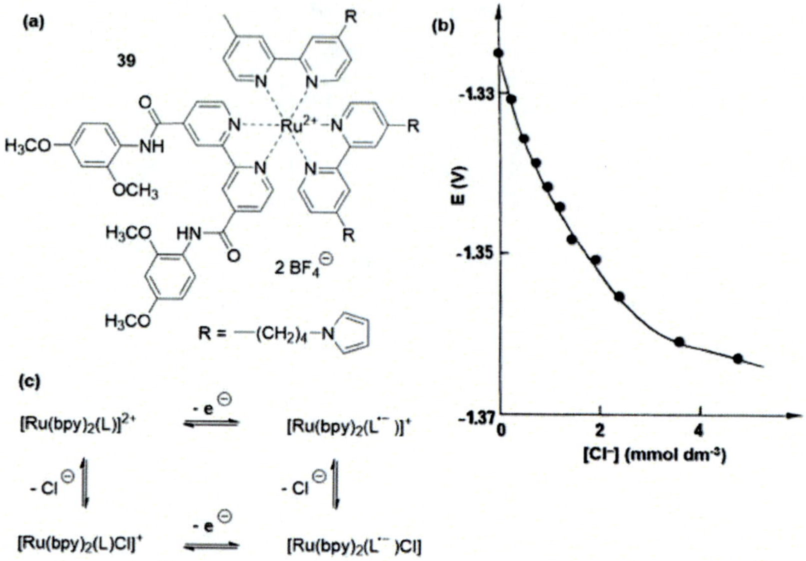

Fig. 3.28 **a** Schematic representation of the complex monomer **39**. **b** Evolution of the potential of the first ligand-based redox couple of poly-**33** as a function of the chloride concentration in MeCN containing 0.1 M TABP (Γ = 1.5 × 10–8 mol cm^{-2}, E vs. Ag/AgCl). **c** Proposed redox and binding equilibria for **39** involving the amide-functionalized ligand (denoted as L) [113]. Figure reproduced with kind permission; © 1996 The Royal Chemical Society

Cl_2, as an oxidative quencher [119]. Moreover, an amperometric immunosensor for the detection of West Nile virus (WNV) antibodies was reported by the same group of authors [120]. This biodevice was prepared by, firstly, electrochemical co-polymerization of **33c** and pyrrole-functionalized benzophenone and, subsequently, grafting of T7 bacteriophages, which exhibit a specific WNV epitope, onto the benzophenone moieties. The resulting electrodes were utilized for the amperometric detection of the WNV antibody at a potential of 0 V via a secondary peroxidase-labeled antibody [e.g. horseradish peroxidase (HRP)]; the measurements were conducted in water containing H_2O_2 and hydroquinone, as enzyme substrate (Fig. 3.29b). The electrochemical responses of various WNV-antibody dilutions from 10 down to 10^{-6} gave a calibration curve; the response time of the device after the substrate injection was in the range of *ca.* 30 s (Fig. 3.29c).

As for the vinyl-functionalized Ru^{II} *tris*bidentate complexes (e.g., **2** in Fig. 3.2), the ECL activity of the pyrrole-functionalized analogs has been a subject of research [121]. The ECL behavior of poly-**33b** was rationalized by two characteristic mechanisms [122]: Redox cycling and oxidative reduction. For the former, no co-reactant was required; whereas, the latter required an aliphatic amine and ammonium oxalate, as co-reactants, in organic and aqueous media, respectively. Besides potential applications in luminance/lighting devices or for imaging, the

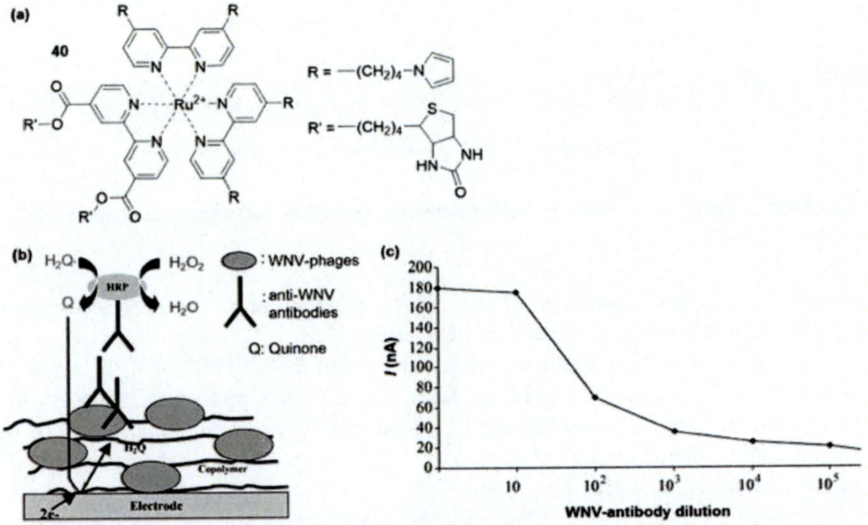

Fig. 3.29 **a** Schematic representation of the biotinylated Ru^{II} complex **40**. **b** Schematic representation of the working principle of the amperometric WNV-immunosensor. **c** Calibration curve for the detection of the WNV-antibody obtained via an electro-enzymatic immunoassay on a metallopolymers-modified electrode (applied potential of 0 V in buffered water at a pH value of 7.2) [120]. Figure reproduced with kind permission; © 2007 Elsevier B.V

ECL in aqueous medium might be of particular relevance to develop ECL-based biosensors for medical and clinical purposes [122]. For example, a sensitive ECL biosensing platform for glucose detection was fabricated. In general, ECL was quenched by adsorbing Au nanoparticles onto the luminescent ruthenopolymer film; such Au nanoparticles might be grown from a H_2O_2-containing solution of $AuCl_4^-$. The reducing agent could also be bio-generated by an oxidase and its corresponding substrate. Ding et al. utilized glucose oxidase, as a model enzyme, to catalyze the glucose transformation and, simultaneously, produce H_2O_2 [123]. The limit for detecting glucose, based on ECL quenching, was as low as 5.0×10^{-6} M. These authors proposed that the biosensor should also be suitable to other biologically produced H_2O_2 or NADH bio-systems (NADH: nicotinamide adenine dinucleotide).

In a related study, a biosensor for EtOH detection was prepared by casting the composite of alcohol dehydrogenase (ADH) and laponite colloids onto a Pt electrode, which was pre-coated with poly-**33b** [124]. This bilayer structure featured a robust performance due to electrostatic interactions between the negatively charged biocomposite and the underlying polycationic metallopolymer film. Ethanol sensing was tested in the presence of NADH, as co-reactant; the latter was required to enhance the ECL according to the proposed mechanism (Fig. 3.30); the detection limit for NADH was as low as 2.0×10^{-6} M. As summarized by Lorenzo et al.

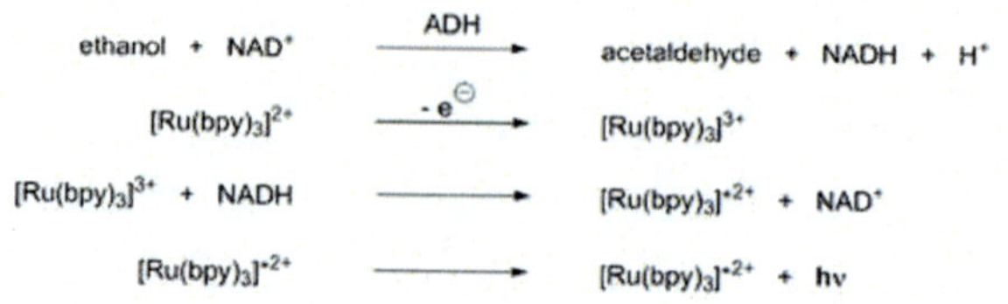

Fig. 3.30 Schematic representation of the proposed mechanism for ethanol detection by ECL

elsewhere, this study might be of interest for applications in nutrition technology, e.g., the quantification of ethanol in beverages [125].

Cosnier et al. coated multi-walled carbon nanotubes (MWCNTs), as electrode materials, with thin metallopolymer films via the electropolymerization of Ru^{II} complexes **41** (Fig. 3.31a) [126, 127]. Due to the electron-donating nature of the substituents, a negative shift of the $Ru^{II/III}$ redox couple compared to $[Ru(bpy)_3]^{2+}$ or **33** was observed (**41a**: $E_{1/2}{}^{ox}$ = 0.59 V). The polymer-coated electrodes were investigated by SEM imaging and revealed a very high surface area compared to those observed on planar electrodes (Fig. 3.31b). The nanostructured electrodes, based on MWCNTs, led also to an increased polymerization yield and a decreased charge-trapping phenomenon, which is typically involved in the oxidation and reduction of the PPy backbone. Moreover, the 3D MWCNT electrodes had an important impact on the electrochemical properties of the generated ruthenopolymers, in particular when concerning the permeability of their films (anodic photocurrents in the range of 19.5 mA cm^{-2} were measured). The latter is beneficial regarding the accessibility of the photo-excited Ru^{II} centers within the polymer film by sacrificial donors/acceptors and, hence, is of relevance for the development of photoelectrochemical devices.

These authors also applied nanosphere lithography (NSL) for the fabrication of highly organized nanostructured polymer films of **41b** on GC electrodes [128]. For this purpose, polystyrene (PS) beads were drop-casted onto the electrode. Subsequently, **41b** was electropolymerized around the deposited nanobeads by CV scanning and, finally, the PS template was removed by rinsing the electrode with THF. As confirmed by 3D laser-scanning confocal microscopy and SEM imaging, poly-**41b** with a honeycomb-like or macroporous nanostructure were obtained utilizing PS template beads with diameters of 900 and 100 nm, respectively (Fig. 3.32). Electrochemical and photo-electrochemical investigations revealed the enhancement of redox-probe diffusion into the patterned metallopolymer layer. This effect was accompanied by an enhancement of the photocurrent for the NSL-generated nanostructured poly-**41b**, in particular for the macroporous structure (0.25 μA cm^{-2} vs. 0.09 μA cm^{-2} for non-patterned poly-**41b**).

The Ru^{II} complex **42** was also electropolymerized anodically onto MWCNTs [129]. In comparison to GC electrodes, a significantly higher apparent surface concentration of Ru^{II} centers was obtained—the value of 80 nmol cm^{-2} was equivalent to *ca.* 1000 compact monolayers of **42**. Moreover, nanostructured poly-

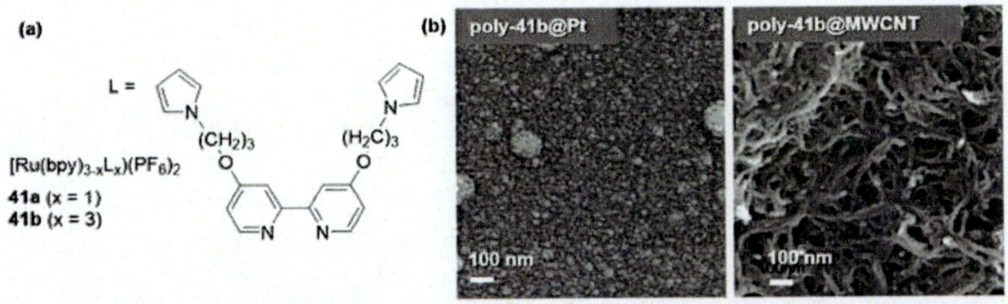

Fig. 3.31 a Schematic representation of the Ru^{II} complexes **41**. **b** Comparison of the surface morphology of electrodes coated with poly-**41** (left: Pt, right: MWCNTs) [126]. Figure reproduced with kind permission; © 2011 Elsevier B.V

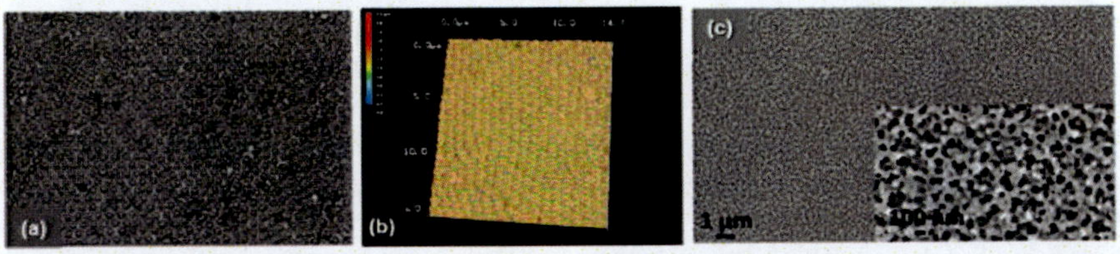

Fig. 3.32 Confocal 3D microscope images of a GC electrode modified with poly-**41b** electrogenerated with 900-nm-diameter PS beads: **a** Laser image and **b** 3D image. **c** SEM images of a GC electrode modified with poly-**41b** electrogenerated with 100-nm-diameter PS beads [128]. Figure reproduced with kind permission; © 2014 Elsevier B.V

42 exhibited efficient electrocatalytic properties towards the oxidation of NADH (1.1 mA cm^{-2} compared to 0.04 mA cm^{-2} for poly-**42**@GC). Since poly-**42** could electrochemically be generated in water from the surface-adsorbed monomer, these authors also targeted the immobilization of enzymes by the co-adsorption of **42** and an enzyme, followed by an electropolymerization step. This methodology allowed the entrapping of 70 to 90% of the deposited enzyme, i.e., glucose dehydrogenase (GDH), within the ruthenopolymer films. In the presence of NAD$^+$, the enzyme-containing electrode featured high current densities for glucose oxidation of 1.04 mA cm^{-2} at low overpotentials of −0.1 V with a detection limit for glucose of 1 μM (Fig. 3.33).

The anodic electropolymerization of various Ru$^{II/III}$ complexes containing the tpy-based ligand **43** was investigated by Sauvage et al. (Fig. 3.34a) [130]; from these, the homoleptic Ru^{II} complex was electropolymerized in a highly controlled fashion (Fig. 3.34b); whereas, Ru(**43**)Cl_3 did not give any stable polymer film. The aquo-ligand of the mixed-ligand complex [Ru(**43**)(bpy)(OH_2)]$^{2+}$ was rapidly exchanged in MeCN by a solvent molecule; the thusly formed [Ru(**43**)(bpy)(MeCN)]$^{2+}$ species could also be electropolymerized, though only one polymerizable group per complex molecule was present. The electrodeposition of the

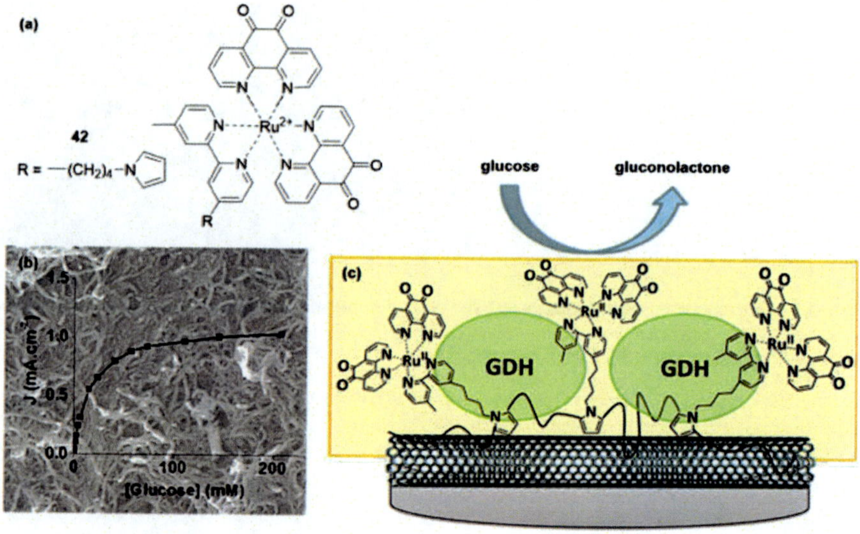

Fig. 3.33 a Schematic representation of the electropolymerizable Ru^{II} complex **42**. **b** Plot of the current density as a function of the glucose concentration (a representative SEM image of the coated MWCNTs is shown as the background). **c** Schematic representation of the work-principle of the hybrid biosensor for glucose detection [129]. Figure reproduced with kind permission; © 2014 American Chemical Society

aquo-complex itself induced a coalescence of the Ru$^{IV/III}$ and Ru$^{III/II}$ redox couples; however, adjusting the pH value allowed to shift these redox potentials and enabled the electropolymerization from a buffered aqueous solution. The resultant ruthenopolymer film exhibited two processes: Two electrons were exchanged with the electrode and two protons with the solution (Fig. 3.34c). The complex [Ru(tpy)(bpy)(OH_2)]$^{2+}$ is long-known for its ability to catalyze the electro-oxidation of benzylic alcohols—either in solution [131] or when incorporated into a Nafion™ membrane [132]. In contrast, the electropolymerized film did not show any activity in this respect. However, when the polymer films were formed in polar, weakly-coordinating solvents (i.e., CH_2Cl_2 or acetone) and when a further polymerizable pyrrole-substituent was present on the bpy ligand, stable and electrocatalytically active films were obtained—the oxidation of allylic and benzylic alcohols could be performed even on a preparative scale [133]. A similar oxidation behavior was observed for the electrogenerated films of the mixed-ligand complex [Ru(tpy)(L)(OH_2)]$^{2+}$ [L = pyrrole-functionalized 2,2'-di(pyridine-2-yl)amine] [134]. The metallopolymer film obtained from the heteroleptic complex **44** (Fig. 3.34a) represents one further example where charge separation has been realized in integrated systems [135]. Upon simultaneous potentiostating at 0 V and irradiation of the Ru^{II} sensitizer in the presence of TEOA, as sacrificial donor, a

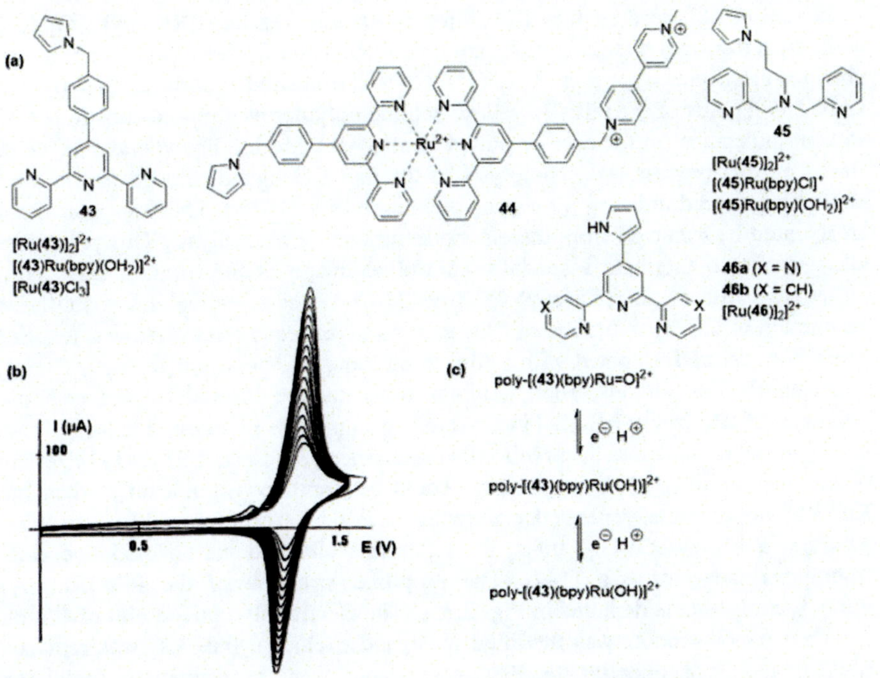

Fig. 3.34 **a** Schematic representation of $Ru^{II/III}$ complexes containing tridentate pyrrole-functionalized ligands. **b** CV measurements of complex [Ru(**43**)$_2$]$^{2+}$ upon continuous anodic scanning revealed the continuous growth of the polymer film. **c** Redox/pH equilibrium of poly-[Ru(**43**)(bpy)(OH$_2$)]$^{2+}$ [130]. Figure reproduced with kind permission; © 1990 Elsevier B.V

significant photocurrent was detected. The light-to-energy conversion was by far more efficient as, e.g., in an analogous copolymer or in multi-layer films.

Besides the classical tpy-based ligands also the more flexible tridentate **45** was utilized (Fig. 3.34a). The homoleptic [Ru(**45**)$_2$]$^{2+}$ complex with *cis,fac*-configuration was electropolymerized to give a remarkably (electro-)chemical stable metallopolymer film [136]. Due to a decrease of π-back bonding and an increase of the σ-bonding capacity of **45**, compared to conventional tpy ligands, the $Ru^{II/III}$ redox potential was shifted by 300 mV to higher potentials {relative to the polymer film obtained from [Ru(**43**)$_2$]$^{2+}$}. Similar to [(**43**)Ru(bpy)(OH$_2$)]$^{2+}$ also the polymer films obtained from the mixed-ligand complexes containing **45** were catalytically active in the oxidation of benzylic alcohols [137]. The tridentate (pyrrol-2-yl)-substituted ligands **46** were employed for the synthesis of homoleptic Ru^{II} complexes and these, in turn, were electropolymerized anodically giving redox-active films on graphite electrodes [138, 139]. The direct linkage of the pyrrole unit to the chelating ligand, thus enabling a strong electronic coupling between the Ru^{II} centers and the polymerizable group, was found to be slightly beneficial for increasing the excited-state lifetime.

Ziessel et al. prepared thin PPy films containing pending [Ru(bpy)$(CO)_2Cl_2$] units by electropolymerizing the monomers **47** (Fig. 3.35a) [140]. The thusly obtained electrodes were immersed in MeCN and irradiated at 366 nm; thereby, the labile CO ligands of the initially yellow-colored polymer were photo-substituted by solvent molecules (in the case of poly-**47a** also one chloride ligand was replaced). Based on this process (i.e., photolysis of the first CO ligand), images in the film could be prepared utilizing a masking technique (Fig. 3.35b): The non-photolyzed areas could be removed from the electrode surface by electrolysis. This procedure allowed one to fabricate a spatially controlled image of the original mask as a red-colored film of poly[Ru(bpy)(CO)(MeCN)Cl_2]. The selective and quantitative production of CO by reduction of CO_2 in a purely aqueous electrolyte was realized on carbon electrodes coated with a film of an inorganic polymer [Ru(bpy)$(CO)_2]_n$ featuring Ru^0-Ru^0 bonds; such polymer films can be formed by the cathodic reduction of [Ru(bpy)$(CO)_2Cl_2$] in a two-electron process involving the loss of two Cl^- ligands (for more details on this type of polymers, see Sect. 1.5) [141–143]. The operational stability of these materials could be improved significantly when the Ru^0-Ru^0 bonds were generated electrochemically in a preformed PPy film with pending [Ru(bpy)$(CO)_2Cl_2$] units; the latter was obtained via an oxidative electropolymerization of **47a** [142]. The proposed structure of the PPy-[Ru(bpy)$(CO)_2]_n$ composite is depicted in Fig. 3.35c. The electropolymerized film of **47a** on a carbon paper support was modified by ligand exchange (i.e., Cl^- was replaced MeCN) and, subsequently, the PF_6^- counterions were exchanged by metal ion complexes. Electroreduction of the dispersed and fixed metal salts in aqueous KCl, as electrolyte, provided the corresponding metallic nanoparticles. Investigation of the surfaces of the resulting composite materials was performed by TEM imaging, XRD analysis, and XPS measurements; the size of the formed metal particle depended on the employed metal (i.e., Pd, Ru, Rh) and ranged from 1 to 6 nm [144]. Representatively, images of the obtained Pd particles are depicted in Fig. 3.36.

More recently, Oyarzúna, Pérez et al. analyzed the stability of [Ru(bpy)$(CO)_2Cl_2$] complexes anchored to nanoporous TiO_2. For this purpose, [Ru(bpy)$(CO)_2Cl_2$] and [Ru(dcbpy)$(CO)_2Cl_2$] were attached to the semiconductor surface via electrostatic interactions and carboxylate binding, respectively (dcbpy: 2,2'-bipyridine-4,4'-dicarboxylate) [146]. Moreover, the oxidative electropolymerization of **47a** was attempted; however, the nanoporous TiO_2 did not offer sufficient conductivity in the required potential regime to enable this process. Thus, these authors opted for a different strategy in which the Ru^0-based organometallic polymer **48** was formed by a two-electron reductive (though poorly controlled) electropolymerization of **47a**. Field-emission SEM imaging revealed a cloudy, inhomogeneous film morphology (Fig. 3.37). The ruthenopolymer **48** was found to be highly unstable—monomeric Ru^{II} species that formed instantaneously under aerobic conditions. Therefore, the application of this hybrid electrode in the photoelectrochemical CO_2 reduction could not be tested. From the other two candidates, the carboxylate-anchored complex was significantly more stable against hydrolysis and offered a better photoelectrocatalytic activity.

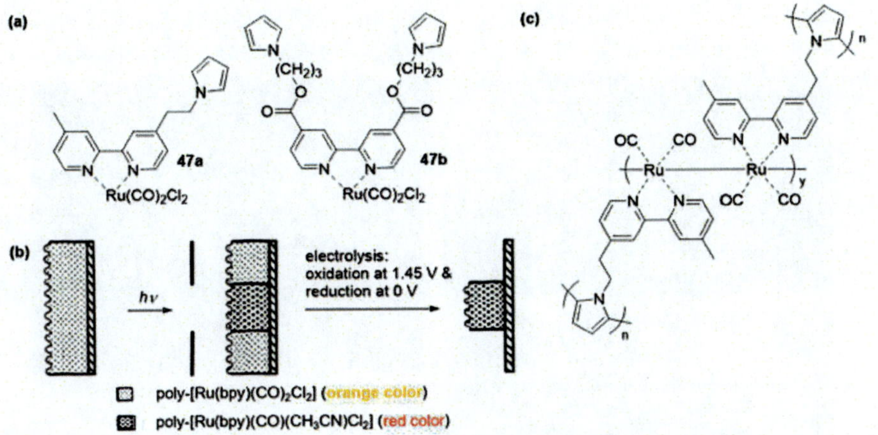

Fig. 3.35 Schematic representation of (**a**) the Ru^{II}-carbonyl complexes **47**, of (**b**) the fabrication of a patterned surface via two-step photolysis-electrolysis methodology, and of (**c**) the PPy-[Ru $(bpy)_2(CO)_2]_n$ composite material [140]. Figure reproduced with kind permission; © 1993 American Chemical Society

Deronzier et al. employed the dinuclear Ru^{I} complexes **49** (Fig. 3.38), as precursor material, for the electropolymerization in two different ways [146]: Firstly, the two-electron reduction process gave access to Ru^{0}-Ru^{0} organometallic films deposited onto electroactive surfaces (see also Sect. 1.5). Alternatively, the oxidative electropolymerization of the pending pyrrole moieties was applied to generate the functionalized PPy films; however, the exact composition of the Ru complexes after the electropolymerization step could not be clarified (the Ru^{I} centers and pyrrole had almost the same oxidation potentials). Nonetheless, the reductive formation of the Ru^{0}-based inorganic chain was achieved from the functionalized PPy films.

Although relying on a different Ru^{II}-containing polymer, i.e., poly-**33c**, the application as anion-exchange membrane (AEM) is worth mentioning in this context [147, 148]. The metallopolymer-coated carbon electrode was emerged into an aqueous K_2RuO_4 solution in order to load the electrode with RuO_4^{2-} anions. The potentiostatic reduction of the anions at 0.2 V versus SCE generated polymer-bound RuO_2 nanoparticles (the amount of RuO_2 was calculate from the charge used for this reduction, Fig. 3.39). This composite material was employed as an electrocatalyst for the oxidation of benzylic alcohols: Thereby, the polymer-bound Ru^{II} complexes acted as an electron relay for the electrochemical generation of the oxidizing RuO_4^{2-} species at a potential of 1.1 V. A variety of substrates was oxidized on a preparative scale in a $LiClO_4$/MeCN electrolyte containing 4% water; for the reactions, TON values in the range of 10^4 were found. When using carboxy-ester-substituted bpy ligands in the coordination sphere of the electropolymerized Ru^{II} complexes, their $Ru^{II/III}$ redox couple at 1.24 V enabled the

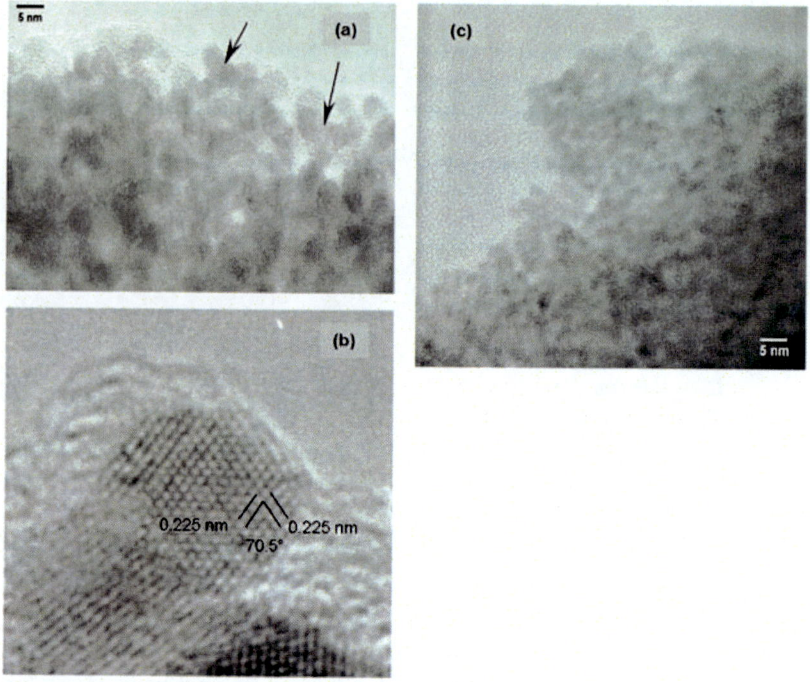

Fig. 3.36 TEM image of the ruthenopolymer film containing Pd particles. **b** High-resolution TEM image of a Pd particle in [1,0,−1]-orientation and exhibiting its (1,1,1) and (1,−1,1) planes at 0.225 nm and separated by 70.5°. **c** TEM image of the metallopolymer film containing Ru particles [144]. Figure reproduced with kind permission; © 2004 The Royal Chemical Society

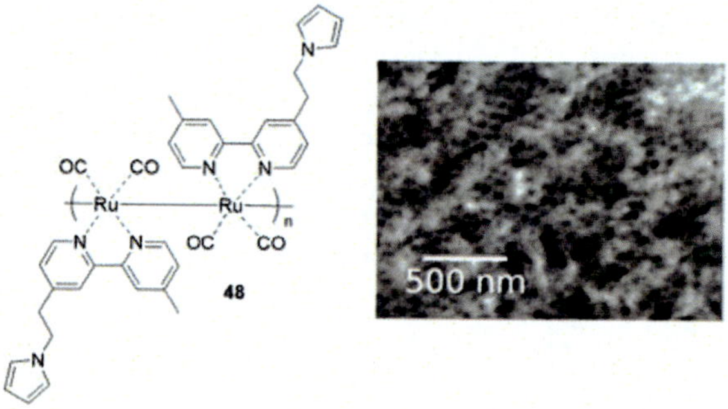

Fig. 3.37 Schematic representation of the Ru^0-based polymer **48**. A representative field-emission SEM image of **48** grown electrochemically onto nanoporous TiO_2 is also depicted [145]. Figure reproduced with kind permission; © 2018 Elsevier B.V

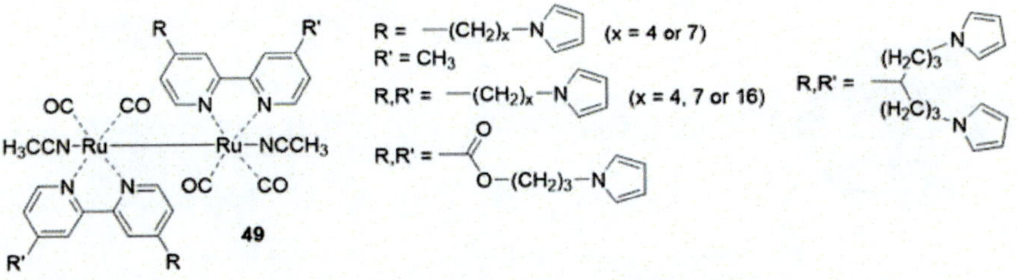

Fig. 3.38 Schematic representation of the dinuclear Ru^{I} complexes **49**

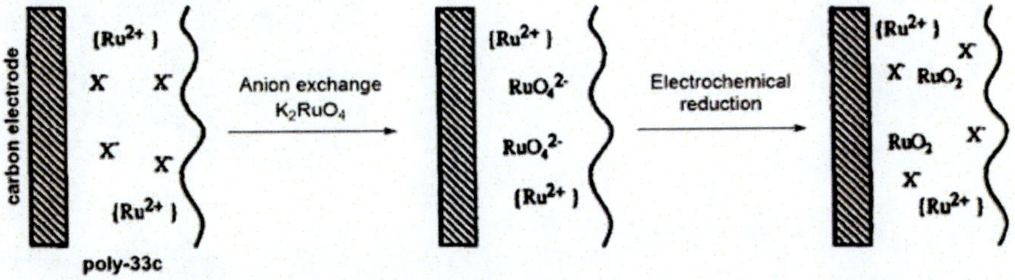

Fig. 3.39 Schematic representation of the preparation of a RuO_2-loaded, polymer-coated carbon electrode [89]. Figure redrawn according the literature, © 1996 Elsevier B.V

formation of the even stronger oxidant RuO_4^-. This hybrid system featured a higher catalytic activity and TONs of >14,000: For example, cinnamyl alcohol was efficiently transformed, in an overall six-electron oxidation reaction, into glyoxal and benzaldehyde.

Fujihara et al. reported on the synthesis of functionalized Au nanoparticles (Fig. 3.40) [149]. The reduction of $AuCl_4^-$ with $NaBH_4$ in the presence of (2,2':6',2''-terpyridin-4'-yl)-octanethiol afforded Au nanoparticles with a corona of tpy moieties. The complexation of these conjugates with $Ru(tpy)Cl_3$ gave nanoparticles decorated with Ru^{II}–tpy *bis*-complex (thereby, an increase in the average core size from 2.0 to 5.5 nm was observed); UV/vis absorption spectroscopy confirmed the identity of these assemblies by the occurrence of both the characteristic MLCT band at 483 nm and the Au plasmon resonance at *ca.* 510 nm. Subsequent surface modification of the nanoparticles was achieved by a ligand exchange using *ω*-(*N*-pyrrolyl)decanethiol; these nanoparticles were immobilized onto a GC electrode by anodic electropolymerization (Fig. 3.40).

The *tris*-ruthenium-μ-oxo-carboxylate clusters with a formula $[Ru_3O(OOCCH_3)_6L_3]^+$ (L denotes a pyridine ligand) feature a remarkable reversible redox chemistry (i.e., five well-separated redox couples in the potential window from +1.70 and to −2.16 V). Besides their mixed-valence properties in the partially oxidized states and their electrochromism also their ability to serve as redox catalyst and mediator

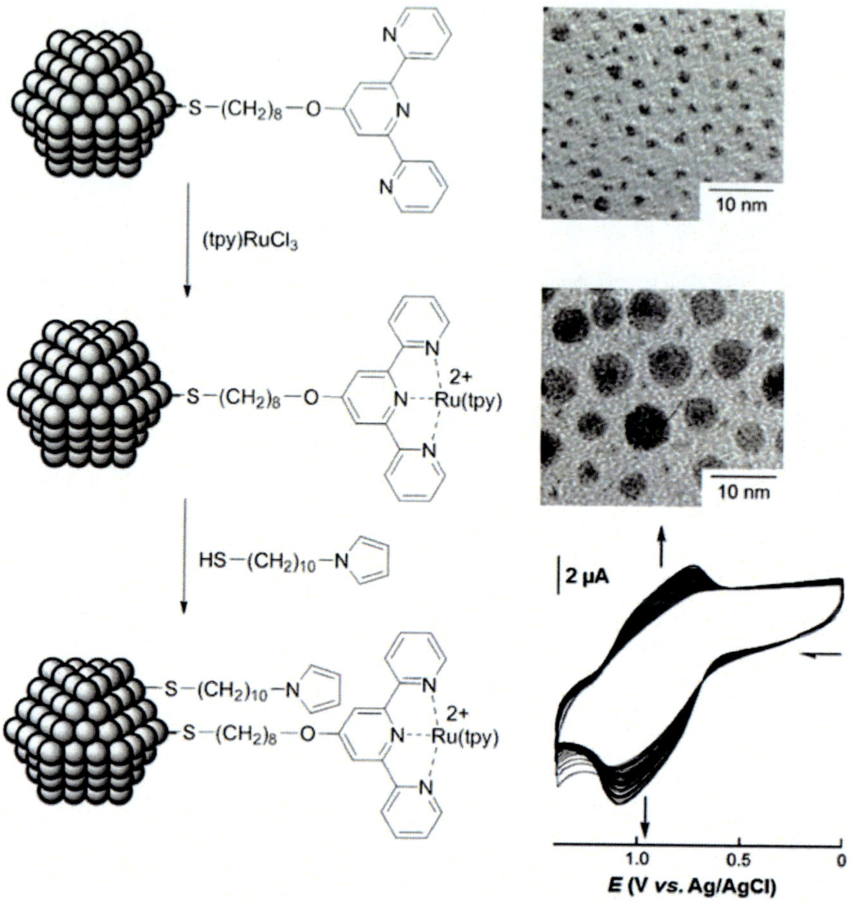

Fig. 3.40 Schematic representation of the synthesis of functionalized Au nanoparticles by complexation and ligand-exchange. The corresponding TEM images for the systems Au–tpy and Au–Ru(tpy)$_2$ are also depicted. Moreover, the oxidative electropolymerization of the pyrrole-functionalized Au–Ru(tpy)$_2$ assembly on a GC electrode by a potentiodynamic method in 0.1 M Bu_4NPF_6 in MeCN, as electrolyte, is also shown (scan rate: 100 mV s^{-1}) [149]. Figure reproduced with kind permission; © 2005 The Royal Chemical Society

in multiple electron-transfer reactions are of interest. Against this background, Cosnier et al. immobilized the clusters **50** onto a Pt electrode by potentiostating the electrode in range of 0.7 to 0.9 V, thereby achieving an average electropolymerization yield of *ca.* 67% [150]. For instance, a polymer film grown by potentiostating at 0.80 V until a total charge of 1.02 × 10^3 C had passed showed a surface coverage of 4.50 × 10^{-9} mol cm^{-2} and exhibited a high stability: Negligible decomposition was observed after scanning 50 times over a potential range from −2.0 to +2.0 V (Fig. 3.41).

Xie et al. followed a completely different strategy [151]; the ruthenopolymer **51** was prepared by mixing pyrrole and $RuCl_3$, thereby forming a Ru-pyrrole adduct, which, in turn, was electropolymerized onto a carbon-paper substrate (Fig. 3.42a). The molecular structure of **51** was concluded from molecular-modelling studies—the tetrahedral coordination geometry and the all-*trans*-orientation of the pyrrole units within the chain was considered to be the most thermodynamically stable architecture. Energy-dispersive X-ray (EDX) analysis of the films gave a relative Ru-to-N-to-Cl ratio of 1:2:2, thus corroborating the proposed molecular composition. These authors investigated the morphology of the obtained metallopolymer films. The presence of the metal centers had a profound impact on the microstructure of the PPy material (a looser packing of the particles was found for **51** compared to bare PPy prepared under the same experimental conditions, Fig. 3.42b). Detailed photophysical and electrochemical studies on the hybrid electrode were performed and suggested its applicability, as an active electrode material, in super capacitors. It was shown that the capacitance of **51** decreased from 618 F g^{-1} at 1.0 A g^{-1} to 301 F g^{-1} at 20.0 A g^{-1}, corresponding to a capacitance retention of 48.7%, higher as for bare PPy under the same conditions (*ca.* 41%). The improved cycling stability of **51** was further shown by continuous cycling (i.e., 2000 cycles at 20.0 A g^{-1}) where a capacity retention ratio of 91.4% was reached (PPy: 79.4%). The enhanced performance of **51** was attributed to the presence of bipolarons, induced by the coordinated metal centers, which facilitated electrochemical conductivity and capacitance. For comparison, the authors also coordinated $RuCl_3$ to a preformed PPy film. The resulting ruthenopolymer was structurally less defined, swellable, and significantly less stable under

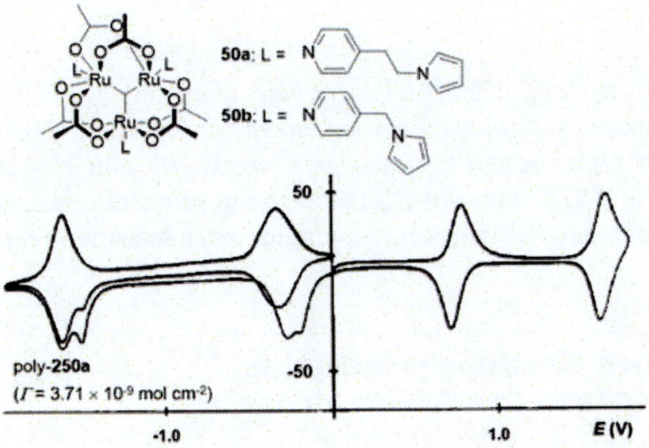

Fig. 3.41 Schematic representation of the pyrrole-functionalized *tris*-ruthenium-clusters **50**. The CV curves of a Pt electrode coated with poly-**50a** are also depicted (positive potentials: 10th consecutive scan; negative potentials: the subsequent 11th and 12th scans) [150]. Figure reproduced with kind permission; © 1990 Elsevier B.V

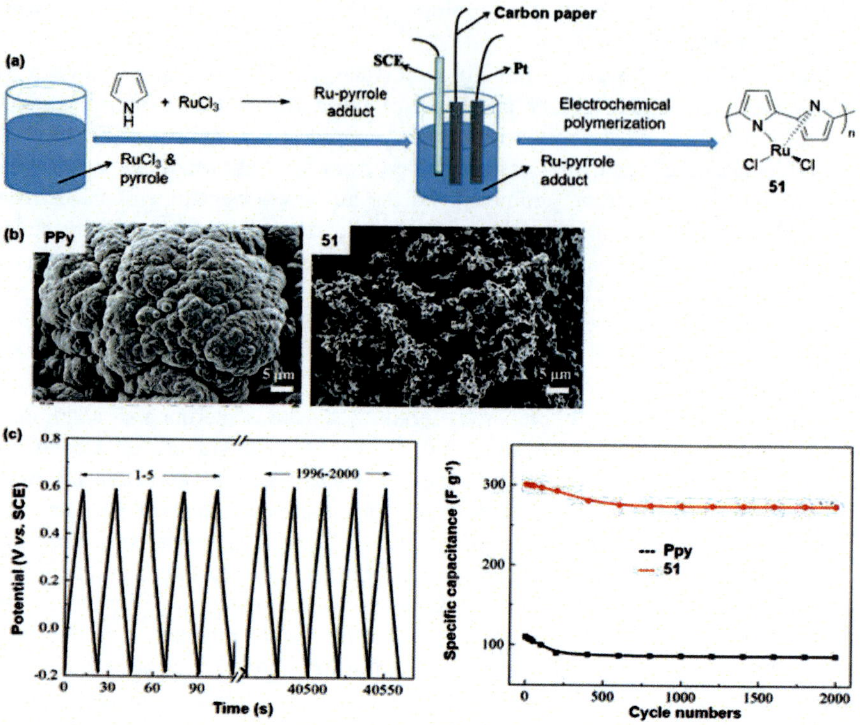

Fig. 3.42 a Schematic representation of the preparation of **51**. **b** Representative SEM image of a carbon-paper electrode coated with PPy (left) and **51** (scale bar: 5 μm). **c** Cycle galvanostatic charge/discharge (GCD) curve of **51** at 20.0 A g^{-1} and corresponding capacitance curves [151]. Figure reproduced with kind permission; © 2017 The Royal Chemical Society

electrochemical cycling. The straightforward preparation of **51**, as capacitor material, represents a significant improvement to the alternative concept where RuO_2, having a high specific capacitance, is combined with PPy into stable composite electrodes [152]. Moreover, the preparation protocol can readily be adopted to fabricate analogous electrodes incorporating other transition-metal ions [153].

3.5 Oxidative Electropolymerization of Thiophene-Containing Monomers

Polythiophene (PT) represents one of most widely used conducting polymers and its synthesis can basically be realized via two synthetic routes [154]: Chemical and electrochemical synthesis. The former is the method of choice when aiming for oligomers of defined structure and various protocols have been established for this purpose. Similar to pyrrole, thiophene can be polymerized oxidatively in the presence of $FeCl_3$ [155]; in addition to this, a range of metal-catalyzed polycondensations is available, which basically use 2-halo or 2,5-dihalothiophenes, as starting materials [156]. These methods offer the advantage of a regioregular coupling of the thiophene units, if unsymmetrical 3-alkylthiophenes are used. However, PTs featuring the highest degree-of-π-conjugation and the highest conductivities have been obtained from oxidative electrochemical polymerizations. As for the electropolymerization of pyrrole derivatives (see the previous Section), the anodic polymerization of thiophene monomers is highly attractive for numerous reasons: the absence of catalysts (as required in all chemical alternatives), direct deposition of the polymer onto the electrode surface, ease in controlling the film thickness via the deposition charge and, finally, possibility to perform in-situ characterization of the growing polymer film by electrochemical and/or spectroscopic means (see Sect. 3.1).

Basically, applying a certain potential generates an activated species, such as a radical cation or anion that subsequently undergoes dimerization. In the following, oligomers and, eventually, polymers are obtained [154]. The accepted mechanism for the anodic polymerization of thiophene is depicted in Fig. 3.43. As for all electron-rich heteroles, coupling may occur via the α- or β-positions of the ring with the former being more favored (however, for most thiophene derivatives, the actual ratio between α,α- and α,β-coupling remains unknown). In general, strictly α,α-coupled systems exhibit the longer effective π-conjugation lengths.

Since the first reports on the electropolymerization of 2,2'-bithiophene [157] and, two years later, of thiophene itself [158], a wide range of thiophene derivatives has been utilized, as monomers, to generate PT-type materials—including metal-containing ones [1, 2, 159]. In recent years, the heavier polychalcogenophenes, i.e., materials based on selenophene or tellurophene, have been established as new classes of π-conjugated polymers offering some advantages

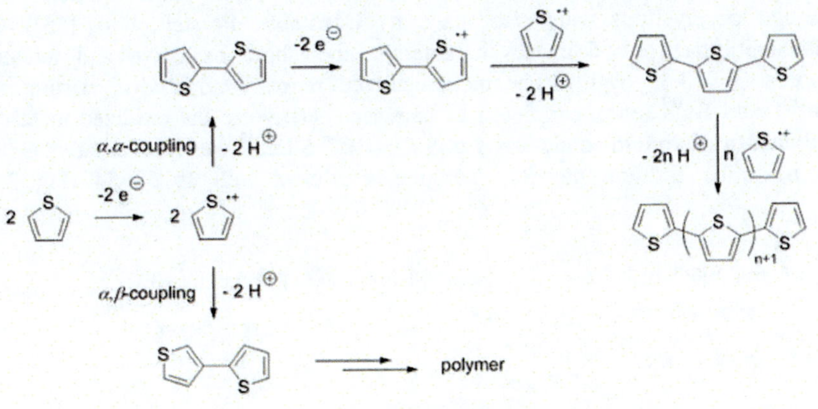

Fig. 3.43 Schematic representation of the oxidative thiophene electropolymerization

compared to the PTs [160, 161]. These polymers possess a more quinoid character, feature lower energy-band gaps, and are more rigid in structure. Accordingly, selenophene- and even tellurophene-containing polymers represent emerging materials that use these beneficial properties. However, only scattered reports dealing with metal-containing selenophenes and tellurophenes and their application, as monomers, in electropolymerizations are known so far, presumably owing to the remaining synthetic restrictions [162, 163].

The first thiophene-functionalized Ru^{II}-complexes, employed in the anodic electropolymerization, were reported in 1996 by Keene et al. (**52**, Fig. 3.44) [164]. The electrochemical properties of poly-**52** were studied by CV and chronoamperometry measurements as well as impedance spectroscopy. Besides an influence of the supporting electrolyte, a strong dependence of the electrochemical behavior of the polymers on the length of the alkyl chains was observed: Longer chains resulted in a (partially) irreversible voltametric behavior and increased charge-transfer resistances in the cathodic processes. The authors suggested that the electrochemical processes of poly-**52** in the anodic region were controlled by both kinetic and diffusion factors; whereas, the primary influence in the cathodic region was assigned to kinetic factors arising from the flexible alkyl chains. In a similar fashion, the heteroleptic complex **53** (Fig. 3.44) was electropolymerized into a stable coating on a Pt electrode; the orange Ru^{II}-containing film showed a reversible $Ru^{II/III}$ oxidation wave at +1.20 V [165].

In contrast to the previous examples, where the bpy and thiophene units were linked via a flexible alkyl spacer, also Ru^{II} complexes, having the thiophene entities directly attached to the ligands, were prepared and electropolymerized. In such polymer architectures, the metal centers are in close proximity to the conjugated backbone enabling a strong electronic coupling. For example, Swager et al. showed that the homoleptic Ru^{II} complex **54b** as well as the metal-free ligand could be electropolymerized (Fig. 3.45, left) [166]; in contrast, the *bis*(thiophene)-containing systems (i.e., complex **54a** and corresponding ligand) were not electropolymerizable, presumably due to the highly labile nature of the initially formed radical cation (similar observations were also made by Constable, Forster et al. [55]). The ruthenopolymer poly-**54b** was conductive upon both oxidative and reductive doping (Fig. 3.45, right). The metal-centered redox conductivity, arising from $Ru^{II/III}$ and $Ru^{0/I}$ redox couples, was increased relative to the oxidized metal-free polymer (a maximum conductivity of 9.4×10^{-4} S cm^{-1} was reached at −1.37 V). In a similar fashion, the Ru^{II} complexes **54c** as well as **55–56** (Fig. 3.46)

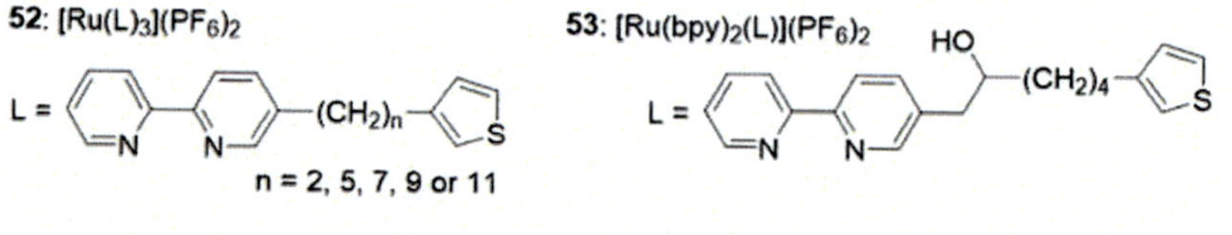

Fig. 3.44 Schematic representation of the thiophene-functionalized Ru^{II} complexes **52** and **53**

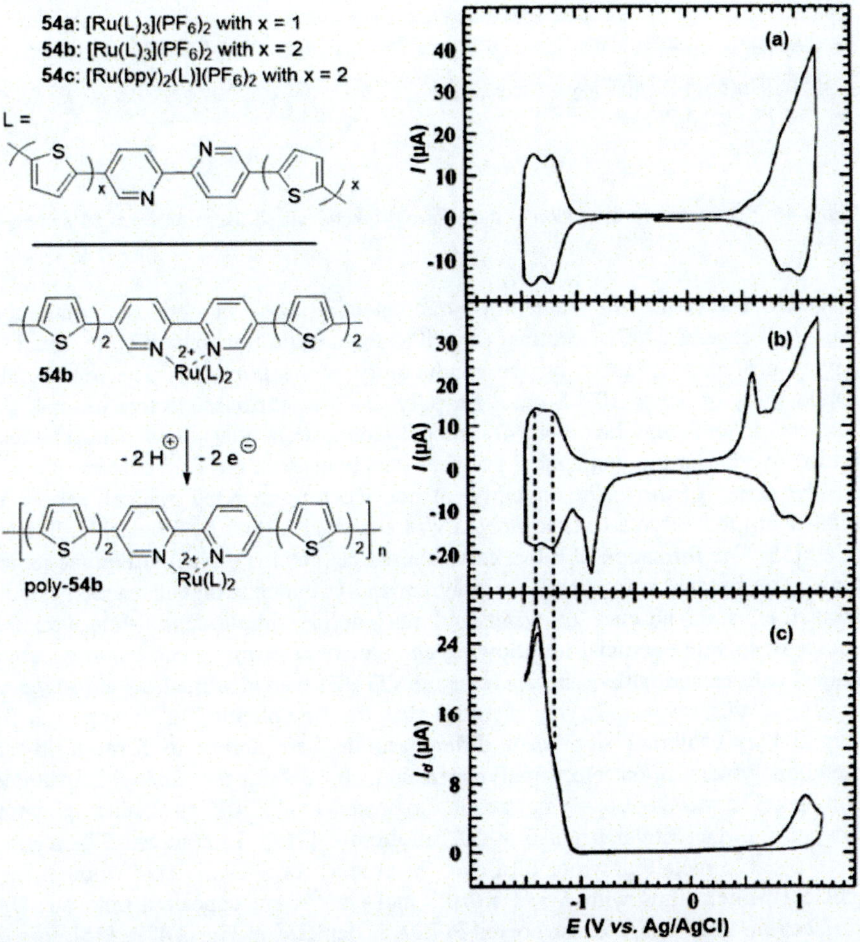

Fig. 3.45 Left: Schematic representation of the homoleptic Ru^{II} complexes **54** and electropolymerization of **54b**. The electrochemistry of poly-**54b**, as observed in the CV measurements is also shown: **a** Continuous cycling over separate cathodic and anodic regions and **b** over the entire potential window (scan rate of 100 mV s^{-1}). **c** Drain current (I_d) measured with an offset potential of 40 mV between alternate microelectrodes (sweep rate of 5 mV s^{-1}) [166]. Figure reproduced with kind permission; © 1996 Wiley–VCH

were employed, as monomers, for the deposition of metallopolymer films on Pt interdigitated electrode arrays (IDAs) [167]. The electrochemical analysis of all films revealed the characteristic Ru^{II}-centered and thiophene-based electroactivity. In particular, high redox conductivity was observed for poly-**55a** and poly-**56** for both the thiophene-based oxidation and the Ru^{II}-based reduction processes. Thus, charge localization for the Ru^{II} complexes and the tetrathienyl fragments was

55a: $[Ru(L^1)_3](PF_6)_2$
55b: $[Ru(bpy)_2(L^1)](PF_6)_2$
56: $[Ru(L^2)_3](PF_6)_2$

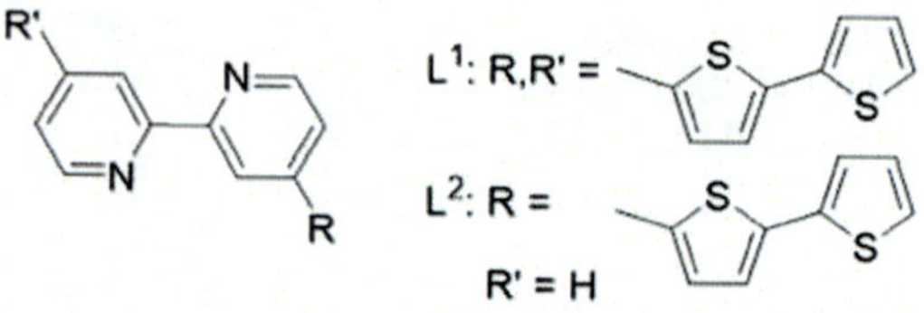

Fig. 3.46 Schematic representation of the bithienyl-functionalized Ru^{II} complexes **55** and **56**

assumed. Moreover, the degree-of-interconnection (i.e., number of tetrathienyl linkages between the Ru^{II} centers) as well as the substitution pattern were found to have a profound influence on the conductivity of the polymers. The remarkable conductivity of 3.3×10^{-3} S cm^{-1} for poly-**55a** was attributed to the, in total, six connection units and favorable 4,4'-substitution pattern that allowed an efficient orbital overlap of the conjugated polymer backbone with the Ru^{II} centers.

The issue of how the stability of an electro-generated radical cation is influencing the polymerization process was analyzed further by Constable, Forster et al. [55] The formation of rather unstable radical cations typically requires rather high onset potentials to enable the polymerization under stringent reaction conditions (i.e., strict absence of water and nucleophilic impurities). Moreover, the choice of solvent is crucial for allowing an efficient electropolymerization of labile radical cations and trifluoroborane etherate (TFBE) was identified, as a good candidate, in this respect. It was reported that the homoleptic Ru^{II} complexes **57** (Fig. 3.47a) exhibited significant differences in their ability to form rigid-rod metallopolymers upon electropolymerization. In detail, the thienyl-substituted complex **57a** was electropolymerized in the presence of TFBE, as stabilizing agent, to give a surface coverage of 5×10^{-8} mol cm^{-2} [168]. In contrast, **57b** readily gave a stable metallopolymer film even in absence of a Lewis acid (though less efficient); thick films with $\Gamma = 1 \times 10^{-6}$ mol cm^{-2} were deposited onto an IDA surface (the SEM image of the coated Pt IDA is depicted in Fig. 3.47b) [55]. In-situ conductivity measurements with the IDAs coated with poly-**257b** and the Os^{II}-containing poly-**57c**, respectively, were performed (Fig. 3.47d). A remarkably high absolute DC conductivity of 1.6×10^{-3} S cm^{-1} was found, irrespective of the nature of the employed metal center. Since the maximum conductivity occurred at the formal potential of each redox process, these authors proposed redox conduction to be the predominant charge-transport pathway.

As mentioned above, due to the presence of the less electronegative and more polarizable Se atom, the oxidation potential of selenophene monomers is decreased relative to the corresponding thiophenes; therefore, a better control of the electro-optical properties of the resulting polymer can be achieved [160, 169, 170]. Additionally, polyselenophenes possess photophysical characteristics different from those of their PT- or polyfuran-type counterparts, such as significantly lower energy-band gaps. This can be utilized, for example, to prepare electrochromic devices, based on polyselenophenes. Though a wide range of

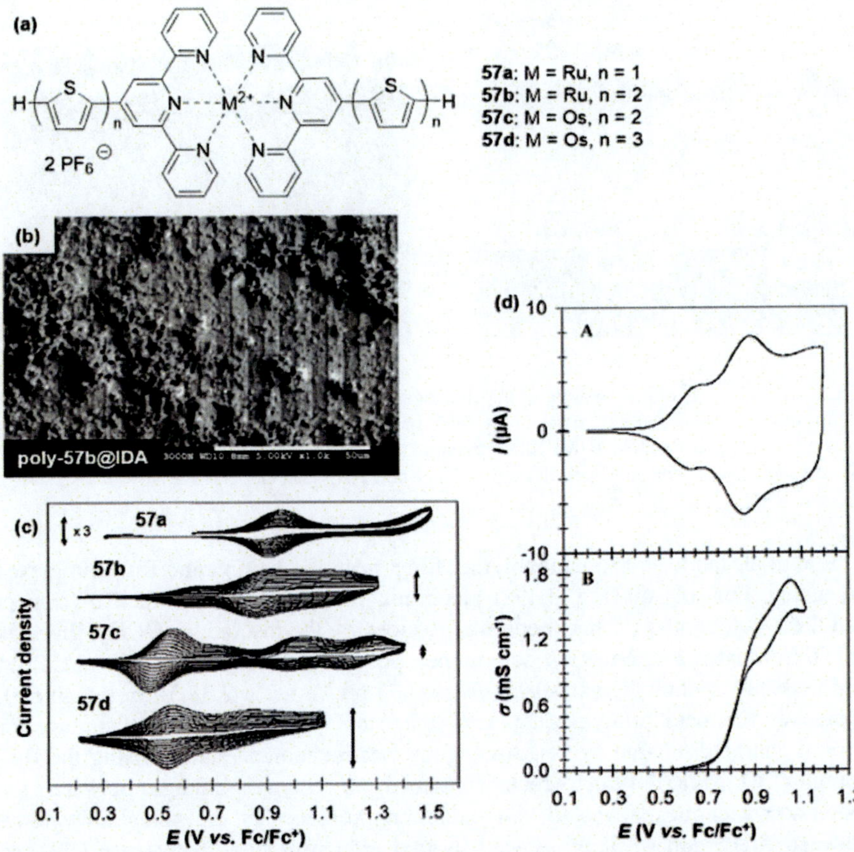

Fig. 3.47 **a** Schematic representation of the homoleptic Ru^{II} or Os^{II} complexes **57**. **b** SEM image of Pt IDAs coated with poly-**57b** (scale bar: 50 μm). **c** Oxidative electropolymerization of complexes **57** onto a Pt disk electrode (the bars represent a current density of 1 mA cm^{-2}; the scan rate was 0.1 V s^{-1} in all cases.). **d** CV (top) and in-situ conductivity profile of poly-**57b** on a Pt IDA at 0.1 V s^{-1} (bottom, the forward and backward scans are shown, the offset voltage was 20 mV) [55]. Figure reproduced with kind permission; © 2005 American Chemical Society

selenophene-containing polymers has been reported in recent years [160], poly-**58** represents the exclusive example for a metal-containing material so far [162]. The electropolymerization of **58** was achieved from a TFBE solution onto a FTO electrode. The polymer film exhibited electrochromism, which could be switched reversibly by scanning between +1.0 and +1.6 V (Fig. 3.48a). These same authors also prepared the first tellurophene-substituted Ru^{II} complex **59** and realized its controlled electropolymerization onto a Pt electrode via oxidative coupling of the tellurophene rings [163]. Thusly, the obtained ruthenopolymer films were analyzed by EDX and SEM imaging regarding their elemental composition and morphology, respectively (Fig. 3.48b).

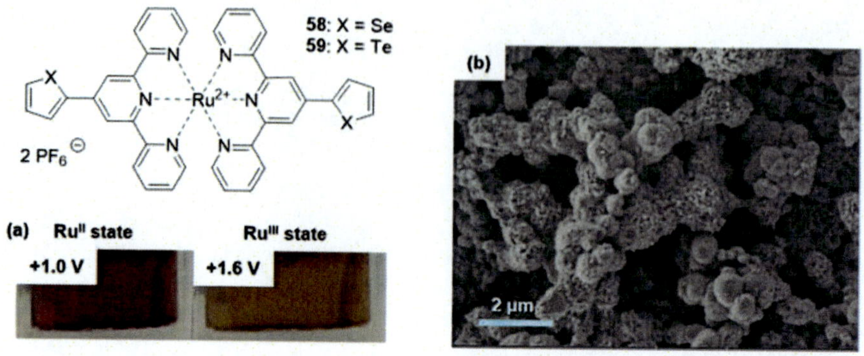

Fig. 3.48 Schematic representation of the chalcogenophene-equipped Ru^{II} complexes **58** and **59**. **a** The colors of the electropolymerized poly-**58** films in the reduced (left) and oxidized state (right) are shown [162]. **b** Representative SEM image of poly-**58** deposited onto a Pt electrode [163]. Figures reproduced with kind permission; © 2014 and 2019 Elsevier B.V

Pilo et al. fabricated ruthenopolymer films from the homoleptic Ru^{II} complexes **60** and **61**. For poly-**60** (Fig. 3.49a), electronic interaction between the Ru^{II} centers and the π-conjugated PT backbone was reduced by the insulating -OCH_2- linkages [171]. Compared to the metal-free polymer, poly-**60** was both easier to be oxidized and reduced (+0.65 V vs. +0.75 V and −1.66 V vs. −2.38 V, respectively). Moreover, the metallopolymer appeared to be stable over a wide potential range for several charge/discharge cycles. Analogous electrochemical studies using the Ru^{II} complex **61** (Fig. 3.49b), which contained an ethynyl linkage between the redox-active entities, revealed two oxidation processes in the potential window between +1.15 and +1.51 V as well as two reduction steps between −1.11 and −1.28 V (the former were attributed to the oxidation of the terthiophene and the Ru^{II} centers, respectively; the latter were arising from reductions on the tpy moiety). In CH_2Cl_2, the anodic electropolymerization gave the expected poly-**61** film on the electrode surface. Changing the solvent to MeCN, however, allowed an untypical cathodic electropolymerization to proceed (poly-**61'**). Both materials were characterized with respect to their photophysical and electrochemical properties. The electronic processes for poly-**61'** appeared to be more complex in nature than those observed for poly-**61** and monomer **61**. Furthermore, a smaller energy gap (E_g = 1.95 eV) was determined for poly-**61** when compared to poly-**61'** (E_g = 2.15 eV), still the E_g values were higher than the one observed for the conducting polymer, derived from the metal-free ligand. The authors proposed that, depending on the applied electropolymerization conditions, either an outer- (poly-**61**) or inner-sphere material (poly-**61'**) was generated. Holliday and Swager suggested two different mechanisms, i.e., the inner-sphere and the outer-sphere electron transfer, to rationalize the electronic properties, as observed in these types of conducting polymers (Fig. 3.49b) [159].

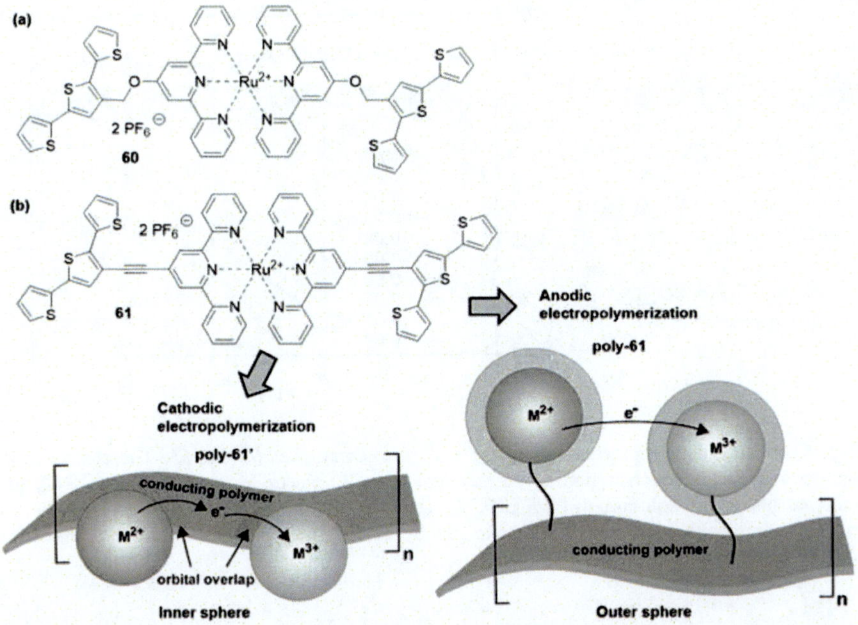

Fig. 3.49 **a** Schematic representation of the Ru^{II} complex **60**. **b** Schematic representation of the Ru^{II} complex **61**, which showed a dual capability for electropolymerization and, thus gave two different types of conducting materials [159]. Figure reproduced with kind permission; © 2005 The Royal Chemical Society

The Ru^{II}–phen complex **62** was electropolymerized on a GC electrode via its 2,5-di(thien-2-yl)-1*H*-pyrrole moiety [172]. The thusly obtained electrochemically-active material was utilized for the direct oxidation of paracetamol to 4-iminocyclohexa-2,5-dien-1-one in a buffered solution at a pH of 7.0. The reaction, as mediated by the Ru$^{II/III}$ redox couple, was studied by CV and chronoamperometry measurements (Fig. 3.51); thereby, a low detection limit of 7.9×10^{-6} M with an accuracy of 99.8% was reached. However, the detection selectivity of the device was rather poor, since other clinically important species were oxidized at a similar potential (e.g., ascorbic acid or dopamine) [173]; thus, a reliable differentiation between acetaminophen and these molecules should be difficult. The same authors also electropolymerized the related Ru^{II} complex **63** and employed the resulting polymer, as catalyst, in a hydrogen-transfer reaction (i.e., reduction of ketones by isopropanol under basic conditions) [174] (Fig. 3.50).

Schmittel et al. reported a series of Ru^{II}–phen *tris*-complexes equipped with azacrown-ether recognition sites [175]. The complexes were electropolymerized via their pending 4*H*-cyclopenta[2,1-*b*:3,4-*b*′]dithiophene (CPDT) moieties onto Pt and ITO electrodes. The obtained hybrid electrodes were employed for the electrochemical sensing of metal ions. It was demonstrated that Ba^{2+} and Pb^{2+} ions, in

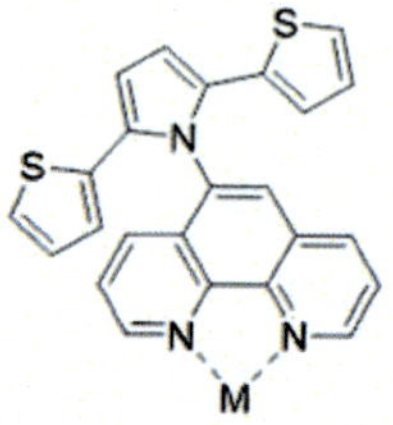

62 (M = $Ru(DMSO)_2Cl_2$)
63 (M = $Ru(bpy)_2^{2+}$)

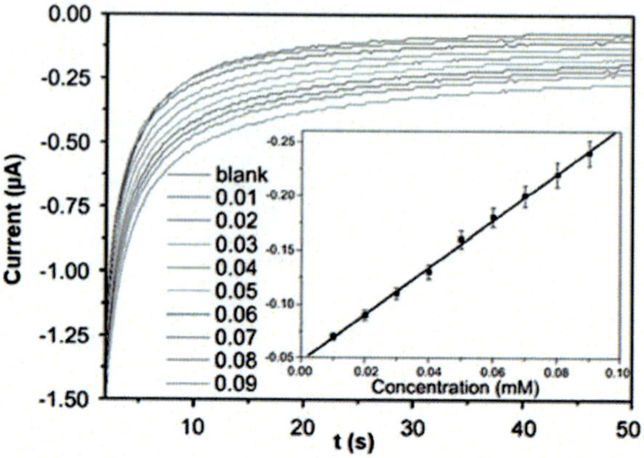

Fig. 3.50 Schematic representation of the Ru^{II}-phen complexes **62** and **63**. The chronoamperograms of various concentration of paracetamol at a GC electrode, coated with poly-**62**, in a buffered phosphate solution (0.1 M, pH 7.0) at 0.4 V are also depicted; the inset shows the corresponding calibration plot for acetaminophen [172]. Figure reproduced with kind permission; ©2011 KCS Publications

Fig. 3.51 Schematic representation of the Ru^{II} complexes **64** and **65**

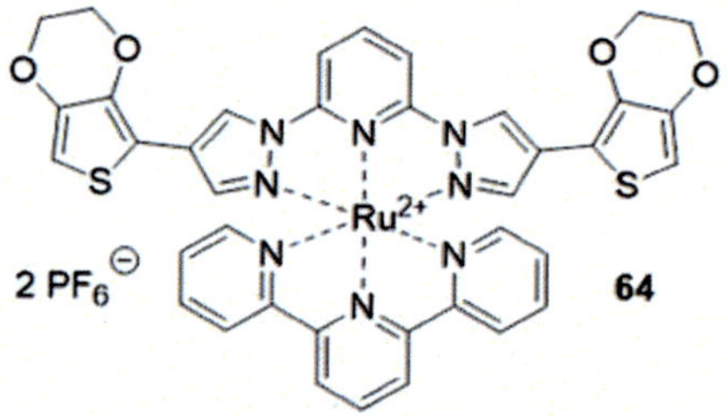

64

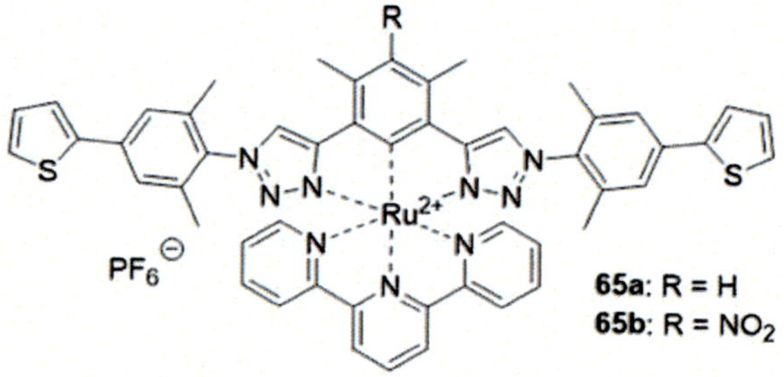

65a: R = H
65b: R = NO_2

contrast to the various alkali metal ions, could be detected (with a higher selectivity for the latter). More recently, Wang et al. fabricated a photo-active electrode by electropolymerizing a homoleptic Ru^{II}–phen *tris*-complex onto an ITO-coated substrate [176]. The electrode gave a photocurrent density of 1.52 μA cm^{-2} upon irradiation with white light at − 0.4 V versus SCE in an aqueous environment; in the presence of O_2, the response increased to 2.30 μA cm^{-2}. Thus, the deposited metallopolymer film was able to act as both, a sensitizer and a catalyst for the oxygen-reduction reaction.

Roncali et al. introduced terminal ethylene dioxythiophene (EDOT) units, as electropolymerizable groups, in order to lower the potential for the anodic electropolymerization. Various bpy-type ligands as well as their Ru^{II} complexes were readily electropolymerized to generate materials, which displayed electroactivity due to the extended PT substructure (at 0.60 V) and the immobilized Ru^{II} complexes (at 0.35 V) [177]. The idea of lowering the electropolymerization potential by using EDOT-functionalized complexes was taken up by Holliday et al., who prepared a linear, conducting ruthenopolymer by the electrochemical polymerization of **64** (Fig. 3.51) [178]. The molecular structure of the deep-red polymer films was corroborated by high-resolution XPS measurements revealing an atomic ratio of Ru-to-S of 1:2.4, being in fairly good agreement with the expected 1:2 ratio. The material featured a broad absorption range from 310 to 700 nm – a behavior that is beneficial for a potential application, as sensitizer, in photovoltaic devices. The first example for the successful anodic electropolymerization of a cyclometallated Ru^{II} complex was given by Schubert et al. (Fig. 3.51) [179]. The oxidative decomposition upon anodic scanning, as observed for **65a**, however, was overcome by introducing an electron-withdrawing substituent on the cyclometallating ring. As a result, **65b** could be electropolymerized at potentials below the critical decomposition-related redox potential of the cyclometallating ring; poly-**65b** featured UV/vis absorption up to 700 nm as well as stable redox switchability associated with electrochromism. Moreover, poly(**65b**-*co*-EDOT) was obtained at various **65b**-to-EDOT ratios; due to the incorporated oligo-/poly-EDOT segments, the absorption was even expanded into the NIR regime. The electrochemical and UV/vis/NIR-spectroelectrochemical behavior of the copolymers reflected the mixed characteristics of the Ru^{II} and EDOT moieties.

Very recently, Yucel, Camurlu et al. reported that the electrochromic properties of electropolymerized ruthenopolymers can be adjusted via the structure of the polymer backbone [180]. In this context, the Ru^{II} complexes **66**, bearing one click-derived (1*H*-1,2,3-triazol-4-yl)-pyridine ligand, were investigated (Fig. 3.52). In general, all three polymer films, poly-**66**, exhibited a multichromatic and ambipolar behavior, which is absent in the respective bare π-conjugated polymers. The photophysical and electrochemical characterization gave E_g values of 2.44, 194, and 1.84 eV for poly-**66a**, poly-**66b**, and poly-**66c**, respectively. The ruthenopolymer films had different colors in their neutral states (yellow, orange, and green, respectively). The electrochromic properties were studied by sequential anodic and cathodic polarization and distinct coloration schemes were found (Fig. 3.52). The CE values of the polymers were 69.42, 298.70, 365.22 cm^2 C^{-1}, respectively; in the case of poly-**66c**, an optical contrast ratio of 46.24% and a switching time of 2.52 s were obtained.

The isomeric Ru^{II} complexes **67** and **68** (Fig. 3.53a) were employed by Jäger, Schubert et al. for the synthesis of type-II and type-III metallopolymer films, respectively, according to Wolf's classification scheme (Fig. 3.1) [181]. Similar to

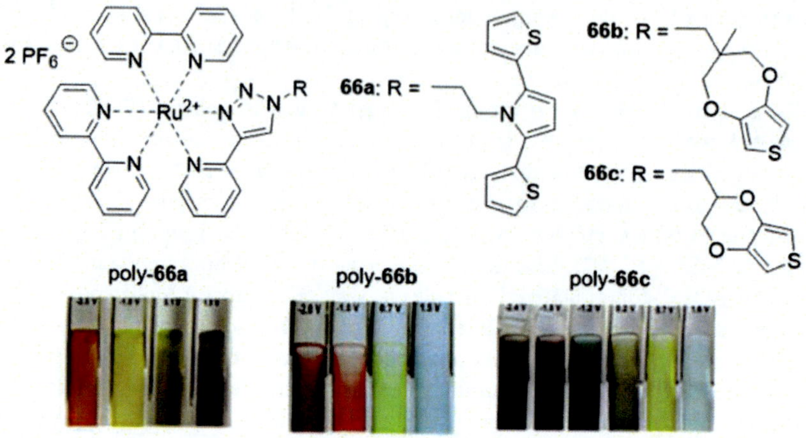

Fig. 3.52 Schematic representation of the Ru^{II} complexes **66**. The different color schemes, as observed in electrochromic measurements of the respective ruthenopolymer films, are also depicted [180]. Figure reproduced with kind permission; © 2020 Elsevier B.V

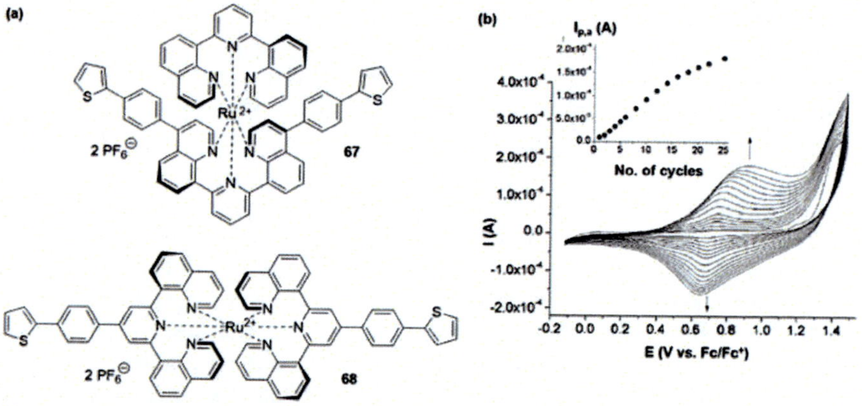

Fig. 3.53 **a** Schematic representation of two isomeric Ru^{II} complexes **67** and **68**. **b** Representative example for the evolution of the CV signal during the electrochemical co-polymerization of **68** and thiophene (monomer ratio of 1:20) [184]. Figure reproduced with kind permission; © 2013 The Royal Chemical Society

their tpy relatives, the dqp-type ligands allow a strictly linear growth of the polymer chain when using monomer **68**, but offer advanced photophysical properties, such as increased excited-state lifetimes in the μs regime (unlike Ru^{II}–tpy *bis*-complexes, the dqp-containing analogs possess ideal octahedral geometries) [182, 183]. The potentiodynamic electropolymerization in the presence of a Lewis acid on GC or

ITO electrodes gave polymer films, which were investigated by (spectro-)electrochemical measurements. For poly-**68**, as a type-III material, a faster electrochemical response was observed, which was attributed to an increased charge-carrier mobility. This assumption was supported by the enhanced polymerization rate as well as by DFT calculations, which revealed a more efficient π-conjugation within poly-**68**. Moreover, copolymers of **67** and **68**, respectively, were prepared with thiophene, as co-monomer (monomer ratios ranging from 0 to 95% thiophene content) [184]. With the increasing content of Ru^{II} complex, the electrochemical properties became more behaved, i.e., reversible switching of the conductivity in combination with a remarkably enhanced stability upon oxidation. Electrochemical impedance spectroscopy revealed conductivities up to 10^{-5} S cm^{-1}; all metallopolymer films featured the characteristic Ru^{II}-based luminescence at *ca.* 750 nm —even under ambient conditions, *i.e.*, in the presence of oxygen.

Mirkin et al. aimed for combining the concepts of redox-switchable hemilabile ligands with conducting polymers. Thereby, the polymer backbone was designed to electrochemically modulate the electronic properties and coordination spheres of coordinated transition metal moieties. For example, a redox-active Ru^{II} complex was covalently attached to the PT backbone through an alkyldiphenylphosphine tether and a labile Ru–S coordinative bond (poly-**69**, Fig. 3.54) [185–187]. It was anticipated that the electron donation to the Ru^{II} center could be sequentially varied by adjusting the redox state of the π-conjugated polymer backbone. FT-IR spectroelectrochemistry was utilized to determine the electron density on the Ru^{II} center through changes in the characteristic CO stretching-band frequency. Only two discrete bands, separated by *ca.* 18 cm^{-1} were observed suggesting that holes were present along the PT chain preventing the tuning of electronic properties over a continuous range. Still, the binding and release of small molecules could be controlled via the oxidation state: The Ru–S bond was cleaved at the cost of an added ligand; whereas, in the oxidized state, no such substitution reaction was observed in NMR experiments (Fig. 3.54b). Later, Wolf et al. prepared some Ru^{II} complexes with terthiophene or pentathiophene entities, similar to Mirkin's complexes; however, an anodic electropolymerization was unrealized for these complexes.[188] More recently, González et al. successfully electropolymerized the half-sandwich complexes **70** (Fig. 3.54a) having two bulky PTA-type stabilizing ligands (PTA: 1,3,5-triaza-7-phosphaadamantane) [188]. In particular, poly-**70b** was suggested to be a promising material for battery or analytical sensor applications: Regarding the former, a reversible p-type doping/undoping response in the potential range between −0.3 and +0.7 V versus SCE might be beneficial; whereas, the response as a reversible redox couple between −0.3 and −1.0 V versus SCE is of interest concerning the latter.

In the 1990s, Rauchfuss et al. investigated sandwich complexes of the general formula $[Ru(\eta^5\text{-thiophene})(\eta^6\text{-arene})]^{2+}$, also in view of their electrochemical properties [189]. Gonzaléz et al. reported on the anodic electropolymerization of such a complex (**71**) [190]. Due to the moisture-sensitivity of the involved intermediates, stringent reaction conditions had to be applied; the obtained poly-**71** was characterized by SEM imaging and CV measurements (Fig. 3.55a). A remarkable

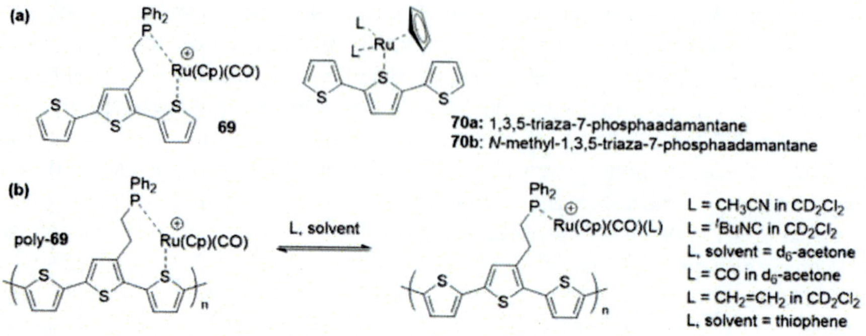

Fig. 3.54 Schematic representation (a) of the Ru^{II} complexes **69** and **70**. **b** Schematic representation of the reversible coordination of small molecules to the not-oxidized poly-**69**

stability in the charging/discharging process of the polymer-coated electrode with regard to the number of scanning cycles was reported—a property that is important for potential implementation, e.g., as cathode in battery applications.

Wolf et al. studied the electropolymerization of the Ru^{II}–oligothienylacetylide complexes **72** (Fig. 3.55b); from these, only **72a** did not give a polymer film on the electrode (presumably due to the aforementioned low stability of the initial radical cation) [191]. Noteworthy, a soluble and, thus, solution-processable derivative of not-accessible poly-**72a** was reported recently (**73**, Fig. 3.55) [192]. For poly-**72b** and poly-**72c**, photophysical and electrochemical measurements suggested an electronic interaction of the π-conjugated oligothienyl units with the Ru^{II} centers

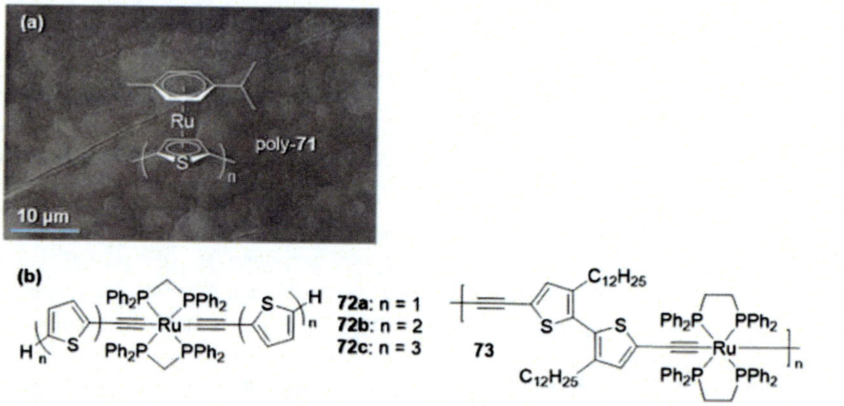

Fig. 3.55 **a** Schematic representation of poly-**71**; the SEM image of the polymer on a Pt disk electrode is shown, as the background [190]. **b** Schematic representation of the Ru^{II}-oligothienylacetylide complexes **72** and **73**. Figure reproduced with kind permission; © 2016 Wiley–VCH

[191]: These electron-donating centers decreased the energy of the π-π* transition in the oligothienyl bridge and the energy of this transition increased upon oxidation of the metal to the +III state. Moreover, the reversibility of the $Ru^{II/III}$ oxidation wave was improved when the length of the oligothienyl unit was increased. Accordingly, stabilization of the Ru^{III} species occurred via delocalization of holes from the metal to the conjugated bridge.

Liu et al. employed a metal-templated [2 + 3]-type cage synthesis to obtain a series of cage-type molecules (**74**), which formed conductive polymers upon subsequent electropolymerization (Fig. 3.56a) [193]. Despite having essentially *iso*-structural organic scaffolds, the materials displayed distinct metal-dependent electrochemical properties, as evidenced by the different $M^{II/III}$ redox windows (M = Fe, Co or Ru). The consecutive electropolymerization using two different monomers gave bilayer devices with different metals in each layer. Beyond simply serving as heavy-atom markers in cross-sectional analysis by, *e.g.*, focused ion beam (FIB) milling and EDX, the redox-active metal centers participated in voltage-dependent inter-layer EIT processes, which gave rise to CVs with distinctively different signatures than those observed for single-layer films or random copolymers (Fig. 3.56b–d).

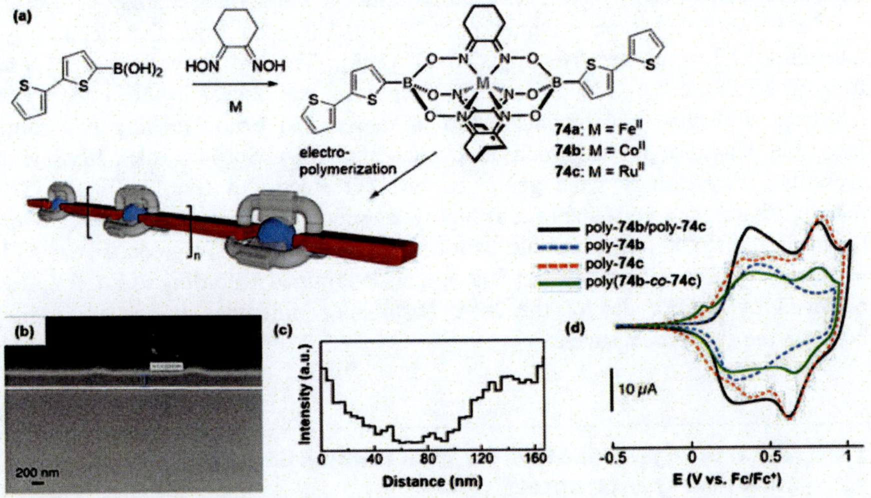

Fig. 3.56 **a** Schematic representation of the synthesis of the cage-type molecules **74**, as monomers for electropolymerization. **b** Cross-sectional field-emission SEM image of poly-**74b**/poly-**74c** on an ITO-coated glass milled by FIB. The bright line under the polymer layer corresponds to the ITO coating on the glass substrates. **c** Pixel-average profile of the poly-**74b**/poly-**74c** bilayer generated by measuring field-emission SEM and averaging 300 pixels along a direction perpendicular to the electrode surface from top to bottom (x = 0: top surface of the material). **d** CVs of the poly-**74b**/poly-**74c** bilayer (black solid line) and the random copolymer poly(**74b**-*co*-**74c**) (gray solid line) on Pt electrodes (CH_2Cl_2 with 0.1 M *n*-Bu_4NPF_6 as supporting electrolyte, scan rate of 25 mV s^{-1}); the CVs of poly-**74b** and poly-**74c** obtained under similar conditions are shown for comparison (dotted lines) [193]. Figure reproduced with kind permission; © 2010 American Chemical Society

Metallophthalocyanines represent organic semiconductor-type materials with significant relevance for various opto-electronic applications: Thin-film transistors, light-emitting devices and, in particular, photovoltaics [194]. Ruthenophthalocyanines have attracted particular interest due to their broad absorption, efficient triplet-state generation, and ability to undergo photo-induced charge-transfer processes; moreover, the binding of axial ligands enabled the coupling to other functional moieties (such metallo-supramolecular Shish-Kebab-type polymers are discussed in Chap. 2) [195]. For example, ruthenophthalocyanines bearing axial pyridine ligands decorated with dendritic oligothiophenes were employed, as electron-donor materials, in BHJ-type solar-cells; these devices reached efficiencies as high as 1.6% [196]. Against this background, Łapkowski et al. presented the ruthenophthalocyanine **75** in which polymerizable 3,5-*bis*(bithienyl)pyridine ligands were coordinated to the Ru^{II} center in both axial positions (Fig. 3.57a) [197]. The electropolymerization of this compound was conducted in various solvents; the resulting polymer networks showed the electrochemical and spectral properties of both subunits—the ruthenophthalocyanines and oligothiophene fragments. In CH_2Cl_2, poly-**75** could be doped/undoped reversibly; whereas, an unusual hysteresis of the redox properties was observed in MeCN. This behavior was attributed to redox matching of the singly oxidized ruthenophthalocyanine moieties with the π-conjugated pyridine-*alt*-quarterthiophene copolymer chains.

The Shish-Kebab-type polymer poly-**76** was obtained from the electropolymerization of monomer **76** (Fig. 3.57a) [198]. The macrocyclic porphycene framework, as a constitutional isomer to porphyrin, was chosen by the authors due to its more intense and tunable nature of absorption bands making it a more favorable candidate for electrochromic pigments [199]. Sub-μm thick films with crystalline morphology were generated on ITO electrodes (Fig. 3.57b, c). The material featured a very stable and highly reproducible electrochromic behavior, based on the $Ru^{II/III}$ redox couple (color scheme: blue ↔ olive-green, Fig. 3.57d, e); the high CE of 178 $cm^2\ C^{-1}$ at 609 nm, as determined according to Eq. 3.2, was achieved by making use of the large absorbance difference in the vis region between the two redox states.

3.6 Electropolymerization of Miscellaneous Ru^{II}-containing Monomers

Beyond the three major classes of electropolymerized films derived from vinyl-, pyrrole-, or thiophene-functionalized Ru^{II} monomers (see the three previous Sections), also other functionalities have—to some extent—been employed basically for the same purpose.

The oxidative electropolymerization of Ru^{II} complexes containing NH_2-functionalized polypyridyl ligands was already discussed by Murray, Meyer et al. in 1983 [200]. A variety of complexes (**77** and **78**) was employed in that study and fairly stable metallopolymer films on Pt, SnO_2 or vitreous carbon electrodes were

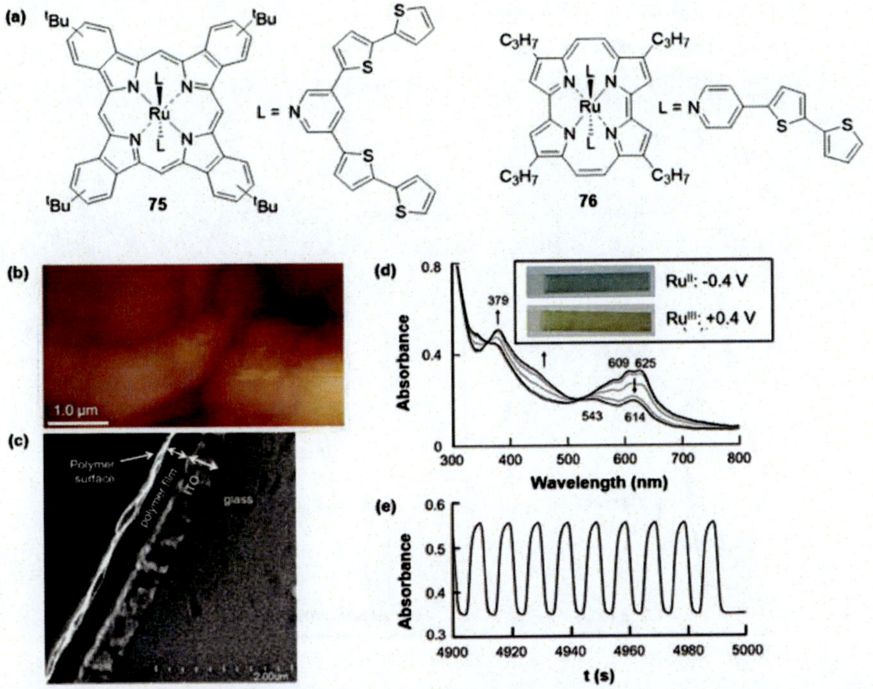

Fig. 3.57 **a** Schematic representation of the Ru^{II}-MAC complexes **75** and **76** (MAC: macrocyclic ligand); **b** tapping-mode AFM image of poly-**76**; **c** cross-sectional SEM image of poly-**76**; **d** UV/vis absorption spectral change of poly-**76** from Ru^{II} (−0.40 V) to Ru^{III} (+0.40 V); electrolyte: 0.1 M n-Bu_4NPF_6 in CH_2Cl_2. The inset shows photographs of the corresponding Ru^{II} and Ru^{III} polymers. **e** Repetitive switching of the absorption intensity of poly-**76** between −0.40 and +0.40 V [198]. Figure reproduced with kind permission; © 2015 American Chemical Society

formed (Fig. 3.58a, c). However, complex **78a**, having only a single NH_2-group, could only be copolymerized with **78c**. According to IR and UV/vis absorption spectroscopic measurements, the polymerization process was ligand-based and highly susceptible towards bases and traces of water. In analogy to the oxidative polymerization of anilines, the first step was suggested to be a ligand-base oxidation forming a labile radical cation which, in turn, was deprotonated to afford a radical intermediate; subsequently, either of its mesomeric species attacks the next monomer at the NH_2-moiety (Fig. 3.58b). Replacing the conventional organic or aqueous solvents by an ionic liquid, i.e., 1-butyl-2,3-dimethylimidazolium *bis* [(trifluoromethyl)sulfonyl]imide, accelerated the deposition rate of poly-**78c** by an order of magnitude [201]. Though the spectral properties of poly-**78c** prepared in ionic liquids, MeCN or H_2SO_4 were almost identical, significant differences were identified regarding the surface morphology: SEM imaging revealed a very rough surface for the ruthenopolymer prepared in the ionic liquid; accordingly, the

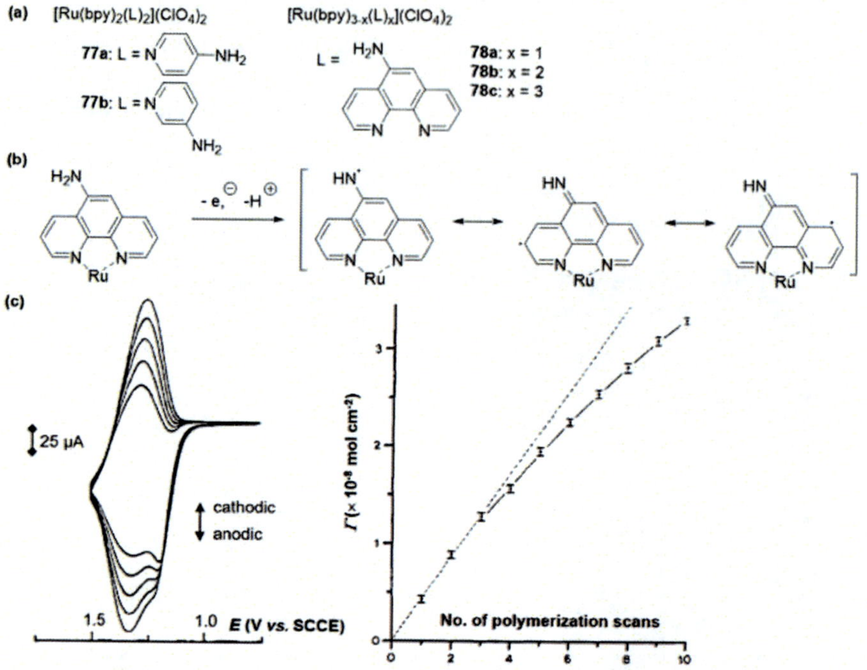

Fig. 3.58 **a** Schematic representation of the NH_2-functionatized Ru^{II} complexes **77** and **78**. **b** Schematic representation of the proposed formation of the intermediate radical, as the initiating species. (**c**, left) Representation of the oxidative electropolymerization of **78c** at a Pt electrode by repeated potential scans (third to seventh cycle, 200 mV s-1). (**c**, right) Plot of the apparent surface coverage (Γ) of poly-**78c** versus the number of electropolymerization scans under the same scan conditions (Γ values were determined after each scan in a fresh electrolyte solution) [200]. Figure reproduced with kind permission; © 1983 American Chemical Society

homogeneous charge transport through that film occurred by two orders of magnitude faster than observed for the smooth films deposited from MeCN (*ca.* 6.5×10^{-9} cm^2 s^{-1}). Moreover, the polymer films showed an ECL behavior when tri(*n*-propyl)amine was added, as sacrificial reactant. The films produced from sulfuric acid were very thin when compared to those produced from the ionic-liquid solution; however, the ECL signal was of similar intensity. The electropolymerization of **78c** (as well as of the analogous Fe^{II} complex) from an non-buffered aqueous solution gave thin films on GC or ITO electrodes [202]. The deposited films were electronically inactive, i.e., an insulator-type behavior was observed in the potential window from 0 to +1 V; this finding was attributed to a (partial) demetallation in the course of polymerization. Electrochemical investigations in the presence of redox-active analyte molecules showed that the polymer films functioned as molecular sieves in a highly selective manner: Neutral and hydrophobic molecules penetrated the polymeric electrode coating much faster than charged or hydrophilic ones. AFM imaging, UV/vis absorption spectroscopy, and XPS were employed to clarify how the molecular structure influenced the apparent electrochemical behavior. The molecular-sieving behavior was proposed to follow a classical membrane-transport model, since the transport rates also varied inversely with the membrane thickness, *i.e.*, thinner films gave larger rates.

Polyaniline (PANI) represents one of the most intensively studied materials due to its notable properties, such as manifold and controlled redox chemistry, environmental stability, and possibility for reversible doping [203]. Remarkably, the discovery of PANI dates back to 1834 when Runge performed oxidation reactions with aniline (named kyanol at that time) [204]. It has, thus, been argued that PANI was the first full synthetic polymer reported in the literature [205, 206]. Noteworthy, the name PANI came into use only in the 1960s; before this, various other names were commonly used which referred to the colors of the different occurring oxidation states of PANI and their salts (i.e., emeraldine, azurine, aniline black, etc.) [206]. Basically, one can distinguish between five forms of PANI, which all display distinct photophysical and electrochemical properties. The idealized structures of these species are summarized in Fig. 3.59; however, besides the depicted head-to-tail connection, also the tail-to-tail phenyl–phenyl linkage significantly contributes to the polymeric structure [206]. The fully reduced form, the so-called leucoemeraldine base, is characteristically transparent; whereas, the fully oxidized species, referred to as pernigraniline, has a dark purple coloration. The half-oxidized, violet-blue-colored emeraldine base form of PANI can be protonated to give the green emeraldine salt, which is electrically conducting. The reversible interconversion between these different species can be achieved either electrically or chemically (i.e., by treatment with an appropriate reducing or oxidizing agent). Thus, PANI has been utilized, as active material, in energy-storage and energy-conversion applications [207]. Moreover, the distinct color changes between the various forms of PANI are of importance for sensing or electrochromic applications [208, 209].

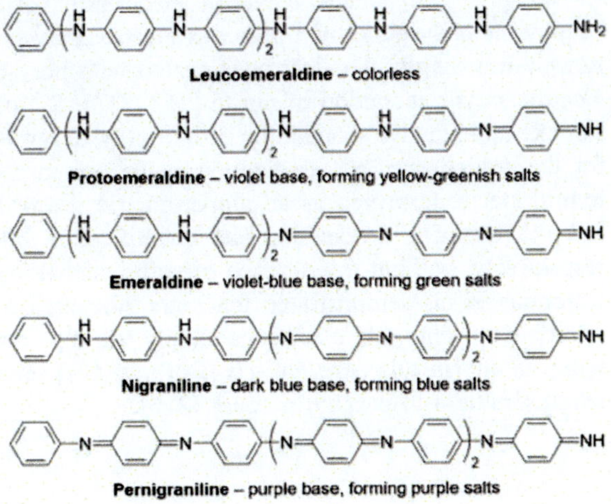

Fig. 3.59 Schematic representation of the various forms of PANI

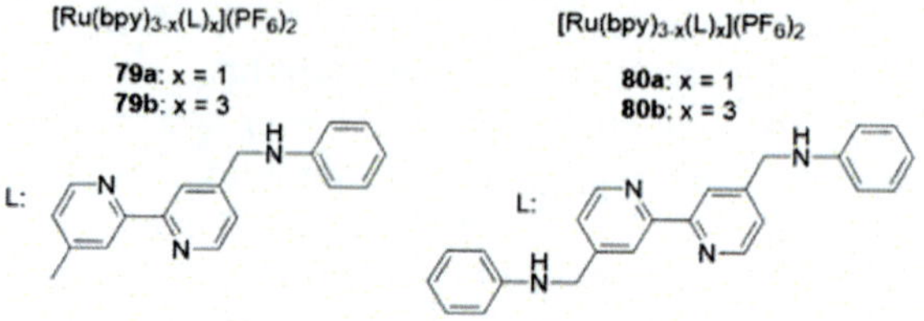

Fig. 3.60 Schematic representation of the aniline-equipped Ru^{II} complexes **79** and **80**

The first controlled electrochemical synthesis of PANI was reported by Diaz and Logan, who obtained PANI, as an adhesive and electroactive film on a Pt electrode, by the potentiodynamic electro-oxidation of aniline [210]. Later, Horwitz et al. attached the aniline group to metal complexes and studied its oxidative electropolymerization. In this context, the Ru^{II} complexes **79** and **80** were also employed (Fig. 3.60) [211]. In contrast to **79a**, this complex possessed only one aniline moiety, **79b** and **80** gave stable polymer films on GC and Pt electrodes. Besides the expected tail-to-tail coupling, also head-to-tail coupling of the monomers was demonstrated by CV measurements. Though these authors gave a proof-of-concept for the electropolymerization of aniline-equipped complexes, further studies in this respect have—to the best of our knowledge—not been published (presumably due to the fact that, that vinyl-, pyrrole- or thiophene-functionalized monomers can be electropolymerized in a much more controlled fashion).

Remarkably, the tetra(pyridin-4-yl)porphyrins, to which four $[Ru(5\text{-}NO_2\text{-}phen)_2Cl]^+$ moieties were coordinated (**81**, Fig. 3.61a), could be electropolymerized reductively to generate polymer coatings on GC electrodes [212]. Presumably, reduction at −0.9 V formed an initial radical anion on the $5\text{-}NO_2$-phen ligand which, in turn, initiated the polymerization. However, the mechanism could not be fully clarified—neither the metal-free $5\text{-}NO_2$-phen ligand nor the fragment $[Ru(5\text{-}NO_2\text{-}phen)_2Cl]^+$ alone could be electropolymerized. Tetraruthenated metalloporphyrins are well-known to act as electrocatalysts in a range of reduction reactions. Accordingly, the electrodes coated with poly-**81** were tested with respect to electrocatalytic reduction of nitrite [212, 213], S^{IV} oxoanions [214] or CO_2 [215]. For example, the Ni^{II}-containing film of poly-**81b** was employed, as electrocatalyst, for the reduction of S^{IV} oxoanions in the presence of gallic acid. Moreover, the hybrid electrode served as an amperometric sensor towards sulfite in white-wine samples featuring a detection and quantification limit of 1.40 and 4.68 mg L^{-1}, respectively, without interference of gallic acid (Fig. 3.61b) [214]. More recently, Calufmán et al. demonstrated that tetraruthenated metalloporphyrins can also be immobilized onto GC electrodes simply via dip-coating when combined with an ionic liquid {in this case, the *cis*-$[Ru(bpy)_2Cl_2]$ precursor was coordinated to the tetra(pyridine-4-yl)porphyrin core} [216].

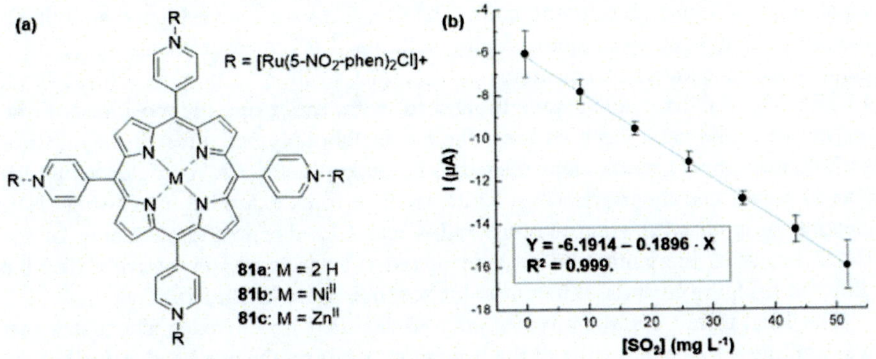

Fig. 3.61 a Schematic representation of the tetraruthenated porphyrins **81**. **b** Calibration curve obtained for the determination of sulfite in Chilean white wine using a GC electrode coated with poly-**81b** (0.1 KCl was used as supporting electrolyte) [214]. Figure reproduced with kind permission; © 2012 Wiley–VCH

In addition to the electrochromic Ru^{II}-containing metallopolymers obtained from the oxidative electropolymerization of vinyl-functionalized complexes (see Sect. 1.2) [217, 218], Zhong et al. also used the dinuclear Ru^{II} complexes **82** (Fig. 3.62a), equipped with two terminal diphenylamino moieties, in a head-to-tail-type electropolymerization [219]. Accordingly, ruthenopolymer films were grown on electrode surfaces by the formation of *bis*(diphenylamine)-

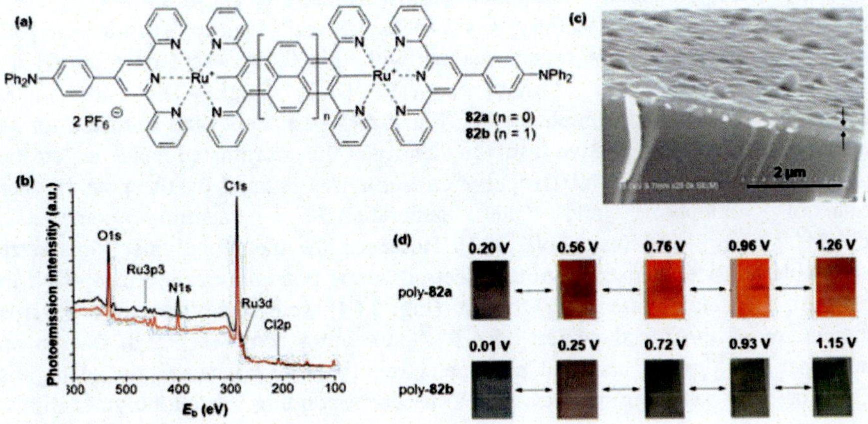

Fig. 3.62 a Schematic representation of the dinuclear complexes **82**. **b** XPS spectra of poly-**82a** (black curve) and poly-**82b** (red curve) on ITO electrodes. **c** SEM image of poly-**82b** on an ITO electrode ($\Gamma = 2.8 \times 10^{-9}$ mol cm^{-2}); the polymeric layer (thickness of *ca.* 100 nm) is indicated by arrows. **d** Pictures of the individual film colors at the respective potentials [219]. Figure reproduced with kind permission; © 2013 American Chemical Society

functionalized biphenyl bridging units. The thin films were investigated applying a range of analytical techniques; high-resolution XPS measurements revealed a slightly lower content of Ru atoms in poly-**82a** relative to poly-**82b** (0.86% vs. 1.03%, Fig. 3.62b), which was in accord with the atom composition of the respective monomers. Each of the adherent metallopolymer films displayed four well-defined anodic redox steps related to the incorporated Ru^{II} complexes and *bis* (triarylamine) chromophores (Fig. 3.62c, d). The films exhibited a five-stage NIR electrochromism with good contrast ratios and CEs due to the switching of the IVCT bands of inorganic and organic mixed-valence units, respectively (for the different bridging ligands, different color schemes were obtained).

The head-to-tail metallopolymer poly-**83** featured a rich redox chemistry that was attributed to the reduction of the bridging ligands on the one hand as well as the oxidation of the tetrapyridylpyrazine and Ru^{II} segments on the other (Fig. 3.63a) [220]. The polymer film was investigated by various surface-characterization techniques, such as XPS and AFM imaging; the latter revealed a homogeneous surface with a mean roughness of 5 nm (Fig. 3.63b). The electrochemical device of the configuration ITO|poly-**83**|Al (Fig. 3.63c) revealed a high resistive memory switching in concert with a low operational voltage, a large ON/OFF current ratio (100 to 1000, Fig. 3.63d), good stability (i.e., 500 cycles were tested), and long retention time. In contrast to this, the polymer derived from the related mononuclear complex **84** showed only a poor memory performance. DFT calculation supported the assumption, that the field-induced conductivity of poly-**83** was a result of the population of a stable charge-transfer state.

In contrast to the oxidative electropolymerization of Ru^{II} complexes with amino-substituents, their OH-functionalized counterparts have attracted far less attention. Though already mentioned briefly by Ellis et al. in 1983 (i.e., electropolymerization of $\{[Ru(bpy)_2(py\text{-}3\text{-}OH)_2]^{2+}$ and $[Ru(bpy)_2(py\text{-}3\text{-}OH)_2]^{2+}\}$ [200], the first hand-tight report dealing with this topic was later reported by Hanabusa et al. and Abruña et al. [221, 222]. The homoleptic catechol-functionalized complexes **85** (Fig. 3.64) were deposited onto GC or Pt electrodes, almost irrespective from the identity of the coordinated metal center; the potential window for the electropolymerization was defined by the potential for catechol oxidation and the generation of neutral complexes: $E^{ox}(M^{0/I}) < E_{dep} < E^{ox}(\text{catechol})$ [223]. However, the metallopolymer films were not stable and were rinsed from the electrode upon potential cycling in MeCN. In contrast, the heteroleptic complexes **86** (Fig. 3.64) were highly sensitive towards the nature of the metal center: The Ru^{II}-containing complex could, due to an improper $Ru^{II/III}$ redox potential, not be electropolymerized; whereas, the analogous Os^{II}- or Co^{II}-based complexes gave the corresponding metallopolymer films. Hanabusa et al. reported the electropolymerization of the OH- and NH_2-functionalized complexes **87** and **88** onto GC or ITO-coated glass electrodes [221]. From these, the Ru^{II} complex **88** gave a metallopolymer film on the electrode, which functioned as a photocathode (as demonstrated by photoelectrochemical reduction of MV^{2+}).

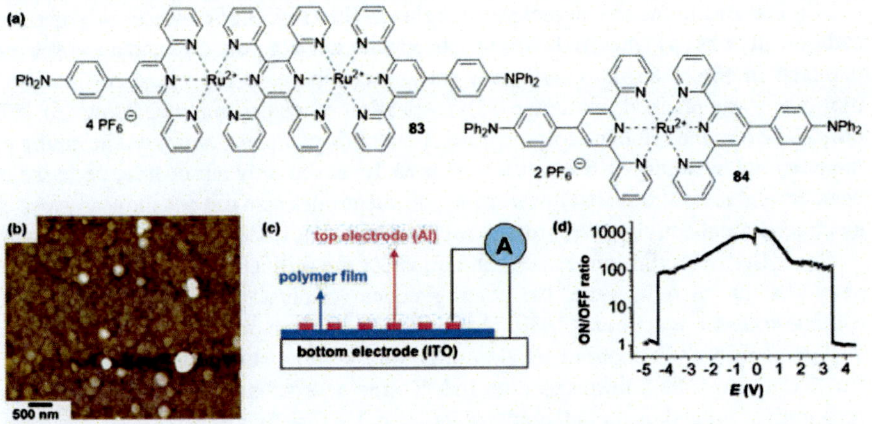

Fig. 3.63 **a** Schematic representation of the triphenylamino-functionalized Ru^{II} complexes **83** and **84**. **b** AFM height image of poly-**83** on an ITO electrode (size: 5 × 5 mm). **c** Schematic representation of the configuration of the memory device. **d** Plot of the ON/OFF current ratio *vs.* voltage [220]. Figure reproduced with kind permission; © 2015 The Royal Chemical Society

The rather uncommon electropolymerization of Ru^{II} complexes with acetylenic moieties was first communicated by Calvert et al., who investigated the ability to reductively electropolymerize $[Ru(bpy)_2(L)_2]^{2+}$ and $[Ru(tpy)(L)_3]^{2+}$ (L = *bis* (pyridin-4-yl)acetylene) [224]. The resulting films showed similar properties as those obtained from analogous vinyl-bearing complexes in terms of charge-transport rate, film morphology, and dry-film conductivity. The structure of the materials resembled that of doped polyacetylene—however, the unique properties of the parent PA could not be reached. A more detailed study by Ziessel et al. employed a series of mono- and binuclear Ru^{II} complexes with bpy or tpy ligands

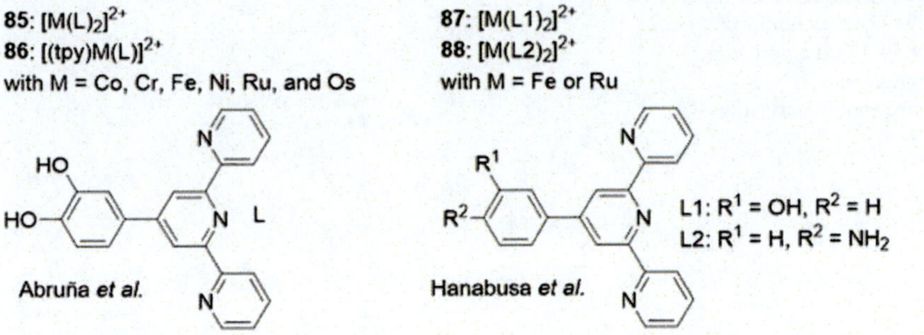

Fig. 3.64 Schematic representation of the OH- or NH_2-functionalized complexes **85**–**88**

[225]. For example, the dinuclear complexes **89** (Fig. 3.65a) were polymerized cathodically onto a Pt electrode (the proposed structures of poly-**89a** and poly-**89b** are depicted in Fig. 3.65b). A sufficient surface coverage in the range of 1×10^{-8} mol cm^{-2} was reached, while the polymerization efficiency remained low (*ca.* 5% only). The obtained ruthenopolymer films were highly stable: Continuous scanning of the potential between +1.6 and 0 V versus SCE caused only a moderate decrease in peak current after 100 scans (by *ca.* 15%); the electrodes also did not show any loss of electrochemical activity after storage for over one year under ambient conditions.

Rault-Berthelot, Simonneaux et al. aimed for the utilization of metalloporphyrin networks as a new type of heterogeneous catalyst, e.g., in oxidation or carbene-transfer reactions [226]. The ruthenoporphyrin monomer **90** (Fig. 3.66) was prepared by treatment of tetra-*meso*-spirobifluorene-substituted porphyrin with $Ru_3(CO)_{12}$ in *o*-dichlorobenzene at 160 °C and was subsequently employed in a preparative potentiostatic electrolysis at +1.4 V. Finally, poly-**90** was removed from the anode, dried under vacuum, and ground to obtain a fine and fully insoluble powder. The polymer was tested as heterogeneous catalyst, in the carbene-transfer reaction with ethyl diazoacetate, as a designated carbene precursor, and styrene or methyl allyl sulfide, as substrates. The reactions proceeded in 87% and 85% yield, respectively, and poly-**90** could be recovered and reused in seven catalytic runs without significant loss-of-activity. In order to reduce the steric crowding in the polymer film and, thereby, make the Ru^{II} center more accessible, these same authors also oxidatively polymerized the tetra-*meso*-fluorene-substituted ruthenoporphyrin **91** (Fig. 3.67) [227]. After removal from the electrode, poly-**91** was utilized to catalyze both the heterogeneous cyclopropanation as well as the 2,3-sigmatropic rearrangements with ethyl diazoacetate even more efficiently than poly-**90** (also the recyclability of the heterogeneous catalyst was improved, Fig. 3.67).

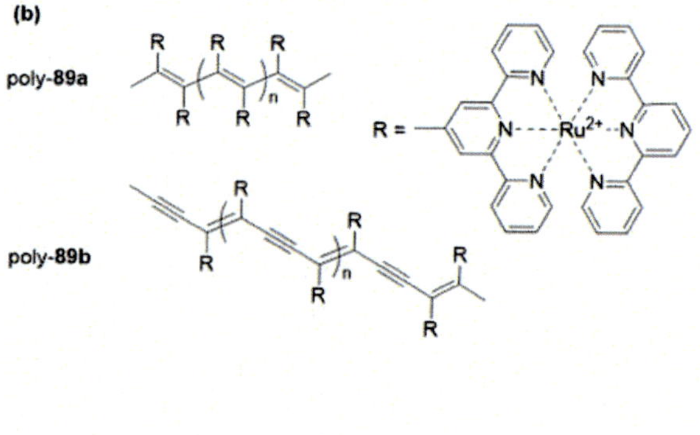

Fig. 3.65 Schematic representation of the dinuclear Ru^{II} complexes **89** (**a**) and of the corresponding polymers poly-**89** obtained from reductive electropolymerization (**b**)

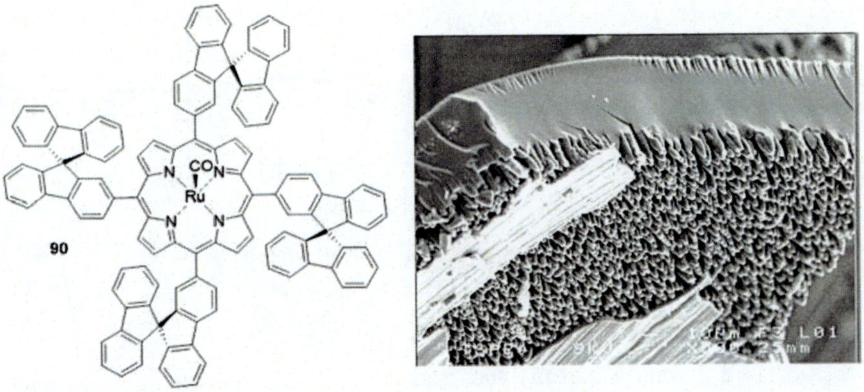

Fig. 3.66 Schematic representation of the Ru^{II} monomer **90** used for reductive electropolymerization via its spirobifluorene substituents. A SEM image of poly-**90** after removal from the electrode is also shown [226]. Figure reproduced with kind permission; © 2003 The Royal Chemical Society

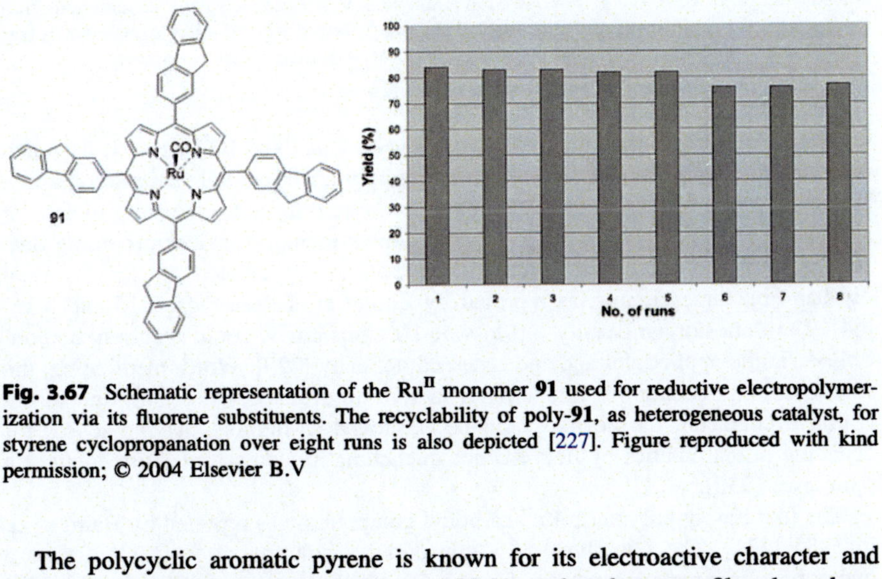

Fig. 3.67 Schematic representation of the Ru^{II} monomer **91** used for reductive electropolymerization via its fluorene substituents. The recyclability of poly-**91**, as heterogeneous catalyst, for styrene cyclopropanation over eight runs is also depicted [227]. Figure reproduced with kind permission; © 2004 Elsevier B.V

The polycyclic aromatic pyrene is known for its electroactive character and ability to be electropolymerized at *ca.* +1.30 V; such polypyrene films have been reported to be highly conductive despite their rather brittle morphology [228]. Zharnikov, Gupta et al. reported on the homoleptic pyrene-equipped Ru^{II} complex **92** (Fig. 3.68a) and its utilization in the electrochemical fabrication of a memory-storage device featuring up to four electrically addressable states (Fig. 3.68b–e) [229]. The obtained ruthenopolymer films, deposited onto ITO-coated glass slides, were found to be highly adherent and robust, since they

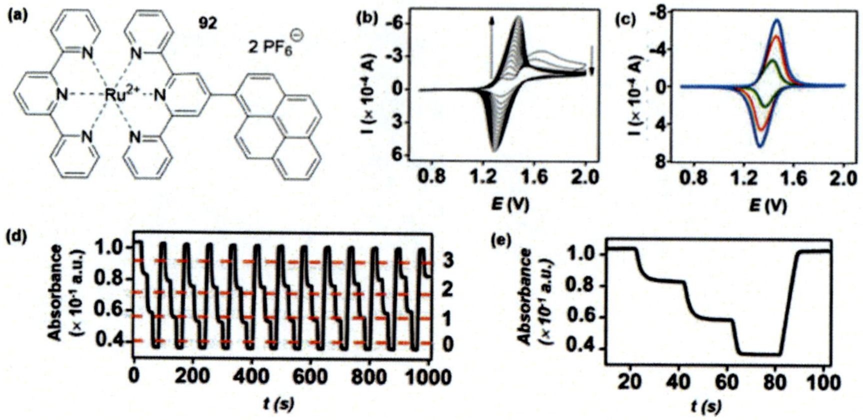

Fig. 3.68 **a** Schematic representation of the pyrene-equipped Ru^{II} complex **92**. **b** Oxidative electropolymerization of **92** (0.5×10^{-3} M, vs. Ag/AgCl, 0.1 M Bu_4NPF_6 in MeCN) on ITO electrode by ten successive potential scans at 50 mV s^{-1}. **c** CV profiles of poly-**92** on ITO obtained after 5 (olive line), 10 (red line), and 20 (blue line) electropolymerization cycles (scan rate of 50 mV s^{-1}). **d** Quaternary states "0", "1", "2" and "3" of poly-**92** upon applying quadruple potential steps of +1.6 V, +1.45 V, +1.35 V, and +1.2 V, respectively; the dashed red lines represent the threshold levels of distinct optical states. **e** Enlarged view of a single quaternary cycle [229]. Figure reproduced with kind permission; © 2016 American Chemical Society

withstood a large number of read–write cycles (*ca.* 5×10^2) as well as harsh temperature conditions (200 °C) which is of importance for facile chip-lithography. These authors proposed their metallopolymer system to be a promising molecular alternative for silicon-based static random access memory devices providing data storage density, which was even higher than that generated by flip-flop and flip-flap-flop logic circuits, as reported by Zhong et al. (see Figs. 3.15 and 3.16) [64]. The data storage density (up to 4×10^{15} bits cm^{-2}) could precisely be controlled via the applied voltage and accessed optically [229]. Worth mentioning, the analogous Os^{II}-containing film offered in five absorbance states and, due to the emissive character, the device could be employed as highly sensitive sensor for detecting nitroaromatics by fluorescence quenching in aqueous medium down to a ppm level [230].

The first binary polymeric Ru^0-carbonyl compound was reported by Baird et al. in 1985 [231]. The photolysis of a concentrated solution of $Ru_3(CO)_{12}$ gave a microcrystalline material with the formula $[Ru(CO)_4]_n$ (**93**), according to elemental analysis. Due to the poor solubility of the material, any structure determination (*e.g.*, estimation of the molar mass) became obsolete. Later, the structure of **93** was solved by detailed XRD analysis (Fig. 3.69a) [232]. The molecular structure consisted of a polymeric array of staggered *trans*-D_{4h}-$Ru(CO)_4$ fragments with a mean Ru–Ru distance of *ca.* 2.86 Å. Computational DFT calculations, using the PBE1PBE and MPW1K hybrid functionals, suggested a Ru–Ru bonding scenario

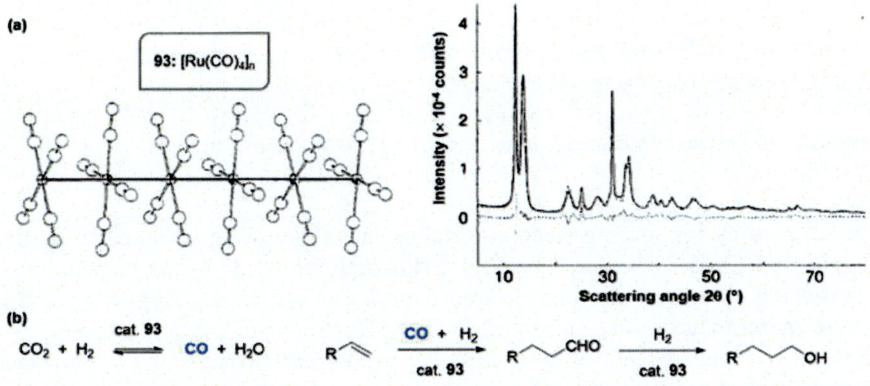

Fig. 3.69 **a** Left: ORTEP representation of the infinite linear Ru^0–Ru^0 chain in **93** with a *trans*-D_{4h} configuration. Right: Experimental (bold line) and calculated XRD pattern (dashed line) of **93** (the dotted line represents the difference plot) [232]. **b** Schematic representation of the reductive hydroformylation of terminal alkenes catalyzed by **93** [235]. Figure reproduced with kind permission; © 1996 The Royal Chemical Society

similar to that of the parent carbonyl complex $Ru_3(CO)_{12}$ [233]. Thermal treatment of **93** gave Ru nanoparticles with a faulted hexagonal close packing (hcp) mode [234]. Haukka et al. employed **93**, as pre-catalyst, for the hydroformylation of terminal alkenes with CO_2 (DMF and LiCl were employed as solvent and promoter, respectively) [235]. These authors proposed, that **93** promoted both—the initial reversed water–gas shift reaction as well as subsequent hydroformylation step. Moreover, **93** also supported the in-situ hydrogenation of the aldehyde intermediates; thus, primary alcohols were formed in high selectivity. Similar results were obtained when using $Ru_3(CO)_{12}$ for the same purpose; however, different (presumably catalytically active) intermediates were identified by mass spectrometric analysis.

A related type-IV material comprising of a polymer backbone built from Ru^0–Ru^0 bonds was reported by Ziessel et al. [141–143]. The potential-controlled electrolysis of the *trans*-[Ru(bpy)$(CO)_2Cl_2$] precursor gave a dark blue, air-sensitive, and strongly adherent electroactive film on the working electrode. FT-IR and UV/vis spectroscopy, elemental analysis as well as electrochemical measurements were conducted to characterize this material. Formation of the polymer with the proposed formula $[Ru^0(bpy)(CO)_2]_n$ (**94**) by cathodic reduction was believed to proceed via a two-electron-addition to the Ru^{II} centers and simultaneous loss of the two Cl^- ligands (Fig. 3.70). The ruthenopolymer **94** featured a high activity in the electrocatalytic CO_2 reduction: At a potential of −1.2 V versus SCE, CO was obtained almost quantitatively—even in an aqueous environment. As already mentioned, the operational stability of these materials could be improved significantly when the Ru^0 chains were established electrochemically in a preformed PPy matrix (Fig. 3.38) [146]. The electrochemical properties of **94** could

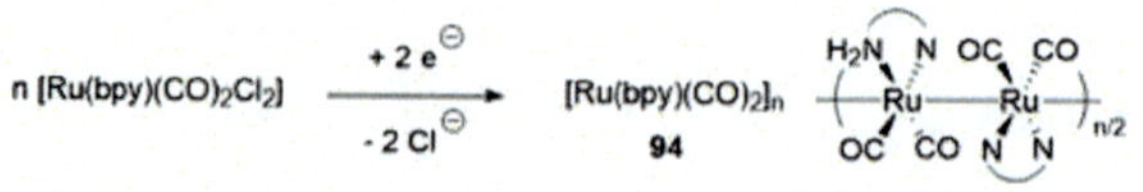

Fig. 3.70 Schematic representation of the synthesis of **94** by cathodic reduction

be adjusted by introducing electron-donating or -withdrawing substituents on the bpy ligand [236]. However, when using 4,4'-dinitro-bpy, as ligand, metallopolymers could not be formed—the electron-poor ligand had a stabilizing effect on the two-electron reduced Ru centers and, thus, the electropolymerization was retarded. Extensive electrochemical studies corroborated that the proposed two-electron-reduction pathway was valid for most of the bpy derivatives studied—except for the electron-rich and/or very bulky ligands. Also the dimeric species **49** (Fig. 3.38) gave the corresponding Ru^0-based polymers, though apparently via different mechanistic routes [237, 238]. The first attempts to validate the molecular structure of various polymers of the general formula $[Ru(L)(CO)_2]_n$ (L = bpy (**94**), dmbpy or phen) by XRD analyses suffered from poor diffraction patterns [239]. However, ab initio XRD studies, based on the known structure of $[Ru(CO)_4]_n$ and various dinuclear Ru^I complexes, suggested that the flat $[Ru(L)(CO)_2]$ fragments were staggered randomly in the polymer chain, with rotations of either 45 or 135° around the Ru^0–Ru^0 bonds. For **94**, these values could be confirmed by using solid-state ^{13}C NMR spectroscopy in concert with DFT calculations [240]. The accuracy of this approach even allowed an estimation of the molar mass from the CO resonances.

Moreover, electrical conductivity in the range of 10^{-1} S cm^{-1} was reported for **94** [241, 242]. The conductivity could be correlated with the degree-of-oxidation of the initially Ru^0 centers to the +I state leading to an increased conductivity, e.g., by four orders of magnitude upon 5% of oxidative doping. Thereby, a linear correlation between conductance and doping level was found at low levels and the conductivity of the films could be repeatedly switched on and off in the presence of charge-compensating inert anions, such as PF_6^- or ClO_4^- (Fig. 3.71). The conduction process was ascribed to a charge delocalization along the metallopolymer chains in analogy to that of other similar chains formed by stacking of planar transition metal complexes with d^8-configuration. Spectroscopic evidence for this process was obtained in the near-IR region, with the absorbance showing a proportional relation to the doping level.

The photoelectrochemical CO_2 reduction, affording higher-value chemicals, represents a very active field of current research addressing the issues of global warming and limited fossil fuel reserves [243]. In this context, hybrid materials—basically combining the properties of semiconductor photoelectrodes with those of absorbed metal complexes—have gained particular interest. One remarkable example for such a photoelectrochemical cell (PEC) was reported by Arai et al., who modified a *p*-type InP photocathode with **94** following Deronzier's protocol for

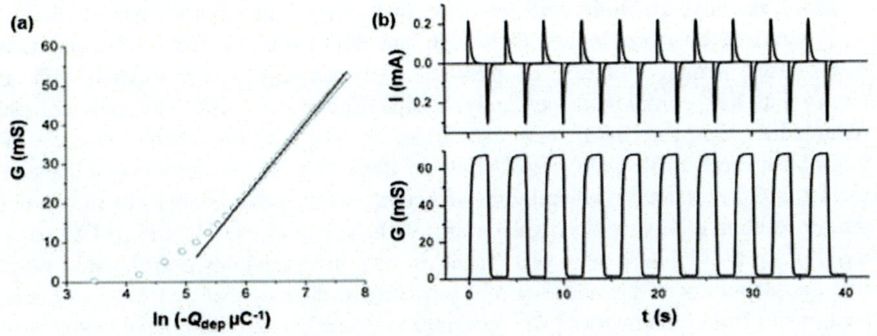

Fig. 3.71 **a** Film conductance at −0.6 V versus ln(-Q_{dep}) with linear regression line. **b** Redox current on a Pt double band electrode induced by potential stepping from −1.1 to −0.6 V (top) and change in film conductance (bottom) as a function of time for a film of **94** in an aq. electrolyte (E_{dep} = −1.25 V, g ≈ 7 μm, last ten cycles of *ca.* 1000 are shown) [241]. Figure reproduced with kind permission; © 2009 Elsevier B.V

the electropolymerization of [Ru(bpy)$(CO)_2Cl_2$] [244]. The working scheme for the resulting PEC is depicted in Fig. 3.72; upon irradiation of the modified photoelectrode (400 nm < λ < 800 nm), the formation of formate occurred at significantly lower applied potentials than required in conventional electrocatalytic systems (efficiency of *ca.* 34%). The performance could be improved further to *ca.* 62% by utilizing a pyrrole-substituted Ru^{I} precursor complex (according to Fig. 3.38); in this case, a two-step electropolymerization was performed in order to deposit the PPy-protected Ru^{0} film onto the semiconductor.

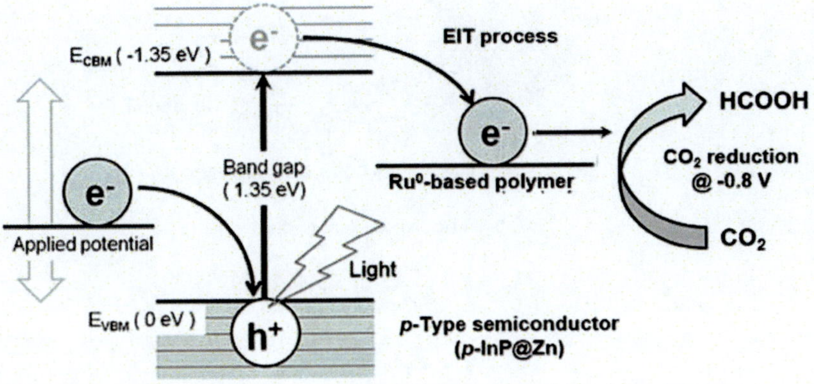

Fig. 3.72 Schematic representation of the energy diagram of a **94**/*p*-InP@Zn photocathode upon irradiation with visible light [244]. Figure reproduced with kind permission; © 2010 The Royal Chemical Society

More recently, graphitic carbon nitride (g-C_3N_4), as an organic semiconductor, was found to be even more efficient in the photocatalytic CO_2 reduction than metal-based semiconductors, in particular when g-C_3N_4 was loaded with an appropriate Ru^{II} complex, as co-catalyst [245]. Either –COOH or –PO_3H_2 moieties (denoted as X in Fig. 3.73) were used to anchor the Ru^{II} complexes onto the flat g-C_3N_4 surface [246]. CO or HCO_2H were identified, as the products of the photoelectric CO_2 reduction, whereby the selectivity of the process could be modulated via the anchor groups: At a high loading with X = -$CH_2PO_3H_2$ a high CO selectivity of up to 70% was achieved. Spectroscopic investigations corroborated that a Ru^0-based polymer—presumably of a structure similar to that of **94**—was generated in situ from the anchored Ru^{II} complexes under irradiation in a solvent of high donor strength (e.g., DMAc, DMF or DMSO); this system was believed to be the active catalysts involved in the CO formation.

Finally, some soluble analogs to **94**, as an archetypical solid-state material, were reported by Deronzier et al. (Fig. 3.74) [247]. UV/vis and IR spectroelectrochemical measurements combined with CV and electrolysis/coulometry experiments showed that the soluble, electron-rich polymer $\{[Ru^0(bpy)(CO)Cl]^-\}_n$ **95a** was assembled from reduced *mer*-$[Ru^{II}Cl_3(CO)(bpy)]^-$ in absence of strongly coordinating solvents. The presence of coordinating solvents, such as propionitrile, enabled the formation of $[Ru(bpy)(CO)(PrCN)]_n$ (**95b**) due to a fast ligand exchange. These two Ru^0-based polymers were well soluble in the electrolyte solution and could be reduced reversibly to the corresponding [bpy•⁻]-containing species. However, a characterization of the molecular structure utilizing solution-based techniques was not performed. Moreover, **95b** was also obtained from *cis*-$[Ru(bpy)(CO)(MeCN)Cl_2]$ in PrCN, while its derivative $[Ru(CO)(MeCN)(bpy)]_n$ (**95c**) was synthesized from the same precursor in THF.

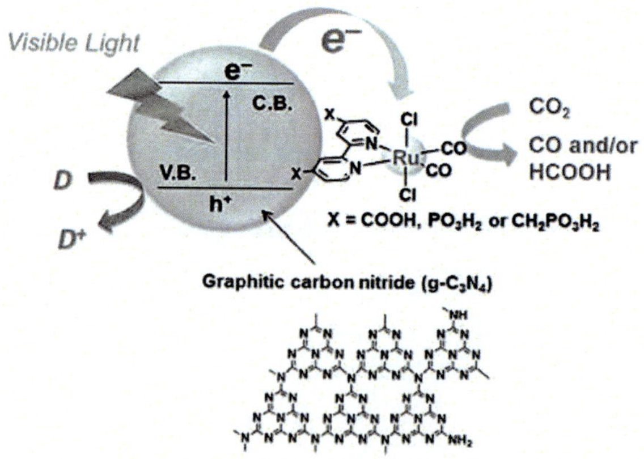

Fig. 3.73 Schematic representation of the hybrid photocatalyst comprising a semiconducting g-C_3N_4 electrode loaded with Ru^{II} complexes [246]. Figure reproduced with kind permission; © 2016 American Chemical Society

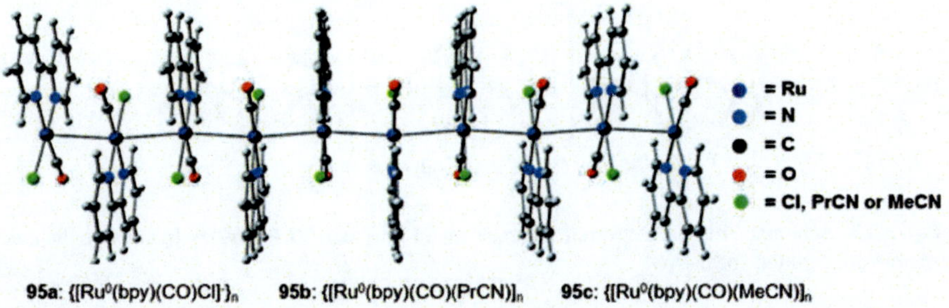

Fig. 3.74 Schematic representation of the soluble linear Ru^0-based polymers **95** [247]. Figure reproduced with kind permission; © 2009 American Chemical Society

The polymerization of fullerenes can basically be achieved via two approaches—chemically or electrochemically; depending on the targeted application either of these methods have been utilized [248]. According to Balch and Olmstead, transition metal complexes can be added to the outer surface of fullerenes to give the so-called fullerides of the general formula M_xC_{60} (where M denotes a main group or transition metal ion) [249]. Consequently, both approaches have been combined giving rise to the formation of metallopolymers **96** comprising of fullerene moieties and transition metal centers in a strictly alternating fashion (Fig. 3.75). Typically, the polymerization gives 1D polymer chains which, in turn, undergo cross-linking towards 2D/3D species. Thermal recombination of these affords more defined and compact 3D arrays that can be metallated even further (Fig. 3.76). Thus, the ratio of transition-metal precursor to C_{60} as well as the reaction conditions strongly influence the composition (and morphology) of the formed polymeric fullerides, potentially ranging from chain-like $-(M\text{-}C_{60})_n-$ assemblies to compact 3D materials (e.g., nanospheres or nanoparticles).

A broad variety of materials, based on Ru and fullerenes, is known from the literature [250]; however the first in-depth investigation on Ru-C_{60} polymers has just been reported rather recently [251]. The chemical synthesis from an appropriate Ru^0 precursor complex and C_{60} was conducted under high-pressure conditions; thereby the composition and morphology of the resulting hybrid polymers could be adjusted by variation of the solvent and feed ratio of educts. An electrochemical synthesis has, to the best of our knowledge, not yet been reported.

3.7 Concluding Remarks

The formation of metallopolymers via electrochemical methods represents a versatile alternative to the chemical synthesis of such materials. The latter typically generates materials, which are (at least partially) soluble in organic solvents; whereas, the former offers the advantage to directly deposit metallopolymers in

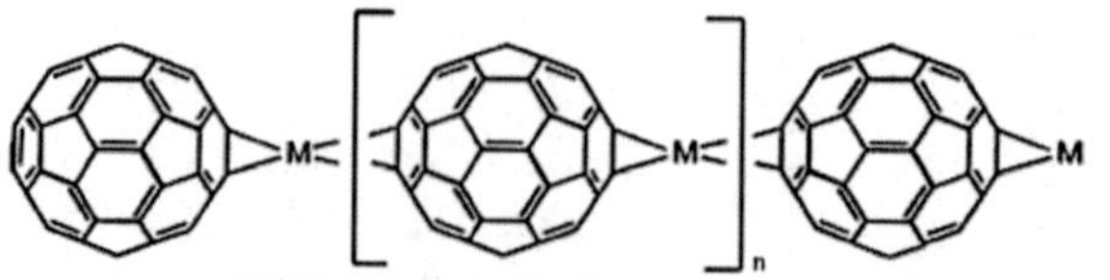

96 (M = Pd, Pt, Ir, Rh, Au, Ag or Ru)

Fig. 3.75 Schematic representation of fullerene-metal-polymers **96** (ancillary ligands on the metal centers are omitted for clarity)

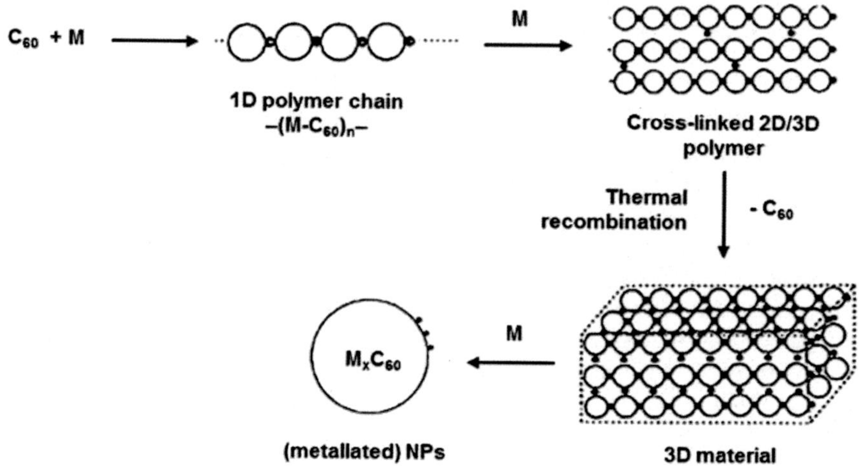

Fig. 3.76 Schematic representation of the formation of fullerene-metallopolymers [248]. Figure redrawn according to the literature; © 2006 Springer Int. Publ

course of their formation onto conductive substrates, such as electrodes. Various functional groups can be electropolymerized—anodically as well as cathodically—without disturbance of the metal complex they are attached to: polypyridyl complexes, metallocenes, metalloporphyrins, etc. Thusly, fabricated ruthenopolymer films have been utilized, as active layers, in various opto-electronic, sensing or catalytic applications. Thereby, the photophysical and electrochemical properties of the metal complexes are combined with those of the (in many) cases conductive polymer backbone. In particular, the electrochromic polymer films are worth mentioning in this context, since their implementation into smart windows is highly aspired. Electrodes coated with redox-active polymer films were found to be highly accurate and sensitive detectors for a range of analytes, inorganic as well as organic ones. These examples demonstrate the enormous potential of electrochemically generated ruthenopolymers. Moreover, the films can be prepared using a single electrochemical set-up—independent from the chemical structure of the complex to be electropolymerized. Taking the limit of the natural abundance and the relatively high price of Ru into account, one further benefit of this approach becomes apparent: Thin, yet highly active films can already be prepared from small amounts of Ru-containing monomers.

References

1. Friebe, C., Hager, M.D., Winter, A., Schubert, U.S.: Metal-containing polymers *via* electropolymerization. Adv. Mater. **24**(3), 332 (2012)
2. Wolf, M.O.: Transition-metal–polythiophene hybrid materials. Adv. Mater. **13**(8), 545 (2001)
3. Abruña, H.D.: Coordination chemistry in two dimensions: chemically modified electrodes. Coord. Chem. Rev. **86**, 135 (1988)
4. Moses, P.R., Wier, L., Murray, R.W.: Chemically modified tin oxide electrode. Anal. Chem. **47**(12), 1882 (1975)
5. Oyama, N., Anson, F.C.: Electrostatic binding of metal complexes to electrode surfaces coated with highly charged polymeric films. J. Electrochem. Soc. **127**(1), 249 (1980)
6. Oyama, N.: Electrostatic binding of metal complexes to electrode surfaces coated with highly charged polymeric films. J. Electrochem. Soc. **127**(1), 247 (1980)
7. Oyama, N., Shimomura, T., Shigehara, K., Anson, F.C.: Electrochemical responses of multiply-charged transition metal complexes bound electrostatically to graphite electrode surfaces coated with polyelectrolytes. J. Electroanal. Chem. **112**(2), 271 (1980)
8. Rubinstein, I., Bard, A.J.: Polymer films on electrodes. Part 4. Nafion-coated electrodes and electrogenerated chemiluminescence of surface-attached tris(2,2'-bipyridine)ruthenium(II). J. Am. Chem. Soc. **102**(21), 6641 (1980)
9. Murray, R.W.: Polymer modification of electrodes. Ann. Rev. Mater. Sci. **14**(1), 145 (1984)
10. Abruna, H.D., Denisevich, P., Umana, M., Meyer, T.J., Murray, R.W.: Rectifying interfaces using two-layer films of electrochemically polymerized vinylpyridine and vinylbipyridine complexes of ruthenium and iron on electrodes. J. Am. Chem. Soc. **103**(1), 1 (1981)
11. Ruiz, V., Colina, Á., Heras, A., López-Palacios, J.: Electropolymerization under potentiodynamic and potentiostatic conditions: spectroelectrochemical study on electrosynthesis of poly[4,4′-bis(2-methylbutylthio)-2,2′-bithiophene]. Electrochim. Acta **50**(1), 59 (2004)
12. Otero, T.F., De Larreta, E.: Electrochemical control of the morphology, adherence, appearance and growth of polypyrrole films. Synth. Metals **26**(1), 79 (1988)
13. Zhou, M., Heinze, J.: Electropolymerization of pyrrole and electrochemical study of polypyrrole: Evidence for structural diversity of polypyrrole. Electrochim. Acta **44**(11), 1733 (1999)
14. Granqvist, C.G.: Transparent conductive electrodes for electrochromic devices: A review. Appl. Phys. A **57**(1), 19 (1993)
15. Wang, J.: Carbon-nanotube based electrochemical biosensors: A review. Electroanalysis **17** (1), 7 (2005)
16. Li, J., Wu, N.: Semiconductor-based photocatalysts and photoelectrochemical cells for solar fuel generation: a review. Catal. Sci. Technol. **5**(3), 1360 (2015)
17. Zhong, Y.-W., Yao, C.-J., Nie, H.-J.: Electropolymerized films of vinyl-substituted polypyridine complexes: synthesis, characterization, and applications. Coord. Chem. Rev. **257**(7), 1357 (2013)
18. Bandeira, M.C.E., Crayston, J.A., Gonçalves, N.S., Noda, L.K., Glidle, A., Franco, C.V.: Electropolymerization of *trans*-[$RuCl_2(vpy)_4$] complex - EQCM and Raman studies. J. Solid State Electrochem. **11**(2), 231 (2007)
19. Bandeira, M.C.E., Crayston, J.A., Franco, C.V., Glidle, A.: Electrochemical deposition of poly(*trans*-[$RuCl_2$(4-vinylpyridine)$_4$]) and its reductive desorption: cyclic voltammetry and

electrochemical quartz crystal microbalance studies. Phys. Chem. Chem. Phys. **9**(8), 1003 (2007)
20. Calvert, J.M., Schmehl, R.H., Sullivan, B.P., Facci, J.S., Meyer, T.J., Murray, R.W.: Synthetic and mechanistic investigations of the reductive electrochemical polymerization of vinyl-containing complexes of iron(II), ruthenium(II), and osmium(II). Inorg. Chem. **22**(15), 2151 (1983)
21. Denisevich, P., Abruna, H.D., Leidner, C.R., Meyer, T.J., Murray, R.W.: Electropolymerization of vinylpyridine and vinylbipyridine complexes of iron and ruthenium: Homopolymers, copolymers, reactive polymers. Inorg. Chem. **21**(6), 2153 (1982)
22. Oyama, N., Anson, F.C.: Facile attachment of transition metal complexes to graphite electrodes coated with polymeric ligands. Observation and control of metal-ligand coordination among reactants confined to electrode surfaces. J. Am. Chem. Soc. **101**(3), 739 (1979)
23. Oyama, N.: Factors affecting the electrochemical responses of metal complexes at pyrolytic graphite electrodes coated with films of poly(4-vinylpyridine). J. Electrochem. Soc. **127**(3), 640 (1980)
24. Murray, R.W.: Chemically modified electrodes. Acc. Chem. Res. **13**(5), 135 (1980)
25. Abruña, H.D., Bard, A.J.: Electrogenerated chemiluminescence. Part 40. A chemiluminescent polymer based on the tris(4-vinyl-4'-methyl-2,2'-bipyridyl)ruthenium(II) system. J. Am. Chem. Soc. **104**(9), 2641 (1982)
26. Tokel-Takvoryan, N.E., Hemingway, R.E., Bard, A.J.: Electrogenerated chemiluminescence. Part XIII. Electrochemical and electrogenerated chemiluminescence studies of ruthenium chelates. J. Am. Chem. Soc. **95**(20), 6582 (1973)
27. Abruña, H.D., Meyer, T.J., Murray, R.W.: Chemical and electrochemical properties of 2,2'-bipyridyl complexes of ruthenium covalently bound to platinum oxide electrodes. Inorg. Chem. **18**(11), 3233 (1979)
28. Maness, K.M., Terrill, R.H., Meyer, T.J., Murray, R.W., Wightman, R.M.: Solid-state diode-like chemiluminescence based on serial, immobilized concentration gradients in mixed-valent poly[Ru(vbpy)$_3$](PF$_6$)$_2$ films. J. Am. Chem. Soc. **118**(43), 10609 (1996)
29. Kaufman, F.B., Schroeder, A.H., Engler, E.M., Kramer, S.R., Chambers, J.Q.: Ion and electron transport in stable, electroactive tetrathiafulvalene polymer coated electrodes. J. Am. Chem. Soc. **102**(2), 483 (1980)
30. Gould, S., O'Toole, T.R., Meyer, T.J.: Photochemical ligand loss as a basis for imaging and microstructure formation in a thin polymeric film. J. Am. Chem. Soc. **112**(26), 9490 (1990)
31. Leasure, R.M., Ou, W., Moss, J.A., Linton, R.W., Meyer, T.J.: Spatial electrochromism in metallopolymeric films of ruthenium polypyridyl complexes. Chem. Mater. **8**(1), 264 (1996)
32. Nie, H.-J., Yao, J., Zhong, Y.-W.: Synthesis of vinyl-substituted polypyridyl ligands through Suzuki-Miyaura cross-coupling of potassium vinyltrifluoroborate with bromopolypyridines. J. Org. Chem. **76**(11), 4771 (2011)
33. Nie, H.-J., Shao, J.-Y., Wu, J., Yao, J., Zhong, Y.-W.: Synthesis and reductive electropolymerization of metal complexes with 5,5′-divinyl-2,2′-bipyridine. Organometallics **31**(19), 6952 (2012)
34. Nallas, G.N.A., Brewer, K.J.: Chemically modified electrodes incorporating ruthenium complexes of polyazine bridging ligands. Inorg. Chim. Acta **257**(1), 27 (1997)
35. Sanjuán, A.M., Reglero Ruiz, J.A., García, F.C., García, J.M.: Recent developments in sensing devices based on polymeric systems. React. Funct. Polym. **133**, 103 (2018)
36. Potts, K.T., Usifer, D.A., Guadalupe, A., Abruna, H.D.: 4-Vinyl-, 6-vinyl-, and 4'-vinyl-2,2':6',2"-terpyridinyl ligands: Their synthesis and the electrochemistry of their transition-metal coordination complexes. J. Am. Chem. Soc. **109**(13), 3961 (1987)
37. Sende, J.A.R., Arana, C.R., Hernandez, L., Potts, K.T., Keshevarz-K, M., Abruña, H.D.: Electrocatalysis of CO_2 reduction in aqueous media at electrodes modified with electropolymerized films of vinylterpyridine complexes of transition metals. Inorg. Chem. **34**(12), 3339 (1995)

38. Takada, K., Storrier, G.D., Pariente, F., Abruña, H.D.: EQCM Studies of the redox processes during and after electropolymerization of films of transition-metal complexes of vinylterpyridine. J. Phys. Chem. B **102**(8), 1387 (1998)
39. Rakowski Dubois, M., Dubois, D.L.: Development of molecular electrocatalysts for CO_2 reduction and H_2 production/oxidation. Acc. Chem. Res. **42**(12), 1974 (2009)
40. Winter, A., Schubert, U.S.: Metal-terpyridine complexes in catalytic application – a spotlight on the last decade. ChemCatChem **12**(11), 2890 (2020)
41. Moss, J.A., Argazzi, R., Bignozzi, C.A., Meyer, T.J.: Electropolymerization of molecular assemblies. Inorg. Chem. **36**(5), 762 (1997)
42. Arana, C.R., Abruña, H.D.: Monomeric and oligomeric complexes of ruthenium and osmium with tetra-2-pyridyl-1,4-pyrazine (TPPZ). Inorg. Chem. **32**(2), 194 (1993)
43. Yao, C.-J., Zhong, Y.-W., Nie, H.-J., Abruña, H.D., Yao, J.: Near-IR electrochromism in electropolymerized films of a biscyclometalated ruthenium complex bridged by 1,2,4,5-tetra (2-pyridyl)benzene. J. Am. Chem. Soc. **133**(51), 20720 (2011)
44. Yao, C.-J., Yao, J., Zhong, Y.-W.: Metallopolymeric films based on a biscyclometalated ruthenium complex bridged by 1,3,6,8-tetra(2-pyridyl)pyrene: applications in near-infrared electrochromic windows. Inorg. Chem. **51**(11), 6259 (2012)
45. Banasz, R., Wałęsa-Chorab, M.: Polymeric complexes of transition metal ions as electrochromic materials: Synthesis and properties. Coord. Chem. Rev. **389**, 1 (2019)
46. Mortimer, R.J.: Electrochromic materials. Chem. Soc. Rev. **26**(3), 147 (1997)
47. Sonmez, G.: Polymeric electrochromics. Chem. Commun. (42), 5251 (2005)
48. Beaujuge, P.M., Reynolds, J.R.: Color control in π-conjugated organic polymers for use in electrochromic devices. Chem. Rev. **110**(1), 268 (2010)
49. Zhong, Y.-W.: Electrochromism within transition-metal coordination complexes and polymers. In: Mortimer, R.J., Rosseinsky, D.R., Monk, P.M.S. (eds.) Electrochromic Materials and Devices, p. 185. Wiley-VCH, Weinheim (2013)
50. Maier, A., Cheng, K., Savych, J., Tieke, B.: Double-electrochromic coordination polymer network films. ACS Appl. Mater. Interfaces **3**(7), 2710 (2011)
51. Motiei, L., Lahav, M., Freeman, D., van der Boom, M.E.: Electrochromic behavior of a self-propagating molecular-based assembly. J. Am. Chem. Soc. **131**(10), 3468 (2009)
52. Higuchi, M.: Electrochromic organic-metallic hybrid polymers: Fundamentals and device applications. Polym. J. **41**(7), 511 (2009)
53. Chen, B.-H., Kao, S.-Y., Hu, C.-W., Higuchi, M., Ho, K.-C., Liao, Y.-C.: Printed multicolor high-contrast electrochromic devices. ACS Appl. Mater. Interfaces **7**(45), 25069 (2015)
54. Chen, X.-Y., Yang, X., Holliday, B.J.: Photoluminescent europium-containing inner sphere conducting metallopolymer. J. Am. Chem. Soc. **130**(5), 1546 (2008)
55. Hjelm, J., Handel, R.W., Hagfeldt, A., Constable, E.C., Housecroft, C.E., Forster, R.J.: Conducting polymers containing in-chain metal centers: electropolymerization of oligothienyl-substituted M(tpy)2 complexes and in situ conductivity studies, M = Os(II), Ru(II). Inorg. Chem. **44**(4), 1073 (2005)
56. Milum, K.M., Kim, Y.N., Holliday, B.J.: Pt-[NCN] pincer conducting metallopolymers that display redox-attenuated metal-ligand interactions. Chem. Mater. **22**(8), 2414 (2010)
57. Powell, A.B., Bielawski, C.W., Cowley, A.H.: Design, synthesis and study of main chain poly(*N*-heterocyclic carbene) complexes: Applications in electrochromic devices. J. Am. Chem. Soc. **132**(29), 10184 (2010)
58. Qi, Y.H., Desjardins, P., Meng, X.S., Wang, Z.Y.: Electrochromic ruthenium complex materials for optical attenuation. Opt. Mater. **21**(1–3), 255 (2003)
59. Han, F.S., Higuchi, M., Kurth, D.G.: Metallosupramolecular polyelectrolytes self-assembled from various pyridine ring-substituted bisterpyridines and metal ions: Photophysical, electrochemical, and electrochromic properties. J. Am. Chem. Soc. **130**(6), 2073 (2008)
60. Nie, H.-J., Zhong, Y.-W.: Near-infrared electrochromism in electropolymerized metallopolymeric films of a phen-1,4-diyl-bridged diruthenium complex. Inorg. Chem. **53**(20), 11316 (2014)

61. Cui, B.-B., Nie, H.-J., Yao, C.-J., Shao, J.-Y., Wu, S.-H., Zhong, Y.-W.: Reductive electropolymerization of bis-tridentate ruthenium complexes with 5,5''-divinyl-4'-tolyl-2,2':6',2''-terpyridine. Dalton Trans. **42**(39), 14125 (2013)
62. Tang, J.-H., Yao, C.-J., Cui, B.-B., Zhong, Y.-W.: Ruthenium-amine conjugated organometallic materials for multistate near-IR electrochromism and information storage. Chem. Rec. **16**(2), 754 (2016)
63. Cui, B.-B., Yao, C.-J., Yao, J., Zhong, Y.-W.: Electropolymerized films as a molecular platform for volatile memory devices with two near-infrared outputs and long retention time. Chem. Sci. **5**(3), 932 (2014)
64. Cui, B.-B., Tang, J.-H., Yao, J., Zhong, Y.-W.: A molecular platform for multistate near-infrared electrochromism and flip-flop, flip-flap-flop and ternary memory. Angew. Chem. Int. Ed. **54**(32), 9192 (2015)
65. Kajita, T., Leasure, R.M., Devenney, M., Friesen, D., Meyer, T.J.: Electropolymerized films of macromeric assemblies. Inorg. Chem. **37**(19), 4782 (1998)
66. Ashford, D.L., Gish, M.K., Vannucci, A.K., Brennaman, M.K., Templeton, J.L., Papanikolas, J.M., Meyer, T.J.: Molecular chromophore–catalyst assemblies for solar fuel applications. Chem. Rev. **115**(23), 13006 (2015)
67. Wee, K.-R., Brennaman, M.K., Alibabaei, L., Farnum, B.H., Sherman, B., Lapides, A.M., Meyer, T.J.: Stabilization of ruthenium(II) polypyridyl chromophores on nanoparticle metal-oxide electrodes in water by hydrophobic PMMA overlayers. J. Am. Chem. Soc. **136** (39), 13514 (2014)
68. Eberhart, M.S., Wee, K.-R., Marquard, S., Skinner, K., Wang, D., Nayak, A., Meyer, T.J.: Fluoropolymer-stabilized chromophore–catalyst assemblies in aqueous buffer solutions for water-oxidation catalysis. ChemSusChem **10**(11), 2380 (2017)
69. Hanson, K., Losego, M.D., Kalanyan, B., Ashford, D.L., Parsons, G.N., Meyer, T.J.: Stabilization of $[Ru(bpy)_2(4,4'-(PO_3H_2)bpy)]^{2+}$ on mesoporous TiO_2 with atomic layer deposition of Al_2O_3. Chem. Mater. **25**(1), 3 (2013)
70. Hanson, K., Losego, M.D., Kalanyan, B., Parsons, G.N., Meyer, T.J.: Stabilizing small molecules on metal oxide surfaces using atomic layer deposition. Nano Lett. **13**(10), 4802 (2013)
71. Lapides, A.M., Sherman, B.D., Brennaman, M.K., Dares, C.J., Skinner, K.R., Templeton, J. L., Meyer, T.J.: Synthesis, characterization, and water oxidation by a molecular chromophore-catalyst assembly prepared by atomic layer deposition. The "mummy" strategy. Chem. Sci. **6**(11), 6398 (2015)
72. Moss, J.A., Yang, J.C., Stipkala, J.M., Wen, X., Bignozzi, C.A., Meyer, G.J., Meyer, T.J.: Sensitization and stabilization of TiO_2 photoanodes with electropolymerized overlayer films of ruthenium and zinc polypyridyl complexes: A atable aqueous photoelectrochemical cell. Inorg. Chem. **43**(5), 1784 (2004)
73. Lapides, A.M., Ashford, D.L., Hanson, K., Torelli, D.A., Templeton, J.L., Meyer, T.J.: Stabilization of a ruthenium(II) polypyridyl dye on nanocrystalline TiO_2 by an electropolymerized overlayer. J. Am. Chem. Soc. **135**(41), 15450 (2013)
74. Fang, Z., Keinan, S., Alibabaei, L., Luo, H., Ito, A., Meyer, T.J.: Controlled electropolymerization of ruthenium(II) vinylbipyridyl complexes in mesoporous nanoparticle films of TiO_2. Angew. Chem. Int. Ed. **53**(19), 4872 (2014)
75. Ashford, D.L., Lapides, A.M., Vannucci, A.K., Hanson, K., Torelli, D.A., Harrison, D.P., Templeton, J.L., Meyer, T.J.: Water oxidation by an electropolymerized catalyst on derivatized mesoporous metal oxide electrodes. J. Am. Chem. Soc. **136**(18), 6578 (2014)
76. Wu, L., Brennaman, M.K., Nayak, A., Eberhart, M., Miller, A.J.M., Meyer, T.J.: Stabilization of ruthenium(II) polypyridyl chromophores on mesoporous TiO_2 electrodes: surface reductive electropolymerization and silane chemistry. ACS Cent. Sci. **5**(3), 506 (2019)
77. Yang, J., Sykora, M., Meyer, T.J.: Electropolymerization of vinylbipyridine complexes of ruthenium(II) and osmium(II) in SiO_2 sol-gel films. Inorg. Chem. **44**(10), 3396 (2005)

78. Ashford, D.L., Sherman, B.D., Binstead, R.A., Templeton, J.L., Meyer, T.J.: Electro-assembly of a chromophore-catalyst bilayer for water oxidation and photocatalytic water splitting. Angew. Chem. Int. Ed. **54**(16), 4778 (2015)
79. Sherman, B.D., Ashford, D.L., Lapides, A.M., Sheridan, M.V., Wee, K.-R., Meyer, T.J.: Light-driven water splitting with a molecular electroassembly-based core/shell photoanode. J. Phys. Chem. Lett. **6**(16), 3213 (2015)
80. Li, F., Fan, K., Wang, L., Daniel, Q., Duan, L., Sun, L.: Immobilizing Ru(bda) catalyst on a photoanode via electrochemical polymerization for light-driven water splitting. ACS Catal. **5** (6), 3786 (2015)
81. Naota, T., Takaya, H., Murahashi, S.-I.: Ruthenium-catalyzed reactions for organic synthesis. Chem. Rev. **98**(7), 2599 (1998)
82. Nestler, O., Severin, K.: A ruthenium porphyrin catalyst immobilized in a highly cross-linked polymer. Org. Lett. **3**(24), 3907 (2001)
83. Elliott, C.M., Dunkle, J.R., Paulson, S.C.: Electrochemical olefin epoxidation employing an iron(IV)porphyrin/trisbipyridineruthenium(III) polymer catalyst. Langmuir **21**(19), 8605 (2005)
84. Elliott, C.M., Redepenning, J.G.: Stability and response studies of multicolor electrochromic polymer modified electrodes prepared from tris(5,5′-dicarboxyester-2,2′-bipyridine)ruthenium(II). J. Electroanal. Chem. Interfacial Electrochem. **197**(1), 219 (1986)
85. Mayo, F.R.: The dimerization of styrene. J. Am. Chem. Soc. **90**(5), 1289 (1968)
86. Puodziukynaite, E., Oberst, J.L., Dyer, A.L., Reynolds, J.R.: Establishing dual electrogenerated chemiluminescence and multicolor electrochromism in functional ionic transition-metal complexes. J. Am. Chem. Soc. **134**(2), 968 (2012)
87. Sadki, S., Schottland, P., Brodie, N., Sabouraud, G.: The mechanisms of pyrrole electropolymerization. Chem. Soc. Rev. **29**(5), 283 (2000)
88. Vernitskaya, T.Y.V., Efimov, O.N.: Polypyrrole: A conducting polymer; its synthesis, properties and applications. Russ. Chem. Rev. **66**(5), 443 (1997)
89. Deronzier, A., Moutet, J.-C.: Polypyrrole films containing metal complexes: Syntheses and applications. Coord. Chem. Rev. **147**, 339 (1996)
90. Nicolau, Y.F., Davied, S., Genoud, F., Nechtschein, M., Travers, J.P.: Polyaniline, polypyrrole, poly(3-methylthiophene) and polybithiophene layer-by-layer deposited thin films. Synth. Metals **42**(1), 1491 (1991)
91. Dall'Olio, A., Dascola, Y., Varacca, V., Bocchi, V.: Résonance paramagnétique électronique et conductivité d'un noir d'oxypyrrol électrolytique. C. R. Hebd. Seances Acad. Sci., Ser. C **267**, 433 (1968)
92. Kanazawa, K.K., Diaz, A.F., Geiss, R.H., Gill, W.D., Kwak, J.F., Logan, J.A., Rabolt, J.F., Street, G.B.: 'Organic metals': Polypyrrole, a stable synthetic 'metallic' polymer. J. Chem. Soc., Chem. Commun. (19), 854 (1979)
93. Diaz, A.F., Kanazawa, K.K., Gardini, G.P.: Electrochemical polymerization of pyrrole. J. Chem. Soc., Chem. Commun. (14), 635 (1979)
94. Bidan, G., Deronzier, A., Moutet, J.-C.: Redox polymeric film electrode by electropolymerization of a pyrrole-containing complex of ruthenium(II). Nouv. J. Chim. **8**, 501 (1984)
95. Cosnier, S., Deronzier, A., Moutet, J.-C.: Oxidative electropolymerization of polypyridinyl complexes of ruthenium(II)-containing pyrrole groups. J. Electroanal. Chem. Interfac. Electrochem. **193**(1), 193 (1985)
96. Eaves, J.G., Munro, H.S., Parker, D.: Electroactive thin films from the anodic electropolymerisation of iron(II) and ruthenium(II) tris(N-bipyridylpyrrole) complexes. J. Chem. Soc., Chem. Commun. (11), 684 (1985)
97. Eaves, J.G., Munro, H.S., Parker, D.: Anodic electrodeposition of thin films of polypyrrole functionalized with metal bipyridyl redox centers. Inorg. Chem. **26**(5), 644 (1987)
98. Franco, C.V., Bodanese Prates, P., de Moraes, V.N., Paula, M.M.S.: Oxidative formation of electroactive film from polypyridinyl complexes of ruthenium(II)-containing 3-(pyrrol-1-ylmethyl) pyridine. Synth. Metals **90**(2), 81 (1997)

99. Cosnier, S., Deronzier, A., Moutet, J.C.: Photoresponse of platinum electrodes coated by electropolymerized polypyridyl complexes of ruthenium(II)-containing pyrrole groups in the presence of an external quencher. Film thickness effect. J. Phys. Chem. **89**(23), 4895 (1985)
100. Deronzier, A., Jardon, P., Martre, A., Moutet, J.-C., Santato, C., Balzani, V., Credi, A., Paolucci, F., Roffia, S.: Photoredox pathways for the polymerization of a pyrrole-substituted ruthenium tris(bipyridyl) complex. New J. Chem. **22**(1), 33 (1998)
101. Deronzier, A., Eloy, D., Jardon, P., Martre, A., Moutet, J.-C.: Electroreductive coating of electrodes from soluble polypyrrole-ruthenium(II) complexes: Ion modulation effects on their electroactivity. J. Electroanal. Chem. **453**(1–2), 179 (1998)
102. Laguitton-Pasquier, H., Martre, A., Deronzier, A.: Photophysical properties of soluble polypyrrole-polypyridyl-ruthenium(II) complexes. J. Phys. Chem. B **105**(21), 4801 (2001)
103. Martre, A., Laguitton-Pasquier, H., Deronzier, A., Harriman, A.: Preparation and properties of a soluble polypyrrole-polypyridyl-ruthenium(II)-viologen dyad. J. Phys. Chem. B **107**(12), 2684 (2003)
104. Ochmanska, J., Pickup, P.G.: Redox conduction in poly-[Ru(2,2'-bipyridine)2(3-{pyrrol-1-ylmethyl}pyridine)$_2$](ClO_4)$_2$ and enhancement of electron transport rates. J. Electroanal. Chem. Interfac. Electrochem. **297**(1), 197 (1991)
105. Foster, K., McCormac, T.: Synthesis, characterization and electrochemical polymerization of a Ru^{2+} functionalized pyrrole monomer. Electroanalysis **19**(14), 1509 (2007)
106. Cosnier, S., Deronzier, A., Roland, J.F.: Controlled permeability of functionalized polypyrrole films by use of different electrolyte anion sizes in the electropolymerization step. J. Electroanal. Chem. Interfac. Electrochem. **310**(1), 71 (1991)
107. Downard, A.J., Surridege, N.A., Meyer, T.J., Cosnier, S., Deronzier, A., Moutet, J.-C.: Electropolymerized multilayer and copolymeric structures based on substituted pyrroles. J. Electroanal. Chem. Interfac. Electrochem. **246**(2), 321 (1988)
108. Downard, A.J., Surridge, N.A., Gould, S., Meyer, T.J., Deronzier, A., Moutet, J.C.: Photocurrents in thin polymeric films: chromophore/quencher assemblies based on polypyrrole. J. Phys. Chem. **94**(17), 6754 (1990)
109. Deronzier, A., Essakalli, M.: A covalently linked ruthenium tris(bipyridine)-viologen complex-polypyrrole film as a molecular photoelectrode. J. Chem. Soc., Chem. Commun. (3), 242 (1990)
110. Marfurt, J., Zhao, W., Walder, L.: Photocurrents at polmeric triads: Sensitized redox cascades under forward and reverse bias. J. Chem. Soc., Chem. Commun. (1), 51 (1994)
111. Ochmanska, J., Pickup, P.G.: Conducting polypyrrole films containing [Ru(2,2'-bipyridine)$_2$(3-{pyrrol-1-ylmethyl} pyridine)Cl]$^+$. J. Electroanal. Chem. Interfac. Electrochem. **271**(1), 83 (1989)
112. Cosnier, S., Deronzier, A., Roland, J.F.: Polypyridinyl complexes of ruthenium(II) having 4,4'-dicarboxy ester-2,2'-bipyridine ligands attached covalently to polypyrrole films. J. Electroanal. Chem. Interfac. Electrochem. **285**(1), 133 (1990)
113. Lopez, C., Moutet, J.-C., Saint-Aman, E.: Electrochemical recognition of chloride ions by a poly [tris-(2,2'-bipyridine)ruthenium(II)] modified electrode. J. Chem. Soc., Faraday Trans. **92**(9), 1527 (1996)
114. Moutet, J.-C., Popescu, A., Saint-Aman, E., Tomaszewski, L.: Electrochemical recognition of alkali metal cations by electrodes modified withpoly[tris-(2,2′-bipyridine) ruthenium(II)] films. Electrochim. Acta **43**(16–17), 2257 (1998)
115. Beer, P.D., Kocian, O., Mortimer, R.J., Ridgway, C.: Cyclic voltammetry of benzo-15-crown-5 ether-vinyl-bipyridyl ligands, their ruthenium(II) complexes and bismethoxyphenyl-vinyl-bipyridyl ruthenium(II) complexes. Electrochemical polymerization studies and supporting electrolyte effects. J. Chem. Soc., Faraday Trans. **89**(2), 333 (1993)
116. Beer, P.D., Kocian, O., Mortimer, R.J., Ridgway, C.: Syntheses, coordination, spectroscopy and electropolymerisation studies of new alkynyl and vinyl linked benzo- and aza-crown

ether-bipyridyl ruthenium(II) complexes. Spectrochemical recognition of group IA/IIA metal cations. J. Chem. Soc., Chem. Commun. (20), 1460 (1991)
117. Beer, P.D., Kocian, O., Mortimer, R.J., Ridgway, C., Stradiotto, N.R.: Electrochemical polymerisation studies of aza-15-crown-5 vinyl-2,2′-bipyridine ruthenium(II) complexes. J. Electroanal. Chem. **408**(1), 61 (1996)
118. Haddour, N., Cosnier, S., Gondran, C.: Electrogeneration of a biotinylated poly (pyrrole-ruthenium(II)) film for the construction of photoelectrochemical immunosensor. Chem. Commun. (21), 2472 (2004)
119. Haddour, N., Chauvin, J., Gondran, C., Cosnier, S.: Photoelectrochemical immunosensor for label-free detection and quantification of anti-cholera toxin antibody. J. Am. Chem. Soc. **128** (30), 9693 (2006)
120. Ionescu, R.E., Cosnier, S., Herzog, G., Gorgy, K., Leshem, B., Herrmann, S., Marks, R.S.: Amperometric immunosensor for the detection of anti-West Nile virus IgG using a photoactive copolymer. Enzyme Microb. Technol. **40**(3), 403 (2007)
121. Ding, S.-N., Cosnier, S., Shan, D., Sun, Y.-M., Wang, Y.: Electrogenerated chemiluminescence of poly[(2,2′-bipyridyl)(4-(2-pyrrol-1-ylethyl)-4′-methyl-2,2′-bipyridyl)2]ruthenium (II) film. Electrochem. Commun. **12**(7), 905 (2010)
122. Hu, L., Xu, G.: Applications and trends in electrochemiluminescence. Chem. Soc. Rev. **39** (8), 3275 (2010)
123. Xia, J., Ding, S.-N., Gao, B.-H., Sun, Y.-M., Wang, Y.-H., Cosnier, S., Guo, X.: A biosensing application based on quenching the enhanced electrochemiluminescence of poly [tris(N-bipyridylethyl)pyrrole] ruthenium(II) film by Au nanoparticles. J. Electroanal. Chem. **692**, 60 (2013)
124. Gao, B.-H., Ding, S.-N., Wang, Y.-H., Dai, Y., Xia, J., Sun, Y.-M., Cosnier, S.: A solid-state electrochemiluminescence ethanol biosensor based on electrogenerated poly(pyrrole-tris (2,2′-bipyridyl)ruthenium(II)) film/alcohol dehydrogenase/laponite composite. Electroanalysis **25**(3), 697 (2013)
125. Lorenzo, J.M., Munekata, P.E., Muchenje, V., Saraiva, J.A., Pinto, C.A., Barba, F.J., Santos, E.M.: Biosensors applied to quantification of ethanol in beverages. In: Grumezescu, A.M., Holban, A.M. (eds.) Engineering Tools in the Beverage Industry, p. 447. Woodhead Publishing, Duxford (2019)
126. Le Goff, A., Holzinger, M., Cosnier, S.: Characterization of multi-walled carbon nanotube electrodes functionalized by electropolymerized tris(pyrrole-ether bipyridine) ruthenium(II). Electrochim. Acta **56**(10), 3633 (2011)
127. Le Goff, A., Cosnier, S.: Photocurrent generation by MWCNTs functionalized with bis-cyclometallated Ir(III)- and trisbipyridyl ruthenium(II)-polypyrrole films. J. Mater. Chem. **21**(11), 3910 (2011)
128. Fritea, L., Haddache, F., Reuillard, B., Le Goff, A., Gorgy, K., Gondran, C., Holzinger, M., Săndulescu, R., Cosnier, S.: Electrochemical nanopatterning of an electrogenerated photosensitive poly-[trisbipyridinyl-pyrrole ruthenium(II)] metallopolymer by nanosphere lithography. Electrochem. Commun. **46**, 75 (2014)
129. Reuillard, B., Le Goff, A., Cosnier, S.: Polypyrrolic bipyridine bis(phenantrolinequinone) Ru(II) complex/carbon nanotube composites for NAD-dependent enzyme immobilization and wiring. Anal. Chem. **86**(9), 4409 (2014)
130. Collin, J.P., Jouaiti, A., Sauvage, J.P.: Anodic electropolymerization of films of polypyrrole functionalized with metal terpyridyl redox centres. J. Electroanal. Chem. Interfac. Electrochem. **286**(1), 75 (1990)
131. Thompson, M.S., Meyer, T.J.: Kinetics and mechanism of oxidation of aromatic hydrocarbons by $Ru(trpy)(bpy)O_2^+$. J. Am. Chem. Soc. **104**(19), 5070 (1982)
132. McHatton, R.C., Anson, F.C.: Electrochemical behavior of $Ru(trpy)(bpy)(OH_2)^{3+}$ in aqueous solution and when incorporated in Nafion coatings. Inorg. Chem. **23**(24), 3935 (1984)

133. De Giovani, W.F., Deronzier, A.: Electrochemical preparation and behaviour of poly {pyrrole-$[Ru(trpy)(bpy)(OH_2)]^{2+}$} (trpy = 2,2′: 6′,2″-terpyridine) (bpy = 2,2′-bipyridine) films. J. Electroanal. Chem. **337**(1), 285 (1992)
134. Cheung, K.-C., Guo, P., So, M.-H., Zhou, Z.-Y., Lee, L.Y.S., Wong, K.-Y.: Ruthenium terpyridine complexes containing a pyrrole-tagged 2,2′-dipyridylamine ligand—synthesis, crystal structure and electrochemistry. Inorg. Chem. **51**(12), 6468 (2012)
135. Collin, J.P., Deronzier, A., Essakalli, M.: Photoelectrodes based on electropolymerized molecular ruthenium and osmium diads. J. Phys. Chem. **95**(15), 5906 (1991)
136. Romero, I., Rodriguez, M., Llobet, A., Collomb-Dunand-Sauthier, M.-N., Deronzier, A., Parella, T., Stoeckli-Evans, H.: Synthesis, structure and redox properties of a new ruthenium (II) complex containing the flexible tridentate ligand N,N-bis(2-pyridylmethyl)ethylamine, cis-fac-$Ru(bpea)_2^{2+}$, and its homologue attached covalently to a polypyrrole film. J. Chem. Soc., Dalton Trans. (11), 1689 (2000)
137. Rodríguez, M., Romero, I., Sens, C., Llobet, A., Deronzier, A.: Redox and catalytic properties of new polypyrrole modified electrodes functionalized by $[Ru(bpea)(bpy)H_2O]^{2+}$ complexes; bpea=N,N′-bis(2-pyridylmethyl)ethylamine, bpy=2,2′-bipyridine. Electrochim. Acta **48**(8), 1047 (2003)
138. Beley, M., Delabouglise, D., Houppy, G., Husson, J., Petit, J.P.: Preparation and properties of ruthenium (II) complexes of 2,2′:6′,2″-terpyridines substituted at the 4′-position with heterocyclic groups. Inorg. Chim. Acta **358**(11), 3075 (2005)
139. Abboud, M., Kalinina, D., Potvin, P.G.: Pyrrole-substituted tridentate complexes of Ru(II): spectroscopy, electrochemistry, photosensitization and the role of orbital mixing. Inorg. Chim. Acta **362**(14), 4953 (2009)
140. Collomb-Dunand-Sauthier, M.N., Deronzier, A., Ziessel, R.: Electrochemical elaboration of thin films of poly$[Ru^{II}(L)(CO)_2Cl_2]$ (L = 4-(2-pyrrol-1-ylethyl)-4’-methyl-2,2’-bipyridine or 4,4’-bis((3-pyrrol-1-ylpropyloxy)carbonyl)-2,2’-bipyridine): photochemical properties and photoimaging. J. Phys. Chem. **97**(22), 5973 (1993)
141. Collomb-Dunand-Sauthier, M.-N., Deronzier, A., Ziessel, R.: Electrocatalytic reduction of CO_2 in water on a polymeric $[\{Ru^0(bpy)(CO)_2\}]$ (bpy = 2,2'-bipyridine) complex immobilized on carbon electrodes. J. Chem. Soc., Chem. Commun. (2), 189 (1994)
142. Collomb-Dunand-Sauthier, M.-N., Deronzier, A., Ziessel, R.: Electrocatalytic reduction of carbon dioxide with mono(bipyridine)carbonylruthenium complexes in solution or as polymeric thin films. Inorg. Chem. **33**(13), 2961 (1994)
143. Chardon-Noblat, S., Collomb-Dunand-Sauthier, M.N., Deronzier, A., Ziessel, R., Zsoldos, D.: Formation of polymeric $[\{Ru^0(bpy)(CO)_2\}_n]$ films by electrochemical reduction of $[Ru(bpy)_2(CO)_2](PF_6)_2$: its implication in CO_2 electrocatalytic reduction. Inorg. Chem. **33**(19), 4410 (1994)
144. Sigaud, M., Li, M., Chardon-Noblat, S., Aires, F.J.C.S., Soldo-Olivier, Y., Simon, J.P., Renouprez, A., Deronzier, A.: Electrochemical preparation of nanometer sized noble metal particles into a polypyrrole functionalized by a molecular electrocatalyst precursor. J. Mater. Chem. **14**(17), 2606 (2004)
145. Oyarzún, D.P., Chardon-Noblat, S., Linarez Pérez, O.E., López Teijelo, M., Zúñiga, C., Zarate, X., Schott, E., Carreño, A., Arratia-Perez, R.: Comparative study of the anchorage and the catalytic properties of nanoporous TiO_2 films modified with ruthenium (II) and rhenium (I) carbonyl complexes. Chem. Phys. Lett. **694**, 40 (2018)
146. Chardon-Noblat, S., Pellissier, A., Cripps, G., Deronzier, A.: Electrochemical behavior of binuclear complexes (L = 2,2′-bipyridine substituted by pyrrole groups). Their reductive and oxidative electropolymerization. J. Electroanal. Chem. **597**(1), 28 (2006)
147. Cosnier, S., Deronzier, A., Moutet, J.C.: A poly[tris(N-(bipyridylylbutyl)pyrrole)ruthenium (II)]-RuO_2 catalytic modified electrode for organic oxidations. Inorg. Chem. **27**(14), 2389 (1988)

148. Cosnier, S., Deronzier, A., Roland, J.F.: Electrocatalytic oxidation of alcohols on carbon electrodes modified by functionalized polypyrrole-RuO_2 films. J. Mol. Catal. **71**(3), 303 (1992)
149. Ito, M., Tsukatani, T., Fujihara, H.: Preparation and characterization of gold nanoparticles with a ruthenium-terpyridyl complex and electropolymerization of their pyrrole-modified metal nanocomposites. J. Mater. Chem. **15**(9), 960 (2005)
150. Cosnier, S., Deronzier, A., Llobet, A.: Triruthenium cluster-polypyrrole films: A remarkably stable immobilized relay at highly positive potentials. J. Electroanal. Chem. Interfac. Electrochem. **280**(1), 213 (1990)
151. Zhou, Y., Xie, Y.: Capacitive performance of ruthenium-coordinated polypyrrole. New J. Chem. **41**(18), 10312 (2017)
152. Li, X., Wu, Y., Zheng, F., Ling, M., Lu, F.: Preparation and characterization of RuO_2/polypyrrole electrodes for supercapacitors. Solid State Commun. **197**, 57 (2014)
153. Xie, Y.: Electrochemical performance of transition metal-coordinated polypyrrole: A mini review. Chem. Rec. **19**(12), 2370 (2019)
154. Roncali, J.: Conjugated poly(thiophenes): synthesis, functionalization and applications. Chem. Rev. **92**(4), 711 (1992)
155. Yoshino, K., Hayashi, S., Sugimoto, R.-I.: Preparation and properties of conducting heterocyclic polymer films by chemical method. Jpn. J. Appl. Phys. **23**(Part 2, No. 12), L899 (1984)
156. McCullough, R.D.: The chemistry of conducting polythiophenes. Adv. Mater. **10**(2), 93 (1998)
157. Diaz, A.F.: Electrochemical preparation and characterization of conducting polymers. Chem. Scr. **17**(1–5), 145 (1981)
158. Tourillon, G., Garnier, F.: New electrochemically generated organic conducting polymers. J. Electroanal. Chem. Interfac. Electrochem. **135**(1), 173 (1982)
159. Holliday, B.J., Swager, T.M.: Conducting metallopolymers: the roles of molecular architecture and redox matching. Chem. Commun. (1), 23 (2005)
160. Patra, A., Bendikov, M.: Polyselenophenes. J. Mater. Chem. **20**(3), 422 (2010)
161. Jahnke, A.A., Seferos, D.S.: Polytellurophenes. Macromol. Rapid Commun. **32**(13), 943 (2011)
162. Et Taouil, A., Husson, J., Guyard, L.: Synthesis and characterization of electrochromic [Ru(terpy)$_2$ selenophene]-based polymer film. J. Electroanal. Chem. **728**, 81 (2014)
163. Husson, J., Abdeslam, E.T., Guyard, L.: A missing member in the family of chalcogenophene-substituted 2,2′:6′,2″-terpyridine: 4′-(Tellurophen-2-yl)-2,2′:6′,2″-terpyridine, its Ru(II) complex and its electropolymerization as a thin film. J. Electroanal. Chem. **855**, 113594 (2019)
164. Wang, J., Keene, F.R.: Synthesis and electrochemical properties of poly[3-{{ω-[4-2,2′-bipyridyl)]alkyl}}thiophenes], P{{B(n)T}}, and of poly[Ru(II){{B(n)T}}$_3$$^{2+}$]. J. Electroanal. Chem. **405**(1), 71 (1996)
165. Crayston, J.A., Iraqi, A., Morrison, J.J., Walton, J.C.: Synthesis of thiophene subtituted ruthenium and rhenium bipyridyl complexes. Synth. Metals **84**(1), 441 (1997)
166. Zhu, S.S., Swager, T.M.: Design of conducting redox polymers: A polythiophene-Ru(bipy)$_3$$^{2n+}$ hybrid material. Adv. Mater. **8**(6), 497 (1996)
167. Zhu, S.S., Kingsborough, R.P., Swager, T.M.: Conducting redox polymers: Investigations of polythiophene-Ru(bpy)$_3$ hybrid materials. J. Mater. Chem. **9**(9), 2123 (1999)
168. Hjelm, J., Constable, E.C., Figgemeier, E., Hagfeldt, A., Handel, R., Housecroft, C.E., Mukhtar, E., Schofield, E.: A rod-like polymer containing Ru(terpy)$_2$ units prepared by electrochemical coupling of pendant thienyl moieties. Chem. Commun. (3), 284 (2002)
169. Rhoden, C.R.B., Zeni, G.: New development of synthesis and reactivity of seleno- and tellurophenes. Org. Biomol. Chem. **9**(5), 1301 (2011)

170. Dong, B., Xing, Y., Xu, J., Zheng, L., Hou, J., Zhao, F.: Electrosyntheses of free-standing and highly conducting polyselenophene films in an ionic liquid. Electrochim. Acta **53**(19), 5745 (2008)
171. Scanu, R., Manca, P., Zucca, A., Sanna, G., Spano, N., Seeber, R., Zanardi, C., Pilo, M.I.: Homoleptic Ru(II) complex with terpyridine ligands appended with terthiophene moieties: synthesis, characterization and electropolymerization. Polyhedron **49**(1), 24 (2013)
172. Kannan, S., Son, J.-I., Yang, J.-E., Shim, Y.-B.: Electrochemical polymerization of ruthenium(II) complex and application to acetaminophen analysis. Bull. Kor. Chem. Soc. **32**(4), 1341 (2011)
173. Eddy, S., Warriner, K., Christie, I., Ashworth, D., Purkiss, C., Vadgama, P.: The modification of enzyme electrode properties with non-conducting electropolymerised films. Biosensors Bioelectron. **10**(9), 831 (1995)
174. Son, J.-I., Kim, A.-R., Noh, H.-B., Lee, H.-J., Shim, Y.-B., Park, K.-H.: Synthesis and catalytic hydrogen-transfer reaction of ruthenium(II) complex. Bull. Kor. Chem. Soc. **33**, 319 (2012)
175. Schmittel, M., Lin, H.: Electropolymerisable ruthenium(II) phenanthrolines carrying azacrown ether receptors: Metal ion recognition in thin film redox sensors. J. Mater. Chem. **18**(3), 333 (2008)
176. Yang, T., Zhang, C.-X., Li, Y.-J., Fu, Y.-H., Yin, Z.-H., Gao, L.-H., Wang, K.-Z.: A 3D electropolymerized thin film based on a thiophene-functionalized Ru(II) complex: Electrochemical and photoelectrochemical insights. Inorg. Chem. Front. **6**(12), 3518 (2019)
177. Jousselme, B., Blanchard, P., Ocafrain, M., Allain, M., Levillain, E., Roncali, J.: Electrogenerated poly(thiophenes) derivatized by bipyridine ligands and metal complexes. J. Mater. Chem. **14**(3), 421 (2004)
178. Zhu, X.J., Holliday, B.J.: Electropolymerization of a ruthenium(II) bis(pyrazolyl)pyridine complex to form a novel Ru-containing conducting metallopolymer. Macromol. Rapid Commun. **31**(9–10), 904 (2010)
179. Friebe, C., Schulze, B., Görls, H., Jäger, M., Schubert, U.S.: Designing cyclometalated ruthenium(II) complexes for anodic electropolymerization. Chem. Eur. J. **20**(8), 2357 (2014)
180. Guven, N., Sultanova, H., Ozer, B., Yucel, B., Camurlu, P.: Tuning of electrochromic properties of electrogenerated polythiophenes through Ru(II) complex tethering and backbone derivatization. Electrochim. Acta **329**, 135134 (2020)
181. Friebe, C., Görls, H., Jäger, M., Schubert, U.S.: Linear metallopolymers from ruthenium(II)-2,6-di(quinolin-8-yl)pyridine complexes by electropolymerization—formation of redox-stable and emissive films. Eur. J. Inorg. Chem. **2013**(24), 4191 (2013)
182. Jäger, M., Kumar, R.J., Görls, H., Bergquist, J., Johansson, O.: Facile synthesis of bistridentate Ru(II) complexes based on 2,6-di(quinolin-8-yl)pyridyl ligands: sensitizers with microsecond ^{3}MLCT excited state lifetimes. Inorg. Chem. **48**(7), 3228 (2009)
183. Abrahamsson, M., Jäger, M., Österman, T., Eriksson, L., Persson, P., Becker, H.-C., Johansson, O., Hammarström, L.: A 3.0 µs room temperature excited-state lifetime of a bistridentate Ru(II)-polypyridine complex for rod-like molecular arrays. J. Am. Chem. Soc. **128**(39), 12616 (2006)
184. Friebe, C., Jager, M., Schubert, U.S.: Emitting electrode coatings with redox-switchable conductivity: incorporation of ruthenium(II)-2,6-di(quinolin-8-yl)pyridine complexes into polythiophene by electropolymerization. RSC Adv. **3**(29), 11686 (2013)
185. Weinberger, D.A., Higgins, T.B., Mirkin, C.A., Stern, C.L., Liable-Sands, L.M., Rheingold, A.L.: Terthienyl and poly-terthienyl ligands as redox-switchable hemilabile ligands for oxidation-state-dependent molecular uptake and release. J. Am. Chem. Soc. **123**(11), 2503 (2001)
186. Weinberger, D.A., Higgins, T.B., Mirkin, C.A., Liable-Sands, L.M., Rheingold, A.L.: Terthienyl-based redox-switchable hemilabile ligands: Transition metal polymeric complexes with electrochemically tunable or switchable coordination environments? . Angew. Chem. Int. Ed. **38**(17), 2565 (1999)

187. Kuchison, A.M., Wolf, M.O., Patrick, B.O.: Photophysical and electrochemical properties of Ru(II) complexes containing tridentate bisphosphino-oligothiophene ligands. Dalton Trans. **40**(26), 6912 (2011)
188. González, B., del Valle, M.A., Díaz, F.R., Espinosa-Bustos, C., Ramírez, A., Hernández, L. A.: Synthesis and electrochemical characterization of new ruthenium-terthiophene complexes. Polyhedron **89**, 232 (2015)
189. Luo, S., Rauchfuss, T.B., Wilson, S.R.: Arene versus thiophene reduction in the (arene) ruthenium thiophene system $(C_6R_6)Ru(C_4R_4S)_{2+}$ and the protonation of η^4-thiophene ligands. J. Am. Chem. Soc. **114**(22), 8515 (1992)
190. González, B., del Valle, M.A., Díaz, F.R., Espinosa-Bustos, C., Ramírez, A.M.R., Hernández, L.A.: Synthesis, characterization, electropolymerization and possible utilities of a new ruthenium-thiophene complex. J. Appl. Polym. Sci. **133**(25), 43559 (2016)
191. Zhu, Y., Millet, D.B., Wolf, M.O., Rettig, S.J.: Models for conjugated metal acetylide polymers: ruthenium oligothienylacetylide complexes. Organometallics **18**(10), 1930 (1999)
192. Ho, P.-Y., Komber, H., Horatz, K., Tsuda, T., Mannsfeld, S.C.B., Dmitrieva, E., Blacque, O., Kraft, U., Sirringhaus, H., Lissel, F.: Synthesis and characterization of a semiconducting and solution-processable ruthenium-based polymetallayne. Polym. Chem. **11**(2), 472 (2020)
193. Liu, W., Huang, W., Pink, M., Lee, D.: Layer-by-layer synthesis of metal-containing conducting polymers: caged metal centers for interlayer charge transport. J. Am. Chem. Soc. **132**(34), 11844 (2010)
194. Martínez-Díaz, M.V., de la Torre, G., Torres, T.: Lighting porphyrins and phthalocyanines for molecular photovoltaics. Chem. Commun. **46**(38), 7090 (2010)
195. Rawling, T., McDonagh, A.: Ruthenium phthalocyanine and naphthalocyanine complexes: Synthesis, properties and applications. Coord. Chem. Rev. **251**(9–10), 1128 (2007)
196. Fischer, M.K.R., López-Duarte, I., Wienk, M.M., Martínez-Díaz, M.V., Janssen, R.A.J., Bäuerle, P., Torres, T.: Functionalized dendritic oligothiophenes: Ruthenium phthalocyanine complexes and their application in bulk heterojunction solar cells. J. Am. Chem. Soc. **131** (24), 8669 (2009)
197. Krompiec, M., Krompiec, S., Grudzka, I., Filapek, M., Skórka, Ł, Flak, T., Łapkowski, M.: A cross-linked conjugated metallopolymer comprised of bisaxially coordinated ruthenium tetra-t-butyl phthalocyanine connected by quaterthiophene linkers. Electrochim. Acta **56** (19), 6824 (2011)
198. Abe, M., Futagawa, H., Ono, T., Yamada, T., Kimizuka, N., Hisaeda, Y.: An electropolymerized crystalline film Incorporating axially-bound metalloporphycenes: Remarkable reversibility, reproducibility and coloration efficiency of ruthenium(II/III)-based electrochromism. Inorg. Chem. **54**(23), 11061 (2015)
199. Waluk, J.: Spectroscopy and tautomerization studies of porphycenes. Chem. Rev. **117**(4), 2447 (2016)
200. Ellis, C.D., Margerum, L.D., Murray, R.W., Meyer, T.J.: Oxidative electropolymerization of polypyridyl complexes of ruthenium. Inorg. Chem. **22**(9), 1283 (1983)
201. Venkatanarayanan, A., Spehar-Délèze, A.-M., Dennany, L., Pellegrin, Y., Keyes, T.E., Forster, R.J.: Ruthenium aminophenanthroline metallopolymer films electropolymerized from an ionic liquid: Deposition and electrochemical and photonic properties. Langmuir **24** (19), 11233 (2008)
202. Bélanger, S., Stevenson, K.J., Mudakha, S.A., Hupp, J.T.: "Perfect" electrochemical molecular sieving by thin and ultrathin metallopolymeric films. Langmuir **15**(3), 837 (1999)
203. Syed, A.A., Dinesan, M.K.: Review: polyaniline—a novel polymeric material. Talanta **38** (8), 815 (1991)
204. Runge, F.F.: Ueber einige Producte der Steinkohlendestillation. Ann. Phys. **107**(33), 513 (1834)
205. Rasmussen, S.C.: Revisiting the early history of synthetic polymers: Critiques and new insights. Ambix **65**(4), 356 (2018)

206. Rasmussen, S.C.: Conjugated and conducting organic polymers: The first 150 years. ChemPlusChem **85**(7), 1412 (2020)
207. Wang, H., Lin, J., Shen, Z.X.: Polyaniline (PANI) based electrode materials for energy storage and conversion. J. Sci.: Adv. Mater. Devices **1**(3), 225 (2016)
208. Pandey, S.: Highly sensitive and selective chemiresistor gas/vapor sensors based on polyaniline nanocomposite: A comprehensive review. J. Sci.: Adv. Mater. Devices **1**(4), 431 (2016)
209. Mortimer, R.J., Dyer, A.L., Reynolds, J.R.: Electrochromic organic and polymeric materials for display applications. Displays **27**(1), 2 (2006)
210. Diaz, A.F., Logan, J.A.: Electroactive polyaniline films. J. Electroanal. Chem. **111**(1), 111 (1980)
211. Horwitz, C.P., Zuo, Q.: Oxidative electropolymerization of iron and ruthenium complexes containing aniline-substituted 2,2'-bipyridine ligands. Inorg. Chem. **31**(9), 1607 (1992)
212. Dreyse, P.A., Isaacs, M.A., Iturriaga, P.E., Villagra, D.A., Aguirre, M.J., Kubiak, C.P., Glover, S.D., Goeltz, J.C.: Electrochemical preparation of conductive films of tetrapyridylporphyrins coordinated to four [Ru(5-NO_2-phen)$_2$Cl]$^+$ groups. J. Electroanal. Chem. **648**(2), 98 (2010)
213. Dreyse, P., Isaacs, M., Calfumán, K., Cáceres, C., Aliaga, A., Aguirre, M.J., Villagra, D.: Electrochemical reduction of nitrite at poly-[Ru(5-NO_2-phen)$_2$Cl] tetrapyridylporphyrin glassy carbon modified electrode. Electrochim. Acta **56**(14), 5230 (2011)
214. Dreyse, P., Quezada, D., Honores, J., Aguirre, M.J., Mendoza, L., Matsuhiro, B., Villagra, D., Isaacs, M.: Determination of S(IV) oxoanions at poly[Ru(5-NO_2-phen)$_2$Cl] tetrapyridylporphyrin glassy carbon modified electrode. Electroanalysis **24**(8), 1709 (2012)
215. Dreyse, P., Honores, J., Quezada, D., Isaacs, M.: Electrocatalytic transformation of carbon dioxide into low carbon compounds on conducting polymers derived from multimetallic porphyrins. ChemSusChem **8**(22), 3897 (2015)
216. Calfumán, K., Honores, J., Isaacs, M., Quezada, D., Valdebenito, J., Urzúa, M.: Quick and easy modification of glassy carbon electrodes with ionic liquid and tetraruthenated porphyrins for the electrochemical determination of atrazine in water. Electroanalysis **31**(4), 671 (2019)
217. Shao, J.-Y., Yao, C.-J., Cui, B.-B., Gong, Z.-L., Zhong, Y.-W.: Electropolymerized films of redox-active ruthenium complexes for multistate near-infrared electrochromism, ion sensing and information storage. Chin. Chem. Lett. **27**(8), 1105 (2016)
218. Zhong, Y.-W., Yao, C.-J., Nie, H.-J.: Electropolymerized films of vinyl-substituted polypyridine complexes: Synthesis, characterization and applications. Coord. Chem. Rev. **257**(7–8), 1357 (2013)
219. Yao, C.-J., Zhong, Y.-W., Yao, J.: Five-stage near-infrared electrochromism in electropolymerized films composed of alternating cyclometalated bisruthenium and bis-triarylamine segments. Inorg. Chem. **52**(17), 10000 (2013)
220. Cui, B.-B., Mao, Z., Chen, Y., Zhong, Y.-W., Yu, G., Zhan, C., Yao, J.: Tuning of resistive memory switching in electropolymerized metallopolymeric films. Chem. Sci. **6**(2), 1308 (2015)
221. Hanabusa, K., Nakamura, A., Koyama, T., Shirai, H.: Electropolymerization and characterization of terpyridinyl iron(II) and ruthenium(II) complexes. Polym. Int. **35**(3), 231 (1994)
222. Storrier, G.D., Takada, K., Abruña, H.D.: Catechol-pendant terpyridine complexes: electrodeposition studies and electrocatalysis of NADH oxidation. Inorg. Chem. **38**(3), 559 (1999)
223. Storrier, G.D., Takada, K., Abruña, H.D.: Synthesis, characterization, electrochemistry, and EQCM studies of polyamidoamine dendrimers surface-functionalized with polypyridyl metal complexes. Langmuir **15**(3), 872 (1999)
224. Calvert, J.M., Peebles, D.L., Nowak, R.J.: Spectral, electrochemical, and conductivity studies of poly(pyridyl)ruthenium complexes containing polymerizable acetylenic and olefinic ligands. Inorg. Chem. **24**(20), 3111 (1985)

225. Grosshenny, V., Harriman, A., Gisselbrecht, J.-P., Ziessel, R.: A remarkable structural sensitivity toward electrogeneration of metallopolyacetylenes. J. Am. Chem. Soc. **118**(42), 10315 (1996)
226. Poriel, C., Ferrand, Y., le Maux, P., Paul, C., Rault-Berthelot, J., Simonneaux, G.: Poly (ruthenium carbonyl spirobifluorenylporphyrin): A new polymer used as a catalytic device for carbene transfer. Chem. Commun. (18), 2308 (2003)
227. Paul-Roth, C., Rault-Berthelot, J., Simonneaux, G.: New polymers for catalytic carbene transfer: Electropolymerization of tetrafluorenylporphyrinruthenium carbon monoxide. Tetrahedron **60**(52), 12169 (2004)
228. Waltman, R.J., Bargon, J.: Electrically conducting polymers: A review of the electropolymerization reaction, of the effects of chemical structure on polymer film properties and of applications towards technology. Can. J. Chem. **64**(1), 76 (1986)
229. Chhatwal, M., Kumar, A., Awasthi, S.K., Zharnikov, M., Gupta, R.D.: An electroactive metallo-polypyrene film as a molecular scaffold for multi-state volatile memory devices. J. Phys. Chem. C **120**(4), 2335 (2016)
230. Chhatwal, M., Kumar, A., Gupta, R.D., Awasthi, S.K.: A pyrene-based electropolymerized film as a solid-state platform for multi-bit memory storage and fluorescence sensing of nitroaromatics in aqueous solutions. J. Mater. Chem. C **4**(19), 4129 (2016)
231. Hastings, W.R., Baird, M.C.: A new form of ruthenium tetracarbonyl. Inorg. Chem. **25**(16), 2913 (1986)
232. Masciocchi, N., Moret, M., Cairati, P., Ragaini, F., Sironi, A.: Solving simple organometallic structures solely from X-ray powder diffraction data: The case of polymeric [{$Ru(CO)_4$}]. J. Chem. Soc., Dalton Trans. (3), 471 (1993)
233. Niskanen, M., Hirva, P., Haukka, M.: Computational DFT study of ruthenium tetracarbonyl polymer. J. Chem. Theory Comput. **5**(4), 1084 (2009)
234. Cervellino, A., Maspero, A., Masciocchi, N., Guagliardi, A.: From paracrystalline $Ru(CO)_4$ 1D polymer to nanosized ruthenium metal: A case of study through total scattering analysis. Cryst. Growth Des. **12**(7), 3631 (2012)
235. Kontkanen, M.-L., Oresmaa, L., Moreno, M.A., Jänis, J., Laurila, E., Haukka, M.: One-dimensional metal atom chain $[Ru(CO)_4]_n$ as a catalyst precursor—hydroformylation of 1-hexene using carbon dioxide as a reactant. Appl. Catal. A: Gen. **365**(1), 130 (2009)
236. Caix-Cecillon, C., Chardon-Noblat, S., Deronzier, A., Haukka, M., Pakkanen, T.A., Ziessel, R., Zsoldos, D.: Electrochemical formation and spectroelectrochemical characterization of organometallic $[Ru(L)(CO)_2]_n$ polymers; L=disubstituted-2,2′-bipyridine. J. Electroanal. Chem. **466**(2), 187 (1999)
237. Myllynen, S., Wasberg, M., Eskelinen, E., Haukka, M., Pakkanen, T.A.: Voltammetric and EQCM investigation of conductive films formed by electroreduction of $Ru(L)(CO)_2Cl_2$ compounds. J. Electroanal. Chem. **506**(2), 115 (2001)
238. Myllynen, S., Wasberg, M.: Formation and oxidative break-up of organometallic Ru–Ru bonded polymer films in DMSO electrolyte studied using the dimer $[Ru(bpy)(CO)_2Cl]_2$ (bpy = 2,2′-bipyridine). J. Electroanal. Chem. **623**(1), 93 (2008)
239. Masciocchi, N., Sironi, A., Chardon-Noblat, S., Deronzier, A.: X-ray powder diffraction study of organometallic polymers: $[Ru(L)(CO)_2]_n$ (L = 2,2'-Bipyridine or 1,10-Phenanthroline). Organometallics **21**(19), 4009 (2002)
240. Gerbaud, G., Mouesca, J.-M., Hediger, S., Chardon-Noblat, S., Lafolet, F., Deronzier, A., Bardet, M.: Structural characterization of metal-metal bonded polymer $[Ru(L)(CO)_2]_n$ (L = 2,2'-bipyridine) in the solid state using high-resolution NMR and DFT chemical shift calculations. Phys. Chem. Chem. Phys. **12**(47), 15428 (2010)
241. Myllynen, S., Wasberg, M.: High electrical conductivity for the partially oxidised d^8 Ru-Ru bonded chain $[Ru(bpy)(CO)_2]_n$ (bpy = 2,2′-bipyridine) and its wide range redox control. Electrochem. Commun. **11**(7), 1453 (2009)

242. Myllynen, S., Wasberg, M.: On the conducting $[Ru(bpy)(CO)_2]_n$ (bpy = 2,2'-bipyridine) polymer: Doping process, near-IR absorption, T-dependence and film growth. J. Electroanal. Chem. **661**(2), 294 (2011)
243. Zhao, J., Wang, X., Xu, Z., Loo, J.S.C.: Hybrid catalysts for photoelectrochemical reduction of carbon dioxide: A prospective review on semiconductor/metal complex co-catalyst systems. J. Mater. Chem. A **2**(37), 15228 (2014)
244. Arai, T., Sato, S., Uemura, K., Morikawa, T., Kajino, T., Motohiro, T.: Photoelectrochemical reduction of CO_2 in water under visible-light irradiation by a p-type InP photocathode modified with an electropolymerized ruthenium complex. Chem. Commun. **46**(37), 6944 (2010)
245. Ong, W.-J., Tan, L.-L., Ng, Y.H., Yong, S.-T., Chai, S.-P.: Graphitic carbon nNitride ($g\text{-}C_3N_4$)-based photocatalysts for artificial photosynthesis and environmental remediation. Are we a step closer to achieving sustainability? Chem. Rev. **116**(12), 7159 (2016)
246. Kuriki, R., Ishitani, O., Maeda, K.: Unique solvent effects on visible-light CO_2 reduction over ruthenium(II)-complex/carbon nitride hybrid photocatalysts. ACS Appl. Mater. Interfaces **8**(9), 6011 (2016)
247. Hartl, F., Renfrew, A.K., Lafolet, F., Mahabiersing, T., Calhorda, M.J., Chardon-Noblat, S., Haukka, M., Deronzier, A.: Soluble redox-active polymetallic chains $[\{Ru^0(CO)(L)(bpy)\}_m]_n$ (bpy = 2,2′-bipyridine, L = PrCN, Cl^-; m = 0, −1): Electrosynthesis and characterization. Inorg. Chem. **48**(17), 8233 (2009)
248. Winkler, K., Balch, A.L., Kutner, W.: Electrochemically formed fullerene-based polymeric films. J. Solid State Electrochem. **10**(10), 761 (2006)
249. Balch, A.L., Olmstead, M.M.: Reactions of transition metal complexes with fullerenes (C_{60}, C_{70}, etc.) and related materials. Chem. Rev. **98**(6), 2123 (1998)
250. Barthelmes, K., Winter, A., Schubert, U.S.: Hybrid materials based on ruthenium and fullerene assemblies. Dalton Trans. **45**(38), 14855 (2016)
251. Leng, F., Gerber, I.C., Lecante, P., Bacsa, W., Miller, J., Gallagher, J.R., Moldovan, S., Girleanu, M., Axet, M.R., Serp, P.: Synthesis and structure of ruthenium-fullerides. RSC Adv. **6**(73), 69135 (2016)

Check for updates

Ruthenodendrimers

4

Abstract

Dendrimers, i.e., highly branched, 3D macromolecular architectures, have been utilized to incorporate (or encapsulate) Ru complexes or nanoparticles. The different possibilities to reach this goal represent the core subject of this chapter. Competing with ruthenopolymers, ruthenodendrimers have *inter alia* been utilized in the context of catalysis and biomedicine; these as well as further applications (e.g., in the field of opto-electronics) are shown.

4.1 Introduction

Hyperbranched polymers (HPs) and dendrimers are both highly branched 3D macromolecular architectures that can, in short, be distinguished by their preparative methods and resulting properties (Fig. 4.1). Hyperbranched polymers are typically prepared by noniterative one-pot polymerization methods and, thus, their architecture is rather irregular due to a significant number of unreacted branching points. As a result, the observed molar-mass distributions or dispersities (Đ) are very broad. Nonetheless, their straightforward synthesis as well as the ability to control their globular structure and functionality make HPs favorable materials for a wide range of applications from opto-electronics, smart coatings, and adhesives to supramolecular, hybrid or biomimetic materials [1].

Dendrimers, on the other hand, are prepared via iterative protocols. The divergent and convergent construction strategies (or combinations thereof) allow one to synthesize dendrimers, as globular, highly branched, fractal-like polymers, which possess well-defined 3D structures and shapes as well as very narrow dispersities (Fig. 4.2) [3, 4]. As pointed out by Newkome et al., the directional multi-step synthesis results in materials with uniform or discrete functionalities. Dendrimers are known for their tunable internal packing densities as well as the predictable void

© The Author(s), under exclusive license to Springer Nature Switzerland AG 2021

U. S. Schubert et al., *Ruthenium-Containing Polymers*,
https://doi.org/10.1007/978-3-030-75598-0_4

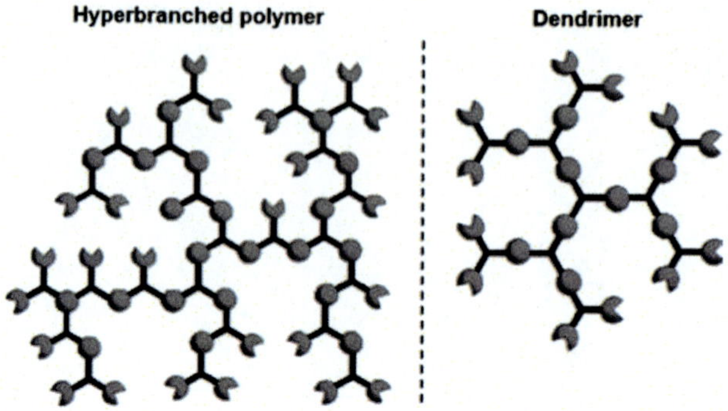

Fig. 4.1 Schematic representation of the topologies of hyperbranched polymers and a typical 1-to-2 branched dendrimers [2]. Figure reproduced with kind permission; © 2001 American Chemical Society

volumes, surface moieties, molar mass, solvent-dependent size, and branching dimensions [3]. However, the potential significant synthetic efforts required to obtain such aesthetic macromolecules represents an obstacle and still hampers large-scale applications, unlike the HPs.

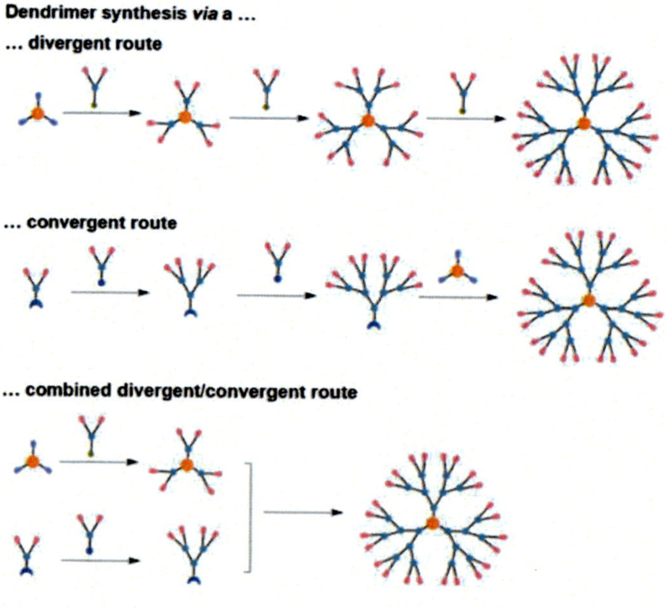

Fig. 4.2 Schematic representation of the different synthetic strategies applied in dendrimer construction [5]. Figure reproduced with kind permission; © 2919 Elsevier B.V.

In 1978, Vögtle et al. reported the first example of a "cascade" polymer, thereby marking the hour-of-birth of dendrimer chemistry [6]. The term "dendrimer" was coined by Tomalia et al. from the Greek term "dendro", for tree-like [7]; contemporaneously, Newkome et al. employed the term "arborol", derived from the Latin term "arbor" for tree [8]. Both dendrimer and arborols represent synonyms for cascade polymers. In order to guard against further misunderstandings, a systematic nomenclature for this class of macromolecules was introduced in the mid-1990s [9, 10]. Since then, a wide range of dendritic architectures has been prepared and the field of research, taking synthesis and applications into account, has been reviewed by various authors [2, 3, 11–16]. Importing supramolecular chemistry, i.e., the usage of non-covalent interactions, opened up new vistas in dendrimer chemistry—the chemistry within the dendrimer or "suprasupermolecular chemistry", according to Newkome [17]. Amongst other types of non-covalent interactions (e.g., host–guest inclusion complexes) metal-to-ligand complexations have—due to its high degree-of-directionality—found widespread application in dendrimer chemistry. This particular field of research is commonly referred to as "metallodendrimers" [18]. In such materials, the metal species, i.e., complexes or nanoparticles, might be included by non-covalent interactions with appropriate binding sites within the dendritic architecture. Alternatively, metal complexes might represent integral parts of the dendrimer—as core, building-block connector, branching unit, and/or terminal functionality (Fig. 4.3) [18]. Similar to metal-containing polymers (see Chaps. 2 and 3), the presence of metal species within the dendrimers evokes properties that are absent for their metal-free counterparts (e.g., photophysical, electro- or magnetochemical properties as well as biological or catalytic activity).

Balzani et al. as well as Newkome et al. were the first to recognize the applicability of Ru^{II}–polypyridyl complexes, as structural unit, in the formation of metallodendrimers. These authors capitalized the directionality of the $Ru^{II/III}$ coordination chemistry, in concert with the remarkable kinetic stability of Ru^{II}–polypyridyl complexes, to obtain "ruthenodendrimers" comprising of such

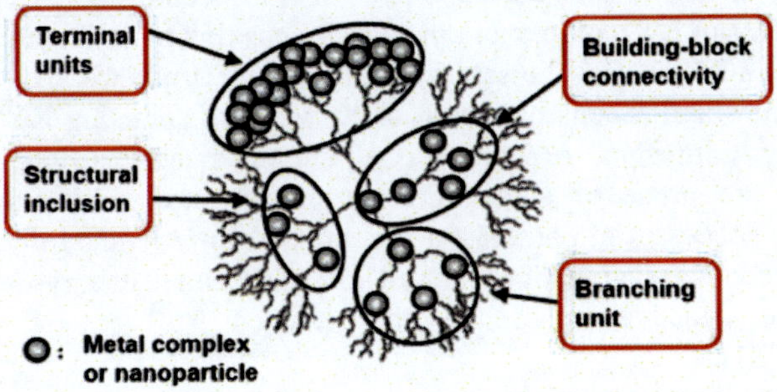

Fig. 4.3 Schematic representation of the potential uses and positioning of metal centers within dendritic architectures [3]. Figure reproduced with kind permission; © 1999 American Chemical Society

complexes as branching unit [19] or building-block connector [20]. In this chapter, ruthenodendrimers are presented; the classification follows their structural role in these materials.

4.2 Dendrimers, Containing Ru^{II} Complexes as the Branching Units

In course of their studies on oligonuclear polypyridyl complexes [21], Balzani et al. prepared a variety of metallodendrimers in which *tris*-bidentate Ru^{II}–polypyridyl complexes were utilized, as branching units, and *tris*-bidentate Ru^{II} or Os^{II}–polypyridyl complexes, as cores [19, 22–28]. These materials, such as the decanuclear Ru^{II} complex **1a** (Fig. 4.4) [22], were assembled by combined divergent/convergent methods (the largest assembly, reported by these authors, comprised of 22 Ru^{II} complexes [24]). 2,4-*Bis*(pyridin-2-yl)pyrazine was identified as a versatile bridging ligand to guarantee the directed construction of the ruthenodendrimer architectures [2,2′-bipyridine (bpy) or 2,2′-*bis*quinoline (biq) were used as outer "shielding" ligands]. Balzani et al. paid special attention to the photophysical and electrochemical properties of their assemblies [29, 30]. The Ru^{II}-centered dendrimers featured the characteristic emission behavior as known from electronically-interacting *tris*-bidentate Ru^{II}–polypyridyl complexes (i.e., emission maximum at *ca.* 770 nm); whereas, the Os^{II}-centered dendrimers, e.g., **1b** (Fig. 4.4), emitted in the spectral range from 850 to 1000 nm. Thus, electronic interactions between neighboring complexes enabled an energy transfer from the photoexcited Ru^{II} centers to the Os^{II} cores. Such a behavior is of significant importance regarding light-harvesting applications, e.g., photocatalysis, where the properties of the natural photosystem of green plants is mimicked by artificial organic or metal–organic structures [30]. The electronic coupling of the Ru^{II} centers within the dendrimers was concluded from electrochemical studies. The ruthenodendrimer containing 22 Ru^{II} centers, for example, showed an intense UV/vis absorption band at 542 nm, characteristic for the metal-to-ligand charge-transfer (MLCT) process of Ru^{II} complexes, an emission maximum at 786 nm, and a $Ru^{II/III}$ redox peak at +1.52 V [*vs.* the saturated calomel electrode (SCE)]; the latter was assigned to the simultaneous oxidation of 12 weakly interacting Ru^{II} centers [24]. More recently, Campagna, Amenitsch et al. demonstrated that ruthenodendrimers, such as **1b**, are prone to aggregation in concentrated solutions; due to this self-assembly process, the photophysics compared to individual dendrimers were strongly altered: An aggregation-induced inter-dendrimer energy-transfer process facilitated the deactivation of the photo-excited Ru^{II} centers even on the ps-timescale [31].

In order to utilize the efficient absorption of visible light, Campagna, Puntereiro et al. employed ruthenodendrimers, such a tetranuclear 1st generation assembly, as photosensitizer, for the photocatalytic water-oxidation reaction (WOR) [32]. In the presence of the persulfate anion, as sacrificial acceptor, and IrO_2, as the actual

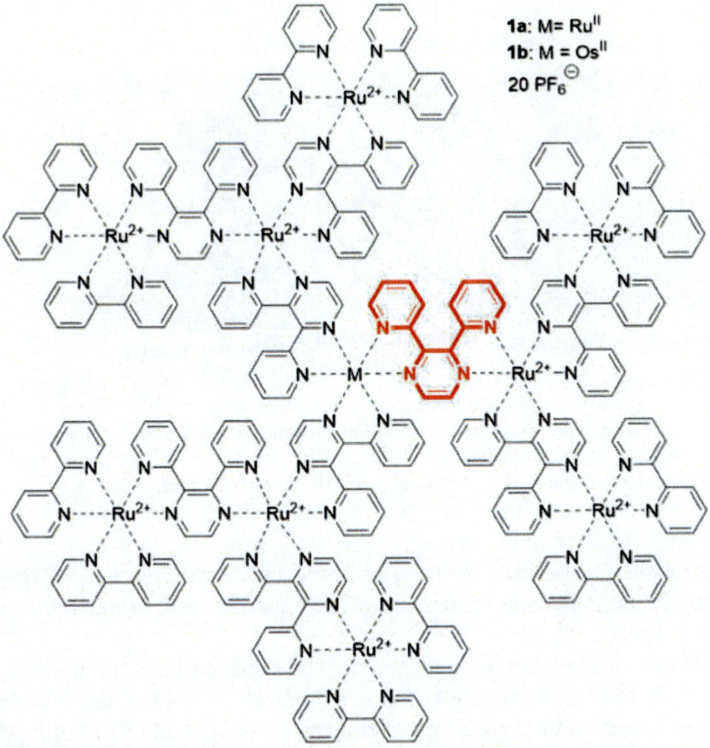

Fig. 4.4 Schematic representation of Balzani's ruthenodendrimers **1** comprising of 10 *bis*-tridentate polypyridyl complexes; the bridging ligand, 2,4-*bis*(pyridine-2-yl)pyrazine, is highlighted in red [22]

water-oxidation catalyst (WOC), the ruthenodendrimer photosensitized the WOC, even when run at a low excitation energy of *ca.* 700 nm. The efficiency of the WOR was increased further by replacing the IrO_2 nanoparticles by a tetraruthenium-bridged polyoxotungstate, as WOC; this system produced O_2 upon irradiation at 550 nm with a remarkable quantum yield of 0.30 (Fig. 4.5) [33].

As for the linear ruthenopolymers containing *tris*-bidentate Ru^{II}–polypyridyl complexes in their backbone, such as Rehahn's polymer containing tetrapyrido[3,2-*a*:2′,3′-*c*:3″,2″-*h*:2‴,3‴-*j*]phenazine, as the bridging ligand (i.e., **77** in Fig. 2.61) [34], metal-centered chirality needs to be addressed. Basically, *tris*-bidentate Ru^{II} complexes might exist in a *Δ*- and *Λ*-form (corresponding to the *fac* and *mer* configuration, respectively). Assuming no control over the assembly process, the synthesis of the decanuclear **1** might give a mixture of 1024 stereoisomers. In 1997, MacDonnell et al. reported one of the very few chiral ruthenodendrimers, i.e., a 1st generation D_3-symmetric assembly. However, chirality doesn't matter. Campagna, MacDonnell et al. compared the photophysical and electrochemical properties of

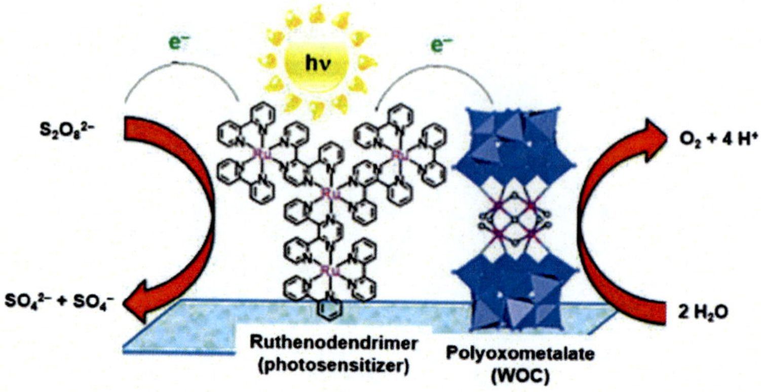

Fig. 4.5 Schematic representation of the photocatalytic WOR using a ruthenodendrimer, as photosensitizer, a polyoxometalate, as WOC, and persulfate, as sacrificial acceptor [33]. Figure reproduced with kind permission; © 2010 The Royal Chemical Society

this chiral ruthenodendrimer to those of its racemic counterpart [35]. These properties were found to be independent of the configuration and, thus, the consideration of chirality became obsolete.

A different synthetic strategy to prepare ruthenodendrimers comprising of Ru^{II} complexes, as branching units, was followed by Juris et al. [36]. These authors used the divergent route and performed "chemistry-on-the-complex": Starting from the homoleptic Ru^{II} or Os^{II} complex of 4-bromo-2,2-bipyridine, as core, bpy units were attached by an aromatic nucleophilic substitution. This bpy units, in turn, were used for the complexation of Ru^{II} precursor complexes, etc. Thus, the ruthenodendrimer was constructed from the core towards the shell. In general, the "chemistry-on-the-complex" concept is highly versatile not only for the functionalization of pre-formed complexes but also regarding their incorporation into macromolecular architectures; this field of research was reviewed recently by Jäger, Schubert et al. [37]. One further example in this respect was reported by Lee, Kim et al.; these authors utilized an esterification under mild conditions to construct their ruthenodendrimers in a divergent fashion [38].

One general limitation when dealing with ruthenodendrimers, is their structural characterization, which was overcome by Arakawa et al. [39] as well as Kirsch-De Mesmaeker et al. [40]. These authors first utilized electrospray ionization mass spectrometry (ESI MS) as a reliable tool to analyze multiple-charged ions; this mild and non-destructive MS technique allowed the accurate determination of the molar mass of the multinuclear complexes. In their first report, Kirsch-De Mesmaeker et al. studied a heptanuclear ruthenodendrimer, similar to Balzani's assemblies [40]. In recent years, ESI MS experienced a tremendous evolution and top-down multi-dimensional techniques are now commonly used [41]. Coupling the ESI set-up with ion-mobility mass spectrometry (IM MS) and energy-resolved (gradient) tandem mass spectrometry (gMS^2) enables one to study the stoichiometries, architectures, and intrinsic stabilities of advanced metallo-supramolecular architectures. Briefly, the mild ionization of a sample by ESI guarantees its structural integrity and IM allows the separation of ionic species by charge and collision cross-section; the latter being a measure for the ions' size and shape. Thusly separated ions are analyzed in the orthogonal MS dimension regarding their mass-to-charge (*m*/*z*) ratios and isotopic patterns. The virtue of this method has been demonstrated by Newkome, Wesdemiotis et al. (as well as various other groups) for advanced metallo-supramolecular 2D and 3D structures formed by transition metal ions and sophisticated 2,2′:6′,2″-terpyridine (tpy) ligands [42, 43]. As one representative example, the analysis of the pentanuclear ruthenodendrimer **2** by multi-dimensional MS is highlighted in Fig. 4.6 [44].

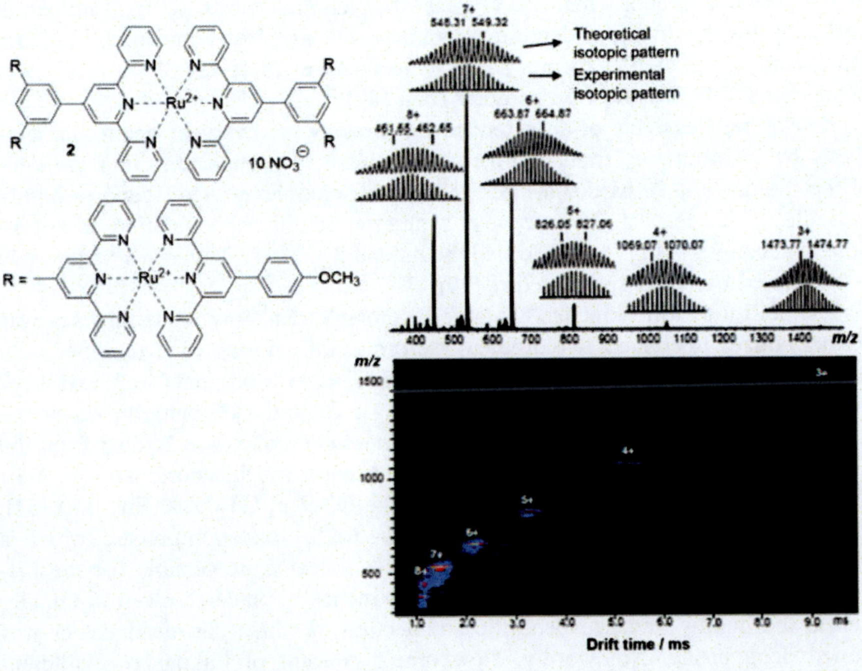

Fig. 4.6 Schematic representation of the pentanuclear ruthenodendrimer **2**, which was analyzed by ESI time-of-flight (TOF) MS, coupled to a travelling-wave IM set-up. The differently charged ions, as observed in the ESI-TOF MS, were separated by their drift time in the IM chamber [44]. Figure reproduced with kid permission; © 2012 The Royal Chemical Society

4.3 Dendrimers, Containing Ru^{II} Complexes as Building-Block Connectors

In 1985, Newkome et al. introduced the concept of "unimolecular micelles" [45]. Unlike a traditional micelle, which is the product of the self-assembly of an amphiphilic molecule in a selective solvent above the critical micellization concentration (cmc), amphiphilic dendrimers are single globular molecules. Such species, typically comprising of a hydrophobic interior and a hydrophilic shell, basically offer the same properties as their supramolecular counterparts: Encapsulation of molecules, reaction "medium" for organic transformations, etc. [46]. In particular, the so-called "micellenoic acid," possessing a 1 → 3 branched purely hydrocarbon framework, was in the focus of pioneering work [47, 48]. The first-generation (G1) and second-generation (G2) dendrimers had (12 × 3) 36 carboxylic-acid groups, respectively, as the terminal units. Using amidation chemistry for the divergent dendrimer synthesis allowed Newkome et al. to obtain giant dendrimers; the G5 species had a molecular weight of 111,373 g mol^{-1} and possessed 972 acid groups on its outer shell [49].

Newkome, Constable et al. extended the strategy of divergent dendrimer synthesis by incorporating transition-metal complexes, as connectors, into the dendrimer framework. In particular, the Ru^{II}–tpy *bis*-complexes were found to highly versatile in this respect. The stepwise synthesis, combining amidation and coordination chemistry (Fig. 4.7), afforded the ruthenodendrimer **3**, whose structure was corroborated by NMR spectroscopy [20].

The ability to form kinetically stable heteroleptic Ru^{II}–tpy *bis*-complexes was further utilized to connect two different preformed dendrimers, each equipped with a tpy unit. This strategy, reminiscent of Emil Fischer's original "lock and key" concept [50], was demonstrated first by linking a G2 and a G3 dendron via such a metallo-supramolecular linkage. First, a Ru^{II} *mono*-complex was formed from the G2-[building block, representing the "key"; subsequently, the coordination to G3-[the "lock", under reductive conditions gave the G2-[Ru]-G3 dyad (Fig. 4.8) [51]. Electrochemical measurements revealed that the $Ru^{II/III}$ oxidation process, which is typically reversible for Ru^{II}–tpy *bis*-complexes, became irreversible for the *bis*-dendrimers. This finding was attributed to a kinetically hindered electron transfer due to the highly branched dendritic architecture in which the metal center was buried deep inside. Noteworthy, Newkome's concept of linking two different dendrimers via directed metal-to-ligand coordination has become powerful tool in today's metallo-supramolecular polymer science. In particular, Schubert et al. have taken up this idea and utilized it for the assembly of pseudo-diblock copolymers containing such complexes, as redox-active and pH-sensitive, bridge between the constituent blocks (see Chap. 2) [52, 53].

In subsequent work, Newkome and He followed a convergent strategy to assemble the octanuclear ruthenodendrimer **4**, which contained two consecutive Ru^{II}-tpy complexes in each of the four dendritic arms (Fig. 4.9) [54]. Due to the presence of the eight Ru^{II} centers an eight-fold increase in the UV/vis absorbance

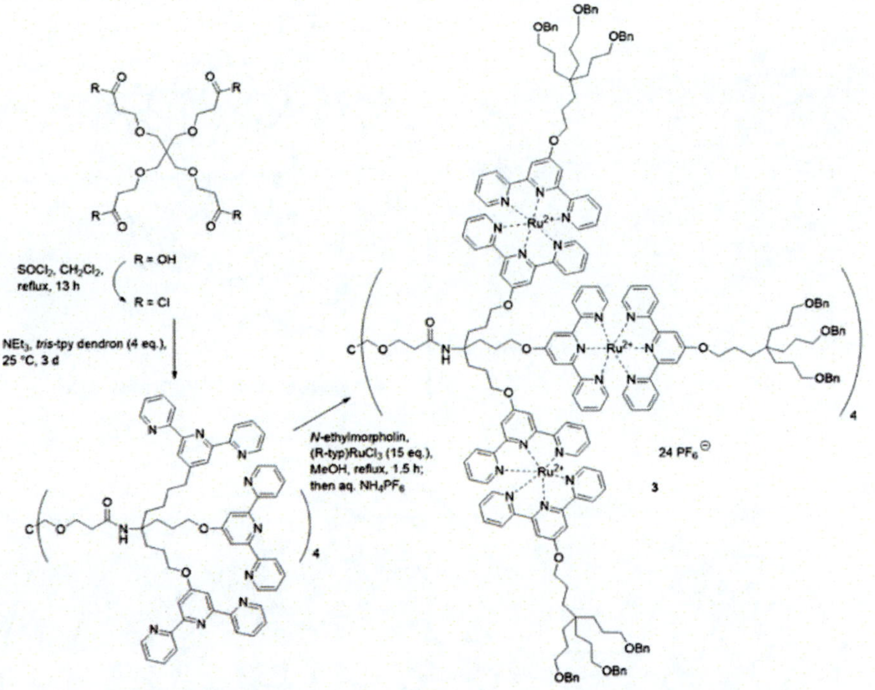

Fig. 4.7 Schematic representation of the synthesis of a ruthenodendrimer containing 12 Ru^{II}–tpy *bis*-complexes, as connector

was observed. In related work, Newkome et al. prepared the tetranuclear ruthenodendrimer **5** in which the charges of the cationic complexes were compensated by eight internal carboxylate sites (Fig. 4.9) [55, 56]. Thus, the overall charge-neutral dendrimer required no external counterions, as present in most of the other reported polycationic structures (e.g., Cl^-, BF_4^- or PF_6^-).

These authors also prepared the isomeric ruthenodendrimers **6** and **7** having virtually the same molecular formula and molar mass ($C_{597}H_{880}F_{48}N_{52}O_{136}P_8Ru_4$, M = 12,526 g mol^{-1}; Fig. 4.10) [55]. The sophisticated structural design allowed to mimic methane's ideal tetrahedral geometry on a nanoscopic level. Various spectroscopic and mass spectrometric analyses, as well as chemical/physical studies (i.e., solubility and thermal stability) gave very similar results for both materials; however, cyclic voltammetry (CV) indicated distinct differences regarding their internal densities and void regions: Dendrimer **7**, possessing the more crowded interior suppressed the electronic interaction between the Ru^{II} centers and, thus, a splitting of the Ru$^{II/III}$ redox wave occurred. This finding suggested that the electronic communication between metal centers within dendritic architectures can be facilitated (or hampered) by their positioning and, thus, environment within the dendrimer.

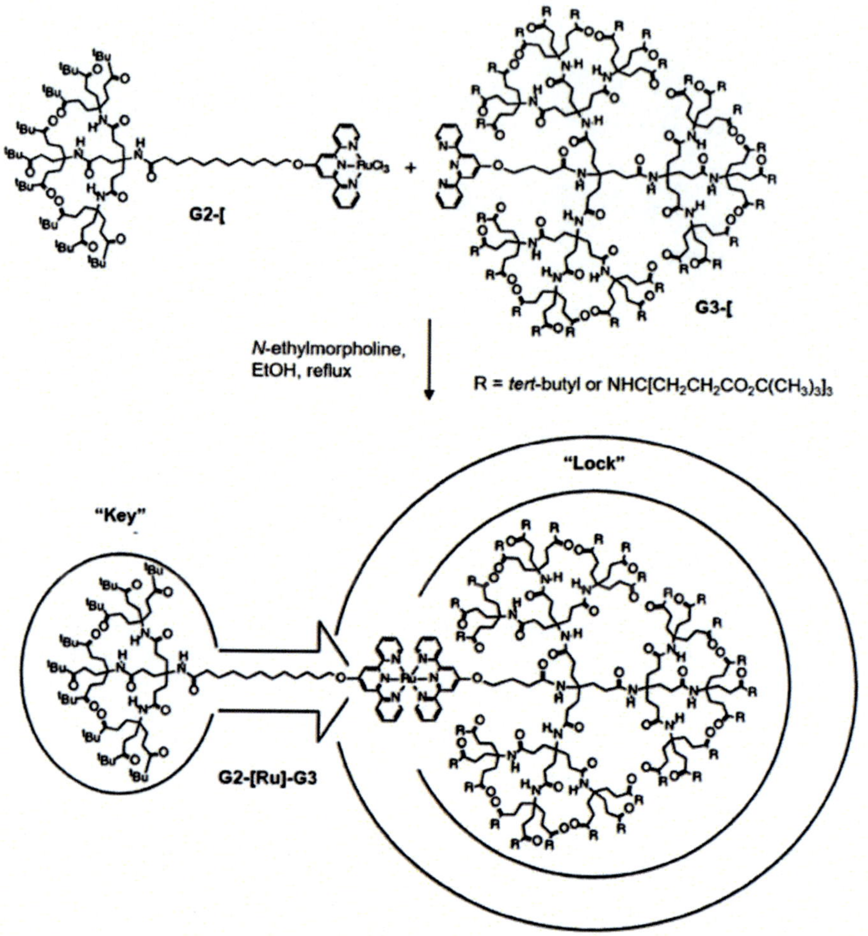

Fig. 4.8 Schematic representation of Newkome's metallo-supramolecular linking of dendrimers according to the "lock and key" concept [3]. Figure reproduced with kind permission; © 1999 American Chemical Society

In one further study, Newkome et al. prepared the four-armed G1 and G2 dendrimers **8** containing four or twelve Ru^{II} complexes, as building-block connectors (Fig. 4.9) [57]. In contrast to the aforementioned dendrimer **4**, additional adamantyl units were present, as terminal units. Adamantane is known for its ability to form inclusion complexes with β-cyclodextrin (β-CD), as macrocyclic host [the association constant (K_a) for the formation of the 1:1 complex is $>10^4$ M^{-1}] [58]. This particular interaction has been widely used in the context of supramolecular materials [59], including polymers [60] and dendrimers [61]. Accordingly, the surface functionalities of the water-soluble dendrimers **8** gave such inclusion

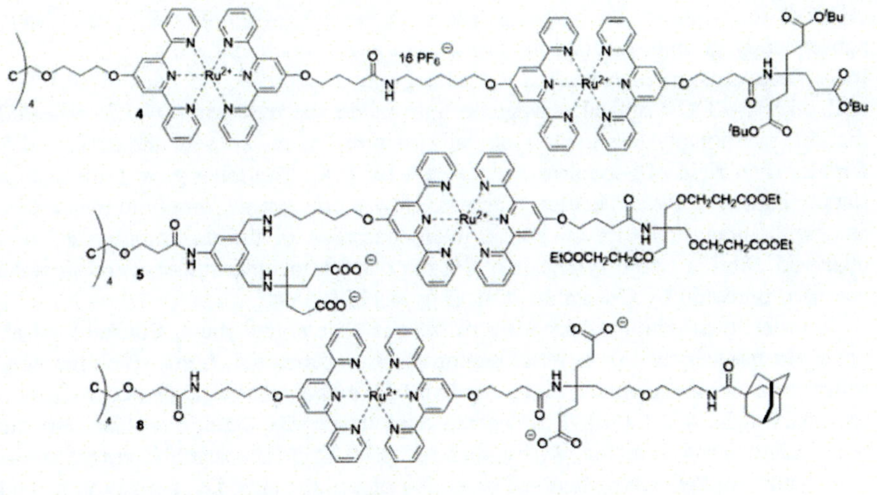

Fig. 4.9 Schematic representation of Newkome's ruthenodendrimers 4, 5 and 8

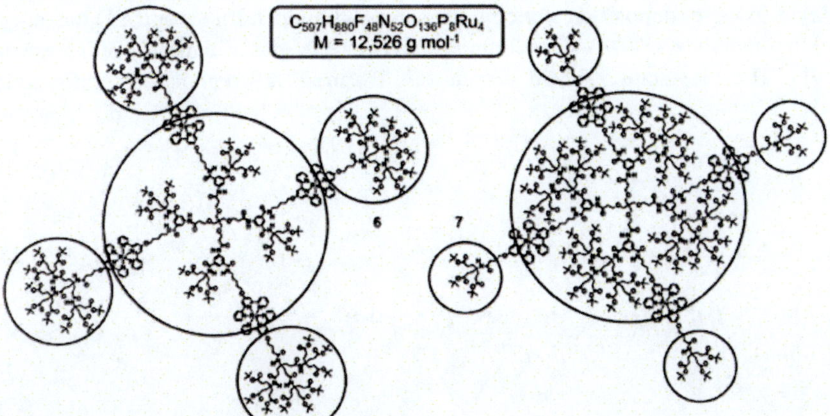

Fig. 4.10 One molecular formula—two dendritic architectures. Schematic representation of Newkome's isomeric ruthenodendrimers [3]. Figure reproduced with kind permission; © 1999 American Chemical Society

complexes with β-CD. This study shows how different types of non-covalent interactions, i.e., metal-to-ligand coordination and host–guest chemistry, can be utilized to construct stimuli-responsive dendrimers.

Constable et al. also targeted the formation of ruthenodendrimers; however, these authors utilized different synthetic strategies compared to Newkome's work [62]. Relying on a convergent strategy, Constable and Harverson developed a protocol to synthesis the oligonuclear back-to-back assemblies **9**, which were subsequently grafted, as dendritic arms, onto a *tris*-functionalized core (Fig. 4.11) [63]. These authors also studied the adsorption of **10** to semiconductor surfaces, such as SiO_2–TiO_2; optical waveguide light-mode spectroscopy (OWLS) revealed that **10**, presumably due to its globular structure, interacted with the surface differently than rigid oligonuclear Ru^{II} complexes [64]. The latter gave films with a flat orientation of the molecules; whereas, **10** was only bound loosely to the surface and, with increasing film thickness, interpenetration of the macromolecules was observed. Similar to the synthesis of **10**, the octadecanuclear ruthenodendrimer **11** was also prepared by Constable et al. (Fig. 4.11) [65].

In order to assemble ruthenodendrimers via a divergent route, Constable et al. opted for pentaerythritol, as tetrafunctional starting material, from which the dendritic arms were constructed in an iterative fashion using organic and coordination chemistry (Fig. 4.12) [66]. Besides the tetranuclear G1 dendrimer **12a** also the hexadecanuclear G2 analog was synthesized [67]. This G1 assembly was tested, as a mediator, for the electrochemical oxidation of amino acids, i.e., L-methionine and L-cystine, as well as As^{III} [68]. Under optimized conditions (potential of +1.1 V *vs.* Ag/AgCl electrode in phosphate-buffered, pH-neutral water), a detection limit of 0.6 μM and 0.5 μM, respectively, was reached for the two amino acids. Following a layer-by-layer deposition strategy, the same ruthenodendrimer and a Dawson-type polyoxometalate (POM) were immobilized onto a glassy carbon (GC) electrode [69]. The resultant hybrid electrode featured a remarkable bifunctional

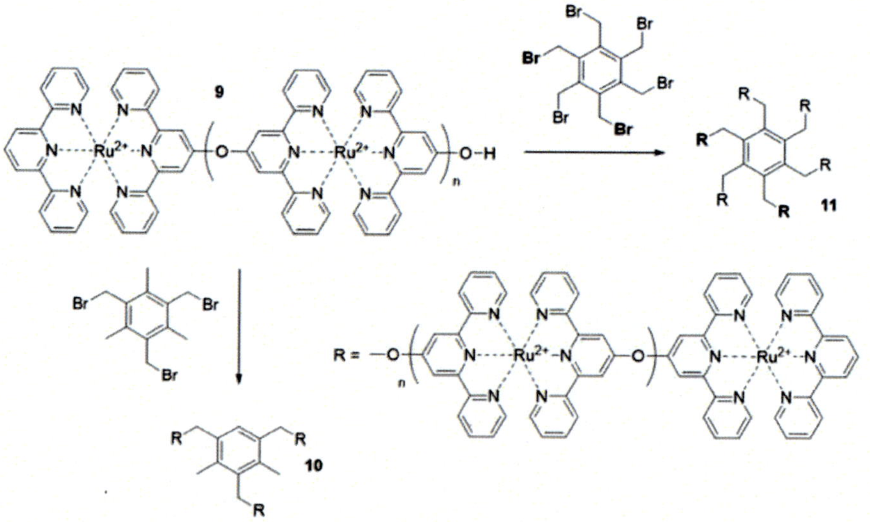

Fig. 4.11 Schematic representation of the synthesis of a three- (**10**) and six-arm ruthenodendrimer (**11**) by grafting the oligonuclear arm (**9**) onto appropriate cores (the counterions are omitted for clarity)

electrocatalytic activity as demonstrated in the reduction of iodate and oxidation of As^{III}. Similar to this, McCormach et al. utilized a different Dawson-type (POM) or a Cu_{20}-tungstophosphate polyanion in the otherwise same electrode configuration for the iodate-reduction reaction in water [70, 71]. The convergent strategy applied in the formation of the G1 ruthenopolymer was extend further to prepare the tetranuclear ruthenodendrimers **12** containing *ortho*-carborane or chiral 1,1-binaphathyl moieties, as terminal units (Fig. 4.12) [66, 72, 73]. The *o*-carborane unit represents—similar to adamantane—an appropriate guest for the inclusion into *β*-CD, as macrocyclic guest (K_a value of 2690 M^{-1}) [74]. Thus, *o*-carborane decorated dendrimers can also be functionalized on their surface via the formation of supramolecular host–guest complexes [75]; however, this strategy was not considered in Constable's work.

It was demonstrated that the emission of the organic chromophores in **12c-f** was quenched by the neighboring Ru^{II} centers, presumably due to an efficient energy transfer of the electron-rich organic domains to the metal centers, which acted as energy-acceptor sites [72, 73]. Moreover, the derivatives containing the Fréchet--type dendrons (**12d-f**) showed a generation-dependent quenching of the emission: The G3-substituted dendrimer (**12f**) exhibited a weak residual emission due to the large distance between the outer dimethoxybenzene units and Ru^{II} complexes close to the dendritic core.

Metal-acetylide polymers have attracted significant interest due to their potential applications, as active materials, in the fields of opto-electronics: Polymer light-emitting diodes, photovoltaic or non-linear optical (NLO) devices as well as low-dimensional conductors ("molecular wires") [76, 77]. Examples for such Ru (II)-containing polymers, in which the linear *trans*-alkynyl*bis*(bidentate phosphine)

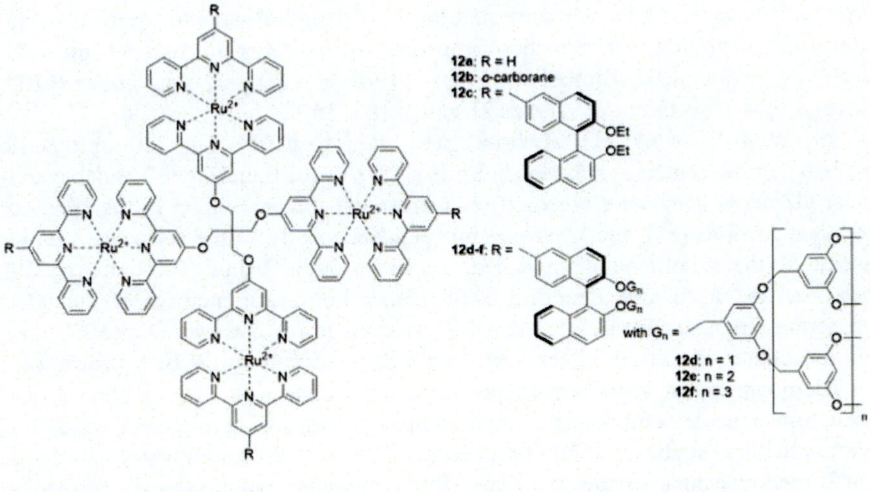

Fig. 4.12 Schematic representation of the tetranuclear, pentaerythritol-centered ruthenodendrimers **12**

ruthenium(II) unit represents the central building block, are discussed in Chap. 2. Humphrey et al. utilized this particular Ru^{II} fragment for the formation of various three-armed, benzene-centered dendrimers. According to the fundamentals of NLO chromophores, octopolar molecules are one class of materials for which rich NLO properties can be expected; such molecules combine the second-order NLO properties with the absence of a dipole moment—in the ground and excited state [78]. In initial studies, Humphrey et al. demonstrated that the NLO response of star-shaped derivatives, such as the D_{3h} symmetric G0 species **13a** (Fig. 4.13), was increased significantly when compared to analogous mononuclear species [79, 80]. The analogous G1-type dendrimer **14a** was prepared by the same authors via a convergent approach (Fig. 4.13) [81]. Transmission-electron microscopy (TEM) was utilized to visualize **14a**; the thus determined diameter of *ca.* 6 nm was in fair agreement with the value derived from molecular-modelling studies [82]. The organo-soluble and highly robust material was analyzed regarding its third-order NLO properties using the so-called Z-scan technique at a wavelength of 800 nm. Both dendrimers, **13a** and **14a**, were optically transparent; however, due to its extended 2D π-conjugated system, **14a** exhibited a strongly increased 2nd hyperpolarizability (**14a**: $\gamma \approx 20{,}700 \times 10^{-36}$ esu; **13a**: $\gamma \approx 3000 \times 10^{-36}$ esu) and an enhanced two-photon absorption (2PA) behavior. A 2PA cross-section of 11,800 GM at 750 nm was calculated for **14a**. Later, the G2 dendrimer, containing 21 Ru^{II} centers, was also reported [83]; however, this material only showed a moderate dendritic effect, i.e., a slight increase in the NLO response relative to that of **14a**. In the same context, also the dendrimers containing electron-donating or electron-withdrawing groups, as terminal units were prepared (**13b/c** and **14b**, Fig. 4.13) [82]. Unlike **14a**, the analogous dendrimer containing vinyl bridges between the phenyl rings extended its 2PA behavior to very low energies (up to 1300 nm), thus even reaching the telecommunication wavelengths [84]. The NO_2-equipped dendrimer **14b** was subjected to a wavelength-dependent analysis of the nonlinear absorption and refraction properties. In the range of 600–800 nm, **14b** acted as two-photon absorber; whereas, at lower energies, i.e., above 1100, three-photon absorption (3PA) was observed [85, 86].

In related work, Humphrey et al. focused on D_{3h}-symmetric triphenylamine-centered ruthenodendrimers [87]. The trinuclear G1 and hexanuclear G2 derivatives were prepared via convergent routes. Similar to the behavior reported for **14b** [85], the wavelength-dependence study of the refractive nonlinearity of these compounds revealed, at wavelengths below 1000 nm, a 2PA behavior and a positive dendritic effect, i.e., a non-linear increase of the NLO properties upon dendrimer growth [87]. At even lower energies ($\lambda > 1000$ nm), photochemistry occurred, which was presumably induced by the 3PA processes.

The quadrupolar, cruciform-shaped dendrimer **15** (Fig. 4.13) exhibited a dual stimuli-responsive behavior [88]. According to linear and non-linear electro- and photo-chemical studies, the NLO properties of **15** were reversibly switched via the $Ru^{II/III}$ redox couple or the pH value (Ru^{II}–vinylidene complexes are formed in acidic media).

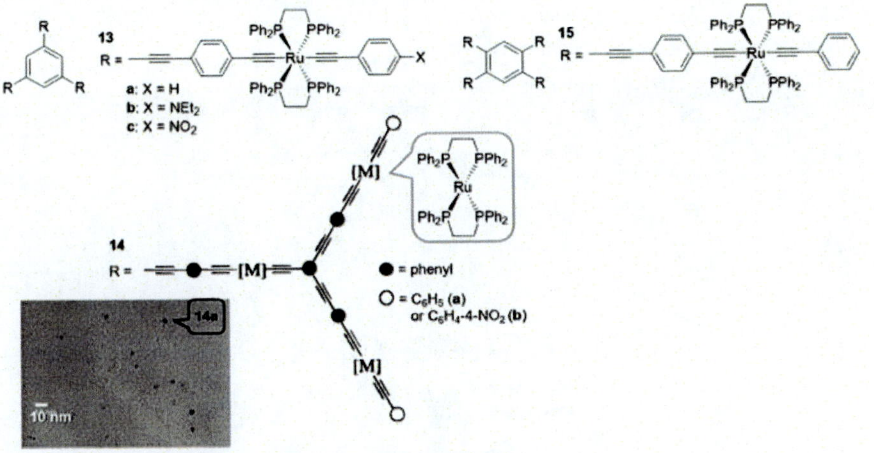

Fig. 4.13 Schematic representation of Humphrey's Ru^{II}–*bis*alkynyl dendrimers **13** and **14**. A representative TEM image of **14a**, deposited onto an alumina substrate, is also depicted [82]. Figure reproduced with kind permission; © 2003 American Chemical Society

Since metallated π-conjugated macrocycles, such as porphyrins and phthalocyanines, represent efficient and robust cubic NLO-phores, Humphrey et al. utilized a Zn^{II}–porphyrin derivative, as core, for their NLO-active ruthenodendrimers. In their first study on such assemblies, the G0 species **16** (Fig. 4.14) showed 2PA in the visible and near-IR spectral range [89]. Their G1 counterparts **17** (Fig. 4.14), containing three Ru^{II} centers per dendritic arm, were analyzed regarding their cubic NLO properties by third-harmonic generation and Z-scan measurements and, in particular, **17b** exhibited a remarkable NLO response in the near-IR range [90, 91]. These results demonstrate that these extended "cross fourchée"-like structures, as in **17b**, are beneficial in improving the 3rd-order NLO properties of these materials. In order to elaborate how the topology of the dendrimers influences their NLO behavior, Humphrey et al. synthesized further dendrimers of, e.g., Y-shape or H-shape [92]. The latter possessed a distorted molecular structure and moderate to large 2PA cross-sections (up to 5500 GM at 880 nm) were reported for these materials.

4.4 Dendrimers, Containing Ru^{II} Complexes as the Core

Similar to Newkome's metallo-supramolecular linking of two dendrimers via a [Ru(tpy)]$^{2+}$ connector (Fig. 4.8) [51], Vögtle, Balzani et al. prepared ruthenodendrimers by coordinating bpy-functionalized dendrons to Ru^{II} centers, thus forming [Ru(bpy)$_3$]$^{2+}$-centered materials; the assembly derived from the G3-dendron possessed 54 peripheral ester units (Fig. 4.15) [93]. Photophysical studies revealed a

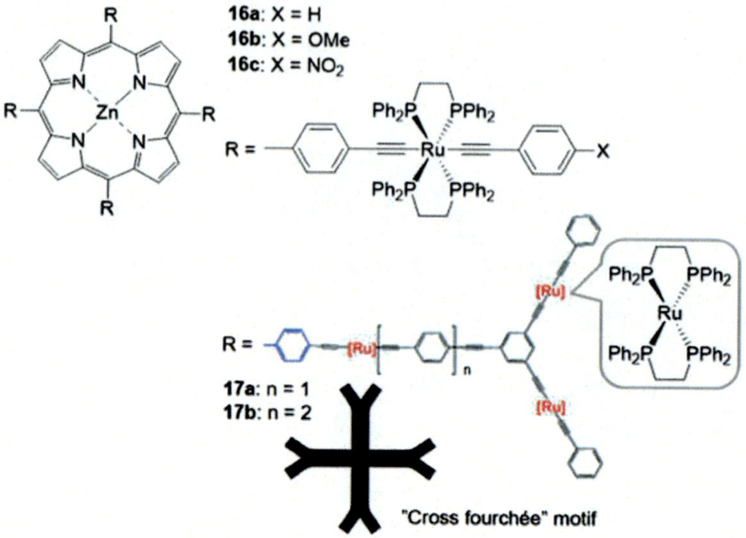

Fig. 4.14 Schematic representation of the Zn^{II}–porphyrin-centered ruthenodendrimers **16** and **17**. The latter represented the heraldic "cross-fourchée" motif

shielding effect of the dendritic shell on the $[Ru(bpy)_3]^{2+}$ complex: With increasing volume of the dendritic "coating", the diffusion rate of potential quenchers (e.g., O_2) decreased, i.e., the rate constant of the emission quenching of the G3 dendrimer was twelve times smaller compared to the parent $[Ru(bpy)_3]^{2+}$ complex. Moreover, the excited-state lifetime of the incorporated complexes was significantly prolonged (>1 μs for the 3rd-generation dendrimer in an aerated solution); presumably, the complex was "solvated" by the dendritic environment, thereby efficiently suppressing deactivating energy-/electron-transfer processes. In subsequent work, these same authors prepared a variety of similar $[Ru(bpy)_3]^{2+}$-centered dendrimers and generalized their proposal of dendritic shielding [94]. For a range of electron-transfer quenchers [i.e., methyl viologen dication (MV^{2+}), tetrathiafulvalene (TTF), and anthraquinone-2,6-disulfonate anion], the rate constant of the quenching process decreased with increasing size of the dendritic shell.

The dendritic effect was further verified for the example of a $[Ru(dpp]_3]^{2+}$ complex (dpp: 4,7-diphenyl-1,10-phenanthroline): With increasing number of dendritic generations, the excited-state lifetime of the Ru^{II} chromophore became elongated in aerated MeCN solution and higher emission quantum yields, compared to the parent $[Ru(dpp)_3]^{2+}$ complex, were found [95].

In subsequent work, these authors also introduced photoactive naphthyl moieties on the outside of the Ru^{II}-centered dendrimers (derivatives with 12 or 24 of such chromophores were prepared) [96]. These assemblies exhibited the characteristic emission band of the central $[Ru(bpy)_3]^{2+}$ unit—irrespective of the excitation

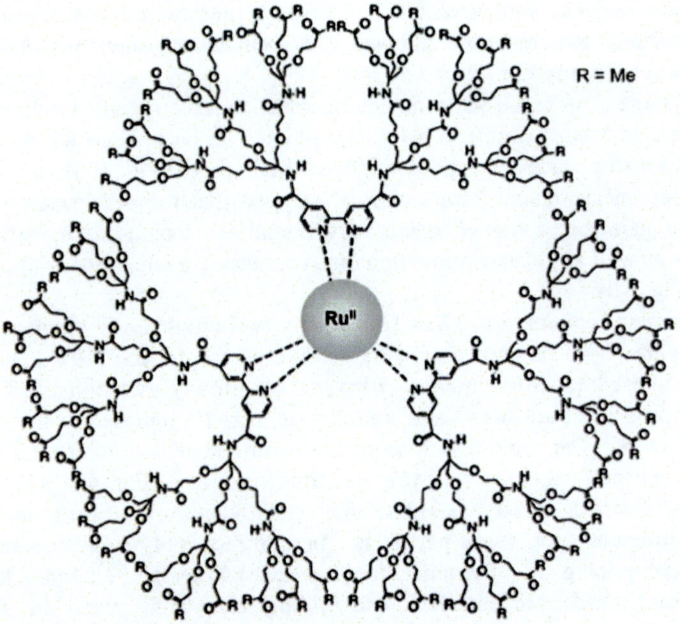

Fig. 4.15 Schematic representation of a Ru^{II}-centered G3 dendrimer

wavelength. This observation was assigned to an inter-dendrimer Förster-type resonance energy-transfer (FRET) process from the photo-excited organic chromophores on the shell to the central Ru^{II} complex, which occurred with >90% efficiency. More recently, similar results were reported by Rivera et al. for a related dendrimer labeled with pyrene units [97]. Similarly, Castellano et al. prepared a Ru^{II}-centered dendrimer with twelve coumarin-450 dyes, as terminal units [98]. Photo-excitation of these dyes, known for their blue ns fluorescence, resulted in the typical red–orange, long-lived Ru^{II}-based emission.

Seeberger et al. followed the same synthetic strategy as Vögtle, Balzani et al. to prepare Ru^{II}-centered dendrimers having peripheral carbohydrate moieties. However, these authors not only focused on the aforementioned stabilization of the ^{3}MLCT state by the dendritic shell and inter-dendrimer energy-transfer processes, but also targeted the application of their rutheno-glycodendrimers, as a biosensor [99, 100]. In particular, the interaction of carbohydrate units with lectins, as model protein substrates, was analyzed. Both the binding affinity as well as the photophysical properties of the dendrimer depended on the number of peripheral sugar sites. Based on these properties, a digital, single-operation analytical method to study glycodendrimer—lectin interactions in a real-time fashion was developed [101]. The rutheno-glycodendrimers were utilized further as fluorescent and electrochemical probes for the detection of lectin—carbohydrate interactions on microarrays or gold substrates [102]. Thereby, prototypical interaction between concanavalin-A and mannose allowed the accurate measurement of saccharide concentrations in solution.

In the same context, Okada et al. demonstrated that the specific binding of lectins or the tetanus toxin to rutheno-glycodendrimers can be utilized for the combinatorial biosensing, based on photo-induced SET processes [103]. Emission and polarization measurements, for example, showed that the dendrimers possessing peripheral galactose moieties specifically bound the tetanus toxin; whereas, the glucose—as well as galactose-functionalized dendrimers interacted with the various lectins (Fig. 4.16).

Dendrimers containing 6, 12 or 18 Ru^{II}–tpy *bis*-complexes in the dendritic arms were prepared by a combined convergent/divergent approach [104]. The Pd^0-catalyzed Sonogashira cross-coupling between the ethynyl-functionalized dendrons and an appropriate transition-metal complex, such as **18**, represented the key step in Osawa's work. The rigid macromolecules were analyzed *inter alia* by TEM imaging and their spherical structure was thus visualized (Fig. 4.17).

In the aforementioned examples, the ruthenodendrimers were, due to their flexible structure, not shape-persistent. In contrast, polyphenylene dendrimers (PPDs), comprising of poly(pentaphenylbenzene) branches, exhibit stiff, mostly radial arms, which prevent their backfolding and, thus, result in an overall shape-persistent macromolecular structure [105, 106]. Therefore, the geometry of the core dictates the shape of the PPD, at least in first approximations [107] In 2008, Müllen et al. synthesized the first example of an octahedral PPD by attaching six polyphenylene arms to a $[Ru(bpy)_3]^{2+}$ core (Fig. 4.18) [108]. The synthesis of the dendrimers was accomplished by a combined convergent/divergent protocol which, remarkably, involved a high-temperature Diels–Alder reaction on the Ru^{II} complex (perfect control over the molecular structure was guaranteed up to the G3 ruthenodendrimer). Despite the dense packing of aromatic rings, exchange of the initial Cl^- counteranion by, e.g., PF_6^-, BF_4^- or the methyl-orange anion, was still possible —at least in the case of the G1 material.

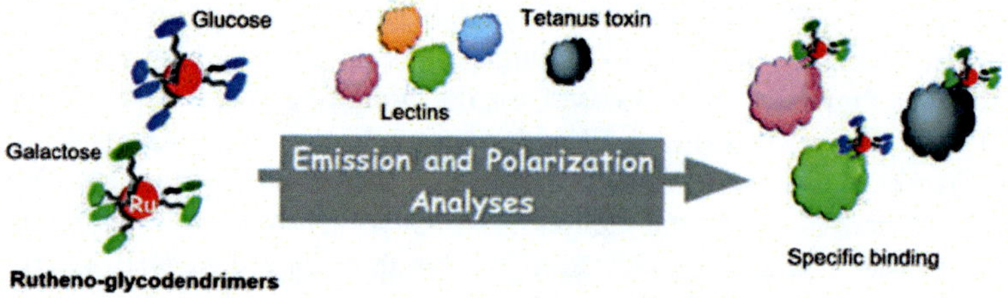

Fig. 4.16 Schematic representation of the specific interaction between carbohydrate-functionalized ruthenodendrimers and protein substrates [103]. Figure reproduced with kind permission; © 2009 American Chemical Society

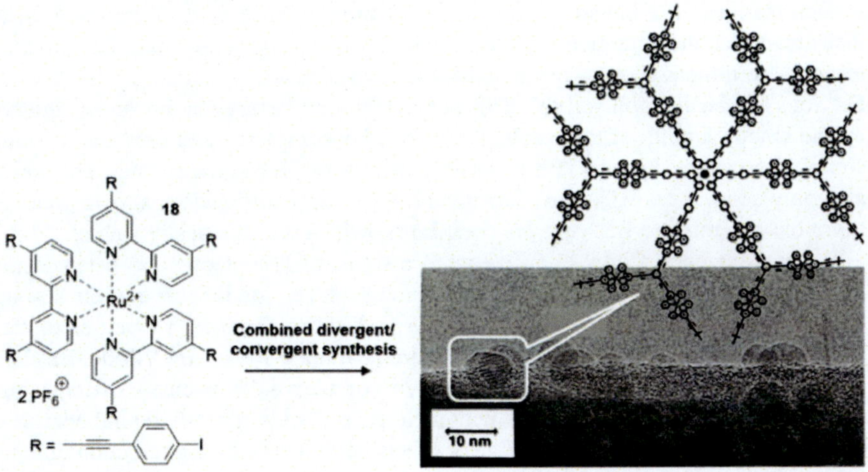

Fig. 4.17 Schematic representation of Osawa's nonadecanuclear ruthenodendrimer **18**. A representative TEM image of the material is also depicted [104]. Figure reproduced with kind permission; © 1999 American Chemical Society

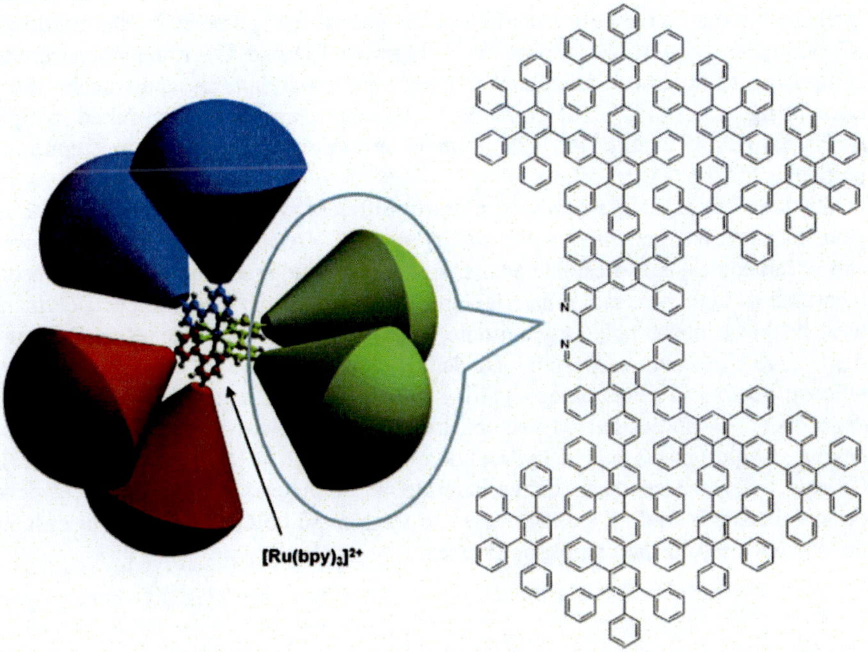

Fig. 4.18 Schematic representation of Müllen's shape-persistent PPD-type ruthenodendrimers [108]. Figure reproduced with kind permission; © 2008, Wiley–VCH

Besides these *tris*-bidentate complexes, commonly used as dendritic core, also their *bis*-tridentate counterparts have been used for the same purpose. In this respect, the dimerization of tpy-functionalized dendrimers by their coordination to Fe^{II} ions has been employed [3]. This straightforward strategy is, however, limited by the kinetic lability of the resultant Fe^{II}–tpy *bis*-complexes and only homoleptic complexes can be formed [109]. As in the case of Newkome's "lock and key" assembly (Fig. 4.8) [51], the directionality of the coordination allows one to assemble heteroleptic Ru^{II}–tpy *bis*-complexes, which are kinetically stable.

As one example, Chow et al. prepared a series of homoleptic and heteroleptic Ru^{II} complexes (**19**) to carve out the dendritic effect, i.e., the influence of the size of the polyether wedges on the photophysical and electrochemical properties of the Ru^{II}-centered dendrimers (similar to the aforementioned studies by Vögtle, Balzani et al., Fig. 4.15) [110]. CV measurements of complexes **19** revealed a decreasing reversibility of the redox processes with increasing shielding of the central complex by the dendritic wedges. These ΔE values were correlated with the retention time in size-exclusion chromatography (SEC) measurements and, thus, with the molecular weight of the dendrimers (Fig. 4.19). In the same context, Bäuerle et al. synthesized dendrimers with a $[Ru(tpy)_2]^{2+}$ core and oligothiophene arms [111]. The photophysical analysis of these assemblies suggested intra-dendrimer energy-transfer processes from the electron-rich oligothiophene dendrons to the Ru^{II} center; however, a dendritic effect, as observed in Chow's dendrimers **19**, was absent (presumably due to the poor shielding of the central Ru^{II} complex by the rigid-rod oligothiophene chains. Newkome et al. reported G0 and G1 ruthenodendrimers comprising of thiophene-functionalized Ru^{II}–tpy *bis*-complexes and truxene moieties, as the nodes [112]. The combination of these chromophores resulted in high molar extinction coefficients and a broad absorption region (250–640 nm), in particular for the G1 dendrimer.

Diederich et al. focused on a completely different type of application, which is gene transfection, i.e., the transport/delivery of DNA to cells. For this purpose, carrier molecules, the so-called vectors, are required and macromolecules have been identified as highly versatile materials in this respect. From the synthetic polymers used in this context, poly(ethylenimine) (PEI) is considered as the "golden standard" [113]. As summarized by Diederich et al., dendrimers can also serve as efficient vectors in gene therapy [114]. These authors assembled the amphiphilic dendrimers **20**, as designated vectors, by coordinating two different tpy ligands, bearing a hydrophilic and a hydrophobic dendron, to a Ru^{II} center (Fig. 4.20). Unexpectedly, the synthesis of the heteroleptic complexes containing to G2 dendrons failed; this finding was attributed to the non-solubility of the tpy ligands in the solvent used in the complexation step.

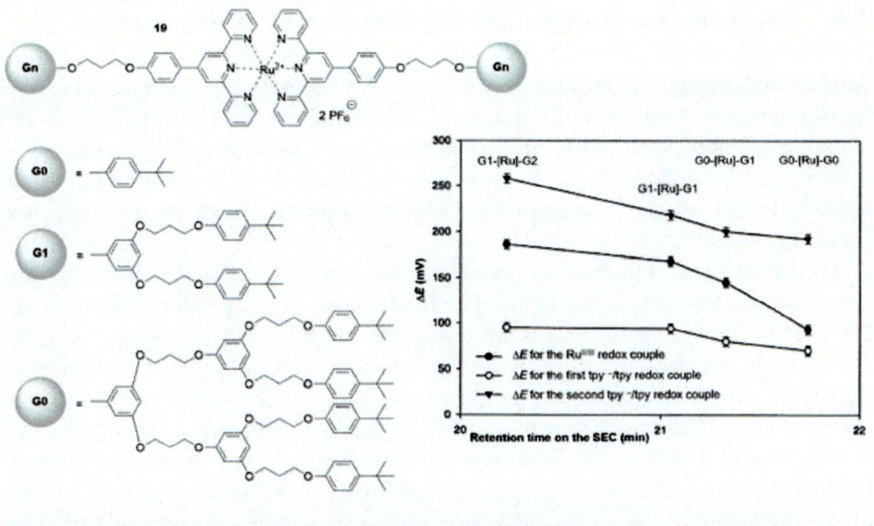

Fig. 4.19 Schematic representation of the Ru^{II}-centered dendrimers **19**. The reversibility of their redox couples could be correlated to the molecular weight of the dendrimers and, thus, with the shielding of the metal center [110]. Figure reproduced with kind permission; © 2001 Elsevier B.V.

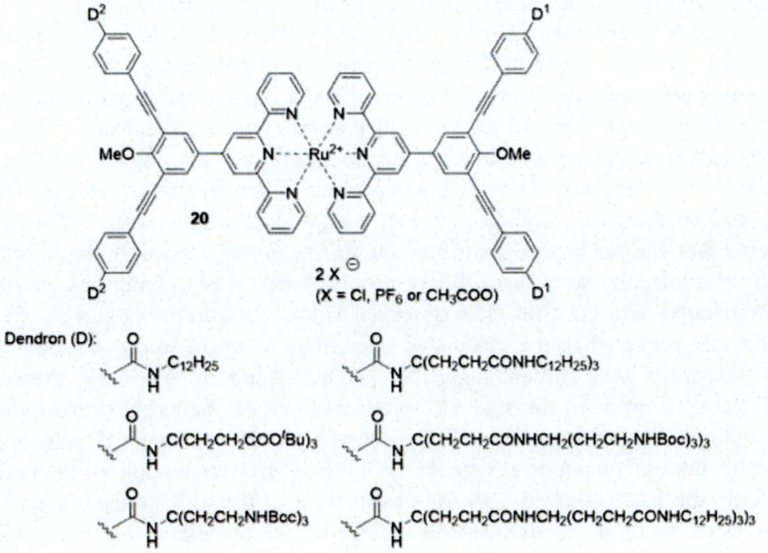

Fig. 4.20 Schematic representation of the amphiphilic ruthenodendrimers **20**

4.5 Dendrimers, Containing Ru^{II} Complexes at the Surface

As depicted in Fig. 4.3, dendrimers can also be decorated on their surface with transition-metal complexes. In this case, the metal centers are exposed to the environment and, thus, easily accessible by foreign substrates. Such metallodendrimers can be utilized, e.g., as heterogeneous catalysts in organic reactions (in analogy to the various examples for polymers equipped with Ru catalysts, see Chap. 2).

Constable et al. reported on metallodendrimers containing six Ru^{II}–tpy *bis*-complexes on the surface and M^{II}–bpy *tris*-complexes, as core unit (M = Fe, Co, or Ru) [115–117]. For synthetic reasons, a convergent approach was followed, i.e., the central complex was assembled in the last step of the synthesis. Such multinuclear ruthenodendrimers were tested, as photosensitizer, in dye-sensitized solar cells (DSSCs) [118]. The attachment to the TiO_2 semiconductor surface was realized via a carboxylate moiety in the periphery of the assembly (thus, offering a stronger binding and more efficient charge injection as when only considering physisorption of the polycationic dye molecules). However, only moderate photocurrents were detected and the ruthenodendrimers were identified as less efficient photosensitizer compared to the established Ru^{II} chromophores used in this context (Chap. 1).

Abruña et al. utilized commercial poly(amidoamine) (PAMAM) dendrimers to attach bpy or tpy units to the terminal NH_2 sites via straightforward peptide-coupling methods [119]. The thus obtained dendrimers with ligand sites, as terminal units, were subsequently used for the complexation of Ru^{II} precursor complexes, i.e., *cis*-Ru(bpy)$_2$Cl$_2$ and Ru(tpy)Cl$_3$. Remarkably, the G4 metallodendrimers possessed 64 terminal Ru^{II} complexes, giving rise to an overall charge of +128 (Fig. 4.21). The electrochemical properties of the dendrimers, deposited onto Pt electrodes, were analyzed by electrochemical quartz-crystal microbalance (EQCM) measurements: The metallodendrimer films displayed charge-trapping peaks and changes in morphology when the potential was changed. The authors proposed that various factors contributed to this behavior: Deposition/dissolution of dendrimer molecules as well as the ejection/absorption of counterions and/or solvent molecules into the film. The observed values for surface coverage (Γ) were significantly higher than the calculated ones, thus compression of the dendrimers upon adsorption was further suggested [120]. In subsequent work, these same authors also determined the size of the dendrimers via their diffusion coefficients [CV, voltammetry at a rotating disk electrode (RDE), and pulsed field gradient spin-echo nuclear magnetic resonance (PFGSE-NMR) were applied] [121]. However, only the latter allowed a realistic estimation of the ruthenodendrimers' sizes, which were, in crude approximation similar to the calculated ones (the electrochemical methods gave inaccurate results due to the multitude of Ru^{II} centers, which could hardly be sampled simultaneously [122]). Though the apparent solvent layer of the ruthenodendrimers attached to a Pt electrode was small (less than one solvent molecule), their solvation number was very high (e.g., the G4 dendrimer with 64 $[Ru(bpy)_3]^{2+}$ sites incorporated 2.5×10^3 MeCN molecules) [123].

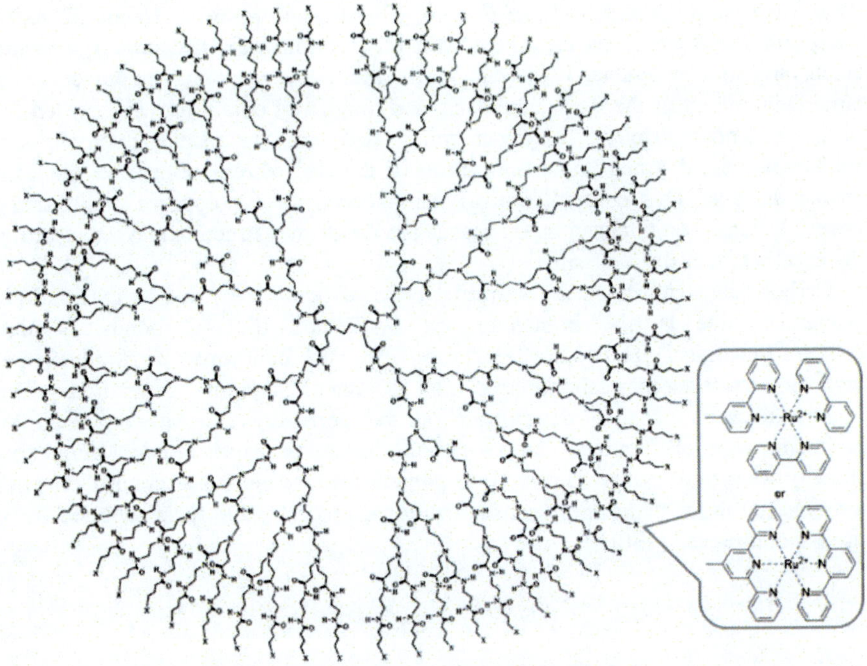

Fig. 4.21 Schematic representation of Abruña's G4 metallodendrimers containing 64 Ru^{II}–polypyridyl complexes, as terminal units (the counterions are omitted for clarity) [120]. Figure reproduced with kind permission; © 1999 American Chemical Society

Finally, Abruña et al. used the emission properties of the dendrimers bearing terminal $[Ru(bpy)_3]^{2+}$ units to fabricate electroluminescent devices [124]. When sandwiched between indium tin oxide (ITO) and Au electrodes, the devices showed a similar behavior as those containing parent $[Ru(bpy)_3]^{2+}$, as active layer [125]. In detail, the ionic and electronic mobility decreased with increasing dendrimer generation (i.e., the turn-on time increased, whilst the steady-state current decreased) [124]. Due to self-quenching effects, the device efficiency also decreased with generation (the G1 ruthenodendrimer was identified as the best emitter within the series). Noteworthy, dendrimers are of relevance regarding their implementation, as emissive material, into light-emitting diodes (LEDs). Besides the various all-organic dendrimers, the Ir^{III}-containing metallodendrimers have attracted considerable interest in this respect [14, 126].

Similar dendrimers, i.e., also possessing up to 64 tpy units on the surface, were reported by Hong et al. [127]. However, these authors utilized a different dendrimer architecture in which a phosphine-substituted polyhedral silsesquioxane served as a cubic core, to which the dendrons were attached via benzylic linkages (thereby creating an internal polycationic phosphonium-salt layer). The G1, G2, and G3 dendrimers were treated with *cis*-Ru(bpy)$_2$Cl$_2$, thus forming 6, 16, or 32 Ru^{II}–polypyridyl complexes on the surface (thereby, the basically tridentate tpy ligand acted only as a bidentate, Fig. 4.22) [128]. Intense UV/vis absorption, due to a cumulative effect of the Ru^{II} complexes, and long-lived emission was reported for these ruthenodendrimers. Electrochemical and time-dependent photophysical studies suggested a negligible interaction of the Ru^{II} chromophores on the dendrimer surface. This contradiction to the photophysical properties of Abruña's materials can be rationalized by the less dense dendrimer structure and the resulting distance between the Ru^{II} sites.

Carbosilane dendrimers are a special class of dendrimers due to their unique properties; due to the branching via Si–C and Si–O–C bonds, a high degree-of-chemical stability is offered; moreover, such dendrimers are fluid, even at high molar masses (corresponding to high numbers of dendritic generations). The functionalization of such dendrimers on the periphery can be conducted by hydrosilylation of terminal carbon–carbon double or triple bonds [129]. This strategy was followed by Kim et al. to prepare carbosilane-based dendrimers with terminal tpy units, which were subsequently used to assemble surface-bound Ru^{II}–tpy *bis*-complexes [130].

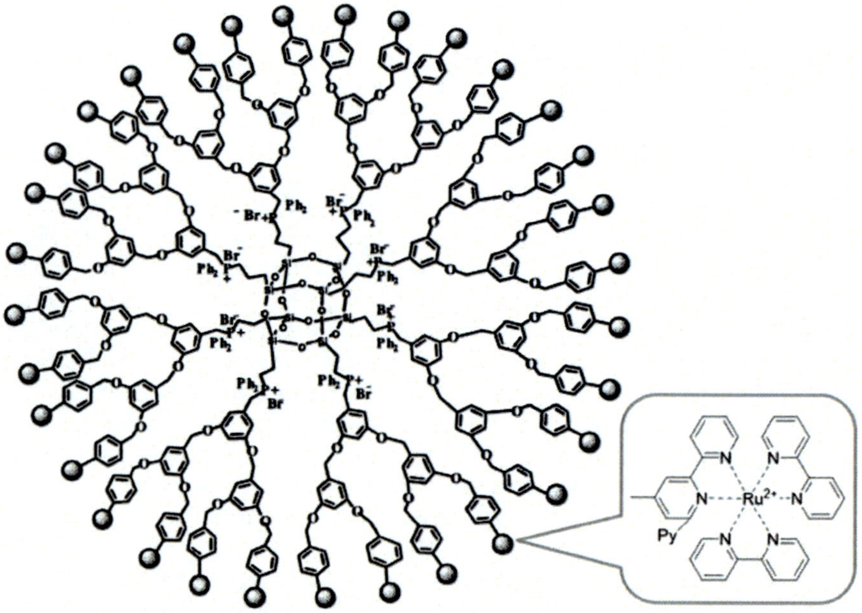

Fig. 4.22 Schematic representation of Hong's silsesquioxane-centered ruthenodendrimers (the G3 assembly is depicted representatively) [128]. Figure reproduced with kind permission; © 2000 American Chemical Society

Besides the aforementioned Ru^{II}–polypyridyl complexes, also a range of other Ru^{II} complexes has been attached to the dendrimer's shell. In one of the first reports in this respect, Liao and Moss prepared ruthenodendrimers, up to the G4 material, containing half-sandwich complexes, as terminal units [131–133]. These dendrimers, based on benzyl-ether linkages, were constructed via Fréchet's convergent approach; remarkably, the Ru^{II} centers were connected to the dendritic frameworks via metal-to-carbon σ-bonds and remained stable throughout the synthesis (Fig. 4.23). The dendrimers, containing up to 48 Ru^{II} centers, were characterized by standard techniques (NMR, elemental analysis, etc.). Though half-sandwich Ru^{II}–arene complexes are long-known for their catalytic activity [134], such applications were not considered by Liao and Moss.

Caminade, Majoral, Chaudret et al. utilized the *bis*-phosphine moieties in the periphery of their *P*-branched dendrimers as bidentate binding sites for the coordination of transition-metal centers (e.g., Pt, Pt, and Rh) [135]. In subsequent work, these authors also attached Ru^{II}–hydride complexes to these binding sites (Fig. 4.24) [136]. The authors used ^{1}H and ^{31}P NMR spectroscopy to analyze the coordination geometry of the dendrimer-bound complexes (in the case of the hydrido-complex, various coordination isomers were identified). The hydrido ligand in these complexes was readily substituted by CO; whereas, the complex missing this ligand preferentially displaced one of its PPh_3 ligands. Catalytic activity of the ruthenodendrimers in the ketone hydrogenation was further reported. Later, the scope of the catalytic applicability was extended to the Knoevenagel condensation and diastereoselective Michael addition of activated methylene

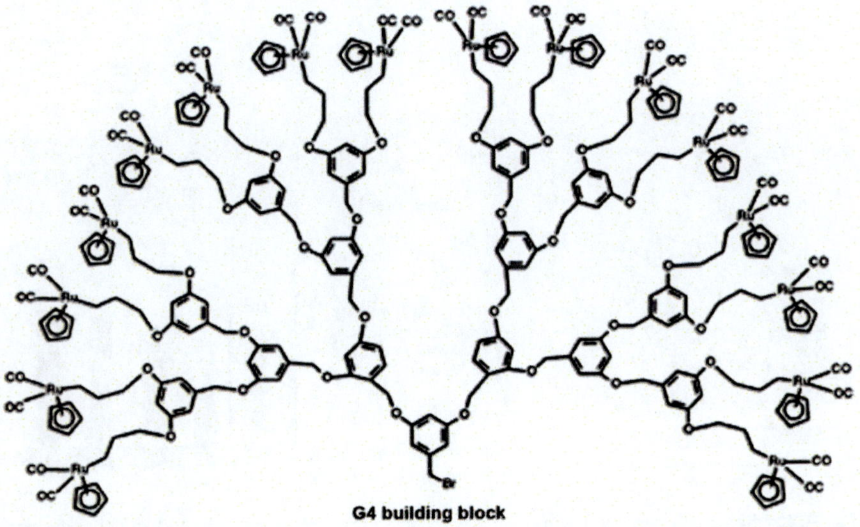

Fig. 4.23 Schematic representation of a dendron-containing half-sandwich Ru^{II}–arene complexes

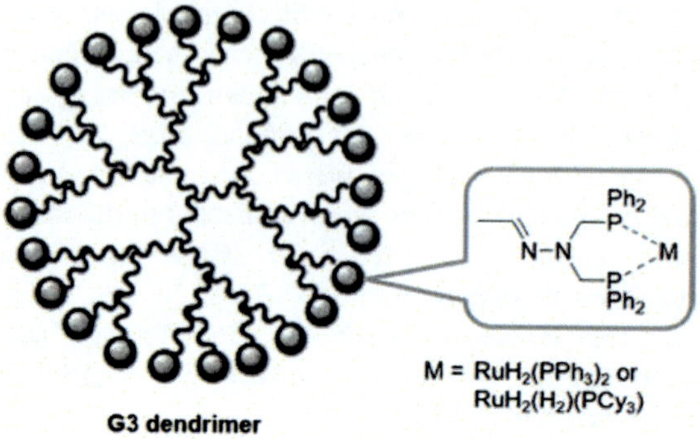

Fig. 4.24 Schematic representation of a dendrimer with terminal catalytically active Ru^{II}–hydride complexes [136]. Figure reproduced with kind permission; 1997 American Chemical Society

compounds; in both cases, the G3 dendrimer showed a high catalytic activity and the recovery/re-use of the catalyst was also demonstrated [137].

The cationic Ru^{II} complex **21** was found to moderately catalyze the isomerization of allylic alcohols to the corresponding ketones in a biphasic H_2O/heptane regime and using Cs_2CO_3, as co-catalyst. The catalytic activity was increased significantly when this complex was attached to the surface of a *P*-branched dendrimer (the G3 dendrimer is depicted in Fig. 4.25) [138]. The polycationic nature of the dendrimers enabled their solubility in water; whereas, reactants and the products remained in the organic phase (strong stirring guaranteed a sufficient intermixing of

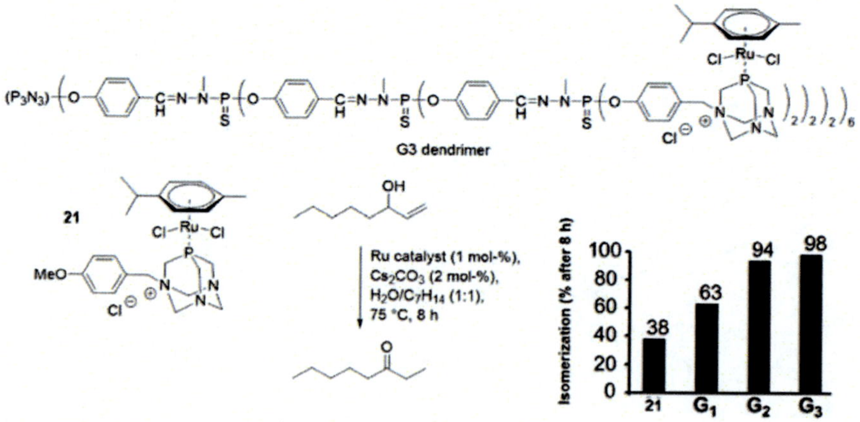

Fig. 4.25 Schematic representation of the Ru^{II}–arene complex **21** and its dendronized counterparts. These materials were used, as catalysts, for the in the isomerization of allylic alcohols [138]. Figure reproduced with kind permission; © 2009 The Royal Chemical Society

the phases and, thus, created the appropriate interface, where the reaction occurred). With increasing generation, the orientation of the dendrimers to the interface became favored and, therefore, their catalytic activity increased (Fig. 4.25). However, such a positive dendritic effect could not be identified in the hydration of terminal alkynes: The regioselectivity of the reaction was high (notably, ketones were formed almost exclusively), but the dendrimers showed no increased catalytic activity compared to **21**. It was shown later that increasing the dendrimer generation had a positive influence on the catalytic activity, but also the density of complex sites on the dendrimer surface was of particular relevance [139].

A related G1 dendrimer containing a redox-active ferrocene (Fc) moiety next to the Ru^{II} complex, was utilized by Hey-Hawkins et al. in the context of redox-switchable catalysis (RSC) [140]. Basically, the RSC concept relies on an oxidation or reduction process within the vicinity of the catalyst site, which can alter its activity and/or selectivity [141]. Similar to the catalytic behavior depicted in Fig. 4.25, the dendrimer was more efficient in catalyzing the aforementioned model reaction (G1 dendrimer: TOF = 600 h^{-1}, **22**: TOF = 400 h^{-1}) [140]. Even more important, in contrast to the non-dendronized complex **22**, its G1 analog showed a pronounced RSC behavior: The oxidation of the Fc units drastically decreased the activity of the G1 dendrimer; also reduction of the Fc^{+} sites regenerated the initial catalyst, which almost completely restored its activity. These authors proposed, that the decreased activity of the oxidized species resulted from its poor solubility in the

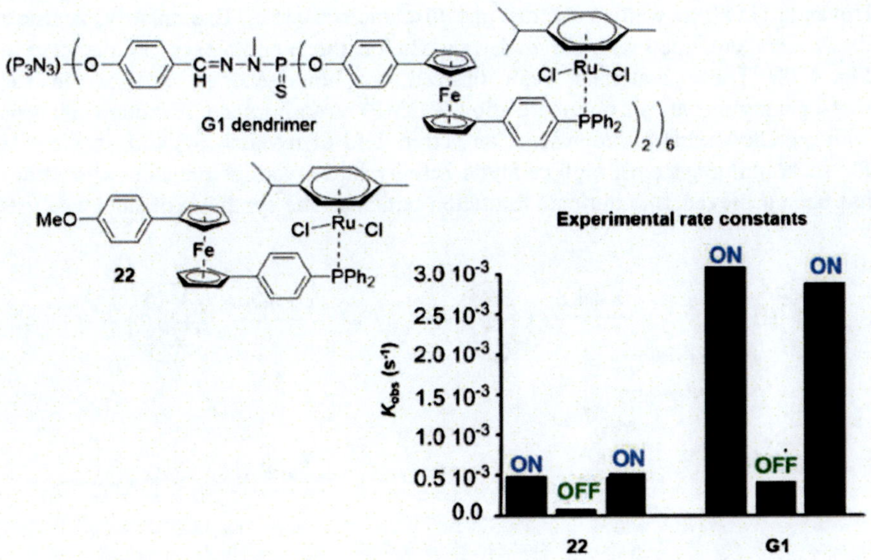

Fig. 4.26 Schematic representation of catalyst **22** and its dendronized G1 analog. Both catalysts showed a RSC behavior in the isomerization of an allylic alcohol (the "off" state corresponds to the oxidized species) [140]. Figure reproduced with kind permission; 2014 Wiley–VCH

reaction medium and electronic interactions between the Fc^+ sites and the neighboring *P* atoms. In subsequent work, it was demonstrated that the catalytic activity can be modulated by adjusting the distance between the Fc^+ sites and the peripheral Ru^{II} complexes (increasing the distance reduced the electronic interactions and, thus, the RSC effect) [142] (Fig. 4.26).

The same Ru(η^6-cymene)Cl_2-fragment, coordinated to the peripheral phosphine units of a carbosilane dendrimer, was used, as catalyst, for an asymmetric hydrogen-transfer reaction (i.e., the reduction of acetophenone to 1-phenylethanol) [143]. Though a high catalytic activity in terms of conversion and selectivity was found, the chiral complexes offered no stereo-control in the reaction.

Van Koten et al. developed a protocol for the peripheral functionalization of dendrimers utilizing polylithiated dendrimers, as the starting material (Fig. 4.27a) [144]. This strategy was employed, e.g., to introduce 2-pyridylcarbinol units, which were subsequently reacted with 1st generation Grubbs catalysts to afford the ruthenodendrimers **23** (Fig. 4.27b). Since such 2-pyridylcarbinol-containing Ru^{II}–carbene complexes are known to be applicable in the various olefin metathesis reactions [145], van Koten et al. tested their dendrimers in this respect [144]. The test reaction, i.e., the ring-closing metathesis (RCM) of a bifunctional olefin, proceeded with high efficiency (full conversion within 30 min). As one further achievement, these authors also established nanofiltration, as a purification method, to separate the soluble dendrimer catalyst from the reaction products and, thus, enabled its recovery [146]. In the following, a variety of dendrimer-based catalysts for olefin metathesis reactions were reported by the groups of Astruc [147–150], Hoveyda [151], as well as Baudry and Fischmeister [152]. Exemplarily, Astruc's dendrimers, equipped with Hoveyda-type Ru^{II}–carbene complexes, are depicted in Fig. 4.28. These materials were utilized, as homogeneous catalyst, in the ring-opening metathesis polymerization (ROMP) of norbornene. Although catalysis with metallodendrimers represents an active field of research [4] and, in view of Ru^{II}-containing systems, high catalytic activity in the various metathesis reactions has been achieved, one intrinsic limitation remains: The synthesis of such catalysts

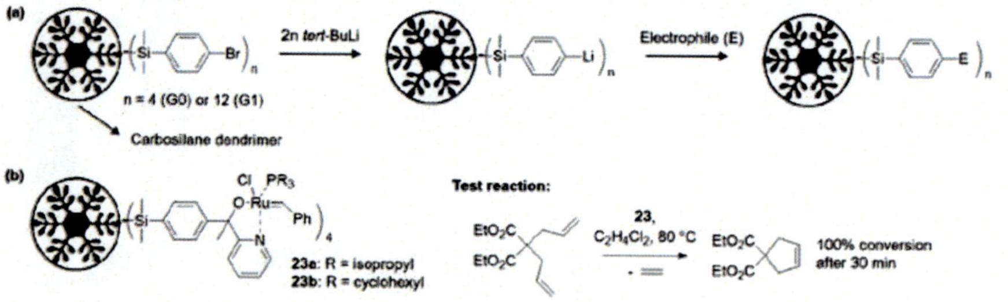

Fig. 4.27 **a** Schematic representation of the synthesis of surface-functionalized dendrimers via a polylithiation approach. **b** Schematic representation of van Koten's dendrimer-based Ru^{II}–carbene catalyst for the RCM reaction [144]. Figure reproduced with kind permission; © 2000 American Chemical Society

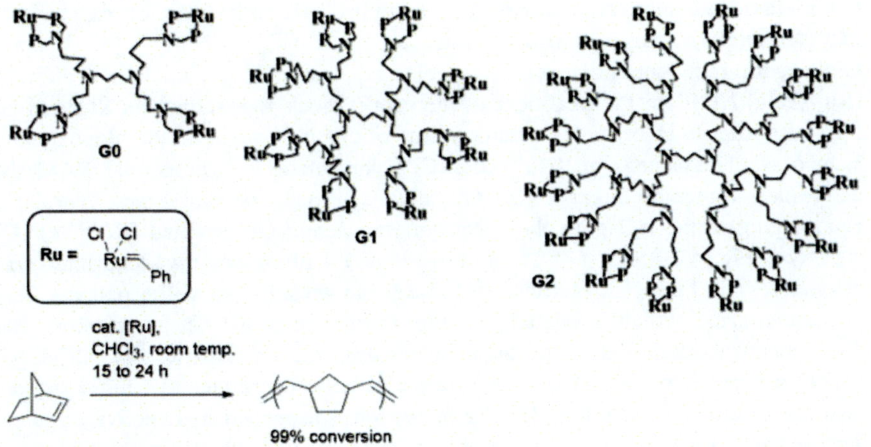

Fig. 4.28 Schematic representation of Astruc's dendrimer-based catalysts for the ROMP of norbornene [149]. Figure reproduced with kind permission; © 2005 Elsevier B.V.

is always related with a significant synthetic effort. The immobilization of Ru catalysts on polymers [153, 154] or even (commercial) polymer resins [154] is easier to realize and, thus, such materials represent the more applicable catalysts, in particular for large-scale reactions. Examples for this are shown in Chap. 2.

The application of ruthenodendrimers for biomedical or biomimetic applications represents one further active field of research. In particular, the development of Ru^{II}-containing potential anti-cancer agents, dyes for photodynamic therapy (PDT) or bio-orthogonal catalysts prompted researchers to immobilize such complexes onto polymeric supports [155]. Thereby their intrinsically poor solubility in water and/or hydrolytic stability under physiological conditions can be overcome [156]. Though polymers are more commonly used in this context, some notable ruthenodendrimers have also been reported.

As a first example in this respect, NO-releasing dendrimers are worth mentioning. Nitric oxide (NO), as relatively short-lived (lifetime in biological media between 10 ms and 1 s) of a readily diffusible molecule, is of significant biological importance; NO has been identified to be involved in signal transduction, neurotransmission, and the host-defense mechanism [157]. In living cells, NO is produced by nitric-oxide synthase (NOS), a heme-protein, which catalyzes the oxidation of L-arginine to NO and citrulline [158]. Due to the prominent role of NO in organisms, the deficiency in NO biosynthesis has been related to the occurrence of some diseases [159]. Thus, the exogeneous delivery of NO has been suggested as a therapeutic approach and carriers for the targeted transport and release of NO have been developed. As summarized by Roveda Jr. and Franco, NO-releasing dendrimers are one class of agents to be employed for this purpose [160]. The $[Ru^{III}(edta)(H_2O)]^-$ complex exhibits a remarkable chemical and biological activity (edta: ethylenediaminetetraacetate). This complex efficiently takes up NO to give [Ru^{III}(edta)(NO)]$^{-}$, as a kinetically stable complex ($K_a = 10^8$ M^{-1}) [161]; upon one-electron reduction, however, NO is released [162]. The potential application of [Ru^{III}(edta)(NO)]$^{-}$ in the treatment of the septic shock was shown by Slade et al. [161]. In order to increase the retention time of this NO-carrier in the bloodstream, Franco et al. attached the [Ru^{III}(edta)(Cl)]$^{-}$ precursor to commercial PAMAM dendrimers via amide linkages (G0, G2, and G3) [163]. The subsequent hydrolysis and purging of the NO gave the corresponding dendrimer-attached [Ru(edta)NO] sites (**24**, Fig. 4.29); for G3-**24**, a storage of 1.4 μmol per mg dendrimer was estimated. The biological activity of G2-**24** was tested against the parasite *Trypanosoma cruzi*, which is related with the occurrence of the Chagas disease. The dendrimer was identified as the more potent drug than parent [Ru^{III}(edta)(NO)]$^{-}$, which only reached 89% of the dendrimer's activity. More recently, these authors also screened the dendrimers **25** (Fig. 4.29) with respect to their activity against *Trypanosoma cruzi* [164]. These dendrimers even showed a dual stimuli-responsive behavior: the controlled release of NO could be triggered either by an one-electron reduction or light irradiation (λ = 355 nm).

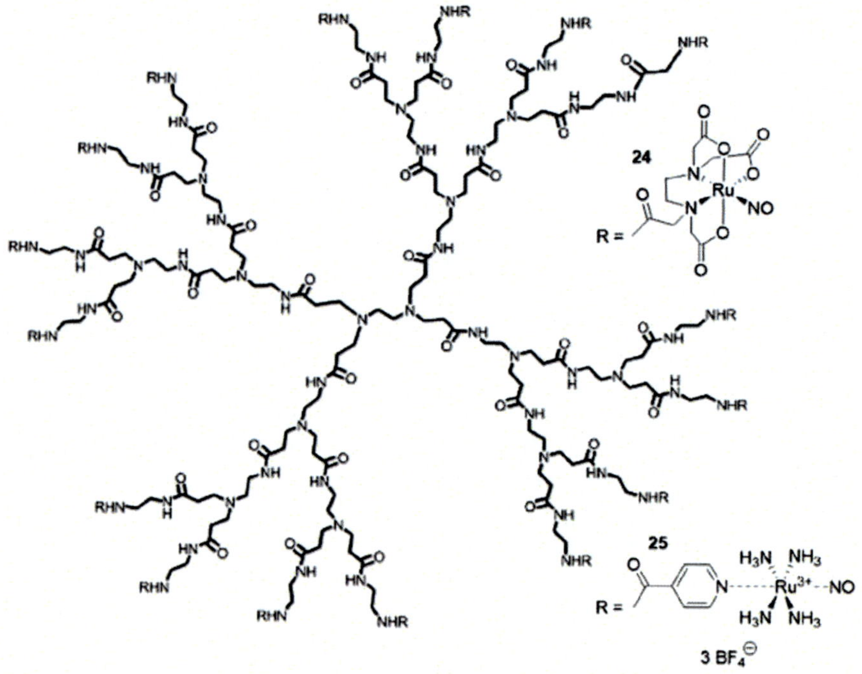

Fig. 4.29 Schematic representation of Franco's ruthenodendrimers **24** and **25**, which were used as NO-releasing agent for the treatment of Chagas disease (representatively, the G2 dendrimer is depicted)

Ru^{II}–arene complexes, such as RAPTA-C, are discussed as novel anti-cancer agents and, for some examples, clinical studies are currently in progress [155]. Similar to the conjugation to polymers (Chap. 2) or the encapsulation in polymer micelles (Chap. 5), Therrien, Smith et al. attached such Ru^{II}–arene complexes to poly(propyleneimine) dendrimer frameworks and studied their activity against the A2780 human ovarian carcinoma cell line [165]. However, the *in-vitro* activity of G1-**26** and G2-**26** as well as of their mononuclear analogs **27** (Fig. 4.30) was much lower compared to that of cisplatin. The tetranuclear und octanuclear dendrimers (G1 and G2) had IC_{50} values in the range of 40 μM and 20 μM, respectively; whereas, for cisplatin an IC_{50} value of 1.5 μM was found (Table 4.1). In subsequent work, these authors improved the antiproliferative activity of their ruthenodendrimers by changing from a monodentate to a bidentate binding mode (the structures of these dendrimers **28** and **30**, along with those of their mononuclear analogs **29** and **31**, are depicted in Fig. 4.30) [166]. In particular, octanuclear almost completely clear G2-**30b** showed a remarkably high activity against the

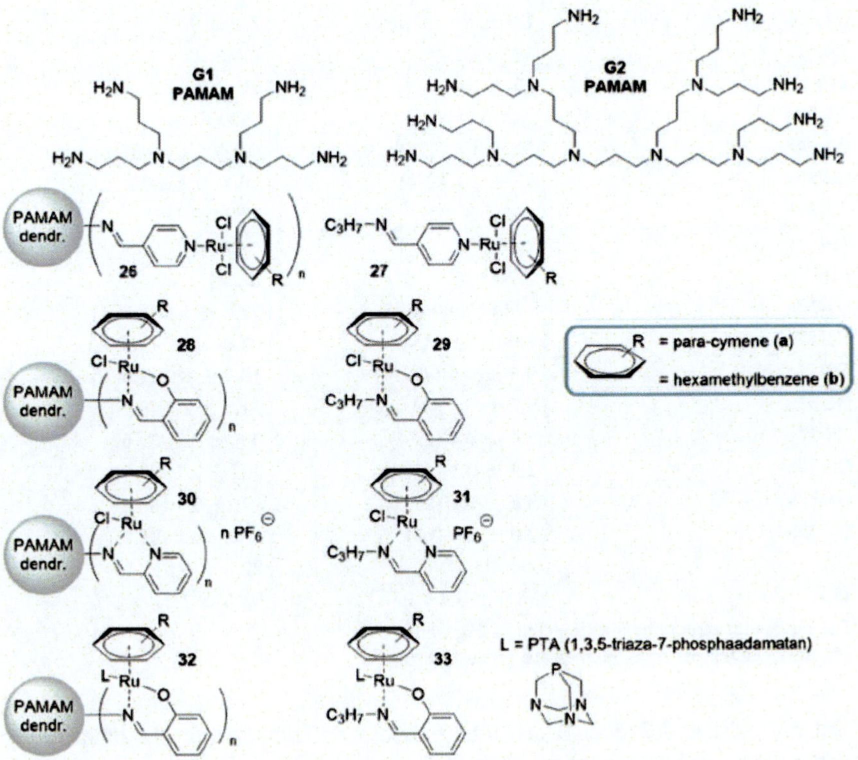

Fig. 4.30 Schematic representation of the various PAMAM-based ruthenodendrimers and their mononuclear counterparts

Table 4.1 Comparison of the antiproliferative activity of various ruthenodendrimers and their mononuclear counterparts [165–167]

Complex or Dendrimer[a]	A2780 (IC_{50}, μM)[b]	A2780cis (IC_{50}, μM)[b]
Cisplatin	1.5	25
G1-**26a**	43.0 ± 5.0 (172)	n/a
G1-**26b**	40.0 ± 5.0 (160)	n/a
G2-**26a**	21.0 ± 5.0 (168)	n/a
G2-**26b**	20.0 ± 5.0 (160)	n/a
27a	98 ± 5.0	n/a
27b	94 ± 5.0	n/a
G1-**28a**	50 ± 1.4 (200)	52 ± 0.8 (208)
G1-**28b**	27 ± 1.3 (108)	25 ± 1.3 (100)
G2-**28a**	22 ± 1.2 (176)	15 ± 1.4 (120)
G2-**28b**	10 ± 0.3 (80)	9 ± 0.3 (72)
G3-**28a**	6 ± 1.0 (96)	13.2 ± 1.4 (211)
G3-**28b**	2.9 ± 0.1 (92.8)	2.1 ± 0.1
G4-**28a**	2.1 ± 0.1 (33.6)	9.9 ± 1.2
G4-**28b**	1.6 ± 0.1 (51.2)	2.1 ± 0.1
29a	49 ± 2.3	47 ± 0.8
29b	19 ± 1.8	18 ± 0.8
G1-**30a**	>200	>200
G1-**30b**	32 ± 1.6 (128)	42 ± 5.1 (168)
G2-**30a**	23 ± 2.1 (184)	76 ± 9.4 (608)
G2-**30b**	4 ± 1.3 (32)	4 ± 0.3 (32)
31a	>200	>200
31b	>200	>200
G1-**32a**	174.0 ± 40 (696)	72.8 ± 1.6 (291.1)
G1-**32b**	8.9 ± 2.8 (35.6)	25.0 ± 5.0 (100)
G2-**32a**	9.3 ± 0.4 (74.4)	19.3 ± 0.2 (154.4)
G2-**32b**	6.2 ± 0.3 (49.6)	11.8 ± 1.1 (94.4)
G3-**32a**	1.4 ± 0.4 (22.4)	3.6 ± 0.2 (57.6)
G3-**32b**	2.9 ± 0.1 (46.4)	2.0 ± 0.1 (32.0)
G4-**32a**	0.8 ± 0.1 (25.6)	27. ± 0.1 (86.4)
G4-**32b**	2.0 ± 01. (64)	1.1 ± 0.1 (35.2)
33a	>200	82.0 ± 6.0
33b	38.0 ± 3.4	93.0 ± 7.0

[a]The molecular structures are depicted in Fig. 4.30
[b]The IC_{50} values per metal center are given in brackets

cisplatin-resistant A2780cisR cell line, thus by far outcompeting the other ruthenodendrimers and the mononuclear ruthenodrugs (i.e., IC_{50} value of 4 μM; Table 4.1). The results revealed a positive dendritic effect, i.e., a correlation between the macromolecular size and the observed cytotoxicity, and suggested an interaction with the DNA strands, which was distinctly different from the one proposed for cisplatin. These authors further showed that the in-vitro cytotoxicity could be increased further by replacing the chloro ligand of **28/29** by a PTA one (thus, forming RAPTA-type analogs attached to PAMAM dendrimers of various generations) [167]. From these ruthenodendrimers, G4-**32a** showed the highest activity, in the high nanomolar range, against the tested cancer cells lines. Moreover, the dendrimers exhibited a significantly higher selectivity towards cancer cells relative to healthy human embryonic kidney (HEK) cells. Similar ruthenodendrimers, up to 4th generation, were also reported by Smith et al. (in this case, 2-phenoxyethanol was used as η^6-coordinative ligand) [168]. Again, the best activity against A2780 cell lines was reported for the G4 material; however, the IC_{50} value (*ca.* 5.0 μM) could not compete with those listed for other the G4 ruthenodendrimers (Table 4.1).

The same Ru^{II} complexes, attached to carbosilane dendrimers (G0 to G2), were utilized by Gómez, de la Mata et al. [169]. For this purpose, the [Ru(η^6-*p*-cymene)Cl_2]$_2$ precursor complex was coordinated to the peripheral monodentate (i.e., NH_2 and pyridin-4-yl) or bidentate binding sites (*N,N*- or *N,O*-chelate). Similar to the aforementioned studies by Therrien, Smith et al., most of the prepared ruthenodendrimers exhibited cytotoxicity in the low μM range against various cancer cell lines; in general, the G1 dendrimers showed the lowest IC_{50} values (since different cancer cell lines were tested in this study, a comparison with the data shown in Table 4.1 cannot be offered). In the same report, also the cell death type was analyzed for selected examples and these authors also tested their reactivity towards some relevant biomolecules [e.g., DNA, human serum albumin (HSA), and Cathepsin-B]. Inhibition of the latter, as observed for the G1 ruthenodendrimers, corroborated their potential antimetastatic behavior. In the same context, Michlewska, Ionov et al. showed that the G1 ruthenodendrimer was also highly cytotoxic against promyelocytic leukemia HL-60 and human leukemia 1301 cells in which apoptosis and necrosis were induced [170, 171]. The same materials were employed further evaluated in the in vitro and in vivo treatment of advanced prostate cancer [172, 173]. The latter, performed on mice, showed a significant decrease in resistant prostate tumor growth, long-term survival of the animals, no indication of toxicity as well as the excretion through urine and feces [inductively coupled plasma (ICP) measurements were conducted to analyze the distribution of the ruthenodrug in the organs]. In order to establish a model how the ruthenodendrimers interact with cell membranes, these dendrimers were labelled with a spin probe, i.e., a 2,2,6,6-tetramethylpiperidine-1-oxyl (TEMPO) unit [174]. Using cetyltrimethylammonium bromide (CTAB) micelles and lecithin liposome, as model membranes, in-situ electron paramagnetic resonance (EPR) spectroscopy indicated a partial insertion of the dendrimers' surface groups into the CTBA micelles via non-covalent polar and hydrophobic interactions. In the case of the lecithin liposome, polar interactions with the ruthenodendrimers were observed and could be correlated to their anticancer activities.

The fluorescein-labeled ruthenodendrimer **34** (Fig. 4.31a) was adsorbed onto graphene-oxide-modified gold nanowires (GO-AuNWs) [175]. The resultant hybrid material was analyzed by energy dispersive X-ray (EDX) spectroscopy; these measurements revealed a homogeneous distribution of 34 over the nanostructure (Fig. 4.31b). The material was applied, as a nanomotor, in the active intracellular transport, which was propelled by sonication (MCF7 breast cancer cells were utilized as test cells in this study). Upon sonication (5 min, 2 V, 2.66 MHz), the assembly was propelled towards the cells; thereby, **34** was detached from the nanomotor's surface. This process was visualized by the recovery of the initial fluorescence of **34**, which was quenched upon binding to the GO-AuNW substrate. On the basis of their experimental findings, these authors proposed, that the sonication-driven cell internalization was significantly faster and, thus, delivery of the ruthenodendrimer payload into the MCF7 cells was higher.

In a different approach, Ortega, de la Mata et al. prepared heterofunctional ruthenodendrimers in which the Ru^{II}–arene complex was attached to the focal point of the dendrimer. The peripheral NMe_2 units accounted for their solubility in an aqueous environment; whereas, allyl groups facilitated solubility in DMSO (Fig. 4.32) [176]. In this recent study, these authors listed the main factors contributing to the stability of their ruthenodendrimers: Nature of the peripheral groups (i.e., NMe_2 *vs.* allyl), number of peripheral groups (i.e., dendritic generation) and, finally, the solvent used in the complexation step.

Nitrile-functionalized low-generation poly(alkylidenamine) dendrimers were employed by Rissanen, Rodrigues et al., as ligand, for the binding of Ru^{II} complexes (Fig. 4.33) [177, 178]. In particular, the ruthenodendrimers **35** as well as the

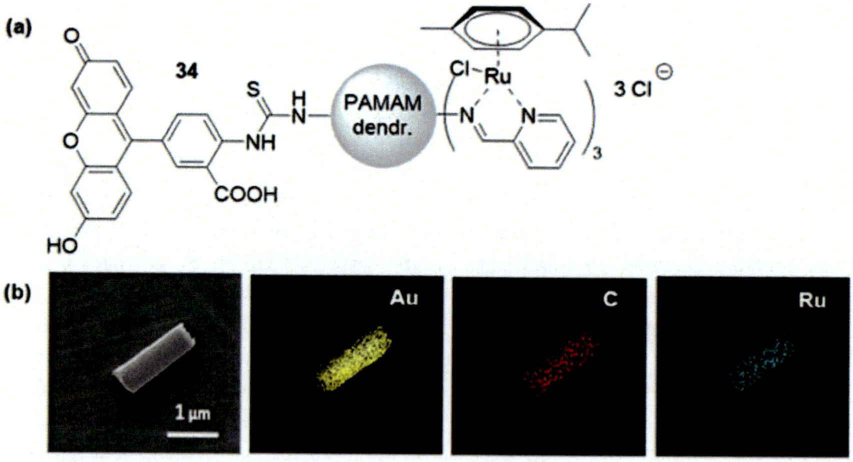

Fig. 4.31 **a** Schematic representation of the dye-labeled ruthenodendrimer **34**. **b** Scanning-electron microscopy (SEM) and element-specific EDX images of **34**@GO-AuNW) [175]. Figure reproduced with kind permission; © 2019 Science Repository

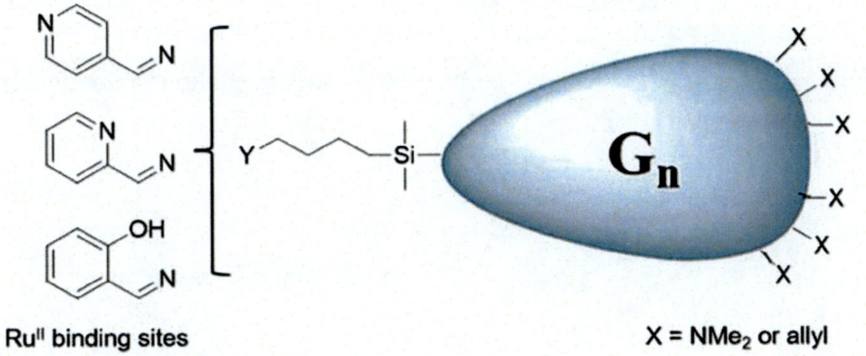

Fig. 4.32 Schematic representation of the heterodifunctional dendrimers, according to Ortega, de la Mata et al. [176]. Figure redrawn according to the literature; © 2019 Elsevier B.V.

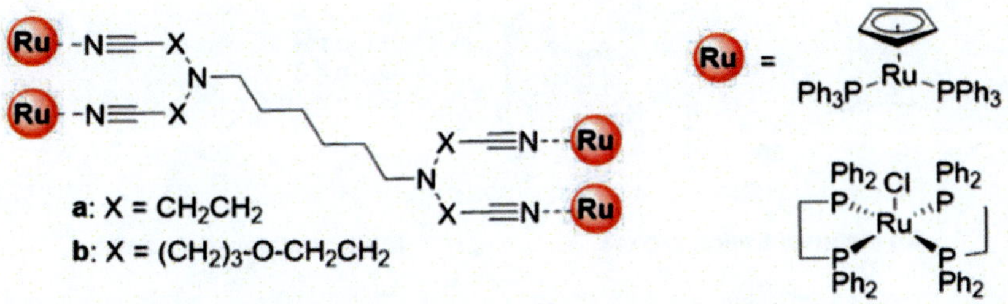

Fig. 4.33 Schematic representation of the ruthenodendrimers **35** and **36**

complex [Ru(η^5-C_5H_5)(PPh_3)$_2$Cl] showed a respectable activity against a range of cancer cell lines (i.e., Caco-2, MCF-7, CAL-72, and A2780, A2780cis) [179]. Regarding the latter IC50 values of 0.3 μM were found for **35a** and **35b**. These materials were also active against human mesenchymal stem cells (hMSCs), which typically occur in the tumor microenvironment and for which a key role in tumor progression and drug resistance has been proposed.

Endo et al. utilized PAMAM dendrimers, in the protonated state, to bind the multifunctional Ru^{II} complex **37** via electrostatic interactions (Fig. 4.34) [180]. The boronic-acid moiety on the ancillary ligand served as a sugar recognition site. The binding of sugars, such as D-fructose, could be monitored electrochemically due to distinct shifts of the Ru$^{II/III}$ redox potential.

Finally, one further aspect associated with the presence of cationic Ru^{II} complexes, as peripheral units in the dendrimers, is worth mentioning: Self-assembly of dendrimers into "megamers". Macchioni et al. utilized 2D NMR spectroscopy to study the concentration-dependent self-assembly of the multicationic ruthenodendrimers **38** (Fig. 4.35a) [181]. With increasing generation or concentration, the

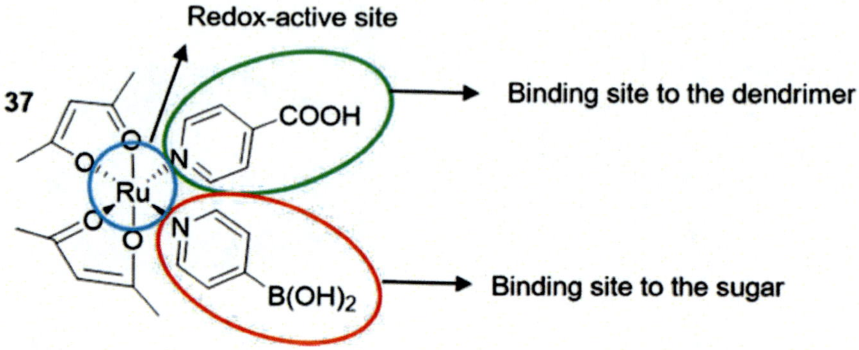

Fig. 4.34 Schematic representation of the multifunctional Ru^{II} complex **37**

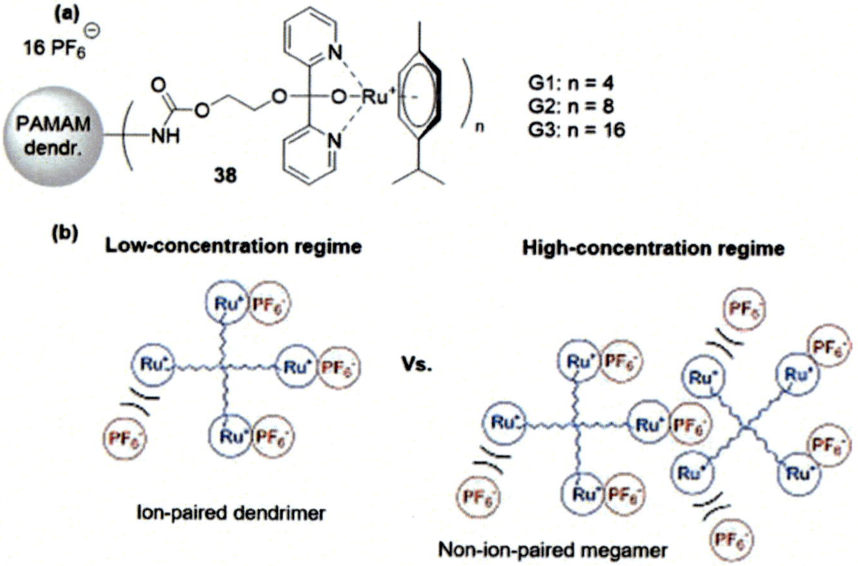

Fig. 4.35 **a** Schematic representation of the G3 ruthenodendrimer **38**. **b** Schematic representation of the formation of megamers due to an imbalance of ion-pairing interactions in the high concentration regime [181]. Figure reproduced with kind permission; © 2009 Wiley–VCH

tendency of **38** to aggregate into dimers, the so-called megamers, increased (Fig. 4.35b). These authors proposed that the solvophobicity of the dendrimers, i.e., the difference in the chemical–physical character of their surface from that of the solvent environment, represented the main driving force for the observed self-assembly behavior. Noteworthy, this behavior was absent when isostructural, but non-charged dendrimers were used.

4.6 Supramolecular Assemblies, Derived from Dendrimers and Ru^{II} Complexes

In this very brief Section, the self-assembly of dendrimers with Ru^{II} complexes into ordered nanostructures will be discussed. According to Maeda et al., ion-pairing interactions between oppositely charged species represents one possibility to prepare such assemblies in a highly dimension-controlled fashion [182]. Basically, three different structural motifs can be derived from such a self-assembly process (Fig. 4.36). In charge-by-charge assemblies, the stacks are comprised of positively and negatively charged species, ordered in a strictly alternating fashion. The charge-segregated mode represents the other extreme case in which stacks of identically charged species are formed. In the latter case, the electrostatic repulsion of the charged species needs to be overcome by strong attractive interactions; whereas, electrostatic attraction of opposite charges account for the formation of the former. However, in most cases, reported in the available literature, intermediate assembly modes were found.

The dimension-controlled ordering via ion-pairing interactions will only occur if some requisites are met: The charged components need appropriate geometries and electronic states. In other words, the cationic and anionic species must have similar shapes and the same absolute charge value. The self-assembly of molecules of complementary shapes into structures of higher complexity is referred to as "molecular automorphogenesis" [183]. This term was coined by Lehn in 1990 and introduced into the field of dendrimers by Newkome et al. In their initial study, these authors showed the self-assembly of a two-directional, amphiphilic dendrimer into single- and multistrand fibers [184].

Later, these same authors utilized the G1 dendrimer **39**$^{12-}$ and the Ru^{II}-based metallomacrocycle **40**$^{12+}$, as oppositely charged components [43, 49], for the ion-promoted, automorphogenic and stoichiometric self-assembly (Fig. 4.37) [185].

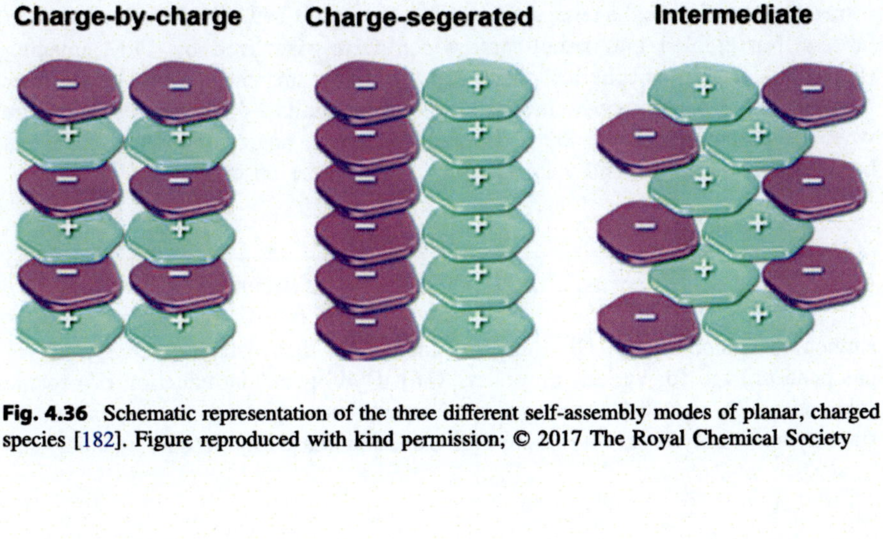

Fig. 4.36 Schematic representation of the three different self-assembly modes of planar, charged species [182]. Figure reproduced with kind permission; © 2017 The Royal Chemical Society

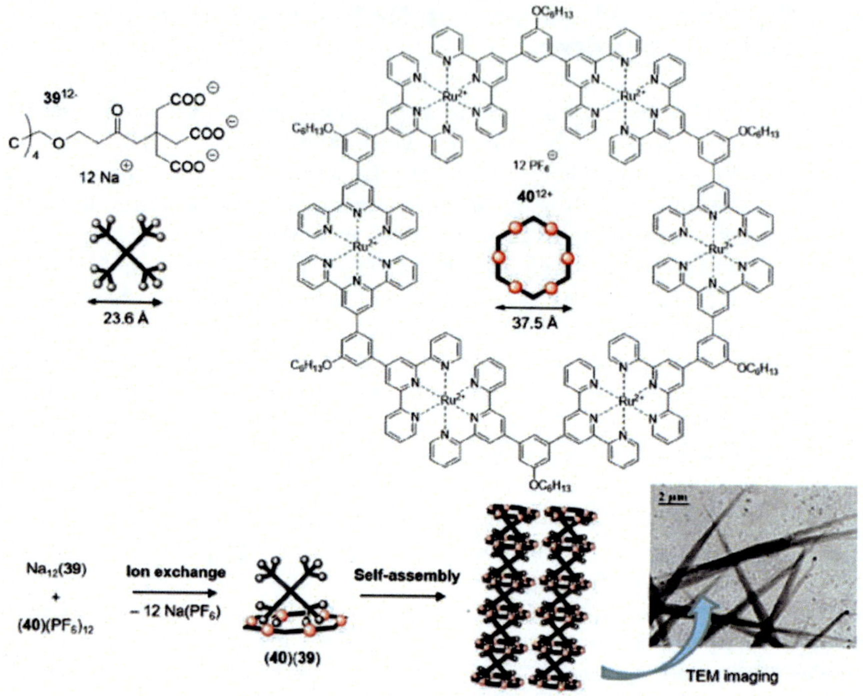

Fig. 4.37 Schematic representation of the hierarchical self-assembly of 39^{12-} and 40^{12+} into nanofibers due to dimension-controlled ion-pairing interactions [185]. Figure reproduced with kind permission; © 2008 Wiley–VCH

The monomers **39** and **40** possessed similar shapes, as concluded from diffusion-ordered NMR spectroscopy (DOSY) and molecular-modelling studies, respectively, and gave the charge-balanced 1:1 complex **(49)(39)**. This assembly, in turn, self-assembled into nanofibers, which were visualized by TEM imaging (Fig. 4.37). Due to its spherical shape and the charge mismatch, the G3 dendrimer —possessing 108 carboxylate moieties, as terminal units—did not give such fibers when co-assembled with $\mathbf{40}^{12+}$. Instead, spherical aggregates with diameters between 25 and 55 nm and varying compositions were observed.

4.7 Ru Species Immobilized Within Dendrimers

Ruthenium nanoparticles (NPs) are well-known for their catalytic activity, which has been utilized for various purposes [186]: Hydrogenation/reduction of organic substrates, CO_2 reduction, hydrogen production, water splitting, etc. In general, the stability and catalytic activity of Ru NPs can be increased by organic ligands, such

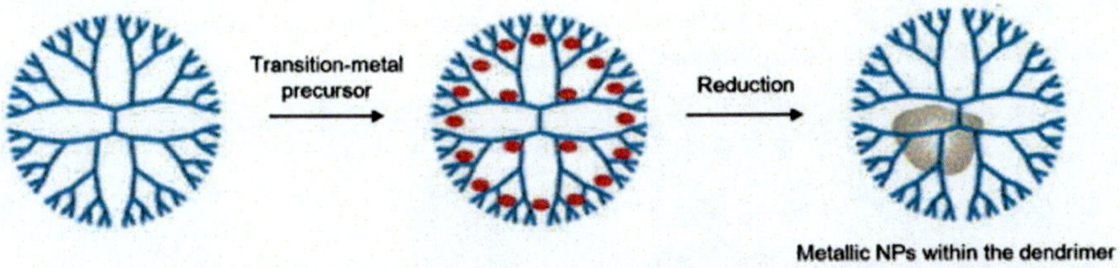

Fig. 4.38 Schematic representation of the preparation of dendrimer-stabilized metallic NPs [188]. Figure reproduced with kind permission; © 2003 Elsevier B.V.

as ionic liquids, CDs or *N*-heterocyclic carbenes (NHCs) [187]. Besides these small-molecule ligands, organic polymers have attracted considerable interest: Poly (vinylpyrrolidone) (PVP) and its derivatives are commonly used, as stabilizer for metallic NPs—this approach is detailed for the Ru NPs in Chap. 7. Similar to this, also dendrimers can be utilized, as matrix, for the binding and stabilization of metallic NPs (Fig. 4.38) [188, 189]. When the dendritic generation is high enough, the 3D spherical structures offer voids in which nanoscopic guests (i.e., ions, molecules or nanoparticles) can be embedded [190].

Crooks et al. pointed out that two types of commercially available dendrimers are particularly suited to host such NPs: PAMAM and poly(propylene imine) (PPI) dendrimers [189]. The latter is thermally very stable (G4 PPI exhibits the onset of weight loss at 450 °C); whereas, the former is prone to decomposition due to retro-Michael addition reaction already at *ca.* 100 °C [191]. These issues have to be considered when aiming for catalytic applications of dendrimer-stabilized NPs. With respect to the formation dendrimer-based host–guest assemblies, a sufficient driving force for the guest encapsulation must be offered—this might be a covalent binding, electrostatic attraction, steric confinement, complexation and/or non-covalent interactions [189, 190, 192]. In order to guarantee the accommodation of a guest in the interior of a dendrimer, its complexation or binding with the terminal units needs to be suppressed (this is of particular relevant when dealing amine-terminated dendrimers). Besides the cautious adjustment of the pH value (the peripheral NH_2-groups are protonated more readily than the internal tertiary amines), the surface functionalization with non-interacting moieties is commonly applied. The latter offers the advantage that a careful control over the pH value during the binding process becomes obsolete.

Crooks et al. demonstrated that transition-metal ions readily entered the interior of OH-terminated dendrimers, where coordination to the tertiary amine sites occurred. For example, $[Ru(NH_2Cl)Cl_2]$, as a precursor for Ru^{III} ions, was complexed in the interior of OH-functionalized PAMAMs [193]. This approach was taken up by Amiridis et al., who reported on the first example for Ru NPs within dendrimers [194]. These authors utilized $RuCl_3 \times xH_2O$, as precursor, and the NP formation was achieved by treatment with H_2. The NP formation was evidenced by EDX analysis and high-resolution TEM imaging was employed to visualize the Ru

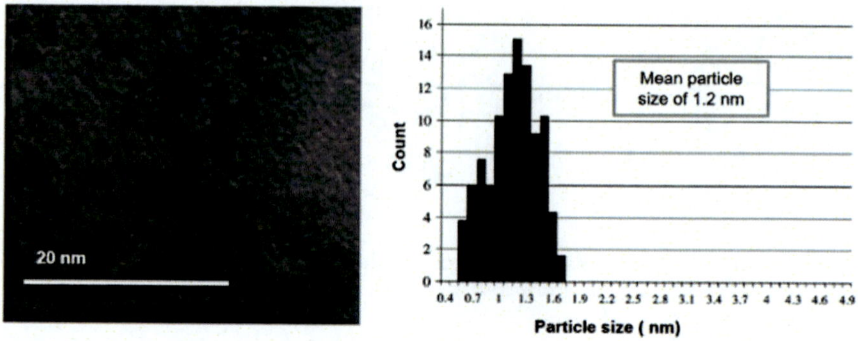

Fig. 4.39 Representative high-resolution TEM image of Ru NPs within an OH-functionalized G4 PAMAM. The corresponding particle-size distribution is also shown [194]. Figure reproduced with kind permission; © 2004 Springer Nature

NPs, for which a mean diameter of 1.2 nm was determined (Fig. 4.39). The obtained material was used for the fabrication of a Ru/γ-Al_2O_3 catalyst. The catalyst particles featured a smaller mean size and a narrower particle-size distribution when compared to the catalyst, which was prepared from $RuCl_3 \times xH_2O$ by the wet-impregnation method. These same authors further showed that the size of the dendrimer-stabilized Ru NPs could simply be controlled via the adjusted Ru^{III}-to dendrimer ratio (mean NP diameters ranging from 0.9 to 1.4 nm) [195].

The hydroxy-functionalized G4 PAMAM was also utilized by Escalante García et al. for the preparation of Ru, Pt, and bimetallic PtRu nanoparticles using $NaBH_4$, as the reducing agent [196]. These authors immobilized the dendrimers encapsulating the NPs onto a GC electrode and tested these hybrid systems regarding their applicability in the electrocatalytic oxygen-reduction reaction (ORR) from an aqueous MeOH environment. However, only in the case of the Pt NPs, a relevant electrocatalytic activity was found; nonetheless, the encapsulation of the NPs in the dendrimers significantly increased their stability and prevented catalyst poisoning by the co-solvent.

However, the complexation of Ru^{III} ions in the interior of OH-terminated G4-PAMAM is intrinsically slow (up to three days), thus the formation of dendrimer-stabilized NPs is a rather time-consuming procedure. In order to accelerate this process, Stevenson et al. introduced a galvanic-displacement method, which starting from dendrimer-stabilized Ni NPs, gave bimetallic NiRu NPs, encapsulated in the dendrimer [197]. The formation of Ni NPs within the PAMAM matrix readily occurred within hours and, according to the reduction potentials, the partial displacement of Ni^0 by Ru^0 was enabled; the bimetallic nanoparticles were then obtained by reduction with $NaBH_4$ (Fig. 4.40). The dendrimer-stabilized nanoalloys, having different Ni-to-Ru ratios, were tested, as catalysts, in the $NaBH_4$-reduction of *p*-nitrophenol. Thereby, the Ni-to-Ru ratio was found to have an impact on the reaction's rate constant; the highest value was found for the

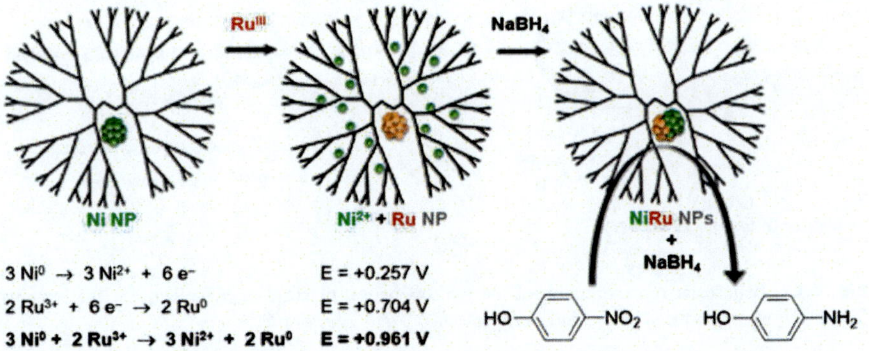

Fig. 4.40 Schematic representation of the synthesis of dendrimer-encapsulated bimetallic NiRu NPs via a galvanic-displacement approach. The resulting materials were used to catalyze the $NaBH_4$-reduction of *p*-nitrophenol [197]. Figure reproduced with kind permission; © 2012 The Royal Chemical Society

$Ru_{10}Ni_{30}$ NPs (for comparison, the dendrimer-stabilized homometallic Ru and Ni NPs showed a negligible and moderate catalytic activity).

Antonels and Meijboom employed the Langmuir–Hinshelwood approach to further analyze the aforementioned reduction, catalyzed by dendrimer-stabilized Ru NPs (G4-, G5-, and G6-PAMAM were used) [198]. Based on this model, these authors correlated the apparent rate constant to the total surface area of the nanoparticle, the kinetic rate constant (representing the rate-determining step), and the adsorption constants of the reactants (i.e., *p*-nitrophenol and BH_4^-). The evaluation of the experimental data indicated a higher catalytic activity for the smaller NPs. These authors proposed that an optimum NP size was required for the catalytic reaction (presumably, Stevenson et al. missed this activity window in their previous study). The same model reaction was also studied by Murugan and Pakrudheen [199]. In this case, an amphiphilic G2-PPI dendrimer was used to host the Ru NPs. A pseudo-first order rate constant of $15.95 \times 10^{-4}\ s^{-1}$ was found, i.e., the reaction was accelerated compared to the previous example. Moreover, the hybrid materials were immobilized onto a GC electrode. This array allowed the electrochemical sensing of H_2O_2 in the absence of enzymes in a concentration regime from 100 μM to 5 mM.

Jess et al. developed the "solid catalyst with ionic liquid layer" (SCILL) concept for hydrogenations. Compared to catalysts dispersed in bulk ionic liquids, the solubility of H_2 and its diffusion to a supported catalyst is strongly increased, when the latter is covered with an ionic-liquid layer [200]. This strategy was followed by Antonels and Meijboom, who prepared SiO_2-supported dendrimer-encapsulated Ru NPs, as heterogeneous catalysts, for the hydrogenation of citral; as in their previous work, OH-terminated PAMAMs of the 4th to 6th generation were used [201]. These authors tested various NP-to-dendrimer ratios and, in particular, focused on the effects of ionic liquids, as catalyst coatings, on the catalytic activity. Various

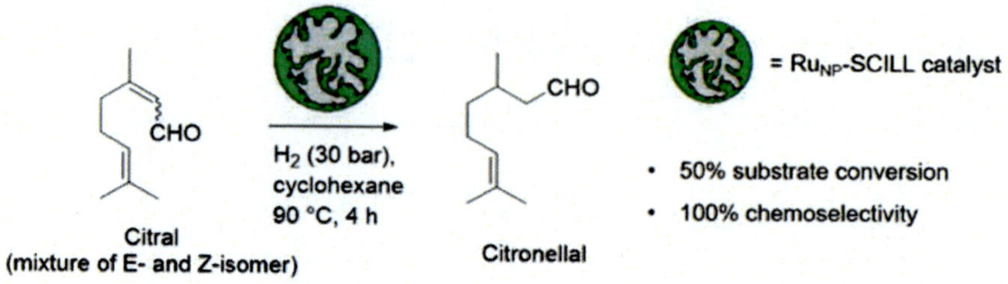

Fig. 4.41 Schematic representation of the hydrogenation of citral using a Ru_{NP}-SCILL catalyst [201]. Figure reproduced with kind permission; © 2014 Elsevier B.V.

ionic liquids were screened in this respect; from these, *N*-butyl-*N'*-methylimidazolium *bis*triflimide [bmim][NTf_2] offered an increased catalytic activity (50% conversion of citral) and an excellent chemoselectivity (citronellal was obtained, as the exclusive product, Fig. 4.41). In absence of ionic liquids, the conversion was significantly lower (29%). Since not all ionic liquids increased the catalytic effect (some even had a detrimental effect), the authors proposed, that the strength of the interactions between the ionic liquid and the NP might be of particular relevance. In subsequent work, these authors also targeted the partial hydrogenation of toluene using Ru_{NP}-SCILL catalysts [202]. In this case, the catalytic system comprising of RuNP within a G5-PAMAM matrix, immobilized onto silica-100 particles and coated with an [emim][NTf_2] layer, showed the highest methylcyclohexene selectivity, i.e., suppressing the full hydrogenation to afford methylcyclohexane (emim: *N*-ethyl-*N'*-methylimidazolium). Independent of the ionic-liquid coating, the selectivity of the reaction was increased compared to the uncoated Ru_{NP}@dendrimer catalyst; whereas, the catalytic activity was decreased in all cases due to catalyst poisoning.

Very recently, a completely different dendrimer architecture was employed by Hadadzadeh et al. [203]. These authors utilized the dendrimers **40** for the coordination of Ru^{III} ions (Fig. 4.42). The subsequent simultaneous $NaBH_4$-reduction and immobilization onto SiO_2 gave hybrid catalysts, which were also utilized for the hydrogenation of citral. Despite their very low loading with Ru (in the range of 0.03%), a high catalytic activity for the formation of 3,7-dimethyloctanol was found. The activity was attributed to the highly porous structure of the hybrid structures and their accessible surface area.

Karakhanov et al. investigated various heterogeneous RuNP@dendrimer systems with respect to their applicability, as catalysts, in the hydrogenation of arenes. For example, these authors prepared Ru and bimetallic PdRu NP dispersed in dendrimer particles (i.e., PPI dendrimers were covalently cross-linked by reaction with hexamethylene diisocyanate, Fig. 4.43) [204]. Under two-phase conditions, the immobilized Ru NPs efficiently catalyzed the hydrogenation of alkylbenzene derivatives [the presence of water increased the turnover frequency (TOF) by a factor of 3–90, depending on the substrate]. The bimetallic NPs were found to be

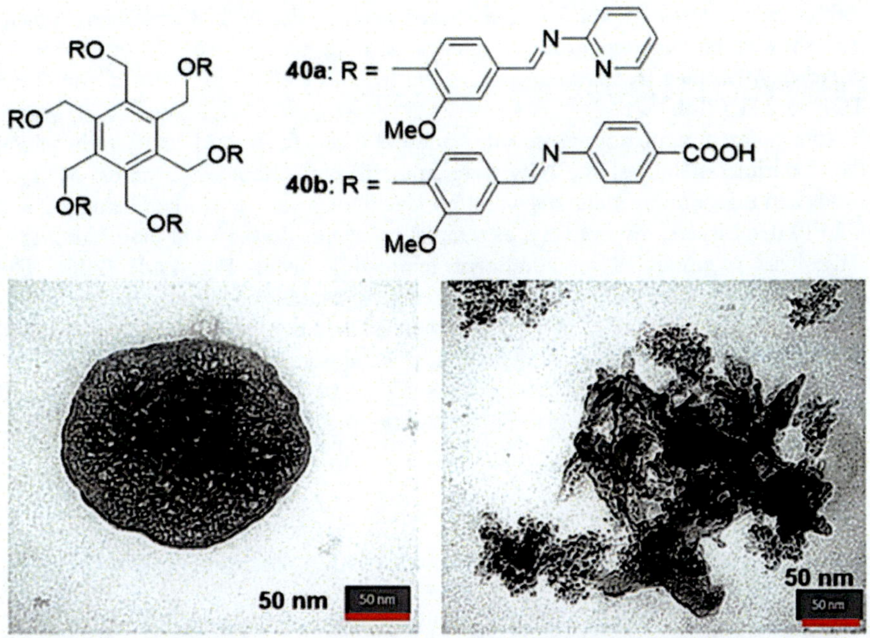

Fig. 4.42 Top: Schematic representation of the dendrimers **40**. Bottom: Representative TEM images of the SiO_2-supported Ru NP catalysts, derived from **40a** and **40b** (left and right, respectively) [203]. Figure reproduced with kind permission; © 2020 Elsevier B.V.

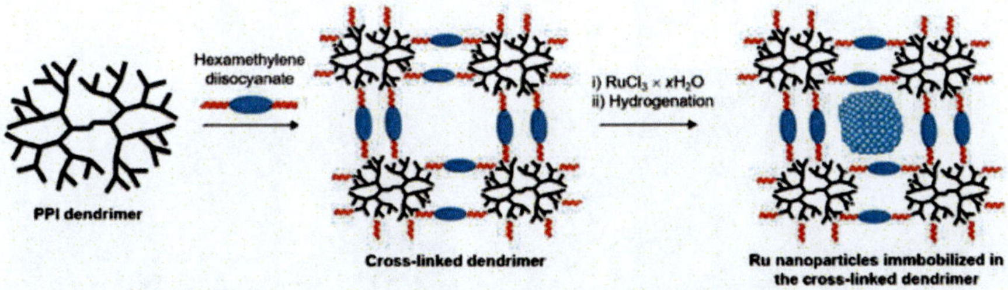

Fig. 4.43 Schematic representation of the preparation of Ru nanoparticles, stabilized by a cross-linked PPI dendrimer network [205]. Figure reproduced with kind permission; © 2015 Wiley–VCH

even more active in catalyzing the conversion of benzene; whereas, the hydrogenation of the alkyl-substituted derivatives was less efficient. The scope of these heterogeneous catalysts was further extended towards the hydrogenation of phenols in aqueous media; a maximum TOF of 2975 h^{-1} was found in these experiments

[205]. Unlike conventional Ru-based heterogeneous catalysts, Karakhanov's catalyst showed an inverted order-of-activity with respect to the hydrogenation of dihydroxybenzenes: Resorcinol was converted most efficiently with a very high TOF of 243,150 h^{-1}.

In a different approach, these same authors opted for hybrid catalysts in which the dendrimer-stabilized Ru NPs were immobilized within silica pores. Microporous and mesoporous materials were fabricated via a sol–gel method using G2 and G3 PPI dendrimers, respectively, as templating agents (in the latter case Pluronic™ P123 was required, as an additional templating agent, Fig. 4.44) [206]. The mesoporous hybrid catalyst exhibited a very high catalytic activity in the hydrogenation of phenol (TOF of 6090 h^{-1}); whereas, the microporous system was found to be even inferior to the aforementioned catalysts, based on cross-linked PPI dendrimers (TOF of 1841 h^{-1} *vs. ca.* 2700 h^{-1}). The same trend was also observed for various substituted phenols. These authors proposed that the structure of the

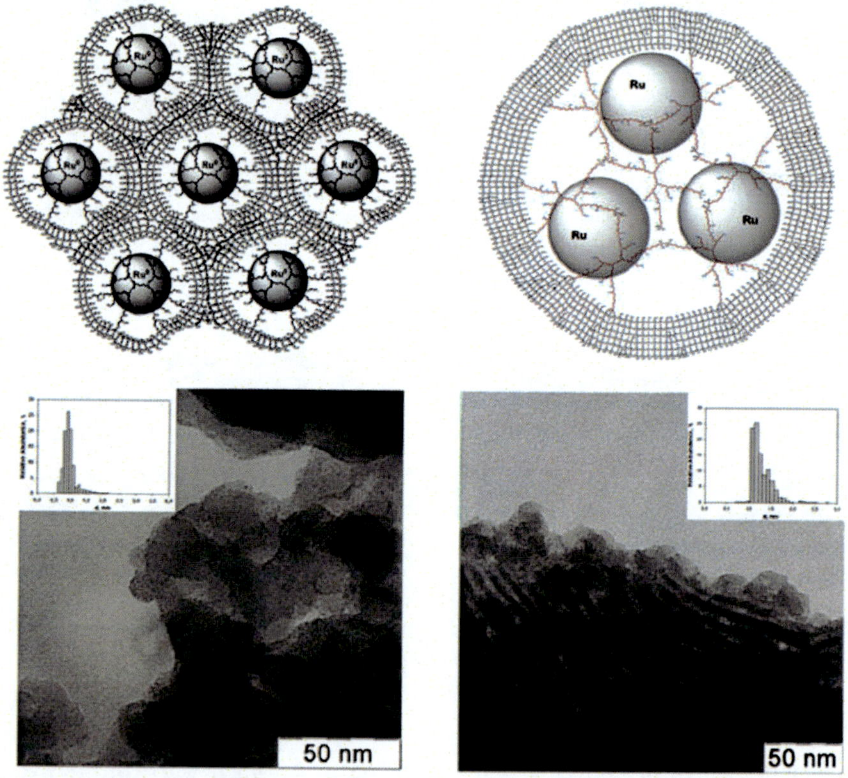

Fig. 4.44 Schematic representation of the microporous and mesoporous hybrid catalysts. Representative TEM images and the corresponding particle-size distributions are also depicted [206]. Figure reproduced with kind permission; © 2017 MDPI

solid support had a significant influence on the chemical and physical properties of embedded nanoparticles, as revealed by their catalytic activities. In subsequent work, the hybrid catalysts were employed for the selective hydrogenation of levulinic acid to γ-valerolactone. This particular reaction is of relevance for the low-cost conversion of biomass-derived compounds into fine chemicals [207]. Karakhanov et al. performed the hydrogenation of levulinic acid (and its esters) in water and γ-valerolactone was formed selectively within 2 h (80 °C, 30 bar of H_2), TOF values of up to 1610 h^{-1} were determined for this process (thereby, the mesoporous catalyst gave higher TOF values than its microporous analog) [208].

In a related study, Meijboom et al. synthesized mesoporous hybrid catalyst [209]. These authors employed pre-formed SiO_2 and TiO_2, as solid support, for the metallic NPs embedded within PAMAM dendrimers. After the deposition the dendrimer-stabilized NPs, calcination at 550 °C was performed to remove the dendrimer and, thus, activate the catalyst. The Ru-containing catalysts, supported or non-supported, showed a higher catalytic activity in the hydrogenation of levulinic acid with 100% selectivity for γ-valerolactone compared to the analogous Pt-containing catalysts.

Phosphorhydrazones represent versatile building blocks for the construction of macromolecular architectures, such as macrocycles and dendrimers [210]. As shown in Sect. 4.5, such dendrimers containing appropriate functional groups, as terminal units, can be used to coordinate transition-metal complexes (Fig. 4.25). Ouali, Redón, Turrin et al. demonstrated that also Ru NPs can be stabilized by such dendritic architectures [211]. These authors utilized a solventless preparation protocol in which $RuCl_3 \times xH_2O$, $NaBH_4$, and dendrimers **41** (Fig. 4.45) were milled under ambient conditions. The resulting materials, containing Ru NP with an average diameter of 2–3 nm, were long-term stable in solution or as solids (this holds particularly true for the larger G1 and G2 dendrimers, independent of the length of the alkyl tail). All six materials were utilized in the side-chain hydrogenation of styrene and gave ethylbenzene in high yields (ethylcyclohexane, as byproduct, was only observed in traces, Fig. 4.45).

Williams et al. compared two different protocol to fabricate dendrimer-stabilized bimetallic PtRu NPs (Fig. 4.46) [212]. Starting from OH-terminated G4-PAMAM, the sequential complexation-reduction method gave these bimetallic NPs within the dendrimer matrix. As an alternative method, the co-complexation of Pt^{II} and Ru^{III} ions by the dendrimer was followed by reduction with $NaBH_4$. However, the latter gave slightly larger particles and a broader size distribution (noteworthy, PtRu NPs prepared via conventional routes had even larger sizes and broader dispersities). Subsequently, the dendrimer-stabilized PtRu nanoparticles were immobilized onto SiO_2. The catalytic activity of these prepared materials was tested in the hydrogenation of 3,4-epoxy-1-butene. In general, the dendrimer-derived nanoparticles showed a higher catalytic activity as the conventionally prepared ones. Moreover, different reaction pathways were followed in the formation of *n*-butanol, as the final hydrogenation product: The catalyst prepared via the co-complexation route favored crotonaldehyde, as an intermediate; whereas, the material prepared via the

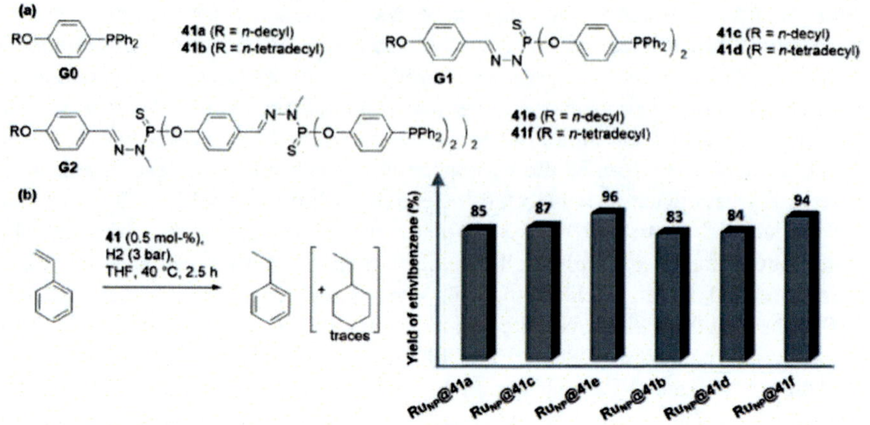

Fig. 4.45 a Schematic representation of the alkyl-functionalized polyphosphorhydrazones **41**, which were utilized to host and stabilize Ru NPs. **b** Schematic representation of the side-chain hydrogenation of styrene, catalyzed by Ru_{NP}@**41** [211]. Figure reproduced with kind permission; © 2016 The Royal Chemical Society

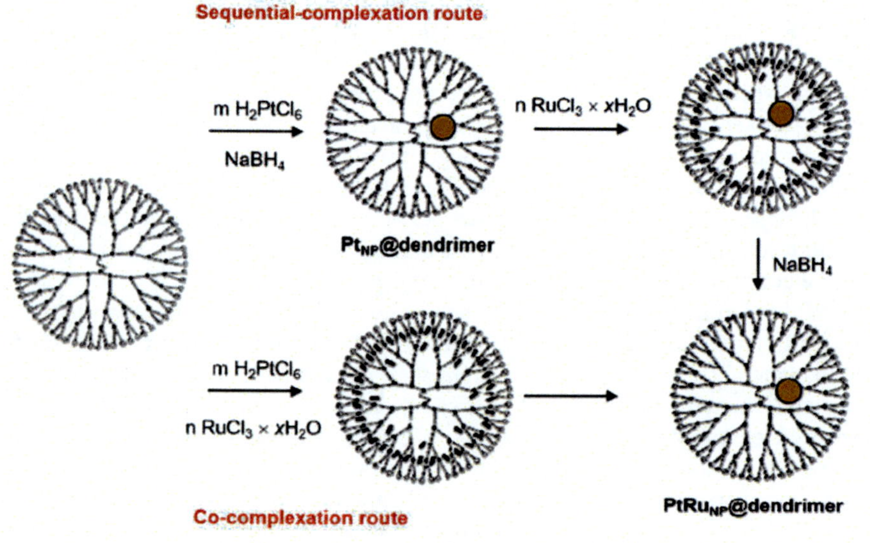

Fig. 4.46 Schematic representation of dendrimer-stabilized bimetallic PtRu NPs via two different preparation routes [212]. Figure reproduced with kind permission; © 2010 Elsevier B.V.

sequential complexation method involved 3-buten-1-ol, as the intermediate. These authors rationalized this finding by relative affinities of the catalysts towards these intermediates.

Dendrimer-encapsulated bimetallic PtRu and RhRu nanoparticles were also prepared by Peng et al. [213]. These authors utilized a partially quarternized G5-PAMAM dendrimer and the nanoparticle synthesis was conducted according to the aforementioned co-complexation method; thereby, the Pt-to-Ru and Rh-to-Ru ratios could be varied over a wide range. Dynamic-light scattering measurements and TEM imaging revealed an uniform size and shape of the nanoparticles as well as their homogeneous dispersion within the dendrimer matrix (Fig. 4.47a). All materials were employed, as catalysts, in the hydrosilylation using poly(methylhydro)siloxane and 1-hexene, as substrates. The functionalization of the siloxane proceeded with high regioselective, as shown by ^{1}H and ^{29}Si NMR spectroscopy (i.e., the addition of the Si–H group to the C–C double bond occurred strictly in an anti-Markovnikov-type fashion, Fig. 4.47b). Noteworthy, the bimetallic nanoparticles were catalytically more active than the monometallic ones (i.e., Ru, Rh, or Pt NPs); this finding was attributed to a synergistic electronic effect between the metals (such a synergy was first proposed by Crooks et al. for dendrimer-stabilized bimetallic PdPt NPs [214]). Fierro, Rojas et al. demonstrated that the dendrimer-stabilized PtRu nanoparticles, when immobilized onto a GC electrode, functioned as electro-catalyst for the oxidation of CO [215].

Similar to this, bimetallic PtRu nanoparticles, stabilized by OH-terminated G4-PAMAM, were immobilized onto carbon-black powder [216]. The subsequent thermal treatment at 400 °C under a H_2 atmosphere gave small-sized nanoparticles, which were well dispersed in the carbon support; in contrast to this

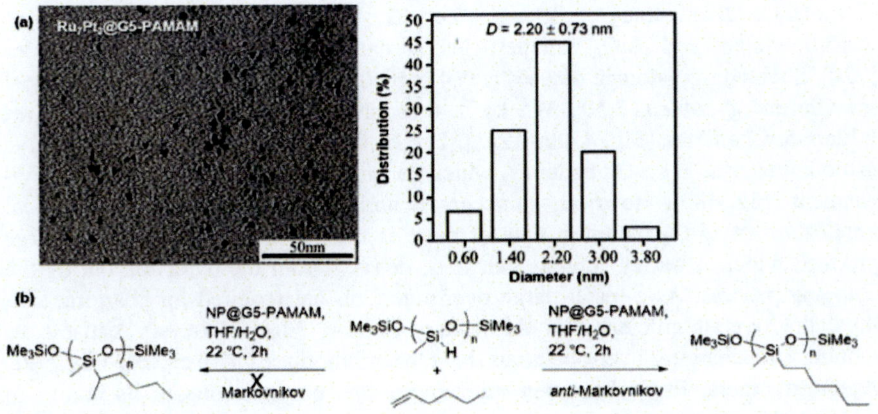

Fig. 4.47 a Representative TEM image of dendrimer-stabilized bimetallic PtRu nanoparticles, the corresponding size distribution is also depicted. **b** Schematic representation of the regioselective hydrosilylation, catalyzed by the bimetallic nanoparticles [213]. Figure reproduced with kind permission; 2011 Wiley–VCH

dendrimer-mediated method, the conventional nanoparticle synthesis afforded, after thermal treatment, larger and partially agglomerated PtRu NPs. The thus prepared materials were utilized, as catalysts, for the electro-oxidation of MeOH, a reaction of particular relevance in view of fuel-cell applications. The highest catalytic activity was found for the nanoparticles with an equal Pt-to-Ru ratio. Based on these results, Pastor et al. utilized carbon nanofibers, as solid support, for their dendrimer-encapsulated bimetallic PtRu NPs (in this study the generation number of the PAMAM dendrimers was varied) [217]. Carbon nanofibers enable a better dispersion of PtRu NPs; as a result, a higher catalytic activity—compared to carbon black—is generally observed [218]. After thermal treatment, active catalysts for the electro-oxidation of CO and MeOH were obtained [217]. Compared to commercial PtRu@C, the materials derived, from G2- and G3-PAMAM, exhibited a significantly higher catalytic activity in this respect. These same authors later showed that the formation of RuO_2 upon thermal treatment could not fully be avoided; beneficially, the presence of RuO_2 even increased the catalytic activity, as revealed by in-depth electrochemical studies [219]. Depending on the applied voltage, deprotonation or CO diffusion were identified as rate-determine steps in the MeOH oxidation process; thereby, the generation number was also identified as crucial parameter for fabricating bimetallic PtRu NPs for fuel-cell applications. In addition to this Fierro, Rojas et al. paid attention to the structure of the bimetallic nanoparticles by synthesizing dendrimer-encapsulated core–shell NPs comprising of a Ru core and Pt shell or the other way around [215]. The sequential-complexation allowed one to prepare these two types of NPs simply via the order-of-complexation/reduction steps. When utilized, as catalyst, in the CO electro-oxidation, the NPs with a Pt-rich shell showed a higher catalytic activity, which was attributed to the presence of catalytically-active Pt atoms on the outside in concert with a stabilizing effect offered by the Ru-rich core.

Finally, Bhattacharya, Zhang et al. focused the application of dendrimer-stabilized RuO_2 nanoparticles, as catalyst, in lithium-oxygen batteries [220]. Such energy-storage devices are of particular interest due to their theoretical specific energy of *ca.* 3,500 Wh kg^{-1}, thus much higher as the state-of-the art lithium-ion batteries (Fig. 4.48a) [221]. Unlike the latter, O_2 represents the active cathode material in Li-O_2 batteries, which can readily be absorbed from the environment (Fig. 48b). However, some grave limitations still hamper the practical application of Li-O_2 batteries. One of these is related to Li_2O_2, as the discharge product, which is barley soluble and, thus, detracts from the oxidation during the charging process. As a result, large over-potentials are required for charging and, thus, the overall efficiency of the battery is low. Electrocatalysts, capable of reducing the charging over-potentials, have been introduced to overcome this issue. Amongst others, RuO_2 has been utilized for this purpose, due to its ability to catalyze the oxygen-evolution reaction [222].

In this context, dendrimer-stabilized Ru nanoparticles were synthesized (OH-terminated G4-PAMAM was used in this study). However, under the given conditions (aqueous environment, air), the metallic NP were readily oxidized to RuO_2 species. Scanning-transmission electron microscopy (STEM) imaging and

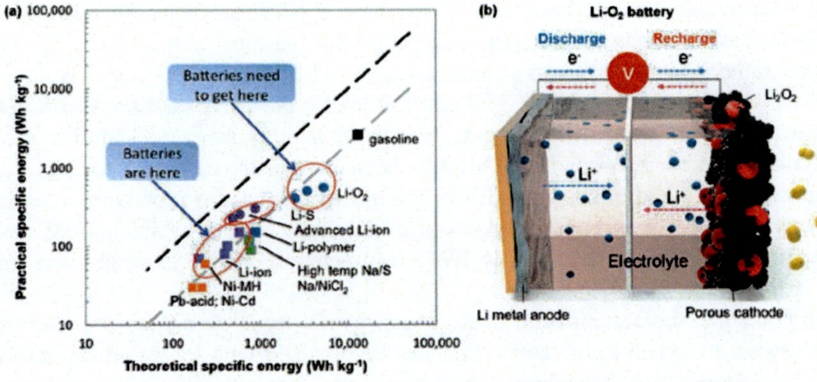

Fig. 4.48 **a** Representation of the current position of various batteries on a plot of practical specific energy versus theoretical energy density. **b** Schematic representation of a Li-O_2 battery [221]. Figure reproduced with kind permission; © 2020 American Chemical Society

X-ray photoelectron spectroscopy (XPS) revealed the formation of RuO_2 NPS, which possessed a narrow size distribution and an average particle diameter of 2.1 nm. When employed, as electrocatalyst, in Li-O_2 batteries, the charging over-potential was lowered by 0.5 V, at a much lower catalyst loading as in batteries using conventional RuO_2 NPs; moreover, also the cycling lifetime was increased. These authors rationalized the high catalytic activity by the large available surface area of the NPs, which were protected against agglomeration by the dendrimer matrix.

4.8 Concluding Remarks

Dendrimers represent highly branched, 3D macromolecular architectures which are, due to their unique molecular structures, already interesting from an aesthetic point-of-view. Metallodendrimers have emerged, as a sub-class, containing transition-metal complexes at various positions of the dendritic structure: Core, branching point, bridging unit or terminal moieties. In the case of Ru^{II} complexes, examples for all cases can be found in the available literature. In contrast to the dendrimers containing Ru^{II} complexes, as peripheral units, all other systems have to be synthesized via multi-step convergent, divergent or combined convergent/divergent routes. This significant synthetic effort hampers any realistic application and such ruthenodendrimers are mainly studied in a scientific content, such as intra-dendrimer electron—and/or energy transfer processes. The surface-functionalized dendrimers, however, are much easier accessible—the terminal functional groups of commercially available dendrimers, such as PAMAM or PPI, can readily be functionalized, e.g., by condensation and complexation reactions.

This relative ease in synthesis facilitated their application in various fields of research, with catalysis and biomedicine being the two most notable ones. In this respect, ruthenodendrimers can at least compete with their ruthenopolymer counterparts, which were the subject of Chap. 2. Finally, the encapsulation of metallic nanoparticles within dendrimers represents a versatile strategy to stabilize such catalytically active nanospecies. This has been demonstrated at the example of homometallic Ru and bimetallic PtRu or RhRu nanoparticles, which were utilized, as catalyst, mainly in hydrogenation reactions. In parallel to the stabilization of such nanoparticles by polymers, such as PVP, commercially available dendrimers are here of particular virtue.

To conclude, the combination of Ru species with dendrimers becomes relevant for applications, when stable assemblies can be obtained with low synthetic effort. Thus, appropriate Ru-containing dendrimers, as alternative to Ru-containing polymers, can be expected to play a future role in catalytic and/or biomedical applications.

References

1. Zheng, Y., Li, S., Weng, Z., Gao, C.: Hyperbranched polymers: Advances from synthesis to applications. Chem. Soc. Rev. **44**(12), 4091 (2015)
2. Grayson, S.M., Fréchet, J.M.J.: Convergent dendrons and dendrimers: From synthesis to applications. Chem. Rev. **101**(12), 3819 (2001)
3. Newkome, G.R., He, E., Moorefield, C.N.: Suprasupermolecules with novel properties: Metallodendrimers. Chem. Rev. **99**(7), 1689 (1999)
4. Astruc, D., Boisselier, E., Ornelas, C.: Dendrimers designed for functions: From physical, photophysical, and supramolecular properties to applications in sensing, catalysis, molecular electronics, photonics, and nanomedicine. Chem. Rev. **110**(4), 1857 (2010)
5. Lyu, Z., Ding, L., Huang, A.Y.T., Kao, C.L., Peng, L.: Poly(amidoamine) dendrimers: Covalent and supramolecular synthesis. Mater. Today Chem. **13**, 34 (2019)
6. Buhleier, E., Wehner, W., Vögtle, F.: "Cascade"- and "nonskid-chain-like" syntheses of molecular cavity topologies. Synthesis (2), 155 (1978)
7. Tomalia, D.A., Baker, H., Dewald, J., Hall, M., Kallos, G., Martin, S., Roeck, J., Ryder, J., Smith, P.: A new class of polymers: Starburst-dendritic macromolecules. Polym. J. **17**(1), 117 (1985)
8. Zweifel, G., Leung, T., Najafi, M.R., Najdi, S.: Stereoselective syntheses of alkenyl substituted 1,3-dioxolanes or 4,7-dihydro-1,3-dioxepins or an (*E*)-α,β-unsaturated aldehyde from (*Z*)-2-butene-1,4-diols. J. Org. Chem. **50**(11), 2004 (1985)
9. Newkome, G.R., Baker, G.R., Young, J.K., Traynham, J.G.: A systematic nomenclature for cascade polymers. J. Polym. Sci. Part A: Polym. Chem. **31**(3), 641 (1993)
10. Baker, G.R., Young, J.K.: A systematic nomenclature for cascade (dendritic) polymers. In: Newkome, G.R. (Ed.) Advances in Dendritic Macromolecules, 169 pp. Greenwich, JAI (1994)
11. Newkome, G.R., Shreiner, C.: Dendrimers derived from 1→3 branching motifs. Chem. Rev. **110**(10), 6338 (2010)
12. Newkome, G.R., Shreiner, C.D.: Poly(amidoamine), polypropylenimine, and related dendrimers and dendrons possessing different 1→2 branching motifs: an overview of the divergent procedures. Polymer **49**(1), 1 (2008)

13. Astruc, D., Deraedt, C., Djeda, R., Ornelas, C., Liu, X., Rapakousiou, A., Ruiz, J., Wang, Y., Wang, Q.: Dentromers, a family of super dendrimers with specific properties and applications. Molecules **23**(4), 966 (2018)
14. Hwang, S.-H., Moorefield, C.N., Newkome, G.R.: Dendritic macromolecules for organic light-emitting diodes. Chem. Soc. Rev. **37**(11), 2543 (2008)
15. Chirag, M., Gowda, D.V., Sathish, B., Famna Roohi, N.K.: A comprehensive review on dendrimers in current advanced drug delivery. Int. J. Pharm. Sci. Res. **11**, 1055 (2020)
16. Boas, U., Heegaard, P.M.H.: Dendrimers in drug research. Chem. Soc. Rev. **33**(1), 43 (2004)
17. Newkome, G.R.: Suprasupermolecular chemistry: The chemistry within the dendrimer. Pure Appl. Chem. **70**(12), 2337 (1998)
18. Gorman, C.: Metallodendrimers: Structural diversity and functional behavior. Adv. Mater. **10**(4), 295 (1998)
19. Campagna, S., Denti, G., Serroni, S., Ciano, M., Juris, A., Balzani, V.: A tridecanuclear ruthenium(II)-polypyridine supramolecular species: Synthesis, absorption and luminescence properties and electrochemical oxidation. Inorg. Chem. **31**(13), 2982 (1992)
20. Newkome, G.R., Cardullo, F., Constable, E.C., Moorefield, C.N., Thompson, A.M.W.C.: Metallomicellanols: Incorporation of ruthenium(II)–2,2′: 6′,2″-terpyridine triads into cascade polymers. J. Chem. Soc. Chem. Commun. (11), 925 (1993)
21. Sauvage, J.P., Collin, J.P., Chambron, J.C., Guillerez, S., Coudret, C., Balzani, V., Barigelletti, F., De Cola, L., Flamigni, L.: Ruthenium(II) and osmium(II) bis(terpyridine) complexes in covalently-linked multicomponent systems: Synthesis, electrochemical behavior, absorption spectra, and photochemical and photophysical properties. Chem. Rev. **94**(4), 993 (1994)
22. Denti, G., Campagna, S., Serroni, S., Ciano, M., Balzani, V.: Decanuclear homo- and heterometallic polypyridine complexes: Syntheses, absorption spectra, luminescence, electrochemical oxidation, and intercomponent energy transfer. J. Am. Chem. Soc. **114** (8), 2944 (1992)
23. Denti, G., Serroni, S., Campagna, S., Ricevuto, V., Balzani, V.: Directional energy transfer in a luminescent tetranuclear Ru(II) polypyridine complex that contains two different types of bridging ligands. Inorg. Chim. Acta **182**(2), 127 (1991)
24. Campagna, S., Denti, G., Serroni, S., Juris, A., Venturi, M., Ricevuto, V., Balzani, V.: Dendrimers of nanometer size based on metal complexes: Luminescent and redox-active polynuclear metal complexes containing up to twenty-two metal centers. Chem. Eur. J. **1**(4), 211 (1995)
25. Issberner, J., Vögtle, F., Cola, L.D., Balzani, V.: Dendritic bipyridine ligands and their tris (bipyridine)ruthenium(II) chelates—syntheses, absorption spectra, and photophysical properties. Chem. Eur. J. **3**(5), 706 (1997)
26. Serroni, S., Denti, G., Campagna, S., Juris, A., Ciano, M., Balzani, V.: Arborols based on luminescent and redox-active transition metal complexes. Angew. Chem. Int. Ed. **31**(11), 1493 (1992)
27. Serroni, S., Juris, A., Venturi, M., Campagna, S., Resino, I.R., Denti, G., Credi, A., Balzani, V.: Polynuclear metal complexes of nanometre size. A versatile synthetic strategy leading to luminescent and redox-active dendrimers made of an osmium(II)-based core and ruthenium (II)-based units in the branches. J. Mater. Chem. **7**(7), 1227 (1997)
28. Balzani, V., Ceroni, P., Juris, A., Venturi, M., Campagna, S., Puntoriero, F., Serroni, S.: Dendrimers based on photoactive metal complexes. Recent advances. Coord. Chem. Rev. **219–221**, 545 (2001)
29. Puntoriero, F., Serroni, S., La Ganga, G., Santoro, A., Galletta, M., Nastasi, F., La Mazza, E., Cancelliere, A.M., Campagna, S.: Photo- and redox-active metal dendrimers: A journey from molecular design to applications and self-aggregated systems. Eur. J. Inorg. Chem. **2018**(35), 3887 (2018)

30. Balzani, V., Campagna, S., Denti, G., Juris, A., Serroni, S., Venturi, M.: Designing dendrimers based on transition-metal complexes. Light-harvesting properties and predetermined redox patterns. Acc. Chem. Res. **31**(1), 26 (1998)
31. Arrigo, A., Puntoriero, F., La Ganga, G., Campagna, S., Burian, M., Bernstorff, S., Amenitsch, H.: Aggregation-induced energy transfer in a decanuclear Os(II)/Ru(II) polypyridine light-harvesting antenna dendrimer. Chem **3**(3), 494 (2017)
32. La Ganga, G., Nastasi, F., Campagna, S., Puntoriero, F.: Photoinduced water oxidation sensitized by a tetranuclear Ru(II) dendrimer. Dalton Trans. (45), 9997 (2009)
33. Puntoriero, F., La Ganga, G., Sartorel, A., Carraro, M., Scorrano, G., Bonchio, M., Campagna, S.: Photo-induced water oxidation with tetra-nuclear ruthenium sensitizer and catalyst: A unique 4 × 4 ruthenium interplay triggering high efficiency with low-energy visible light. Chem. Commun. **46**(26), 4725 (2010)
34. Kelch, S., Rehahn, M.: High-molecular-weight ruthenium(II) coordination polymers: Synthesis and solution properties. Macromolecules **30**(20), 6185 (1997)
35. Campagna, S., Serroni, S., Bodige, S., MacDonnell, F.M.: Absorption spectra, photophysical properties, and redox behavior of stereochemically pure dendritic ruthenium(II) tetramers and related dinuclear and mononuclear complexes. Inorg. Chem. **38**(4), 692 (1999)
36. Börje, A., Köthe, O., Juris, A.: A new bridging ligand for the synthesis of luminescent polynuclear Ru(II) and Os(II) polypyridine complexes. New J. Chem. **25**(2), 191 (2001)
37. Mede, T., Jäger, M., Schubert, U.S.: "Chemistry-on-the-complex": Functional Ru^{II} polypyridyl-type sensitizers as divergent building blocks. Chem. Soc. Rev. **47**(20), 7577 (2018)
38. Lee, D.N., Soh, B.K., Kim, S.H., Jun, Y.M., Yoon, S.H., Lee, W.-Y., Kim, B.H.: Star-shaped electrochemiluminescent metallodendrimers with central polypyridyl Ru(II) complexes: Synthesis and their photophysical and electrochemical properties. J. Organomet. Chem. **693**(4), 655 (2008)
39. Arakawa, R., Matsuo, T., Nozaki, K., Ohno, T., Haga, M.-A.: Analysis of multiply charged Ions of ruthenium(II) tetranuclear complexes by electrospray ionization mass spectrometry. Inorg. Chem. **34**(9), 2464 (1995)
40. Moucheron, C., Kirsch-De Mesmaeker, A., Dupont-Gervais, A., Leize, E., Van Dorsselaer, A.: Synthesis and characterization by electrospray mass spectrometry of a novel dendritic heptanuclear complex of ruthenium(II). J. Am. Chem. Soc. **118**(50), 12834 (1996)
41. Wesdemiotis, C.: Multidimensional mass spectrometry of synthetic polymers and advanced materials. Angew. Chemie Int. Ed. **56**(6), 1452 (2017)
42. Guo, K., Guo, Z., Ludlow, J.M., III., Xie, T., Liao, S., Newkome, G.R., Wesdemiotis, C.: Characterization of metallosupramolecular polymers by top-down multidimensional mass spectrometry methods. Macromol. Rapid Commun. **36**(17), 1539 (2015)
43. Chakraborty, S., Newkome, G.R.: Terpyridine-based metallosupramolecular constructs: Tailored monomers to precise 2D-motifs and 3D-metallocages. Chem. Soc. Rev. **47**(11), 3991 (2018)
44. Wang, J.-L., Li, X., Shreiner, C.D., Lu, X., Moorefield, C.N., Tummalapalli, S.R., Medvetz, D.A., Panzner, M.J., Fronczek, F.R., Wesdemiotis, C., et al.: Shape-persistent, ruthenium (II)- and iron(II)-bisterpyridine metallodendrimers: Synthesis, traveling-wave ion-mobility mass spectrometry, and photophysical properties. New J. Chem. **36**(2), 484 (2012)
45. Newkome, G.R., Yao, Z., Baker, G.R., Gupta, V.K.: Micelles. Part 1. Cascade molecules—a new approach to micelles: a [27]-arborol. J. Org. Chem. **50**(11), 2003 (1985)
46. Moorefield, C.N., Newkome, G.R.: Unimolecular micelles: Supramolecular use of dendritic constructs to create versatile molecular containers. Compt. Rend. Chim. **6**(8), 715 (2003)
47. Newkome, G.R., Moorefield, C.N., Baker, G.R., Johnson, A.L., Behera, R.K.: Alkane cascade polymers possessing micellar topology: Micellanoic acid derivatives. Angew. Chem. Int. Ed. **30**(9), 1176 (1991)
48. Newkome, G.R., Moorefield, C.N., Baker, G.R., Saunders, M.J., Grossman, S.H.: Unimolecular micelles. Angew. Chem. Int. Ed. **30**(9), 1178 (1991)

49. Newkome, G.R., Young, J.K., Baker, G.R., Potter, R.L., Audoly, L., Cooper, D., Weis, C. D., Morris, K., Johnson, C.S.: Cascade polymers. Part 35. pH Dependence of hydrodynamic radii of acid-terminated dendrimers. Macromolecules **26**(9), 2394 (1993)
50. Lichtenthaler, F.W.: 100 Years "Schlüssel-Schloss-Prinzip": What made Emil Fischer use this analogy? Angew. Chem. Int. Ed. **33**(23–24), 2364 (1995)
51. Newkome, G.R., Güther, R., Moorefield, C.N., Cardullo, F., Echegoyen, L., Pérez-Cordero, E., Luftmann, H.: Routes to dendritic networks: Bis-dendrimers by coupling of cascade macromolecules through metal centers. Angew. Chem. Int. Ed. **34**(18), 2023 (1995)
52. Winter, A., Schubert, U.S.: Synthesis and characterization of metallo-supramolecular polymers. Chem. Soc. Rev. **45**(19), 5311 (2016)
53. Winter, A., Schubert, U.S.: Block copolymers with element blocks: The metal-bisterpyridine linkage. In: Chujo, Y. (Ed.) New Polymeric Materials Based on Element-Blocks, 307 pp. Springer, Singapore (2019)
54. Newkome, G.R., He, E.: Nanometric dendritic macromolecules: stepwise assembly by double(2,2′:6′,2″-terpyridine)ruthenium(I) connectivity. J. Mater. Chem. **7**(7), 1237 (1997)
55. Newkome, G.R., He, E., Godínez, L.A., Baker, G.R.: Neutral highly branched metallo-macromolecules: Incorporation of a (2,2′:6′,2″-terpyridine)ruthenium(II) complex without external counterions. Chem. Commun. (1), 27 (1999)
56. Newkome, G.R., He, E., Godínez, L.A., Baker, G.R.: Electroactive metallomacromolecules via tetrabis(2,2′:6′,2″-terpyridine)ruthenium(II) complexes: Dendritic nanonetworks toward constitutional isomers and neutral species without external counterions. J. Am. Chem. Soc. **122**(41), 9993 (2000)
57. Newkome, G.R., Kim, H.J., Choi, K.H., Moorefield, C.N.: Synthesis of neutral metallo-dendrimers possessing adamantane termini: Supramolecular assembly with β-cyclodextrin. Macromolecules **37**(17), 6268 (2004)
58. Eftink, M.R., Andy, M.L., Bystrom, K., Perlmutter, H.D., Kristol, D.S.: Cyclodextrin inclusion complexes: Studies of the variation in the size of alicyclic guests. J. Am. Chem. Soc. **111**(17), 6765 (1989)
59. Del Valle, E.M.M.: Cyclodextrins and their uses: A review. Proc. Biochem. **39**(9), 1033 (2004)
60. Harada, A., Takashima, Y., Yamaguchi, H.: Cyclodextrin-based supramolecular polymers. Chem. Soc. Rev. **38**(4), 875 (2009)
61. Arima, H., Motoyama, K., Higashi, T.: Potential use of polyamidoamine dendrimer conjugates with cyclodextrins as novel carriers for siRNA. Pharmaceuticals **5**(1), 61 (2012)
62. Constable, E.C.: Metallodendrimers: metal ions as supramolecular glue. Chem. Commun. (12), 1073 (1997)
63. Constable, E.C., Harverson, P.: A new convergent strategy for high-nuclearity metallodendrimers. Chem. Commun. (1), 33 (1996)
64. Constable, E.C., Harverson, P., Ramsden, J.R.: Adsorption of ruthenadendrimers to silica–titania surfaces studied by optical waveguide lightmode spectroscopy (OWLS). Chem. Commun. (17), 1683 (1997)
65. Constable, E.C., Harverson, P.: Convergent synthesis of an octadecaruthenium metallodendrimer. Inorg. Chim. Acta **252**(1), 9 (1996)
66. Armspach, D., Cattalini, M., Constable, E.C., Housecroft, C.E., Phillips, D.: Boron-rich metallodendrimers—mix-and-match assembly of multifunctional metallosupramolecules. Chem. Commun. (15), 1823 (1996)
67. Constable, E.C., Housecroft, C.E., Cattalini, M., Phillips, D.: Pentaerythritol-based metallodendrimers. New J. Chem. **22**(2), 193 (1998)
68. Holmstrom, S.D., Cox, J.A.: Electrocatalysis at a conducting composite electrode doped with a ruthenium(II) metallodendrimer. Anal. Chem. **72**(14), 3191 (2000)
69. Cheng, L., Cox, J.A.: Nanocomposite multilayer film of a ruthenium metallodendrimer and a Dawson-type polyoxometalate as a bifunctional electrocatalyst. Chem. Mater. **14**(1), 6 (2002)

70. Ali, B., Imar, S., Laffir, F., McCormac, T.: Electrochemical, surface and electrocatalytic properties of electrode multilayer assemblies composed of a ruthenium metallodendrimer and a wheel-shaped Cu-20 tungstophosphate. J. Electroanal. Chem. **836**, 77 (2019)
71. Zynek, M., Serantoni, M., Beloshapkin, S., Dempsey, E., McCormac, T.: Electrochemical and surface properties of multilayer films based on a Ru^{2+} metallodendrimer and the mixed addenda Dawson heteropolyanion. Electroanal. **19**(6), 681 (2007)
72. Jiang, H., Joong Lee, S., Lin, W.: Chiral ruthenium–terpyridine based metallodendrimers: Facile synthesis, characterization, and photophysical studies. J. Chem. Soc. Dalton Trans. (18), 3429 (2002)
73. Jiang, H., Lee, S.J., Lin, W.: Chiral hybrid metal–organic dendrimers. Org. Lett. **4**(13), 2149 (2002)
74. Sadrerafi, K., Moore, E.E., Lee, M.W.: Association constant of β-cyclodextrin with carboranes, adamantane, and their derivatives using displacement binding technique. J. Incl. Phenom. Macrocycl. Chem. **83**(1), 159 (2015)
75. Grimes, R.N.: Carborane polymers and dendrimers. In: Grimes, R.N. (Ed.) Carboranes, 2nd edn., 1015 pp. Academic Press, Oxford (2011)
76. Ho, C.-L., Yu, Z.-Q., Wong, W.-Y.: Multifunctional polymetallaynes: Properties, functions and applications. Chem. Soc. Rev. **45**(19), 5264 (2016)
77. Grelaud, G., Cifuentes, M.P., Paul, F., Humphrey, M.G.: Group-8 metal alkynyl complexes for nonlinear optics. J. Organomet. Chem. **751**, 181 (2014)
78. Verbiest, T., Houbrechts, S., Kauranen, M., Clays, K., Persoons, A.: Second-order nonlinear optical materials: Recent advances in chromophore design. J. Mater. Chem. **7**(11), 2175 (1997)
79. Whittall, I.R., Humphrey, M.G., Houbrechts, S., Maes, J., Persoons, A., Schmid, S., Hockless, D.C.R.: Organometallic complexes for nonlinear optics. Part 14. Syntheses and second-order nonlinear optical properties of ruthenium, nickel and gold σ-acetylides of 1,3,5-triethynylbenzene: X-Ray crystal structures of 1-(HC≡C)-3,5-C_6H_3(trans-C≡CRuCl$(dppm)_2)_2$ and 1,3,5-C_6H_3(C≡CAu$(PPh_3))_3$. J. Organomet. Chem. **544**(2), 277 (1997)
80. McDonagh, A.M., Humphrey, M.G., Samoc, M., Luther-Davies, B., Houbrechts, S., Wada, T., Sasabe, H., Persoons, A.: Organometallic complexes for nonlinear optics. Part 16. Second and third order optical nonlinearities of octopolar alkynylruthenium complexes. J. Am. Chem. Soc. **121**(6), 1405 (1999)
81. McDonagh, A.M., Humphrey, M.G., Samoc, M., Luther-Davies, B.: Organometallic complexes for nonlinear optics. Part 17. Synthesis, third-order optical nonlinearities, and two-photon absorption cross section of an alkynylruthenium dendrimer. Organometallics **18** (25), 5195 (1999)
82. McDonagh, A.M., Powell, C.E., Morrall, J.P., Cifuentes, M.P., Humphrey, M.G.: Convergent synthesis of alkynylbis(bidentate phosphine)ruthenium dendrimers. Organometallics **22**(7), 1402 (2003)
83. Green, K.A., Simpson, P.V., Corkery, T.C., Cifuentes, M.P., Samoc, M., Humphrey, M.G.: Divergent synthesis of ruthenium alkynyl dendrimers and a two-photon absorption cross-section dendritic effect. Macromol. Rapid Commun. **33**(6–7), 573 (2012)
84. Jeffery, C.J., Cifuentes, M.P., Dalton, G.T., Corkery, T.C., Randles, M.D., Willis, A.C., Samoc, M., Humphrey, M.G.: Organometallic complexes for nonlinear optics. Part 47. Synthesis and cubic optical nonlinearity of a stilbenylethynylruthenium dendrimer. Macromol. Rapid Commun. **31**(9–10), 846 (2010)
85. Samoc, M., Morrall, J.P., Dalton, G.T., Cifuentes, M.P., Humphrey, M.G.: Two-photon and three-photon absorption in an organometallic dendrimer. Angew. Chem. Int. Ed. **46**(5), 731 (2007)
86. Samoc, M., Corkery, T.C., McDonagh, A.M., Cifuentes, M.P., Humphrey, M.G.: Organometallic complexes for non-linear optics. Part 49. Third-order non-linear optical spectral dependence studies of arylalkynylruthenium dendrimers. Austr. J. Chem. **64**(9), 1269 (2011)

87. Roberts, R.L., Schwich, T., Corkery, T.C., Cifuentes, M.P., Green, K.A., Farmer, J.D., Low, P.J., Marder, T.B., Samoc, M., Humphrey, M.G.: Organometallic complexes for nonlinear optics. Part 45. Dispersion of the third-order nonlinear optical properties of triphenylamine-cored alkynylruthenium dendrimers. Adv. Mater. **21**(22), 2318 (2009)
88. Dalton, G.T., Cifuentes, M.P., Petrie, S., Stranger, R., Humphrey, M.G., Samoc, M.: Independent switching of cubic nonlinear optical properties in a ruthenium alkynyl cruciform complex by employing protic and electrochemical stimuli. J. Am. Chem. Soc. **129**(39), 11882 (2007)
89. Drouet, S., Merhi, A., Yao, D., Cifuentes, M.P., Humphrey, M.G., Wielgus, M., Olesiak-Banska, J., Matczyszyn, K., Samoc, M., Paul, F., et al.: Cubic nonlinear optical properties of new zinc tetraphenyl porphyrins peripherally functionalized with electron-rich Ru(II) alkynyl substituents. Tetrahedron **68**(50), 10351 (2012)
90. Merhi, A., Grelaud, G., Morshedi, M., Abid, S., Green, K.A., Barlow, A., Groizard, T., Kahlal, S., Halet, J.-F., Ngo, H.M., et al.: Synthesis, characterization and third-order nonlinear optical properties of a dodecaruthenium organometallic dendrimer with a zinc(II) tetraphenylporphyrin core. Dalton Trans. **47**(32), 11123 (2018)
91. Merhi, A., Grelaud, G., Green, K.A., Minh, N.H., Reynolds, M., Ledoux, I., Barlow, A., Wang, G., Cifuentes, M.P., Humphrey, M.G., et al.: A hybrid ruthenium alkynyl/zinc porphyrin "cross fourchée" with large cubic NLO properties. Dalton Trans. **44**(17), 7748 (2015)
92. Morshedi, M., Kodikara, M.S., Corkery, T.C., Hurst, S.K., Chavan, S.S., Kulasekera, E., Stranger, R., Samoc, M., Van Cleuvenbergen, S., Asselberghs, I., et al.: Quadratic and cubic optical nonlinearities of Y-shaped and distorted-H-shaped arylalkynylruthenium complexes. Chem. Eur. J. **24**(61), 16332 (2018)
93. Issberner, J., Vögtle, F., De Cola, L., Balzani, V.: Dendritic bipyridine ligands and their tris (bipyridine)ruthenium(II) chelates: Syntheses, absorption spectra, and photophysical properties. Chem. Eur. J. **3**(5), 706 (1997)
94. Vögtle, F., Plevoets, M., Nieger, M., Azzellini, G.C., Credi, A., De Cola, L., De Marchis, V., Venturi, M., Balzani, V.: Dendrimers with a photoactive and redox-active $[Ru(bpy)_3]^{2+}$-type core: Photophysical properties, electrochemical behavior, and excited-state electron-transfer reactions. J. Am. Chem. Soc. **121**(26), 6290 (1999)
95. Hahn, U., Vögtle, F., De Paoli, G., Staffilani, M., De Cola, L.: Long-lived luminescent dendrimers with a $[Ru(dpp)_3]^{2+}$-type core: Synthesis and photophysical properties. Eur. J. Inorg. Chem. **2009**(18), 2639 (2009)
96. Plevoets, M., Vögtle, F., De Cola, L., Balzani, V.: Supramolecular dendrimers with a $[Ru(bpy)_3]^{2+}$ core and naphthyl peripheral units. New J. Chem. **23**(1), 63 (1999)
97. Vonlanthen, M., Cevallos-Vallejo, A., Aguilar-Ortíz, E., Ruiu, A., Porcu, P., Rivera, E.: Synthesis, characterization and photophysical studies of novel pyrene labeled ruthenium(II) trisbipyridine complex cored dendrimers. Polymer **99**, 13 (2016)
98. Zhou, X., Tyson, D.S., Castellano, F.N.: First generation light-harvesting dendrimers with a $[Ru(bpy)_3]^{2+}$ core and aryl ether ligands functionalized with coumarin-450. Angew. Chem. Int. Ed. **39**(23), 4301 (2000)
99. Kikkeri, R., García-Rubio, I., Seeberger, P.H.: Ru(II)–carbohydrate dendrimers as photoinduced electron transfer lectinbiosensors. Chem. Commun. (2), 235 (2009)
100. Kikkeri, R., Liu, X., Adibekian, A., Tsai, Y.-H., Seeberger, P.H.: Facile synthesis of size dependent Ru(II)–carbohydrate dendrimers *via* click chemistry. Chem. Commun. **46**(13), 2197 (2010)
101. Kikkeri, R., Grünstein, D., Seeberger, P.H.: Lectin biosensing using digital analysis of Ru (II)-glycodendrimers. J. Am. Chem. Soc. **132**(30), 10230 (2010)
102. Kikkeri, R., Kamena, F., Gupta, T., Hossain, L.H., Boonyarattanakalin, S., Gorodyska, G., Beurer, E., Coullerez, G., Textor, M., Seeberger, P.H.: Ru(II) glycodendrimers as probes to study lectin−carbohydrate interactions and electrochemically measure monosaccharide and oligosaccharide concentrations. Langmuir **26**(3), 1520 (2010)

103. Okada, T., Makino, T., Minoura, N.: Fluorescence emission and polarization for analyzing binding of ruthenium metalloglycocluster to lectin and tetanus toxin C-fragment. Bioconjug. Chem. **20**(7), 1296 (2009)
104. Osawa, M., Hoshino, M., Horiuchi, S., Wakatsuki, Y.: Palladium-mediated one-step coupling between polypyridine metal complexes: preparation of rigid and dendritic nano-sized ruthenium complexes. Organometallics **18**(2), 112 (1999)
105. Morgenroth, F., Reuther, E., Müllen, K.: Polyphenylene dendrimers: From three-dimensional to two-dimensional structures. Angew. Chem. Int. Ed. **36**(6), 631 (1997)
106. Rosenfeldt, S., Dingenouts, N., Pötschke, D., Ballauff, M., Berresheim, A.J., Müllen, K., Lindner, P.: Analysis of the spatial dimensions of fully aromatic dendrimers. Angew. Chem. Int. Ed. **43**(1), 109 (2004)
107. Wiesler, U.M., Berresheim, A.J., Morgenroth, F., Lieser, G., Müllen, K.: Divergent synthesis of polyphenylene dendrimers: The role of core and branching reagents upon size and shape. Macromolecules **34**(2), 187 (2001)
108. Haberecht, M.C., Schnorr, J.M., Andreitchenko, E.V., Clark, C.G., Jr., Wagner, M., Müllen, K.: Tris(2,2′-bipyridyl)ruthenium(II) with branched polyphenylene shells: A family of charged shape-persistent nanoparticles. Angew. Chem. Int. Ed. **47**(9), 1662 (2008)
109. Schubert, U.S., Winter, A., Newkome, G.R.: Terpyridine-based materials. Wiley-VCH, Weinheim (2011)
110. Chow, H.-F., Chan, I.Y.K., Fung, P.-S., Mong, T.K.K., Nongrum, M.F.: The 'dendritic effect' on the electrochemical properties of an encapsulated redox center: correlation of size exclusion chromatography data and redox reversibility of electrochemically active homoleptic and heteroleptic ruthenium(II)-bis(terpyridine) metallodendrimers. Tetrahedron **57**(8), 1565 (2001)
111. Mishra, A., Mena-Osteritz, E., Bäuerle, P.: Synthesis, photophysical and electrochemical characterization of terpyridine-functionalized dendritic oligothiophenes and their Ru(II) complexes. Beilstein J. Org. Chem. **9**, 866 (2013)
112. Wang, J.-L., Chan, Y.-T., Moorefield, C.N., Pei, J., Modarelli, D.A., Romano, N.C., Newkome, G.R.: Shape-persistent, truxene-based, nano-sized bisterpyridine ruthenium(II) complexes: Synthesis and photophysical properties. Macromol. Rapid Commun. **31**(9–10), 850 (2010)
113. Godbey, W.T., Wu, K.K., Mikos, A.G.: Poly(ethylenimine) and its role in gene delivery. J. Control. Release **60**(2), 149 (1999)
114. Guillot-Nieckowski, M., Eisler, S., Diederich, F.: Dendritic vectors for gene transfection. New J. Chem. **31**(7), 1111 (2007)
115. Constable, E.C., Harverson, P., Oberholzer, M.: Convergent and divergent approaches to metallocentric metallodendrimers. Chem. Commun. (15), 1821 (1996)
116. Constable, E.C., Harverson, P.: A convergent approach to heteroheptanuclear star complexes. Polyhedron **18**(14), 1891 (1999)
117. Constable, E.C., Housecroft, C.E., Poleschak, I.: Convergent synthesis of a heptaruthenium metallostar. Inorg. Chem. Commun. **2**(11), 565 (1999)
118. Bozic-Weber, B., Constable, E.C., Figgemeier, E., Housecroft, C.E., Kylberg, W.: Evaluation of polynuclear dendrons as photosensitizers for dye-sensitized solar cells. Energy Environ. Sci. **2**(3), 299 (2009)
119. Storrier, G.D., Takada, K., Abruña, H.D.: Synthesis, characterization, electrochemistry, and EQCM studies of polyamidoamine dendrimers surface-functionalized with polypyridyl metal complexes. Langmuir **15**(3), 872 (1999)
120. Takada, K., Storrier, G.D., Morán, M., Abruña, H.D.: Thermodynamics and kinetics of adsorption of poly(amido amine) dendrimers surface functionalized with ruthenium(II) complexes. Langmuir **15**(21), 7333 (1999)
121. Goldsmith, J.I., Takada, K., Abruña, H.D.: Probing diffusional transport in redox-active dendrimers. J. Phys. Chem. B **106**(34), 8504 (2002)

122. Fealy, R.J., Goldsmith, J.I.: Investigating electron transfer in macromolecular ruthenium tris (bipyridyl) complexes using collection experiments at a rotating ring–disk electrode. J. Phys. Chem. C **116**(24), 13133 (2012)
123. Takada, K., Abruña, H.D.: Determination of solvation numbers of $[Ru^{II}(bpy)_3]$ and $[Ru^{II}(tpy)_2]$ functionalized-PAMAM dendrimers adsorbed onto platinum electrode surfaces. J. Electroanal. Chem. **567**(2), 249 (2004)
124. Barron, J.A., Bernhard, S., Houston, P.L., Abruña, H.D., Ruglovsky, J.L., Malliaras, G.G.: Electroluminescence in ruthenium(II) dendrimers. J. Phys. Chem. A **107**(40), 8130 (2003)
125. Lee, K.W., Slinker, J.D., Gorodetsky, A.A., Flores-Torres, S., Abruña, H.D., Houston, P.L., Malliaras, G.G.: Photophysical properties of tris(bipyridyl)ruthenium(II) thin films and devices. Phys. Chem. Chem. Phys. **5**(12), 2706 (2003)
126. Burn, P.L., Lo, S.-C., Samuel, I.D.W.: The development of light-emitting dendrimers for displays. Adv. Mater. **19**(13), 1675 (2007)
127. Hong, B., Thoms, T.P.S., Murfee, H.J., Lebrun, M.J.: Highly branched dendritic macromolecules with core polyhedral silsesquioxane functionalities. Inorg. Chem. **36**(27), 6146 (1997)
128. Murfee, H.J., Thoms, T.P.S., Greaves, J., Hong, B.: New metallodendrimers containing an octakis(diphenylphosphino)-functionalized silsesquioxane core and ruthenium(II)-based chromophores. Inorg. Chem. **39**(23), 5209 (2000)
129. Kim, C., Hong, J.H.: Carbosilane and carbosiloxane dendrimers. Molecules **14**(9), 3719 (2009)
130. Kim, C., Kim, H.: 2,2′:6′,2″-Terpyridine and bis(2,2′:6′,2″-terpyridine)ruthenium(II) complex on the dendritic periphery. J. Organomet. Chem. **673**(1), 77 (2003)
131. Liao, Y.-H., Moss, J.R.: Ruthenium-containing organometallic dendrimers. J. Chem. Soc. Chem. Commun. (23), 1774 (1993)
132. Liao, Y.-H., Moss, J.R.: Synthesis of very large organoruthenium dendrimers. Organometallics **14**(5), 2130 (1995)
133. Liao, Y.-H., Moss, J.R.: Organoruthenium dendrimers. Organometallics **15**(20), 4307 (1996)
134. Kumar, P., Gupta, R.K., Pandey, D.S.: Half-sandwich arene ruthenium complexes: synthetic strategies and relevance in catalysis. Chem. Soc. Rev. **43**(2), 707 (2014)
135. Bardaji, M., Kustos, M., Caminade, A.-M., Majoral, J.-P., Chaudret, B.: Phosphorus-containing dendrimers as multidentate ligands: Palladium, platinum, and rhodium complexes. Organometallics **16**(3), 403 (1997)
136. Bardají, M., Caminade, A.-M., Majoral, J.-P., Chaudret, B.: Ruthenium hydride and dihydrogen complexes with dendrimeric multidentate ligands. Organometallics **16**(15), 3489 (1997)
137. Maraval, V., Laurent, R., Caminade, A.-M., Majoral, J.-P.: Phosphorus-containing dendrimers and their transition metal complexes as efficient recoverable multicenter homogeneous catalysts in organic synthesis. Organometallics **19**(20), 4025 (2000)
138. Servin, P., Laurent, R., Gonsalvi, L., Tristany, M., Peruzzini, M., Majoral, J.-P., Caminade, A.-M.: Grafting of water-soluble phosphines to dendrimers and their use in catalysis: Positive dendritic effects in aqueous media. Dalton Trans. **23**, 4432 (2009)
139. Servin, P., Laurent, R., Dib, H., Gonsalvi, L., Peruzzini, M., Majoral, J.-P., Caminade, A.-M.: Number of terminal groups versus generation of the dendrimer, which criteria influence the catalytic properties? Tetrahedron Lett. **53**(30), 3876 (2012)
140. Neumann, P., Dib, H., Caminade, A.-M., Hey-Hawkins, E.: Redox control of a dendritic ferrocenyl-based homogeneous catalyst. Angew. Chem. Int. Ed. **54**(1), 311 (2015)
141. Allgeier, A.M., Mirkin, C.A.: Ligand design for electrochemically controlling stoichiometric and catalytic reactivity of transition metals. Angew. Chem. Int. Ed. **37**(7), 894 (1998)
142. Neumann, P., Dib, H., Sournia-Saquet, A., Grell, T., Handke, M., Caminade, A.-M., Hey-Hawkins, E.: Ruthenium complexes with dendritic ferrocenyl phosphanes: Synthesis, characterization, and application in the catalytic redox isomerization of allylic alcohols. Chem. Eur. J. **21**(17), 6590 (2015)

143. Rodríguez, L.-I., Rossell, O., Seco, M., Muller, G.: Rhodium or ruthenium units peripherally coordinated to carbosilane dendrimers functionalized with P-stereogenic monophosphines. J. Organomet. Chem. **692**(4), 851 (2007)
144. Wijkens, P., Jastrzebski, J.T.B.H., van der Schaaf, P.A., Kolly, R., Hafner, A., van Koten, G.: Synthesis of periphery-functionalized dendritic molecules using polylithiated dendrimers as starting material. Org. Lett. **2**(11), 1621 (2000)
145. Fürstner, A.: Recent advancements in ring closing olefin metathesis. Top. Catal. **4**(3), 285 (1997)
146. Dijkstra, H.P., van Klink, G.P.M., van Koten, G.: The use of ultra- and nanofiltration techniques in homogeneous catalyst recycling. Acc. Chem. Res. **35**(9), 798 (2002)
147. Gatard, S., Kahlal, S., Méry, D., Nlate, S., Cloutet, E., Saillard, J.-Y., Astruc, D.: Synthesis, chemistry, DFT calculations, and ROMP activity of monomeric benzylidene complexes containing a chelating diphosphine and of four generations of metallodendritic analogues. Positive and negative dendritic effects and formation of dendritic ruthenium−polynorbornene stars. Organometallics **23**(6), 1313 (2004)
148. Gatard, S., Nlate, S., Cloutet, E., Bravic, G., Blais, J.-C., Astruc, D.: Dendritic stars by ring-opening-metathesis polymerization from ruthenium–carbene initiators. Angew. Chem. Int. Ed. **42**(4), 452 (2003)
149. Méry, D., Astruc, D.: Synthesis of monomeric and dendritic ruthenium benzylidene cis-bis-tertiobutyl phosphine complexes that catalyze the ROMP of norbornene under ambient conditions. J. Mol. Catal. A **227**(1), 1 (2005)
150. Astruc, D., Diallo, A.K., Gatard, S., Liang, L., Ornelas, C., Martinez, V., Méry, D., Ruiz, J.: Olefin metathesis in nano-sized systems. Beilstein J. Org. Chem. **7**, 94 (2011)
151. Garber, S.B., Kingsbury, J.S., Gray, B.L., Hoveyda, A.H.: Efficient and recyclable monomeric and dendritic Ru-based metathesis catalysts. J. Am. Chem. Soc. **122**(34), 8168 (2000)
152. Keraani, A., Nasser, G., Shahane, S., Renouard, T., Bruneau, C., Rabiller-Baudry, M., Fischmeister, C.: Syntheses and characterization of molecular weight enlarged olefin metathesis pre-catalysts. Compt. Rend. Chim. **20**(7), 717 (2017)
153. Lu, J., Toy, P.H.: Organic polymer supports for synthesis and for reagent and catalyst immobilization. Chem. Rev. **109**(2), 815 (2009)
154. Buchmeiser, M.R.: Polymer-supported well-defined metathesis catalysts. Chem. Rev. **109** (2), 303 (2008)
155. Zeng, L., Gupta, P., Chen, Y., Wang, E., Ji, L., Chao, H., Chen, Z.-S.: The development of anticancer ruthenium(II) complexes: from single molecule compounds to nanomaterials. Chem. Soc. Rev. **46**(19), 5771 (2017)
156. Villemin, E., Ong, Y.C., Thomas, C.M., Gasser, G.: Polymer encapsulation of ruthenium complexes for biological and medicinal applications. Nat. Rev. Chem. **3**(4), 261 (2019)
157. Levine, A.B., Punihaole, D., Levine, T.B.: Characterization of the role of nitric oxide and its clinical applications. Cardiology **122**(1), 55 (2012)
158. Roy, S., Sharma, S., Sharma, M., Aggarwal, R., Bose, M.: Induction of nitric oxide release from the human alveolar epithelial cell line A549: An *in vitro* correlate of innate immune response to Mycobacterium tuberculosis. Immunology **112**(3), 471 (2004)
159. Luiking, Y.C., Engelen, M.P., Deutz, N.E.: Regulation of nitric oxide production in health and disease. Curr. Opin. Clin. Nutr. Metab. Care **13**(1), 97 (2010)
160. Roveda, A.C., Jr., Franco, D.W.: Nitric oxide releasing-dendrimers: An overview. Braz. J. Pharm. Sci. **49**, 1 (2013)
161. Davies, N.A., Wilson, M.T., Slade, E., Fricker, S.P., Murrer, B.A., Powell, N.A., Henderson, G.R.: Kinetics of nitric oxide scavenging by ruthenium(III) polyaminocarboxylates: novel therapeutic agents for septic shock. Chem. Commun. (1), 47 (1997)
162. Zanichelli, P.G., Miotto, A.M., Estrela, H.F.G., Soares, F.R., Grassi-Kassisse, D.M., Spadari-Bratfisch, R.C., Castellano, E.E., Roncaroli, F., Parise, A.R., Olabe, J.A., et al.: The

[Ru(Hedta)NO]^{0,1}−system: structure, chemical reactivity and biological assays. J. Inorg. Biochem. **98**(11), 1921 (2004)
163. Benini, P.G.Z., McGarvey, B.R., Franco, D.W.: Functionalization of PAMAM dendrimers with [Ru^{III}(edta)(H_2O)]$^-$. Nitric Oxide **19**(3), 245 (2008)
164. Roveda, A.C., Papa, T.B.R., Castellano, E.E., Franco, D.W.: PAMAM dendrimers functionalized with ruthenium nitrosyl as nitric oxide carriers. Inorg. Chim. Acta **409**, 147 (2014)
165. Govender, P., Antonels, N.C., Mattsson, J., Renfrew, A.K., Dyson, P.J., Moss, J.R., Therrien, B., Smith, G.S.: Anticancer activity of multinuclear arene ruthenium complexes coordinated to dendritic polypyridyl scaffolds. J. Organomet. Chem. **694**(21), 3470 (2009)
166. Govender, P., Renfrew, A.K., Clavel, C.M., Dyson, P.J., Therrien, B., Smith, G.S.: Antiproliferative activity of chelating N, O- and N, N-ruthenium(II) arene functionalised poly(propyleneimine) dendrimer scaffolds. Dalton Trans. **40**(5), 1158 (2011)
167. Govender, P., Sudding, L.C., Clavel, C.M., Dyson, P.J., Therrien, B., Smith, G.S.: The influence of RAPTA moieties on the antiproliferative activity of peripheral-functionalised poly(salicylaldiminato) metallodendrimers. Dalton Trans. **42**(4), 1267 (2013)
168. Govender, P., Riedel, T., Dyson, P.J., Smith, G.S.: Higher generation cationic *N*, *N*-ruthenium(II)-ethylene-glycol-derived metallodendrimers: Synthesis, characterization and cytotoxicity. J. Organomet. Chem. **799–800**, 38 (2015)
169. Maroto-Díaz, M., Elie, B.T., Gómez-Sal, P., Pérez-Serrano, J., Gómez, R., Contel, M., Javier de la Mata, F.: Synthesis and anticancer activity of carbosilane metallodendrimers based on arene ruthenium(II) complexes. Dalton Trans. **45**(16), 7049 (2016)
170. Michlewska, S., Ionov, M., Maroto-Díaz, M., Szwed, A., Ihnatsyeu-Kachan, A., Abashkin, V., Dzmitruk, V., Rogalska, A., Denel, M., Gapinska, M., et al.: Ruthenium dendrimers against acute promyelocytic leukemia: *In vitro* studies on HL-60 cells. Fut. Med. Chem. **11**(14), 1741 (2019)
171. Michlewska, S., Ionov, M., Szwed, A., Rogalska, A., Sanz del Olmo, N., Ortega, P., Denel, M., Jacenik, D., Shcharbin, D., de la Mata, F.J., et al.: Ruthenium dendrimers against human lymphoblastic leukemia 1301 cells. Int. J. Mol. Sci. **21**(11), 4119 (2020)
172. Maroto-Diaz, M., Sanz del Olmo, N., Muñoz-Moreno, L., Bajo, A.M., Carmena, M.J., Gómez, R., García-Gallego, S., de la Mata, F.J.: *In vitro* and *in vivo* evaluation of first-generation carbosilane arene Ru(II)-metallodendrimers in advanced prostate cancer. Eur. Polym. J. **113**, 229 (2019)
173. Sanz del Olmo, N., Bajo, A.M., Ionov, M., García-Gallego, S., Bryszewska, M., Gómez, R., Ortega, P., de la Mata, F.J.: Cyclopentadienyl ruthenium(II) carbosilane metallodendrimers as a promising treatment against advanced prostate cancer. Eur. J. Med. Chem. **199**, 112414 (2020)
174. Carloni, R., Sanz del Olmo, N., Ortega, P., Fattori, A., Gómez, R., Ottaviani, M.F., García-Gallego, S., Cangiotti, M., de la Mata, F.J.: Exploring the interactions of ruthenium (II) carbosilane metallodendrimers and precursors with model cell membranes through a dual spin-label spin-probe technique using EPR. Biomolecules **9**(10), 540 (2019)
175. Garaiová, Z., Bolat, G., de Ávila, B.E.-F., Gong, H., del Olmo, N.S., Ortega, P., de la Mata, F.J., Michlewska, S., Wang, J., Hianik, T.: Combination of ruthenium dendrimers and acoustically propelled gold nanowires as a platform for active intracellular drug delivery towards breast cancer therapy. Clin. Ocol. Res. **2**(4), 1 (2019)
176. Maroto-Diaz, M., Sanz del Olmo, N., Garcia-Gallego, S., Gómez, R., Ortega, P., de la Mata, F.J.: Synthesis and structural characterization of carbosilane ruthenium(II) metallodendrons containing cymene units. J. Organomet. Chem. **901**, 120942 (2019)
177. Jardim, M.G., Rissanen, K., Rodrigues, J.: Preparation and characterization of novel poly (alkylidenamine) nitrile ruthenium metallodendrimers. Eur. J. Inorg. Chem. **2010**(11), 1729 (2010)

178. Rodrigues, J., Jardim, M.G., Figueira, J., Gouveia, M., Tomás, H., Rissanen, K.: Poly (alkylidenamines) dendrimers as scaffolds for the preparation of low-generation ruthenium based metallodendrimers. New J. Chem. **35**(10), 1938 (2011)
179. Gouveia, M., Figueira, J., Jardim, M.G., Castro, R., Tomás, H., Rissanen, K., Rodrigues, J.: Poly(alkylidenimine) dendrimers functionalized with the organometallic moiety $[Ru(\eta^5\text{-}C_5H_5)(PPh_3)_2]^+$ as promising drugs against cisplatin-resistant cancer cells and human mesenchymal stem cells. Molecules **23**(6), 1471 (2018)
180. Endo, A., Minesaka, H., Hashimoto, T., Hayashita, T.: Electrochemical sugar recognition using a ruthenium complex with boronic acid assembled on polyamidoamine (PAMAM) dendrimer. Anal. Meth. **4**(9), 2657 (2012)
181. Pettirossi, S., Bellachioma, G., Ciancaleoni, G., Zuccaccia, C., Zuccaccia, D., Macchioni, A.: Diffusion and NOE NMR studies on multicationic DAB-organoruthenium dendrimers: Size-dependent noncovalent self-assembly to megamers and ion pairing. Chem. Eur. J. **15** (21), 5337 (2009)
182. Haketa, Y., Maeda, H.: Dimension-controlled ion-pairing assemblies based on π-electronic charged species. Chem. Commun. **53**(20), 2894 (2017)
183. Lehn, J.-M.: Perspectives in supramolecular chemistry—from molecular recognition towards molecular information processing and self-organization. Angew. Chem. Int. Ed. **29**(11), 1304 (1990)
184. Newkome, G.R., Moorefield, C.N., Baker, G.R., Behera, R.K., Escamillia, G.H., Saunders, M.J.: Supramolecular self-assemblies of two-directional cascade molecules: Automorphogenesis. Angew. Chem. Int. Ed. **31**(7), 917 (1992)
185. Wang, P., Moorefield, C.N., Jeong, K.-U., Hwang, S.-H., Li, S., Cheng, S.Z.D., Newkome, G.R.: Dendrimer–metallomacrocycle composites: Nanofiber formation by multi-ion pairing. Adv. Mater. **20**(7), 1381 (2008)
186. Axet, M.R., Philippot, K.: Catalysis with colloidal ruthenium nanoparticles. Chem. Rev. **120** (2), 1085 (2020)
187. Karakhanov, E.A., Maksimov, A.L., Zolotukhina, A.V., Kardasheva, Y.S.: Hydrogenation catalysts based on metal nanoparticles stabilized by organic ligands. Russ. Chem. Bull. **62** (7), 1465 (2013)
188. Niu, Y., Crooks, R.M.: Dendrimer-encapsulated metal nanoparticles and their applications to catalysis. Compt. Rend. Chim. **6**(8), 1049 (2003)
189. Crooks, R.M., Zhao, M., Sun, L., Chechik, V., Yeung, L.K.: Dendrimer-encapsulated metal nanoparticles: Synthesis, characterization, and applications to catalysis. Acc. Chem. Res. **34** (3), 181 (2001)
190. Jansen, J.F.G.A., de Brabander-van den Berg, E.M.M., Meijer, E.W.: Encapsulation of guest molecules into a dendritic box. Science **266**(5188), 1226 (1994)
191. Zhao, M., Liu, Y., Crooks, R.M., Bergbreiter, D.E.: Preparation of highly impermeable hyperbranched polymer thin-film coatings using dendrimers first as building blocks and then as *in situ* thermosetting agents. J. Am. Chem. Soc. **121**(5), 923 (1999)
192. Bosman, A.W., Janssen, H.M., Meijer, E.W.: About dendrimers: Structure, physical properties, and applications. Chem. Rev. **99**(7), 1665 (1999)
193. Crooks, R.M., Lemon, B.I., Sun, L., Yeung, L.K., Zhao, M.: Dendrimer-encapsulated metals and semiconductors: Synthesis, characterization, and applications. Top. Curr. Chem. **212**, 81 (2001)
194. Lafaye, G., Williams, C.T., Amiridis, M.D.: Synthesis and microscopic characterization of dendrimer-derived Ru/Al_2O_3 catalysts. Catal. Lett. **96**(1), 43 (2004)
195. Lafaye, G., Siani, A., Marécot, P., Amiridis, M.D., Williams, C.T.: Particle size control in dendrimer-derived supported ruthenium catalysts. J. Phys. Chem. B. **110**(15), 7725 (2006)
196. Escalante García, I.L., Ledesma García, J., Rodríguez, F.J., Chapman, T.W., Godinez, L.A.: Synthesis and electrochemical characterization of platinum and ruthenium nanoparticles encapsulated in PAMAM-G4OH dendrimer for oxygen reduction in the presence of methanol. ECS Trans. **3**(30), 1 (2019)

197. Marvin, K.A., Thadani, N.N., Atkinson, C.A., Keller, E.L., Stevenson, K.J.: Preparation and catalytic evaluation of ruthenium–nickel dendrimer encapsulated nanoparticles via intradendrimer redox displacement of nickel nanoparticles. Chem. Commun. **48**(50), 6289 (2012)
198. Antonels, N.C., Meijboom, R.: Preparation of well-defined dendrimer encapsulated ruthenium nanoparticles and their evaluation in the reduction of 4-nitrophenol according to the Langmuir-Hinshelwood approach. Langmuir **29**(44), 13433 (2013)
199. Murugan, E., Pakrudheen, I.: Efficient amphiphilic poly(propylene imine) dendrimer encapsulated ruthenium nanoparticles for sensing and catalysis applications. Sci. Adv. Mater. **7**(5), 891 (2015)
200. Korth, W., Jess, A.: Solid catalysts with ionic liquid layer (SCILL). In: Fehrmann, R., Riisager, A., Haumann, M. (Eds.) Supported Ionic Liquids, 279 pp. Wiley-VCH, Weinheim (2014)
201. Antonels, N.C., Meijboom, R.: Preparation of well-defined dendrimer encapsulated ruthenium nanoparticles and their application as catalyst and enhancement of activity when utilised as SCILL catalysts in the hydrogenation of citral. Catal. Commun. **57**, 148 (2014)
202. Antonels, N.C., Benjamin Williams, M., Meijboom, R., Haumann, M.: Well-defined dendrimer encapsulated ruthenium SCILL catalysts for partial hydrogenation of toluene in liquid-phase. J. Mol. Catal. A **421**, 156 (2016)
203. Keshtiara, P., Hadadzadeh, H., Daryanavard, M., Mousavi, N., Dinari, M.: New dendrimers containing ruthenium nanoparticles as catalysts for hydrogenation of citral to 3,7-dimethyloctanol. Mater. Chem. Phys. **249**, 122962 (2020)
204. Karakhanov, E.A., Maximov, A.L., Zolotukhina, A.V., Terenina, M.V., Vutolkina, A.V.: Nanoheterogeneous ruthenium-containing catalysts based on dendrimers in the hydrogenation of aromatic compounds under two-phase conditions. Pet. Chem. **56**(6), 491 (2016)
205. Maximov, A., Zolotukhina, A., Murzin, V., Karakhanov, E., Rosenberg, E.: Ruthenium nanoparticles stabilized in cross-linked dendrimer matrices: hydrogenation of phenols in aqueous media. ChemCatChem **7**(7), 1197 (2015)
206. Karakhanov, E., Maximov, A., Zolotukhina, A., Mamadli, A., Vutolkina, A., Ivanov, A.: Dendrimer-stabilized Ru nanoparticles immobilized in organo-silica materials for hydrogenation of phenols. Catalysts **7**(3), 86 (2017)
207. Palkovits, R.: Pentenoic acid pathways for cellulosic biofuels. Angew. Chem. Int. Ed. **49** (26), 4336 (2010)
208. Maximov, A.L., Zolotukhina, A.V., Mamedli, A.A., Kulikov, L.A., Karakhanov, E.A.: Selective levulinic acid hydrogenation in the presence of hybrid dendrimer-based catalysts. Part I. Monometallic. ChemCatChem **10**(1), 222 (2018)
209. Nemanashi, M., Noh, J.-H., Meijboom, R.: Hydrogenation of biomass-derived levulinic acid to γ-valerolactone catalyzed by mesoporous supported dendrimer-derived Ru and Pt catalysts: An alternative method for the production of renewable biofuels. Appl. Catal. A: Gen. **550**, 77 (2018)
210. Majoral, J.-P., Caminade, A.-M.: Phosphorhydrazones as useful building blocks for special architectures: macrocycles and dendrimers. Eur. J. Inorg. Chem. **2019**(11–12), 1457 (2019)
211. García-Peña, N.G., Caminade, A.-M., Ouali, A., Redón, R., Turrin, C.-O.: Solventless synthesis of Ru(0) composites stabilized with polyphosphorhydrazone (PPH) dendrons and their use in catalysis. RSC Adv. **6**(69), 64557 (2016)
212. Liu, D., López-De Jesús, Y.M., Monnier, J.R., Williams, C.T.: Preparation, characterization, and kinetic evaluation of dendrimer-derived bimetallic Pt–Ru/SiO_2 catalysts. J. Catal. **269** (2), 376 (2010)
213. Peng, X., Pan, Q., Lu, X.: Regioselective catalyzed modification of poly(methylhydro) siloxane using RuRh and RuPt bimetallic dendrimer-encapsulated nanoparticles. J. Appl. Polym. Sci. **122**(1), 334 (2011)
214. Scott, R.W.J., Datye, A.K., Crooks, R.M.: Bimetallic palladium−platinum dendrimer-encapsulated catalysts. J. Am. Chem. Soc. **125**(13), 3708 (2003)

215. Bernechea, M., García-Rodríguez, S., Terreros, P., de Jesús, E., Fierro, J.L.G., Rojas, S.: Synthesis of core−shell PtRu dendrimer-encapsulated nanoparticles. Relevance as electrocatalysts for CO oxidation. J. Phys. Chem. C **115**(4), 1287 (2011)
216. Gu, Y., Wu, G., Hu, X.F., Chen, D.A., Hansen, T., zur Loye, H.-C., Ploehn, H.J.: PAMAM-stabilized Pt–Ru nanoparticles for methanol electro-oxidation. J. Power Sources **195**(2), 425 (2010)
217. Calderón, J.C., Calvillo, L., Lázaro, M.J., Pastor, E.: Use of dendrimers during the synthesis of Pt-Ru electrocatalysts for PEM fuel cells: Effects on the physical and electrochemical properties. Int. J. Electrochem. **2011**, 564828 (2011)
218. Park, I.-S., Park, K.-W., Choi, J.-H., Park, C.R., Sung, Y.-E.: Electrocatalytic enhancement of methanol oxidation by graphite nanofibers with a high loading of PtRu alloy nanoparticles. Carbon **45**(1), 28 (2007)
219. Calderón, J.C., Calvillo, L., Lázaro, M.J., Rodríguez, J.L., Pastor, E.: Effect of the dendrimer generation used in the synthesis of Pt-Ru nanoparticles supported on carbon nanofibers on the catalytic activity towards methanol oxidation. Energies **10**(2), 159 (2017)
220. Bhattacharya, P., Nasybulin, E.N., Engelhard, M.H., Kovarik, L., Bowden, M.E., Li, X.S., Gaspar, D.J., Xu, W., Zhang, J.-G.: Dendrimer-encapsulated ruthenium oxide nanoparticles as catalysts in lithium-oxygen batteries. Adv. Funct. Mater. **24**(47), 7510 (2014)
221. Kwak, W.-J., Rosy, Sharon, D., Xia, C., Kim, H., Johnson, L.R., Bruce, P.G., Nazar, L.F., Sun, Y.-K., Frimer, A.A., et al.: Lithium–oxygen batteries and related systems: potential, status, and future. Chem. Rev. **120**(14), 6626 (2020)
222. Shi, L., Zhao, T., Xu, A., Wei, Z.: Unraveling the catalytic mechanism of rutile RuO_2 for the oxygen reduction reaction and oxygen evolution reaction in Li–O_2 batteries. ACS Catal. **6**(9), 6285 (2016)

Polymers and Small-Molecule Ru Species

5

Abstract

The blending of Ru complexes in polymers represents a straightforward alternative to the covalent or coordinative incorporation into polymer architectures. Though applications in the fields of opto-electronics (e.g., light-emitting electrochemical cells or diodes) or catalysis have been reported, an intrinsic limitation remains, which is bleaching. These aspects are discussed in this chapter. Furthermore, polymer staining is worth mentioning. The reaction between Ru species and polymers increases the electron-density contrast and, thus, facilitates the microscopy imaging of polymer samples. The history and important examples for this application are shown in the second half of the chapter.

5.1 Introduction

The blending of transition-metal complexes into polymers represents a versatile alternative to the metallopolymer approach, i.e., the covalent attachment of such species to polymers. Regarding the latter, diverse architectures have been reported over the last few decades and applications in various fields of research have been envisioned [1–6]. The different types of ruthenopolymers have been summarized in Chap. 2; however, the synthesis of such metallopolymers can be demanding, in particular when compared to the "mixing" of polymers with transition-metal complexes or salts. This strategy is, for example, commonly used for the fabrication of polymer-stabilized nanoparticles (NPs); there, well-defined NPs are formed from appropriate precursor species, which have been dispersed in a polymer matrix (the field of polymer-stabilized Ru NPs is covered in Chap. 7) [7]. Moreover, the ability of polymer matrices to stabilize sensitive organometallic guest molecules was recognized in the 1980s. In the case of transparent, inert polymer films, even the

© The Author(s), under exclusive license to Springer Nature Switzerland AG 2021
U. S. Schubert et al., *Ruthenium-Containing Polymers*,
https://doi.org/10.1007/978-3-030-75598-0_5

spectroscopic analysis of photochemical or photomechanical reactions of the embedded guests was enabled. In particular, polystyrene (PS) [8, 9], poly(methyl methacrylate) (PMMA) [8], poly(tetrafluoroethylene) (PTFE) [10] as well as various copolymers [8] have been utilized in this context. Due to negligible absorption of the polymer matrix between 2100 and 1750 cm^{-1}, metal–carbonyl complexes, as guests, can be readily characterized by infrared (IR) spectroscopy [9].

Solid-state opto-electronic devices, such light-emitting electrochemical cells (LECs) or light-emitting diodes (LEDs), represent an important application where polymers have been combined, as matrix, with transition-metal complexes, as emitters [11]. Initially, phosphorescent Ru^{II} or Ir^{III} complexes were used in this context [12]; however, emitters based on first-row transition-metal ions have recently emerged, as an interesting low-cost alternative [13–15]. In a similar fashion, also the formation of heterogeneous catalysts for organic reactions has been targeted. However, the non-covalent immobilization of a transition-metal complex, as catalyst, in a polymer matrix bears the risk of bleaching. As a result, such heterogeneous catalysts lose their activity and, in many cases, cannot be easily recycled. On the other hand, the covalent attachment of catalytic complexes to a solid polymer resin or their incorporation into polymer networks has widely been used to generate stable and efficient heterogeneous catalysts for diverse purposes. The family of metathesis reactions represents one example where immobilized Ru^{II} complexes have been applied successfully [16]. Nonetheless, a few examples have been reported where Ru^{II} complexes were embedded into a polymer matrix. Their application as stable, reusable catalysts in organic reactions has been demonstrated [17].

Finally, polymer staining needs to be mentioned as one further important application. With respect to the transmission electron microscopy (TEM) imaging of heterogeneous polymer samples, the electron-density contrast can be enhanced by treating the sample with an appropriate staining agent. Not only OsO_4, but also RuO_4, as an even stronger oxidant, have proven their applicability as selective staining agents [18, 19].

Though the aforementioned applications are of highly different character, they share one common feature: Ru species are imbedded into polymer matrices. In the following, these fields of research will be summarized.

5.2 Ru Complexes Embedded in Polymer Matrices: Catalytic Applications

Shaver et al. pointed out that the mixing of complexes into polymers represents a significant simplification for the preparation of “metallopolymers” compared to the covalent routes shown in Chap. 2 [9]. In order to study chemical processes between transition-metal complexes and gases in a rigid or semi-rigid environment, these authors developed a straightforward protocol to fabricate polymer films with embedded complexes [8]. It has been demonstrated that PS and PMMA, homopolymers as well as copolymers, are versatile matrices in this respect—mainly for two reasons: Firstly, these polymers possess a regime of low absorptivity and, thus, embedded complexes can be studied by, e.g., vis/IR spectroscopy (the latter is of relevance for studying carbonyl complexes via the stretching vibration of the CO ligands). Moreover, light-emitting complexes can thusly be immobilized—a strategy, which has been followed to fabricate light-emitting devices (e.g., LECs or OLEDs; see Sect. 1.4). Secondly, the matrices are considered as appropriate mimics for typical solvents, such as toluene (in the case of PS) or ethyl acetate (in the case of PMMA).

For example, [CpRu(cod)Cl] (Cp: cyclopentadiene, cod: 1,5-cyclooctadiene) was dispersed in a PS matrix, thereby mimicking a toluene solution [9]. This particular complex is a versatile catalyst precursor, which is utilized in various alkyne transformations; thereby, the catalytically active species is formed by displacement of the cod ligand [20]. Substitution of this labile ligand in solution might be induced thermally or photochemically [21, 22]. Against this background, Shaver, Butler et al. studied the ligand-substitution reaction of [CpRu(cod)Cl] using CO, as the nucleophile [9]. This reaction represents one example where the sorption of CO by coordination to transition-metal centers was realized. More recently, Tsukada et al. employed the Ru^{II} complex **1** for the same purpose (Fig. 5.1a) [23]. Such dithiolene complexes, integrated in metal–organic frameworks (MOFs) or polymers, are widely used in the context of olefin addition/purification, adsorption/chemisorption of small molecules (e.g., CO or NO), hydrogen production, and CO_2 reduction [24]. The formation of the CO adduct of **1** occurred readily, in THF solution as well as in the solid state; the initial state could be restored upon treatment with an oxidant (e.g., Me_3NO or O_2) [23]. These authors also prepared a free-standing hybrid film, which is comprised of a polysiloxane matrix (M_w = 18,000 g mol^{-1}) with imbedded **1**. This material exhibited a significant color change from black to orange-brown upon exposure to CO (Fig. 5.1b, c). Later, these same authors presented some structural variants of **1** containing halogen-substituents on the dithiolate ligand [25]. These substituents were important to (a) control the electron density of the metal center and (b) to modulate the solubility of the complex in the polymer matrix, i.e., prevent its aggregation.

The latter aspect, i.e., the compatibility of Ru^{II} complexes with the polymer matrix, is of utmost importance also with respect to potential applications in the field of catalysis. Flexible polymer membranes in which transition-metal catalysts are dispersed have been utilized for a wide range of organic reactions [26, 27]. This straightforward, low-cost strategy is, however, limited by the thermal and solvent stability of the membrane material itself. As a consequence, considerable catalyst bleaching and disintegration of the membrane might occur under the reaction conditions. The issue of compatibility was, for example, addressed by Jacobs et al., who fabricated a bifunctional polydimethylsiloxane (PDMS) membrane containing a Brønsted acid and Ru-BINAP, as an enantioselective hydrogenation catalyst [28]. Stable membranes were only obtained at low additive concentrations; nonetheless, the heterogeneous catalyst was highly active in the hydrogenation of β-keto esters. The hydrogenation of methyl acetoacetate occurred at 40 °C with a turnover frequency (TOF) of 4.6 h^{-1} and offered a good enantioselectivity (*ee* value of 92). The

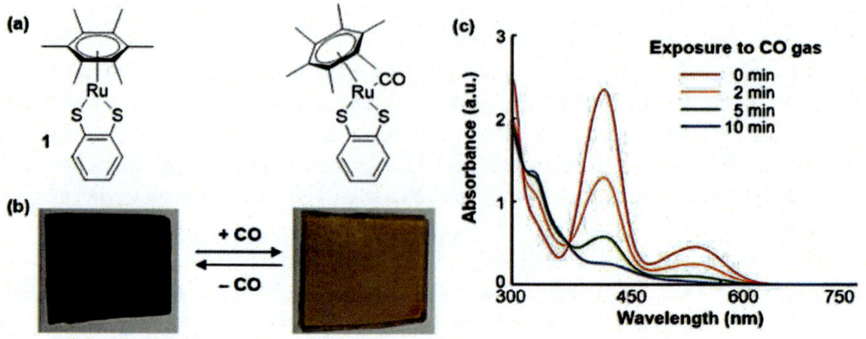

Fig. 5.1 **a** Schematic representation of the Ru^{II}–dithiolene complex **1**, which reversibly formed an adduct with CO. **b** Pictures of the free-standing polymer film incorporating **1** before and after chemisorption of CO. **c** Temporal evolution of the UV/vis absorption spectrum of the hybrid film upon exposure to gaseous CO [23]. Figure reproduced with kind permission; © 2015 Wiley–VCH

superiority of the heterogeneous catalyst compared to the homogeneous variant was shown by its recovery by filtration and reuse in consecutive experiments.

Charge-neutral ruthenoporphyrins are highly compatible with hydrophobic polymers. This was shown by Buonomenna et al. for **2**, which afforded homogeneous dispersions when immobilized in poly(vinylidene fluoride) (PVDF) or PDMS membranes (Fig. 5.2a) [17]. These authors performed leaching tests with these membranes in aromatic solvents, representing the optimal solvents for aziridination. The lowest degree-of-leaching was found for **2a**, owing to the low solubility in such solvents (Fig. 5.2b). In subsequent studies, polyethersulfone (PES), polysulfone (PSf), and Hyflon™ AD60X, a robust perfluorinated copolymer of tetrafluoroethylene and 2,2,4-trifluoro-5-trifluoromethoxy-1,3-dioxole, were employed as polymeric hosts for **2c** [17, 29]. From these, the catalyst immobilized in the Hyflon™ polymer exhibited a promising synthetic applicability due to the ease of catalyst separation, recovery, and recycling. Moreover, the heterogeneous catalyst allowed the utilization of *n*-hexane, as a more convenient solvent, for the aziridination of olefins. When tested in aziridination reactions, high substrate conversion and product selectivity values were obtained.

Ragaini, Buonomenna et al. reported a polymer membrane containing the Ru-BIAN complexes **3** [BIAN: *bis*(arylimino)acenaphthene; Fig. 5.3a] [30]. Under homogeneous reaction conditions, **3a** represents a highly active catalyst for both the reduction of nitroarenes to anilines by CO/H_2O and the allylic amination of unfunctionalized olefins by nitroarenes and CO. Since the rather forcing reaction conditions (temperatures in the range of 160–170 °C), these authors opted for polymers with high glass-transition temperatures (T_g). In particular, polyethersulfone, polysulfone, and **4**, which is basically a modified poly(ether ketone), all offer high thermal, mechanical, and chemical robustness below their T_g values. In order to increase the compatibility of the catalyst with the polymer matrix, flexible alkyl

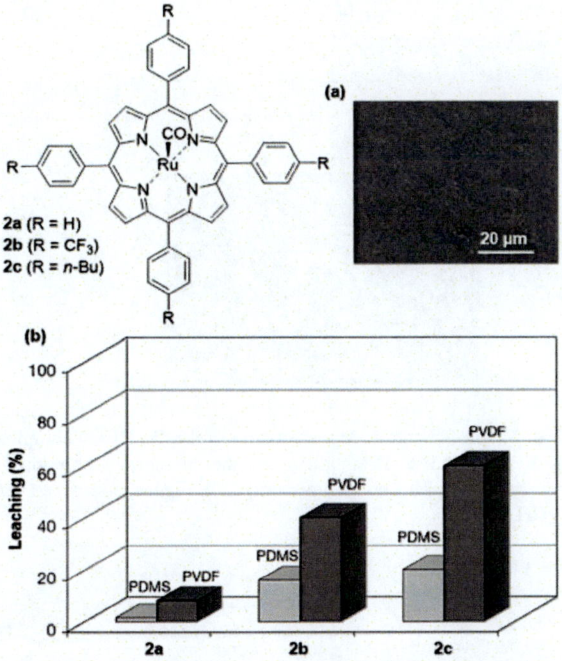

Fig. 5.2 Schematic representation of the $Ru^{II}(CO)$-porphyrin catalysts **2**. **a** EDX surface analysis in the secondary-electron mode of the PVDF membrane containing 8 wt-% of **2a**. **b** Leaching tests in toluene (at 25 °C) from PVDF- and PDMS-based membranes containing 8 wt-% of **2** [17]. Figure reproduced with kind permission; © 2006 Elsevier B.V

chains were introduced (**3b**). The polymer membranes were prepared via the phase-inversion technique [31]; however, aggregates of **3b** were still detectable by scanning-electron microscopy (SEM) imaging of the obtained polymer membranes (Fig. 5.3b) [30]. This hybrid membrane, comprising of **3b** and **4**, was found to be as efficient in catalyzing the reduction of nitrobenzene by CO/H_2O as **3a** under homogeneous conditions. Remarkably, the heterogeneous catalyst was able to deal with the applied forcing high-temperature and high-pressure conditions of the test reactions; very little catalyst leaching was observed for this hybrid material.

In a different approach, Chen et al. prepared polymeric ruthenoporphyrin-functionalized CNTs and reduced graphene oxide (RGO), as heterogeneous catalysts for the transformation of ethyl levulinate (Fig. 5.4) [32]. The synthesis involved a two-step protocol in which a porphyrin network was obtained from the condensation of terephthaldehyde and pyrrole in the presence of the respective carbon allotrope; the subsequent metalation with $Ru_3(CO)_{12}$ gave the hybrid materials, denoted as Ru-PP/CNTs and Ru-PP/RGO, which were thoroughly characterized by microscopy imaging, XRD analysis, solid-state NMR spectroscopy, and XPS. Both hybrid materials efficiently catalyzed the hydrogenation of ethyl levulinate into γ-valerolactone (full conversion, >99%). Moreover, Ru-PP/CNTs also supported the reductive amination of the ester to give pyrrolidone derivatives in moderate to high yields. Noteworthy, the transformation of ethyl levulinate into γ-valerolactone was also catalyzed by Ru clusters immobilized

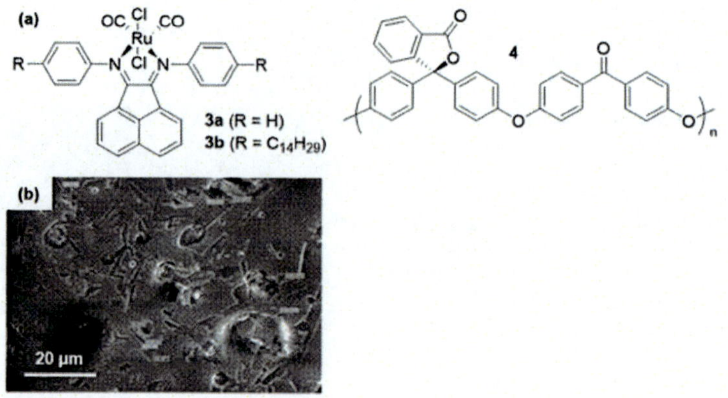

Fig. 5.3 **a** Schematic representation of the Ru-BIAN complexes **3** and the poly(ether ketone) **4**. **b** Representative SEM image of the membrane comprising of polymer **4** and catalyst **3b** (crystallites of **3b** were still present) [30]. Figure reproduced with kind permission; © 2010 Wiley–VCH

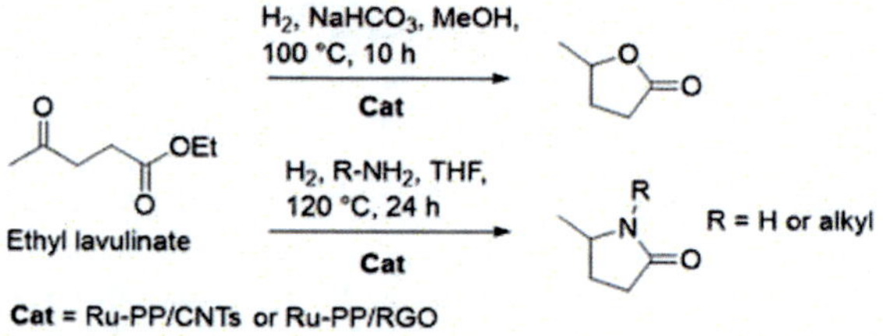

Fig. 5.4 Schematic representation of the transformation of ethyl levulinate into γ-valerolactone or pyrrolidone derivatives, catalyzed by the hybrid materials Ru-PP/CNTs or Ru-PP/RGO

within a cross-linked polyethersulfone network [33]. Both types of heterogeneous catalysts showed high degrees-of-stability and -recyclability.

Nafion™ is a sulfonated tetrafluoroethylene-based polymer, which was introduced in the late 1960s by the DuPont company (Fig. 5.5a) [34]. Due to its perfluorinated structure, Nafion™ is highly resistant to chemical attack (even against oxidants at elevated temperatures) and has, therefore, found widespread use, as a membrane material, and was first employed by Rubinstein and Bard, as an electrode-coating material [35]. The morphology of hydrated Nafion™, according to the generally accepted cluster-network model, is depicted in Fig. 5.5b [34]. The pronounced cation-exchange ability of Nafion™ was, in particular, employed to immobilize $[Ru(bpy)_3]^{2+}$ via electrostatic interactions (Sect. 1.4). Beyond this, also other Ru salts and complexes were combined with this remarkable polymer.

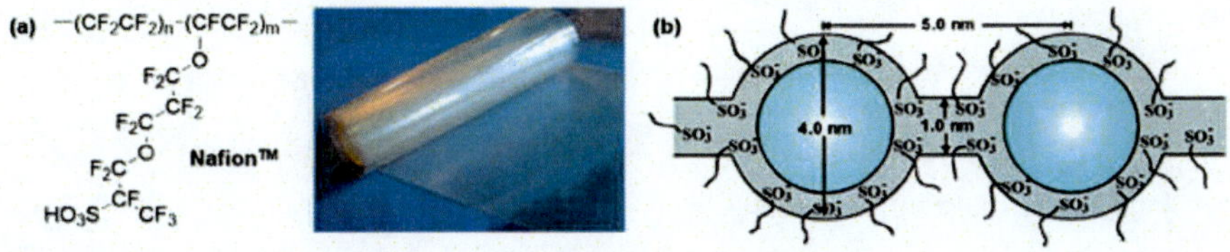

Fig. 5.5 **a** Schematic representation of Nafion™ (the H^+ form is shown) and picture of a commercial Nafion™ foil [36]. **b** Schematic representation of the morphology of hydrated NafionTM (according to the cluster-network model) [34]. Figure reproduced with kind permission; © 2004 American Chemical Society

The immobilization of Ru^{III} ions within the pores of Nafion™ by a straightforward anion exchange was reported by Prakash, Mathew et al. [37]. The material, which was not structurally characterized, was employed as a heterogeneous catalyst for the hydration of aromatic nitriles. The corresponding primary amides were obtained in good to high yields (23 to 97%) and a high functional-group tolerance was observed (in total, 20 nitriles were screened). Moreover, the catalyst could be recovered and used in up to five consecutive runs without significant loss-of-activity. Against the same background, also the immobilization of Ru^{III} onto biopolymers, such as chitosan [38] or chitin [39], was investigated. However, in these cases the catalytically active species, polymer-bound Ru nanoparticles, were generated by *in-situ* reduction with $NaBH_4$. Similar to $RuCl_3$, also the immobilization of $Ru(DMSO)Cl_2$ onto Nafion™ afforded a hybrid material which, when deposited on a GC electrode, functioned as electrochemical detector for hypoxanthine, xanthine, and uric acid [40] (Fig. 5.6).

The ability of Ru^{III}–ammonia complexes to catalyze the oxidation of water by Ce^{IV}, as the oxidant, was first recognized by Kaneko et al. [41]. These authors utilized the mixed-valence, trinuclear complex "ruthenium red" (Ru-red, Fig. 5.7a), as catalyst, under homogeneous as well as heterogeneous conditions (Kaolin, as the

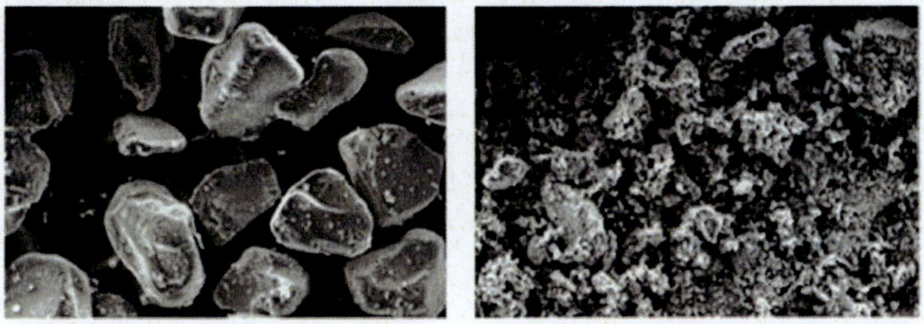

Fig. 5.6 Representative SEM images of Nafion™ before (left) and after cation exchange with Ru^{III} ions (for both images × 100 magnification) [37]. Figure reproduced with kind permission; © 2012 Wiley–VCH

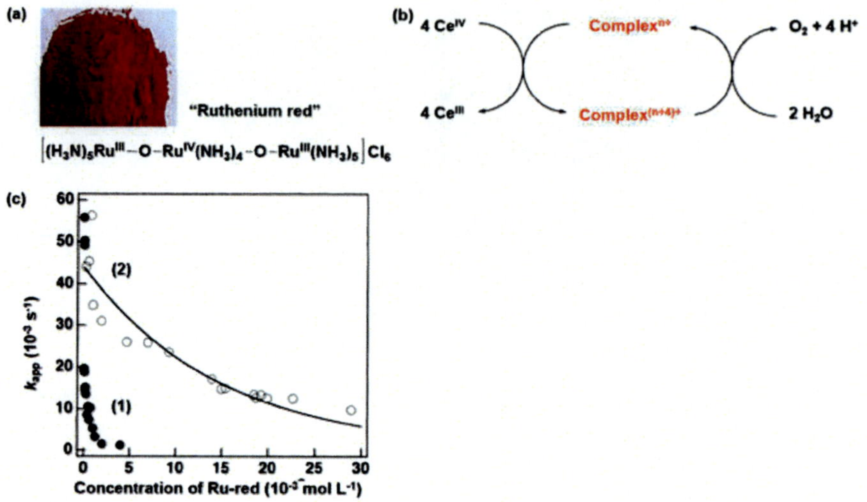

Fig. 5.7 **a** Chemical formula and picture of the dye "ruthenium red" [51]. **b** Schematic representation of the chemical water-oxidation reaction, catalyzed by a redox-active transition-metal complex. **c** Relationship between the apparent catalytic activity (k_{app}) and the concentration of Ru-red in (1) an aqueous environment and (2) a Nafion™ membrane; the solid line represents the theoretical curve determined with $k_{app} = 44\ s^{-1}$ and $r_d = 1.23$ nm [43]. Figure redrawn according to the literature; © 1996 The Royal Chemical Society

solid support). Noteworthy, this particular dye is well-known and commonly used as a pharmacological agent to study specific cellular mechanisms [42]. Later, Nafion™ has also been used for the immobilization of Ru-red [43]. This material acted as heterogeneous catalyst for the water-oxidation reaction (WOR) with Ce^{IV}, as oxidant, even when a high concentration of Ru-red in the membrane was established; whereas, in aqueous medium, at high concentrations, the N_2 evolution rate increased significantly at the cost of the O_2 evolution rate. It was concluded that the bimolecular decomposition of the catalyst was efficiently suppressed by the polymer matrix. A critical decomposition distance (r_d) between two adjacent Ru-red molecules was determined to be 1.23 nm. The heterogeneous catalyst featured an apparent catalytic activity (k_{app}) of 44 s^{-1}. In the following, various related bi- and mononuclear complexes were studied, after immobilization in Nafion™, as the heterogeneous water-oxidation catalysts (WOCs), the thusly determined k_{app} and r_d values are summarized in Table 5.1 [44–50]. Moreover for the mononuclear complexes, an optimum intermolecular distance for the catalytic activity was identified (r_{co}). In these cases, the distance between two metal centers within the membrane is limited by r_d and r_{co} to allow cooperative catalysis (i.e., the coordination of water generates the catalytically active species possessing a Ru–O–O–Ru bond). For further mechanistic considerations of the WOR, catalyzed by the aforementioned complexes, the reader is referred to Kaneko's review article on this topic [44].

Table 5.1 Summary of the catalytic activity of various heterogeneous WOC/Nafion™ systems[a]

WOC	k_{app} (s^{-1})	r_d (nm)	r_{co} (nm)	Reference
Ru-red	44	1.23		43
$[(H_3N)_5Ru\text{-}(\mu\text{-}O)\text{-}Ru(NH_3)_5]^{4+}$	13	1.06		45
$[(H_3N)_3Ru\text{-}(\mu\text{-}Cl)_3\text{-}Ru(NH_3)_3]^{2+}$	67	1.22		46
$[Ru(NH_3)_6]^{3+}$	0.035	0.82	1.21	47
$[Ru(NH_3)_5Cl]^{2+}$	2.7	0.84	1.21	48
cis-$[Ru(NH_3)_4Cl_2]^+$	15	0.88	n/a	49
$[Ru(en)_3]^{3+}$ [b]	0.085	1.07	1.44	50

[a]Chemical WOR with Ce^{IV}, as oxidant;
[b]en: ethylenediamine

Besides the chemical WOR, the electrochemical variant has also been investigated [52]. For this purpose, Ru-red was immobilized in a Nafion™ layer, which was coated on a Pt electrode [53]. Kaneko et al. identified an optimal concentration for immobilized Ru-red, where a maximum value of the turnover number (TON) for the O_2 evolution was reached. These authors proposed that, at low concentrations, the charge transfer between catalyst sites via a hopping mechanism was facilitated; whereas, bimolecular catalyst decomposition became predominant at higher Ru-red concentrations (Fig. 5.8b). In subsequent work, these authors demonstrated that the electrocatalytic activity can be increased by adding mediators for the charge-transfer (CT) process (in proteins, amino-acid residues are known to have this function). In particular, *p*-cresol and 4-methylimidazole, as mimics for tyrosine and histidine, were found to be potent mediators in this respect, which allowed the CT process to proceed over a much longer distance (Fig. 5.8a) [54, 55]. Noteworthy, also other complexes from Table 5.1 as well as $RuCl_3$ were utilized for the electrocatalytic WOR [56–58]. These studies showed that the catalytic activity increased with the number of Ru centers present in the complex, i.e., trinuclear Ru-red was found to be the most powerful electrocatalyst for the WOR within the series.

In contrast to the aforementioned complexes, where the molecular structure of the catalytically active species could not (yet) be clarified, the μ-oxo-bridged dinuclear Ru^{III} complex **5**, commonly referred to as the "blue dimer", represents the first structurally well-defined molecular WOC for which in-depth mechanistic studies are available (Fig. 5.9) [59]. Introduced by Meyer et al. in the early 1980s, **5** was found to catalyze water oxidation by Ce^{IV}, as the oxidant [60–63]. These authors also incorporated **5** via cation metathesis into anionic polymers coatings on glassy-carbon (GC) electrodes; Nafion™ and poly(styrene sulfonate) (PSS) were utilized in these studies [64, 65]. These hybrid electrodes were employed for the catalytic oxidation of Cl^- ions. In the case of the (**5**/PSS)-coated electrode, the oxidation of an 1 M LiCl solution occurred with a current density of 100 mA cm^{-2} and gave a TON of 26,000 per Ru^{III} sites on the electrode before the catalyst was deactivated [64]. The deactivation was believed to be result of an oxidative adnation, which presumably involved the binding of the sulfonate moieties of the polymer to the metal centers.

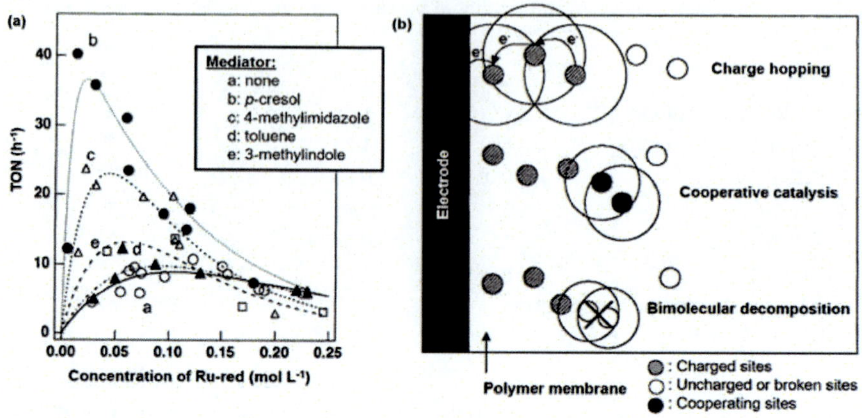

Fig. 5.8 **a** Relationships between the TON of the O_2 evolution and the concentration of Ru-red in the membrane in the absence and presence of a CT mediator [55]. **b** Schematic representation of the three relevant processes in the electrocatalytic WOR using immobilized complexes [56]. Figure reproduced with kind permission; © 1998 American Chemical Society and 1999 Springer Nature

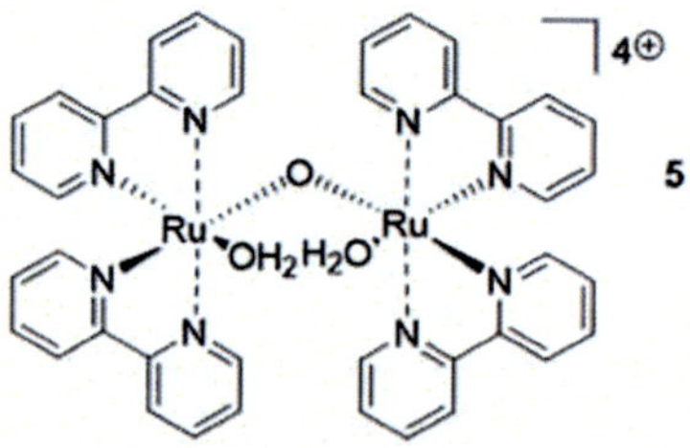

5

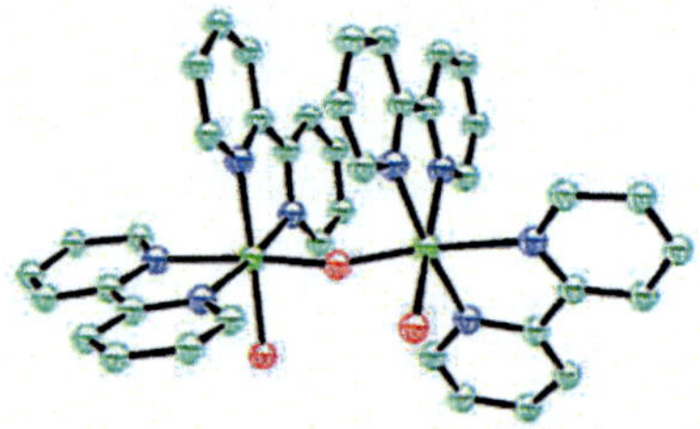

Fig. 5.9 Schematic representation (left) and solid-state structure (right) of the "blue dimer", *cis*, *cis*-$[(bpy)_2(H_2O)Ru^{III}ORu^{III}(H_2O)(bpy)_2]^{4+}$ (**5**) [59]. Figure reproduced with kind permission; © 2008 American Chemical Society

A different electrode configuration was reported by Casella et al. [66]. These authors prepared a Ru^{III}-loaded electrode by emerging a Nafion™-coated GC electrode into a $RuCl_3$ solution. The electrochemically stable electrode was used for the oxidation of hydrazine (or its derivatives) at an operational voltage of +1.0 V. It was further demonstrated that the hybrid electrode could be utilized for the amperometric detection of hydrazine in water samples; this sensor offered a very high accuracy, sensitivity (in the range of 1–13 μM), and recovery values of >95%. The main advantage of this setup, compared to other hydrazine sensors reported in the literature, was the ease of preparation and long-term activity, as revealed by continuous-flow experiments.

One further chemically-modified electrode was fabricated by Zen and Chen, who immobilized lead-ruthenate pyrochlore in a Nafion™-coated GC electrode. The pyrochlore particles were formed inside the polymer layer by oxidative treatment of the Ru^{3+}/Pb^{2+}-loaded electrode, immersed into an alkaline aqueous solution (Fig. 5.10) [67]. This hybrid electrode served as a reliable sensor for dissolved O_2 in both acidic and alkaline environments. Moreover, the selective electrochemical determination of dopamine, in the presence of a high concentration of ascorbic acid, was demonstrated; in this case, a lower detection limit of 0.1 μM was reached [68]. In the following, Zen et al. extended the scope of the electrode towards the amperometric detection of a wide range of inorganic, organic, and biological molecules: hydrazine [69], amitrole [70], acetaminophen [71], caffeine [71, 72], theophylline [73], codeine [74], dopamine [68], serotonin [75], adenine [75], guanine [75], cysteine [76], hypoxanthine [77], xanthine [78], and uric acid [78]. As summarized by Zen and Kumar, square-wave voltammetry or flow-injection analysis were utilized to probe the various analytes. Not only pure analyte solutions were investigated, the applicability to "real" samples (e.g., commercial drugs and beverages, blood or urine) was also shown [79].

Zen et al. also fabricated a hybrid membrane, lead-ruthenate pyrochlore incorporated into Nafion™, via the aforementioned two-step procedure [80]. Microscopy imaging as well as X-ray diffraction (XRD) analysis was conducted; the latter revealed the presence of metal-oxide clusters with an average diameter of 4 nm. The hybrid membrane had a pancake-like agglomerate morphology with the metal-oxides crystallites preferentially exposed to the outer surface (Fig. 5.11a). Remarkably, the sulfonate sites within the membrane remained accessible for a further cation exchange, as shown by the loading of the hybrid membrane with $[Ru(bpy)_3]^{2+}$ complexes. The incorporation of these cations could be proven by solid-state UV/vis absorption spectroscopy—the characteristic metal-to-ligand

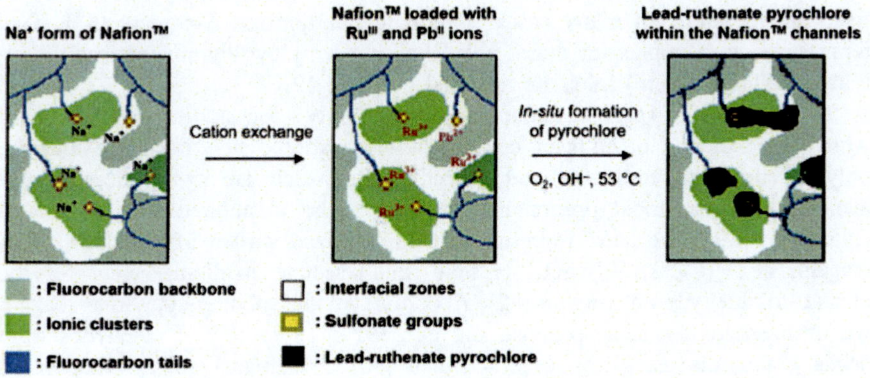

Fig. 5.10 Schematic representation of the immobilization of lead-ruthenate pyrochlore in a Nafion™-coated membrane [79]. Figure reproduced with kind permission; © 2001 American Chemical Society

charge-transfer (MLCT) band of the complex, at *ca.* 475 nm, was detected. Moreover, these authors also tested the catalytic activity of the hybrid membrane in the oxidation of benzyl alcohol with H_2O_2, as oxidant. This test reaction afforded benzaldehyde with >99% chemoselectivity (i.e., over-oxidation to benzoic acid was not observed). The good recyclability of this material was demonstrated by its recovery and reuse in four consecutive runs without a loss of catalytic activity (there was also no indication for leaching of heavy-metal ions into the solution). Later, these authors employed their polymeric membrane reactor for the general alcohol-to-carbonyl oxidation using (NaOCl, as oxidant) [81]. In the case of primary alcohols, an excellent chemoselectivity for the formation of aldehydes was again found (>99%, 16 substrates were tested). Electrochemical studies suggested that the oxidation was mediated by the *in-situ* generated RuO_4^- anions (thus, the mechanism of the reaction might be comparable to that of the Ley oxidation [82, 83]). Furthermore, Zen et al. also used molecular oxygen (or even air), as oxidant, for the room-temperature oxidation of alcohols—primary or secondary as well as aliphatic, allylic or benzylic ones. In general, high yields (>85%, with the exception of some fatty alcohols) and excellent chemoselectivity (>99%) were reported (in total, 22 substrates were screened) [84]. This strategy was later adapted to the oxidation of primary amines—the corresponding nitriles were obtained in moderate to very high yields (30 to 99%, 7 examples) [85]. In a similar fashion, organic sulfides were converted into the corresponding sulfoxides. This selective oxidation proceeded only photocatalytically with O_2, as oxidant; $[Ru(bpy)_3]^{2+}$ was co-incorporated into the hybrid membrane and acted as a photosensitizer [86]. A summary of the various catalytic applications related to the pyrochlore@Nafion™ membrane has been published by Zen et al. [87].

Nafion™ and PSS represent anionic polyelectrolytes, which have been combined with various cationic Ru species (vide supra). However, the orthogonal combination—anionic Ru species and cationic polyelectrolytes—has attracted significantly less attention. A broad range of such synthetic polymers is known from the literature, in many cases containing quarternized *N*-centers [88]. Two representative examples for these, poly(*N*-alkyl-4-vinylpyridinium) and poly(diallyldimethylammonium) salts, are depicted in Fig. 5.12.

Since anionic Ru species are comparably rare, only a few studies are available in which such anions have been combined with cationic polymers. Presumably, polymer-supported ruthenate and perruthenate, which are both commercially available, are the most prominent examples to be mentioned in this context (Fig. 5.13) [89]. The latter represents an immobilized variant of the Ley-Griffith reagent, *i.e.*, $[(C_3H_7)_4N][RuO_4]$, a mild and selective oxidizing agent for the alcohol-to-aldehyde conversion [82]. Accordingly, the polymer-supported reagent was utilized for the same purpose and allowed the synthesis of aldehydes from primary alcohols using O_2, as co-oxidant [90]. Disadvantageously, Ley et al. observed the poor recyclability of the reagent and, thus, the focus-of-interest was shifted to inorganic supports for the $[RuO_4]^-$ anion [91]. Related to this, Friedrich and Singh employed cross-linked poly(4-vinylpyridine) (P4VP), as solid support, for the immobilization of ruthenate [92]. In contrast to perruthenate, non-supported

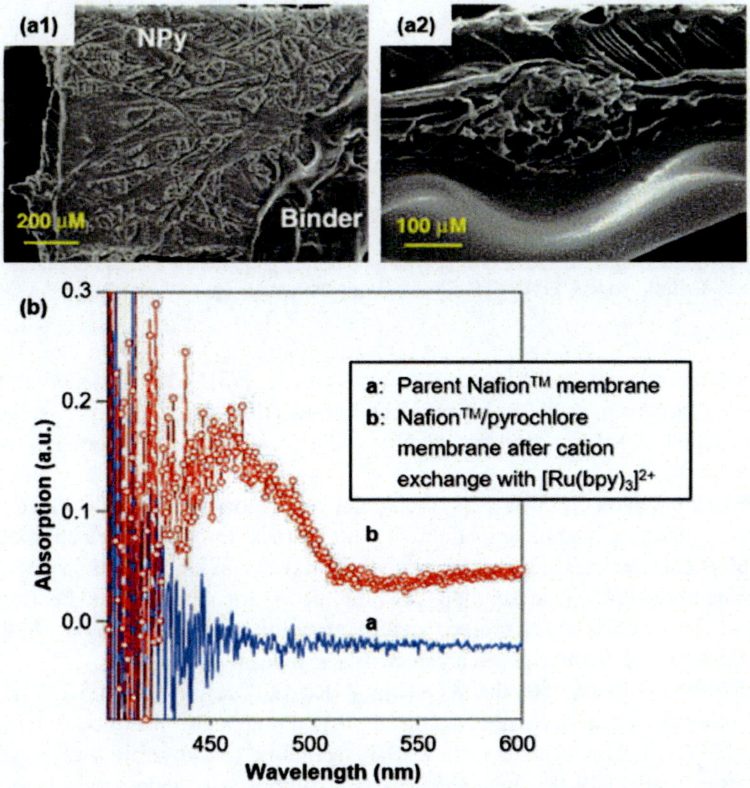

Fig. 5.11 **a** Representative SEM images of the hybrid membrane (a1: top view, a2: cross-sectional view). **b** Solid-state UV/vis absorption spectra of the parent Nafion™ membrane and the hybrid membrane after loading with $[Ru(bpy)_3]^{2+}$ cations [80]. Figure reproduced with kind permission; © 2005 Elsevier B.V

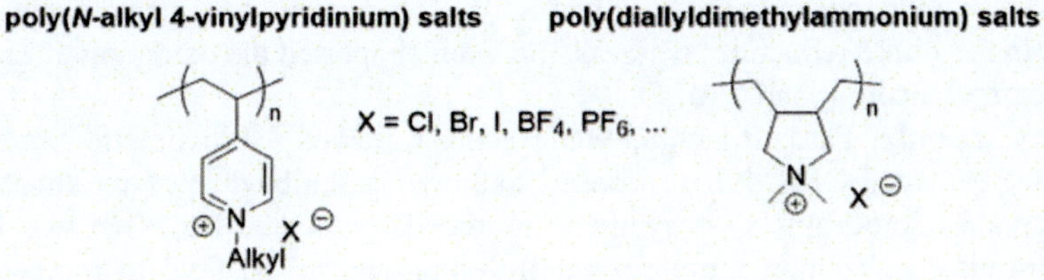

Fig. 5.12 Schematic representation of typical cationic polyelectrolytes

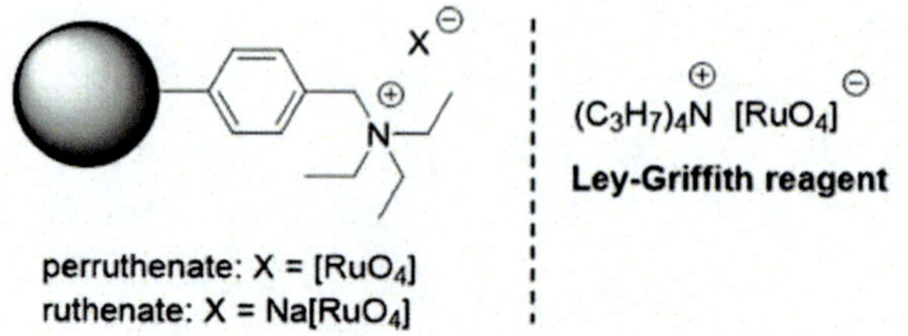

Fig. 5.13 Schematic representation of polymer-supported perruthenate and ruthenante as well as of the Ley-Griffith reagent TPAP (TPAP: tetrapropylammonium perruthenate)

ruthenate is a non-selective oxidant for alcohols [93]. The polymer-supported ruthenate, however, acted as an efficient two-electron oxidant, which selectively formed aldehydes from primary alcohols (PhIO or periodate were the optimal co-oxidants).

Polyoxometalates (POMs), as molecular transition-metal oxide clusters, are currently attracting significant attention due to their remarkable electrochemical, photophysical, and magnetic properties, which have been employed in a wide range of applications [94]. Due to their polyanionic nature, POMs can be bound to cationic polyelectrolytes via electrostatic attraction and, accordingly, functional POM films can be fabricated via an electrostatic layer-by-layer (LBL) self-assembly approach [95]. For example, the WOC $Na_{10}[\{Ru_4O_4(OH)_2(H_2O)_4\},(\gamma\text{-}SiW_{10}O_{36})_2]$, commonly referred to as Ru_4POM, was deposited onto a GC electrode or an indium tin oxide (ITO) glass substrate via a LBL technique (polypyrrole was utilized, as conductive matrix) [96]. Thus, the generated hybrid electrode was tested in the electrocatalytic WOR from a buffered pH-neutral environment (Fig. 5.14). In a related study, Ma et al. fabricated a multilayered film, comprising of Ru_4POM and polyethyleneimine (PEI), by the LBL self-assembly onto quartz or ITO-coated glass substrates [97]. The controlled growth of the film was monitored by UV/vis absorption spectroscopy. Remarkably, this assembly functioned as a bifunctional electrocatalyst, since both subunits of the R_4POM—the $[SiW_{10}]$ and the Ru_4 fragment—could be addressed individually: The former was electrocatalytically active in the nitrite reduction; whereas, the latter supported the oxidation of ascorbic acid, benzyl alcohol, and DMSO.

Very recently, Patil, Bonchio, Mann et al. followed a biomimetic strategy to immobilize Ru_4POM [98]. These authors assembled polymer/nucleotide co-acervate microdroplets, as synthetic protocells, to which Ru_4POM was bound electrostatically. Though more commonly employed in the field of biomedicine, such systems have also been tested as nanoreactors for catalytic applications [99]. It was demonstrated that the membrane-free coacervate microdroplets, assembled from poly(diallyldimethylammonium chloride) and adenosine 5'-triphosphate (ATP), experienced a transformation into membrane-bounded coacervate vesicles

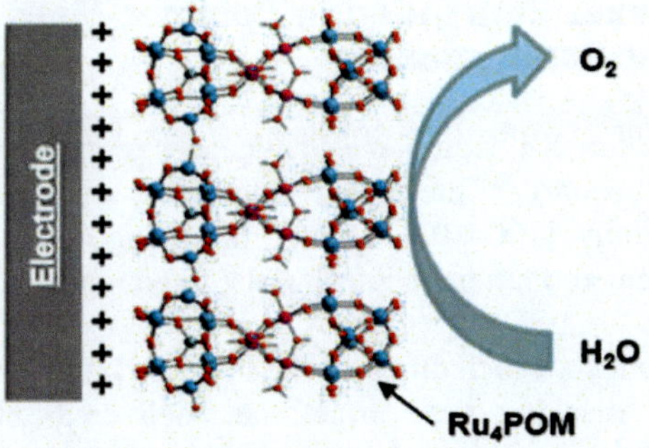

Fig. 5.14 Schematic representation of a hybrid electrode, coated with a bilayer of polypyrrole and Ru_4POM, for the electrocatalytic water-oxidation reaction [96]. Figure reproduced with kind permission; © 2014 American Chemical Society

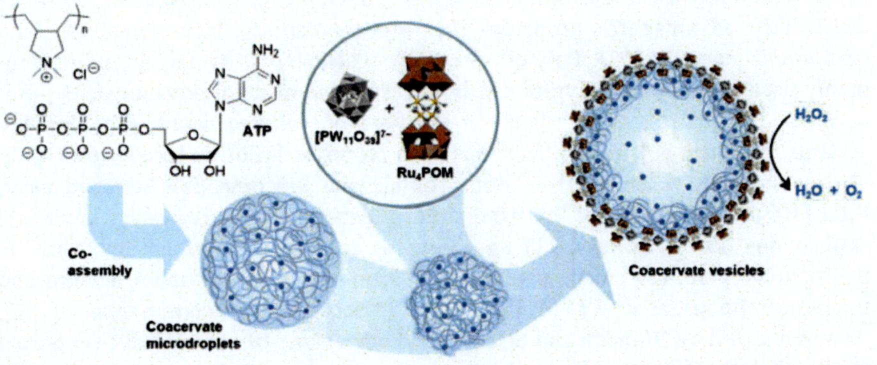

Fig. 5.15 Schematic representation of the two-step self-assembly of functionalized coacervate vesicles, which were utilized, as catalyst, for the oxidation of peroxide [98]. Figure reproduced with kind permission; © 2020 Springer Nature

upon simultaneous addition of the polyanions Ru_4POM $[PW_{11}O_{39}]^{7-}$ (Fig. 5.15) [98]. These vesicles were utilized, as catalyst, for the production of O_2 from H_2O_2. Mainly owing to diffusion effects, the 2nd order rate constant of this reaction was, however, significantly lower compared to Ru_4POM under homogeneous conditions (43.0×10^{-3} M^{-1} s^{-1} vs. 36.8 M^{-1} s^{-1}).

5.3 Ru Complexes Embedded in Polymer Matrices: Biomedical Applications

For a variety of Ru^{III} or Ru^{II} complexes, anticancer properties have been reported and, thus, these "ruthenodrugs" are considered as alternatives to the established Pt^{II}- or Pt^{IV}-containing drugs [100–104]. Though not being commercialized yet, some ruthenodrugs have entered human clinical trials; three representative candidates are shown in Fig. 5.16.

The therapeutic effect of such drugs can be increased if they are protected against degradation before reaching their target, i.e., delivered into the cancer cells (Fig. 5.17). Against this background, polymeric micelles, as carrier, have attracted tremendous interest. Such species, typically comprising of a hydrophobic core and a hydrophilic corona, are formed by the self-assembly of amphiphilic copolymers in aqueous medium above the critical micellization concentration (cmc, Fig. 5.18). Amongst others, poly(L-lactic acid) (PLA), poly(L-lactic-*co*-glycolic acid) (PLGA), poly(ε-caprolactone) (PCL), polyanhydrides, and poly(amino acid)s (PAAs) represent commonly used hydrophobic segments used for the amphiphilic polymers due to their advantageous properties (i.e., biodegradability, biocompatibility, and pH-responsiveness) [105]. Poly(ethylene glycol) (PEG) is, by far, the most commonly used hydrophilic polymer and is—irrespective of its disadvantageous—still considered as the "gold standard" in the field of polymer-based, drug-delivery systems for which a "stealth effect" has been proposed [106]. In recent years, poly (2-oxazoline)s have emerged as water-soluble and biocompatible alternatives to PEG [107]. The benefits of the nano-sized polymer micelles regarding medicinal applications are as follows: (1) an ability to solubilize non-polar drugs in the hydrophobic core, (2) biocompatibility, (3) long-term stability under in vitro and in vivo conditions, as well as (4) the aptitude to accumulate in certain tissues [108]. As summarized by Blunden and Stenzel, also other types of carriers have been used in the context of ruthenodrugs: Liposomes, polymer particles, (polymer-coated) inorganic nanoparticles, single polymer chains, and dendrimers [109].

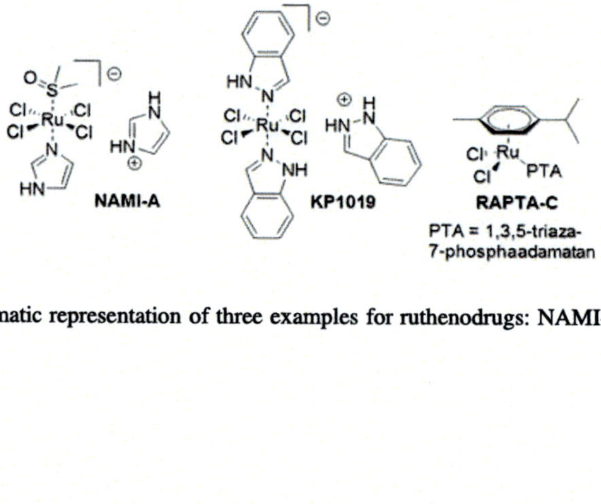

Fig. 5.16 Schematic representation of three examples for ruthenodrugs: NAMI-A, KP1019, and RAPTA-C

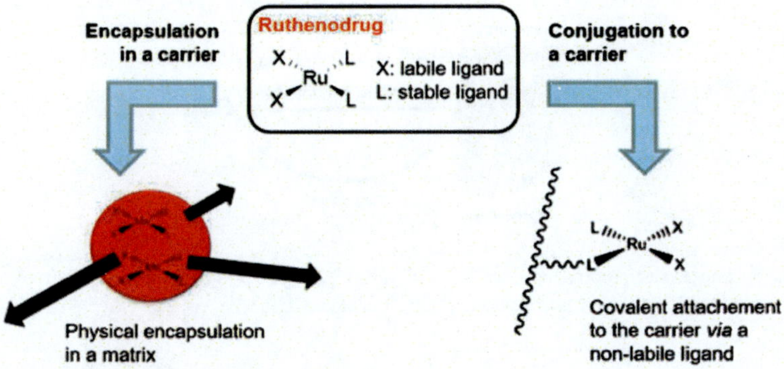

Fig. 5.17 Schematic representation of the two alternative strategies for the immobilization of ruthenodrugs: Covalent attachment to a carrier or encapsulation in a carrier [109]. Figure reproduced with kind permission; © 2014 Wiley–VCH

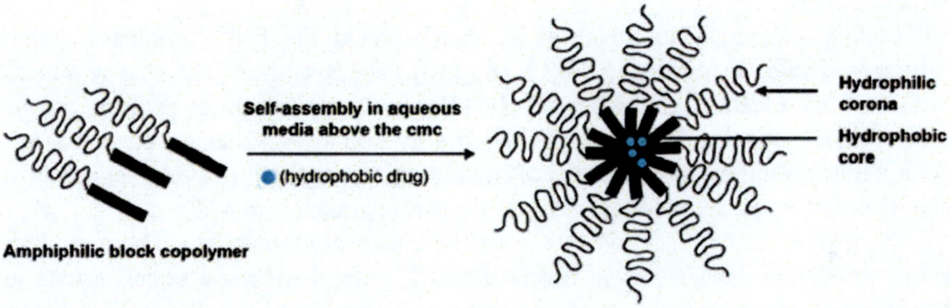

Fig. 5.18 Schematic representation of the formation of a polymeric core–shell micelle, as nanocarrier for drug molecules [108]. Figure reproduced with kind permission; © 2014 Frontiers Media SA

Besides this non-covalent approach, drug molecules have also been attached to polymers via covalent, yet hydrolyzable bonds, thereby forming polymer–drug conjugates. Many examples for such conjugates using biocompatible or biodegradable polymers have already been reported and even successfully utilized in clinical research [110–113]. As shown in Chap. 2, this strategy has also been applied for some ruthenodrugs. These examples include RAPTA-type half-sandwich complexes, which were attached to the side chains of polymers, e.g., of the polyphosphazene, PMMA, and PLA families [109, 114]. Moreover, Wu et al. reported on photoresponsive side-chain polymers for potential applications in photodynamic therapy (PDT) [115].

In the PDT, a photosensitizer is stimulated by irradiation with visible light. Thus, the generated excited state can be deactivated via two different pathways: an O_2-dependent and an O_2-independent one. The former produces highly reactive species

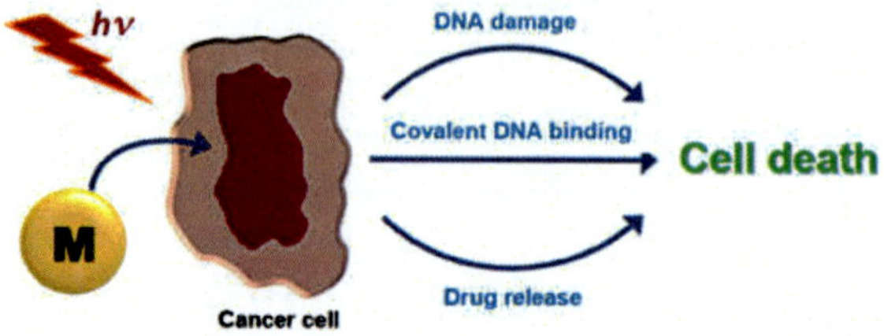

Fig. 5.19 Schematic representation of the principle of phototherapeutic therapy [116]. Figure reproduced with kind permission; © 2015 Elsevier B.V

within the cells, such as $^1O_2$. In the other case, also referred to as photoactivated chemotherapy, electron-transfer processes, covalent binding to the DNA or photoinduced drug delivery may occur (Fig. 5.19) [116]. Against this background, Ru (II) complexes, as potential photosensitizers in PDT, have attracted considerable interest—ruthenoporphyrins and polypyridyl complexes have to be named in this context [117].

Recently, Wang, Wang, Hao et al. encapsulated the Ru^{II} complex **6** into a polymer micelle, according to Fig. 5.18 [118]. These authors used the reversible addition/fragmentation chain-transfer (RAFT) polymerization to prepare amphiphilic diblock copolymers **7** comprising of a hydrophobic PS and a brush-shaped, hydrophilic poly[poly(ethylene glycol) methyl ether acrylate] (PEGA) segment (a). The copolymer with an equal length of the segments (m = 26, n = 28, M_n = 15,480 g mol^{-1}, Đ = 1.29) was used for the encapsulation studies. Dynamic-light scattering (DLS) measurements revealed that **7** formed spherical micelles with an average diameter of 21 nm—with or without encapsulated dye **6** (b). The encapsulation of **6** by **7** was further corroborated by UV/vis absorption and emission measurements, which showed the characteristic features of the Ru^{II} complex within the micelles (c). Furthermore, cell-toxicity studies were performed using human hepatoma cells (SK-HEP-1). Pure **6** exhibited a poor cytotoxicity under dark conditions (IC_{50} value of 1237 ± 124 μM); whereas, the toxicity was increased significantly under light conditions (IC_{50} value of 340 ± 39 μM). These authors ascribed the elevated activity to the light-induced formation of reactive oxygen species (e.g., $^1O_2$). Polymer-encapsulated **6** even showed a higher toxicity in dark conditions, which was strongly enhanced under light conditions (IC_{50} values of 292 ± 21 μM and 92 ± 21 μM, respectively). The results of the cell-toxicity tests are summarized in (d). The aggregation of **6** in aqueous media was suppressed by its encapsulation, thus smaller particles were formed in the latter case. These particles were taken up by the cells more readily and, as a result from this, the activity in the PDT was increased (Fig. 5.20).

The pros and cons of the two alternative strategies—covalent conjugation of ruthenodrugs to polymers or their encapsulation in nanocarriers—were analyzed by Blunden and Stenzel [109]. Regarding the encapsulation of such complexes in polymers, one obstacle is worth mentioning: Many ruthenodrug, such as the

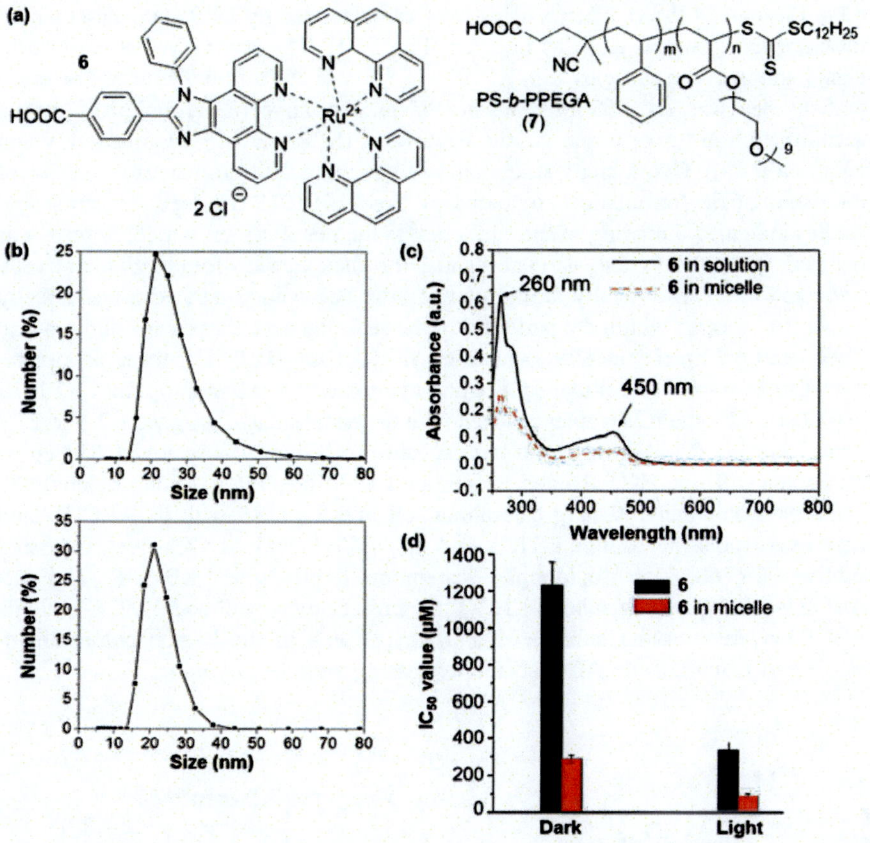

Fig. 5.20 **a** Schematic representation of the Ru^{II} complex **6** and the amphiphilic deblock copolymer **7**. **b** Particle size distributions of the micelles according to DLS measurements (top: **7** alone, bottom: **7** encapsulating **6**). **c** UV/vis absorption spectra of **6** in solution (black curve) and encapsulated in **7** (red curve). **d** Results of the cell-toxicity studies (at the example of SK-HEP-1 cells) [118]. Figure reproduced with kind permission; © 2019 Elsevier B.V

examples shown in Fig. 5.16, are hydrophilic and, thus, their encapsulation in a hydrophobic matrix, e.g., in the core of a micelle, is difficult. Despite this shortcoming various formulations, where ruthenodrugs have been encapsulated in (organic) carriers, have been reported.

A nanoparticle formulation was, for example, prepared with the Ru^{III} drug KP1019 [119]. KP1010 itself is of low stability in aqueous solution, i.e., under physiological condition—a property that is related to the high reactivity of the Ru–Cl bonds. In order to overcome this limitation, Kowol et al. immobilized the ruthenodrug in PLA nanoparticles. Nanoparticle formation was conducted in a single oil-in-water emulsion in the presence of a non-ionic surfactant (i.e., Tween™

80 or Pluronic™ F-68). The particles were characterized by DLS and zeta-potential measurements as well as TEM imaging (Fig. 5.21). The latter revealed an insufficient loading of the particles with KP1019 in the case of Pluronic™ F-68. The mean particle size was *ca.* 165 nm, thus in the proper range (<300 nm) to allow an accumulation in tumor tissue via the EPR effect (EPR: enhanced permeability and retention [120]). Cytotoxicity studies showed an up to 20-fold increased activity of the nanoparticle formulation compared to "free" KP1019. In depth spectroscopic studies indicated a binding of the surfactant to the metal center, which, in turn, was reduced from +III to the +II state. Thus, the nanoparticle formulation not only stabilized the hydrolytically labile ruthenodrug, but even generated a biologically active Ru^{II} species within the particles. In the same context, these same authors also PEGylated polymeric micelles, as carrier, for KP1019 [121]. For these nanomaterials also an improved cellular uptake and cytotoxicity compared to "free" KP1019 was observed—the IC_{50} values reached with the encapsulated drug were 1.6 to 22.7 times lower. In detail, a variety of human cancer cell lines was screened: The colon carcinoma cell line HCT116 and its subline HCT116 (p53/ko) with a deleted p53 gene, the non-small cell lung carcinoma cell line SW1573 with its ABCC1- and LRP-overexpressing subline 2R120, and its ABCB1- and ABCC1-overexpressing subline 2R160, the epidermal carcinoma cell line KB-3–1 and its ABCB1-overexpressing subline KBC-1, the breast cancer cell model MCF-7 and its ABCB1-overexpressing subline MCF-7/adr, as well as the human colon adenocarcinoma Lovo and its ABCB1-overexpressing subline Lovo/dox. However, the

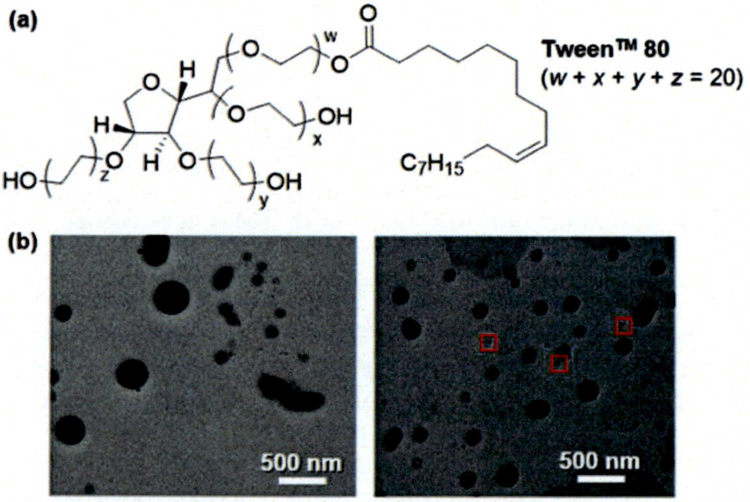

Fig. 5.21 **a** Schematic representation of the non-ionic surfactant Tween™ 80. **b** Representative TEM images of the fabricated nanoparticles using Tween™ 80 (left) and Pluronic™ F-68 (right), as surfactant. Based on a contrast analysis, the marked nanoparticles were not loaded with the ruthenodrug [119]. Figure reproduced with kind permission; © 2014 The Royal Chemical Society

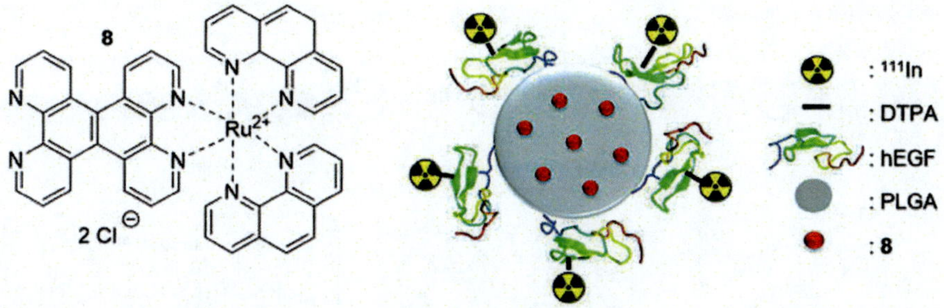

Fig. 5.22 **a** Schematic representation of the radiosensitizer **8** and the multifunctional polymer nanoparticles encapsulating this dye. **b** Representative TEM image of the multifunctional nanoparticles. **c** Top: EGFR levels in normal human fibroblasts (HFF-1) as well as OE21, OE33, and FLO-1 oesophageal cancer cells (determined by immunoblotting of whole-cell lysates with anti-EGFR antibodies. Bottom: Normalized EGFR-to-(β-actin) [122]. Figure reproduced with kind permission; © 2018 The Royal Chemical Society

micelles contained a significant amount of DMSO and possessed a rather low ruthenodrug-to-polymer ratio; thus, this formulation was, despite its promising cytotoxicity, not considered for further in vivo or in vitro studies.

In order to combine the theragnostic properties of radionuclides, the controlled release of chemotherapeutic drugs and the targeting of cancer cells, Vallis et al. designed a radiolabeled, drug-loaded nanoparticle formulation [122]. These authors focused on the DNA-intercalating and radiosensitizing Ru^{II} complex **8**, which was found to be as active as cisplatin towards oesophageal cancer cells, but with reduced cytotoxicity towards healthy cells [123]. Complex **8** was encapsulated in PLGA nanoparticles, which were functionalized on their surface with DTPA binding sites and hEGF moieties (DTPA: diethylenetriaminepentaacetic acid, hEGF: human epidermal growth factor). The former enabled the complexation of the ^{111}In radionuclide to achieve synergies between the therapeutic properties of ionizing radiation (IR) and controlled release of **8**, as chemotherapeutic radiosensitizer [124]. The hEGF moieties were introduced to enable a selective binding to cell-surface EGF receptor (EGFR) sites, which are overexpressed for cancer cells; thus, the delivery of the multifunctional nanoparticles to cancerous cells was targeted [125]. The structure of the particles, for which a mean diameter of *ca.* 133 nm (polydispersity of 0.236) was determined by TEM imaging, is depicted in Fig. 5.22a, b [122]. The radiolabeled nanoparticles were efficiently taken up only by oesophageal cancer cells, exhibiting an overexpression of EFGR; there, radiotoxicity due to cellular DNA damage was observed (Fig. 5.22c). Control experiments demonstrated the synergy of the co-delivery of (^{111}In and **8**) as revealed by an increased cell death compared to nanoparticles containing either ^{111}In or **8**. These results show how the radiolabeling of nanostructures can potentially be utilized in chemotherapy; further, the possibility of multimodality imaging is given—an important technique in modern nuclear medicine and cancer diagnosis [126].

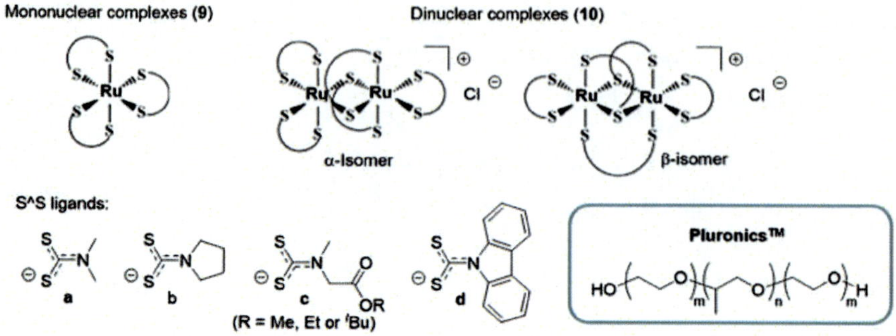

Fig. 5.23 a Schematic representation of the mono- and binuclear Ru^{III} complexes (**9** and **10**) as well as of the amphiphilic triblock copolymer Pluronics™

Nargon, Fregona et al. reported a variety of mono- and dinuclear Ru^{III} complexes bearing aromatic or non-aromatic dithiocarbamate ligands (**9** and **10**, Fig. 5.23). Though these complexes exhibited considerable anti-cancer activity, their solubility under physiological conditions represented a drawback [127–130]. In order to circumvent this restraint, these same authors encapsulated their complexes in the hydrophobic core of micelles, as nanocarriers (according to Fig. 5.18); the triblock copolymer Pluronic™ F127, as a non-ionic surfactant, was used in this study (M_n = 12,600 g mol^{-1}; Fig. 5.23) [131]. Complexes **9b/d** and **10b/d** were successfully encapsulated within the micelles and, upon storage in an aqueous medium mimicking physiological condition, leaching of Ru^{III} species was not observed. The "free" complexes as well as their nanoformulations were tested regarding their in vitro cytotoxicity against HeLa (cervix adenocarcinoma) and HCT116 (colon carcinoma) cancer cell lines (cisplatin was used as reference). In particular, encapsulated **9d** and **β-10d** exhibited a significant increase in activity compared to their "free" versions; in general, all these formulations were more cytotoxic against the HeLa cells; however, encapsulated **9b** and **10b** did not show any relevant activity against the tested cell lines.

Very recently, de Oliveira Siolva et al. reported the encapsulation of dinuclear mixed-valence $Ru^{II/III}$ complexes in a terpolymer formulation [132]. The paddlewheel-shaped complexes, for which anti-tumor activity was shown earlier, **11** contained non-steroidal anti-inflammatory drugs, as bidentate ligands (Fig. 5.24) [133]. The terpolymer comprised of poly(methacrylate) (PMA) and Tween™ 80 chain, which were grafted onto starch—nanoparticles, prepared from this polymer, were found to play an important role in in vivo experiments, such as the suppression of the surfactant displacement in blood proteins or the encapsulation of drugs and imaging agents with respect to cancer therapy and diagnostics [134]. The fabricated nanoparticles, which also contained ethyl arachidate, as non-ionic surfactant, were and characterized by DLS and zeta-potential measurements as well as AFM imaging (Fig. 5.24) [132]. The thusly determined parameters, such as the

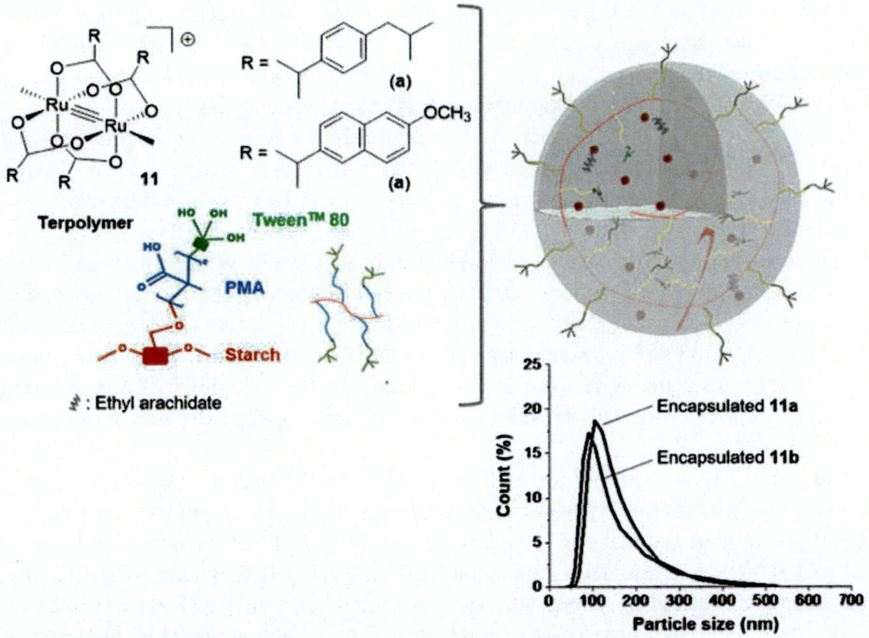

Fig. 5.24 Schematic representation of the terpolymer polymer nanoparticles with contained complexes **11** and ethyl arachidate. The size distribution, as observed in DLS measurements, is also shown [132]. Figure reproduced with kind permission; © 2020 Elsevier B.V

average particle size of *ca.* 130 nm and zeta potential of *ca.* −40 mV, were in the proper range to enable a delivery into tumor cells. Moreover, their colloidal stability under physiological conditions suggested a potential administration via intravenous injection. For both nanoformulations, an enhanced antiproliferative effects in U87MG cells was observed; whereas, only the nanoparticles containing **11a** also exhibited an activity against the cisplatin-resistant T98G cancer cells.

5.4 Ru Complexes Embedded in Polymer Matrices: Opto-Electronic Applications

Similar to the fundamental studies dealing with the embedding of charge-neutral Ru^{II} complexes in apolar (or weakly polar) polymer matrices (see Sect. 1.2), also cationic polypyridyl complexes combined with anionic polyelectrolytes has been investigated.

The (presumably) first report in this respect dates back to 1959, when Jensen et al. analyzed the binding of various cationic complexes to PSS by dialysis methods; *inter alia*, $[Ru(phen)_3]^{2+}$ was studied in this context (phen:

1,10-phenanthroline) [135]. For this particular system, an intrinsic binding constant of *ca.* $2 \times 10^3\ M^{-1}$ was estimated. Later, Kurimura et al. established a model for the binding of $[Ru(bpy)_3]^{2+}$ to PSS [136]. These authors proposed a cooperativity between electrostatic and hydrophobic interactions to account for the strong binding of the complexes to the polymer. It was demonstrated that the adsorption of $[Ru(bpy)_3]^{2+}$ from an aqueous solution by cation-exchange resins was decreased drastically by its binding to the anionic polyelectrolyte. However, in the presence of NaCl, the binding cooperativity was repealed and, consequently, $[Ru(bpy)_3]^{2+}$ could be loaded onto the resins. A commercial cation-exchange resin, loaded with $[Ru(bpy)_3]^{2+}$ and MV^{2+} cations, enabled the light-induced reduction of molecular oxygen to H_2O_2 in water (MV^{2+}: methyl viologen) [137]. Later, Tuite et al. applied molecular-dynamics simulations to show that the PSS chains adopted an open-chain conformation in water with local curvatures around the complexes. Thus, the $[Ru(bpy)_3]^{2+}$ moieties were still exposed to water, though significantly less than in bulk solution [138].

The luminescence properties of $[Ru(bpy)_3]^{2+}$, in the presence of anionic polyelectrolytes, was studied by Meisel and Matheson [139]. These authors showed that the photo-excited state of the complex, though only in the presence of poly(vinyl sulfate) (PVS), was efficiently quenched by Cu^{II} ions. It was concluded that the quenching proceeded by energy-transfer processes between $[Ru(bpy)_3]^{2+}$ and Cu^{II} sites, coordinated to the polymer in close proximity to each other. This fundamental study already forestalled the assumptions made later by, e.g., Fréchet et al. [140] or Meyer et al. [141] regarding their polymers with covalently attached Ru^{II} complexes and redox-active chromophores (see Chap. 2). However, the presence of PVS alone did not affect the emission behavior of $[Ru(bpy)_3]^{2+}$ in aqueous solution [139]. Similar observations were also made in the presence of other anionic polyelectrolytes [i.e., poly(acrylic acid) and poly(methacrylic acid)]; whereas, the emission intensity of $[Ru(bpy)_3]^{2+}$ was increased when PSS was added to an aqueous solution [136]. This observation was rationalized by the rather confined environment around the complex established by the aforementioned cooperative non-covalent interactions (Fig. 5.25). The strength of the hydrophobic interactions was later quantified by Slama-Schwok and Rabani, who performed laser-pulse studies on the system $[Ru(bpy)_3]^{2+}$/PSS. The determined change in the free energy (ΔG^0) was in the same range as reported for the formation of polymer micelles [142]. According to Miyashita et al., microdomains comprising of up to 43 monomer-repeat units per $[Ru(bpy)_3]^{2+}$ complex were formed, reminiscent of a typical micelle domain [143].

Turro, Barton et al. elaborated that the luminescence enhancement of Ru^{II} complexes in the presence of anionic polyelectrolytes correlated with the strength of their interactions; in particular, the hydrophobic forces were identified as the main contributor in this respect [144]. Three heteroleptic complexes {i.e., $[Ru(bpy)_3]^{2+}$, $[Ru(phen)_3]^{2+}$, and $[Ru(dip)_3]^{2+}$} were combined with the sodium salts of PSS and PAA—from these, the systems $[Ru(dip)_3]^{2+}$/PSS, $[Ru(dip)_3]^{2+}$/PAA and $[Ru(phen)_3]^{2+}$/PSS gave the highest emission enhancement (dip: 4,7-diphenyl-1,10-phenanthroline). In these environments, the complexes were also

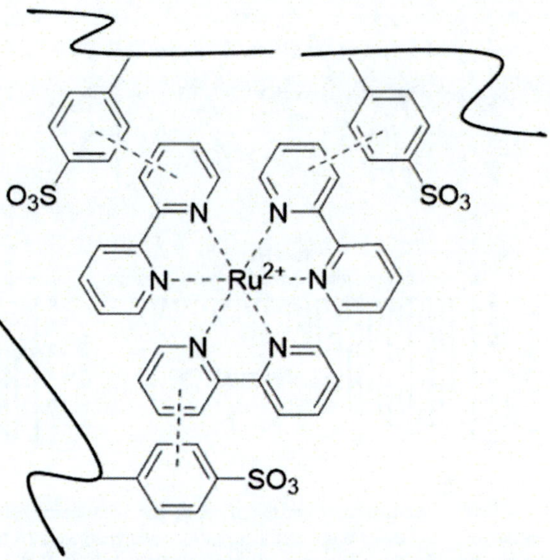

Fig. 5.25 Schematic representation of the proposed binding of $[Ru(bpy)_3]^{2+}$ to PSS via electrostatic and hydrophobic interactions

protected against quenching, at least in the case of anionic and neutral quenchers {e.g., $[Fe(CN)_6]^{4-}$ and O_2, respectively}.

More recently, Moore et al. demonstrated that anionic polyelectrolytes not only stabilized the excited states of Ru^{II} complexes in water, but also enhanced their photocatalytic performance in the hydrogen-evolution reaction (HER) from water under ambient conditions (TON values of *ca.* 30 were reached in these experiments using complex **12**, as the photocatalyst; Fig. 5.26a) [145]. Aggregation of **12** and PSS, as shown by cryo-TEM imaging (Fig. 5.26b), was proposed to increase the ^{3}MLCT excited-state lifetime and protect the photocatalyst from quenching by O_2 due to the strong ionic field established around the charged polymer. These same authors applied isothermal titration calorimetry (ITC) to gain further insight into the binding between Ru^{II} complexes and anionic polyelectrolytes [146]. The rather weak interaction between $[Ru(bpy)_3]^{2+}$ and PVS, as already proposed by Meisel and Matheson on the basis of their photophysical studies, was verified by ITC, which gave only low exothermic heats; whereas, in the case of PSS larger values were obtained ($\Delta H = -21 \pm 1$ kJ mol^{-1}, Fig. 5.26c). The thermogram was fitted using an independent site model; from this, a binding constant (K_a) of $4.3 \pm 0.6 \times 10^5$ M^{-1} was calculated {stoichiometric ratio of *ca.* 3.8 monomer-repeat units per $[Ru(bpy)_3]^{2+}$ site}. The remarkably high K_a value compared to the one obtained for the system $[Ru(bpy)_3]^{2+}$/DNA was rationalized by the polyelectrolyte's higher hydrophobicity and chain mobility. For the isothermal titration of **12** into an aqueous solution of PSS, an additional binding event occurred at high complex concentrations: Beyond the point-of-equilibrium of the titration, charge neutralization induced a conformational change of the PSS chains, which eventually gave the aggregates that were visualized by TEM imaging (Fig. 5.26b).

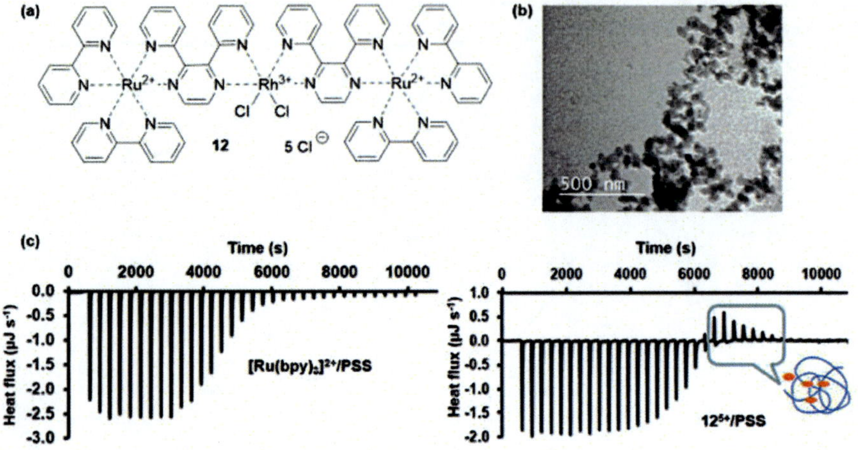

Fig. 5.26 a Schematic representation of the trinuclear, heterobimetallic complex **12**. **b** Representative cryo-TEM image of the aggregates formed by **12**/PSS in water [145]. **c** Thermograms of the raw heat for the isothermal titration of $[Ru(bpy)_3]^{2+}$ (left) or **12** (right) into a solution of PSS in water (25 °C, corrected for dilution effects) [146]. Figure reproduced with kind permission; © 2015 Wiley–VCH and 2017 American Chemical Society, respectively

This process was accompanied by a change from an exothermic to an endothermic heat flow in the thermogram—a feature, which was absent for $[Ru(bpy)_3]^{2+}$. The data were fitted using a two-stage model to account for the two sequential binding events and the aggregation process. For the binding of **12** to the elongated PSS chains via strong electrostatic and hydrophobic interactions, a binding enthalpy of -29 ± 1 kJ mol^{-1} and stoichiometry of *ca.* 6.5 repeat units per complex were determined. Due to the stronger non-covalent interactions in the system **12/PSS**, a significantly higher K_a value of *ca.* 1.3×10^6 M^{-1} compared to $[Ru(bpy)_3]^{2+}$/PSS was found.

To sum up the aforementioned considerations, the general conclusions, to be drawn when combining Ru^{II} complexes with anionic polyelectrolytes, are as follows [147]:

(a) The presence of hydrophobic groups (such as the phenyl rings in PSS) enables a strong binding of the complexes via electrostatic and hydrophobic interactions.
(b) The more hydrophobic ligands of the complexes are, the stronger is the interaction with the hydrophobic units in the polymers.
(c) Polymers in a densely coiled conformation bind the complexes stronger than polymers in an extended conformation (e.g., PMA at pH values <5 vs. PMA at pH values >5).

(d) The stronger the bound complexes are imbedded within the polymer structure, the stronger their emission intensity is increased and tendency to be quenched (e.g., by MV^{2+}, O_2 or Cu^{II} ions) is reduced.

A summary of the early contributions in the field of polyelectrolytes with respect to potential applications in the fields of photochemical energy conversion or storage was presented by Rabani in 1991 [148]. The usage of Ru^{II}–polypyridyl complexes, in particular of the $[Ru(bpy)_3]^{2+}$ family, in solid-state electroluminescent devices was, for the first time, reviewed by Malliaras et al. in 2003 [11]. These authors pointed out that such an application requires the efficient transport of electrons and holes in the solid state, i.e., within the thin films. As important, intense luminescence from the films is highly desirable. Basically, the ElT in such films follows an electron hopping via a sequence of redox reactions; in this process, the LUMOs of the complexes are involved (the same holds true for the hole migration, which proceeds through the HOMOs). The luminescence results from $\pi^* \rightarrow t_{2g}$ ligand-to-metal transitions [149]. Moreover, the counterions accompanying the cationic Ru^{II} complexes provide ionic conductivity due to their mobility within the films at ambient temperatures; thus, the anions, such as PF_6^- or ClO_4^-, also have a profound impact on the device characteristics. When compared to the original light-emitting electrochemical cells (LECs), in which salts are dispersed in an organic semiconductor [150], the devices resulting from the fabrication of Ru^{II} complexes into solid-state exhibit a significantly higher stability (Fig. 5.27). However, crystallization of the material needs to be suppressed either by introducing appropriate substituents (e.g., bulky or flexible alkyl residues), dispersing the complexes in polymer matrices or—as discussed in the previous Sections—by immobilization onto polymers. Advantageously, the synthetic effort is rather low when the Ru^{II} complex and the polymer are prepared separately and mixed in the desired ratio prior to the processing from solution by spin-coating or inkjet-printing techniques. On the other hand, segregation and aggregation of the charged complexes in the polymer matrix might still occur—as a consequence, the performance of the device will be decreased significantly due to concentration quenching. Many examples for using blended systems in LECs, as one of the simplest type of electroluminescent device comprising an active layer of the emitter and of an electrolyte, can be found in the literature [11, 14].

The first example for LECs, based on the Ru^{II} complex **13**, were reported by Rubner et al. in 1996 [151]. Single-layer as well as multilayer LECs were fabricated with ITO-coated glass and Al, as the electrodes. In the case of the multilayer LEC, negatively charged **13** was combined with the polycation PEI. However, the single-layer LEC of the composition ITO|**13**|Al gave a very poor quantum efficiency of 0.005%. A significant increase in the device performance was achieved by changing to the heterostructure design ITO|PPV|**13**|Al in which the additional PPV layer facilitated the hole injection into the light-emitting layer. This set-up gave a luminance of 70–110 cd m^{-2} at current densities in the range of 400 mA cm^{-2}; the efficiency for the orange-red emitting device was 0.05% (Fig. 5.28). In subsequent work, Rubner et al. investigated the influence of various types of polymers, with

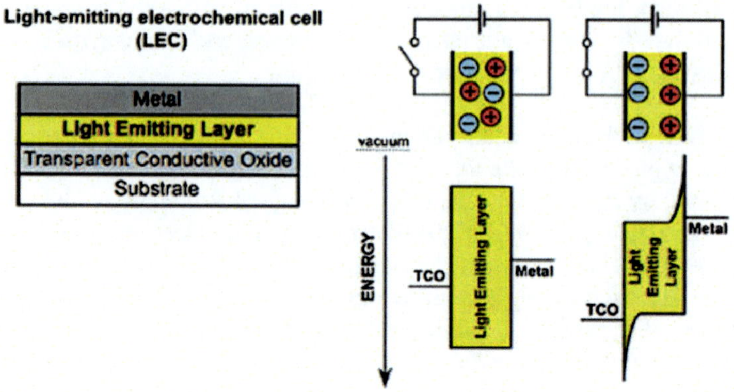

Fig. 5.27 Left: Schematic representation of a single-layer LEC. Right: The movement of the ions in the active layer under an applied bias enables an efficient charge-carrier injection from air-stable electrodes [14]. Figure reproduced with kind permission; © 2014 Elsevier B.V

glassy or polyanionic character, on the performance of LECs based on the luminescent $[Ru(bpy)_3]^{2+}$ complex [152–155]. It was shown that a LEC fabricated from $[Ru(bpy)_3](PF_6)_2$ with an Al cathode gave an external quantum efficiency (EQE) in the range of 1.2–1.5% at 100 to 1000 cd m^{-2} [154]. Degradation in time was observed when the LEC was kept in the off-mode even under inert conditions. Blending with glassy polymers, such as PMMA, polycarbonate (PC) or PS, led to an improved film quality and the degradation was significantly slowed. Moreover, the EQE reached values of up to 2.5% with a light output of 200 cd m^{-2}; the EQE could however be enhanced to 3% when metallizing an Ag layer, as the cathode. Noteworthy, flexible devices can be obtained in similar fashion, when ITO-coated polymer foils, e.g., polyester substrates, are employed [156].

A different type of polymer matrix was employed by De Cola et al. [157]. A polyphenylenevinylene (PPV) derivative, as a semiconducting polymer, was combined with dinuclear Ru^{II} complexes, which acted both as triplet emitter and electron-transfer mediator (e.g., **14**, Fig. 5.29). The LEC, containing a blend of these two components, as active layer, displayed a switchable emission color: At a forward bias of +4.0 V, the excited state of the complex was populated, giving rise to its characteristic red emission; whereas, reversing the bias to −4.0 V led to the emission of green light (attributed to the population of the lowest excited singlet state of the PPV host (Fig. 5.29). About the same time, Chao et al. reported on a flexible light emitting device in which a hexa*kis*phenyl-substituted Ru^{II}–bpy *tris*-complex, as emissive dye, was doped into a blend of poly(*N*-vinylcarbazole) (PVK) and 2-*tert*-butylphenyl-5-biphenyl-1,3,4-oxadizole (PBD). These two host materials facilitated the hole- and electron-transfer processes within the fabricated device [158]. In a similar fashion, Yang and Gordon employed a heteroleptic Ru^{II} complex, as dopant in a PVK matrix [159]. The initially poor performance of the device incorporating this material, as emissive layer, was increased by one order of magnitude by

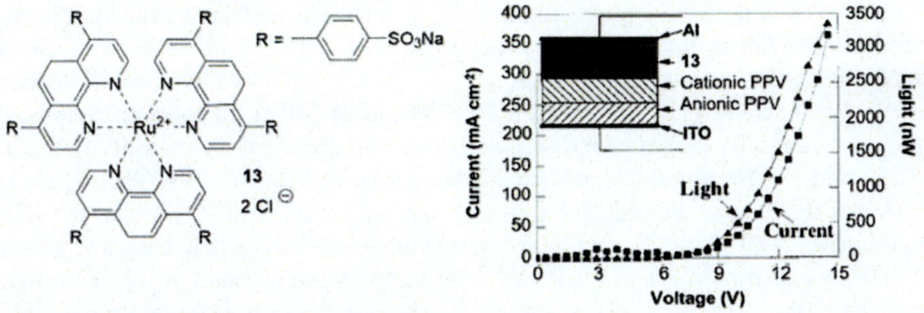

Fig. 5.28 Schematic representation of the Ru^{II} complex **13**, which was incorporated, as an emissive material, into LECs. The light-voltage and current–voltage characteristics of the LEC with the architecture ITO|**13**|PPV|Al are also shown [151]. Figure reproduced with kind permission; © 1996 AIP Publishing

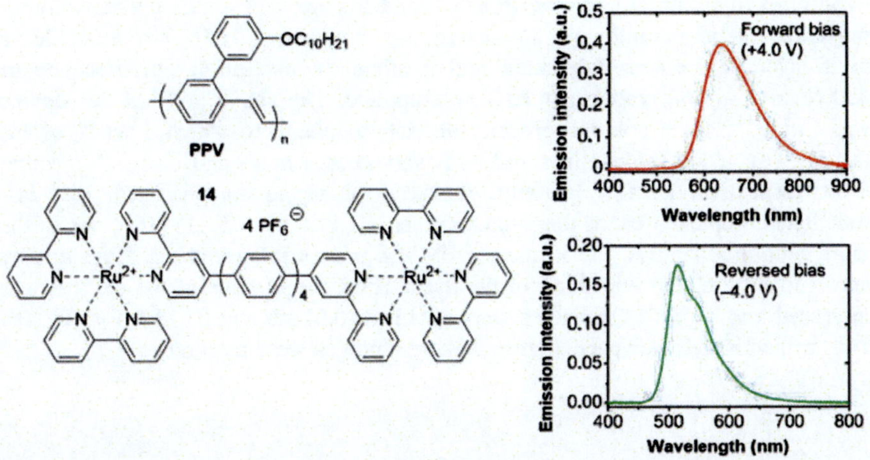

Fig. 5.29 Schematic representation of the dinuclear Ru^{II} complex **14** and the PPV, which were used, as a blended active layer, in a LEC. The electroluminescence spectra of this LEC are also depicted (top: forward bias, bottom: reversed bias [157]. Figure reproduced with kind permission; © 2003 Nature Publishing Group

employing an additional charge-balancing layer of *tris*(8-hydroxyquinolinato)aluminum and 2,9-dimethyl-4,7-diphenyl-1,10-phenanthroline. A maximum efficiency of 0.14% was reached under optimized conditions. When dispersing the dye in a mixture of PVK and PBD, the charge-trapping effect enabled the direct exciton formation from the Ru^{II} complex [160]. A single-layer electroluminescent device of the configuration ITO|(Ru^{II} dye/PVK/PBD)|Ba was fabricated (EL maximum at 640 nm) [161]. In this study, an ester-substituted $[Ru(bpy)_3]^{2+}$ complex was employed, as a pure red emitter, and an EQE of 3.0% was obtained with a luminous efficiency of 2.4 cd A^{-1} and a maximum brightness of 935 cd m^{-2}. Ma et al. tested further semiconductor-type polymers, as potential host materials (e.g., polyfluorenes and polyphenylenes) [162]. However, PVK was identified as the most efficient host polymer due to its longer excited-state lifetime and chemical compability with the Ru^{II} emitter. The multilayer device of the configuration ITO|PVK/Ru^{II} dye|PBD|Alq_3|LiF|Al gave red light with an emission maximum of 612 nm and the CIE coordinates of (0.62, 0.37); for this setup, maximum luminance efficiency and power efficiency of 8.6 cd A^{-1} and 2.1 lm W^{-1}, respectively, were reported [163]. Varlikli et al. emphasized the necessity of long alkyl substituents to enhance the compatibility of the complexes with the polymer matrix, i.e., suppression of dye aggregation [164]. The Ru^{II} complexes **15** were incorporated into a multi-layer device in which 2,2',2"-(1,3,5-benzinetriyl)-*tris*(1-phenyl-1*H*-benzimidazole) (TPBi) was employed as additional electron-transport material (Fig. 5.30).

Not only Ru^{II} complexes of the *tris*-bpy or *tris*-phen family have been employed in light-emitting devices. In 1998, Tripathy et al. demonstrated that the performance of polymer light-emitting diodes (PLEDs) can be increased by adding a transition or rare-earth metal complex to the active polymer layer [165]. For example, a multi-layer device was fabricated and exhibited a maximum luminescence of 430 cd m^{-2} when a voltage of 10 V was applied. The active layer of this device was formed from a solution-processable polythiophene to which 1 wt-% of Ru$(acac)_3$ was added (polyaniline and Alq_3 served as a hole- and electron-injecting layer, respectively). More recently, Heinze et al. tested the push–pull Ru^{II} *bis*-tridentate complexes **16**, as deep-red emitters in LECs (Fig. 5.31) [166]. As in the aforementioned studies, the insulating PMMA matrix improved the film's quality and reduced the self-quenching of the dyes. Thus, the lifetime of the devices was increased and higher EQE values were obtained (0.028% and 0.013% for **16a** and **16b**, respectively); however, higher driving voltages were required.

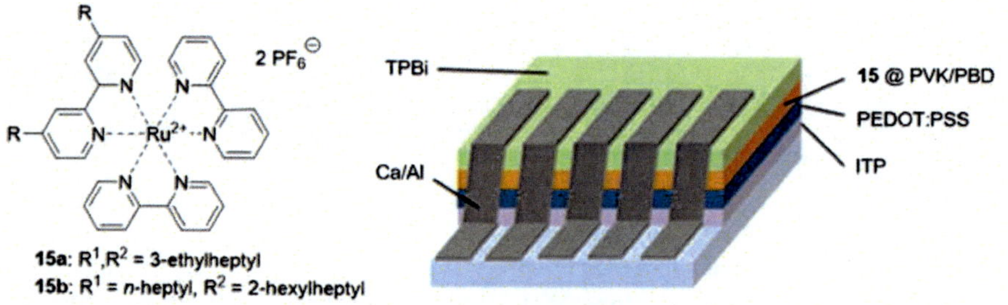

Fig. 5.30 Schematic representation of the Ru^{II} complexes **15**, which were incorporated, as orange-red emitters, into a multi-layer EL device [164]. Figure reproduced with kind permission; © 2012 Elsevier B.V

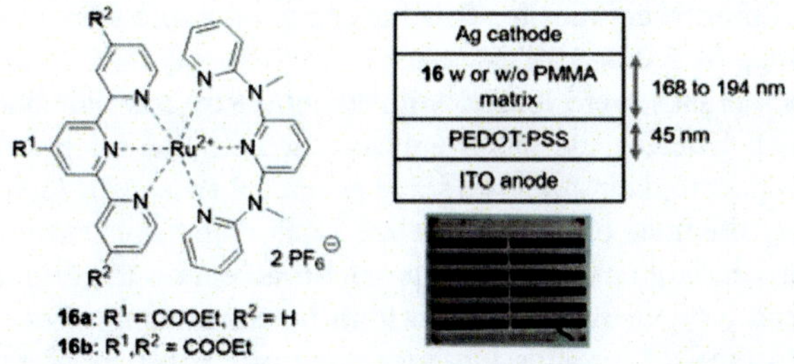

Fig. 5.31 Schematic representation of the Ru^{II} *bis*-tridentate complexes **16**. A schematic representation and a picture of the fabricated LEC are also depicted [166]. Figure reproduced with kind permission; © 2014 Wiley–VCH

As mentioned above, Nafion™ represents an important anionic polyelectrolyte and cation-exchange membrane, which has been combined with a broad range of cationic species—ions as well as complexes (Fig. 5.5) [34]. In this context, the pronounced cation-exchange ability of Nafion™ was also employed to immobilize $[Ru(bpy)_3]^{2+}$ via electrostatic interactions (Fig. 3.7). As already noted in Sect. 3.2, in the presence of oxalate, electrogenerated chemiluminescence (ECL) was observed upon oxidation of the Ru^{II} centers within the membrane. ECL occurred over a wide concentration range with 1 μmol of oxalate representing the lower limit (alternatively, NADH or alkylamines could be used for the same purpose) [167]. Bard et al. investigated the mechanism for the involved charge-transport and concluded from the observed diffusional behavior that both physical diffusion of the redox-active species as well as ElT processes contributed to this [168, 169]. Remarkably, cathodic ECL was observed from a similarly modified graphite oxide electrode in water [170]. Although the reduction of $[Ru(bpy)_3]^{2+}$ is basically disabled in an aqueous environment (even when using a graphite electrode featuring a large overpotential for hydrogen evolution) [35], the immobilization into Nafion™ circumvented this limitation—ECL of the complex at −1.80 V, in presence of peroxodisulfate, was attributed to the occurrence of an one-electron reduced $[Ru(bpy)_3]^{+}$ species. Starting from these pioneering studies on the system Nafion™/$[Ru(bpy)_3]^{2+}$, a broad range of analytical applications has been developed targeting the detection of diverse analytes by ECL stimulation or quenching [171]: Oxalate and organic acids, amines, amino acids and proteins, pharmaceuticals as well as various inorganic compounds. For example, the highly sensitive detection of dopamine and noradrenaline, based on ECL inhibition, even in the presence of interfering compounds (such as ascorbic acid), was enabled using a $[Ru(bpy)_3]^{2+}$-loaded hybrid CNT/Nafion™ film adsorbed on a GC electrode [172]. A highly stable and sensitive ECL biosensor for the detection of EtOH was prepared by Lee et al. [173]. These authors assembled a two-layer material in which the inner layer ionic-liquid doped Nafion™/$[Ru(bpy)_3]^{2+}$ film, was surrounded by a film of alcohol dehydrogenase within a TiO_2-Nafion™ composite. This biosensor offered a very low detection for EtOH in the range of 5.0×10^{-6} M.

Nafion™ membranes were loaded with various Ru^{II} *tris*-bidentate polypyridyl complexes [174]. These materials were screened in the dried-state as well in solution with respect to their ability to serve as optical sensor for molecular oxygen. The excited-state lifetimes of the immobilized Ru^{II} complexes ranged from 0.03 to 4.9 μs, reflecting the influence of the microenvironment on the luminescent probes (i.e., location and oxygen accessibility within the membrane). For the membrane loaded with $[Ru(dip)_3]^{2+}$ an emission-quenching rate constant of *ca.* 1.7 M^{-1} s^{-1} was determined (dip: 4,7-diphenyl-1,10-phenanthroline). Utilizing this particular Ru^{II} dye, a highly O_2-sensitive device (thickness of 178 μm), suitable for the continuous monitoring in solution or in the gas phase, was fabricated and featured response times below 1.5 min. Beyond the various analytical application, also an opto-electronic application is worth mentioning. Mitchell et al. fabricated a PLED in which two polymer layers were incorporated, the Nafion™/$[Ru(bpy)_3]^{2+}$ systems and PPV [175]. The significant role of Ru^{II} polypyridyl complexes, immobilized in Nafion™ membranes, is revealed by the enormous amount of scientific literature dealing with this topic (>450 publications by March 1st 2021 according to SciFinder™). Thus, a detailed evaluation of these goes far beyond the scope of this article.

Recently, Comton et al. reported on the preparation of Nafion™ nanoparticles, doped with $[Ru(bpy)_3]^{2+}$, and their potential application, as electrocatalysts [176]. The nanoparticles were synthesized by a re-precipitation procedure according to Fig. 5.32a and analyzed by UV/vis absorption spectroscopy, CV measurements as well as SEM imaging. The latter revealed a broad size distribution and an average particle diameter of 0.40 ± 0.46 μm (Fig. 5.32b). These authors also applied the Ru^{II}-loaded nanoparticles, deposited onto a carbon electrode, for the electrocatalytic oxidation of oxalate at a potential of +1.1 V (Fig. 5.32c). This reaction was, by a factor of 50, more efficient as when using non-modified electrodes.

Elabd et al. demonstrated that Nafion™, though only in the presence of PAA, can be electrospun into nanofibers [177]. It was proposed that the addition of PAA to a Nafion™ solution would change the ionic strength of the solvent; as a result, the typical aggregation of the polymer chains was suppressed, the viscosity of the solution was increased, and finally, the polymer-chain entanglement became enhanced. Thus, proper conditions for nanofiber formation via electrospinning were established. Xiao et al. utilized this strategy to fabricate Nafion™ nanofibers in which $[Ru(bpy)_3]^{2+}$ complexes were immobilized (Fig. 5.33a) [178]. SEM imaging revealed the formation of a 3D porous membrane in which the nanofibers were randomly oriented (Fig. 5.33b). The nanofibers possessed an increased surface-area-to-volume ratio and were able to accommodate a larger number of complexes when compared to conventional Nafion™ membranes. The strong binding of $[Ru(bpy)_3]^{2+}$ gave rise to an enhanced (by a factor of *ca.* 15) and sustainable ECL behavior in solution (the signal was stable over >50 continuous CV scanning cycles). The nanofibers were also deposited onto electrodes (i.e., Au, Pt, GC, and ITO); these hybrid electrodes also showed intense ECL. These authors

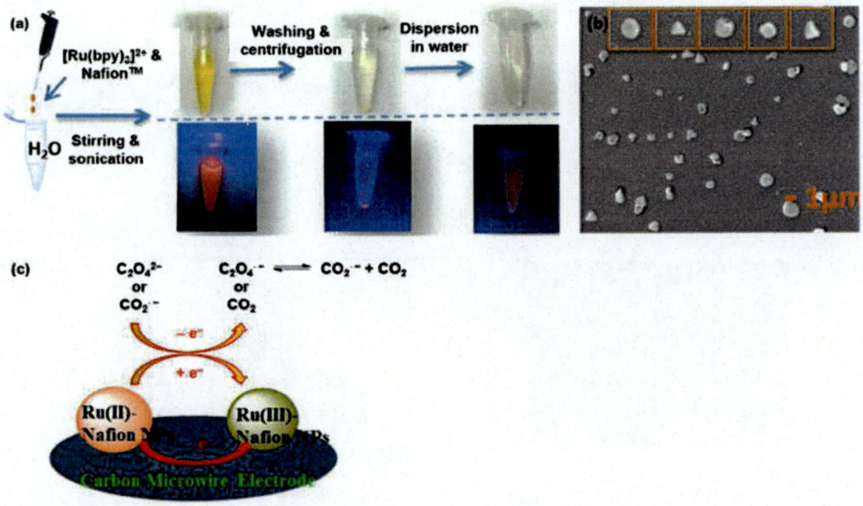

Fig. 5.32 **a** Schematic representation of the synthesis of $[Ru(bpy)_3]^{2+}$-doped Nafion™ nanoparticles. Images of the samples under white light (top) and UV light (bottom) are also depicted. **b** Representative SEM image of the $[Ru(bpy)^3]^{2+}$-doped Nafion™ nanoparticles. **c** Schematic representation of the electrocatalytic oxalate oxidation [176]. Figure reproduced with kind permission; © 2017 Wiley–VCH

used their nanofibers for the detection of phenol derivatives, which acted as efficient ECL quenchers—in the case of phenol, a detection limit of 1.0 nM in solution was determined. In similar fashion, Shan et al. deposited nanofibers onto a GC electrode. In this study, where poly(acrylonitrile-*co*-acrylic acid) was used to host $[Ru(bpy)_3]^{2+}$ complexes, even a 100 times higher ECL intensity compared to an aqueous solution of $[Ru(bpy)_3]^{2+}$ and the co-polymer was observed [179].

Very recently, You et al. reported an ultrasensitive self-enhanced ECL sensor, based on a nanofiber network [180]. These authors fabricated the nanofibers from a mixture of polyacrylonitrile (PAN), $[Ru(bpy)_3]^{2+}$, poly(ethylenimine) (PEI), and Nafion™. PEI is known to function as an ECL enhancing co-reactant, a property that has been ascribed to the presence of the various amine groups along its backbone; the co-assemblies of $[Ru(bpy)_3]^{2+}$ and PEI were employed in various chemo- or immunosensor applications [181–184]. Noteworthy, a similar behavior was also reported for the system PAMAM/$[Ru(bpy)_3]^{2+}$ (PAMAM: polyamidoamine) [185, 186]. In You's multi-component nanofibers, the electronic transmission distance between the luminophore and PEI was strongly decreased and, consequently, the ECL intensity was enhanced [180]. These authors tested the applicability of the nanofibers, as sensor, also for the detection of phenol derivatives e.g., α-naphthol. For this analyte, a linear detector response over a wide concentration range from 1 to 0.1 μM down to 1.0 pM was found (the latter represented the lower detection limit).

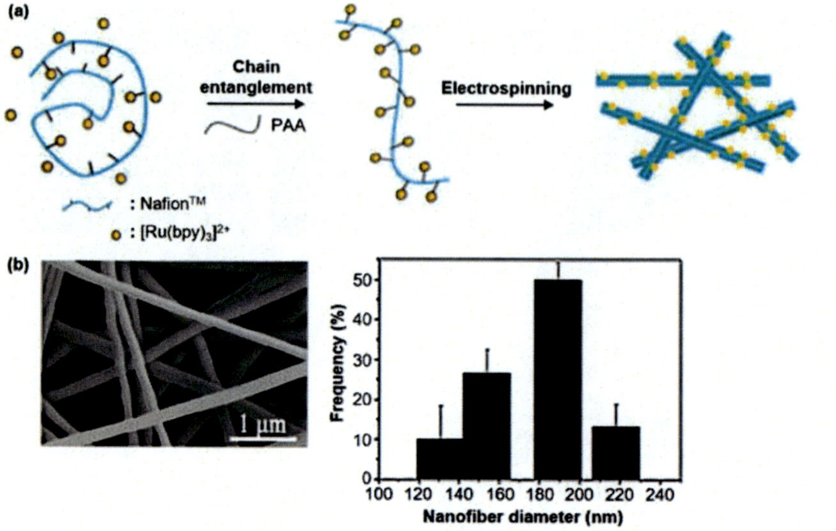

Fig. 5.33 **a** Schematic representation of the fabrication of Nafion™/$[Ru(bpy)_3]^{2+}$ nanofibers. **b** Representative SEM image of the Nafion™/$[Ru(bpy)_3]^{2+}$ nanofibers and the resultant diameter distribution [178]. Figure reproduced with kind permission; © 2010 The Royal Chemical Society

As already pointed out in Chap. 2, ruthenoporphyrins represent versatile catalysts for a range of organic reactions (e.g., epoxidation, cyclopropanation or aziridation of olefins) [187]. However, due to their non-charged character, their immobilization in charged Nafion™ membranes can only be realized if charge-carrying moieties are introduced on the porphyrin unit. To the best to our knowledge, the exclusive example in this respect was reported by Yamamoto et al., who utilized a dinuclear porphyrin stack, which was formed from a cobalto^{III}porphyrin and a rutheno^{II}porphyrin due to electrostatic attraction of their oppositely charged substituents (K_a value for the dimerization of 5×10^{-7} mol L^{-1}) [188, 189]. When incorporated into a Nafion™ membrane coated onto a GC electrode, the metalloporphyrin stacks exhibited a high activity for the four-electron reduction of oxygen.

5.5 Polymer-Staining Using Ru Reagents

Electron microscopy (EM) imaging represents a versatile technique to visualize the structure and morphology of heterogeneous (polymer) systems in the solid state with a high level of spatial resolution. However, when dealing with polymer samples, it is often necessary to enhance the image contrast by utilizing a so-called staining agent, typically a transition-metal compound of high electron density

[190]. In this respect, OsO_4 represents a versatile reagent, which can be applied for many types of polymers to enhancing the EM image contrast, though excluding the saturated ones. This limitation has been overcome by employing RuO_4, as a low-boiling (40 °C) and highly reactive staining agent. Due to its intrinsic instability, RuO_4 is favorably formed in situ from $Ru^{III/IV}$ precursors, such as $RuCl_3$ xH_2O or $RuO_2 \times xH_2O$, by oxidation with metaperiodate [191]. This compound represents an important specie, since it's a distillable intermediate for the refining of elemental Ru and its basic compounds (e.g., $RuCl_3 \times xH_2O$) [192, 193]. In 1980, Vitali and Montani reported the improved image contrast for polybutadiene lattices (i.e., the block copolymers of acrylonitrile, butadiene, and styrene), as achieved with RuO_4 [19]. Almost by the same time, Trent et al. demonstrated the polymer staining with RuO_4 for TEM imaging—PS-PMMA blends and high-impact PS films were used as examples in this study [194]. In a more detailed study, the scope and limitations of RuO_4 were evaluated [18]. A wide range of polymer systems, saturated and unsaturated, were treated with RuO_4. Besides the rather reactive polymers, which contained ether, alcohol, aromatic or amide moieties, also high-density polyethylene (PE), linear PE wax, poly(viny1 methyl ketone) and isotactic as well as atactic polypropylene (iPP and aPP, respectively) were successfully stained. However, PMMA, poly(viny1 chloride) (PVC), poly(viny1idene fluoride), and polyacrylonitrile did not show any staining effect. In contrast to OsO_4, where the staining is directly related to a chemical reaction with unsaturated sites, RuO_4 preferentially forms clusters. The diverse behavior of RuO_4 to the various polymers was mainly attributed to the diffusion and accumulation rates in the polymers microdomains (thus, chemical selectivity is considered to play a minor role only). In general, the diffusion into low-density amorphous domains is faster and, thus, staining preferentially occurs there; consequently, the crystalline, hence more compact domains are not stained [195]. As one example, the staining experiments on linear low-density polyethylene (LLDPE) and iPP are highlighted. Sao et al. studied their melt-crystallized samples by TEM imaging [196]. The lamellar crystalline phase of LLDPE appeared as stripes; whereas, the crystalline phase of iPP showed a cross-hatched lamellar texture (Fig. 5.34). These authors proposed a staining mechanism, which involved the partial oxidation of the amorphous polyolefin chains to afford carboxylate units and RuO_2. Noteworthy, also for Ru-red (Fig. 5.7a) a long-term history as staining agent is recorded: This agent was already been used in the 1890s to stain various plant materials, such as pectins, mucilages, and gums. In the case of pectic acid, as a carboxylated polysaccharide, a remarkable stereoselectivity of the staining reaction was found: According to Steeling, staining involved "each monomer unit and its adjacent neighbor" [197]. The application of Ru-red, as staining reagent, for EM imaging was introduced by Luft in the 1960s [198–201]. This author investigated animal material and, in general, observed fine grain, high resolution, and good contrast. More recently, Ru-red has attracted significant interest in pharmacological research, in particular to gain insight into specific cellular mechanisms (in this respect, selectivity represents a considerable issue, since Ru-red will interact with a broad range of proteins) [202]. Despite the common usage of Ru-red, as a staining agent,

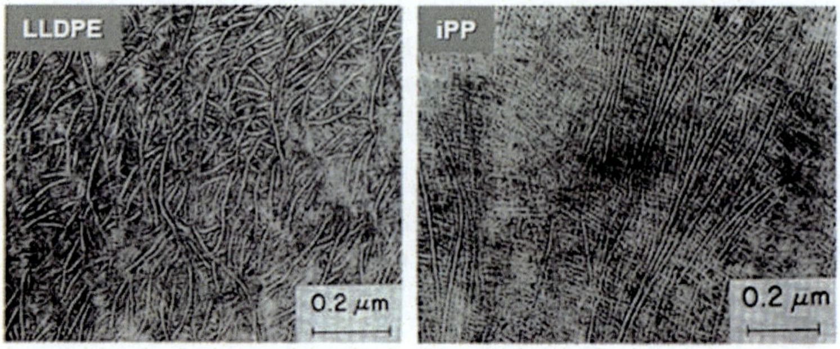

Fig. 5.34 Representative TEM images of melt-crystallized LLDPE and iPP samples, after staining with RuO_4 [196]. Figure reproduced with kind permission; © 1986 Elsevier B.V

in the EM imaging of natural materials, its application for synthetic polymers has not been demonstrated.

Starting from these pioneering contributions, the investigation of RuO_4-stained polymers by EM imaging has become a state-of-the-art method and, accordingly, a plethora of publications are available where this technique has been applied. Beyond polymer blends, also semi-crystalline homopolymers were investigated and RuO_4-staining generated a contrast between crystalline lamellar and amorphous regions [203]. In a similar fashion, also semi-crystalline or amphiphilic block copolymers were treated. For example, Gido et al. investigated the morphology of PE-*b*-aPP by TEM imaging [204]. Staining of the samples enhanced the contrast between amorphous PE and amorphous aPP regions as well as between amorphous PE regions and crystalline PE regions. *Avowedly, RuO_4 is classified as a hazardous compound and precautions must be taken when handling this sensitive reagent.* Addressing this issue, Montezinos et al. introduced a convenient way for polymer staining, which is based on the in-situ formation of RuO_4 by oxidation of $RuCl_3$ with hypochlorite. The efficiency of this method was shown exemplarily at various PE samples (Fig. 5.35) [205].

With respect to polyesters, such as poly(butylene terephthalate) (PBT) [206] or polycarbonate (PC) [207, 208], staining with OsO_4 or RuO_4 readily occurred; whereas, the staining of poly(ethylene terephthalate) (PET) is rather difficult. Haubruge et al. cautiously optimized the staining conditions and could visualize the crystalline morphology of PET at the lamellar scale by TEM imaging of thin-solid films or section from the bulk (pure PET as well as PET-containing blends were analyzed) [209].

Going beyond polymer staining for contrast enhancement between different polymer domains in EM imaging, this technique has also attracted considerable interest due to the ability to fabricate patterned surfaces. Ansari and Hamley showed that Au nanoparticles could be selectively deposited onto the stained PI block of a PS-*b*-PI-*b*-PS triblock copolymer [210]. In a more sophisticated approach, arrays of magnetic nanodots were prepared from a thin cylinder-forming

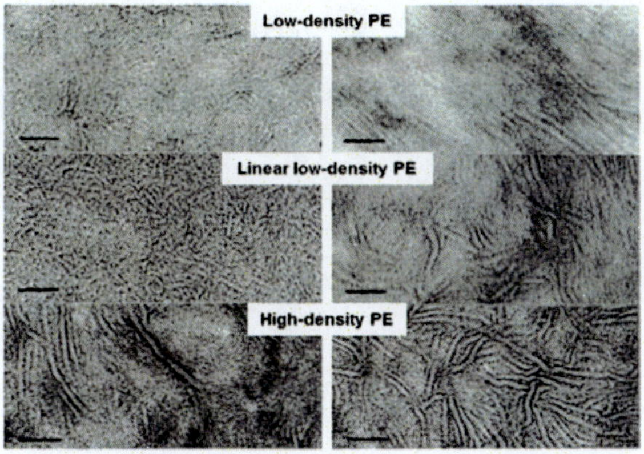

Fig. 5.35 TEM images of various polymer samples stained with $RuCl_3$/NaOCl (left) and RuO_4 (right). The scale bar represents 100 nm [205]. Figure reproduced with kind permission; © 1985 John Wiley & Sons

PS-*b*-PLA film (Fig. 5.36) [211]. Annealing of the polymer film gave PLA cylinders with perpendicular orientation to the surface, subsequently the PS phase was stained with RuO_4 vapors. Removal of the cylinders by O_2-reactive ion etching generated a mask onto which the metal was deposited via molecular-beam epitaxy; in the final step of the procedure, the mask was removed by sonication, followed by sintering at 600 °C under Ar. The thusly obtained $Ni_{80}Fe_{20}$ or Co nanodots had an average diameter of 49 ± 14 nm, thereby revealing the initial template features. The enhancement of etching resistance of the stained domains was investigated in more detail by Vayer et al. [212]. When short staining periods were applied, two-layered films were formed; thereby, the surface layer, which was less than 20 nm thick, featured mainly Ru–O–C bonds. This surface layer had a relatively high atomic concentration of Ru of up to 9%, which accounted for the low etching rate of *ca.* 3 nm min^{-1} of the stained film.

The selective staining of the PS segments within PS-*b*-PMMA or PS-*b*-PLA diblock copolymers was not only used for morphology investigation via microscopy imaging but, even more important, to prepare a precursor surface for the subsequent generation of defined arrays of RuO_2 by calcination at 350 °C (Fig. 5.37) [213]. The electrical and electrochemical properties of RuO_2 have attracted interest in view of potential applications in electrochemical energy storage/conversion devices (*e.g.*, the high capacitance of up to 1300 F g^{-1} makes RuO_2 a good candidate for usage in supercapacitors) [214]. Since conductive polymers, such as PPy or poly(3,4-ethylenedioxythiophene) (PEDOT), also exhibit high specific capacitances in the range of 100–250 F g^{-1}, their combination with RuO_2 was targeted by Lee et al. [215]. Following a multi-step assembly protocol, RuO_2-loaded conductive nanowires were prepared and characterized by EDX and XPS (Fig. 5.38). Different heterogeneous nanocomposites were formed depending on the polymer properties: For the rather compact PPy nanowires, a shell of RuO_2 was observed; whereas, RuO_2 nanoparticles were detected inside the porous PEDOT nanowires. CV and galvanostatic charge–discharge experiments were conducted to estimate the electrochemical performance of the materials—high

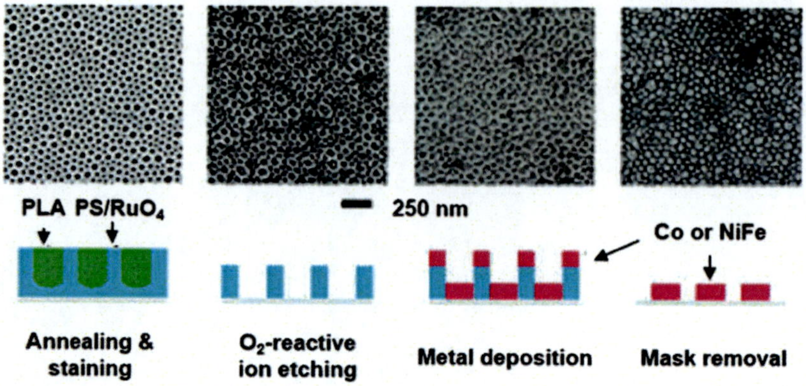

Fig. 5.36 Schematic representation of the multi-step preparation of metal nanodots starting from a thin PS-*b*-PLA film. The SEM images, as obtained after each single production step, are depicted above [211]. Figure reproduced with kind permission; © 2004 The Royal Chemical Society

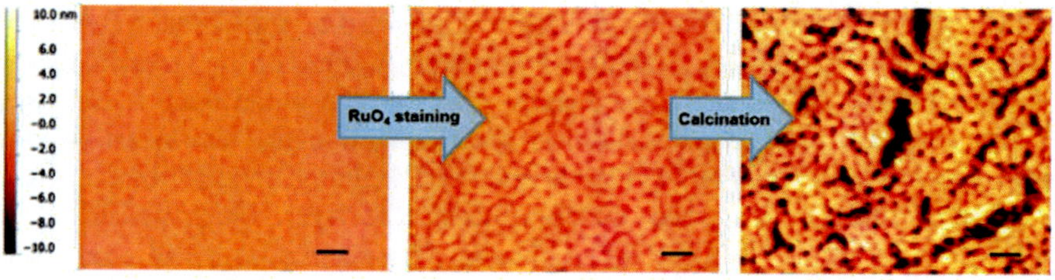

Fig. 5.37 AFM height images of the initial PS-*b*-PLA film (left), after staining with RuO_4 (middle) and subsequent calcination (right). The scale bar corresponds always to 100 nm [213]. Figure reproduced with kind permission; © 2014 Elsevier B.V

specific capacitance of 371 and 500 F g^{-1} were measured for the PEDOT- and PPy-derived composite nanowires, respectively. These results are of particular relevance regarding potential supercapacitor applications—RuO_2-based composites have attracted considerable interest in this respect [216].

Recently, Spontak et al. demonstrated that metallic species can selectively be incorporated into block ionomers which, similar to neutral block copolymers, are prone to phase separation into microphase-ordered morphologies [217]. These authors studied the phase-separation behavior of the midblock-sulfonated triblock ionomer **17** by TEM imaging. Using two different staining agents, RuO_4 and Pd $(OAc)_2$, allowed one to visualize the coexistence of lamellar and cylindrical domains in the solid state. However, the mode-of-action of these reagents was rather different and enabled a complementary metalation of the polymer (Fig. 5.39): Upon exposure to RuO_4 vapor, reaction of the staining agent preferentially occurred

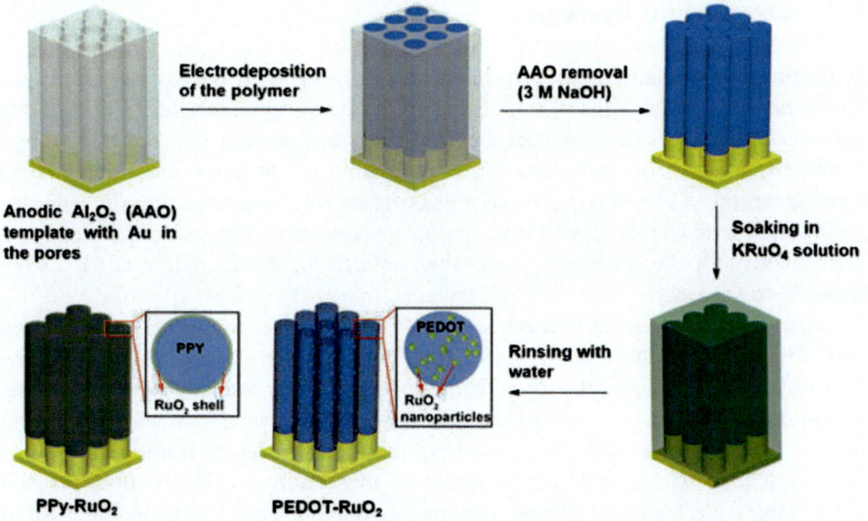

Fig. 5.38 Schematic representation of the step-wise fabrication of conductive polymer-RuO_2 nanowires [215]. Figure reproduced with kind permission; © 2014 The Royal Chemical Society

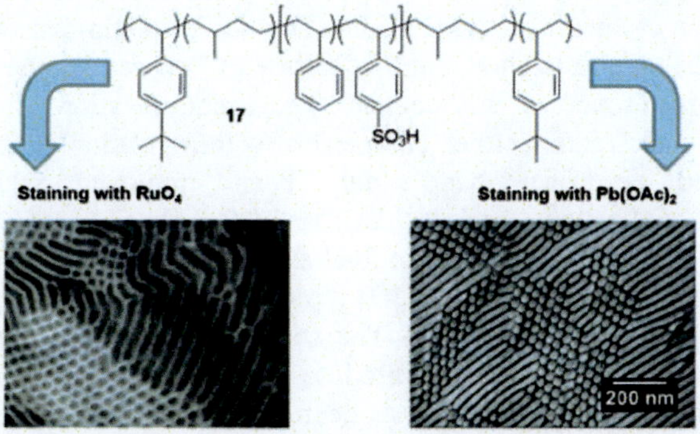

Fig. 5.39 Schematic representation of the block ionomer **17.** Representative TEM images of the stained polymer samples are also shown [217]. Figure reproduced with kind permission; © 2018 Wiley–VCH

with the phenyl rings of the styrenic blocks, thus Ru species were enriched in the apolar domains. Due to an ion-exchange reaction with the sulfonic-acid groups, the Pb^{II} ions were trapped in the hydrophilic segment. In a similar fashion, metalation of the central block was also possible with $AgNO_3$, $CuSO_4$, $ZnCl_2$, and $AlCl_3$.

5.6 Concluding Remarks

A range of Ru-containing species has been combined with polymers, thereby targeting highly diverse fields of applications: Catalysis, opto-electronic, biomedicine and electron microscopy. Starting from fundamental studies devoted to homogeneous dispersion of Ru ions or complexes in polymer matrices these applications have emerged. This strategy represents a versatile alternative for the covalent incorporation of such Ru species into polymers. However, the ease of preparation is confronted with the, in many cases, low stability of these mixtures, e.g., when exposed to solvents.

Regarding catalytic applications, numerous heterogeneous catalysts showing promising activity and stability have been reported—this holds true for various organic reactions (e.g., oxidation or hydrogenation) as well as the cutting-edge energy-related purposes, such as the (electro)chemical water oxidation. However, the corresponding homogeneous catalysts or their covalently immobilized counterparts (Chap. 2) are typically superior to these non-covalent hybrid systems. Presumably, the most significant achievements have been made in the field of biomedical applications where cytotoxic ruthenodrugs were encapsulated in core–shell micelles, assembled from amphiphilic copolymers. These nanostructures have been used in the context of drug delivery and enabled both the stabilization of the ruthenodrugs under physiological conditions and the targeted delivery into cancer cells. Further progress in this field has to be expected, since these ruthenodrugs are known for their cytotoxic activity against various cisplatin-resistant cancer cell lines. By far, the combination Nafion™/$[Ru(bpy)_3]^{2+}$ represents the most common system to be mentioned in this context. The immobilization of the prominent light-emitting and redox-active complexes within Nafion™ membranes or Nafion™-coated electrodes has widely been employed with respect to opto-electronic applications, such as LECs or ECL sensors.

The field of polymer staining is the final aspect, which has been covered in this chapter. Due to its high reactivity, RuO_4 represents a versatile alternative to OsO_4 or uranyl acetate, as staining agent. The staining of polymer samples, i.e., the contrast enhancement, is often required to visualize phase separation by EM imaging. Accordingly, this strategy has been applied to a wide range of polymer architectures and blends. As a significant achievement, the development of *in-situ* protocols avoids the direct handling of harmful RuO_4. Moreover, surface patterning and the production of RuO_2-polymer composites have been demonstrated. In particular, the latter might be of relevance for fabricating the next generation supercapacitor materials.

References

1. Whittell, G.R., Hager, M.D., Schubert, U.S., Manners, I.: Functional soft materials from metallopolymers and metallosupramolecular polymers. Nat. Mater. **10**(3), 176 (2011)
2. Abd-El-Aziz, A.S., Shipman, P.O., Boden, B.N., McNeil, W.S.: Synthetic methodologies and properties of organometallic and coordination macromolecules. Prog. Polym. Sci. **35**(6), 714 (2010)
3. Winter, A., Schubert, U.S.: Synthesis and characterization of metallo-supramolecular polymers. Chem. Soc. Rev. **45**(19), 5311 (2016)
4. Ho, C.-L., Yu, Z.-Q., Wong, W.-Y.: Multifunctional polymetallaynes: properties, functions and applications. Chem. Soc. Rev. **45**(19), 5264 (2016)
5. Wong, W.-Y., Ho, C.-L.: Organometallic photovoltaics: A new and versatile approach for harvesting solar energy using conjugated polymetallaynes. Acc. Chem. Res. **43**(9), 1246 (2010)
6. Callari, M., Aldrich-Wright, J.R., de Souza, P.L., Stenzel, M.H.: Polymers with platinum drugs and other macromolecular metal complexes for cancer treatment. Prog. Polym. Sci. **39** (9), 1614 (2014)
7. Axet, M.R., Philippot, K.: Catalysis with colloidal ruthenium nanoparticles. Chem. Rev. **120** (2), 1085 (2020)
8. Shaver, A., Butler, I.S., Eisenberg, A., Gao, J.P., Xu, Z.H., Fong, B., Uhm, H., Klein, D.: Complexes in polymers: FT-IR spectra and photochemistry of some monomeric organometallic carbonyl complexes in polystyrene, poly(methyl methacrylate), polystyrene—poly(methyl methacrylate) and polystyrene—polyacrylonitrile copolymers. Appl. Organomet. Chem. **1**(5), 383 (1987)
9. Shaver, A., Gao, J.P., Butler, I.S.: Complexes in polymers. Part III: Addition reactions of *trans*-$Ir(PPh_3)_2(CO)Cl$ embedded in polystyrene with hydrogen, oxygen, sulfur dioxide, carbon monoxide and gaseous iodine and of $(\eta\text{-}C_5H_5)Ru(\eta^4\text{–}1,5\text{-cyclooctadiene})Cl$ in polystyrene with carbon monoxide. Appl. Organomet. Chem. **3**(5), 425 (1989)
10. De Paoli, M.-A., De Oliveira, S.M., Galembeck, F.: The use of polymers as matrices for organometallic photochemical reactions. J. Organomet. Chem. **193**(1), 105 (1980)
11. Slinker, J., Bernards, D., Houston, P.L., Abruña, H.D., Bernhard, S., Malliaras, G.G.: Solid-state electroluminescent devices based on transition metal complexes. Chem. Commun. (19), 2392 (2003)
12. Holder, E., Langeveld, B.M.W., Schubert, U.S.: New trends in the use of transition metal–ligand complexes for applications in electroluminescent devices. Adv. Mater. **17**(9), 1109 (2005)
13. Costa, R.D., Ortí, E., Bolink, H.J., Monti, F., Accorsi, G., Armaroli, N.: Luminescent ionic transition-metal complexes for light-emitting electrochemical cells. Angew. Chem. Int. Ed. **51**(33), 8178 (2012)
14. Meier, S.B., Tordera, D., Pertegás, A., Roldán-Carmona, C., Ortí, E., Bolink, H.J.: Light-emitting electrochemical cells: Recent progress and future prospects. Mater. Today **17**(5), 217 (2014)
15. Fresta, E., Costa, R.D.: Beyond traditional light-emitting electrochemical cells—a review of new device designs and emitters. J. Mater. Chem. C **5**(23), 5643 (2017)
16. Buchmeiser, M.R.: Polymer-supported well-defined metathesis catalysts. Chem. Rev. **109** (2), 303 (2008)
17. Buonomenna, M.G., Gallo, E., Ragaini, F., Cenini, S., Drioli, E.: New ruthenium porphyrin polymeric membranes: Preparation, characterization and catalytic activity in alkenes aziridination. Desalination **199**(1), 167 (2006)
18. Trent, J.S., Scheinbeim, J.I., Couchman, P.R.: Ruthenium tetraoxide staining of polymers for electron microscopy. Macromolecules **16**(4), 589 (1983)
19. Vitali, R., Montani, E.: Ruthenium tetroxide as a staining agent for unsaturated and saturated polymers. Polymer **21**(10), 1220 (1980)

20. Dérien, S., Dixneuf, P.H.: The versatility of molecular ruthenium catalyst RuCl(cod)(C_5Me_5). J. Organomet. Chem. **689**(8), 1382 (2004)
21. Albers, M.O., Robinson, D.J., Shaver, A., Singleton, E.: The chemistry of cyclopentadienyl-ruthenium and -osmium complexes. Part 1. The preparation and substitution reactions of (η^4-cycloocta-1,5-diene)(η^5-cyclopentadienyl)haloruthenium(II) complexes. Versatile new synthetic precursors to open-face cyclopentadienylruthenium(II) chemistry. Organometallics **5**(11), 2199 (1986)
22. Blackmore, T., Cotton, J.D., Bruce, M.I., Stone, F.G.A.: Chemistry of the metal carbonyls. Part LIII. Some dicarbonyl-(π-cyclopentadienyl)ruthenium complexes containing metal–metal bonds. J. Chem. Soc. A 2931 (1968)
23. Tsukada, S., Sagawa, T., Gunji, T.: Carbon monoxide addition to ruthenium–dithiolene complex and polysiloxane hybrid film formation. Chem. Asian J. **10**(9), 1881 (2015)
24. Pitchaimani, J., Ni, S.-F., Dang, L.: Metal dithiolene complexes in olefin addition and purification, small molecule adsorption, H_2 evolution and CO_2 reduction. Coord. Chem. Rev. **420**, 213398 (2020)
25. Tsukada, S., Sagawa, T., Yamamoto, K., Gunji, T.: Preparation of ruthenium dithiolene complex/polysiloxane films and their responses to CO gas. Molecules **23**(4), 845 (2018)
26. Vankelecom, I.F.J.: Polymeric membranes in catalytic reactors. Chem. Rev. **102**(10), 3779 (2002)
27. Vankelecom, I.F.J., Jacobs, P.A.: Dense organic catalytic membranes for fine chemical synthesis. Catal. Today **56**(1), 147 (2000)
28. Tas, D., Thoelen, C., Vankelecom, I.F.J., Jacobs, P.A.: Bifunctional catalytic membrane containing Brønsted acids and sites for enantioselective hydrogenation. Chem. Commun. (23), 2323 (1997)
29. Gallo, E., Buonomenna, M.G., Viganò, L., Ragaini, F., Caselli, A., Fantauzzi, S., Cenini, S., Drioli, E.: Heterogenization of ruthenium porphyrin complexes in polymeric membranes: Catalytic aziridination of styrenes. J. Mol. Catal. A: Chem. **282**(1–2), 85 (2008)
30. Viganò, M., Ragaini, F., Buonomenna, M.G., Lariccia, R., Caselli, A., Gallo, E., Cenini, S., Jansen, J.C., Drioli, E.: Catalytic polymer membranes under forcing conditions: reduction of nitrobenzene by CO/H_2O catalyzed by ruthenium bis(arylimino)acenaphthene complexes. ChemCatChem **2**(9), 1150 (2010)
31. Buonomenna, M.G., Choi, S.H., Galiano, F., Drioli, E.: Membranes prepared via phase inversion. In: Basile, A., Gallucci, F. (eds.) Membranes for Membrane Reactors, p. 475. Wiley, Chichester (2011)
32. Zhang, T., Ge, Y., Wang, X., Chen, J., Huang, X., Liao, Y.: Polymeric ruthenium porphyrin-functionalized carbon nanotubes and graphene for levulinic ester transformations into γ-valerolactone and pyrrolidone derivatives. ACS Omega **2**(7), 3228 (2017)
33. Yao, Y., Wang, Z., Zhao, S., Wang, D., Wu, Z., Zhang, M.: A stable and effective Ru/polyethersulfone catalyst for levulinic acid hydrogenation to γ-valerolactone in aqueous solution. Catal. Today **234**, 245 (2014)
34. Mauritz, K.A., Moore, R.B.: State of understanding of Nafion. Chem. Rev. **104**(10), 4535 (2004)
35. Rubinstein, I., Bard, A.J.: Polymer films on electrodes. Part 4. Nafion-coated electrodes and electrogenerated chemiluminescence of surface-attached tris(2,2'-bipyridine)ruthenium(II). J. Am. Chem. Soc. **102**(21), 6641 (1980)
36. www.fuelcellstore.com. Last accessed 01 Nov 2020
37. Prakash, G.K.S., Munoz, S.B., Papp, A., Masood, K., Bychinskaya, I., Mathew, T., Olah, G. A.: Nafion-Ru: A sustainable catalyst for selective hydration of nitriles to amides. Asian J. Org. Chem. **1**(2), 146 (2012)
38. Baig, R.B.N., Nadagouda, M.N., Varma, R.S.: Ruthenium on chitosan: a recyclable heterogeneous catalyst for aqueous hydration of nitriles to amides. Green Chem. **16**(4), 2122 (2014)

39. Matsuoka, A., Isogawa, T., Morioka, Y., Knappett, B.R., Wheatley, A.E.H., Saito, S., Naka, H.: Hydration of nitriles to amides by a chitin-supported ruthenium catalyst. RSC Adv. **5** (16), 12152 (2015)
40. Kumar, A.S., Swetha, P.: Ru(DMSO)$_4$Cl$_2$ nano-aggregated Nafion membrane modified electrode for simultaneous electrochemical detection of hypoxanthine, xanthine and uric acid. J. Electroanal. Chem. **642**(2), 135 (2010)
41. Ramaraj, R., Kira, A., Kaneko, M.: O_2-generation by oxidation of water with di- and trinuclear ruthenium complexes as homogeneous and heterogeneous catalysts. Angew. Chem. Int. Ed. **25**(11), 1009 (1986)
42. Luft, J.H.: Ruthenium red and violet. Part II. Fine structural localization in animal tissues. Anatomical Rec. **171**(3), 369 (1971)
43. Yagi, M., Tokita, S., Nagoshi, K., Ogino, I., Kaneko, M.: Activity analysis of a water oxidation catalyst immobilized in a polymer membrane. J. Chem. Soc., Faraday Trans. **92** (13), 2457 (1996)
44. Yagi, M., Kaneko, M.: Molecular catalysts for water oxidation. Chem. Rev. **101**(1), 21 (2001)
45. Kentaro, N., Masayuki, Y., Masao, K.: Analysis of water oxidation catalysis by dinuclear ruthenium complex $[(NH_3)_5Ru\text{-}O\text{-}Ru(NH_3)_5]^{4+}$ incorporated in a Nafion® membrane. Bull. Chem. Soc. Jpn. **73**(10), 2193 (2000)
46. Yagi, M., Osawa, Y., Sukegawa, N., Kaneko, M.: Catalytic activity of tri(μ-chloro)-bridged dinuclear ruthenium complex confined in a polymer membrane as an artificial model of photosynthetic oxygen evolving center. Langmuir **15**(21), 7406 (1999)
47. Yagi, M., Sukegawa, N., Kasamastu, M., Kaneko, M.: Cooperative catalysis and critical decomposition distances in water oxidation by the mononuclear ammineruthenium(III) complex in a Nafion membrane. J. Phys. Chem. B **103**(12), 2151 (1999)
48. Yagi, M., Nagoshi, K., Kaneko, M.: Cooperative catalysis and critical decomposition distances between molecular water oxidation catalysts incorporated in a polymer membrane. J. Phys. Chem. B **101**(26), 5143 (1997)
49. Yagi, M., Sukegawa, N., Kaneko, M.: Analysis of catalytic water oxidation by *cis*-tetraamminedichlororuthenium(III) complex incorporated in a polymer membrane. J. Phys. Chem. B **104**(17), 4111 (2000)
50. Yagi, M., Kasamastu, M., Kaneko, M.: Cooperative catalysis and critical decomposition distances in water oxidation by tris(ethylenediamine)ruthenium (III) complex confined in a Nafion membrane. J. Mol. Catal. A **151**(1), 29 (2000)
51. https://www.metal-powder-dust.com. Last accessed 01 Nov 2020
52. Yagi, M., Kaneko, M.: Charge transport and catalysis by molecules confined in polymeric materials and application to future nanodevices for energy conversion. Appl. Polym. Sci. 143 (2006)
53. Yagi, M., Kinoshita, K., Kaneko, M.: Activity analysis of electrochemical water oxidation catalyst confined in a coated-polymer membrane. J. Phys. Chem. **100**(26), 11098 (1996)
54. Yagi, M., Kinoshita, K., Kaneko, M.: Enhancing effect of an amino acid residue model for the electrochemical water oxidation catalyst confined in a polymer membrane. J. Phys. Chem. B **101**(20), 3957 (1997)
55. Kinoshita, K., Yagi, M., Kaneko, M.: Enhancing effects of amino acid residue model compounds on catalytic activity of an artificial oxygen evolving center. Macromolecules **31**(18), 6042 (1998)
56. Shiroishi, H., Yamashita, S., Kinoshita, K., Kaneko, M.: Activity analysis of $[(NH_3)_5RuORu(NH_3)_5]^{5+}$, confined in a coated-polymer membrane for electrochemical water oxidation. Polym. J. **31**(11), 1175 (1999)
57. Kinoshita, K., Yagi, M., Kaneko, M.: Activity analysis of mononuclear ruthenium ammine complex dispersed in a polymer membrane as an electrochemical water oxidation catalyst. J. Mol. Catal. A **142**(1), 1 (1999)

58. Shiroishi, H., Sayama, H., Moroi, T., Kaneko, M.: Electrocatalytic water oxidation using Ru moieties incorporated into a Nafion® coated electrode. J. Electroanal. Chem. **502**(1), 132 (2001)
59. Liu, F., Concepcion, J.J., Jurss, J.W., Cardolaccia, T., Templeton, J.L., Meyer, T.J.: Mechanisms of water oxidation from the blue dimer to photosystem II. Inorg. Chem. **47**(6), 1727 (2008)
60. Gersten, S.W., Samuels, G.J., Meyer, T.J.: Catalytic oxidation of water by an oxo-bridged ruthenium dimer. J. Am. Chem. Soc. **104**(14), 4029 (1982)
61. Gilbert, J.A., Eggleston, D.S., Murphy, W.R., Geselowitz, D.A., Gersten, S.W., Hodgson, D.J., Meyer, T.J.: Structure and redox properties of the water-oxidation catalyst $[(bpy)_2(OH_2)RuORu(OH_2)(bpy)_2]^{4+}$. J. Am. Chem. Soc. **107**(13), 3855 (1985)
62. Chronister, C.W., Binstead, R.A., Ni, J., Meyer, T.J.: Mechanism of water oxidation catalyzed by the μ-oxo dimer $[(bpy)_2(OH_2)Ru^{III}ORu^{III}(OH_2)(bpy)_2]^{4+}$. Inorg. Chem. **36**(18), 3814 (1997)
63. Hurst, J.K., Zhou, J., Lei, Y.: Pathways for water oxidation catalyzed by the (μ-oxo)bis [aquabis(bipyridine)ruthenium]$^{4+}$ ion. Inorg. Chem. **31**(6), 1010 (1992)
64. Vining, W.J., Meyer, T.J.: Redox properties of the water oxidation catalysts diaquatetrakis (2,2'-bipyridine)oxoderuthenium(4+) in thin polymeric films. Electrocatalytic oxidation of chloride to chlorine. Inorg. Chem. **25**(12), 2023 (1986)
65. Vining, W.J., Meyer, T.J.: Direct evidence for chemically distinct regions within Nafion films on electrodes. J. Electroanal. Chem. Interfacial Electrochem. **237**(2), 191 (1987)
66. Casella, I.G., Guascito, M.R., Salvi, A.M., Desimoni, E.: Catalytic oxidation and flow detection of hydrazine compounds at a Nafion/ruthenium(III) chemically modified electrode. Anal. Chim. Acta **354**(1), 333 (1997)
67. Zen, J.M., Wang, C.B.: Determination of dissolved oxygen by catalytic reduction on a Nafion® / ruthenium-oxide pyrochlore chemically modified electrode. J. Electroanal. Chem. **368**(1), 251 (1994)
68. Zen, J.-M., Chen, I.-L.: Voltammetric determination of dopamine in the presence of ascorbic acid at a chemically modified electrode. Electroanalysis **9**(7), 537 (1997)
69. Zen, J.-M., Tang, J.-S.: Flow injection amperometric detection of hydrazine by electrocatalytic oxidation at a perfluorosulfonated ionomer/ruthenium oxide pyrochlore chemically modified electrode. Anal. Chem. **67**(1), 208 (1995)
70. Zen, J.M., Kumar, A.S., Chang, M.R.: Electrocatalytic oxidation and trace detection of amitrole using a Nafion/lead–ruthenium oxide pyrochlore chemically modified electrode. Electrochim. Acta **45**(10), 1691 (2000)
71. Zen, J.-M., Ting, Y.-S.: Simultaneous determination of caffeine and acetaminophen in drug formulations by square-wave voltammetry using a chemically modified electrode. Anal. Chim. Acta **342**(2), 175 (1997)
72. Zen, J.-M., Ting, Y.-S., Shih, Y.: Voltammetric determination of caffeine in beverages using a chemically modified electrode. Analyst **123**(5), 1145 (1998)
73. Zen, J.-M., Yu, T.-Y., Shih, Y.: Determination of theophylline in tea and drug formulation using a Nafion®/lead–ruthenium oxide pyrochlore chemically modified electrode. Talanta **50**(3), 635 (1999)
74. Zen, J.-M., Chang, M.-R., Chung, H.-H., Shih, Y.: Determination of codeine in human plasma and drug formulation using a chemically modified electrode. Electroanalysis **10**(8), 536 (1998)
75. Zen, J.-M., Chen, I.L., Shih, Y.: Voltammetric determination of serotonin in human blood using a chemically modified electrode. Anal. Chim. Acta **369**(1), 103 (1998)
76. Zen, J.-M., Kumar, A.S., Chen, J.-C.: Electrocatalytic oxidation and sensitive detection of cysteine on a lead ruthenate pyrochlore modified electrode. Anal. Chem. **73**(6), 1169 (2001)
77. Zen, J.-M., Lai, Y.-Y., Ilangovan, G., Kumar, A.S.: Electrocatalytic oxidation of hypoxanthine on a Nafion/lead-ruthenium oxide pyrochlore modified electrode. Electroanalysis **12**(4), 280 (2000)

78. Zen, J.-M., Tang, J.-S.: Square-wave voltammetric determination of uric acid by catalytic oxidation at a perfluorosulfonated ionomer/ruthenium oxide pyrochlore chemically modified electrode. Anal. Chem. **67**(11), 1892 (1995)
79. Zen, J.-M., Kumar, A.S.: A mimicking enzyme analogue for chemical sensors. Acc. Chem. Res. **34**(10), 772 (2001)
80. Ke, J.-H., Kumar, A.S., Sue, J.-W., Venkatesan, S., Zen, J.-M.: Catalysis and characterization of a rugged lead ruthenate pyrochlore membrane catalyst. J. Mol. Catal. A **233**(1), 111 (2005)
81. Venkatesan, S., Kumar, A.S., Zen, J.-M.: A rugged lead-ruthenate pyrochlore membrane catalyst for highly selective oxidation of alcohols. J. Mol. Catal. A **250**(1), 87 (2006)
82. Ley, S.V., Norman, J., Griffith, W.P., Marsden, S.P.: Tetrapropylammonium perruthenate, $Pr_4N^+RuO_4^-$, TPAP: a catalytic oxidant for organic synthesis. Synthesis (7), 639 (1994)
83. Ley, S.V., Norman, J., Wilson, A.J.: Tetra-n-propylammonium Perruthenate. In: Bode, J., Rovis, T., Shenvi, R. (eds.) e-EROS. John Wiley & Sons, Chichester (2011)
84. Venkatesan, S., Kumar, A.S., Zen, J.M.: Oxidation of alcohols with molecular oxygen promoted by Nafion ionomer anchored pyrochlore composite at room temperature. Tetrahedron Lett. **49**(28), 4339 (2008)
85. Venkatesan, S., Kumar, A.S., Lee, J.-F., Chan, T.-S., Zen, J.-M.: Room temperature aerobic oxidation of amines by a nanocrystalline ruthenium oxide pyrochlore Nafion composite catalyst. Chem. Eur. J. **18**(20), 6147 (2012)
86. Zen, J.-M., Liou, S.-L., Kumar, A.S., Hsia, M.-S.: An efficient and selective photocatalytic system for the oxidation of sulfides to sulfoxides. Angew. Chem. Int. Ed. **42**(5), 577 (2003)
87. Zen, J.-M., Kumar, A.S., Venketesan, S.: Lead-ruthenate pyrochlore modified Nafion® membrane for tunable heterogeneous catalytic oxidation reaction. In: Gunther, M.B. (ed.) Heterogeneous catalysis research progress, p. 345. Nova Science Publishers Inc., U.S.A. (2008)
88. Jaeger, W., Bohrisch, J., Laschewsky, A.: Synthetic polymers with quaternary nitrogen atoms—synthesis and structure of the most used type of cationic polyelectrolytes. Prog. Polym. Sci. **35**(5), 511 (2010)
89. Hinzen, B., Ley, S.V.: Polymer supported perruthenate (PSP): a new oxidant for clean organic synthesis. J. Chem. Soc., Perkin Trans. 1 **1997**(13), 1907 (1997)
90. Hinzen, B., Lenz, R., Ley, S.V.: Polymer supported perruthenate (PSP): clean oxidation of primary alcohols to carbonyl compounds using oxygen as cooxidant. Synthesis (7), 977 (1998)
91. Piccialli, V.: Ruthenium tetroxide and perruthenate chemistry. Recent advances and related transformations mediated by other transition metal oxo-species. Molecules **19**(5), 6534 (2014)
92. Friedrich, H.B., Singh, N.: A study of poly(4-vinylpyridine)-supported ruthenate in the oxidation of alcohols. Catal. Lett. **110**(1), 61 (2006)
93. Green, G., Griffith, W.P., Hollinshead, D.M., Ley, S.V., Schröder, M.: Oxo complexes of ruthenium(VI) and (VII) as organic oxidants. J. Chem. Soc., Perkin Trans. 1 681 (1984)
94. Roy, S., Crans, D.C., Parac-Vogt, T.N.: Polyoxometalates in catalysis, biology, energy and materials science. Front. Chem. **7**(646) (2019)
95. Liu, S., Volkmer, D., Kurth, D.G.: Functional polyoxometalate thin films via electrostatic layer-by-layer self-assembly. J. Cluster Sci. **14**(3), 405 (2003)
96. Anwar, N., Sartorel, A., Yaqub, M., Wearen, K., Laffir, F., Armstrong, G., Dickinson, C., Bonchio, M., McCormac, T.: Surface immobilization of a tetra-ruthenium substituted polyoxometalate water oxidation catalyst through the employment of conducting polypyrrole and the layer-by-layer (LBL) technique. ACS Appl. Mater. Interfaces **6**(11), 8022 (2014)
97. Ma, H., Shi, S., Zhang, Z., Pang, H., Zhang, Y.: An thin film of the di-decatungstosilicate with a tetra-ruthenium(IV)-oxo core and its electrochemical properties. J. Electroanal. Chem. **648**(2), 128 (2010)

98. Gobbo, P., Tian, L., Pavan Kumar, B.V.V.S., Turvey, S., Cattelan, M., Patil, A.J., Carraro, M., Bonchio, M., Mann, S.: Catalytic processing in ruthenium-based polyoxometalate coacervate protocells. Nature Commun. **11**(1), 41 (2020)
99. Xu, J., Sigworth, F.J., LaVan, D.A.: Synthetic protocells to mimic and test cell function. Adv. Mater. **22**(1), 120 (2010)
100. Bergamo, A., Gaiddon, C., Schellens, J.H.M., Beijnen, J.H., Sava, G.: Approaching tumour therapy beyond platinum drugs: status of the art and perspectives of ruthenium drug candidates. J. Inorg. Biochem. **106**(1), 90 (2012)
101. Mjos, K.D., Orvig, C.: Metallodrugs in medicinal inorganic chemistry. Chem. Rev. **114**(8), 4540 (2014)
102. Bergamo, A., Sava, G.: Ruthenium anticancer compounds: Myths and realities of the emerging metal-based drugs. Dalton Trans. **40**(31), 7817 (2011)
103. Zeng, L., Gupta, P., Chen, Y., Wang, E., Ji, L., Chao, H., Chen, Z.-S.: The development of anticancer ruthenium(II) complexes: From single molecule compounds to nanomaterials. Chem. Soc. Rev. **46**(19), 5771 (2017)
104. Antonarakis, E.S., Emadi, A.: Ruthenium-based chemotherapeutics: Are they ready for prime time? Cancer Chemother Pharmacol. **66**(1), 1 (2010)
105. Jones, M.-C., Leroux, J.-C.: Polymeric micelles—a new generation of colloidal drug carriers. Eur. J. Pharm. Biopharm. **48**(2), 101 (1999)
106. Knop, K., Hoogenboom, R., Fischer, D., Schubert, U.S.: Poly(ethylene glycol) in drug delivery: Pros and cons as well as potential alternatives. Angew. Chem. Int. Ed. **49**(36), 6288 (2010)
107. Adams, N., Schubert, U.S.: Poly(2-oxazolines) in biological and biomedical application contexts. Adv. Drug Delivery Rev. **59**(15), 1504 (2007)
108. Jhaveri, A.M., Torchilin, V.P.: Multifunctional polymeric micelles for delivery of drugs and siRNA. Front. Pharmacol. **5**(77) (2014)
109. Blunden, B.M., Stenzel, M.H.: Incorporating ruthenium into advanced drug delivery carriers —an innovative generation of chemotherapeutics. J. Chem. Technol. Biotechnol. **90**(7), 1177 (2015)
110. Khandare, J., Minko, T.: Polymer–drug conjugates: Progress in polymeric prodrugs. Prog. Polym. Sci. **31**(4), 359 (2006)
111. Kopeček, J.: Polymer–drug conjugates: Origins, progress to date and future directions. Adv. Drug Delivery Rev. **65**(1), 49 (2013)
112. Greco, F., Vicent, M.J.: Combination therapy: Opportunities and challenges for polymer–drug conjugates as anticancer nanomedicines. Adv. Drug Delivery Rev. **61**(13), 1203 (2009)
113. Li, C., Wallace, S.: Polymer-drug conjugates: recent development in clinical oncology. Adv. Drug Delivery Rev. **60**(8), 886 (2008)
114. Aderibigbe, B.A.: Polymeric prodrugs containing metal-based anticancer drugs. J. Inorg. Organomet. Polym. Mater. **25**(3), 339 (2015)
115. Zhou, H., Chen, M., Liu, Y., Wu, S.: Stimuli-responsive ruthenium-containing polymers. Macromol. Rapid Commun. **39**(22), 1800372 (2018)
116. Knoll, J.D., Turro, C.: Control and utilization of ruthenium and rhodium metal complex excited states for photoactivated cancer therapy. Coord. Chem. Rev. **282–283**, 110 (2015)
117. Mari, C., Pierroz, V., Ferrari, S., Gasser, G.: Combination of Ru(II) complexes and light: New frontiers in cancer therapy. Chem. Sci. **6**(5), 2660 (2015)
118. Wang, D., Wang, J., Huang, H., Zhao, Z., Gunatillake, P.A., Hao, X.: Brush-shaped RAFT polymer micelles as nanocarriers for a ruthenium(II) complex photodynamic anticancer drug. Eur. Polym. J. **113**, 267 (2019)
119. Fischer, B., Heffeter, P., Kryeziu, K., Gille, L., Meier, S.M., Berger, W., Kowol, C.R., Keppler, B.K.: Poly(lactic acid) nanoparticles of the lead anticancer ruthenium compound KP1019 and its surfactant-mediated activation. Dalton Trans. **43**(3), 1096 (2014)
120. Portney, N.G., Ozkan, M.: Nano-oncology: drug delivery, imaging, and sensing. Anal. Bioanal. Chem. **384**(3), 620 (2006)

121. Heffeter, P., Riabtseva, A., Senkiv, Y., Kowol, C.R., Körner, W., Jungwith, U., Mitina, N., Keppler, B.K., Konstantinova, T., Yanchuk, I., et al.: Nanoformulation improves activity of the (pre)clinical anticancer ruthenium complex KP1019. J. Biomed. Nanotechnol. **10**(5), 877 (2014)
122. Gill, M.R., Menon, J.U., Jarman, P.J., Owen, J., Skaripa-Koukelli, I., Able, S., Thomas, J.A., Carlisle, R., Vallis, K.A.: ^{111}In-labelled polymeric nanoparticles incorporating a ruthenium-based radiosensitizer for EGFR-targeted combination therapy in oesophageal cancer cells. Nanoscale **10**(22), 10596 (2018)
123. Gill, M.R., Jarman, P.J., Halder, S., Walker, M.G., Saeed, H.K., Thomas, J.A., Smythe, C., Ramadan, K., Vallis, K.A.: A three-in-one-bullet for oesophageal cancer: Replication fork collapse, spindle attachment failure and enhanced radiosensitivity generated by a ruthenium(II) metallo-intercalator. Chem. Sci. **9**(4), 841 (2018)
124. Wang, A.Z., Yuet, K., Zhang, L., Gu, F.X., Huynh-Le, M., Radovic-Moreno, A.F., Kantoff, P.W., Bander, N.H., Langer, R., Farokhzad, O.C.: ChemoRad nanoparticles: A novel multifunctional nanoparticle platform for targeted delivery of concurrent chemoradiation. Nanomedicine **5**(3), 361 (2010)
125. Bae, Y.H., Park, K.: Targeted drug delivery to tumors: Myths, reality and possibility. J. Control. Release **153**(3), 198 (2011)
126. Ge, J., Zhang, Q., Zeng, J., Gu, Z., Gao, M.: Radiolabeling nanomaterials for multimodality imaging: New insights into nuclear medicine and cancer diagnosis. Biomaterials **228**, 119553 (2020)
127. Giovagnini, L., Sitran, S., Castagliuolo, I., Brun, P., Corsini, M., Zanello, P., Zoleo, A., Maniero, A., Biondi, B., Fregona, D.: Ru(III)-based compounds with sulfur donor ligands: Synthesis, characterization, electrochemical behaviour and anticancer activity. Dalton Trans. **47**, 6699 (2008)
128. Nagy, E.M., Nardon, C., Giovagnini, L., Marchiò, L., Trevisan, A., Fregona, D.: Promising anticancer mono- and dinuclear ruthenium(III) dithiocarbamato complexes: Systematic solution studies. Dalton Trans. **40**(44), 11885 (2011)
129. Giovagnini, L., Mancinetti, E., Ronconi, L., Sitran, S., Marchiò, L., Castagliuolo, I., Brun, P., Trevisan, A., Fregona, D.: Preliminary chemico-biological studies on Ru(III) compounds with S-methyl pyrrolidine/dimethyl dithiocarbamate. J. Inorg. Biochem. **103**(5), 774 (2009)
130. Nagy, E.M., Pettenuzzo, A., Boscutti, G., Marchiò, L., Dalla Via, L., Fregona, D.: Ruthenium(II/III)-based compounds with encouraging antiproliferative activity against non-small-cell lung cancer. Chem. Eur. J. **18**(45), 14464 (2012)
131. Scintilla, S., Brustolin, L., Gambalunga, A., Chiara, F., Trevisan, A., Nardon, C., Fregona, D.: Ru(III) anticancer agents with aromatic and non-aromatic dithiocarbamates as ligands: loading into nanocarriers and preliminary biological studies. J. Inorg. Biochem. **165**, 159 (2016)
132. Alves, S.R., Colquhoun, A., Wu, X.Y., de Oliveira Silva, D.: Synthesis of terpolymer-lipid encapsulated diruthenium(II, III)-anti-inflammatory metallodrug nanoparticles to enhance activity against glioblastoma cancer cells. J. Inorg. Biochem. **205**, 110984 (2020)
133. Ribeiro, G., Benadiba, M., Colquhoun, A., de Oliveira Silva, D.: Diruthenium(II, III) complexes of ibuprofen, aspirin, naproxen and indomethacin non-steroidal anti-inflammatory drugs: Synthesis, characterization and their effects on tumor-cell proliferation. Polyhedron **27** (3), 1131 (2008)
134. Shalviri, A., Chan, H.K., Raval, G., Abdekhodaie, M.J., Liu, Q., Heerklotz, H., Wu, X.Y.: Design of pH-responsive nanoparticles of terpolymer of poly(methacrylic acid), polysorbate 80 and starch for delivery of doxorubicin. Colloids Surf. B: Biointerfaces **101**, 405 (2013)
135. Jensen, A., Basolo, F., Neumann, H.M.: The binding of metal complexes by polystyrene sulfonate ion. J. Am. Chem. Soc. **81**(3), 509 (1959)
136. Yoshimi, K., Hiroshi, Y., Kiyotaka, S., Eishun, T.: The interaction of the tris(2,2′-bipyridine) ruthenium(II) ion with poly(p-styrenesulfonate). Bull. Chem. Soc. Jpn. **55**(1), 55 (1982)

137. Yoshimi, K., Kimiyo, K.: Tris(2,2′-bipyridine)ruthenium(II)-photosensitized reductions of methyl viologen and molecular oxygen in a network of water-swollen cation-exchange resin. Bull. Chem. Soc. Jpn. **55**(8), 2560 (1982)
138. Tuite, E.M., Rose, D.B., Ennis, P.M., Kelly, J.M.: Influence of polystyrenesulfonate on electron transfer quenching of ruthenium trisbipyridine luminescence by viologens: non-covalent assembly and covalent tethering of the ruthenium complex. Phys. Chem. Chem. Phys. **14**(10), 3681 (2012)
139. Meisel, D., Matheson, M.S.: Quenching and quenching reversal of tris(2,2'-bipyridine) ruthenium(II) emission in polyelectrolyte solutions. J. Am. Chem. Soc. **99**(20), 6577 (1977)
140. Schultze, X., Serin, J., Adronov, A., Fréchet, J.M.J.: Light harvesting and energy transfer in a ruthenium-coumarin-2 copolymer. Chem. Commun. (13), 1160 (2001)
141. Younathan, J.N., McClanahan, S.F., Meyer, T.J.: Synthesis and characterization of soluble polymers containing electron- and energy-transfer reagents. Macromolecules **22**(3), 1048 (1989)
142. Slama-Schwok, A., Rabani, J.: Free energy change in hydrophobic interactions involving a polyelectrolyte and ruthenium tris(bipyridine). A pulsed laser study. Macromolecules **21**(3), 764 (1988)
143. Miyashita, T., Arito, Y., Matsuda, M.: Kinetic study on electron-transfer quenching of a photoexcited ruthenium complex in polyelectrolyte solutions. Macromolecules **24**(4), 872 (1991)
144. Duveneck, G.L., Kumar, C.V., Turro, N.J., Barton, J.K.: Spectroscopic investigations of the interaction between polypyridyl-ruthenium complexes and anionic polyelectrolytes. J. Phys. Chem. **92**(7), 2028 (1988)
145. Canterbury, T.R., Arachchige, S.M., Moore, R.B., Brewer, K.J.: Increased water reduction efficiency of polyelectrolyte-bound trimetallic [Ru, Rh, Ru] photocatalysts in air-saturated aqueous solutions. Angew. Chem. Int. Ed. **54**(43), 12819 (2015)
146. Canterbury, T.R., Arachchige, S.M., Brewer, K.J., Moore, R.B.: Probing co-assembly of supramolecular photocatalysts and polyelectrolytes using isothermal titration calorimetry. J. Phys. Chem. B **121**(25), 6238 (2017)
147. Park, J.W., Kim, M.H., Ko, S.H.; Paik, Y.H.: Binding, photophysical, and photochemical properties of tris(α,α'-diimine)ruthenium(II) complexes in polyelectrolyte solutions: Dependence on the nature of the ligands of the metal complexes and polyelectrolytes. J. Phys. Chem. **97**(20), 5424 (1993)
148. Rabani, J.: Polyelectrolytes: assessment of their possible role in future solar energy storage. In: Photochemical Conversion and Storage of Solar Energy, p. 103. Dordrecht (1991)
149. Kalyanasundaram, K.: Photochemistry of Polypyridine and Porphyrin Complexes. Academic Press, London (1992)
150. Pei, Q., Yu, G., Zhang, C., Yang, Y., Heeger, A.J.: Polymer light-emitting electrochemical cells. Science **269**(5227), 1086 (1995)
151. Lee, J.K., Yoo, D.S., Handy, E.S., Rubner, M.F.: Thin film light emitting devices from an electroluminescent ruthenium complex. Appl. Phys. Lett. **69**(12), 1686 (1996)
152. Wu, A., Lee, J., Rubner, M.F.: Light emitting electrochemical devices from sequentially adsorbed multilayers of a polymeric ruthenium(II) complex and various polyanions. Thin Solid Films **327**, 663–667 (1998)
153. Wu, A., Yoo, D., Lee, J.K., Rubner, M.F.: Solid-state light-emitting devices based on the tris-chelated ruthenium(II) complex: Part 3. High efficiency devices via a layer-by-layer molecular-level blending approach. J. Am. Chem. Soc. **121**(20), 4883 (1999)
154. Rudmann, H., Rubner, M.F.: Single layer light-emitting devices with high efficiency and long lifetime based on tris(2,2′ bipyridyl) ruthenium(II) hexafluorophosphate. J. Appl. Phys. **90**(9), 4338 (2001)
155. Lyons, C.H., Abbas, E.D., Lee, J.K., Rubner, M.F.: Solid-state light-emitting devices based on the trischelated ruthenium(II) Complex. Part 1. Thin film blends with poly(ethylene oxide). J. Am. Chem. Soc. **120**(46), 12100 (1998)

156. Santos, G., Fonseca, F.J., Andrade, A.M., Patrocínio, A.O.T., Mizoguchi, S.K., Iha, N.Y.M., Peres, M., Simões, W., Monteiro, T., Pereira, L.: Opto-electrical properties of single layer flexible electroluminescence device with ruthenium complex. J. Non-Cryst. Solids **354**(19), 2571 (2008)
157. Welter, S., Brunner, K., Hofstraat, J.W., De Cola, L.: Electroluminescent device with reversible switching between red and green emission. Nature **421**(6918), 54 (2003)
158. Wang, T.-J., Chen, C.-H., Chang, S.-M., Tzeng, Y.-J., Chao, Y.-C.: Flexible polymer light-emitting devices based on ruthenium complexes. Microw. Opt. Technol. Lett. **38**(5), 406 (2003)
159. Yang, J., Gordon, K.C.: Light emitting devices from blended films of ruthenium(II)bis(2,2′-bipyridine)(4,7-dimethyl-1,10-phenanthroline) complex with poly(*N*-vinylcarbazole). Chem. Phys. Lett. **372**(3), 577 (2003)
160. Yang, J., Gordon, K.C.: Electroluminescence of ruthenium(II)(4,7-diphenyl-1,10-phenanthroline)$_3$ from charge trapping by doping in carrier-transporting blend films. Chem. Phys. Lett. **385**(5), 481 (2004)
161. Xia, H., Zhu, Y., Lu, D., Li, M., Zhang, C., Yang, B., Ma, Y.: Ruthenium(II) complexes with the mixed ligands 2,2'-bipyridine and 4,4'-dialkyl ester-2,2'-bipyridine as pure red dopants for a single-layer electrophosphorescent device. J. Phys. Chem. B **110**(37), 18718 (2006)
162. Xia, H., Zhang, C., Qiu, S., Lu, P., Zhang, J., Ma, Y.: Highly efficient red phosphorescent light-emitting diodes based on ruthenium(II)-complex-doped semiconductive polymers. Appl. Phys. Lett. **84**(2), 290 (2004)
163. Xia, H., Zhang, C., Liu, X., Qiu, S., Lu, P., Shen, F., Zhang, J., Ma, Y.: Ruthenium(II) complex as phosphorescent dopant for highly efficient red polymers light-emitting diodes. J. Phys. Chem. B **108**(10), 3185 (2004)
164. Oner, I., Sahin, C., Varlikli, C.: Electroluminescence from two new ruthenium(II) complexes as phosphorescent dopant: Positive effect of swallow-tail bipyridyl ligand. Dyes Pigments **95**(1), 23 (2012)
165. Cazeca, M.J., Chittibabu, K.G., Kim, J., Kumar, J., Jain, A., Kim, W., Tripathy, S.K.: Enhanced performance of polythiophene derivative based light emitting diodes by addition of europium and ruthenium complexes. Synth. Metals **98**(1), 45 (1998)
166. Breivogel, A., Park, M., Lee, D., Klassen, S., Kühnle, A., Lee, C., Char, K., Heinze, K.: Push-pull design of bis(tridentate) ruthenium(II) polypyridine chromophores as deep red light emitters in light-emitting electrochemical cells. Eur. J. Inorg. Chem. **2014**(2), 288 (2014)
167. Downey, T.M., Nieman, T.A.: Chemiluminescence detection using regenerable tris(2,2'-bipyridyl)ruthenium(II) immobilized in Nafion. Anal. Chem. **64**(3), 261 (1992)
168. White, H.S., Leddy, J., Bard, A.J.: Polymer films on electrodes. Part 8. Investigation of charge-transport mechanisms in Nafion polymer modified electrodes. J. Am. Chem. Soc. **104** (18), 4811 (1982)
169. Martin, C.R., Rubinstein, I., Bard, A.J.: Polymer films on electrodes. Part 9. Electron and mass transfer in Nafion films containing tris(2,2'-bipyridine)ruthenium(II). J. Am. Chem. Soc. **104**(18), 4817 (1982)
170. Cui, H., Zhao, X.-Y., Lin, X.-Q.: Cathodic electrochemiluminescence of $Ru(bpy)_3^{2+}$/Nafion coated on graphite oxide electrode in purely aqueous solution. Luminescence **18**(4), 199 (2003)
171. Gerardi, R.D., Barnett, N.W., Lewis, S.W.: Analytical applications of tris(2,2′-bipyridyl) ruthenium(III) as a chemiluminescent reagent. Anal. Chim. Acta **378**(1–3), 1 (1999)
172. Guo, Z., Dong, S.: Electrogenerated chemiluminescence determination of dopamine and epinephrine in the presence of ascorbic acid at carbon nanotube/Nafion-$Ru(bpy)_3^{2+}$ composite film modified glassy carbon electrode. Electroanalysis **17**(7), 607 (2005)

173. Yang, Y., Nam, S., Lee, W.-Y.: Tris(2,2′-bipyridyl)ruthenium(II) electrogenerated chemiluminescence ethanol biosensor based on ionic liquid doped titania-Nafion composite film. Microchem. J. **142**, 62 (2018)
174. García-Fresnadillo, D., Marazuela, M.D., Moreno-Bondi, M.C., Orellana, G.: Luminescent Nafion membranes dyed with ruthenium(II) complexes as sensing materials for dissolved oxygen. Langmuir **15**(19), 6451 (1999)
175. Susac, D., Kono, M., Wong, K.C., Mitchell, K.A.R.: XPS Study of interfaces in a two-layer light-emitting diode made from PPV and Nafion with ionically exchanged $Ru(bpy)_3^{2+}$. Appl. Surface Sci. **174**(1), 43 (2001)
176. Yang, H., Li, X., Batchelor-McAuley, C., Sokolov, S.V., Kätelhön, E., Compton, R.G.: Immobilised electrocatalysts: Nafion particles doped with ruthenium(II) tris(2,2′-bipyridyl). Chem. Eur. J. **23**(69), 17605 (2017)
177. Chen, H., Snyder, J.D., Elabd, Y.A.: Electrospinning and solution properties of Nafion and poly(acrylic acid). Macromolecules **41**(1), 128 (2008)
178. Zhou, C., Liu, Z., Dai, J., Xiao, D.: Electrospun $Ru(bpy)_3^{2+}$-doped nafion nanofibers for electrochemiluminescence sensing. Analyst **135**(5), 1004 (2010)
179. Shan, D., Qian, B., Ding, S.-N., Zhu, W., Cosnier, S., Xue, H.-G.: Enhanced solid-state electrochemiluminescence of tris(2,2′-bipyridyl)ruthenium(II) incorporated into electrospun nanofibrous mat. Anal. Chem. **82**(13), 5892 (2010)
180. Li, L., Zhou, L., Liu, X., You, T.: Ultrasensitive self-enhanced electrochemiluminescence sensor based on novel PAN@Ru@PEI@Nafion nanofiber mat. J. Mater. Chem. B **8**(16), 3590 (2020)
181. Zhuo, Y., Liao, N., Chai, Y.-Q., Gui, G.-F., Zhao, M., Han, J., Xiang, Y., Yuan, R.: Ultrasensitive aurinic/apyrimidinic endonuclease 1 immunosensing based on self-enhanced electrochemiluminescence of a Ru(II) complex. Anal. Chem. **86**(2), 1053 (2014)
182. Zhou, L., Huang, J., Yu, B., You, T.: A novel self-enhanced electrochemiluminescence immunosensor based on hollow Ru-SiO_2@PEI nanoparticles for NSE analysis. Sci. Rep. **6** (1), 22234 (2016)
183. Chen, M.-M., Cheng, S.-B., Ji, K., Gao, J., Liu, Y.-L., Wen, W., Zhang, X., Wang, S., Huang, W.-H.: Construction of a flexible electrochemiluminescence platform for sweat detection. Chem. Sci. **10**(25), 6295 (2019)
184. Xiong, C., Liang, W., Zheng, Y., Zhuo, Y., Chai, Y., Yuan, R.: Ultrasensitive assay for telomerase activity via self-enhanced electrochemiluminescent ruthenium complex doped metal-organic frameworks with high emission efficiency. Anal. Chem. **89**(5), 3222 (2017)
185. Wang, H., Yuan, Y., Chai, Y., Yuan, R.: Self-enhanced electrochemiluminescence immunosensor based on nanowires obtained by a green approach. Biosens. Bioelectron. **68**, 72 (2015)
186. Zhang, L., Xiong, C., Wang, H., Yuan, R., Chai, Y.: A sensitive electrochemiluminescence immunosensor for cardiac troponin I detection based on dual quenching of the self-enhanced Ru(II) complex by folic acid and in situ generated oxygen. Sens. Actuators B: Chem. **241**, 765 (2017)
187. Barona-Castaño, J.C., Carmona-Vargas, C.C., Brocksom, T.J., De Oliveira, K.T.: Porphyrins as catalysts in scalable organic reactions. Molecules **21**(3), 310 (2016)
188. Imaoka, T., Yamamoto, K.: Electrochemistry of Co–Ru hetero-dinuclear porphyrin complex in a Nafion matrix. Phys. Chem. Chem. Phys. **3**(19), 4462 (2001)
189. Imaoka, T., Nakazawa, S., Yamamoto, K.: Electroreduction of dioxygen using Co–Ru dinuclear porphyrin catalyst in Nafion film. Chem. Lett. **30**(5), 412 (2001)
190. Smith, R.W., Bryg, V.: Staining polymers for microscopical examination. Rubber Chem. Technol. **79**(3), 520 (2006)
191. Grube, H.L.: Section 29—The Platinum Metals In Handbook of Preparative Inorganic Chemistry, 2nd edn, p. 1560. Academic Press, New York and London (1965) (G. Brauer, Ed.)

192. Bernardis, F.L., Grant, R.A., Sherrington, D.C.: A review of methods of separation of the platinum-group metals through their chloro-complexes. React. Funct. Polym. **65**(3), 205 (2005)
193. Swain, P., Mallika, C., Srinivasan, R., Mudali, U.K., Natarajan, R.: Separation and recovery of ruthenium: A review. J. Radioanal. Nucl. Chem. **298**(2), 781 (2013)
194. Trent, J.S., Scheinbeim, J.I., Couchman, P.R.: Electron microscopy of PS/PMMA and rubber-modified polymer blends: Use of ruthenium tetroxide as a new staining agent. J. Polym. Sci.: Polym. Lett. Ed. **19**(6), 315 (1981)
195. Tencé-Girault, S., Woehling, V., Oikonomou, E.K., Karpati, S., Norvez, S.: About the art and science of visualizing polymer morphology using transmission electron microscopy. Macromol. Chem. and Phys. **219**(3), 1700483 (2018)
196. Sano, H., Usami, T., Nakagawa, H.: Lamellar morphologies of melt-crystallized polyethylene, isotactic polypropylene and ethylene-propylene copolymers by the RuO_4 staining technique. Polymer **27**(10), 1497 (1986)
197. Steeling, C.: Crystal-structure of ruthenium red and sterochemistry of its pectic stain. Am. J. Bot. **57**(2), 172 (1970)
198. Luft, J.H.: Electron microscopy of cell extra-neous coats as revealed by ruthenium red staining. J. Cell Biol. **23**, 54A (1964)
199. Luft, J.H.: Fine structure of hyaline cartlitage matrix following ruthenium red fixative and staining. J. Cell Biol. **27**, 61A (1965)
200. Luft, J.H.: Ruthenium red and violet. Part I. Chemistry, purification, methods of use for electron microscopy and mechanism of action. Anat. Rec. **171**(3), 347 (1971)
201. Luft, J.H.: Ruthenium red and violet. Part II. Fine structural localization in animal tissues. Anat. Rec. **171**(3), 369 (1971)
202. Fabien, V., Matthew, A.J.D.: TRPV4 agonists and antagonists. Curr. Top. Med. Chem. **11** (17), 2216 (2011)
203. Li, J.X., Ness, J.N., Cheung, W.L.: Lamellar structure of POM spherulites imaged by a two-stage RuO_4 staining technique. J. Appl. Polym. Sci. **59**(11), 1733 (1996)
204. Hong, S., Bushelman, A.A., MacKnight, W.J., Gido, S.P., Lohse, D.J., Fetters, L.J.: Morphology of semicrystalline block copolymers: Polyethylene-*b*-atactic-polypropylene. Polymer **42**(13), 5909 (2001)
205. Montezinos, D., Wells, B.G., Burns, J.L.: The use of ruthenium in hypochlorite as a stain for polymeric materials. J. Polym. Sci.: Polym. Lett. Ed. **23**(8), 421 (1985)
206. Janik, H., Walch, E., Gaymans, R.J.: Ruthenium tetroxide staining of polybutylene terephthalate (PBT) and polyisobutylene-*b*-PBT segmented block copolymers. Polymer **33** (16), 3522 (1992)
207. Hobbs, S.Y., Dekkers, M.E.J., Watkins, V.H.: Toughened blends of poly(butylene terephthalate) and BPA polycarbonate. J. Mater. Sci. **23**(4), 1219 (1988)
208. Delimoy, D., Goffaux, B., Devaux, J., Legras, R.: Thermal and morphological behaviours of bisphenol-A polycarbonate/poly(butylene terephthalate) blends. Polymer **36**(17), 3255 (1995)
209. Haubruge, H.G., Jonas, A.M., Legras, R.: Staining of poly(ethylene terephthalate) by ruthenium tetroxide. Polymer **44**(11), 3229 (2003)
210. Ansari, I.A., Hamley, I.W.: Templating the patterning of gold nanoparticles using a stained triblock copolymer film surface. J. Mater. Chem. **13**(10), 2412 (2003)
211. Olayo-Valles, R., Lund, M.S., Leighton, C., Hillmyer, M.A.: Large area nanolithographic templates by selective etching of chemically stained block copolymer thin films. J. Mater. Chem. **14**(18), 2729 (2004)
212. Vital, A., Vayer, M., Sinturel, C., Tillocher, T., Lefaucheux, P., Dussart, R., Boufnichel, M.: Modification of poly(styrene) thin films and enhancement of cryogenic plasma etching resistance by ruthenium tetroxide vapor staining. Polymer **76**, 123 (2015)
213. Vayer, M., Nguyen, T.H., Sinturel, C.: Ruthenium staining for morphological assessment and patterns formation in block copolymer films. Polymer **55**(4), 1048 (2014)

214. Lokhande, C.D., Dubal, D.P., Joo, O.-S.: Metal oxide thin film based supercapacitors. Curr. Appl. Phys. **11**(3), 255 (2011)
215. Gui, Z., Duay, J., Hu, J., Lee, S.B.: Redox-exchange induced heterogeneous RuO_2-conductive polymer nanowires. Phys. Chem. Chem. Phys. **16**(24), 12332 (2014)
216. Majumdar, D., Maiyalagan, T., Jiang, Z.: Recent progress in ruthenium oxide-based composites for supercapacitor applications. ChemElectroChem **6**(17), 4343 (2019)
217. Deng, J., Yan, J., Tilly, J.C., Deng, L., Mineart, K.P., Spontak, R.J.: Incorporation of metallic species into midblock-sulfonated block ionomers. Macromol. Rapid Commun. **39** (22), 1800427 (2018)

Metal–organic and Covalent Organic Frameworks Incorporating Ru Species

6

Abstract

The 3D metal–organic frameworks (MOFs) or 2D covalent organic frameworks (COFs) represent highly ordered, solid-state materials, which are employed in various fields of research. Hybrid materials in which Ru complexes or nanoparticles are caged or incorporated are discussed in the contexts of photovoltaics or catalysis. In particular, the energy-related applications (e.g., CO_2 reduction of the hydrogen-evolution reaction) have emerged as central fields of research.

6.1 Introduction

As pointed out in the Introduction, metal–organic-frameworks (MOFs) represent porous 3D coordination architectures of, typically, crystalline nature, which are discussed in the context of gas adsorption/separation, magnetism, catalysis, (optical) sensing, drug delivery, etc. [1]. There are still no strict rules regarding the nomenclature of MOFs. Beside the usage of the formula unit (defined by the stoichiometry of the metal centers and organic ligands in the MOF), acronyms are most commonly used. These either refer to the place of origin (e.g., research institutes and universities, as in the case of MIL, HKUST or UiO) or describe the material (e.g., MOF or ZIF). The most prominent MOF structures are depicted in Fig. 6.1. In general, metal ions (or small metal-oxide clusters) are combined with organic linkers—most commonly bi- or multidentate carboxylates—to afford highly-ordered MOFs via hydro- or solvothermal, microwave- or sonication-assisted as well as electro- or mechanochemical reactions; in addition, also diffusion methods, solvent evaporation, and ionothermal approaches have been applied to assembled such materials. Thus, MOFs have to be distinguished from covalent organic frameworks (COFs), which are exclusively based on the light main-group

© The Author(s), under exclusive license to Springer Nature Switzerland AG 2021
U. S. Schubert et al., *Ruthenium-Containing Polymers*,
https://doi.org/10.1007/978-3-030-75598-0_6

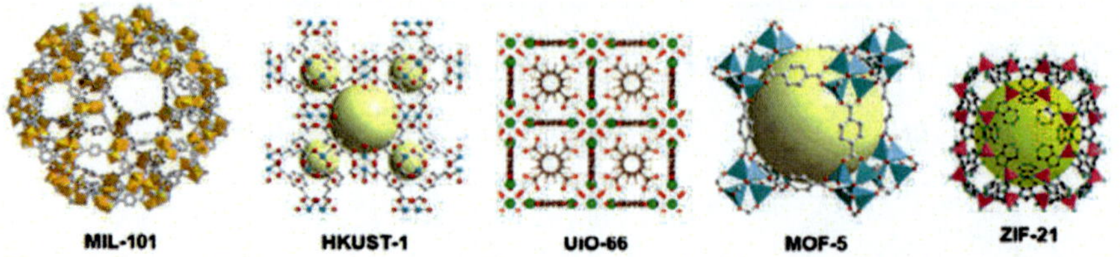

Fig. 6.1 Schematic representation of representative MOF structures: (MIL: Matériaux de l'Institut Lavoisier; HKUST: Hong Kong University of Technology; UiO: Universitetet i Oslo; ZIF: zeolitic imidazole frameworks [7, 8]. Figure reproduced with kind permission; © 2010 The Royal Chemical Society and 2016 Elsevier B.V

elements (i.e., C, N, O, B, S, P). These materials, which are commonly prepared by dynamic covalent self-assembly processes (Fig. 6.2), are discussed in the context of gas-storage, catalytic, sensing, and opto-electronic applications [2–6].

Making use of the aforementioned synthetic strategies, a plethora of MOF architectures have been assembled and, when focusing onto ruthenium *ca.* 450 literature reports can be found (search terms "metal–organic-framework" and "ruthenium", according to Scifinder™ by Dec 31, 2020) [9]. With respect to Ru-containing MOFs, two different types of architectures can be distinguished: MOFs in which small-molecule Ru species—complexes or nanoparticles—are "dispersed" or "caged" and MOFs in which Ru^{II} complexes represent integral parts of the 3D architecture. The same holds true for COFs, where, e.g., Ru nanoparticles have been dispersed in COFs.

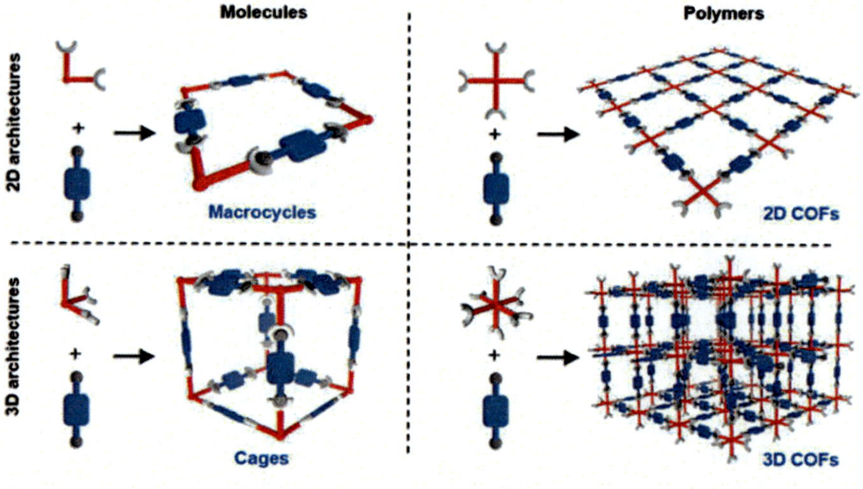

Fig. 6.2 Schematic representation of the various 2D and 3D COF nanostructures, which can be prepared via dynamic covalent self-assembly procedures using appropriate building blocks [2]. Figure reproduced with kind permission; © 2018 Wiley–VCH

Compared to the broad range of Ru-containing polymers, which were discussed in the previous Chapters, the combination of Ru chemistry with MOFs and COFs represents a rather fledgling field of research. The first Ru-containing polymers were already studied in 1970s regarding catalytic applications [10]; whereas, MOFs and COFs have emerged as support or lead structure to host Ru species quite recently.

6.2 Metal–organic-Frameworks with Embedded Ru Nanoparticles or Ru^{II} Complexes

Regarding the "dispersion" of ligand-free Ru nanoparticles in metal–organic frameworks, the application of ZIFs, as solid support, has attracted considerable interest. ZIFs are porous, crystalline architectures formed from imidazole derivatives and transition metal ions, which prefer the tetrahedral coordination geometry. Tailor-made ZIFs can be assembled by proper choice of building blocks and, thus, the porosity and surface can be adjusted over a broad range (Fig. 6.3a) [11, 12]. As the first example, Zahmakiran et al. immobilized $RuCl_3$ into the cages of $[Zn(2\text{-mim})_2]_n$, commonly referred to as ZIF-8 (2-mim: 2-methylimidazolate) [13]. Subsequent treatment with $NaBH_4$ generated Ru_{NP}@ZIF-8—a material that was characterized by X-ray diffraction analysis (XRD, Fig. 6.3b), X-ray photoelectron spectroscopy (XPS), N_2 adsorption–desorption measurements, and transmission electron microscopy (TEM) imaging. Ru_{NP}@ZIF-8 was employed, as a heterogeneous catalyst, in the transfer hydrogenation of olefins with *N,N*-dimethylamine-borane, as the hydrogen source. Throughout the catalytic runs, leaching or sintering of the nanoparticles was not observed. In the same context, ultrafine Ru NPs were

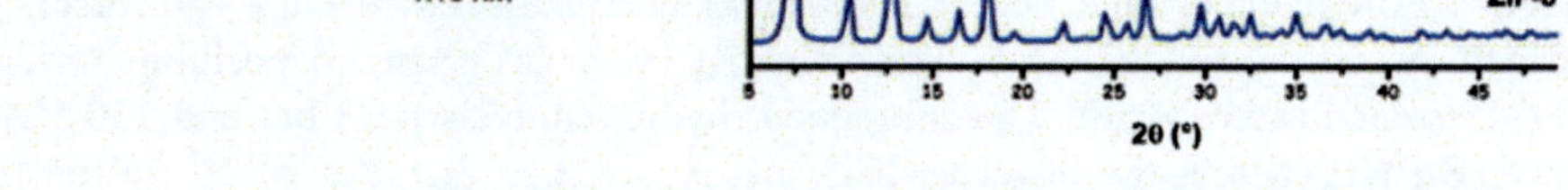

Fig. 6.3 **a** Representation of the solid-state structure of ZIF-8, a framework assembled from Zn^{II} ions and 2-methyl-imidazolate [the blue polyhedral represent the Zn^{II} centers, N and C atoms are shown as spheres and lines, respectively] [11]. **b** Comparison of the powder XRD patterns of ZIF-8 and Ru_{NP}@ZIF-8 [13]. Figure reproduced with kind permission; © 2010 American Chemical Society and 2014 Elsevier B.V

embedded into the pores of MIL-101, as an example for a zeolite-type MOF, via a straightforward liquid impregnation method [14]. This hybrid material offered a high catalytic activity for the hydrolysis of NH_3BH_3 [turnover frequency (TOF) of 178 min^{-1}].

As discussed in Chap. 7, Ru NPs can also be prepared via the so-called polyol process [15] or by $NaBH_4$ reduction; however, in either case stabilization by surfactants or the dispersion in polymer matrix [e.g., polyvinylalcohol (PVA) or cellulose] represents a prerequisite. The immobilization of Ru NPs within MOF architectures is an attractive alternative approach in which caging effects limit the NPs growth and their agglomeration.

More recently, the dispersion of atomic Ru species in a ZIF-8 template was reported by Wang, Li et al. [16] These authors utilized $Ru(acac)_3$ and $Ru_3(CO)_{12}$, as precursors, which were in-situ encapsulated within the framework during its assembly and crystallization process. The thermal reduction of these frameworks gave atomic Ru_1 or small Ru_3 clusters, respectively, dispersed in the ZIF-8 environment. The presence of these species, trapped within the framework's cages, was corroborated by high-angle annular dark-field (HAADF) scanning-tunneling electron microscopy imaging (STEM, Fig. 6.4). A size-dependent catalytic activity was identified: Ru_{NP}@ZIF-8 and Ru1@ZIF-8 showed a similar performance in the hydrogenation of phenylacetylene, as model substrate (conversion of 13% and 18%, respectively); whereas, Ru_3@ZIF-8 offered a conversion of 48% under the otherwise identical experimental conditions. Noteworthy, the confined pyrolysis of $Ru_3(CO)_{12}$@ZIF-8 (800 °C, 3 h, H_2/Ar atmosphere) gave Ru_3 clusters, which were stabilized by carbon–nitrogen species [17]. These clusters, in turn, were highly active and selective catalysts for the oxidation of 2-aminobenzaldehyde [turn-over frequency (TOF) of 4,320 h^{-1}].

ZIF-67 represents an isostructural analogue to ZIF-8, which can be readily assembled from Co^{II} and 2-mim at ambient temperature. Following Zahmakiran's protocol, Tuan and Lin prepared Ru_{NP}@ZIF-67 and tested the material, as a heterogeneous catalyst, for the hydrogen production from $NaBH_4$ under basic conditions [18]. Compared to Ru_{NP}@ZIF-8, a significantly higher catalytic activity was found—presumably due to the Co^{II} centers within the ZIF-67, which actively contributed to the H_2 production (according to Kim et al. [19] Co^{II}-based ZIFs themselves represent catalysts for the hydrolysis of $NaBH_4$).

A different strategy to prepare Ru nanoparticles within a pre-assembled MOF was reported by Fischer et al. [20] These authors employed [Ru(cod)(cot)] (cod = 1,5-cyclooctadiene, cot = 1,3,5-cyclooctatriene), as volatile precursors, which were embedded into $[Zn_4O(bdc)_3]$ via gas-phase deposition (bdc: 1,4-benzenedicarboxylate). The subsequent hydrogenolysis (at 3 bar and 150 °C) gave Ru NPs within the intact MOF-5 matrix (average diameter of *ca.* 1.6 nm). Solid-state NMR spectroscopy indicated a weak interaction between the particles and the surrounding MOF structure. The colloidal deposition of preformed Ru, P, or bimetallic RuPt nanoparticles onto MIL-101, was reported by Li et al.; the RuPt@MIL-1010 hybrid material acted as a heterogeneous catalyst for the hydrogenation of benzene under solventless conditions (TOF of 0.71 s^{-1} at 60 °C

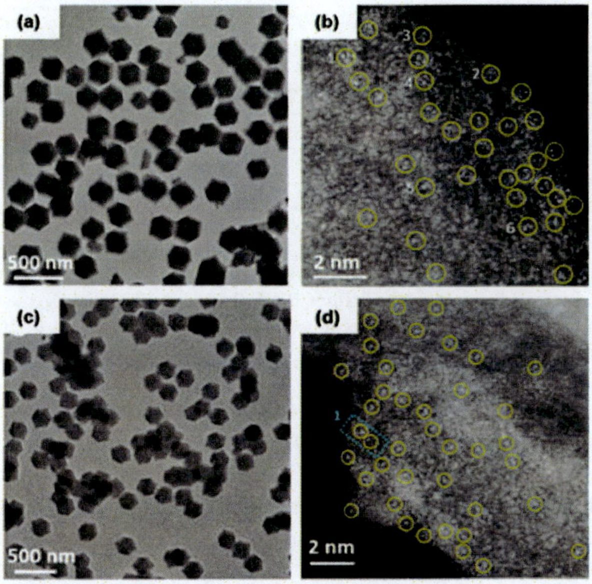

Fig. 6.4 Representative TEM (**a** and **c**) and HAADF-STEM images (**b** and **d**) of Ru_1@ZIF-8 (**a** and **b**) and Ru_3@ZIF-8 (**c** and **d**) [16]. Figure reproduced with kind permission; © 2019 Wiley–VCH

and 1 MPa of H_2) [21]. The related RuB@MIL-53 catalyst gave a TOF of 6.4 s^{-1} for the same reaction; remarkably the catalyst functioned even at 30 °C and a low H_2 pressure of 1 MPa [22].

Han et al. employed a supercritical [CO_2-MeOH] solution of $RuCl_3 \times xH_2O$ to deposit Ru NPs onto a MOF, assembled from La^{III} ions and btc (1,3,5-benzenetricarboxylate) under solvothermal conditions [23]. This particular technique was introduced by Watkins et al. in the late 1990s and allows the controlled deposition of films or nanoparticles into planar and porous substrates [24]. This hybrid MOF was analyzed by microscopy imaging, XPS, and infrared (IR) spectroscopy (Fig. 6.5) [23]. On the basis of these experiments, the authors proposed a binding of the NPs to the MOF via its outward facing carboxylate moieties. It was further shown that the hybrid material was a potent catalyst for the hydrogenation of cyclohexene or benzene (a significantly higher activity compared to Ru/C composites or Ru_{NP}@MOF, derived from aqueous suspensions, was found). In related work, these same authors utilized the same reaction conditions to embed Ru NPs within a MOF structure, which was assembled from $ZrOCl_2$ and bdc [25]. This MOF with embedded Ru NPs was found to be a superior heterogeneous catalyst for the hydrogenation of benzene, which proceeded at 60 °C and a H_2 pressure of 6 MPa (TOF of 5260 h^{-1}); for comparison, the aforementioned La^{III}-based MOF, which carried a layer of Ru NPs, gave a much lower TON of 2475 h^{-1} under the

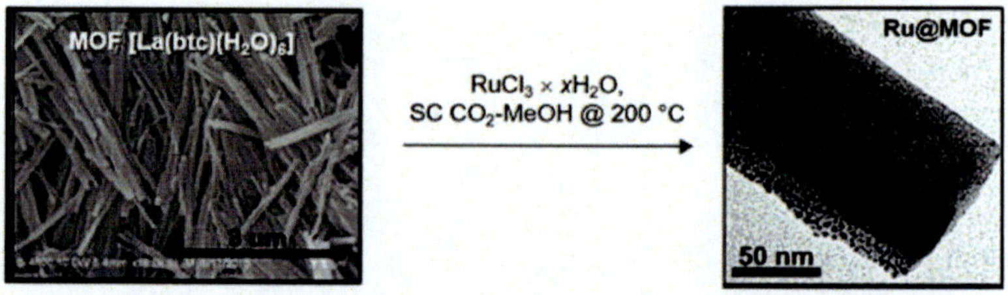

Fig. 6.5 Deposition of Ru NP onto the surface of a Ln^{III}-based MOF. A representative scanning electron microscopy (SEM) image of the pristine MOF (left) and a TEM image the obtained hybrid MOF (right) are shown [23]. Figure reproduced with kind permission; © 2011 The Royal Chemical Society

otherwise same experimental conditions. Noteworthy, the supercritical CO_2 has also been employed, as reaction medium, to deposit Ru NPs onto various inorganic materials: Mesoporous SiO_2 [26], carbon nanotubes (CNTs) [27], or Si wafers [28].

Besides the embedding of Ru NPs, the caging of Ru^{II} complexes, in particular of the prominent $[Ru(bpy)_3]^{2+}$ photosensitizer and photocatalyst, within appropriate MOF architectures represents a highly active field of research. From a synthetic point-of-view, this approach is more straightforward as an alternative way to directly incorporate such complexes into the MOF architecture (thus, making them integral parts of the MOF structure). As discussed in Sect. 6.3, such systems are often prone to catalyst leaching due to photodegradation. The main benefits of the caging approach were summarized by Grzybowski et al. [29] The cavities of a proper MOF are able to accommodate the Ru^{II} dye; whereas, its channels are too narrow to allow leaching. Moreover, the confined environment suppresses photodegradation, even upon prolonged irradiation.

These requirements are met, for example, by MOFs assembled from γ-cyclodextrin (γ-CD), a naturally occurring cyclic oligosaccharide, and heavy alkali-metal ions (i.e., Rb^{I} or Cs^{I}). MOFs derived from CD ("CD-MOFs") have found widespread application in highly diverse areas of research: Drug delivery, sorption, separation and purification of gases, sensing, food packaging, electrical conductors, memristors, photocatalysis, polymer science, etc. [30]. Grzybowski et al. prepared CD-MOFs from the co-crystallization of γ-CD and Rb(OH)/Cs(OH) in the presence of $[Ru(bpy)_3]Cl_2$ (Fig. 6.6a) [29]. The resultant materials, reminiscent of Calzaferri's Y-zeolites with entrapped $[Ru(bpy)_3]^{2+}$ dyes [31], were analyzed by powder XRD analysis, TEM imaging, and UV/vis absorption spectroscopy (Fig. 6.6b). Besides the anticipated benefits, i.e., photostability and suppression of dye leaching, the CD-MOF provided OH^- ions, which served, as electron donors, in the photoreduction of metal salts to afford nanoparticles, such as Pd, Ag, or Au NPs. Very recently, Zhu and Liu et al. investigated the utilization of the chiral environment within CD-MOFs to induce circularly polarized luminescence (CPL) [32]. These authors tested a wide range of chromophores, including the $[Ru(bpy)_3]^{2+}$ and hexa*kis*phenyl-substituted $[Ru(phen)_3]^{2+}$ ion (phen: 1,10-phenanthroline).

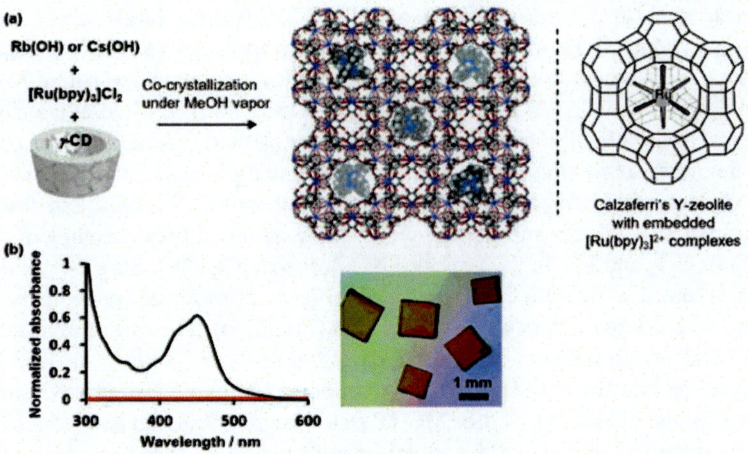

Fig. 6.6 a Schematic representation of the synthesis of a $[Ru(bpy)_3]^{2+}$-containing CD-MOF via a co-crystallization strategy [29]; a related Y-zeolite structure is also shown for comparison [31]. **b** UV/vis absorption spectrum and photograph of the orange-brown nanocrystals of the [Ru (bpy)$_3$]$^{2+}$-containing CD-MOF [29]. Figure reproduced with kind permission; © 2013 Wiley–VCH and 1996 American Chemical Society

However, only the latter gave a distinct CPL signal, which was attributed to the stronger interaction of the bulkier dye with the surrounding CD-MOF structure.

The intrinsic metal-centered chirality of $[Ru(bpy)_3]^{2+}$ was capitalized by Train et al. to obtain 3D oxalate-bridged networks of the formula $\{[Ru(bpy)_3][Cu_{2x}Ni_{2(1-x)}(C_2O_4)_3]\}_n$ ($0 \leq x \leq 1$) [33]. Thereby, the enantiomeric pure Ru^{II} dye acted as templating agent to translate chirality into the MOF architecture, as evidenced by circular dichroism (CD) spectroscopy. This work represents one of the earliest examples, where $[Ru(bpy)_3]^{2+}$ was employed, as a template, for the assembly of MOF structures. Vice versa, the MOFs also had an impact on the properties of the $[Ru(bpy)^3]^{2+}$ chromophore around which they were assembled. For example, Koner et al. prepared anionic, sheet-like coordination networks comprising of $[M(dca)_3]^-$ units (dca: dicyanamide; M: Mn^{II} or Fe^{II}) in which $[Ru(bpy)_3]^{2+}$ ions were entrapped [34]. These assemblies featured dual photoluminescence at room temperature, unlike for pristine $[Ru(bpy)_3]^{2+}$; this finding was attributed to a decoupling of the excited states due to subtle differences in the microenvironment around the bpy ligands within the network. Larsen and Wojtas reported on the stabilization of the triplet metal-to-ligand charge-transfer (^{3}MLCT) state of $[Ru(bpy)_3]^{2+}$ when caged in a MOF, i.e., the USF-2 framework [35]. Remarkably, the excited-state lifetime (τ) was, presumably due to the deactivation of non-radiative relaxation pathways, increased from 614 ns (in EtOH) to 1.2 μs. The same USF-2 motif, comprised of Zn^{II} ions and btc ligands, was also used to host $[Ru(phen)_3]^{2+}$ or, very recently, $[Os(bpy)_3]^{2+}$ ions; in both cases, the stabilization of the ^{3}MLCT state resulted in prolonged emission lifetimes [36, 37].

Larsen et al. also prepared two isomeric MOFs from a single set of building blocks, i.e., $Zn(NO_3)_2$, the btc ligand, and $[Ru(bpy)_3]X_2$ (X: Cl^- or PF_6^-), as template [38]. Noteworthy, in the absence of the Ru^{II} template, the parent MOF-39 structure was formed. RWLC-1 and RWLC-2 were formed simultaneously in a single reaction vial and their crystals had to be separated from each other by hand under the microscope (Fig. 6.7a). XRD structure analysis suggested the location of the Ru^{II} dyes within the cavities (RWLC-1) or channels (RWLC-2) of the negatively charged MOF. In subsequent work, these same authors assembled a MOF from $Zn(NO_3)_2$ and bdc in the presence of $[Ru(bpy)_3]Cl_2$ [39]. Single-crystal XRD studies revealed a twofold interpenetrated pillared honeycomb network with the Ru^{II} dyes distributed within the crystal lattice (Fig. 6.7b). This material, referred to as RWLC-3, exhibited two emission decay lifetimes (τ of 120 ns and 453 ns). It was proposed that the Ru^{II} complexes were located in two different environments. In one of these, quenching of the ^{3}MLCT state occurred from an exogenous donor; whereas, the other modulated the triplet ligand field (^{3}LF) state and, thus, allowed the population of either a ^{1}MLCT or a fourth ^{3}MLCT state {different from those accessible from pristine $[Ru(bpy)_3]^{2+}$}. A similar behavior was observed for the $[Ru(bpy)_3]^{2+}$ complexes embedded into a MOF comprised of Cd^{II} ions and bdc ligands; this material, named RWLC-5, had a different structure than the related Zn^{II}-based

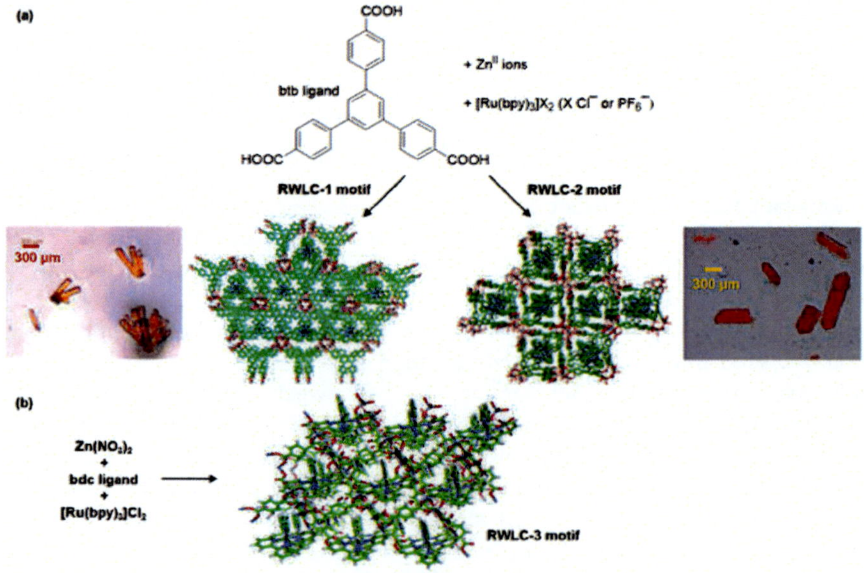

Fig. 6.7 **a** Schematic representation of the formation of the isomeric MOFs RWLC-1 and RWLC-2 from a single set of building blocks. The microscopy images of the two materials are also depicted [38]. **b** Schematic representation of the synthesis of the MOF RWLC-3 [39]. Figure reproduced with kind permission; © 2014 American Chemical Society and 2015 The Royal Chemical Society

RWLC-3 assembly with crystallographic resolvable and non-disordered complex sites [40]. In recent years, the concept of encapsulating $[Ru(bpy)_3]^{2+}$ in the cavities of MOFs was then extended to a range of other MOF architectures: classical UiO-66 [41], amine-modified UiO-66 (UiO-66-NH_2) [42], HKUST-1 [43], and RWLC-6 [44]. In contrast to other frameworks, UiO-66-NH_2 did not stabilize the excited state of the encapsulated Ru^{II} complexes; detrimentally, the amino groups accelerated the decay, presumably due to perturbations of lower lying ^{1,3}MLCT states [42].

A remarkable improvement of the photophysical properties of $[Ru(bpy)_2(py)_2]^{2+}$ by caging this dye in the cavities of the anionic Zn^{II}–oxalate MOF was reported by Lin et al. [45] The monodentate py ligands of $[Ru(bpy)_2(py)_2]^{2+}$ were prone to rapid photo-substitution by, e.g., solvent molecules, thus photophysical studies on this particular complex were very difficult to perform. However, the ligand-exchange processes were fully suppressed when the complex was encapsulated in a confined MOF environment; molecular-modelling studies indicated that the cavities of the MOF were fully occupied by the Ru^{II} complex, thus not leaving any space for competitive solvent molecules. As a result, the photoluminescence lifetime and quantum yield as well as the photostability were increased tremendously. This study highlights that not only the modulation of photophysical properties is possible when Ru^{II} dyes are encapsulated within MOF architectures, but even sensitive complexes, which diminish themselves from photophysical characterization by fast photo-decomposition, can be studied when caged by a rigid MOF environment.

Yamashita et al. prepared an amino-modified MIL-125 (MIL-125-NH_2), comprised of 2-aminoterephthalic acid, linkers, and $[Ti_8O_8(OH)_4]$ SBUs, in which $[Ru(bpy)_3]^{2+}$ complexes were encapsulated [46]. This assembly was highly active in the photocatalytic oxidation of benzyl alcohol to benzaldehyde under aerobic conditions. For comparison, neither the Ru^{II}-free MIL-125-NH_2, parent MIL-125, nor MIL-125 with encapsulated Ru^{II} complexes offered such a high catalytic activity. Thus, absorption of visible light by both subunits, the MIL-125-NH_2 framework and Ru^{II} complexes, was found to be crucial for the observed photocatalytic activity, in particular regarding the fast formation of the reactive $[O_2^{\cdot-}]$ and $[^{\cdot}OH]$ intermediates. However, MIL-125, encapsulating $[Ru(bpy)_3]^{2+}$ complexes within its framework, was found to be a versatile photocatalyst for the selective oxidative coupling of amines under ambient conditions [47] (Fig. 6.8).

The heterogeneous photocatalyst FJI-Y2 was prepared by Kang, Yuan et al. under solvothermal conditions from $MnCl_2$, a tetracarboxylated NDI derivative and $[Ru(bpy)_3]Cl_2$ (FJI: Fujian Institute, NDI: naphthalenediimide; Fig. 6.9) [48]. Remarkably, the lattice comprised of pillared NDI radicals, which strongly interacted with the encapsulated Ru^{II} complexes via multiple CH-π and slipped π-π interactions. These non-covalent interactions, in turn, facilitated the light-induced single electron-transfer (SET) process between the NDI and Ru^{II} sites and, thus, boosted the overall photocatalytic activity of the assembly compared to a combination of NDI and $[Ru(bpy)_3]^{2+}$ under homogeneous conditions. This was demonstrated as an example of the aerobic cross-dehydrogenative coupling of 2-phenyl-1,2,3,4-tetrahydro-isoquinoline derivatives with phosphite esters. Very

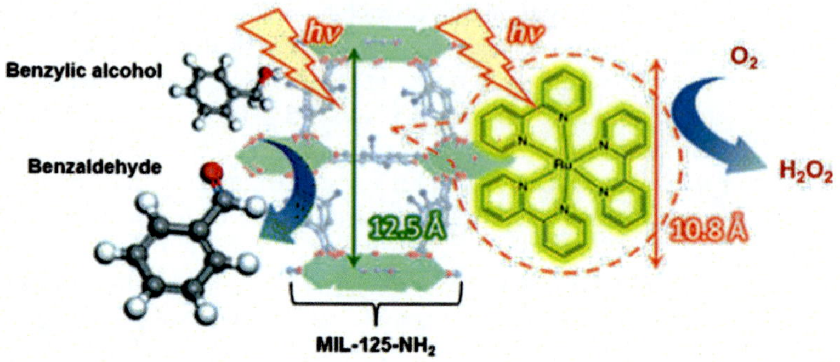

Fig. 6.8 Schematic representation of MIL-125-NH_2 with encapsulated Ru^{II} complexes. This hybrid material was used, as photocatalyst, for the aerobic oxidation of benzylic alcohol to benzaldehyde [46]. Figure reproduced with kind permission; © 2019 The Royal Chemical Society

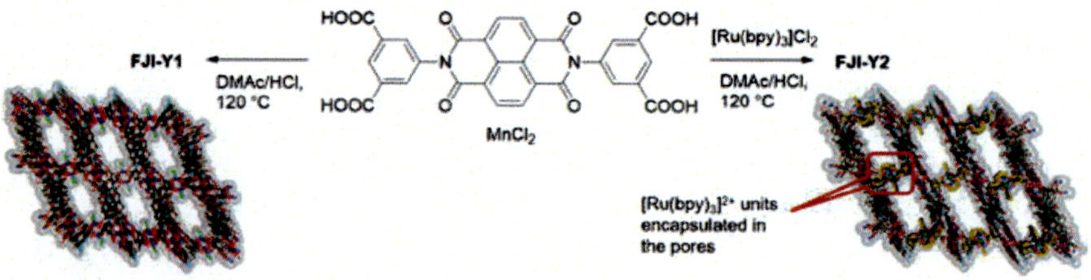

Fig. 6.9 Schematic representation of the synthesis of the MOFs FJI-Y1 and FJI-Y2, the latter encapsulating $[Ru(bpy)_3]^{2+}$ moieties. The solid-state structures of these MOFs are also shown [48]. Figure reproduced with kind permission; © 2018 Elsevier B.V

recently, Zhang et al. encapsulated $[Ru(bpy)_3]^{2+}$ within the UiO-67 framework [49]. This assembly was used, as a heterogeneous photocatalyst, in an oxidative cross-coupling reaction between enamides and aryl sulfonates in the presence of nitrobenzene, as oxidant (25 examples, >60% yield) (Fig. 6.10).

The PCN-99 framework, featuring cuboctahedral cages and 1D channels, was designed as a matrix MOF for the encapsulation of $[Ru(bpy)_3]^{2+}$. A topological analysis guided Zhou et al. to use a sophisticated tripodal linker for the solvothermal synthesis of PCN-99 with $In(NO_3)_3$ (PCN: porous coordination network, Fig. 6.11) [50]. Due to the anionic nature of PCN-99, the uptake of $[Ru(bpy)_3]^{2+}$ cations from solution readily occurred; whereas, anionic organic dyes, such as Congo red, were not embedded into the framework. The thusly obtained material was tested, as a heterogeneous photocatalyst, in the oxidative hydroxylation of arylboronic acids. However, a slightly decreased catalytic activity, compared to $[Ru(bpy)_3]^{2+}$ under homogeneous conditions, was found. This finding was attributed to an impeded access of the substrate to the catalytic sites within the framework.

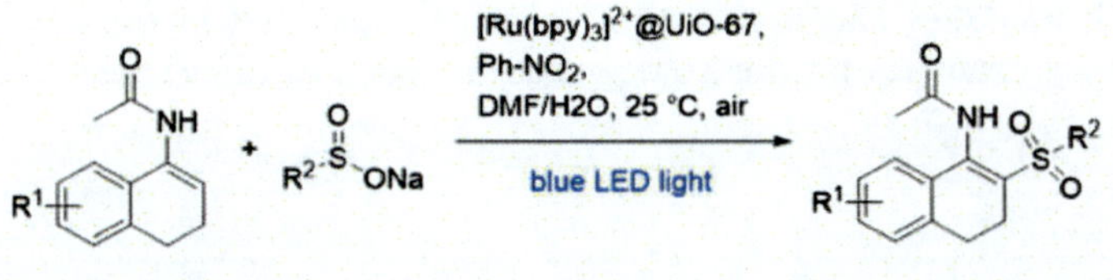

Fig. 6.10 Schematic representation of the synthesis of *β*-acetylamino acryl sulfones via an oxidative cross-coupling procedure, using $[Ru(bpy)_3]^{2+}$ encapsulated within UiO-67, as photocatalyst [49]

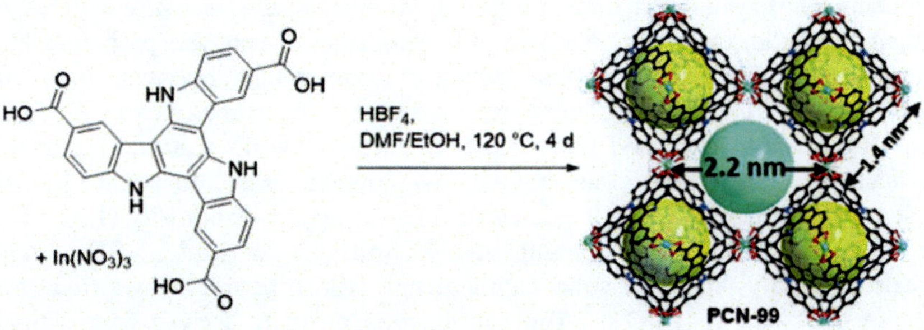

Fig. 6.11 Schematic representation of the solvothermal synthesis of PCN-99. The solid-state structure of this MOF is also depicted [50]. Figure reproduced with kind permission; © 2016 The Royal Chemical Society

Meyer, Lin et al. performed luminescence-quenching studies on MOFs with caged $[Ru(bpy)_3]^{2+}$ complexes (Zn^{II}–oxalate and mixed-metal Na^{I}/Al^{III}–oxalate frameworks were used in this study) [51]. It was demonstrated that the trace amounts of O_2, trapped within the MOF structure, significantly reduced the excited-state lifetime by almost one order of magnitude. Moreover, $[Ru(bpy)_3]^{2+}$ and $[Os(bpy)_3]^{2+}$ (0.2 to 1 mol-%) were co-encapsulated in the $[NaAl(oxalate)_3]$ structure. As in the case of the MOF incorporating these complexes within their framework structure (see Fig. 6.15), the photo-excited Ru^{II} dyes were rapidly deactivated by energy transfer to the Os^{II} sites. Thus, light harvesting and energy transfer, reminiscent of the processes occurring in natural photosystems, were found to be "in action" in this assembled hybrid structure.

The encapsulation of $[Ru(bpy)_3]^{2+}$ within an anionic porous coordination cage enabled the efficient degradation of methylene blue [52]. In solution, the degradation of this cationic dye by $[Ru(bpy)_3]^{2+}$, as photocatalyst, is hampered by the electronic repulsion of the chromophores. The anionic framework, however, attracted the dye, thus bringing it into close proximity to the encapsulated photocatalyst and, thus, improved the catalytic performance. Porous coordination cages, as in this example, are currently considered as artificial enzymes, which offer distinct binding sites and/or microenvironments where certain reactions are promoted; many examples for this strategy can be found in Zhou's recent review on this topic [53].

6.3 Metal–organic-Frameworks Assembled from Ru^{II} Species

The diruthenium–tetracarboxylate motif has been widely used to assemble coordination polymers comprised of mixed-valence $Ru_2^{II/III}$ and/or reduced $Ru_2^{II/II}$ fragments. These solid-state materials have been investigated *inter-alia* with respect to their magnetic, electronic, mesomorphic, occlusion or catalytic properties [54]. For example, Takamizawa et al. prepared coordination polymers from the mixed-valence $Ru_2(\mu\text{-OAc})_4Cl$ precursor and various *bis*-carboxylates (i.e., fumarate, all-*trans*-muconate, and bdc) [55, 56]. These networks featured remarkable gas-occlusion properties; the highest N_2 uptake was observed for the bdc-containing network, which offered the highest degree-of-porosity (1.3 mol N_2 per mole of Ru atoms). Regarding the $[Ru_2(bdc)_2Cl]$ network, XRD analysis revealed a highly organized solid-state structure with supermicropores (total pore sizes of 0.24 $cm^3\ g^{-1}$) [57].[57] The coordination polymer derived from adipate, however, was mesoporous structure (total pore sizes of 0.47 $cm^3\ g^{-1}$). Both assemblies were thermally stable up to 300 °C and had surface areas of 235 $m^2\ g^{-1}$ and 281 $m^2\ g^{-1}$, respectively, according to BET analyses. In a similar fashion, coordination polymers exclusively containing $Ru_2^{II/II}$–dicarboxylate units were studied with respect to their gas-uptake ability and catalytic activity in the hydrogenation of ethylene [58]. The first phase-pure crystalline MOF containing $Ru^{II/III}$ fragments was reported by Fischer et al. [59] These authors used $RuCl_3 \times xH_2O$ and btc to grow thin MOF films on alumina or silica surfaces under solvothermal conditions (Fig. 6.12). According to XRD analysis, applying the Rietveld's refinement, the MOF had a HKUST-1 structure, thus being isostructural to $[Cu_3(btc)_2]$. An improved synthesis, using the $Ru_2(\mu\text{-OPiv})_4Cl$ precursor (Piv: pivalate), was reported by Wade and Dincă [60]. The thusly obtained $[Ru_3(btc)_2]$ MOF had a significantly increased BET surface compared to Fischer's material (1180 ± 5 vs. 704 $m^2\ g^{-1}$) and showed a higher CO_2-adsorption enthalpy than other MOFs of the HKUST-1 family. This finding was attributed to the higher formal charge of the dimetal units, which enabled a stronger electrostatic interaction with the CO_2 molecules.

It is well-known that the structure, physical properties, and chemical reactivity of MOFs can be modified by varying the structure of the organic ligand [61]. In particular, the creation of structural defects, i.e., vacant coordination sites on the metal centers, was found to have a profound impact on the catalytic activity of MOFs [62]. Such defective MOFs can be prepared by solvothermal synthesis using a mixture of organic ligands, i.e., a combination of the "regular" ligand and a "defect-introducing" one [63]. The latter typically possess the same size as the

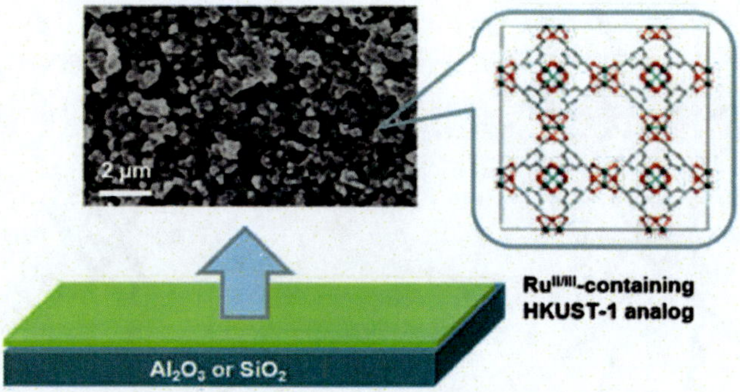

Fig. 6.12 Schematic representation of the $Ru^{II/III}$-containing MOF film assembled on Al_2O_3 or SiO_2 substrates. A representative SEM image of the MOF film on Al_2O_3 is also shown [59]. Figure reproduced with kind permission; © 2011 The Royal Chemical Society

regular ligand, but has a decreased number of binding sites. Against this background, Wang, Fischer et al. prepared defective MOFs using btc and pydc, as organic ligands (pydc: 3,5-pyridinedicarboxylate; Fig. 6.13) [64]. Remarkably, the pydc ligand did not only introduce vacant coordination sites, but also induced the partial reduction of the Ru centers. These defective Ru-containing MOFs were active in the $-CO_2$ to $-CO$ reduction via chemisorption of CO_2 as well as in the catalytic hydrogenation of olefins. The latter involved formation of a Ru–H species, as evidenced from FT-IR spectroscopy. It was demonstrated later that these defective MOFs also enabled the ethylene dimerization with high selectivity (99% production of 1-butene) and durability (>120 h), even in the absence of activators and solvents [65]. The catalytic activity of the defective MOFs was strongly enhanced by applying a hydrogen pre-treatment; both the Meerwein-Ponndorf-Valerey reduction of ketones and the isomerization of allylic alcohol to ketones were catalyzed by the H_2-activated MOFs [66].

One further defective MOF, using isophthalate, as a defect-introducing co-ligand, was reported by these same authors [67]. As in the previous case, this material also showed a good capability in the sorption of gaseous small molecules (e.g., CO_2, CO, H_2) and the catalytic activity. The latter was demonstrated as the examples of the ethylene dimerization and Paal–Knorr reaction.

Recently also mixed-metal HKUST-1 analogs, combining Cu^{II} and Ru^{III} ions or $Ru^{II/III}$ and Rh^{II} ions have been reported [68, 69]. Such mixed-metal MOFs can be prepared by direct preparative routes or by performing a cation exchange on a pre-formed monometallic MOF. The increasing interest in mixed-metal MOFs results from the ability to modify the catalytic, gas-sorption or opto-electronic properties of the related monometallic MOFs by introducing different metals within a single framework [70]. For example, a direct preparative route was utilized by Bauer, Kleist et al. to prepare a bimetallic MOF in which 8% of the Cu^{II} centers of the parent HKUST-1 framework were replaced by Ru^{III} sites (thus, giving rise to a

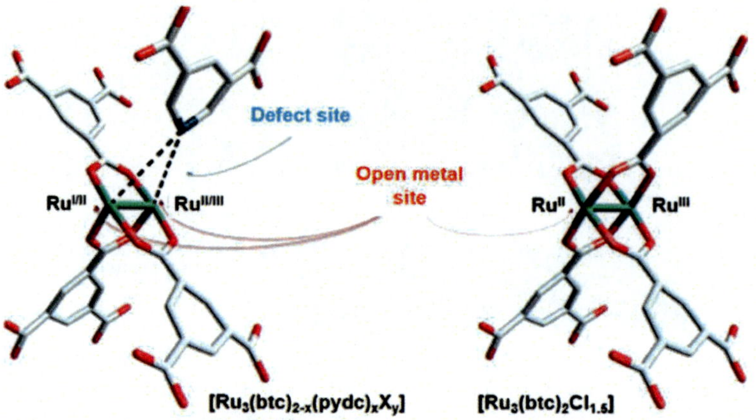

Fig. 6.13 Schematic representation of the defective MOF $[Ru_3(btc)_{2-x}(pydc)_xX_y]$ (left) and its "regular" counterpart $[Ru_3(btc)_2Cl_{1.5}]$ (right) [64]. Figure reproduced with kind permission; © 2014 Wiley–VCH

stoichiometric formula of $[Cu_{2.75}Ru_{0.25}(btc)_2]$) [68]. This bimetallic MOF was used to fabricate Ru-HPC electrodes (Ru-HPC: Ru-decorated hierarchically porous carbon) for the electrocatalytic hydrogen-evolution reaction (HER) [71]. For this purpose, the mixed-metal MOF was pyrolyzed to afford Cu- and Ru-doped carbon (CuRu-C) from which the Cu nanoparticles were removed by washing with a Fe^{III} solution (Fig. 6.14). Thusly, they obtained Ru-HPC that was subsequently combined with Nafion™ and immobilized onto a glassy carbon (GC) electrode. This hybrid electrode functioned at an overpotential of 22.7 mV (current density of 25 mA cm^{-2}) and produced H_2 with a very high turnover frequency (TOF) of 1.79 s^{-1} at 25 mV, thus outcompeting most electrocatalysts performing in an alkaline environment. Noteworthy, the hybrid electrodes fabricated from the respective monometallic MOFs were, by far, less efficient in producing H_2. Targeting the water-splitting reaction, a two-electrode device was constructed in which the Ru-HPC electrode was combined with a porous RuO_2 electrode (the latter was prepared by oxidizing Ru-HPC). This device gave a current density of 10 mA cm^{-2} at only 1.53 V and was more productive than a water-splitting cell comprising commercial Pt/C and RuO_2 electrodes.

Fischer et al. prepared a full series of mixed-metal MOFs of the general formula $[Ru_xRh_{3-x}(btc)_2]$; thereby, the Rh content was varied from 0 to 100% [69]. The crystalline and highly porous products were thoroughly characterized by powder XRD and thermogravimetric analysis (TGA), gas-sorption studies, various spectroscopic methods (i.e., IR, Raman, UV/vis absorption, NMR, and XPS) as well by high-resolution TEM imaging with elemental mapping (HAADF-EDS). The experimental data suggested a homogeneous distribution of the $Ru_2^{II/III}$ and Rh_2^{II}

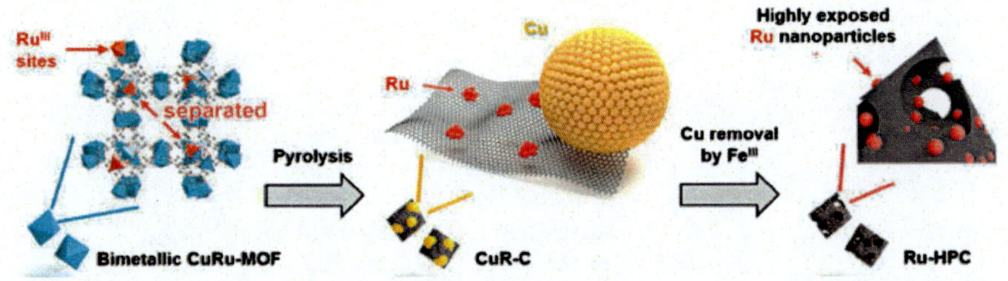

Fig. 6.14 Schematic representation of the preparation of Ru-decorated hierarchically porous carbon (Ru-HPC) from a bimetallic CuRu-MOF; Cu- and Ru-doped carbon (CuRu-C) was obtained as intermediate material [71]. Figure reproduced with kind permission; © 2019 Elsevier B.V

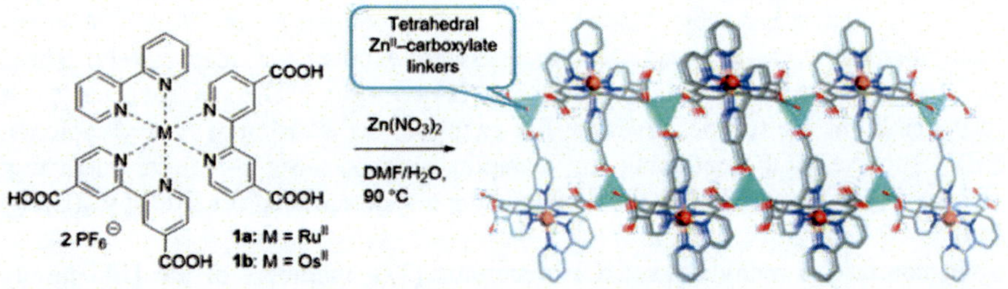

Fig. 6.15 Schematic representation of the synthesis of MOFs containing photoactive Ru^{II} and/or Os^{II} complexes [73]. Figure reproduced with kind permission; © 2010 American Chemical Society

units within the framework, which was isoreticular to the HKUST-1 motif (acetate anions were incorporated into the MOF to compensate structural defects, arising from missing btc ligands, and some charge-balancing chloride ions were coordinated to the Ru$^{II/III}$ centers).

Similar to their polymers with Ru^{II} and/or Os^{II} *tris*-bpy complexes in the side chains (for details, see Sects. 2.1 and 2.2) [72], Papanikolas, Meyer, Lin et al. assembled MOFs from the photoactive complexes **1** and $Zn(NO_3)_2$ (Fig. 6.15) [73]. When only one type of metal center was incorporated, the MOFs exhibited photophysical properties similar to the small-molecule complexes, i.e., $[Ru(bpy)_3]^{2+}$ and $[Os(bpy)_3]^{2+}$. However, the mixed-metal MOFs (containing 0.3 to 2.6 mol-% of Os) revealed a fast, long-distance energy transfer (EnT) from photo-excited Ru^{II} centers to Os^{II} sites. In-depth photophysical studies suggested that the EnT process followed a Dexter-type mechanism, i.e., a triplet-to-triplet EnT involving a multistep incoherent exciton hopping [74]. Different pathways, such as an 1D and 3D hopping, were evidenced on the basis of experimental and computational work. These results were consistent with those obtained from the investigation of the aforementioned MOFs with caged $[Ru(bpy)_3]^{2+}$ chromophores (see Sect. 6.3) [51].

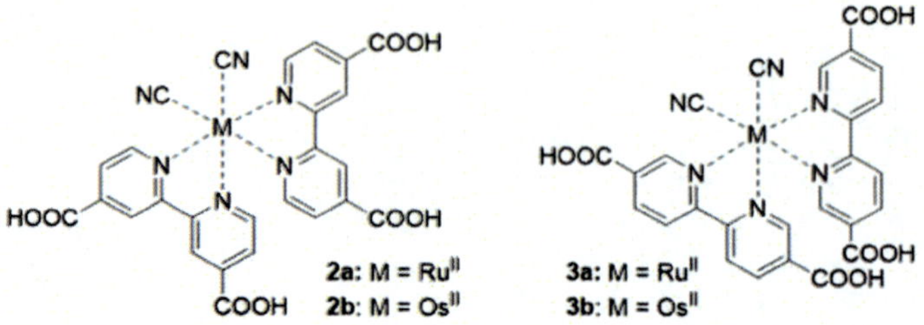

Fig. 6.16 Schematic representation of the Ru^{II} and Os^{II} complexes **2** and **3**, which were used, as building blocks, for the formation of photoactive MOFs

In the same context, Meyer et al. also reported MOFs containing the photoactive complexes **2** and **3** (Fig. 6.16) [75, 76]. The phosphorescence of these MOFs was efficiently quenched by adding, e.g., benzoquinone or *N,N,N′,N′*-tetramethylbenzidine, as an oxidative or a reductive quencher, respectively. These authors proposed non-covalent interactions between the MOF and quencher to be responsible for the fast deactivation. For example, the addition of methyl viologen (MV^{2+}) to the MOF containing Ru^{II} complex **3a** gave a Stern–Volmer quenching constant, which was 7,000 times higher as the one determined for $[Ru(bpy)_2(CN)_2]$ [75].

In contrast to many other MOF structures [77], members of the UiO family exhibit a reasonable hydrolytic stability in aqueous environments. The UiO-type MOFs contain $[Zr_6(\mu_3\text{-}O)_4(\mu_3\text{-}OH)_4]^{12+}$ units, as secondary building units (SBUs), which are linked dicarboxylate ligands [78]. The most prominent representatives are the MOFs in which bdc (UiO-66) or bpdc ligands (bpdc: 4,4′-biphenyldicarboxylate, UiO-67) are incorporated. It was demonstrated that UiO-67 was stable against a variety of polar solvents; however, disassembly occurred in 0.1 M aqueous HCl and NaOH solutions [79]. This property was utilized by Lin et al. to dope UiO-67 with transition-metal complexes for which (photo-)catalytic activity in water has been reported [80]. Besides various Ir^{III}-based water-oxidation catalysts (WOCs) and a Rh^{I}-based catalyst for the photoreduction of CO_2, also the photocatalysts **4** were used, as a building block, to assemble doped UiO-67 derivatives (Fig. 6.17a). These assemblies were tested, as heterogeneous photocatalysts, in organic model reactions: the aza-Henry reaction, aerobic amine coupling, aerobic thioanisole oxidation, and debromination of α-bromoketones (Fig. 6.17b) [78, 81]. In all cases, a very high catalytic activity was found; in control experiments, the parent UiO-67 was catalytically inactive. Moreover, the stability of the MOFs under the catalytic conditions was evidenced by comparing the powder XRD patterns before and after catalysis. The highly porous MOFs, having BET surfaces of 1277 and 2292 $m^2\ g^{-1}$, respectively, were also employed for the sensing of O_2 via phosphorescence quenching [82]. The materials showed a quenching efficiency of 65% at an O_2 pressure of 0.8 atm. Maza and Morris studied the photophysical properties of UiO-67 doped with $[Ru(bpy)_2(dcbpy)]^{2+}$ units (dcbpy: 5,5′-

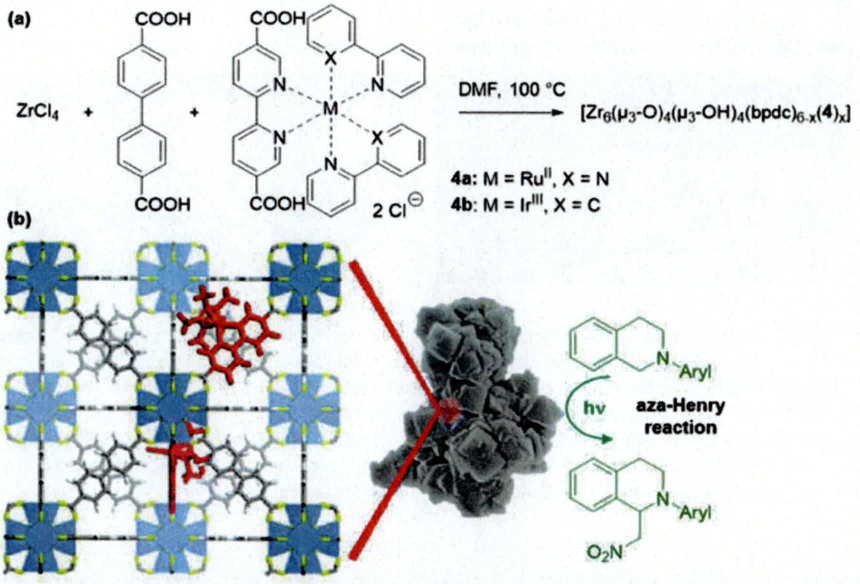

Fig. 6.17 **a** Schematic representation of the synthesis of UiO-67, doped with Ir^{III} or Ru^{II} photocatalysts. **b** Schematic representation of the usage of the doped UiO-67, as photocatalyst, in organic transformations (e.g., in the aza-Henry reaction) [80]. Figure reproduced with kind permission; © 2011 American Chemical Society

dicarboxy-2,2'-bipyridine) [83–85]. These authors identified a concentration dependency of the emission lifetime (τ) and energy of the phosphorescence band (λ_{ph}): Both were found to decrease with an increasing degree-of-doping. This finding was attributed to EnT processes between neighboring Ru^{II} complexes within the MOF, presumably via a resonance EnT, and a poor solvation of the complex sites due to the occupancy of pores at high degrees-of-doping.

Morris et al. also raised the question whether the Ru^{II}-doped UiO-frameworks (or related structures) could even be employed, as sensitizer, in solar-cell applications [86, 87]. It was reported that the MOF-based solar cells (MOFSCs) had significantly lower efficiencies (below 1%) compared to conventional dye-sensitized solar cells (DSSCs) containing Ru^{II} photosensitizers. In more detail, light harvesting was strongly dependent on the distribution of the Ru^{II} sites in the thin MOF film; the film thickness as well as the boundary between the MOF and the semiconductor were identified as crucial parameters. The device structure and the relevant processes are summarized in Fig. 6.18.

These same authors also fabricated a solid-state electrochemoluminescent (ECL) device containing the Ru^{II}-doped UiO-67, as the emissive active material [88]. In ECL devices, luminescence basically originates from excited states, which result from the electrochemical formation of intermediates and subsequent

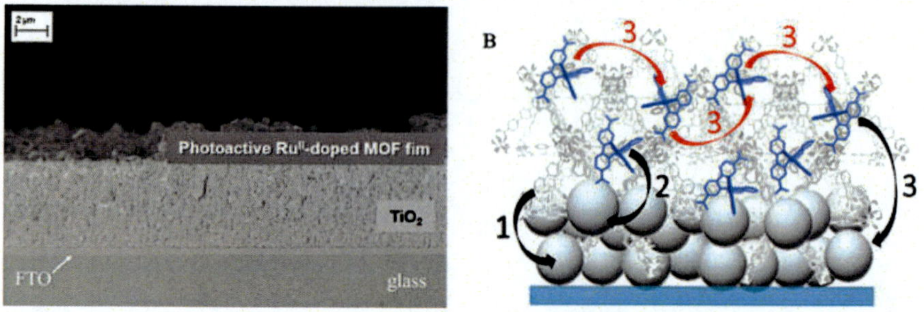

Fig. 6.18 Left: Side-on SEM image of the solar-cell device with the configuration Ru^{II}-doped UiO-67|TiO_2|FTO-coated glass; the MOF films extends into the semiconductor layer by *ca.* 2 µm. Right: Schematic representation of the aforementioned device configuration. Three different elemental steps are depicted: (1) Charge separation between the organic ligands and the TiO_2; (2) Charge separation between the Ru^{II} centers and the TiO_2; (3) Non-directional EnT processes within the MOF (red arrows) and subsequent electron injection into the TiO_2 (black arrows) [86]. Figure reproduced with kind permission; © 2017 Elsevier B.V

electron-transfer reactions [89]. It has been demonstrated that solid-state devices are superior to solution-based ones, since expensive reagents and/or additives (e.g., Nafion™ membranes, examples for this are shown in Sect. 3.1) are avoided and, even more important, the ECL intensity is increased [90]. An FTO electrode was covered with the Ru^{II}-doped UiO-67 film and gave, in the presence of tripropylamine, bright and sustainable ECL with an emission maximum of *ca.* 660 nm (thus, matching to the characteristic photoluminescence wavelength of the employed Ru^{II} complex). The ECL intensity could be modulated via the applied amine concentration, giving a linear relationship over a concentration range (from 0.04 to 20 mM). The applicability of the device, as sensor for dopamine, was further shown.

Carboni et al. performed a post-synthetic modification of the aforementioned Ru^{II}-doped UiO-67 structure by substituting some of the Zr^{IV} centers in the coordination nodes by isoelectronic Ti^{IV} units [91]. In the resultant assembly, the Ru^{II} complexes served as light-harvesting antennae and the Ti^{IV} sites offered catalytic activity. The applicability of this hybrid MOF design was successfully tested in the degradation of organic dyes, such as methylene blue, upon irradiation with visible light.

Tackling the (electro-)chemical water-oxidation reaction (WOR), Pushkar et al. recently incorporated $[Ru(bpy)(dcbpy)(OH_2)_2]^{2+}$ units, as WOC sites, via carboxylate substituents into the UiO-67 framework [92]. These authors tested both synthetic routes, i.e., the direct synthesis and the post-synthetic modification of UiO-67. However, only the latter was found to give the modified MOF—either as a solid-state material or immobilized onto a fluorine-doped tin oxide (FTO) electrode (Fig. 6.19). It is well-established that mononuclear Ru^{II} WOCs generate Ru^{V}=O species, which are crucial for their catalytic performance {at least in the case of *cis*-

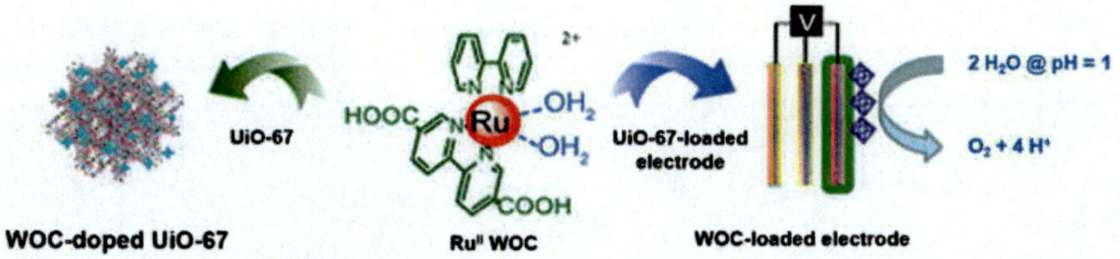

Fig. 6.19 Schematic representation of the post-synthetic modification of UiO-67 and an FTO electrode covered with a thin UiO-67 film. The latter enabled the electrochemical WOR [92]. Figure reproduced with kind permission; © 2020 American Chemical Society

$[Ru(bpy)_2(OH_2)_2]^{2+}$ and related complexes} [93]. These oxo species enable the O–O bond formation by a nucleophilic reaction with water, further oxidation than facilitates the O_2 evolution. As an alternative pathway, μ-oxo-bridged dimers, the so called "blue dimers" might be formed in the presence of Ce^{IV}, as oxidant, which might affect the catalytic activity [94]. However, this deactivation pathway was efficiently suppressed by embedding the WOC into the rigid UiO-67 framework [92]. Under operative conditions, the essential $Ru^{V} = O$ intermediate could be identified by in-depth spectroscopic studies, thus corroborating that the proposed mechanism for the WOR was also valid for the UiO-immobilized catalyst. Moreover, the WOC modified UiO-67@FTO electrode allowed the electrocatalytic production of O_2; a TOF of 0.26 s^{-1} over 6 h was observed for the bulk electrolysis, thereby exceeding the performance of the pristine UiO-67@FTO electrode by two orders of magnitude.

The corresponding hydrogen-evolution reaction (HER) is of similar importance and represents a highly active field of current research; in this context, photocatalytic, electrocatalytic as well as photoelectrocatalytic strategies are discussed. In one approach molecular dyads are used in which a Ru^{II} photosensitizer is covalently linked to a $Pt(bpy)Cl_2$ unit, as the actual HER catalyst [95]. Sakai et al. showed that these dyads efficiently produced H_2 from water upon irradiation with visible light; however, ethylenediaminetetraacetic acid (EDTA) had to be added, as the sacrificial donor. In line with Sakai's work on molecular HER catalysts, Chen et al. prepared a UiO-67 derivative, which was doped with $[Ru(bpy)_2(dcbpy)]^{2+}$ (**4a**) and Pt(dcbpy) Cl_2 units [96]. This doped UiO-67 was characterized thoroughly by a range of analytical methods [e.g., XRD and thermogravimetric analysis, gas-sorption measurements, microscopy imaging (including elemental mapping), etc.; Fig. 6.20]. The incorporation of the Ru^{II} and Pt^{II} complexes units within the MOF facilitated the electron transfer between the metal centers and enabled H_2 production to proceed at a pH of 5.0 upon irradiation with visible light. However, the considerable leaching of the sensitizer led to a drastic decrease of the H_2 evolution within 4 h. Huang et al. replaced the Pt^{II} centers by Co^{II} centers, as HER catalyst sites; this doped UiO-67 functioned as a potent photocatalyst for the HER and gave H_2 from aqueous MeCN over 40 h [turn-over number (TON) of 99 based on Co^{II}] [97]. In

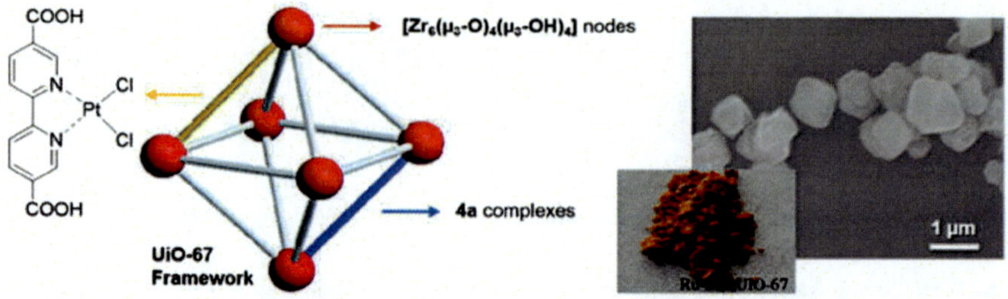

Fig. 6.20 Schematic representation of the UiO-67 derivative containing Ru^{II} and Pt^{II} complexes, as photosensitizer and HER catalyst, respectively. A representative photograph and a SEM image of the material are also depicted [96]. Figure reproduced with kind permission; © 2015 The Royal Chemical Society

related work, Pan et al. introduced electron-deficient Ru^{II} or Rh^{III} half-sandwich complexes into the UiO-67 framework. Such complexes are *inter-alia* known for their catalytically active properties, which include the photocatalytic HER and CO_2 reduction [98]. In particular, Rh^{III}-doped UiO-67 featured a high hydrolytic stability and durability in terms of continuous H_2 production over 174 h without significant decrease in efficiency.

The Ru^{II} *bis*-thiocyanate complex **N3** represents one of most prominent dyes to be used, as a photosensitizer, in DSSCs [99]. It was shown by Liu et al., that **N3** can also be utilized for the colorimetric detection of Hg^{II} ions: The interaction of the heavy-metal ions with the *S*-atoms of the thiocyanate ligands resulted in a distinct color change from red to yellow [100]. This behavior was applied by Gu et al., who incorporated $[Ru(bpy)(dcbpy)(NCS)_2]$ into an UiO-67 framework, thus preparing a robust colorimetric detector for Hg^{2+} ions in water (Fig. 6.21) [101]. UV/vis absorption spectroscopy offered a detection limit (as low as 0.5 μM); Hg^{II} ions were detected selectivity, even in the presence of competitive metal ions at high concentrations. Moreover, the detector could be regenerated by adding I^-, which formed HgI_2, as a barely water-soluble salt (solubility product of 3.20×10^{-29} $mol^3 L^{-3}$).

As one limiting factor regarding the application of modified MOFs of the UiO family in the context of light-driven processes, their photostability must be considered. Recently, García et al. noted that UiO-66 (as well as other prominent MOF structures) are prone to decarboxylation upon prolonged irradiation with UV/vis light—even under inert conditions [102]. As a result, the porosity and surface area of the MOFs are strongly altered. This, in turn, is expected to have considerable influence on the photophysical properties and the resulting applications. Thus, the strategy of using UiO, as lead structure, for preparing Ru-containing MOFs for sustainable photocatalytic or light-harvesting applications is questionable. Regarding the photostability, the zeolitic imidazole frameworks, in particular ZIF-8, seem to be much more promising (such MOFs exhibited no gas evolution upon continuous irradiation).

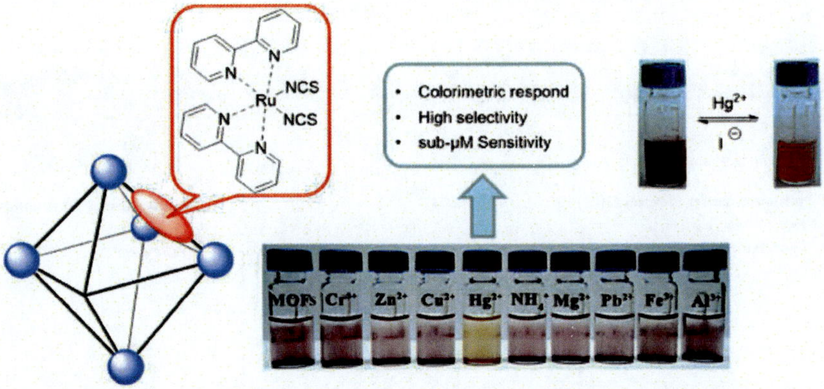

Fig. 6.21 Schematic representation of the UiO-67, doped with [Ru(bpy)$_2$(NCS)$_2$] complexes, which enabled a highly selective, sensitive, and reversible colorimetric detection of Hg^{II} ions [101]. Figure reproduced with kind permission; © 2018 The Royal Chemical Society

The obstacle of MOF's poor stability under photocatalytic conditions was addressed by Luo et al. [103]. These authors assembled a two-fold interpenetrated structure from the heteroleptic Ru^{II} complex **1a** and Cd^{II} ions under solvothermal conditions (similarly, the non-interpenetrated counterpart was obtained from **5** and Cd^{II} ions). Interpenetration, i.e., the entanglement of (at least) two isolated motifs, has been found to increase the stability of MOFs while retaining (or even improving) their gas-storage or catalytic properties [104]. Accordingly, Luo's two-fold interpenetrated MOF showed an increased thermal and photic stability when compared to the non-interpenetrated one (Fig. 6.22). Even more important, the material exhibited a high durability and recyclability under the conditions of the photocatalytic CO_2-to-formate reduction.

Luo et al. also combined complex **5** with In^{III} ions to afford a highly symmetric MOF, which was characterized *inter-alia* by SEM imaging and single-crystal XRD analysis [105]. The microcrystalline material showed a broad absorption of visible light and long-lived emission arising from the Ru^{II} complexes within the confined environment. The photo- and redox-activity of the MOF was used for the photo-degradation of organic dyes, such as methyl orange, and detection of harmful nitroaromatics by fluorescence quenching. For 2,4,6-trinitrotoluene (TNT) a quenching efficiency of *ca.* 25% was reached at a very low concentration of 125 ppm. The same complex, **5**, was employed by Luo et al. to assemble a MOF with a flower-like hierarchical nanostructure [106]. In this MOF, the [$Cd_2(CO_2)_6$] SBUs linked six **5** molecules; whereas, each complex bridged six adjacent SBU units thus giving rise to a regular 3D 6-connected primitive cubic topology (Fig. 6.24a). The remarkable flower-like nanostructure of the assembly was visualized by microscopy imaging (Fig. 6.24b). These authors also employed the Ru^{II}-containing MOF, as a heterogeneous catalyst, in a photocatalytic CO_2 reduction. The experiments showed a high selectivity for the formation of formate, good photostability and recyclability (Fig. 6.23).

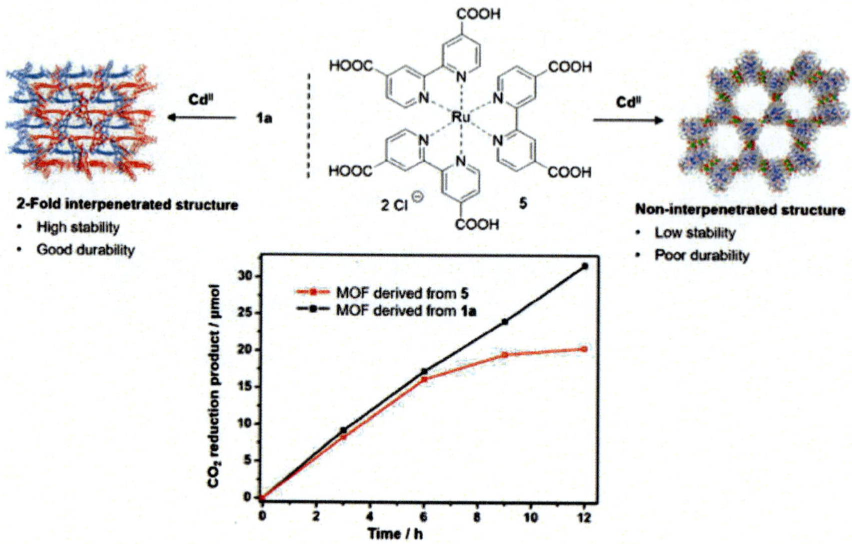

Fig. 6.22 Schematic representation of the synthesis of a two-fold interpenetrated and a non-interpenetrated MOF via the self-assembly of **1a** and **5**, respectively, with Cd^{II} centers. The MOF structure was found to have an impact on the stability and durability under the conditions of the light-induced CO_2 reduction [103]. Figure reproduced with kind permission; © 2015 American Chemical Society

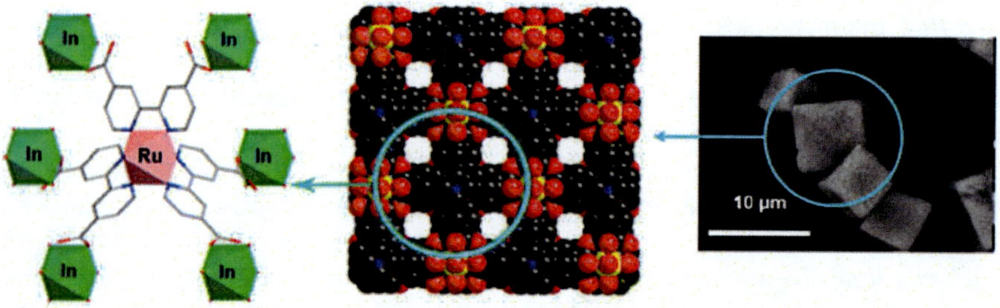

Fig. 6.23 Schematic representation of the connectivity within the synthesized mixed-metal MOF (left). The space-filling model (view along the [010] dimension), as determined by XRD analysis is also shown (middle). The MOF afforded cubic microcrystals, which were visualized by SEM imaging (right) [105]. Figure reproduced with kind permission; © 2013 American Chemical Society

The first isoreticular MOF (IRMOF), $[Zn_4O(bdc)_3]$, was reported by Yaghi et al. in 1999 (MOF-5 or IRMOF-1, Fig. 6.25a) [107]. Since then, the "reticular synthesis" has become a versatile strategy to design functional materials [108]. For example, Xiang, Yan et al. incorporated the phosphonate-functionalized Ru^{II}

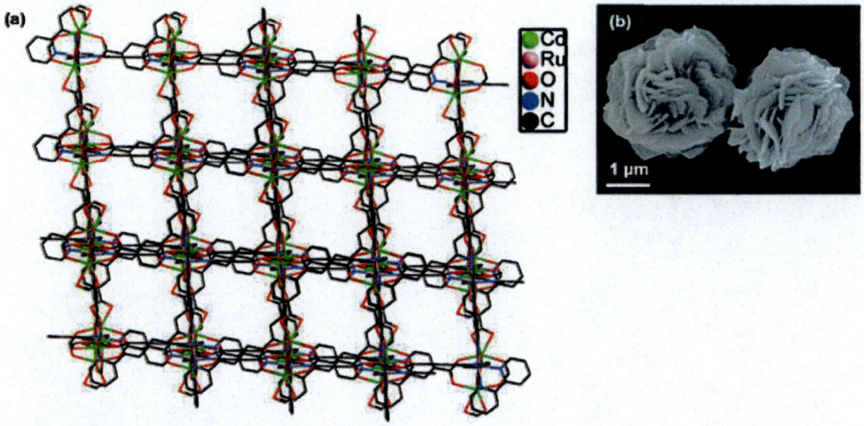

Fig. 6.24 **a** Ball-stick representation of the solid-state structure of the MOF comprised of $[Cd_2(CO_2)_6]$ SBUs and complex WW. **b** Representative SEM image showing the flower-like nanostructure of the MOF [106]. Figure reproduced with kind permission; © 2015 The Royal Chemical Society

complex **6** into the framework of IRMOF-10 without alteration of the overall architecture (Fig. 6.25b) [110]. Remarkably, the doped MOF exhibited a two-color emission behavior, as visualized by 3D confocal fluorescence microscopy; i.e., emission of blue and red light occurred from the interior and exterior of **6**@IRMOF-10, respectively. Moreover, energy up-conversion and an increased emission lifetime, compared to pristine **6**, was also found. The same holds true regarding the photoelectrochemical properties of TiO_2 nanostructures, which differed when loaded with **6**@IRMOF-10, IRMOF-10 or **6**. These authors proposed that host–guest interactions were responsible for the opto-electronic properties of the hybrid assembly (i.e., energy- and/or electron-transfer processes between the MOF and the Ru^{II} complex).

Lin et al. investigated the influence of the charge of isostructural complexes on the overall architecture of the MOF that they were incorporated into [111]. The charge-neutral complex **7** could readily be incorporated into the $[Zn_4(\mu_4\text{-}O)(bpdc)_3]$ framework at a doping level ranging from 0 to 100%. However, the steric demand of **7** induced a phase-transition from the parent interpenetrated IRMOF-9 structure to the non-interpenetrated, charge-balanced IRMOF-10 motif (Fig. 6.26). The same behavior was also observed for charged **1a/b**, though only at low degrees-of-doping due to the charge-mismatch phenomenon. Thus, at higher degrees-of-doping (**1a**: > 6%; **1b**: > 8%), the formation of a new interpenetrated, charge-balanced framework was found (Fig. 6.26). Thus, two factors were identified to control the structure of MOFs: (1) steric demand and (2) charge balance of the organic building blocks.

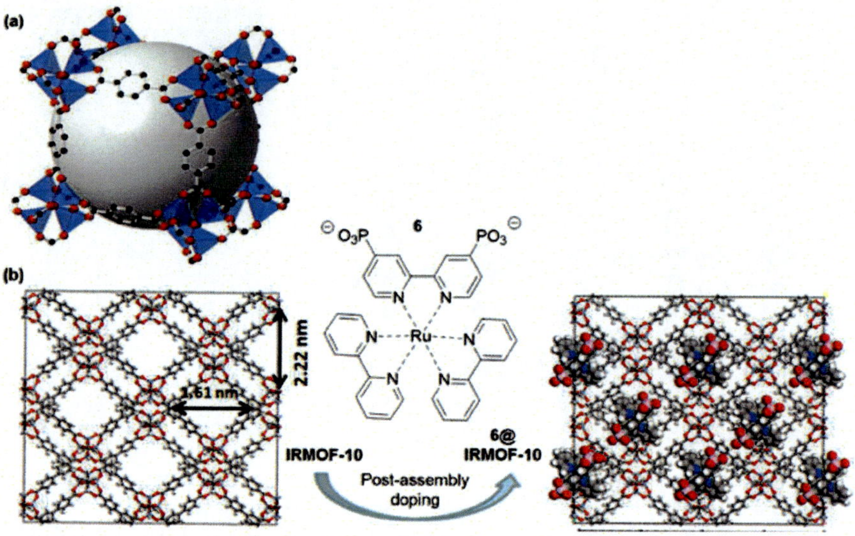

Fig. 6.25 a Schematic representation of the unit cell of MOF-5 *alias* IRMOF-1. The sphere in the middle of the cube represents the size of the cavity, i.e., the maximum size of a guest molecule to fit into the cavity [109]. **b** Schematic representation of the structure of IRMOF-10, which could be doped with Ru^{II} dye **6** without alteration of the MOF architecture [110]. Figure reproduced with kind permission; © 2012 IOP and 2014 American Chemical Society, respectively

Two structurally related MOFs comprising of metalloporphyrin-centered tetracarboxylates and $[Ru_2]$ units were synthesized by Lin et al. (the used metalloporphyrin contained Ru^{II} or Zn^{II} ions) [112]. In these assemblies, the close proximity of these SBU and the metalloporphyrin sites facilitated the multiple ElT from the photoexcited porphyrins to the $\{Ru_2\}$, which acted, as catalyst sites, for the HER in pH-neutral water (Fig. 6.27); the photocatalytic HER activity of the MOFs was, in the presence of triethanolamine (TEOA), as the sacrificial donor, almost 30 times higher compared to a mixture of the metalloporphyrins and $\{Ru_2\}$ clusters under homogeneous conditions (the MOFs gave TONs of 21.2 and 39.4 in the case of incorporated Ru^{II}-porphyrin and Zn^{II}-porphyrin, respectively). Tavernelli, Smit et al. performed computational studies on these MOFs and identified anions, coordinated to the metal centers in the porphyrin units to have a considerable impact on the overall HER activity (this effect was not addressed in Lin's catalytic studies) [113]. In particular, the coordination of OH^- had a detrimental effect due to a promotion of non-adiabatic relaxation pathways and acceleration of the recombination of photo-generated holes and electrons.

Mulfort et al. demonstrated that different MOF topologies are accessible from a single set of molecular components. These authors combined the $[Ru_3(\mu_3\text{-}O)(\mu_2\text{-}OAc)_6(pyCO_2H)_3]^+$ precursor with Co^{II} ions at different temperatures to obtain a stack of 2D sheets (at 50 °C) or a 3D coordination framework (at 80 °C).

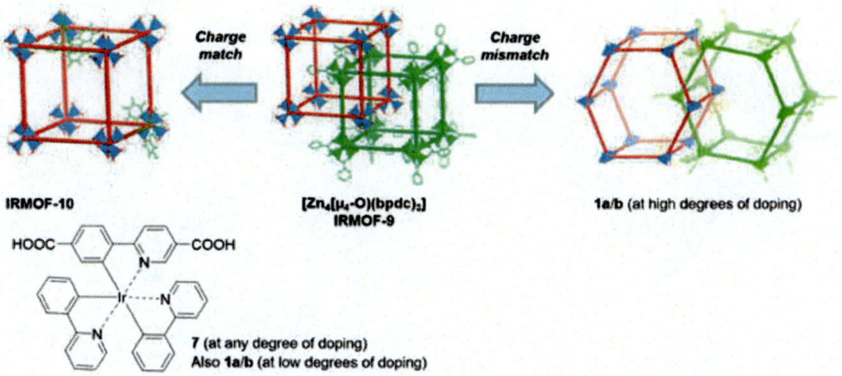

Fig. 6.26 Schematic representation of the phase transition of IRMOF-9 into IRMOF-10 and another interpenetrated MOF upon doping with **7** (or **1a/b** at low degrees-of-doping) and **1a/b** (at higher degrees-of-doping) [111]. Figure reproduced with kind permission; © 2014 American Chemical Society

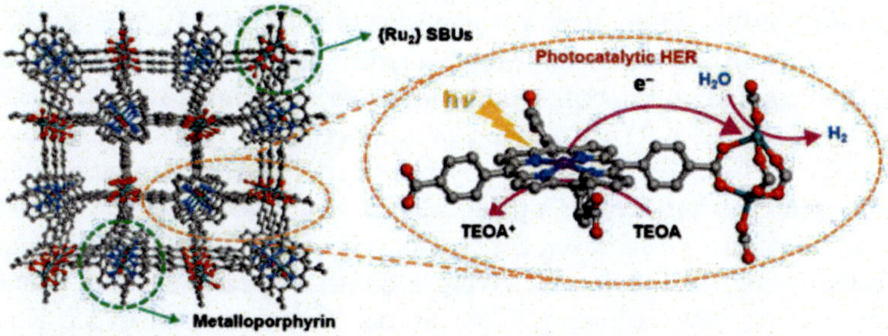

Fig. 6.27 Schematic representation of a MOF assembled from metalloporphyrin-centered tetracarboxylates and $\{Ru_2\}$ secondary building blocks. These MOFs enabled the photocatalytic HER to proceed efficiently [112]. Figure reproduced with kind permission; © 2018 American Chemical Society

The $\{Ru_3\}$ nodes of the latter were formed in situ from the original ones via ligand-exchange reactions. One might speculate that the stack of 2D sheets represents a kinetic product; whereas, the MOF structure obtained at the elevated temperature was thermodynamically favored.

The microporous MOF-253 framework, containing vacant bpy sites, was first reported by Yaghi et al.; the material had a BET surface of 2160 m^2g^{-1} and was found to readily coordinate transition-metal ions to its binding sites [114]. The robustness of MOF-253, as heterogeneous catalyst, under the conditions of various organic reaction was also shown [115]. In line with these studies, Gustafsson, Zou, Martín-Matute et al. prepared a heterogeneous catalyst for the oxidation of alcohols

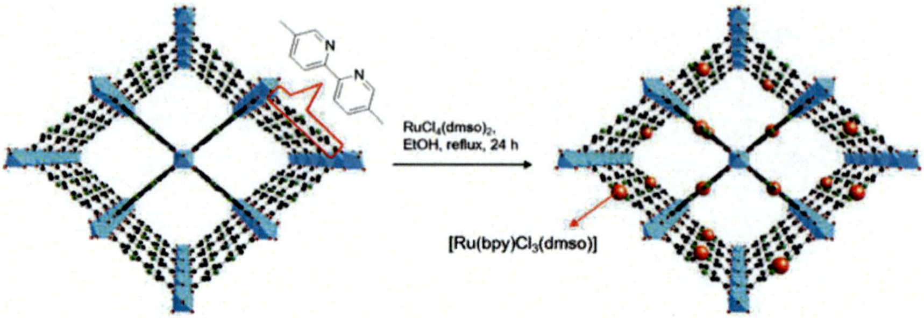

Fig. 6.28 Schematic representation of the modification of MOF-253 by the coordination of Ru^{II} centers [116]. Figure reproduced with kind permission; © 2012 Wiley–VCH

with $PhI(OAc)_2$, as oxidant [116]. For this purpose, MOF-253 was subjected a post-synthetic modification, i.e., Ru^{II} centers were coordinated to the bpy sites (presumably due to steric constraints, only 13% of the bpy sites were complexed). A variety of secondary alcohols was oxidized to the corresponding ketones with high conversions, short reaction times, under mild conditions at low catalyst loadings (i.e., using 0.5 to 1 mol-% of metallated MOF-253); primary benzylic alcohols selectively afforded aromatic aldehydes, as reaction products. A similar strategy was applied by Li et al., who coordinated the $[Ru(CO)_3Cl_2]_2$ precursor to bpy sites of MOF-253. The formation of $Ru(bpy)(CO)_2Cl_2$ complexes was corroborated by infrared IR and X-ray absorption spectroscopy measurements. The resultant assembly was tested, as a photocatalyst, in the CO_2 reduction (a total TON of *ca.* 10 was reached). Moreover, the *cis*-$[Ru(bpy)_2Cl_2]$ precursor was additionally coordinated to the MOF-253, thus giving a framework carrying both photosensitizer and catalyst sites; accordingly, the overall TON increased to 55. However, both Ru^{II}-loaded materials suffered from considerable leaching of metal ions; the authors pointed out, that the bpy sites of MOF-253 did not offer sufficient stability to prevent the Ru^{II} centers to dissociate in the course of the photocatalytic experiments (Fig. 6.28).

6.4 Covalent Organic-Frameworks Incorporating Ru Species

In contrast to the aforementioned MOFs and coordination nanosheets (CONASHs) [117], as their 2D congeners, covalent organic frameworks are exclusively formed by covalent bonds. In other words, non-covalent interactions, such as metal-to-ligand interactions, are not integral parts of their molecular structure. For both types of material, COFs and CONASHs, the design of the organic building blocks and the reversibility in bond formation represent the most important requirements to assemble materials with perfect 2D or 3D periodicity. The dynamic nature of metal-to-ligand interactions accounts for the formation of highly ordered CONASH structures; whereas, dynamic covalent chemistry is responsible for formation of COFs (Fig. 6.2) [3–5]. Regarding the classification of COFs, the structure of the chemical linkage between the monomer units typically serves as a guide. As the most popular types of COFs, the boron-containing, triazine-based and imine-based structures have to be named (a representative example for each of these is depicted in Fig. 6.29).

The first examples for triazine-based COFs, such as CTF-1, were obtained in 2008 from the ionothermal reaction of aromatic di- or tricyano derivatives in molten $ZnCl_2$, which is still the most commonly used for CTF synthesis [118]. Since then, the class of covalent triazine frameworks (CTFs) has attracted widespread attention due to their chemical and thermal robustness, high porosity, and environmentally friendly structure (i.e., absence of toxic elements within the structure) [119]. Accordingly, applications in the fields of gas storage and separation, energy storage, photocatalysis, and heterogeneous catalysis have been proposed [119]. It was demonstrated by Lotsch et al. that the incorporation of *N*-heterocyclic building blocks, such as bpy units, into the CTF structure enabled the binding of transition-metal ions to these binding sites and, due to the heteroatom effect, increased the CO_2 uptake ability (the bpy-containing CTF took up the remarkable amount of 5.58 mmol g^{-1} at 0 °C) [120, 121]. Moreover, the CTF, to which transition-metal complexes were coordinated, were employed, as heterogeneous catalysts in various reactions (e.g., CO_2 reduction, water oxidation, and CH activation) [119].

In line with these studies, Lejaeghere, Van Speybroeck et al. coordinated the *cis*-$[Ru(bpy)_2]^{2+}$ precursor to Lotsch's bpy-containing CTF and studied the photophysical properties of the resultant material [122]. Compared to various small molecule Ru^{II} complexes, the energies of the allowed electronic transitions became significantly red-shifted, thus reaching into the near-IR (NIR) region of the UV/vis/NIR absorption spectrum. Computational studies suggested that increasing the *N* content of the framework, for example by using bipyrimidine or bipyridazine monomers, would shift the absorption range even further towards the NIR regime [123]. These authors concluded that this property could be beneficial for performing near-IR photocatalysis in nanoporous materials; thereby harvesting a broad wavelength regime to activate the coordinated Ru^{II} photocatalysts (Fig. 6.30).

Almost simultaneously, Ren et al. prepared a CTF, named PAF-72, from 9,10-dicyanoantharacene under ionothermal conditions [124]. Gas-sorption studies on the amorphous material indicated small pore sizes, as expected due to the bulky anthracene moieties within the framework. The coordination of Ru^{III} ions to the binding sites within the structure and the in-situ formation of Ru NPs was evidenced from microscopy imaging, XPS, and TGA measurements; the latter revealed a Ru content of *ca.* 2.6 wt.-% within the material. TEM element mapping was employed to show the homogeneous dispersion of the formed NP within the PAF-72 framework (Fig. 6.31). These authors also tested the applicability of

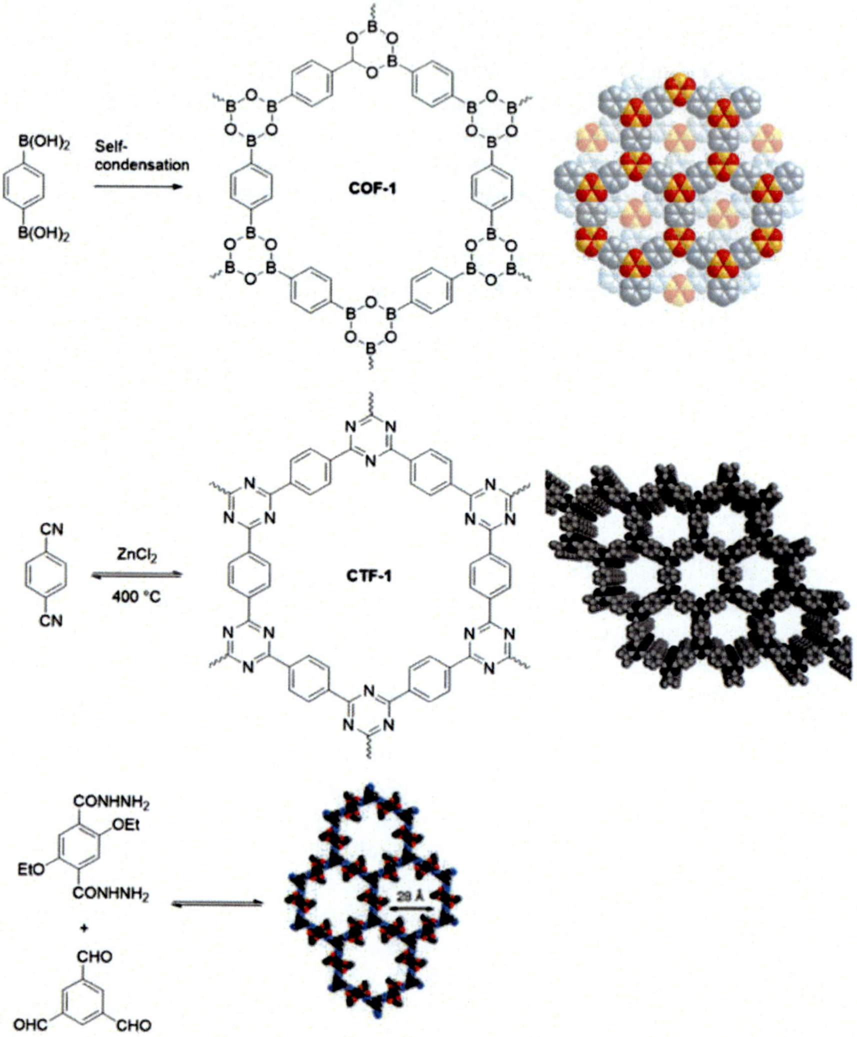

Fig. 6.29 Schematic representation of three representative COFs—a boron-containing, a triazine-based, and an imine-based one (from top to down). The corresponding solid-state structures are also depicted [4]. Figure reproduced with kind permission; © 2013 The Royal Chemical Society

RuNP@PAF-72 to catalyze the hydrolysis of ammonia borane; for the H_2 evolution, a maximum TOF of 294 min^{-1} was detected.

A similar strategy to immobilize Ru nanoparticles within a COF was followed by Chen, Dong et al. [125]. These authors relied on the Schiff base condensation to assemble the framework COF-ASB1. Powder XRD analysis and molecular

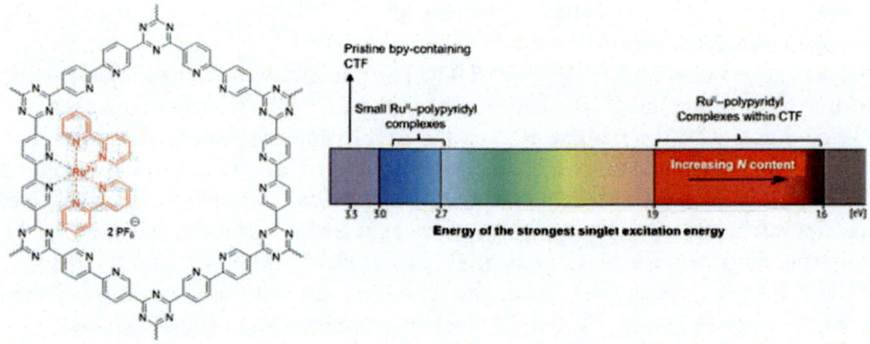

Fig. 6.30 Left: Schematic representation of a bpy-containing CTF to which Ru^{II} complexes were coordinated. Right: Schematic representation of the strongest singlet excitation energies of Ru^{II}–polypyridyl complexes—isolated and embedded into a CTF [122]. Figure reproduced with kind permission; © 2019 American Chemical Society

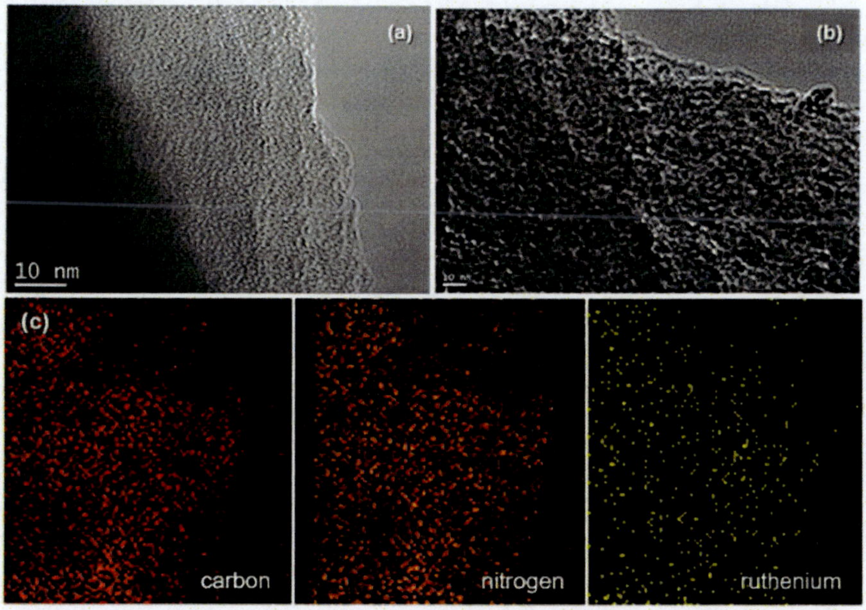

Fig. 6.31 Representative TEM images of pristine PAF-72 (**a**) and Ru_{NP}@PAF-72 (**b**). TEM elemental mapping (**c**) showed the equal distribution of RuNP over the entire CAF material [124]. Figure reproduced with kind permission; © 2019 Frontiers Media SA

modelling suggested an eclipsed stacking of the 2D layer; the diameter of the hexagonal channels was estimated at 3.1 nm, thus offering a proper environment for the accommodation of Ru NPs (Fig. 6.32). The particles were formed by reduction of Ru^{III} ions within the COF-ASB1 structure (the Ru^{III} ions were coordinated to the N-rich domains of the framework, i.e., the acyl hydrazone moieties). The NP formation was accompanied by a distinct color change of the initially yellow colored COF to black (Fig. 6.32). The hybrid COF was used as a robust, efficient, selective, and recyclable catalyst for the solventless oxidation of benzylic alcohols to the corresponding benzaldehyde derivatives in air (80 °C for 24 h, TON = 18 h^{-1}, TON = 0.75 h^{-1}). Moreover, the in-situ formation of imines was realized when the catalytic reaction was run in the presence of an aromatic or benzylic amine.

Kolen'ko et al. prepared a COF via the solvothermal condensation of 1,3,5-triformylphloroglucinol and *o*-toluidine [126]. The thusly obtained COF was loaded with Ru^{III} ions, using the $RuCl_3 \times xH_2O$ precursor; then subsequent treatment with NaOH gave RuO_2 NPs homogeneously dispersed within the COF structure. This hybrid material was used, as a pre-catalyst, for the dehydrogenation of formic acid—a reaction, which is of importance for fuel-cell technology (formic

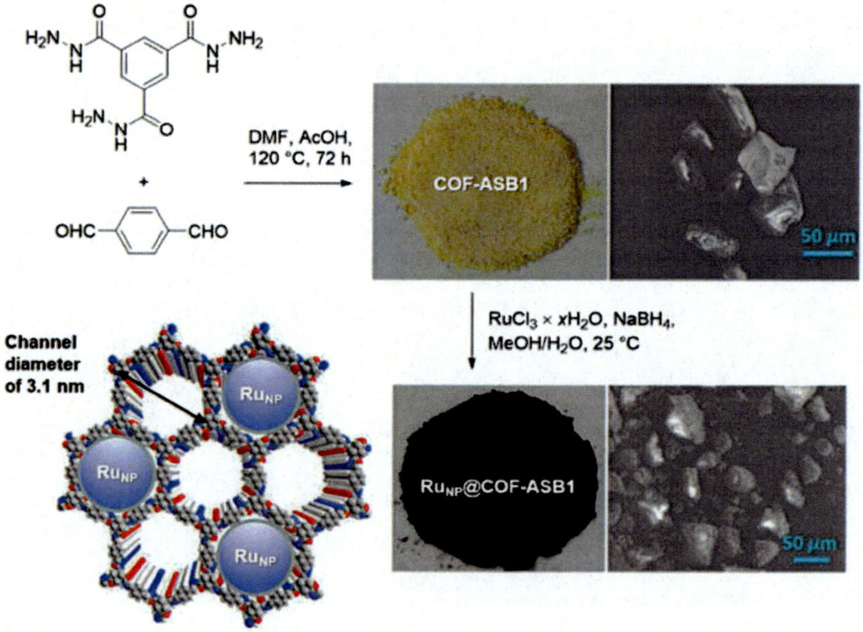

Fig. 6.32 Schematic representation of the formation of COF-ASB1 under solvothermal conditions. The embedding of Ru NPs into the channels of COF-ASB1 by the in-situ reduction of coordinated Ru^{III} ions was visible to the naked eye. The simulated structure of COF-ASB1 with Ru NPs, as guests within the channels, is also shown [125]. Figure reproduced with kind permission; © 2018 American Chemical Society

acid can serve as a carrier for H_2 which, when deliberated, represents the fuel in such cells [127]). The authors proposed that Ru NPs, as the catalytically active species, were formed by in-situ reduction of RuO_2 under the applied reaction conditions. When operated at 150 °C, this heterogeneous catalyst was found to be superior to other catalyst systems (e.g., Ru on a carbon support [128]) in terms of both yield and selectivity (i.e., increase in yield from 15 to 50%, increase in selectivity from 95 to 98%).

The COF derived from the acid-catalyzed Friedel–Crafts-type polymerization of triphenylamine by 1,3,5-triformylphloroglucinol under forcing conditions was, after loading with Ru NPs, studied in the context of "biodiesel" production, i.e., the direct hydrodeoxygenation of long-chain fatty acids to give alkanes [129]. This hybrid material offered a much higher selectivity for the formation of long-chains alkanes compared to other Ru-based heterogeneous catalysts (i.e., Ru NPs on SiO_2, carbon, TiO_2 or ZrO_2 supports).

One further COF was assembled from 3,4-diaminobenzohydrazide and benzene-1,3,5-tricarboxaldehyde and used as a carrier for Ru^{III} ions [130]. This material was immobilized onto a glassy carbon (GC) electrode and employed for the electrochemical HER. Basically, the efficiency of the electrochemical HER can be correlated to the overpotential value (η)—the lower a material's η value, the higher it's catalytic activity in the HER. Das' hybrid electrode showed an overpotential of 159 mV for the HER from water at low pH values, thus being more efficient than GC electrodes loaded with Ru^{III} ions or the metal-free COF. A broad range of Ru-based electrocatalysts for the HER are already known [131]; however, many of these require binders to ensure the immobilization of the catalyst onto the electrode and account for its stability during the electrochemical runs. Remarkable, the hybrid electrode reported by Das et al. was highly stable even in the absence of such a binder.

6.5 Concluding Remarks

The metal–organic and covalent frameworks, possessing highly ordered 2D or 3D structures, have emerged as prominent solid-state materials for a variety of cutting-edge applications: Gas storage/separation, sensing, catalysis, drug delivery, opto-electronic, etc. The structures of these assemblies, i.e., their porosity and cavity size, can be tuned by the proper choice of molecular building blocks. Today, modelling studies and computational ligand design allow one to synthesize tailor-made frameworks in which functional guests can be encapsulated. Ru nanoparticles and photoactive Ru^{II} complexes represent two important examples in this respect. The caging such species or, in the case of the Ru^{II} complexes, their integration into the framework structure gives hybrid materials, which have been tested in photophysical and catalytic applications, such as photovoltaics or the interconversion of organic functional groups. Beyond these, the usage of Ru-containing MOFs or COFs in energy-related applications, i.e., the CO_2

reduction or HER, are of particular interest in the overall context of today's climate crises. Thus, research going into these directions is expected to increase further within the coming years.

References

1. Gangu, K.K., Maddila, S., Mukkamala, S.B., Jonnalagadda, S.B.: A review on contemporary metal–organic framework materials. Inorg. Chim. Acta **446**, 61 (2016)
2. Beuerle, F., Gole, B.: Covalent organic frameworks and cage compounds: Design and applications of polymeric and discrete organic scaffolds. Angew. Chem. Int. Ed. **57**(18), 4850 (2018)
3. Feng, X., Ding, X., Jiang, D.: Covalent organic frameworks. Chem. Soc. Rev. **41**(18), 6010 (2012)
4. Ding, S.-Y., Wang, W.: Covalent organic frameworks (COFs): from design to applications. Chem. Soc. Rev. **42**(2), 548 (2013)
5. Waller, P.J., Gándara, F., Yaghi, O.M.: Chemistry of covalent organic frameworks. Acc. Chem. Res. **48**(12), 3053 (2015)
6. Huang, N., Wang, P., Jiang, D.: Covalent organic frameworks: a materials platform for structural and functional designs. Nat. Rev. Mater. **1**(10), 16068 (2016)
7. Xiang, Z., Cao, D., Lan, J., Wang, W., Broom, D.P.: Multiscale simulation and modelling of adsorptive processes for energy gas storage and carbon dioxide capture in porous coordination frameworks. Energy Environ. Sci. **3**(10), 1469 (2010)
8. Andersson, S., Larsson, P.-E.: Atomic-scale modeling of hydrogen storage in the UiO-66 and UiO-67 metal-organic frameworks. Micropor. Mesopor. Mater. **224**, 349 (2016)
9. Griffin, S.L., Champness, N.R.: A periodic table of metal-organic frameworks. Coord. Chem. Rev. **414**, 213295 (2020)
10. Bailey, D.C., Langer, S.H.: Immobilized transition-metal carbonyls and related catalysts. Chem. Rev. **81**(2), 109 (1981)
11. Phan, A., Doonan, C.J., Uribe-Romo, F.J., Knobler, C.B., O'Keeffe, M., Yaghi, O.M.: Synthesis, structure, and carbon dioxide capture properties of zeolitic imidazolate frameworks. Acc. Chem. Res. **43**(1), 58 (2010)
12. Chen, B., Yang, Z., Zhu, Y., Xia, Y.: Zeolitic imidazolate framework materials: Recent progress in synthesis and applications. J. Mater. Chem. A **2**(40), 16811 (2014)
13. Yurderi, M., Bulut, A., Zahmakiran, M., Gülcan, M., Özkar, S.: Ruthenium(0) nanoparticles stabilized by metal-organic framework (ZIF-8): Highly efficient catalyst for the dehydrogenation of dimethylamine-borane and transfer hydrogenation of unsaturated hydrocarbons using dimethylamine-borane as hydrogen source. Appl. Catal. B: Environ. **160–161**, 534 (2014)
14. Cao, N., Liu, T., Su, J., Wu, X., Luo, W., Cheng, G.: Ruthenium supported on MIL-101 as an efficient catalyst for hydrogen generation from hydrolysis of amine boranes. New J. Chem. **38**(9), 4032 (2014)
15. Yan, X., Liu, H., Liew, K.Y.: Size control of polymer-stabilized ruthenium nanoparticles by polyol reduction. J. Mater. Chem. **11**(12), 3387 (2001)
16. Ji, S., Chen, Y., Zhao, S., Chen, W., Shi, L., Wang, Y., Dong, J., Li, Z., Li, F., Chen, C., et al.: Atomically dispersed ruthenium species inside metal–organic frameworks: Combining the high activity of atomic sites and the molecular sieving effect of MOFs. Angew. Chem. Int. Ed. **58**(13), 4271 (2019)
17. Ji, S., Chen, Y., Fu, Q., Chen, Y., Dong, J., Chen, W., Li, Z., Wang, Y., Gu, L., He, W., et al.: Confined pyrolysis within metal–organic frameworks to form uniform Ru_3 clusters for efficient oxidation of alcohols. J. Am. Chem. Soc. **139**(29), 9795 (2017)

18. Tuan, D.D., Lin, K.-Y.A.: Ruthenium supported on ZIF-67 as an enhanced catalyst for hydrogen generation from hydrolysis of sodium borohydride. Chem. Eng. J. **351**, 48 (2018)
19. Li, Q., Kim, H.: Hydrogen production from $NaBH_4$ hydrolysis via Co-ZIF-9 catalyst. Fuel Process. Technol. **100**, 43 (2012)
20. Schröder, F., Esken, D., Cokoja, M., van den Berg, M.W.E., Lebedev, O.I., Van Tendeloo, G., Walaszek, B., Buntkowsky, G., Limbach, H.-H., Chaudret, B., et al.: Ruthenium nanoparticles inside porous [$Zn_4O(bdc)_3$] by hydrogenolysis of adsorbed [Ru(cod)(cot)]: A solid-state reference system for surfactant-stabilized ruthenium colloids. J. Am. Chem. Soc. **130**(19), 6119 (2008)
21. Liu, H., Fang, R., Li, Z., Li, Y.: Solventless hydrogenation of benzene to cyclohexane over a heterogeneous Ru–Pt bimetallic catalyst. Chem. Emg. Sci. **122**, 350 (2015)
22. Bi, H., Tan, X., Dou, R., Pei, Y., Qiao, M., Sun, B., Zong, B.: Ru–B nanoparticles on metal–organic frameworks as excellent catalysts for hydrogenation of benzene to cyclohexane under mild reaction conditions. Green Chem. **18**(7), 2216 (2016)
23. Zhao, Y., Zhang, J., Song, J., Li, J., Liu, J., Wu, T., Zhang, P., Han, B.: Ru nanoparticles immobilized on metal–organic framework nanorods by supercritical CO_2-methanol solution: highly efficient catalyst. Green Chem. **13**(8), 2078 (2011)
24. Watkins, J.J., McCarthy, T.J.: Polymer/metal nanocomposite synthesis in supercritical CO_2. Chem. Mater. **7**(11), 1991 (1995)
25. Wu, T., Zhang, P., Ma, J., Fan, H., Wang, W., Jiang, T., Han, B.: Catalytic activity of immobilized Ru nanoparticles in a porous metal-organic framework using supercritical fluid. Chin. J. Cat. **34**(1), 167 (2013)
26. Morère, J., Torralvo, M.J., Pando, C., Renuncio, J.A.R., Cabañas, A.: Supercritical fluid deposition of Ru nanoparticles onto SiO_2 SBA-15 as a sustainable method to prepare selective hydrogenation catalysts. RSC Adv. **5**(49), 38880 (2015)
27. Sun, Z., Liu, Z., Han, B., Miao, S., Miao, Z., An, G.: Decoration carbon nanotubes with Pd and Ru nanocrystals via an inorganic reaction route in supercritical carbon dioxide–methanol solution. J. Colloid Interface Sci. **304**(2), 323 (2006)
28. O'Nei, A., Watkins, J.J.: Reactive deposition of conformal ruthenium films from supercritical carbon dioxide. Chem. Mater. **18**(24), 5652 (2006)
29. Han, S., Wei, Y., Grzybowski, B.A.: A metal–organic framework stabilizes an occluded photocatalyst. Chem. Eur. J. **19**(34), 11194 (2013)
30. Rajkumar, T., Kukkar, D., Kim, K.-H., Sohn, J.R., Deep, A.: Cyclodextrin-metal–organic framework (CD-MOF): From synthesis to applications. J. Ind. Eng. Chem. **72**, 50 (2019)
31. Lainé, P., Lanz, M., Calzaferri, G.: Limits of the in situ synthesis of tris(2,2'-bipyridine) ruthenium(II) in the supercages of zeolite Y. Inorg. Chem. **35**(12), 3514 (1996)
32. Hu, L., Li, K., Shang, W., Zhu, X., Liu, M.: Emerging cubic chirality in γCD-MOF for fabricating circularly polarized luminescent crystalline materials and the size effect. Angew. Chem. Int. Ed. **59**(12), 4953 (2020)
33. Pointillart, F., Train, C., Gruselle, M., Villain, F., Schmalle, H.W., Talbot, D., Gredin, P., Decurtins, S., Verdaguer, M.: Chiral templating activity of tris(bipyridine)ruthenium(II) cation in the design of three-dimensional (3D) optically active oxalate-bridged {[$Ru(bpy)_3$] [$Cu_{2x}Ni_{2(1-x)}(C_2O_4)_3$]}$_n$ ($0 \leq x \leq 1$; bpy = 2,2'-bipyridine): structural, optical, and magnetic studies. Chem. Mater. **16**(5), 832 (2004)
34. Sen, R., Koner, S., Bhattacharjee, A., Kusz, J., Miyashita, Y., Okamoto, K.-I.: Entrapment of $[Ru(bpy)_3]^{2+}$ in the anionic metal–organic framework: Novel photoluminescence behavior exhibiting dual emission at room temperature. Dalton Trans. **40**(26), 6952 (2011)
35. Larsen, R.W., Wojtas, L.: Photophysical studies of Ru(II) tris(2,2′-bipyridine) confined within a Zn(II)–trimesic acid polyhedral metal–organic framework. J. Phys. Chem. C **116** (30), 7830 (2012)
36. Larsen, R.W., Wojtas, L.: Photo-physical studies of ruthenium(II) tris(1,10-phenanthroline) confined within a polyhedral zinc(II)-trimesic acid metal organic framework. Inorg. Chim. Acta **466**, 243 (2017)

37. Mayers, J.M., Larsen, R.W.: Modulation of osmium(II) tris(2,2′-bipyridine) photophysics through encapsulation within zinc(II) trimesic acid metal-organic frameworks. Inorg. Chem. **59**(11), 7761 (2020)
38. Whittington, C.L., Wojtas, L., Larsen, R.W.: Ruthenium(II) tris(2,2′-bipyridine)-templated zinc(II) 1,3,5-tris(4-carboxyphenyl)benzene metal-organic frameworks: Structural characterization and photophysical properties. Inorg. Chem. **53**(1), 160 (2014)
39. Whittington, C.L., Wojtas, L., Gao, W.-Y., Ma, S., Larsen, R.W.: A new photoactive Ru(I) tris(2,2′-bipyridine) templated Zn(II) benzene-1,4-dicarboxylate metal organic framework: Structure and photophysical properties. Dalton Trans. **44**(12), 5331 (2015)
40. Larsen, R.W., Mayers, J.M., Wojtas, L.: A novel photo-active Cd:1,4-benzene dicarboxylate metal organic framework templated using $[Ru(II)(2,2'\text{-bipyridine})_3]^{2+}$: synthesis and photophysics of RWLC-5. Dalton Trans. **46**(37), 12711 (2017)
41. Larsen, R.W., Wojtas, L.: Photophysical properties of $[Ru(2,2'\text{-bipyridine})_3]^{2+}$ encapsulated within the Uio-66 zirconium based metal organic framework. J. Solid State Chem. **247**, 77 (2017)
42. Mayers, J.M., Larsen, R.W.: Photophysical study of $[Ru(2,2'\text{-bipyridine})^3]^{2+}$ and $[Ru(1,10\text{-phenanthroline})^3]^{2+}$ encapsulated in the UiO-66-NH_2 metal organic framework. Polyhedron **171**, 382 (2019)
43. McKeithan, C.R., Mayers, J.M., Wojtas, L., Larsen, R.W.: Photophysical studies of Ru(II) tris(2,2′-bipyridine) encapsulated within the ZnHKUST-1 metal organic framework. Inorg. Chim. Acta **483**, 1 (2018)
44. McKeithan, C.R., Wojtas, L., Larsen, R.W.: Photophysical properties of the $[Ru(2,2'\text{-bipyridine})_3]^{2+}$ templated metal organic framework, RWLC-6. Inorg. Chim. Acta **496**, 119034 (2019)
45. Micheroni, D., Lin, Z., Chen, Y.-S., Lin, W.: Luminescence enhancement of *cis*-$[Ru(bpy)_2(py)_2]^{2+}$ *via* confinement within a metal–organic framework. Inorg. Chem. **58**(12), 7645 (2019)
46. Isaka, Y., Kondo, Y., Kuwahara, Y., Mori, K., Yamashita, H.: Incorporation of a Ru complex into an amine-functionalized metal–organic framework for enhanced activity in photocatalytic aerobic benzyl alcohol oxidation. Cat. Sci. Technol. **9**(6), 1511 (2019)
47. Yang, X., Huang, T., Gao, S., Cao, R.: Boosting photocatalytic oxidative coupling of amines by a Ru-complex-sensitized metal-organic framework. J. Catal. **378**, 248 (2019)
48. Tan, Y.-X., Lin, S.-X., Liu, C., Huang, Y., Zhou, M., Kang, Q., Yuan, D., Hong, M.: Boosting photocatalytic cross-dehydrogenative coupling reaction by incorporating $[Ru^{II}(bpy)^3]$ into a radical metal-organic framework. Appl. Catal. B **227**, 425 (2018)
49. Xu, S., Kong, H., Zhang, R.: Visible-light-induced, UiO-67-Ru-catalyzed oxidative cross-coupling for constructing *β*-acetylamino acrylosulfones. Tetrahedron Lett. **61**(11), 151629 (2020)
50. Wang, X., Lu, W., Gu, Z.-Y., Wei, Z., Zhou, H.-C.: Topology-guided design of an anionic bor-network for photocatalytic $[Ru(bpy)_3]_{2+}$ encapsulation. Chem. Commun. **52**(9), 1926 (2016)
51. Kent, C.A., Liu, D., Ito, A., Zhang, T., Brennaman, M.K., Meyer, T.J., Lin, W.: Rapid energy transfer in non-porous metal–organic frameworks with caged $Ru(bpy)_3{}^{2+}$ chromophores: oxygen trapping and luminescence quenching. J. Mater. Chem. A **1**(47), 14982 (2013)
52. Fang, Y., Xiao, Z., Kirchon, A., Li, J., Jin, F., Togo, T., Zhang, L., Zhu, C., Zhou, H.-C.: Bimolecular proximity of a ruthenium complex and methylene blue within an anionic porous coordination cage for enhancing photocatalytic activity. Chem. Sci. **10**(12), 3529 (2019)
53. Fang, Y., Powell, J.A., Li, E., Wang, Q., Perry, Z., Kirchon, A., Yang, X., Xiao, Z., Zhu, C., Zhang, L., et al.: Catalytic reactions within the cavity of coordination cages. Chem. Soc. Rev. **48**(17), 4707 (2019)
54. Aquino, M.A.S.: Recent developments in the synthesis and properties of diruthenium tetracarboxylates. Coord. Chem. Rev. **248**(11), 1025 (2004)

55. Takamizawa, S., Yamaguchi, K., Mori, W.: The gas-occlusion properties of dicarboxylate (fumarate, trans-trans-muconate and terephthalate) ruthenium(II, III) dinuclear complexes. Inorg. Chem. Commun. **1**(5), 177 (1998)
56. Takamizawa, S., Ohmura, T., Yamaguchi, K., Mori, W.: Synthesis and gas-occlusion properties of ruthenium(II,III) dicarboxylates (fumarate, *trans-trans*-muconate and terephthalate) bridged by halogen atoms. Mol. Cryst. Liq. Cryst. Sci. Technol., Sect. A **342**(1), 199 (2000)
57. Ribeiro, G., Vichi, F.M., de Oliveira Silva, D.: Synthesis, characterization and adsorption properties of porous mixed valent diruthenium(II, III)-terephthalate and diruthenium(II, III)-adipate polymers. J. Mol. Struct. **890**(1), 209 (2008)
58. Tetsushi, O., Wasuke, M., Hiroki, H., Masato, O., Yuko, N.: Magnetic and gas-occlusion properties and catalytic activity of microporous materials: Dinuclear ruthenium(II, II) dicarboxylates. Chem. Lett. **32**(5), 468 (2003)
59. Kozachuk, O., Yusenko, K., Noei, H., Wang, Y., Walleck, S., Glaser, T., Fischer, R.A.: Solvothermal growth of a ruthenium metal–organic framework featuring HKUST-1 structure type as thin films on oxide surfaces. Chem. Commun. **47**(30), 8509 (2011)
60. Wade, C.R., Dincă, M.: Investigation of the synthesis, activation, and isosteric heats of CO_2 adsorption of the isostructural series of metal–organic frameworks $M_3(BTC)_2$ (M = Cr, Fe, Ni, Cu, Mo, Ru). Dalton Trans. **41**(26), 7931 (2012)
61. Gascon, J., Corma, A., Kapteijn, F., Llabrés i Xamena, F.X.: Metal organic framework catalysis: Quo vadis? ACS Catal. **4**(2), 361 (2014)
62. Rogge, S.M.J., Bavykina, A., Hajek, J., Garcia, H., Olivos-Suarez, A.I., Sepúlveda-Escribano, A., Vimont, A., Clet, G., Bazin, P., Kapteijn, F., et al.: Metal–organic and covalent organic frameworks as single-site catalysts. Chem. Soc. Rev. **46**(11), 3134 (2017)
63. Marx, S., Kleist, W., Huang, J., Maciejewski, M., Baiker, A.: Tuning functional sites and thermal stability of mixed-linker MOFs based on MIL-53(Al). Dalton Trans. **39**(16), 3795 (2010)
64. Kozachuk, O., Luz, I., Llabrés i Xamena, F.X., Noei, H., Kauer, M., Albada, H.B., Bloch, E. D., Marler, B., Wang, Y., Muhler, M., et al.: Multifunctional, defect-engineered metal–organic frameworks with ruthenium centers: sorption and catalytic properties. Angew. Chem. Int. Ed. **53**(27), 7058 (2014)
65. Agirrezabal-Telleria, I., Luz, I., Ortuño, M.A., Oregui-Bengoechea, M., Gandarias, I., López, N., Lail, M.A., Soukri, M.: Gas reactions under intrapore condensation regime within tailored metal–organic framework catalysts. Nat. Commun. **10**(1), 2076 (2019)
66. Epp, K., Luz, I., Heinz, W.R., Rapeyko, A., Llabrés i Xamena, F.X., Fischer, R.A.: Defect-engineered ruthenium MOFs as versatile heterogeneous hydrogenation catalysts. ChemCatChem, **12**(6), 1720 (2020)
67. Zhang, W., Kauer, M., Halbherr, O., Epp, K., Guo, P., Gonzalez, M.I., Xiao, D.J., Wiktor, C., Llabrés i Xamena, F.X., Wöll, C., et al.: Ruthenium metal–organic frameworks with different defect types: influence on porosity, sorption, and catalytic properties. Chem. Eur. J. **22**(40), 14297 (2016)
68. Gotthardt, M.A., Schoch, R., Wolf, S., Bauer, M., Kleist, W.: Synthesis and characterization of bimetallic metal–organic framework Cu–Ru-BTC with HKUST-1 structure. Dalton Trans. **44**(5), 2052 (2015)
69. Heinz, W.R., Kratky, T., Drees, M., Wimmer, A., Tomanec, O., Günther, S., Schuster, M., Fischer, R.A.: Mixed precious-group metal–organic frameworks: A case study of the HKUST-1 analogue $[Ru_xRh_{3-x}(btc)_2]$. Dalton Trans. **48**(32), 12031 (2019)
70. Abednatanzi, S., Gohari Derakhshandeh, P., Depauw, H., Coudert, F.-X., Vrielinck, H., Van Der Voort, P., Leus, K.: Mixed-metal metal–organic frameworks. Chem. Soc. Rev. **48**(9), 2535 (2019)
71. Qiu, T., Liang, Z., Guo, W., Gao, S., Qu, C., Tabassum, H., Zhang, H., Zhu, B., Zou, R., Shao-Horn, Y.: Highly exposed ruthenium-based electrocatalysts from bimetallic metal-organic frameworks for overall water splitting. Nano Energy **58**, 1 (2019)

72. Happ, B., Winter, A., Hager, M.D., Schubert, U.S.: Photogenerated avenues in macromolecules containing Re(I), Ru(II), Os(II), and Ir(III) metal complexes of pyridine-based ligands. Chem. Soc. Rev. **41**(6), 2222 (2012)
73. Kent, C.A., Mehl, B.P., Ma, L., Papanikolas, J.M., Meyer, T.J., Lin, W.: Energy-transfer dynamics in metal−organic frameworks. J. Am. Chem. Soc. **132**(37), 12767 (2010)
74. Lin, J., Hu, X., Zhang, P., Van Rynbach, A., Beratan, D.N., Kent, C.A., Mehl, B.P., Papanikolas, J.M., Meyer, T.J., Lin, W., et al.: Triplet excitation energy dynamics in metal–organic frameworks. J. Phys. Chem. C **117**(43), 22250 (2013)
75. Kent, C.A., Liu, D., Meyer, T.J., Lin, W.: Amplified luminescence quenching of phosphorescent metal–organic frameworks. J. Am. Chem. Soc. **134**(9), 3991 (2012)
76. Kent, C.A., Liu, D., Ma, L., Papanikolas, J.M., Meyer, T.J., Lin, W.: Light harvesting in microscale metal–organic frameworks by energy migration and interfacial electron-transfer quenching. J. Am. Chem. Soc. **133**(33), 12940 (2011)
77. Low, J.J., Benin, A.I., Jakubczak, P., Abrahamian, J.F., Faheem, S.A., Willis, R.R.: Virtual high throughput screening confirmed experimentally: Porous coordination polymer hydration. J. Am. Chem. Soc. **131**(43), 15834 (2009)
78. Cavka, J.H., Jakobsen, S., Olsbye, U., Guillou, N., Lamberti, C., Bordiga, S., Lillerud, K.P.: A new zirconium inorganic building brick forming metal-organic frameworks with exceptional stability. J. Am. Chem. Soc. **130**(42), 13850 (2008)
79. DeCoste, J.B., Peterson, G.W., Jasuja, H., Glover, T.G., Huang, Y.-G., Walton, K.S.: Stability and degradation mechanisms of metal–organic frameworks containing the $Zr_6O_4(OH)_4$ secondary building unit. J. Mater. Chem. A **1**(18), 5642 (2013)
80. Wang, C., Xie, Z., deKrafft, K.E., Lin, W.: Doping metal–organic frameworks for water oxidation, carbon dioxide reduction, and organic photocatalysis. J. Am. Chem. Soc. **133**(34), 13445 (2011)
81. Santiago-Portillo, A., Baldoví, H.G., Carbonell, E., Navalón, S., Álvaro, M., García, H., Ferrer, B.: Ruthenium(II) tris(2,2′-bipyridyl) complex incorporated in UiO-67 as photoredox catalyst. J. Phys. Chem. C **122**(51), 29190 (2018)
82. Barrett, S.M., Wang, C., Lin, W.: Oxygen sensing *via* phosphorescence quenching of doped metal–organic frameworks. J. Mater. Chem. **22**(20), 10329 (2012)
83. Maza, W.A., Morris, A.J.: Photophysical characterization of a ruthenium(II) tris(2,2′-bipyridine)-doped zirconium UiO-67 metal–organic framework. J. Phys. Chem. C **118**(17), 8803 (2014)
84. Maza, W.A., Ahrenholtz, S.R., Epley, C.C., Day, C.S., Morris, A.J.: Solvothermal growth and photophysical characterization of a ruthenium(II) tris(2,2′-bipyridine)-doped zirconium UiO-67 metal organic framework thin film. J. Phys. Chem. C **118**(26), 14200 (2014)
85. Maza, W.A., Padilla, R., Morris, A.J.: Concentration dependent dimensionality of resonance energy transfer in a postsynthetically doped morphologically homologous analogue of UiO-67 MOF with a ruthenium(II) polypyridyl complex. J. Am. Chem. Soc. **137**(25), 8161 (2015)
86. Zhu, J., Maza, W.A., Morris, A.J.: Light-harvesting and energy transfer in ruthenium(II)-polypyridyl doped zirconium(IV) metal-organic frameworks: A look toward solar cell applications. J. Photochem. Photobiol. A **344**, 64 (2017)
87. Maza, W.A., Haring, A.J., Ahrenholtz, S.R., Epley, C.C., Lin, S.Y., Morris, A.J.: Ruthenium (II)-polypyridyl zirconium(IV) metal–organic frameworks as a new class of sensitized solar cells. Chem. Sci. **7**(1), 719 (2016)
88. Cai, M., Loague, Q.R., Zhu, J., Lin, S., Usov, P.M., Morris, A.J.: Ruthenium(II)-polypyridyl doped zirconium(IV) metal–organic frameworks for solid-state electrochemiluminescence. Dalton Trans. **47**(46), 16807 (2018)
89. Miao, W.: Electrogenerated chemiluminescence and its biorelated applications. Chem. Rev. **108**(7), 2506 (2008)

90. Spehar-Déléze, A.-M., Almadaghi, S., Sullivan, C.K.: Development of solid-state electrochemiluminescence (ECL) sensor based on $Ru(bpy)_3{}^{2+}$-encapsulated silica nanoparticles for the detection of biogenic polyamines. Chemosensors **3**(2), 178 (2015)
91. Navarro Amador, R., Carboni, M., Meyer, D.: Sorption and photodegradation under visible light irradiation of an organic pollutant by a heterogeneous UiO-67–Ru–Ti MOF obtained by post-synthetic exchange. RSC Adv. **7**(1), 195 (2017)
92. Ezhov, R., Karbakhsh Ravari, A., Page, A., Pushkar, Y.: Water oxidation catalyst *cis*-[Ru (bpy)(5,5′-dcbpy)$(H_2O)_2]^{2+}$ and its stabilization in metal–organic framework. ACS Catal. **10** (9), 5299 (2020)
93. Hetterscheid, D.G.H., Reek, J.N.H.: Mononuclear water oxidation catalysts. Angew. Chem. Int. Ed. **51**(39), 9740 (2012)
94. Erdman, D., Pineda-Galvan, Y., Pushkar, Y.: Mechanistic analysis of water oxidation catalyst *cis*-$[Ru(bpy)_2(H_2O)_2]^{2+}$: effect of dimerization. Catalysts **7**(2), 39 (2017)
95. Ozawa, H., Haga, M.-A., Sakai, K.: A photo-hydrogen-evolving molecular device driving visible-light-induced EDTA-reduction of water into molecular hydrogen. J. Am. Chem. Soc. **128**(15), 4926 (2006)
96. Hou, C.-C., Li, T.-T., Cao, S., Chen, Y., Fu, W.-F.: Incorporation of a $[Ru(dcbpy)(bpy)_2]^{2+}$ photosensitizer and a $Pt(dcbpy)Cl_2$ catalyst into metal–organic frameworks for photocatalytic hydrogen evolution from aqueous solution. J. Mater. Chem. A **3**(19), 10386 (2015)
97. Yang, S., Pattengale, B., Lee, S., Huang, J.: Real-time visualization of active species in a single-site metal–organic framework photocatalyst. ACS Energy Lett. **3**(3), 532 (2018)
98. Liao, W.-M., Zhang, J.-H., Hou, Y.-J., Wang, H.-P., Pan, M.: Visible-light-driven CO_2 photo-catalytic reduction of Ru(II) and Ir(III) coordination complexes. Inorg. Chem. Commun. **73**, 80 (2016)
99. Grätzel, M.: Dye-sensitized solar cells. J. Photochem. Photobiol. C **4**(2), 145 (2003)
100. Hao, Y., Xiong, D., Wang, L., Chen, W., Zhou, B., Liu, Y.-N.: A reversible competition colorimetric assay for the detection of biothiols based on ruthenium-containing complex. Talanta **115**, 253 (2013)
101. Wang, Z., Yang, J., Li, Y., Zhuang, Q., Gu, J.: Zr-Based MOFs integrated with a chromophoric ruthenium complex for specific and reversible Hg^{2+} sensing. Dalton Trans. **47** (16), 5570 (2018)
102. Mateo, D., Santiago-Portillo, A., Albero, J., Navalón, S., Alvaro, M., García, H.: Long-term photostability in terephthalate metal–organic frameworks. Angew. Chem. Int. Ed. **58**(49), 17843 (2019)
103. Zhang, S., Li, L., Zhao, S., Sun, Z., Luo, J.: Construction of interpenetrated ruthenium metal–organic frameworks as stable photocatalysts for CO_2 reduction. Inorg. Chem. **54**(17), 8375 (2015)
104. Jiang, H.-L., Makal, T.A., Zhou, H.-C.: Interpenetration control in metal–organic frameworks for functional applications. Coord. Chem. Rev. **257**(15), 2232 (2013)
105. Zhang, S., Han, L., Li, L., Cheng, J., Yuan, D., Luo, J.: A highly symmetric metal-organic framework based on a propeller-like Ru-organic metalloligand for photocatalysis and explosives detection. Cryst. Growth Des. **13**(12), 5466 (2013)
106. Zhang, S., Li, L., Zhao, S., Sun, Z., Hong, M., Luo, J.: Hierarchical metal–organic framework nanoflowers for effective CO_2 transformation driven by visible light. J. mater. Chem. A **3**(30), 15764 (2015)
107. Li, H., Eddaoudi, M., O'Keeffe, M., Yaghi, O.M.: Design and synthesis of an exceptionally stable and highly porous metal-organic framework. Nature **402**(6759), 276 (1999)
108. Yaghi, O.M., O'Keeffe, M., Ockwig, N.W., Chae, H.K., Eddaoudi, M., Kim, J.: Reticular synthesis and the design of new materials. Nature **423**(6941), 705 (2003)
109. Li, Q., Thonhauser, T.: A theoretical study of the hydrogen-storage potential of $(H2)_4CH_4$ in metal organic framework materials and carbon nanotubes. J. Phys.: Condens. Matter **24**(42), 424204 (2012)

110. Tang, Y., He, W., Lu, Y., Fielden, J., Xiang, X., Yan, D.: Assembly of ruthenium-based complex into metal–organic framework with tunable area-selected luminescence and enhanced photon-to-electron conversion efficiency. J. Phys. Chem. C **118**(44), 25365 (2014)
111. Wang, C., Liu, D., Xie, Z., Lin, W.: Functional metal–organic frameworks *via* ligand doping: Influences of ligand charge and steric demand. Inorg. Chem. **53**(3), 1331 (2014)
112. Lan, G., Zhu, Y.-Y., Veroneau, S.S., Xu, Z., Micheroni, D., Lin, W.: Electron injection from photoexcited metal-organic framework ligands to Ru2 secondary building units for visible-light-driven hydrogen evolution. J. Am. Chem. Soc. **140**(16), 5326 (2018)
113. Ortega-Guerrero, A., Fumanal, M., Capano, G., Tavernelli, I., Smit, B.: Insights into the electronic properties and charge transfer mechanism of a porphyrin ruthenium-based metal–organic framework. Chem. Mater. **32**(10), 4194 (2020)
114. Bloch, E.D., Britt, D., Lee, C., Doonan, C.J., Uribe-Romo, F.J., Furukawa, H., Long, J.R., Yaghi, O.M.: Metal insertion in a microporous metal−organic framework lined with 2,2′-bipyridine. J. Am. Chem. Soc. **132**(41), 14382 (2010)
115. Liu, H., Yin, B., Gao, Z., Li, Y., Jiang, H.: Transition-metal-free highly chemo- and regioselective arylation of unactivated arenes with aryl halides over recyclable heterogeneous catalysts. Chem. Commun. **48**(14), 2033 (2012)
116. Carson, F., Agrawal, S., Gustafsson, M., Bartoszewicz, A., Moraga, F., Zou, X., Martín-Matute, B.: Ruthenium complexation in an aluminium metal–organic framework and its application in alcohol oxidation catalysis. Chem. Eur. J. **18**(48), 15337 (2012)
117. Sakamoto, R., Takada, K., Pal, T., Maeda, H., Kambe, T., Nishihara, H.: Coordination nanosheets (CONASHs): Strategies, structures and functions. Chem. Commun. **53**(43), 5781 (2017)
118. Kuhn, P., Antonietti, M., Thomas, A.: Porous, covalent triazine-based frameworks prepared by ionothermal synthesis. Angew. Chem. Int. Ed. **47**(18), 3450 (2008)
119. Liu, M., Guo, L., Jin, S., Tan, B.: Covalent triazine frameworks: Synthesis and applications. J. Mater. Chem. A **7**(10), 5153 (2019)
120. Hug, S., Tauchert, M.E., Li, S., Pachmayr, U.E., Lotsch, B.V.: A functional triazine framework based on N-heterocyclic building blocks. J. Mater. Chem. **22**(28), 13956 (2012)
121. Hug, S., Stegbauer, L., Oh, H., Hirscher, M., Lotsch, B.V.: Nitrogen-rich covalent triazine frameworks as high-performance platforms for selective carbon capture and storage. Chem. Mater. **27**(23), 8001 (2015)
122. Muniz-Miranda, F., De Bruecker, L., De Vos, A., Vanden Bussche, F., Stevens, C.V., Van Der Voort, P., Lejaeghere, K., Van Speybroeck, V.: Optical properties of isolated and covalent organic framework-embedded ruthenium complexes. J. Phys. Chem. A **123**(32), 6854 (2019)
123. De Vos, A., Lejaeghere, K., Muniz Miranda, F., Stevens, Christian V., Van Der Voort, P., Van Speybroeck, V.: Electronic properties of heterogenized Ru(II) polypyridyl photoredox complexes on covalent triazine frameworks. J. Mater. Chem. A **7**(14), 8433 (2019)
124. Cui, P., Ren, H., Zhu, G.: Ruthenium ilaying porous aromatic framework for hydrogen generation from ammonia borane. Front. Mater. **6**(223) (2019)
125. Chen, G.-J., Li, X.-B., Zhao, C.-C., Ma, H.-C., Kan, J.-L., Xin, Y.-B., Chen, C.-X., Dong, Y.-B.: Ru Nanoparticles-loaded covalent organic framework for solvent-free one-pot tandem reactions in air. Inorg. Chem. **57**(5), 2678 (2018)
126. Gonçalves, L.P.L., Christensen, D.B., Meledina, M., Salonen, L.M., Petrovykh, D.Y., Carbó-Argibay, E., Sousa, J.P.S., Soares, O.S.G.P., Pereira, M.F.R., Kegnæs, S., et al.: Selective formic acid dehydrogenation at low temperature over a RuO_2/COF pre-catalyst synthesized on the gram scale. Catal. Sci. Technol. **10**(7), 1991 (2020)
127. Singh, A.K., Singh, S., Kumar, A.: Hydrogen energy future with formic acid: A renewable chemical hydrogen storage system. Catal. Sci. Technol. **6**(1), 12 (2016)
128. Solymosi, F., Koós, Á., Liliom, N., Ugrai, I.: Production of CO-free H_2 from formic acid. A comparative study of the catalytic behavior of Pt metals on a carbon support. J. Catal. **279**(1), 213 (2011)

129. Mondal, S., Singuru, R., Chandra Shit, S., Hayashi, T., Irle, S., Hijikata, Y., Mondal, J., Bhaumik, A.: Ruthenium nanoparticle-decorated porous organic network for direct hydrodeoxygenation of long-chain fatty acids to alkanes. ACS Sustain. Chem. Eng. **6**(2), 1610 (2018)
130. Maiti, S., Chowdhury, A.R., Das, A.K.: Electrochemically facile hydrogen evolution using ruthenium encapsulated two dimensional covalent organic framework (2D COF). ChemNanoMat **6**(1), 99 (2020)
131. Bae, S.-Y., Mahmood, J., Jeon, I.-Y., Baek, J.-B.: Recent advances in ruthenium-based electrocatalysts for the hydrogen evolution reaction. Nanoscale Horiz. **5**(1), 43 (2020)

Polymer-Stabilized Ru Nanoparticles

7

Abstract

Ru nanoparticles are well-known for their high catalytic activity. In order to retain the catalytic performance and prevent agglomeration, such nanoparticles have to be stabilized by appropriate ligands. Besides a range of small molecules, macromolecular architectures have been utilized for this purpose. The different strategies to produced polymer-stabilized nanoparticles, which exhibit catalytic activity, are summarized in this chapter.

7.1 Introduction

The discovery of the so-called quantum-size effect [1] has initiated intense research on nanosized metal particles that feature advanced electronic [2, 3], optical, [3–5] or magnetic characteristics [3, 6–8]—often far beyond the properties of the corresponding bulk metals. Furthermore, the catalytic activity is worth mentioning: Due to their size and shape, metal nanoparticles (NPs) can be considered to be a bridge between homogeneous and heterogeneous catalysis. [9] In general, NPs exhibit high surface-to-volume ratios, in particular when the size of the NPs is in the range of 1 nm (or even below). In this case, >90% of the metal atoms are located on the surface and, thus, a maximum number of (potentially) catalytically active sites is offered [10]. It needs to be pointed out, that the catalytic activity of such NPs depends on numerous parameters. Besides the nature of the metal itself (one metal might be useful for a certain type of catalysis; whereas, the others are not), the stabilizers and, if applicable, the solid support are of significant relevance. With respect to the latter, metal oxides, metal–organic-frameworks (MOFs, see Chap. 6), dendrimers (see Chap. 4), or carbon materials might be used. When aiming for the fabrication of heterogeneous catalysts, a calcination step is required in order to

© The Author(s), under exclusive license to Springer Nature Switzerland AG 2021
U. S. Schubert et al., *Ruthenium-Containing Polymers*,
https://doi.org/10.1007/978-3-030-75598-0_7

remove all organics and to liberate the active sites; however, sintering or oxidation of the NPs might occur and needs to be suppressed (if desired).

Over the last decades, significant effort has been made to enable the synthesis of tailor-made metal NPs, in particular of ligand-stabilized ones in which crucial parameters (e.g., their size and shape as well as surface state) can be adjusted by proper choice of reaction conditions and by adding appropriate stabilizing agents in order to prevent coalescence and aggregation as well as enable solubility in organic solvents (or even water) [10, 11]. In this context, a broad range of stabilizers has already been used: Ionic liquids, polymers, surfactants, polyols, organic ligands [such as *N*-heterocyclic carbenes (NHCs)], etc. These species interact with NPs and thus, affect the surface properties in quite different ways (Fig. 7.1) [12]. Polymers, such as poly(*N*-vinylpyrrolidone) (PVP), only bind loosely to NPs, but offer an efficient steric shielding; whereas, thiols or phosphines enable a very strong interaction which, at high surface loadings, might even result in the loss of catalytic activity. As pointed out by Axet and Philippot, one has to find the balance between efficiently stabilizing the NPs while still guaranteeing substrates have access to the surface and, thereby, retaining the catalytic activity [10].

From the heavy transition metals, the representatives of the platinum and coinage metals have been studied most extensively with respect to NP formation. However, Ru has attracted significantly less attention though Ru nanoparticles represent versatile catalysts, e.g., in reduction or oxidation reactions as well as in CH activation [10, 13]. Moreover, ligand-stabilized Ru nanoparticles can be investigated *inter alia* by solution and solid-state NMR spectroscopy due to the negligible disturbance by magnetic effects (e.g., Knight shift, paramagnetism, ferromagnetism, etc.) [12]. Unlike for many other metal NPs, the chemical processes on Ru NPs can be spectroscopically monitored and the intermediates, occurring in catalytic processes, can be readily identified. This represents a versatile add-on to the typical methods applied to characterize NPs with respect to their size and shape (e.g., by microscopy imaging and light-scattering measurements) or elemental compositions [e.g., by X-ray photoelectron spectroscopy (XPS)].

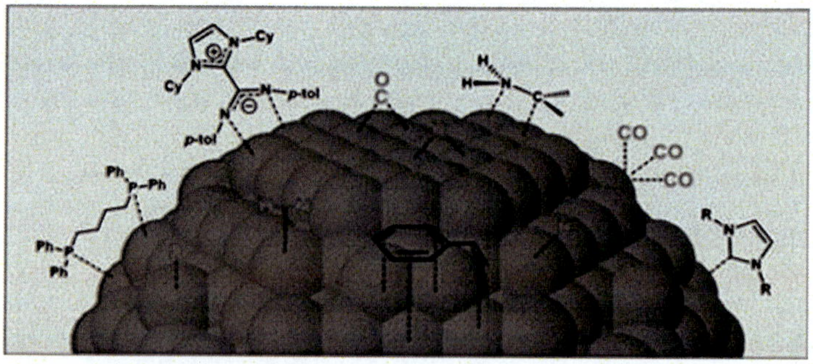

Fig. 7.1 Schematic representation of the interaction of (organic) molecules with a NP surface [12]. Figure reproduced with kind permission; © 2018 American Chemical Society

7.2 Synthesis of Ru Nanoparticles—General Considerations

Before dealing with polymer-stabilized Ru nanoparticles, some general considerations regarding the synthesis of such species have to be made. As for most metal nanoparticles, the liquid-phase synthesis of Ru NPs represents the most versatile strategy [14]; the gas-phase NP synthesis, which is experimentally more demanding, has—to the best of our knowledge—not yet been applied in this context [15].

The commercially available $RuCl_3 \times xH_2O$ represents, by far, the most commonly used precursor for NP synthesis from water [14]. A wide range of reducing agents has been employed for this reaction: Alcohols, amines, CO, hydrides (i.e., $NaBH_4$ and $LiAlH_4$), $H_2N–NH_2$, citrate salts, or hydrogen. In any case, a water-soluble stabilizing agent must be present to suppress agglomeration/separation and control the NP growth [10]. Detrimentally, all these variants lead to a certain degree-of -NP surface contamination, which results from the solvent, salts, and/or reagents used in the reduction step. Moreover, a passivating RuO_2 shell might be formed, which limits the ability of substrates to reach the surface and, thereby, reduces the activity of the NPs. The latter can, however, be circumvent by simply applying a hydrogenation step prior to the use of the Ru NP (Fig. 7.2).

As an alternative approach, organometallic precursor complexes can be utilized, which are decomposed by hydrogenation under mild conditions (i.e., low temperature and H_2 pressure) in the presence of a compatible stabilizer [16]. It has been demonstrated that this method generally gives high-quality NPs with a narrow size distribution; typically, the surface characteristics are well-behaved and these NPs can readily be applied, e.g., in catalysis. However, these organometallic precursors require stringent handling in order to avoid their degradation by water and/or oxygen. This concept was introduced by Bradley et al. [17] and developed further by Philippot et al. [18]. In particular, the olefinic complexes [Ru(cod)(2-methylallyl)$_2$] and [Ru(cod)(cot)] are of high virtue, since their reduction by treatment with H_2 generates NPs with clean surfaces—the ligands are reduced to (cyclo)alkanes, which do not interact with the NPs and can, thus, easily be removed

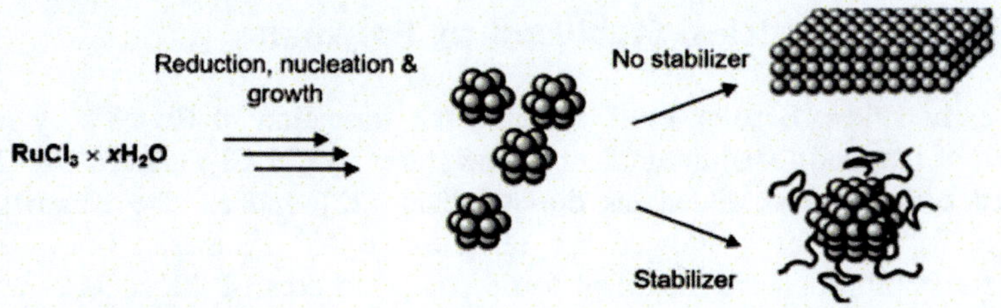

Fig. 7.2 Schematic representation of the synthesis of Ru NPs by reduction of the $RuCl_3 \times xH_2O$ precursor [10]. Figure redrawn according to the literature; © 2020 American Chemical Society

(cod: 1,5-cyclooctadiene; cot: 1,3,5-cyclooctatriene). Moreover, Ru(acac)$_3$ (acac: acetylacetonate) and [$Ru_3(CO)_{12}$] can be employed, as precursors, though only at elevated temperatures. Using the latter, the formed Ru NPs will have CO molecules coordinated to the surface (remarkably, this precursor offers the unique opportunity to perform a shape-controlled NP synthesis) [19].

Finally, the so-called polyol process, working at high temperature, allows the synthesis of metal NPs with good control over their sizes, shapes, compositions, and architectures [10, 20]. As for the aforementioned reduction of $RuCl_3 \times xH_2O$, this process is low-cost, facile, and scalable. Commercially available polyols, i.e., simple 1,2-diols or ether glycols, are utilized as the solvent for the NP precursor and, simultaneously, as a reducing agent; it has been shown, that they even can act as stabilizers for the formed NPs. Nonetheless, reducing agents (e.g., acetates or hydrogen) and stabilizers (e.g., polymers) are sometimes added to further improve the NP formation. Besides $RuCl_3 \times xH_2O$, which is most commonly used, also Ru (acac)$_3$ has been utilized, as the precursor, in the polyol process [20].

Besides these various methods to prepare colloidal (or suspended) nanoparticles, their immobilization onto solid supports is also of particular relevance for catalytic applications [10]. Typically, metal oxides or carbon materials are utilized for this purpose. The main benefits of this approach are the improved stability and recovery of the supported NPs (compared to colloidal systems); moreover, the proper choice of the support enables a tuning of the catalytic activity of the NPs, i.e., the support might play an active role in the targeted application. There are basically two strategies to prepare such supported Ru NPs. A two-step protocol can be applied in which colloidal NPs are first fabricated via one of the aforementioned routes and then immobilized by impregnation. The pre-formed NPs are mixed with an (activated) support and binding occurs via anchoring groups on the support's surface or within its pores. Thereby, the structural characteristics of the NPs are retained, thus enabling a comparison of the catalytic activity of colloidal and supported NPs. Alternatively, NP formation can be conducted in the presence of a (functionalized) support. In this case, however, the structural characteristics of the formed NPs can be strongly affected by the support.

7.3 Ru Nanoparticles, Stabilized by Polymers

In the early 1990s, Bradley et al. reported the formation of Ru NPs by hydrogenation of [Ru(cod)(cot)] at ambient temperatures in CH_2Cl_2; either PVP [21] or modified cellulose was added, as the stabilizer [22]. Indeed, the commercially available Ru^0 precursor allowed the residue-free NP synthesis; however, the polymeric stabilizers were identified to be crucial to control the coalescence and limit particle growth (Fig. 7.3) [23].

These studies demonstrated that crystalline Ru nanoparticles of spherical shape, small size and narrow size distribution could be formed in a medium that offered only a poor electronic interaction with the surface of the nanoparticles but, instead,

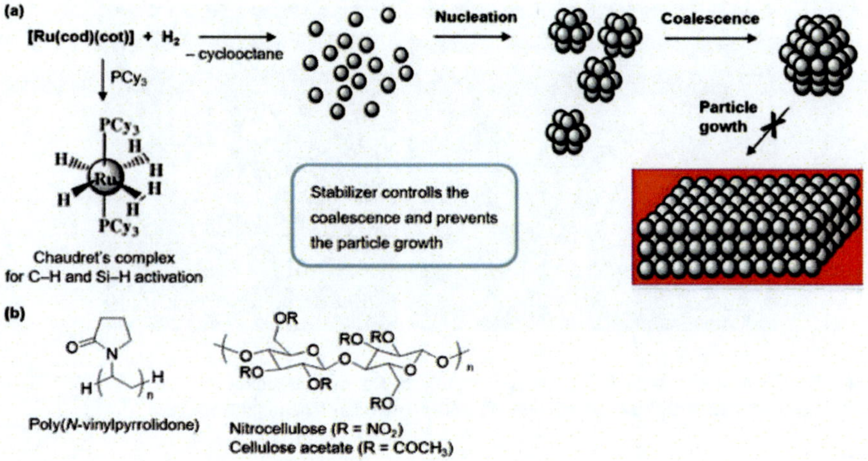

Fig. 7.3 **a** Schematic representation of the controlled synthesis of Chaudret's *bis*-dihydrogen complex and Ru nanoparticles [12]. **b** Schematic representation of polymer stabilizers. Figure redrawn according to the literature; © 2018 American Chemical Society

enabled an efficient steric shielding [24]. For example, at a Ru content of 5 wt% in the polymer stabilizer, crystalline Ru nanoparticles were formed for which a hexagonal close-packing (hcp) was evidenced by high-resolution electron microscopy (HREM) imaging. In the case of PVP, a homogeneous particle dispersion in the polymer matrix was observed (uniform NP mean size of 1.1 nm); whereas, agglomeration of the particles (1.7 nm) occurred in cellulose acetate (Fig. 7.4). Chaudret et al. also prepared PVP-stabilized bimetallic RuPt nanoparticles starting from $Pt(dba)_2$ and Ru(cod)(cot), as precursors; the composition of the nanoparticles was adjusted via the applied precursor ratio (Fig. 7.5a) [25, 26]. With increasing Ru content, a structural change from a face-centered cubic (fcc) to a hcp phase was observed. This transition generated a relative composition of Pt_1Ru_3 (75 wt% of Ru) where the particle size was contracted (ca. 1.1. nm) and the size distribution was narrowed; for this intermediate phase, a model of truncated twinned octahedrons of 79 atoms was proposed (Fig. 7.5b). Later, Jeszka et al. followed a different strategy to fabricate (*inter alia*) Ru nanoparticles [27]. These authors utilized a solid-state synthesis in which a dispersion of the aforementioned Ru^0 precursor in a polymer film was treated with dihydrogen at ambient temperature. The resulting Ru nanoparticles were stabilized by the polymers by π-electronic interactions involving the aromatic rings or the lone-pairs in the case of PS and PMMA, respectively. It was further shown by IR measurements that CO or O_2 could rapidly diffuse through the polymer matrix and coordinated to the particles' surfaces.

Philippot et al. showed that the catalytic activity of Ru NPs, stabilized by PVP, can be modulated by grafting Sn atoms onto their surface (n-Bu_3SnH was used, as precursor, in this study) [28]. Spectroscopic studies (i.e., IR, NMR, and

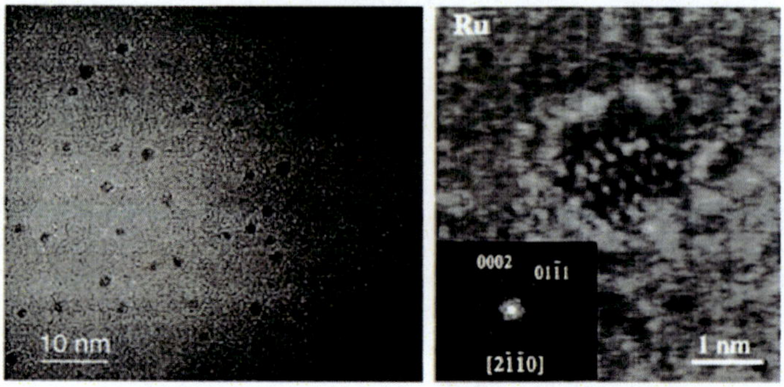

Fig. 7.4 TEM (left) and HREM image (right) of Ru nanoparticles stabilized by PVP [24]. Figure reproduced with kind permission; © 2001 American Chemical Society

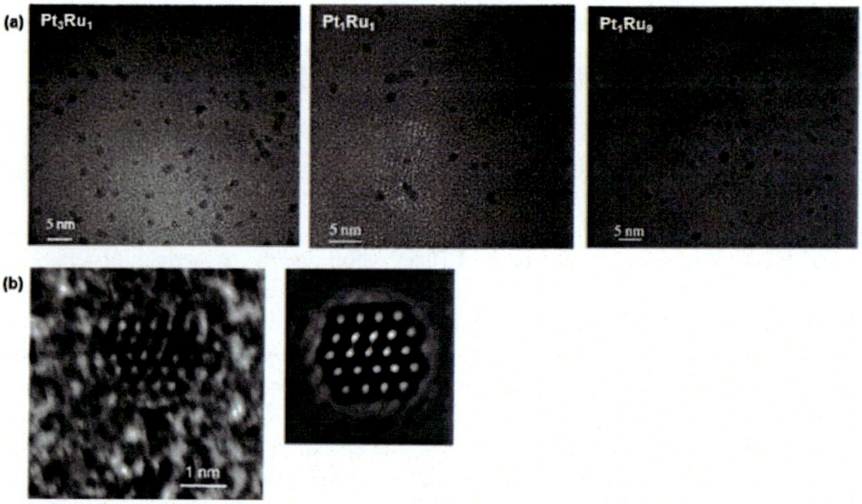

Fig. 7.5 **a** TEM images of the PVP-stabilized bimetallic nanoparticles at various compositions. **b** HREM image (left) of Pt_1Ru_3 showing the twinning and the calculated structure is also depicted (right) [25]. Figure reproduced with kind permission; © 1999 American Chemical Society

Mössbauer), supported by computational methods, suggested the binding of "*n*-BuSn" fragments to the NPs in a μ^3-bridging fashion. In the absence of Sn atoms, the PVP-stabilized the hydrogenation of styrene to give ethylcyclohexane, as the exclusive product; whereas, at a loading with 0.2 eq. of Sn atoms, ethylbenzene was obtained with a high chemoselectivity of 95%. Tao et al. noted that the presence of Sn atoms on the surface of Ru NPs significantly decreased the barrier for H_2 penetration into the particle; presumably, the Sn–H species were involved in this process and could possibly contribute to the different reactivity of the Sn-doped Ru NPs [29].

Liu et al. utilized a different precursor for the preparation of PVP-stabilized Ru nanoparticles [30]. The reduction of $RuCl_3{\cdot}xH_2O$ with $NaBH_4$ in aqueous MeOH at ambient temperature afforded well-defined nanoparticles, their size could be adjusted via the amounts of added stabilizer and reductant. The particles exhibited good catalytic activity in the heterogeneous hydrogenation of various olefins or aromatic nitro compounds; [30, 31] however, considerable drawbacks arose. The related PVP-shielded Pt or Ir NPs were stable even in protic solvents (e.g., water, alcohols or acetic acid) over several months; whereas, the PVP-stabilized Ru nanoparticles decomposed in such solvents within days. Moreover, the small amount of boride associated with Ru as well as a significant amount of surface-adsorbed borate could be identified by XPS analysis. It was shown that the latter had a profound influence on the catalytic performance of the Ru nanoparticles [32]. Dried Ru nanoparticles, prepared by $NaBH_4$ reduction (stabilized and non-stabilized ones), were prone to explode upon exposure to air [33, 34]; thus, this method for preparing these nanosized materials was seldom followed thereafter. Recently, Karakhanov et al. revisited this protocol and prepared well-defined Ru nanoparticles, which were stabilized in a cross-linked network of poly(propylene imine) (PPI) dendrimers (Fig. 7.6) [35]. These nanoparticles were employed, as efficient heterogeneous and reusable catalysts, for the hydrogenation of unsaturated and aromatic compounds, in particular phenols [36]. These authors also demonstrated that the immobilization of the PPI-stabilized Ru particles with mesoporous SiO_2 gave a robust hybrid material, which could be employed, as heterogeneous catalyst, for the hydrogenation of alkyl-substituted phenols [37]. Further examples for the stabilization of Ru NPs within dendrimers can be found in Chap. 4.

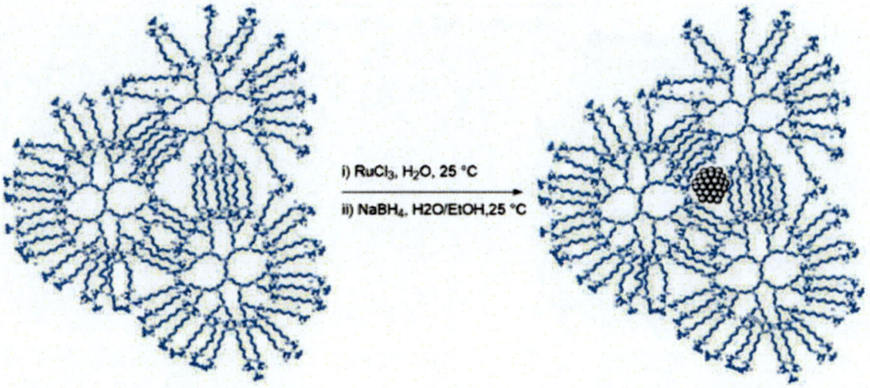

Fig. 7.6 Schematic representation of the synthesis of Ru nanoparticles, stabilized by a cross-linked PPI dendrimer matrix [36]. Figure reproduced with kind permission; © 2015 Wiley-VCH

PVP-stabilized Ru nanoparticles have been widely employed, as catalysts, to reduce nitroarenes or azo derivatives with $NaBH_4$ or hydrazine, as reductants [10]. For example, Song et al. reported the synthesis of Ru NPs by reduction of $RuCl_3 \times xH_2O$ by *n*-propanol in the presence of PVP; thereby, the size of the particles—ranging from 2.6 to 51.5 nm—was controlled via the reaction conditions (i.e., pH value and reaction temperature) [38]. These particles exhibited a size-dependent catalytic activity in the reduction of nitroarenes with $NaBH_4$. The highest activity was found for the NPs with a size up to 8.2 nm; the larger NPs were less active, presumably due to the lower specific surface area as well as the resulting smaller contact area as well as weaker interaction with substrate and reagent. In a related study, the degradation of orange-I, as an example for a diazo dye, by PVP-stabilized Ru NPs was shown [39]. A significantly higher catalytic activity compared to e.g., Pt or Ir NPs was observed. Moreover, the Ru NPs exhibited a pronounced intolerance towards thiols, as revealed by a decrease-of-catalytic-performance. On the basis of this finding, these authors designed a calorimetric sensor for H_2S featuring a low detection limit of 0.6 nM (Fig. 7.7).

The triblock copolymer PEG-*b*-PPO-*b*-PEG is well-known for its self-assembly behavior in an aqueous medium; the resultant micelles offer an environment in which organic reactions can readily proceed [PEG: poly(ethylene glycol) and PPO:

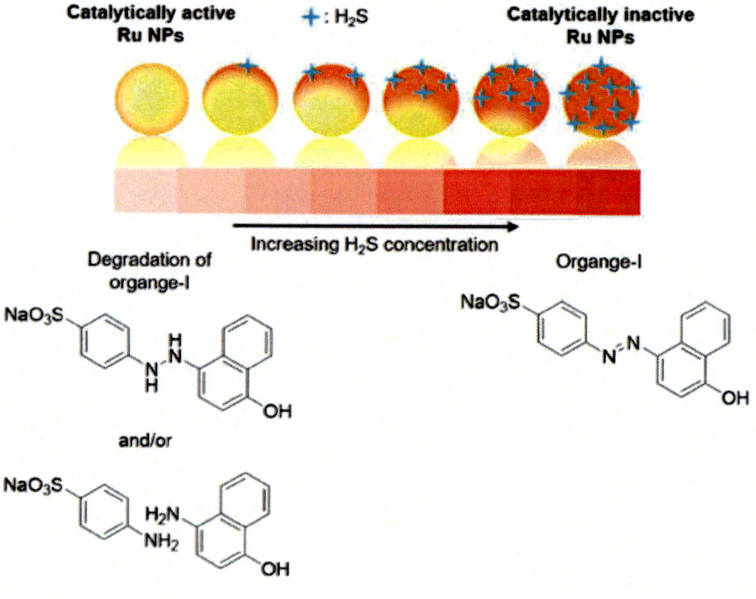

Fig. 7.7 Schematic representation of colorimetric sensing of H_2S based on the catalytic degradation of orange-I by PVP-stabilized Ru NPs [39]. Figure reproduced with kind permission; © 2017 American Chemical Society

poly(propylene oxide)]. Xie et al. demonstrated that Ru NPs, in situ generated from $RuCl_3 \times xH_2O$, were also taken up by these micelles [40, 41]. Micelle-stabilized NPs were utilized, as catalyst, for the hydrogenation of α-pinene to *cis*-pinane under mild conditions (i.e., 40 °C, 3 bar H_2, 3 h); almost full substrate conversion (>99.9%) and a very high chemoselectivity (99%) were observed for this model reaction. These same authors also tested D-α-tocopheryl-PEG1000 succinate to stabilize Ru NPs within a micellar environment [41]. This system efficiently catalyzed the hydrogenation of α-pinene and an improved recyclability of the catalyst was reported. Remarkably, the micelle-stabilized Ru NPs were, by far, the superior catalyst for this particular reaction when compared to cellulose- or PVA-stabilized Ru NPs [PVA: poly(vinyl alcohol)] [42, 43]. The latter required significantly harsher reaction conditions (i.e., 70 °C, 20 bar H_2, 3 h) to reach a comparable substrate conversion.

A water-soluble PEG-cyclodextrin network was used, as stabilizer and size-controlling agent, for in situ generated Ru nanoparticles [44]. Compared to other stabilizers, such as PEG diethers or PVP, significantly smaller particles and a higher catalytic activity in the aqueous hydrogenation of lignocellulose-derived molecules were found: Furfural and phenol derivatives were reacted with high degrees-of-conversion and excellent chemoselectivity, The cross-linking of β-cyclodextrin (β-CD) with citric acid also afforded a polymer network (M_w = 25,000 g mol^{-1}, Đ = 2.2), which stabilized Ru NPs, derived from the $NaBH_4$ reduction of $Ru(NO)(NO_3)_3$, via steric and electronic interactions (the ionizable –CO_2H moieties accounted for the latter) [45]. Though these polymer-stabilized Ru NPs represented highly efficient and recyclable catalysts for the hydrogenation of biomass-derived molecules, their stability in acidic medium remained, however, a drawback. This limitation was overcome by introducing quaternary ammonium sites (Fig. 7.11a) [46]. The Ru NPs, stabilized by this polymer network, were stable over a wide pH regime (Fig. 7.11b) and were used to catalyze the hydrogenation of a wide range of unsaturated compounds, including aromatic and fatty acids. Compared to Ru NPs stabilized by β-CD alone, the β-CD-citrate network or PVP, a significantly higher catalytic activity as revealed by an increased substrate conversion, was observed (Fig. 7.8).

Astruc et al. employed the amphiphile **1** for the stabilization of metal NPs [47]. Remarkably, this particular macromolecular stabilizer enabled the fabrication of very small metal NPs (Fig. 7.9). These stabilized, water-soluble NPs were tested, as catalysts, in a range of organic reactions in aqueous media. *Inter alia*, the Ru NPs were utilized to catalyze the reduction of nitrobenzene, which proceeded within 24 h at 80 °C in the presence of a base and at a low catalyst loading; the turnover number (TON) and turnover frequency (TOF) of this reaction were 2300 and 96 h^{-1}, respectively. Recycling of the catalyst was achieved by simply extracting the products into an organic phase and adding new reactants to the aqueous phase. This straightforward protocol might even allow an implementation in a continuous work-flow.

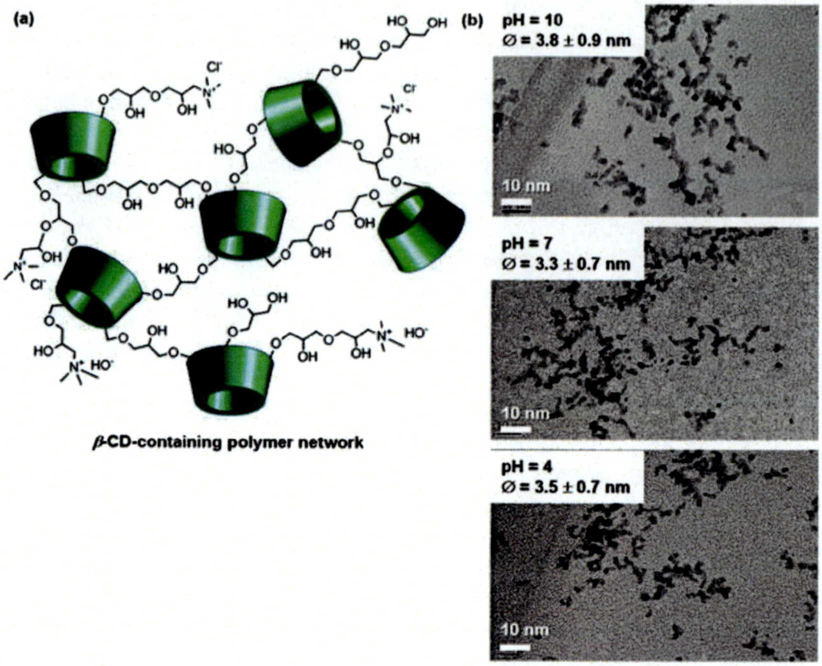

Fig. 7.8 **a** Schematic representation of the polycationic, β-CD-containing polymer network which utilized Ru NPs, as stabilizer. **b** Representative TEM images of Ru NPs, stabilized by the polymer network, in aqueous media of various pH values [46]. Figure reproduced with kind permission; © 2017 The Royal Chemical Society

As detailed in Chap. 6, the metal–organic frameworks (MOFs) and covalent organic frameworks (COFs) represent versatile host materials for the stabilization of transition-metal NPs. However, the former exhibits a considerable sensitivity towards moisture, acids, bases and/or solvents due to the lability of the metal-to-ligand coordinative bonds from which they were assembled. COFs, on the other hand, are chemically robust materials due to the covalent nature of their bonds. Similar to these, porous organic polymers (POPs) have been introduced as macromolecular stabilizers for transition-metal NPs. Though typically exhibiting a lower degree-of-structural order, POPs and COFs still share some important properties: Large surface area, tunable pore size distribution, high hydrothermal and mechanical stability, flexibility in synthesis as well as ability to introduce functional groups [48, 49]. Moreover, catalysts can easily be incorporated into their 3D structure either by covalent bonds or via non-covalent interactions. In the case of NPs, the latter are of significant importance. Noteworthy, appropriate binding sites within the organic architecture facilitate the binding of the precursor species from which the NPs are finally prepared.

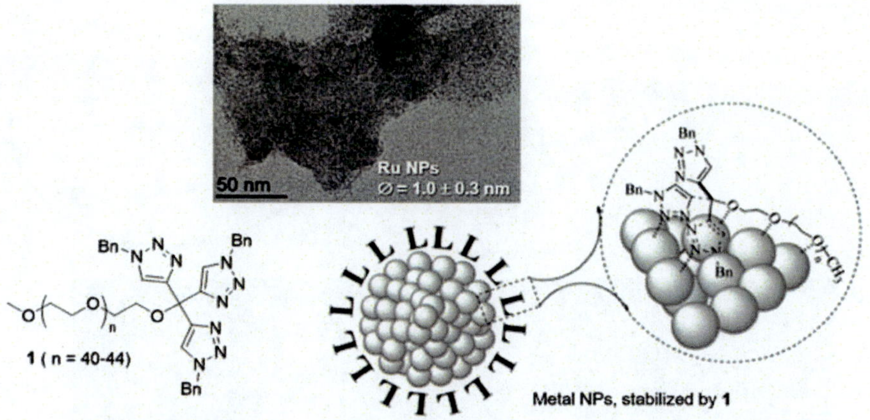

Fig. 7.9 Schematic representation of the amphiphilic stabilizer **1** and its proposed binding to metal nanoparticles. A representative TEM image of the prepared **1**-stabilized Ru NPs is also shown [47]. Figure reproduced with kind permission; © 2016 Wiley-VCH

In one example, a POP was synthesized by a Friedel–Crafts-type reaction from benzylamine and benzene using formaldehyde dimethyl acetal, as a cross-linking agent, and $FeCl_3$, as a Lewis-acid catalyst (Fig. 7.10a) [50]. The $RuCl_3 \times xH_2O$, as precursor, was coordinated to the NH_2-moieties of POP and the subsequent NP formation was achieved via three different methods: Reduction with $NaBH_4$, polyol synthesis, and hydrothermal reduction. The latter two approaches gave isolated Ru NPs dispersed within the POP structure. All these hybrid materials were thoroughly characterized by XRD measurements, electron-microscopy imaging, N_2 adsorption/desorption analysis, and various spectroscopic techniques [i.e., energy-dispersive X-ray, X-ray photoelectron, FT-IR, and CP/MAS ^{13}C NMR; Fig. 7.10b]. All stabilized Ru NPs exhibited a very high catalytic activity in the hydrogenation of nitroarenes under mild conditions (i.e., reduction with $NaBH_4$ at room temperature). For all tested substrates, high conversion and chemoselectivity were observed. The catalysts were stable and their recyclability over several catalytic runs was demonstrated without any indication of NP agglomeration or loss of catalytic activity. These authors proposed that the 3D polymer network not only offered a proper stabilizing environment, but also functioned as an electron donor within the catalytic cycle.

More recently, Bhaumik et al. prepared a POP by the acid-catalyzed condensation of 1,3,5-triformylphloroglucinol and triphenylamine [51]. Ru NPs were formed on the surface of this organic network and utilized, as catalyst, in the hydrodeoxygenation of fatty acids to long-chain alkanes in water—a process, which is of particular relevance in the context of biodiesel production from renewable resources.

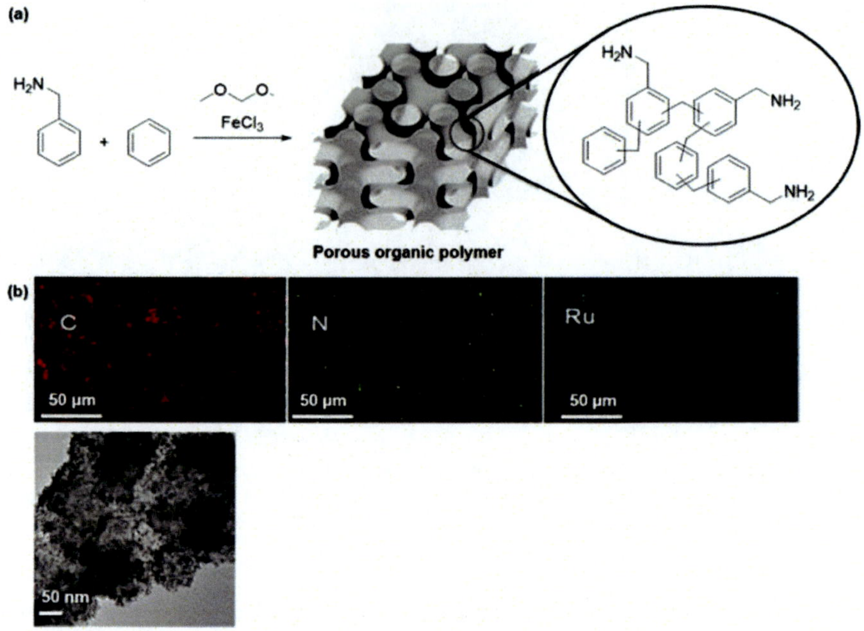

Fig. 7.10 **a** Schematic representation of the synthesis of a POP via a Friedel–Crafts-type cross-linking approach. **b** Characterization of the Ru NPs, prepared via the hydrothermal synthesis —the EDX element mapping and a representative TEM image are shown [50]. Figure reproduced with kind permission; © 2015 Wiley-VCH

A different strategy to disperse catalytically active Ru NPs within polymer matrix was followed by Trapp et al. [52]. These authors prepared a capillary microreactor by cross-linking two different polysiloxanes via a Ru^0-catalyzed hydrosilylation within the capillary; thereby, Ru nanoparticles were formed in situ from the $[Ru(cod)(2\text{-methylallyl})_2]$ precursor. TEM imaging revealed spherical, crystalline Ru NPs dispersed in the polysiloxane matrix (Fig. 7.11). The capillary microreactor was implemented into a GC-MS instrument, thus enabling a high-throughput usage of the catalytically active column. These authors focused on the hydrogenation of carbonyl compounds (i.e., aldehydes, ketones, and pyruvates) and observed a high chemoselectivity for aromatic derivatives. Moreover, the capillary exhibited a remarkable long-term stability under operative conditions, thus being more robust then related homogeneous or heterogeneous catalysts, based on Ru NPs.

The microwave-assisted synthesis of polymer-stabilized nanoparticles (i.e., Pt, Ir, Rh, Pd, Au, and Ru) was investigated by Liu et al. [53]. In particular, the formation of PVP-stabilized colloidal Ru by reduction of $RuCl_3 \times xH_2O$ in high-boiling multivalent alcohols (the so-called "polyol method" represents a versatile method to bypass the major problems involved in the aforementioned $NaBH_4$

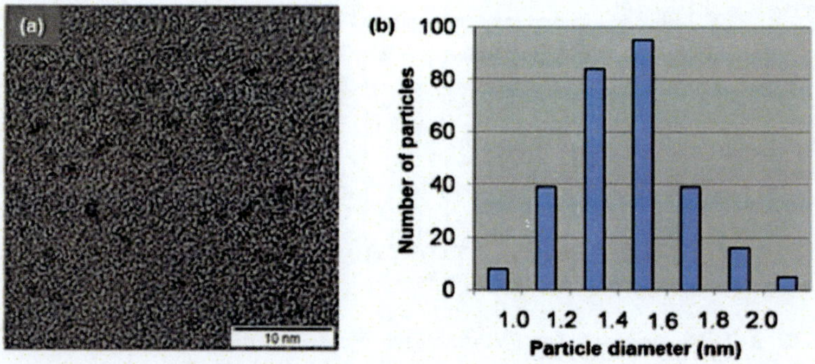

Fig. 7.11 Representative TEM image of Ru nanoparticles dispersed in a polysiloxane matrix, the corresponding particle-size distribution is also depicted [52]. Figure reproduced with kind permission; © 2016 Wiley-VCH

reduction). Starting from the first report on this method in 1989 [54], the polyol synthesis has emerged to what is today one of the standard approaches for synthesizing high-quality nanomaterials and a number of advantages can be listed: (a) Water-comparable solubility of simple metal-salt precursors; (b) high boiling points (up to 320 °C); (c) reducing properties for the immediate synthesis of metals; (d) coordinating properties that guarantee colloidal stabilization of nanoparticles; and (e) broad range of applicable polyols ranging from ethylene glycol to PEGs of high molar masses [55]. By utilizing microwave irradiation, the reaction time of the polyol synthesis could be strongly decreased compared to the related work by Bonet et al., who relied instead on conventional heating (Ru nanoparticles with a diameter of ca. 2 nm were observed) [56].

Tu and Liu obtained nanoparticles with a mean diameter of ca. 1.4 nm and a narrow size distribution from the microwave-assisted protocol with ethylene glycol, as reductant [53]. A TEM image and the corresponding size distribution histogram are depicted in Fig. 7.12a, b. In order to verify the formation of metallic Ru clusters, XPS measurements were conducted: The determined binding energy of 279.9 eV for Ru $3d_{5/2}$ in the clusters was consistent with that for bulk Ru metal (Fig. 7.12c). Avowedly, also in absence of polymeric stabilizers, Ru nanoparticles were formed, but fast agglomeration already occurred during the preparation [57]. In a subsequent study, the reaction parameters were varied regarding the reduction temperature, the solvent, the type of the polyol reductant, and the quantity of PVP [34]. Thereby, particles with sizes ranging from 1.4 to 7.4 nm could be obtained; the reduction rate of Ru^{III} strongly influenced the size of the obtained nanoparticles (i.e., large particles were formed at slow reduction rates). The same issue, i.e., how the reaction conditions of the polyol method control the formation of PVP-stabilized Ru nanoparticles, was also addressed by Simakova et al.; in detail, higher PVP to Ru ratios and high reduction temperatures were found to support the growth of small nanoparticles down to 1.7 nm in size with narrow size distribution

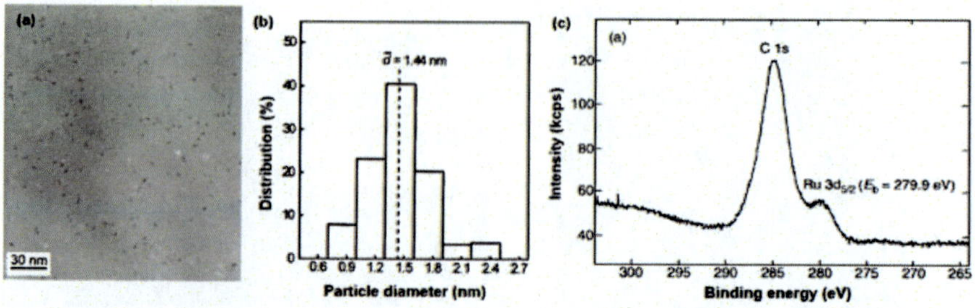

Fig. 7.12 **a** TEM image of the PVP-stabilized Ru nanoparticles. **b** Representation of the corresponding size distribution diagram. **c** XPS spectrum of the Ru nanoparticles particles (only the C 1s and Ru $3d_{5/2}$ region is shown) [53]. Figure reproduced with kind permission; © 2000 The Royal Chemical Society

[58]. Liu et al. paid attention to influence of the polyol reductant on the size of the formed PVP-stabilized Ru nanoparticles [34]. The spherical particles of the largest size were obtained with ethylene glycol, as reductant; whereas, triethylene glycol favored the formation of small particles (diameters of 7.4 vs. 1.8 nm). The authors concluded that the reduction rate plays a considerable role when aiming for larger particles.

Utilizing two different precursors, i.e., $RuCl_3 \times xH_2O$ and $Ru(acac)_3$, Kitagawa et al. synthesized PVP-stabilized Ru NPs with different crystalline structures (triethylene glycol and ethylene glycol were employed, as solvent and reducing agent, respectively) [59]. In the former case, the polyol synthesis gave the characteristic hpc packing motif; whereas, the reduction of $Ru(acac)_3$ afforded fcc Ru NPs (Fig. 7.13). Apparently, the Ru precursor, which could be dissolved in the organic solvent as a neutral species (and not as an ion), facilitated the formation of fcc Ru NPs. The NPs were immobilized onto a γ-Al_2O_3 support and tested regarding their catalytic activity as an example of CO-oxidation. Both a structure- and size-dependency were observed in these experiments (Fig. 7.13).

Gedanken et al. demonstrated that the microwave-assisted polyol synthesis can be applied to prepare anisotropic Ru NPs. These PVP-stabilized NPs were obtained in a crystalline hcp structure and different aspect ratios were observed [60]. Moreover, the seeding process allowed one to control the NP agglomeration and, as a result from this, nanodendritic particles were obtained—presumably due to an agglomeration of single-crystalline domains of different orientations.

Harada and Cong compared the microwave-assisted polyol synthesis of various PVP-stabilized monometallic nanoparticles (i.e., Pd, Rh, Ru, and Pt) by using two different types of microwave reactors, which both facilitated the up-scaling of the nanoparticle production [61]. The single-mode microwave-assisted continuous-flow processing gave well-dispersed nanoparticles at a flow rate of 1 mL min^{-1} (i.e., the residence time in the tubular reactor was ca. 10 min); in the multimode microwave-assisted batch process, the Ru particles formed in glycerol were slightly

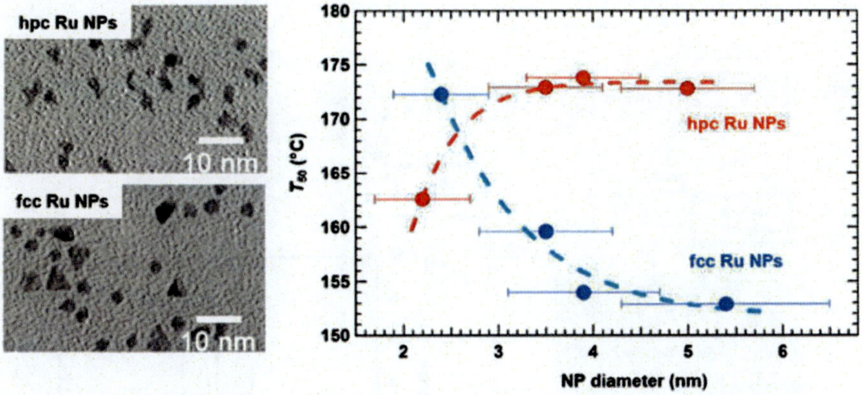

Fig. 7.13 Left: Representative TEM images of hpc and fcc Ru NPs of similar size (average diameter of ca. 3.9 nm. Right: Structure- and size-dependency of the CO-oxidation (expressed by the required temperature to reach a 50% conversion of CO to CO_2, T_{50}) [59]. Figure reproduced with kind permission; © 2013 American Chemical Society

smaller than those obtained when running the reaction in ethylene glycol (the same holds true for the size distribution) (Fig. 7.14).

As a major technical improvement, a so-called microwave applicator was employed [62]. This particular set-up allowed an efficient, cheap, and safe solvothermal synthesis of Ru nanoparticles by in-situ activation of the reaction in a stirred glass vessel without resorting to an oven. Both the preparation of PVP-stabilized as well as γ-Al_2O_3-supported Ru nanoparticles was achieved. Reduction of the Ru^{III} ions could be concluded from the color change from an initially red-orange to a dark brown solution (images of the solutions and their UV/vis absorption spectra are shown in Fig. 7.15a, b). Both the colloidal and γ-Al_2O_3-supported nanoparticles featured an average diameter of ca. 2 nm and narrow size distributions (Fig. 7.15). Such PVP-stabilized Ru/γ-Al_2O_3 nanoparticles were also prepared under solvothermal conditions in an autoclave under N_2 over-pressure [63]. The thusly obtained particles were successfully tested, as catalyst, in the selective hydrogenation of phenol to cyclohexanone at 160 °C under 5 MPa of H_2 (selectivity for this process was 83% with a phenol conversion of 88%).

In a different approach, the synthesis of colloidal Ru/γ-Al_2O_3 was investigated by Zawadzki et al. [64]. A pre-formed PVP-stabilized Ru colloid was immobilized on a γ-Al_2O_3 support; in an alternative method, the colloidal Ru/γ-Al_2O_3 was generated in situ by polyol reduction of $RuCl_3 \times xH_2O$ in the presence of γ-Al_2O_3. Both routes gave Ru nanoparticles, showing a narrow size distribution, which were uniformly distributed over the support (Fig. 7.16). However, only the *in-situ* prepared system was free of chlorine according to XPS analysis and featured a high activity for hydrogen chemisorption (99 µmol H_2 per g Ru/γ-Al_2O_3). The poor

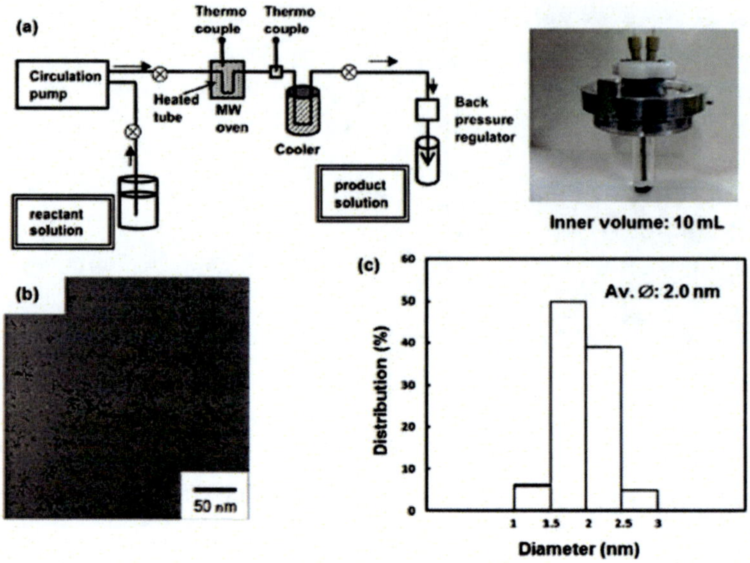

Fig. 7.14 a Schematic representation of the continuous-flow microwave reactor used for the nanoparticle synthesis. The continuous-flow cell is depicted aside. (**b** and **c**) TEM image and the corresponding size distribution histogram of the Ru nanoparticles prepared in the single-mode flow-type reactor [61]. Figure reproduced with kind permission; © 2016 American Chemical Society

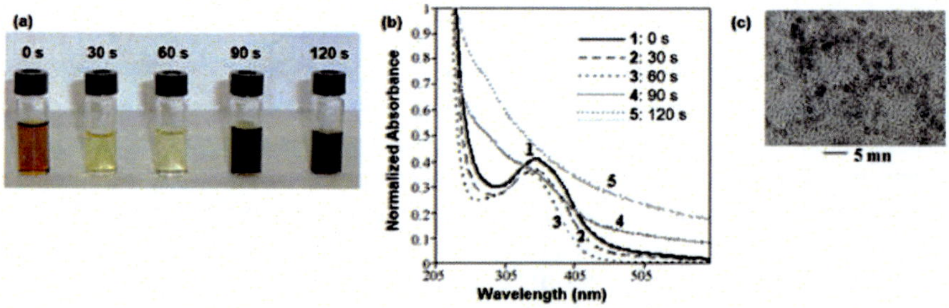

Fig. 7.15 a Image of the samples before and after microwave-assisted heating. **b** UV/vis absorption spectra of the same samples. **c** TEM image of the nanoparticles obtained after a reaction time of 30 s [62]. Figure reproduced with kind permission; © 2008 Elsevier B.V

hydrogen uptake of the immobilized material was attributed to a shielding of the PVP stabilizer. The authors concluded that removal of the PVP from Ru/γ-Al_2O_3 is required when aiming for catalytic applications, i.e., hydrogenation.

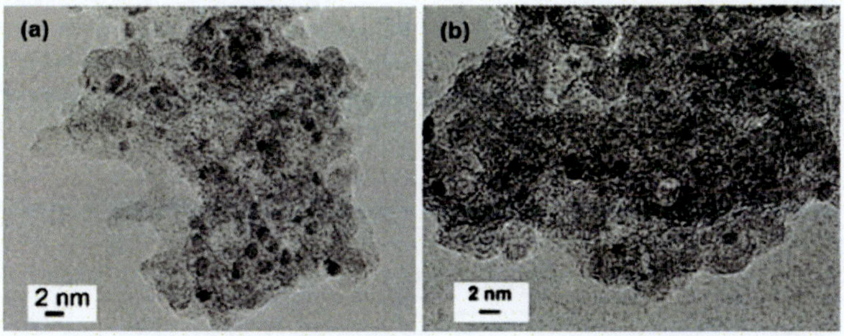

Fig. 7.16 TEM images of Ru/γ-Al_2O_3 obtained via immobilization of PVP-stabilized Ru nanoparticles (**a**) and the in-situ preparation route (**b**) [64]. Figure reproduced with kind permission; © 2008 Elsevier B.V

Zhang et al. utilized a variety of low-boiling alcohols, as reducing agents, in the solvothermal nanoparticle synthesis and investigated the morphology of the obtained materials by UV/vis absorption spectroscopy, TEM imaging, and XRD measurements [65]. Remarkably, in *n*-propanol and ethanol, the shape of the PVP-stabilized nanoparticles was concentration dependent with respect to the employed Ru^{III} precursor: at low concentrations, spherical particles were observed (average diameter of 2.1 nm, Fig. 7.17a). However, with increasing concentration, the formation of shuttle-like particles was favored, these particles had spheroid aspect ratios in the range of 4.50 to 5.02 (Fig. 7.17a). The presence of shuttle-like particles could also be confirmed by UV/vis absorption spectroscopy: With increasing reaction time, the deep-purple solutions featured an absorption band at ca. 500 nm that was attributed to the Ru surface plasmon resonance (Fig. 7.17). In general, the frequency of such a localized resonance is influenced by various parameters (i.e., size, shape, and dielectric environment of the particles); the observed band position was in accordance with those reported for Ag, Pd, Pt, and Au particles of similarly shapes [66]. Later, the PVP-stabilized particles prepared in *n*-butanol were employed, as catalyst, for the room-temperature alcoholysis of $(BH_3)(NH_3)$ and for the gas-phase oxidation of ethanol to acetaldehyde [67, 68].

The size-controlled synthesis of Ru nanoparticles was achieved by Chen et al. making use of two different strategies [69]: one-pot and seeded-growth reactions. In this study, a different type of precursor species, i.e., $Ru(NO)(NO_3)_3$, was utilized. The one-pot method gave particles with sizes of 1.2–6.5 nm, depending on the particular reaction conditions. The seeded-growth approach also allowed the precise control of particle size (3.8–7.3 nm) by varying the seeds to Ru^{III} ratio. Starting from the same precursor, PPV-stabilized Ru nanoparticles were prepared and immobilized onto TiO_2 nanotube arrays [70]. The application of these hybrid assemblies in the photocatalytic CO_2-to-CH_4 conversion was demonstrated (UV as well as high-energy visible light were used for this purpose).

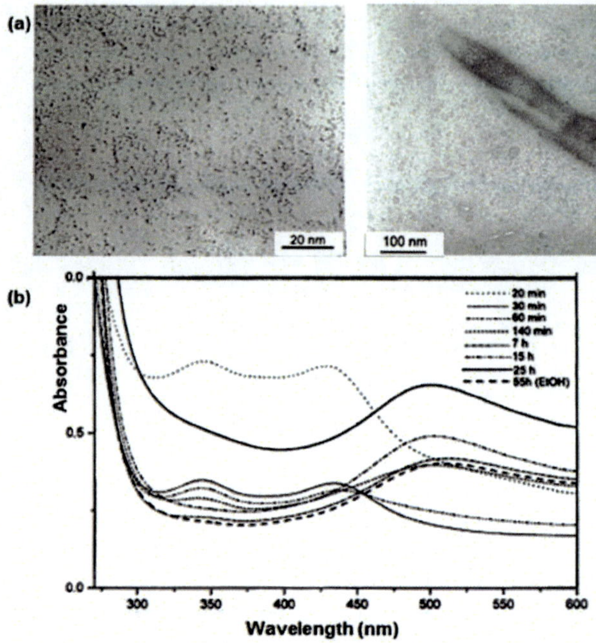

Fig. 7.17 **a** TEM images of Ru particles formed in *n*-propanol; with $[Ru^{III}] = 7.5 \times 10^{-4}$ (left, scale bar: 20 nm) and $[Ru^{III}] = 7.5 \times 10^{-3}$ (right, scale bar: 100 nm). **b** UV/vis absorption spectra measured during the formation of PVP-stabilized Ru nanoparticles in *n*-propanol after different reaction times, the spectrum measured in ethanol after 55 h is also shown [65]. Figure reproduced with kind permission; © 2007 Elsevier B.V

The final methodology allowing the preparation of colloidal dispersions of PVP-stabilized Ru nanoparticles is the photo-reduction of Ru^{III} ions in aqueous ethanol [71]. In comparison to the polyol or hydrothermal methods, this approach—requiring a photo-activator (e.g., benzophenone or benzoin)—is characterized by the convenient and mild reaction conditions. The size and shape of the resultant nanoparticles were analyzed by UV/vis spectroscopy, TEM imaging, and extended X-ray absorption fine structure (EXAFS) measurements (Fig. 7.18). With benzophenone, as photo-activator, metallic Ru nanoparticles with a mean diameter of 1.3 nm were obtained; whereas, a mixture of partially oxidized and metallic Ru nanoparticles was formed, when using benzoin instead. Apparently, the photo-reduction of Ru^{III} to Ru^{0} was promoted more efficiently by the ketyl radicals generated by photo-activation of benzophenone than by the radicals resulting from the photo-fragmentation of benzoin (Fig. 7.18). Irradiation of benzophenone in a Ru^{III}-containing solution represents an efficient and convenient method to produce polymer-stabilized Ru nanoparticles, when compared to conventional refluxing or hydrothermal methods.

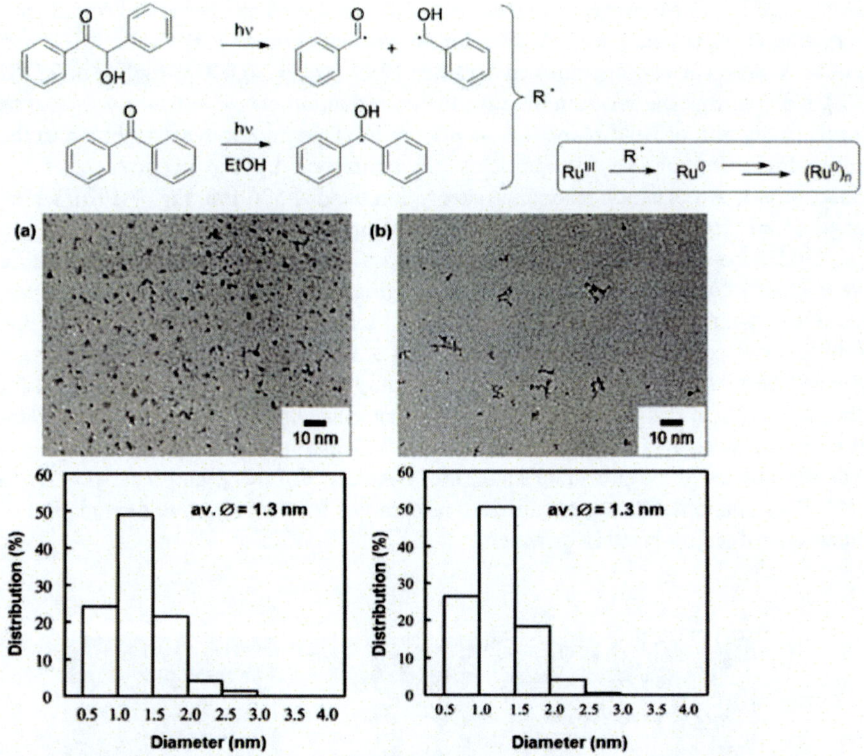

Fig. 7.18 Top: Schematic representation of the photo-fragmentation of the activators (benzoin and benzophenone). The generated radicals (R˙) serve as reducing agent for the Ru^{III} species. Bottom: TEM images and corresponding size distribution histograms of the PVP-stabilized Ru nanoparticles obtained, when using benzoin (**a**) or benzophenone (**b**), as photo-reductant [71]. Figure reproduced with kind permission; © 2008 Elsevier B.V

In the second part of this section, some bimetallic NP systems containing Ru will be summarized. Kumar et al. investigated the applicability of sonochemistry for the synthesis of AuRu nanoparticles [72]. Today, the sonochemical reduction of metal ions represents a versatile and universal protocol for the formation of nanoparticles [73]. The first ultrasound-assisted reduction of Ru^{III} using PEG in a rigid zeolite matrix was published by Gedanken et al. [74]. However, ultrasound did not play a direct role in the reduction process, but rather enabled the interstitial diffusion of the formed Ru nanoparticles into the pores of the zeolite host. A *de-facto* sonochemical reduction of Ru^{III} in a water/alcohol mixture was demonstrated later by using ultrasound with frequencies in the range of 20–1056 kHz. According to UV/vis absorption spectrophotometry, the reduction proceeded sequentially from Ru^{III} via the Ru^{II} state down to Ru^0 within 13 h. According to TEM imaging, the surfactant-stabilized Ru particles had diameters between 10 and 20 nm, as determined by TEM imaging [75]. Kumar et al. employed an ultrasound transducer, operating at a frequency of 355 kHz, for the co-reduction of $HAuCl_4 \times 3H_2O$ and $RuCl_3 \times xH_2O$ at various ratios in acidified PEG (M_w = 10,000 g mol^{-1}) at 25 °C [72]. Noteworthy, the reaction time for the co-reduction was significantly shorter as for the reduction of Ru^{III} alone (i.e., 4 h vs. 7 h). This finding was attributed to the initial formation of Au nanoparticles (as confirmed by the occurrence of the characteristic Au plasmon resonance absorption band at 536 nm, Fig. 7.19b), which acted as an electron sink mediating the subsequent Ru^{III} reduction. According to this model, core–shell-type nanoparticles with a layer of Ru covering the Au core were expected and corroborated by TEM imaging (Fig. 7.19c). A GC electrode was modified with these AuRu nanoparticles and exhibited a decrease in the double-layer current in CV measurements compared to an analogous electrode covered only with Ru nanoparticles. This finding was proposed to be beneficial for potential applications of the bimetallic nanoparticles, e.g., in fuel cells. Later, Patra et al. synthesized PVP-stabilized, trimetallic AuPdRu nanoparticles by a step-wise reduction of the three precursor salts, i.e., $HAuCl_4$, K_2PdCl_4, and $RuCl_3 \times xH_2O$ [76]. These nanoparticles were utilized, as catalyst, for the reductive degradation of diazo-type dyes by $NaBH_4$ in water.

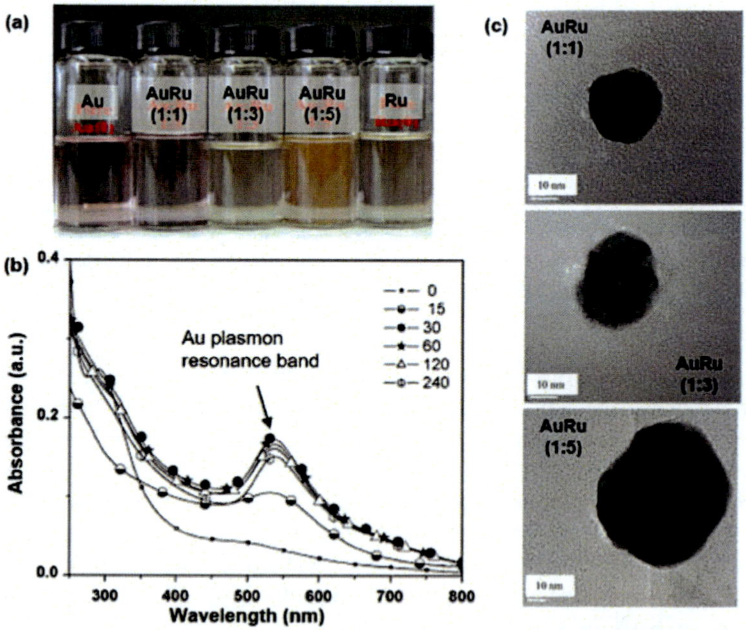

Fig. 7.19 **a** Images of pure Au, pure Ru, and bimetallic AuRu nanoparticles with various Au-to-Ru ratios. **b** Formation of the AuRu nanoparticles (1:1 ratio) as monitored by UV/vis absorption spectroscopy (from 0 to 240 min). **c** TEM images of AuRu nanoparticles of various

A series of PVP-stabilized bimetallic PtRu nanoparticles was obtained from an EtOH co-reduction of $H_2PtCl_6 \times 6H_2O$ and $RuCl_3 \times xH_2O$ [77]. The average diameters of the nanoparticles depended on the applied ratio of precursor salts and decreased with increasing Ru content from 2.8 to 1.9 nm. XPS measurements confirmed the metallic state of both Pt and Ru; as for the aforementioned AuRu nanoparticles, a core–shell-type morphology was found with the particle surfaces being rich in Ru. Furthermore, the PtRu nanoparticles were employed in the selective hydrogenation of *ortho*-chloronitrobenzene. The PtRu catalysts featured a high activity (TOF of 0.8 to 5.3×10^{-2} mol s^{-1}) and selectivity for the formation of *ortho*-chloroaniline (93–99%). Thereby, the selectivity of the reaction increased with increasing Ru content, though the TOF simultaneously decreased (at the Pt-to-Ru ratio of 1:4 a maximum selectivity of 99% was found).

Yan et al. prepared bimetallic PVP-stabilized NiRu, NiRh, and NiPd nanoparticles via a straightforward co-reduction of the respective chloride salts with $NaBH_4$ [78]. Due to subtle differences in the reduction potential, the ultra-small NPs were surface-enriched with Ni atoms; whereas, the core was comprised of mixture of Ni and Ru atoms. Lignin, a biopolymer comprising of cross-linked phenylpropanoid units, represents an important renewable source for the production of value-added aromatic chemicals, such as substituted phenols; however, the cleavage of lignin's C–O bonds by hydrogenolysis requires efficient catalysts [79]. Against this background, the synthesized bimetallic NPs were employed, as catalyst, in the hydrogenolysis of lignin in water and $Ni_{0.85}Ru_{0.15}$ NPs (Fig. 7.20) were identified as the most powerful candidate enabling the selective cleavage of lignin's β-O-4 C–O bonds: Small-molecule arenes were obtained under comparably mild conditions (i.e., 100 °C and 1 bar H_2). These authors performed mechanistic studies and proposed three factors to contribute to the high catalytic activity of the bimetallic NiRu NPs (compared to monometallic Ni and Ru NPs): Increased fraction of Ni surface atoms (compared to Ni NPs), improved H_2 and substrate activation (compared to Ni NPs) and inhibition of aromatic-ring hydrogenation (compared to Ru NPs).

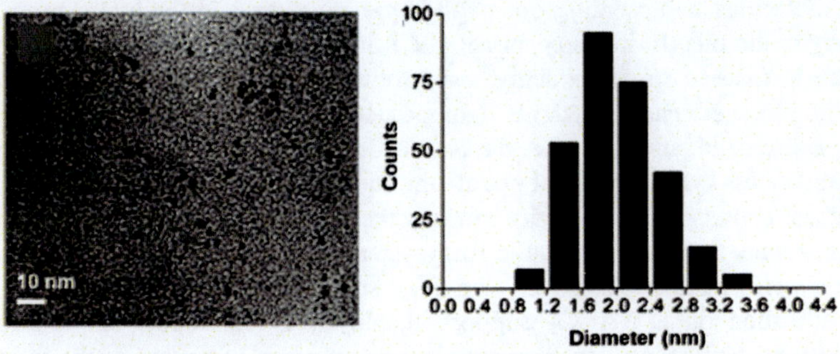

Fig. 7.20 Representative TEM image and corresponding size histogram of $Ni_{0.85}Ru_{0.15}$ nanoparticles [78]. Figure reproduced with kind permission; © 2014 American Chemical Society

It has to be pointed out that Ru and Pd are immiscible at the atomic level. As a consequence, any binary Pd-Ru combination will phase separate into segregated domains in the bulk (this holds true at all temperatures upto the melting points of Ru and Pd) [80]. For this reason, such alloys are basically impossible to prepare. This restriction, however, does not hold true for bimetallic PdRu nanoparticles, which can be considered as pseudo-Rh systems (at least for the $Pd_{0.5}Ru_{0.5}$ composition, Fig. 7.21a). Kitagawa et al. introduced the concept of "interelement fusion," which basically relies on the solid-solution alloying of metals to create new properties by changing the constituent metals composition and/or combination [81]. In the case of Pd and Ru, being the direct neighbors of Rh in the periodic table, the bimetallic PdRu nanoparticles were expected to have similar properties as Rh NPs while being less expensive [82]. These authors reported on a series of such bimetallic Pd_xRu_{1-x} NPs ($x = 0.1$, 0.3., 0.5, 0.7 or 0.9), stabilized by PVP [83]. The NPs were analyzed regarding their size and composition by TEM imaging and EDX element-mapping measurements (Fig. 7.21b); XRD analysis of the particles revealed a transition from a fcc structure (low Ru content) to a hpc motif (high Ru content). The H_2-storage capacity of the NPs was investigated; a transition from an exothermic to an endothermic H_2 uptake was found (with increasing Ru content, the H_2 uptake became thermodynamically less favored). Moreover, the bimetallic NPs catalyzed CO-oxidation; in particular, the pseudo-Rh NPs (i.e., $Pd_{0.5}Ru_{0.5}$ NPs) were most efficient, outcompeting monometallic Rh or Ru NPs as well as a physical mixture of Pd and Ru NPs (Fig. 7.21c). For this application, the NPs were immobilized onto a γ-Al_2O_3 support by the wet-impregnation technique (see Sect. 7.2).

Similar to this, Shen et al. prepared various PVP-stabilized Ru, Ir, and Pd nanoparticles as well as the bimetallic systems RuIr and RuPd by simultaneous reduction of the corresponding precursor salts [84]. Formation of the nanoparticles was monitored by UV/vis absorption spectroscopy (Fig. 7.22a) and TEM imaging was applied to determine the size (distribution) of the obtained materials (in all cases, particles with an average diameter of ca. 2 nm and narrow size distributions were found, Fig. 7.22b); XPS analysis confirmed the bimetallic nature of the mixed particles. The authors addressed the issue of the selective ring-opening of naphthenic molecules, representing one of the upgrading steps in the hydro-processing of heavy crude oils. In industry, bimetallic PtIr catalysts are mainly employed and research is focused on introducing more robust, selective, and efficient systems. From the screened materials, the Pd nanoparticles showed the lowest activity in the ring-opening of indane; whereas, the Ru nanoparticles gave the highest content of undesired *ortho*-xylene and light petroleum; the Ir particles, however, featured both the highest activity and selectivity towards the formation of 2-ethyltoluene and *n*-propylbenzene. From the bimetallic RuIr systems, a higher Ir content enhanced the catalytic performance. Moreover, the PVP-stabilized RuIr particles (1:4 ratio), when adsorbed on a γ-Al_2O_3 support, displayed a remarkable single-cleavage selectivity as well as low cracking activity—even when compared to the industrial state-of-art PtIr catalysts at a comparable indane conversion (Fig. 7.22c). It was

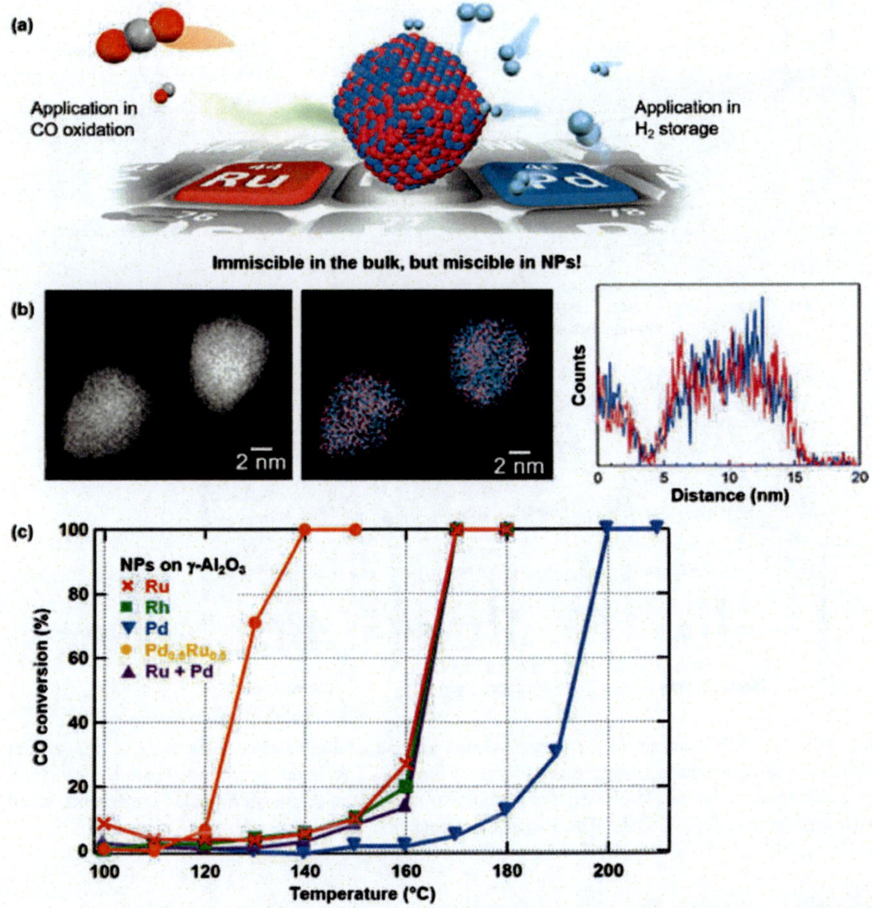

Fig. 7.21 **a** Schematic representation of bimetallic PdRu nanoparticles. **b** Representative TEM image and reconstructed overlay EDX element map (blue: Pd, red: Ru) of $Pd_{0.5}Ru_{0.5}$ NPs; the corresponding compositional line profile is also shown (blue: Pd, red: Ru). **c** Temperature-dependent CO oxidation, catalyzed by mono- or bimetallic NPs [83]. Figure reproduced with kind permission; © 2014 American Chemical Society

proclaimed that the synthesized bimetallic RuIr catalysts would be suited for industrial application, since also a higher tolerance against sulfur would be given as for the Pt-containing catalysts.

Xia et al. utilized PdRu nanoparticles, as precursor, for the fabrication of Ru nanoframes [85]. Such nanostructures have gained significant attention due to their unique properties, which basically result from their highly open structure. Therefore, nanoframes have been employed, as alternative to conventional nanoparticles in catalytic, plasmonic or biomedical applications [86–88]. It is generally accepted

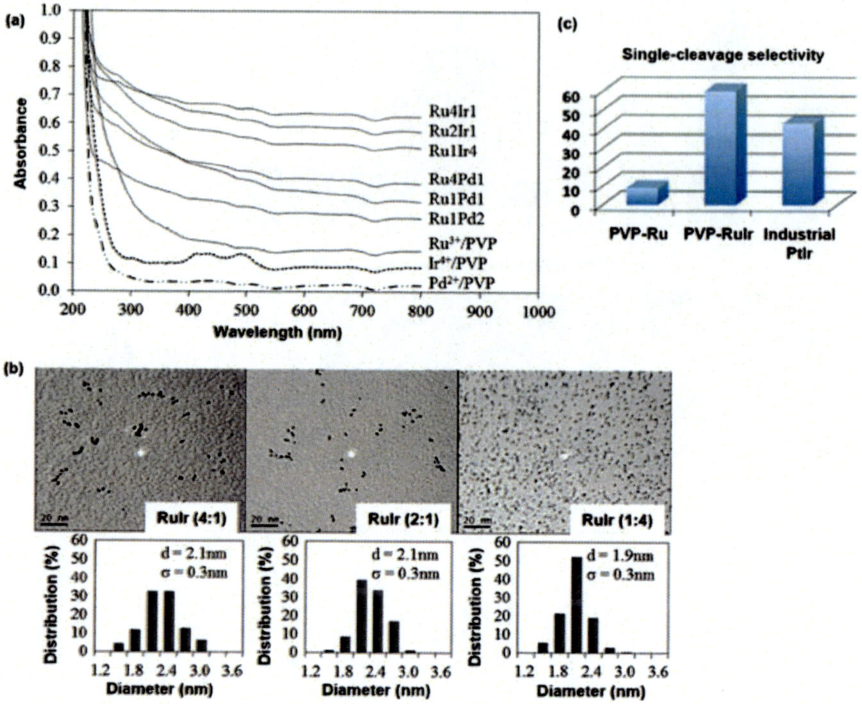

Fig. 7.22 a UV/vis spectra of the various mono- and bimetallic nanoparticles. b TEM images and deduced size distribution histograms of various bimetallic RuIr nanoparticles (scale bar: 20 nm). c Comparison of the single-cleavage selectivity in the indane conversion [84]. Figure reproduced with kind permission; © 2013 The Royal Chemical Society

that the increased catalytic activity of noble-metal nanoframes results from their specific surface areas, high densities of catalytically active sites, and low vulnerability towards sintering [88]. The synthesis of Ru nanoframes was conducted in a three-step procedure [85]. First, PVP-stabilized truncated Pd octahedra were formed. This special geometry allowed the kinetically controlled overgrowth of Ru on the corners and edges of the Pd NP seeds. Finally, chemical etching gave octahedral Ru nanoframes with an average thickness of ca. 2 nm (Fig. 7.23). Remarkably, the fcc crystal structure of the Pd seeds was replicated during the nanoframe formation (i.e., the Ru nanoframes did not adopt their usual hpc packing motif). These authors tested the catalytic activity of these nanomaterials: Both, the reduction of nitrobenzene with $NaBH_4$ as well as the dehydrogenation of $NH_3{\cdot}BH_3$ were catalyzed more efficiently compared to using hcp Ru nanowires of similar thickness. As summarized by Fan and Zhang [89], the catalytic performance of nanostructures can be modulated via their crystal structure and remarkable catalytic activities can be found when such nanostructures are stabilized in unusual packing motifs (as in the previous example).

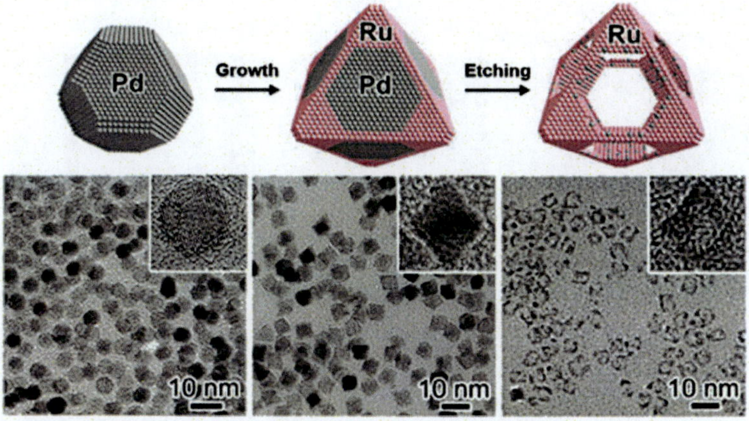

Fig. 7.23 Schematic representation of the fabrication of Ru nanoframes via a step-wise protocol involving a seeded growth and chemical etching. The corresponding TEM images of the three nano-objects are also shown [85]. Figure reproduced with kind permission; © 2016 American Chemical Society

The same templating strategy was also applied by Xia et al. to prepare fcc Ru nanocages with cubic or octahedral shapes (Fig. 7.24) [90, 91]. The former had a wall thickness of only 1.1 nm, corresponding to a layer of six Ru atoms [91]. PVP-stabilized Pd cubic NPs of different dimensions were employed, as seeds (edge length of 6 to 18 nm); from these, the smaller seeds gave the structurally more defined Ru nanocages. These authors predicted, on the basis of self-consistent density-functional theory (DFT) calculations, that the fcc Ru nanocages might be potent catalysts for the NH_3 synthesis due to their lower activation energy for N_2 dissociation and increased binding strength of atomic N (the former represents the rate-determining step in the NH_3 synthesis, catalyzed by conventional hcp Ru catalysts). It was proposed that residual Pd atoms in the nanostructure could facilitate these processes. The same behavior was also expected for the analogous fcc octahedral Ru nanocages, which were fabricated by a similar two-step process, starting from appropriate PVP-stabilized Pd seeds [92]. However, the synthesis of the intermediate Pd@Ru core–shell nanostructures, possessing a conformal, uniform shell and smooth surface, required optimized reaction conditions, such as slow addition of the Ru(III) precursor, presence of KBr, and an elevated reaction temperature. The scope of the methodology was extended further towards fcc icosahedral Ru nanocages, which were thermally stable up to 300 °C [92]. These nanoobjects catalyzed the $NaBH_4$ reduction of nitrobenzene and H_2 evolution from hydrazine more efficiently than their cubic or octahedral counterparts.

Okal et al. applied a microwave-assisted polyol protocol for the synthesis of PVP-stabilized bimetallic RuRe nanoparticles from $(NH_4)ReO_4$ and $RuCl_3 \times xH_2O$, as the precursor salts [93]. The obtained materials were thoroughly

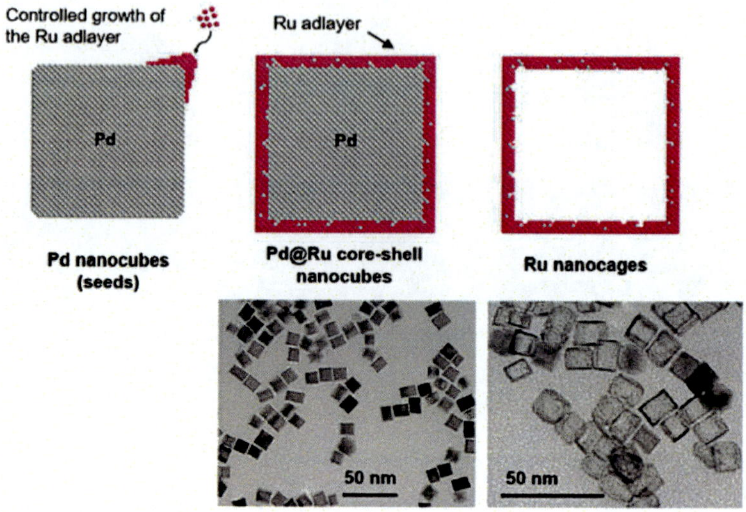

Fig. 7.24 Schematic representation of the stepwise preparation of cubic Ru nanocages. Representative TEM images of the intermediate Pd@Ru core–shell nanocubes and the final Ru nanocages are also shown [90]. Figure reproduced with kind permission; © 2016 American Chemical Society

characterized by UV/vis absorption spectroscopy, EDX, XRD, XPS, and TEM imaging. The latter revealed spherical nanoparticles with a mean size of 1.4 nm and 1.6 nm for Ru_4Re_1 and RuRe, respectively (Fig. 7.25), thusly slightly larger as the PVP-stabilized Ru nanoparticles prepared under the same conditions. In the same context, RuRe nanoparticles supported by SiO_2 or γ-Al_2O_3 were prepared. However, only the latter indicated a potential applicability, as catalyst, in the propane oxidation reaction.

Finally, the fabrication of trimetallic RuNiM nanoparticles was reported by Zhao et al. (M = Fe or Cu) [94]. The particles were characterized by XRD analysis and microscopy imaging; scanning TEM element mapping revealed a homogeneous distribution of the three different metals within the NPs (Fig. 7.26). From the various PVP-stabilized alloy NPs, the system $Ru_{0.44}Ni_{0.28}Fe_{0.28}$ was identified as the most promising candidate to be utilized, as catalyst, for reduction of nitrobenzene with $NaBH_4$, as reductant. The substrate was degraded with an efficiency of ca. 90% within a few seconds (for comparison, bimetallic $Ru_{0.60}Ni_{0.40}$ NPs were inferior catalysts for this reaction, as revealed by longer degradation times and much lower efficiencies). These authors concluded from experimental and computational studies and DFT that the three different metals had a synergistic

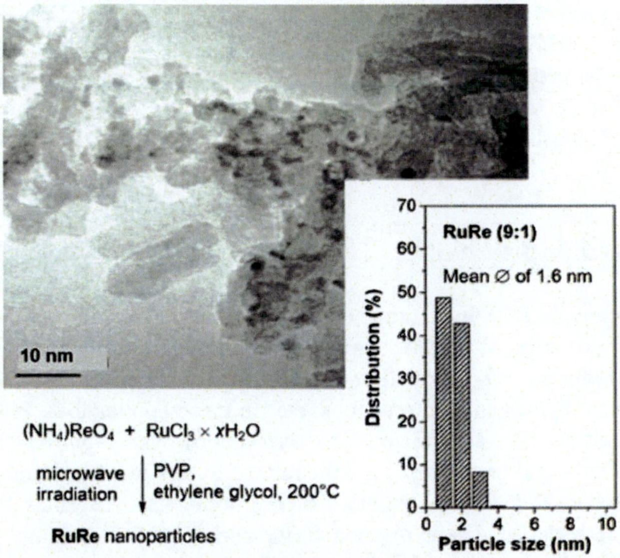

Fig. 7.25 Schematic representation of the synthesis of bimetallic RuRe nanoparticles. A Figure reproduced with kind permission; © 2019 Elsevier B.V

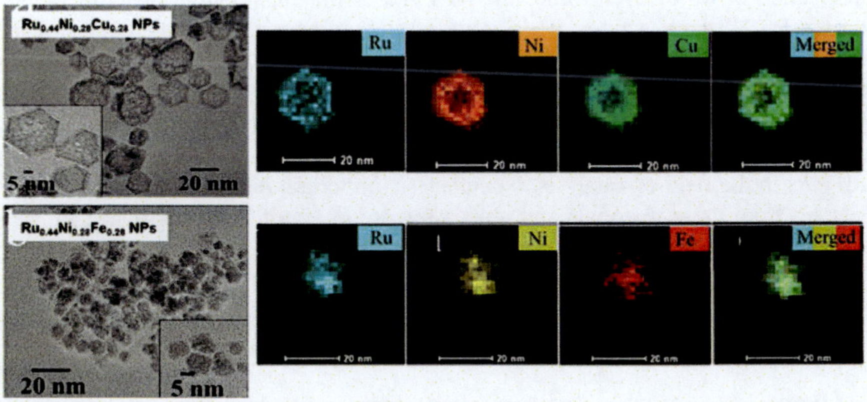

Fig. 7.26 Representative TEM images of the trimetallic $Ru_{0.44}Ni_{0.28}Cu_{0.28}$ (top) and $Ru_{0.44}Ni_{0.28}Fe_{0.28}$ nanoparticles (bottom). The corresponding scanning TEM element mappings are also shown [94]. Figure reproduced with kind permission; © 2019 American Chemical Society

effect on the catalytic activity: The vacant orbitals of the Ru atoms enabled a strong binding of the BH_4^- anion, a precondition for the generation of active H species; whereas, the Ni centers guaranteed the transport of these active species as well as electrons. The Fe atoms were believed to support the Ru and Ni sites in their proposed mechanistic roles.

7.4 Concluding Remarks

Ruthenium nanoparticles have attracted considerable interest due to their ability to catalyze a wide range of organic reactions: Fischer–Tropsch synthesis, C–H activation, CO_2 reduction, H_2 production (e.g., by ammonia borane dehydrogenation or water splitting), hydrogenation reactions, etc. In these applications, either colloidal or immobilized Ru NPs are utilized. The former, however, require stabilization by organic ligands to suppress their agglomeration. In this respect, ionic liquids, surfactants, polyols, NHCs or polymers are commonly used. Besides these, metal–organic frameworks, covalent organic frameworks, porous organic polymers or dendrimers have been employed to host Ru NPs (these systems have been discussed separately in Chaps. 4 and 6). Regarding the stabilization of Ru NPs by polymers, PVP represents the state-of-the-art material, which is most commonly used.

Today, Ru NPs can be prepared via different methods, which all offer good control over size and shape. The focus of current research has shifted towards the fabrication of bi- or even trimetallic NPs and of the construction of more sophisticated Ru NP architectures. The tuning of already existing catalytic cycles or the opening of new catalytic pathways represents the main driving force for research in this field. Many examples for this can be found in Philippot's recent review [10]. However, as for all applications involving Ru, the relatively high price and the low abundance of this metal will retard any large-scale or even industrial application of Ru NPs in the field-of-catalysis. Nonetheless, important lessons in catalysis can be learned from these materials and, thus, they are of significant importance for fundamental studies.

References

1. Halperin, W.P.: Quantum size effects in metal particles. Rev. Mod. Phys. **58**(3), 533 (1986)
2. Matsui, I.: Nanoparticles for electronic device applications: A brief review. J. Chem. Eng. Jpn. **38**(8), 535 (2005)
3. Kruis, F.E., Fissan, H., Peled, A.: Synthesis of nanoparticles in the gas phase for electronic, optical and magnetic applications—a review. J. Aerosol Sci. **29**(5), 511 (1998)
4. Zhang, Y.-X., Wang, Y.-H.: Nonlinear optical properties of metal nanoparticles: A review. RSC Adv. **7**(71), 45129 (2017)
5. Jain, P.K., Huang, X., El-Sayed, I.H., El-Sayed, M.A.: Review of some interesting surface plasmon resonance-enhanced properties of noble metal nanoparticles and their applications to biosystems. Plasmonics **2**(3), 107 (2007)

6. Gul, S., Khan, S. B., Rehman, I.U., Khan, M.A., Khan, M.I.: A comprehensive review of magnetic nanomaterials: Modern day theranostics. Front. Mater. 6(179) (2019)
7. Mohammed, L., Gomaa, H.G., Ragab, D., Zhu, J.: Magnetic nanoparticles for environmental and biomedical applications: A review. Particuology **30**, 1 (2017)
8. Indira, T.K., Lakshmi, P.K.: Magnetic nanoparticles—a review. Int. J. Pharm. Sci. Nanotechnol. **3**(3) (2010)
9. Gates, B.C.: Supported metal clusters: Synthesis, structure, and catalysis. Chem. Rev. **95**(3), 511 (1995)
10. Axet, M.R., Philippot, K.: Catalysis with colloidal ruthenium nanoparticles. Chem. Rev. **120** (2), 1085 (2020)
11. Gual, A., Godard, C., Castillón, S., Curulla-Ferré, D., Claver, C.: Colloidal Ru-, Co- and Fe-nanoparticles. Synthesis and application as nanocatalysts in the Fischer-Tropsch process. Catal. Today **183**(1), 154 (2012)
12. Martínez-Prieto, L.M., Chaudret, B.: Organometallic ruthenium nanoparticles: Synthesis, surface chemistry, and insights into ligand coordination. Acc. Chem. Res. **51**(2), 376 (2018)
13. Lara, P., Philippot, K., Chaudret, B.: Organometallic ruthenium nanoparticles: A comparative study of the influence of the stabilizer on their characteristics and reactivity. ChemCatChem **5** (1), 28 (2013)
14. Cushing, B.L., Kolesnichenko, V.L., O'Connor, C.J.: Recent advances in the liquid-phase syntheses of inorganic nanoparticles. Chem. Rev. **104**(9), 3893 (2004)
15. Grammatikopoulos, P., Steinhauer, S., Vernieres, J., Singh, V., Sowwan, M.: Nanoparticle design by gas-phase synthesis. Adv. Phys.: X **1**(1), 81 (2016)
16. Amiens, C., Ciuculescu-Pradines, D., Philippot, K.: Controlled metal nanostructures: Fertile ground for coordination chemists. Coord. Chem. Rev. **308**, 409 (2016)
17. Duteil, A., Queau, R., Chaudret, B., Mazel, R., Roucau, C., Bradley, J.S.: Preparation of organic solutions or solid films of small particles of ruthenium, palladium, and platinum from organometallic precursors in the presence of cellulose derivatives. Chem. Mater. **5**(3), 341 (1993)
18. Amiens, C., Chaudret, B., Ciuculescu-Pradines, D., Collière, V., Fajerwerg, K., Fau, P., Kahn, M., Maisonnat, A., Soulantica, K., Philippot, K.: Organometallic approach for the synthesis of nanostructures. New J. Chem. **37**(11), 3374 (2013)
19. Lignier, P., Bellabarba, R., Tooze, R.P., Su, Z., Landon, P., Ménard, H., Zhou, W.: Facile synthesis of branched ruthenium nanocrystals and their use in catalysis. Cryst. Growth Design **12**(2), 939 (2012)
20. Fiévet, F., Ammar-Merah, S., Brayner, R., Chau, F., Giraud, M., Mammeri, F., Peron, J., Piquemal, J.Y., Sicard, L., Viau, G.: The polyol process: A unique method for easy access to metal nanoparticles with tailored sizes, shapes and compositions. Chem. Soc. Rev. **47**(14), 5187 (2018)
21. Bradley, J.S., Millar, J.M., Hill, E.W., Behal, S., Chaudret, B., Duteil, A.: Surface chemistry on colloidal metals: Spectroscopic study of adsorption of small molecules. Faraday Discuss. **92**, 255 (1991)
22. Duteil, A., Queau, R., Chaudret, B., Mazel, R., Roucau, C., Bradley, J.S.: Preparation of organic solutions or solid films of small particles of ruthenium, palladium and platinum from organometallic precursors in the presence of cellulose derivatives. Chem. Mater. **5**(3), 341 (1993)
23. Philippot, K., Chaudret, B.: Organometallic approach to the synthesis and surface reactivity of noble metal nanoparticles. C. R. Chimie **6**(8), 1019 (2003)

24. Pan, C., Pelzer, K., Philippot, K., Chaudret, B., Dassenoy, F., Lecante, P., Casanove, M.-J.: Ligand-stabilized ruthenium nanoparticles: Synthesis, organization and dynamics. J. Am. Chem. Soc. **123**(31), 7584 (2001)
25. Pan, C., Dassenoy, F., Casanove, M.-J., Philippot, K., Amiens, C., Lecante, P., Mosset, A., Chaudret, B.: A new synthetic method toward bimetallic ruthenium platinum nanoparticles; composition induced structural changes. J. Phys. Chem. B **103**(46), 10098 (1999)
26. Dassenoy, F., Casanove, M.J., Lecante, P., Pan, C., Philippot, K., Amiens, C., Chaudret, B.: Size and composition effects in polymer-protected ultrafine bimetallic $Pt_x Ru_{1-x}$ ($0<x<1$) particles. Phys. Rev. B **63**(23), 235407 (2001)
27. Wostek-Wojciechowska, D., Jeszka, J.K., Amiens, C., Chaudret, B., Lecante, P.: The solid-state synthesis of metal nanoparticles from organometallic precursors. J. Colloid Interface Sci. **287**(1), 107 (2005)
28. Bonnefille, E., Novio, F., Gutmann, T., Poteau, R., Lecante, P., Jumas, J.-C., Philippot, K., Chaudret, B.: Tin-decorated ruthenium nanoparticles: A way to tune selectivity in hydrogenation reaction. Nanoscale **6**(16), 9806 (2014)
29. Onwudinanti, C., Tranca, I., Morgan, T., Tao, S.: Tin, the enabler—hydrogen diffusion into ruthenium. Nanomaterials **9**(1), 129 (2019)
30. Yu, W., Liu, M., Liu, H., Ma, X., Liu, Z.: Preparation, characterization and catalytic properties of polymer-stabilized ruthenium colloids. J. Colloid Interface Sci. **208**(2), 439 (1998)
31. Liu, M., Yu, W., Liu, H.: Selective hydrogenation of o-chloronitrobenzene over polymer-stabilized ruthenium colloidal catalysts. J. Mol. Catal. A: Chem. **138**(2–3), 295 (1999)
32. Yan, X., Liu, M., Liu, H., Liew, K.Y.: Role of boron species in the hydrogenation of o-chloronitrobenzene over polymer-stabilized ruthenium colloidal catalysts. J. Mol. Catal. A: Chem. **169**(1–2), 225 (2001)
33. Cusumano, J.A.: Safety in the preparation of multimetallic catalysts. Nature **247**(5441), 456 (1974)
34. Yan, X., Liu, H., Liew, K.Y.: Size control of polymer-stabilized ruthenium nanoparticles by polyol reduction. J. Mater. Chem. **11**(12), 3387 (2001)
35. Karakhanov, E.A., Maksimov, A.L., Zolotukhina, A.V., Kardashev, S.V.: Design of dendrimer-based nanostructured catalyst systems and their catalytic activity in hydrogenation: synthesis of ruthenium nanoparticles immobilized in dendrimer networks. Pet. Chem. **50**(4), 290 (2010)
36. Maximov, A., Zolotukhina, A., Murzin, V., Karakhanov, E., Rosenberg, E.: Ruthenium nanoparticles stabilized in cross-linked dendrimer matrices: Hydrogenation of phenols in aqueous media. ChemCatChem **7**(7), 1197 (2015)
37. Karakhanov, E., Maximov, A., Zolotukhina, A., Mamadli, A., Vutolkina, A., Ivanov, A.: Dendrimer-stabilized Ru nanoparticles immobilized in organo-silica materials for hydrogenation of phenols. Catalysts **7**(3), 86 (2017)
38. Zhao, Y., Luo, Y., Yang, X., Yang, Y., Song, Q.: Tunable preparation of ruthenium nanoparticles with superior size-dependent catalytic hydrogenation properties. J. Hazard. Mater. **332**, 124 (2017)
39. Zhao, Y., Luo, Y., Zhu, Y., Sun, Y., Cui, L., Song, Q.: Sensitive colorimetric assay of H_2S depending on the high-efficient inhibition of catalytic performance of Ru nanoparticles. ACS Sustainable Chem. Eng. **5**(9), 7912 (2017)
40. Hou, S., Wang, X., Huang, C., Xie, C., Yu, S.: Highly selective hydrogenation of α-pinene catalyzed by Ru nanoparticles in aqueous micellar microreactors. Catal. Lett. **146**(3), 580 (2016)

41. Hou, S., Xie, C., Zhong, H., Yu, S.: Mild water-promoted ruthenium nanoparticles as an efficient catalyst for the preparation of *cis*-rich pinane. RSC Adv. **5**(109), 89552 (2015)
42. Qu, L., Yu, H., Yu, F., Yuan, B., Xie, C., Yu, S.: Catalytic reduction of α-pinene using Ru nanoparticles stabilized by modified carboxymethyl cellulose. Appl. Surf. Sci. **453**, 271 (2018)
43. Wang, X., Yu, F., Xie, C., Yu, S.: Highly selective hydrogenation of α-pinene in aqueous medium using PVA-stabilized Ru nanoparticles. Mol. Catal. **444**, 62 (2018)
44. Chen, M., Dong, Q., Ni, W., Zhao, X., Gu, Q., Tang, G., Li, D., Ma, W., Hou, Z.: Cyclodextrin-based polymer-assisted Ru nanoparticles for the aqueous hydrogenation of biomass-derived platform molecules. ChemistrySelect **2**(32), 10537 (2017)
45. Herbois, R., Noël, S., Léger, B., Tilloy, S., Menuel, S., Addad, A., Martel, B., Ponchel, A., Monflier, E.: Ruthenium-containing β-cyclodextrin polymer globules for the catalytic hydrogenation of biomass-derived furanic compounds. Green Chem. **17**(4), 2444 (2015)
46. Noël, S., Bourbiaux, D., Tabary, N., Ponchel, A., Martel, B., Monflier, E., Léger, B.: Acid-tolerant cyclodextrin-based ruthenium nanoparticles for the hydrogenation of unsaturated compounds in water. Catal. Sci. Technol. **7**(24), 5982 (2017)
47. Wang, C., Ciganda, R., Salmon, L., Gregurec, D., Irigoyen, J., Moya, S., Ruiz, J., Astruc, D.: Highly efficient transition metal nanoparticle catalysts in aqueous solutions. Angew. Chem. Int. Ed. **55**(9), 3091 (2016)
48. Zhang, Y., Riduan, S.N.: Functional porous organic polymers for heterogeneous catalysis. Chem. Soc. Rev. **41**(6), 2083 (2012)
49. Kaur, P., Hupp, J.T., Nguyen, S.T.: Porous organic polymers in catalysis: opportunities and challenges. ACS Catal. **1**(7), 819 (2011)
50. Mondal, J., Kundu, S.K., Hung Ng, W.K., Singuru, R., Borah, P., Hirao, H., Zhao, Y., Bhaumik, A.: Fabrication of ruthenium nanoparticles in porous organic polymers: Towards advanced heterogeneous catalytic nanoreactors. Chem. Eur. J. **21**(52), 19016 (2015)
51. Mondal, S., Singuru, R., Chandra Shit, S., Hayashi, T., Irle, S., Hijikata, Y., Mondal, J., Bhaumik, A.: Ruthenium nanoparticle-decorated porous organic network for direct hydrodeoxygenation of long-chain fatty acids to alkanes. ACS Sustainable Chem. Eng. **6**(2), 1610 (2018)
52. Gmeiner, J., Behrens, S., Spliethoff, B., Trapp, O.: Ruthenium nanoparticles in high-throughput studies of chemoselective carbonyl hydrogenation reactions. ChemCatChem **8**(3), 571 (2016)
53. Tu, W., Liu, H.: Rapid synthesis of nanoscale colloidal metal clusters by microwave irradiation. J. Mater. Chem. **10**(9), 2207 (2000)
54. Fievet, F., Lagier, J.P., Figlarz, M.: Preparing monodisperse metal powders in micrometer and submicrometer sizes by the polyol process. MRS Bull. **14**(12), 29 (1989)
55. Dong, H., Chen, Y.C., Feldmann, C.: Polyol synthesis of nanoparticles: Status and options regarding metals, oxides, chalcogenides and non-metal elements. Green Chem. **17**(8), 4107 (2015)
56. Bonet, F., Delmas, V., Grugeon, S., Herrera Urbina, R., Silvert, P.Y., Tekaia-Elhsissen, K.: Synthesis of monodisperse Au, Pt, Pd, Ru and Ir nanoparticles in ethylene glycol. Nanostruct. Mater. **11**(8), 1277 (1999)
57. Miyazaki, A., Takeshita, K., Aika, K.-I., Nakano, Y.: Formation of ruthenium colloid in ethylene glycol. Chem. Lett. **27**(4), 361 (1998)
58. Simakova, I.Y.D., Prosvirin, I., Murzin, D.Y., Simakov, A.: Development of polyol method for the synthesis of concentrated colloids of PVP-stabilised Ru nanoparticles. Int. J. Nanotechnol. **13**(1–3), 15 (2016)

59. Kusada, K., Kobayashi, H., Yamamoto, T., Matsumura, S., Sumi, N., Sato, K., Nagaoka, K., Kubota, Y., Kitagawa, H.: Discovery of face-centered-cubic ruthenium nanoparticles: Facile size-controlled synthesis using the chemical reduction method. J. Am. Chem. Soc. **135**(15), 5493 (2013)
60. Harpeness, R., Peng, Z., Liu, X., Pol, V.G., Koltypin, Y., Gedanken, A.: Controlling the agglomeration of anisotropic Ru nanoparticles by the microwave–polyol process. J. Colloid Interface Sci. **287**(2), 678 (2005)
61. Harada, M., Cong, C.: Microwave-assisted polyol synthesis of polymer-protected monometallic nanoparticles prepared in batch and continuous-flow processing. Ind. Eng. Chem. Res. **55** (19), 5634 (2016)
62. Raspolli Galletti, A.M., Antonetti, C., Longo, I., Capannelli, G., Venezia, A.M.: A novel microwave assisted process for the synthesis of nanostructured ruthenium catalysts active in the hydrogenation of phenol to cyclohexanone. Appl. Catal. A: Gen. **350**(1), 46 (2008)
63. Raspolli Galletti, A.M., Antonetti, C., Giaiacopi, S., Piccolo, O., Venezia, A.M.: Innovative process for the synthesis of nanostructured ruthenium catalysts and their catalytic performance. Top. Catal. **52**(8), 1065 (2009)
64. Zawadzki, M., Okal, J.: Synthesis and structure characterization of Ru nanoparticles stabilized by PVP or γ-Al_2O_3. Mater. Res. Bull. **43**(11), 3111 (2008)
65. Zhang, Y., Yu, J., Niu, H., Liu, H.: Synthesis of PVP-stabilized ruthenium colloids with low boiling point alcohols. J. Colloid Interface Sci. **313**(2), 503 (2007)
66. Kreibig, U., Vollmer, M.: Optical Properties of Metal Clusters. Springer, Berlin (1995)
67. Li, Q., Peng, S., Zheng, Q., Niu, Q.: Effect of morphology on catalytic performance of colloid Ru for hydrogen generation from H_3NBH_3 alcoholysis: A comparative study. Integr. Ferroelectr. **170**(1), 73 (2016)
68. Peng, S., Liu, J., Li, Q., Zhang, J., Niu, Q.: PVP-stabilized Ru colloidal nanoparticles by solvothermal synthesis: preparation, characterization and catalytic properties. Integr. Ferroelectr. **170**(1), 83 (2016)
69. Chen, Y., Liew, K.Y., Li, J.: Size-controlled synthesis of Ru nanoparticles by ethylene glycol reduction. Mater. Lett. **62**(6–7), 1018 (2008)
70. Kar, P., Farsinezhad, S., Mahdi, N., Zhang, Y., Obuekwe, U., Sharma, H., Shen, J., Semagina, N., Shankar, K.: Enhanced CH_4 yield by photocatalytic CO_2 reduction using TiO_2 nanotube arrays grafted with Au, Ru, and ZnPd nanoparticles. Nano Res. **9**(11), 3478 (2016)
71. Harada, M., Takahashi, S.: Synthesis of ruthenium particles by photoreduction in polymer solutions. J. Colloid Interface Sci. **325**(1), 1 (2008)
72. Kumar, P.S.S., Manivel, A., Anandan, S., Zhou, M., Grieser, F., Ashokkumar, M.: Sonochemical synthesis and characterization of gold-ruthenium bimetallic nanoparticles. Colloids Surf. A **356**(1–3), 140 (2010)
73. Gedanken, A., Mastai, Y.: Sonochemistry and other novel methods developed for the synthesis of nanoparticles. In: Rao, C.N.R., Müller, A., Cheetham, A.K. (eds.) The Chemistry of Nanomaterials, pp. 113 (2004)
74. Li, H., Wang, R., Hong, Q., Chen, L., Zhong, Z., Koltypin, Y., Calderon-Moreno, J., Gedanken, A.: Ultrasound-assisted polyol method for the preparation of SBA-15-supported ruthenium nanoparticles and the study of their catalytic activity on the partial oxidation of methane. Langmuir **20**(19), 8352 (2004)
75. He, Y., Vinodgopal, K., Ashokkumar, M., Grieser, F.: Sonochemical synthesis of ruthenium nanoparticles. Res. Chem. Intermed. **32**(8), 709 (2006)
76. Sahoo, A., Tripathy, S.K., Dehury, N., Patra, S.: A porous trimetallic Au@Pd@Ru nanoparticle system: Synthesis, characterisation and efficient dye degradation and removal. J. Mater. Chem. A **3**(38), 19376 (2015)

77. Liu, M., Zhang, J., Liu, J., Yu, W.W.: Synthesis of PVP-stabilized Pt/Ru colloidal nanoparticles by ethanol reduction and their catalytic properties for selective hydrogenation of ortho-chloronitrobenzene. J. Catal. **278**(1), 1 (2011)
78. Zhang, J., Teo, J., Chen, X., Asakura, H., Tanaka, T., Teramura, K., Yan, N.: A series of NiM (M = Ru, Rh, and Pd) bimetallic catalysts for effective lignin hydrogenolysis in water. ACS Catal. **4**(5), 1574 (2014)
79. Sanderson, K.: Lignocellulose: A chewy problem. Nature **474**(7352), S12 (2011)
80. Tripathi, S.N., Bharadwaj, S.R., Dharwadkar, S.R.: The Pd-Ru system (palladium-ruthenium). J. Phase Equilib. **14**(5), 638 (1993)
81. Kobayashi, H., Kusada, K., Kitagawa, H.: Creation of novel solid-solution alloy nanoparticles on the basis of density-of-states engineering by interelement fusion. Acc. Chem. Res. **48**(6), 1551 (2015)
82. Sato, K., Tomonaga, H., Yamamoto, T., Matsumura, S., Zulkifli, N.D.B., Ishimoto, T., Koyama, M., Kusada, K., Kobayashi, H., Kitagawa, H., et al.: A synthetic pseudo-Rh: NOx reduction activity and electronic structure of Pd–Ru solid-solution alloy nanoparticles. Sci. Rep. **6**(1), 28265 (2016)
83. Kusada, K., Kobayashi, H., Ikeda, R., Kubota, Y., Takata, M., Toh, S., Yamamoto, T., Matsumura, S., Sumi, N., Sato, K., et al.: Solid solution alloy nanoparticles of immiscible Pd and Ru elements neighboring on Rh: changeover of the thermodynamic behavior for hydrogen storage and enhanced CO-oxidizing ability. J. Am. Chem. Soc. **136**(5), 1864 (2014)
84. Shen, J., Yin, X., Karpuzov, D., Semagina, N.: PVP-stabilized mono- and bimetallic Ru nanoparticles for selective ring opening. Catal. Sci. Technol. **3**(1), 208 (2013)
85. Ye, H., Wang, Q., Catalano, M., Lu, N., Vermeylen, J., Kim, M.J., Liu, Y., Sun, Y., Xia, X.: Ru nanoframes with an fcc structure and enhanced catalytic properties. Nano Lett. **16**(4), 2812 (2016)
86. Fang, Z., Wang, Y., Liu, C., Chen, S., Sang, W., Wang, C., Zeng, J.: Rational design of metal nanoframes for catalysis and plasmonics. Small **11**(22), 2593 (2015)
87. Zhu, C., Du, D., Eychmüller, A., Lin, Y.: Engineering ordered and nonordered porous noble metal nanostructures: Synthesis, assembly, and their applications in electrochemistry. Chem. Rev. **115**(16), 8896 (2015)
88. Wang, X., Ruditskiy, A., Xia, Y.: Rational design and synthesis of noble-metal nanoframes for catalytic and photonic applications. Natl. Sci. Rev. **3**(4), 520 (2016)
89. Fan, Z., Zhang, H.: Template synthesis of noble metal nanocrystals with unusual crystal structures and their catalytic applications. Acc. Chem. Res. **49**(12), 2841 (2016)
90. Zhao, M., Figueroa-Cosme, L., Elnabawy, A.O., Vara, M., Yang, X., Roling, L.T., Chi, M., Mavrikakis, M., Xia, Y.: Synthesis and characterization of Ru cubic nanocages with a face-centered cubic structure by templating with Pd nanocubes. Nano Lett. **16**(8), 5310 (2016)
91. Zhao, M., Elnabawy, A.O., Vara, M., Xu, L., Hood, Z.D., Yang, X., Gilroy, K.D., Figueroa-Cosme, L., Chi, M., Mavrikakis, M., et al.: Facile synthesis of Ru-based octahedral nanocages with ultrathin walls in a face-centered cubic structure. Chem. Mater. **29**(21), 9227 (2017)
92. Zhao, M., Xu, L., Vara, M., Elnabawy, A.O., Gilroy, K.D., Hood, Z.D., Zhou, S., Figueroa-Cosme, L., Chi, M., Mavrikakis, M., et al.: Synthesis of Ru icosahedral nanocages with a face-centered-cubic structure and evaluation of their catalytic properties. ACS Catal. **8** (8), 6948 (2018)

93. Baranowska, K., Okal, J., Tylus, W.: Microwave-assisted polyol synthesis of bimetallic RuRe nanoparticles stabilized by PVP or oxide supports (γ-alumina and silica). Appl. Catal. A: Gen. **511**, 117 (2016)
94. Zhao, Y., Ke, W., Shao, J., Zheng, F., Liu, H., Shi, L.: Rational design of multisite trielement Ru–Ni–Fe alloy nanocatalysts with efficient and durable catalytic hydrogenation performances. ACS Appl. Mater. Interfaces **11**(44), 41204 (2019)